Lecture Notes in Computer Science 6683

Commenced Publication in 1973
Founding and Former Series Editors:
Gerhard Goos, Juris Hartmanis, and Jan van

T0092619

Carlos A. Coello Coello (Ed.)

Learning and Intelligent Optimization

5th International Conference, LION 5
Rome, Italy, January 17-21, 2011
Selected Papers

 Springer

Volume Editor

Carlos A. Coello Coello
Centro de Investigación y de Estudios
Avanzados del Instituto Politécnico Nacional
(CINVESTAV-IPN)
Departmento de Computación
Av. IPN No. 2508, Col. San Pedro Zacatenco
México, D.F. 07360, México
E-mail: ccoello@cs.cinvestav.mx

ISSN 0302-9743
ISBN 978-3-642-25565-6
DOI 10.1007/978-3-642-25566-3
Springer Heidelberg Dordrecht London New York

e-ISSN 1611-3349
e-ISBN 978-3-642-25566-3

Library of Congress Control Number: 2011941277

CR Subject Classification (1998): F.2, F.1, I.2, G.1.6, C.2, J.3

LNCS Sublibrary: SL 1 – Theoretical Computer Science and General Issues

Typesetting: Camera-ready by author, data conversion by Scientific Publishing Services, Chennai, India

Printed on acid-free paper

Springer is part of Springer Science+Business Media (www.springer.com)

Preface

LION 5, the 5th International Conference on Learning and Intelligent OptimizatioN, was held during January 17–21 in Rome, Italy. This meeting, which continues the successful series of LION conferences, aimed at exploring the intersections and uncharted territories between machine learning, artificial intelligence, mathematical programming and algorithms for hard optimization problems. The main purpose of the event was to bring together experts from these areas to discuss new ideas and methods, challenges and opportunities in various application areas, general trends and specific developments.

As in previous years, three different paper categories were available for submission: (1) regular papers on original and unpublished work, (2) short papers on original and unpublished work, and (3) works for oral presentation only. Accepted papers from the first two categories are published in the proceedings. A total of 99 submissions were received, from which 79 fell into the first category, 18 into the second one, and only 2 into the last one. After a thorough review process, 43 regular papers and 6 short papers were accepted for publication in the proceedings (the overall acceptance rate was 49%). None of the submissions from the third category was accepted for presentation.

These 49 contributions that were accepted for presentation cover the general track as well as the following four special sessions that were organized:

- *IMON: Intelligent Multiobjective OptimizatioN*
  Organizers: Dario Landa-Silva, Qingfu Zhang, David Wolfe Corne, Hui Li
- *LION-PP: Performance Prediction*
  Organizers: Kate Smith-Miles, Leo Lopes
- *Self* EAs: Self-Tuning, Self-Configuring and Self-Generating Evolutionary Algorithms*
  Organizers: Gabriela Ochoa, Marc Schoenauer
- *LION-SWAP: Software and Applications*
  Organizers: Mauro Brunato, Youssef Hamadi, Silvia Poles, Andrea Schaerf

The conference program was further enriched by the following tutorials, given by respected scientists in their respective domains. Carlos A. Coello Coello, from CINVESTAV-IPN (México), spoke about "Metaheuristics for Multiobjective Optimization," Yaochu Jin, from the University of Surrey (UK), talked about "A Systems Approach to Evolutionary Aerodynamic Design Optimization," Silvia Poles, from EnginSoft (Italy), gave a tutorial on "Multiobjective Optimization for Innovation in Engineering Design," and Roberto Battiti, from the University of Trento (Italy) spoke about "Reactive Business Intelligence and Data Mining."

The technical program also featured two invited talks by Benjamin W. Wah from the University of Illinois at Urbana-Champaign in USA (title: "Planning

Problems and Parallel Decomposition: A Critical Look") and Edward Tsang from the University of Essex in UK (title: "Intelligent Optimization in Finance and Economics"). Additionally, there was also a steering talk by Xin Yao, from the University of Birmingham in the UK, with the title: "Evolving and Designing Neural Network Ensembles."

Finally, we would like to express our sincere thanks to the authors for submitting their papers to LION 5, and to all the members of the Program Committee for their hard work. The organization of such an event would not be possible without the voluntary work of the Program Committee members. Many thanks also go to the invited speakers and tutorial speakers and to Thomas Stützle, for serving as the scientific liaison with Springer. Special thanks go to Marco Schaerf and Laura Palagi from the Sapienza Università di Roma who dealt with the local organization of this event. Final thanks go to Franco Mascia, the Web Chair of LION 5.

Last but not least, we would also like to acknowledge the contribution of our sponsors: the Associazione Italiana per lIntelligenza Artificiale, IEEE Computational Intelligence Society, Microsoft Research, Sapienza Università di Roma, and University of Trento for their technical co-sponsorship, as well as the industrial sponsor EnginSoft S.P.A.

April 2011   Carlos A. Coello Coello

Organization

Conference General Chair

Xin Yao   The University of Birmingham, UK

Local Organization Co-chairs

Marco Schaerf   Sapienza Università di Roma, Italy
Laura Palagi   Sapienza Università di Roma, Italy

Technical Program Committee Chair

Carlos A. Coello Coello   CINVESTAV-IPN, México

Program Committee

Hernan Aguirre   Shinshu University, Japan
Ethem Alpaydin   Bogazici University, Turkey
Julio Barrera   CINVESTAV-IPN, Mexico
Roberto Battiti   University of Trento, Italy
Mauro Birattari   Université Libre de Bruxelles, Belgium
Christian Blum   Universitat Politècnica de Catalunya, Spain
Juergen Branke   University of Warwick, UK
Mauro Brunato   Università di Trento, Italy
David Corne   Heriot-Watt University, UK
Carlos Cotta   Universidad de Málaga, Spain
Luca Di Gaspero   Università degli Studi di Udine, Italy
Karl F. Doerner   University of Vienna, Austria
Marco Dorigo   Université Libre de Bruxelles, Belgium
Andries Engelbrecht   University of Pretoria, South Africa
Shaheen Fatima   Loughborough University, UK
Antonio J. Fernández Leiva Universidad de Málaga, Spain
Álvaro Fialho   Microsoft Research - INRIA Joint Centre,
                             France
Valerio Freschi   University of Urbino, Italy
Deon Garrett   Icelandic Institute for Intelligent Machines,
                             Iceland
Michel Gendreau   École Polytechnique de Montréal, Canada
Martin Charles Golumbic   CRI Haifa, Israel

Walter J. Gutjahr	University of Vienna, Austria
Youssef Hamadi	Microsoft Research, UK
Jin-Kao Hao	University of Angers, France
Richard Hartl	University of Vienna, Austria
Geir Hasle	SINTEF Applied Mathematics, Norway
Alfredo G. Hernández-Díaz	Pablo de Olavide University, Spain
Francisco Herrera	University of Granada, Spain
Tomio Hirata	Nagoya University, Japan
Frank Hutter	University of British Columbia, Canada
Matthew Hyde	University of Nottingham, UK
Márk Jelasity	University of Szeged, Hungary
Yaochu Jin	University of Surrey, UK
Narendra Jussien	Ecole des Mines de Nantes, France
Zeynep Kiziltan	University of Bologna, Italy
Oliver Kramer	International Computer Science Institute, USA
Dario Landa-Silva	University of Nottingham, UK
Guillermo Leguizamón	Universidad Nacional de San Luis, Argentina
Khoi Le	University of Nottingham, UK
Hui Li	Xi'an Jiaotong University, China
Leo Lopes	Monash University, Australia
Eunice López Camacho	ITESM, México
Manuel López-Ibáñez	Université Libre de Bruxelles, Belgium
Antonio López-Jaimes	CINVESTAV-IPN, México
Vittorio Maniezzo	University of Bologna, Italy
Francesco Masulli	University of Genoa, Italy
Jorge Maturana	Universidad Austral de Chile, Chile
Juan J. Merelo Guervós	University of Granada, Spain
Bernd Meyer	Monash University, Australia
Zbigniew Michalewicz	University of Adelaide, Australia
Nenad Mladenovic	Brunel University, UK
Marco A. Montes de Oca	IRIDIA, Université Libre de Bruxelles, Belgium
Pablo Moscato	University of Newcastle, Australia
Gabriela Ochoa	University of Nottingham, UK
Yew-Soon Ong	Nanyang Technological University, Singapore
Djamila Ouelhadj	University of Portsmouth, UK
Panos M. Pardalos	University of Florida, USA
Andrew Parkes	University of Notthingham, UK
Marcello Pelillo	University of Venice, Italy
Vincenzo Piuri	Università degli Studi di Milano, Italy
Silvia Poles	Enginsoft Srl, Italy
Rong Qu	University of Nottingham, UK
Günther R. Raidl	Vienna University of Technology, Austria
Franz Rendl	Alpen-Adria University Klagenfurt, Austria
Celso C. Ribeiro	Universidade Federal Fluminense, Brazil
María Cristina Riff	Universidad Técnica Federico Santa María, Chile

Andrea Roli	Alma Mater Studiorum Università di Bologna, Italy
Eduardo Rodríguez-Tello	CINVESTAV-Tamaulipas, México
Rubén Ruiz García	Universidad Politécnica de Valencia, Spain
Wheeler Ruml	University of New Hampshire, USA
Ilya Safro	Argonne National Laboratory, USA
Horst Samulowitz	National ICT Australia, Australia
Frédéric Saubion	University of Angers, France
Andrea Schaerf	University of Udine, Italy
Marc Schoenauer	INRIA Saclay, France
Meinolf Sellmann	Brown University, USA
Yaroslav D. Sergeyev	Università della Calabria, Italy
Patrick Siarry	Université Paris-Est Créteil, France
Kate Smith-Miles	Monash University, Australia
Christine Solnon	Université de Lyon, France
Thomas Stützle	Université Libre de Bruxelles, Belgium
Ke Tang	University of Science and Technology of China, China
Hugo Terashima	ITESM - Centre for Intelligent Systems, México
Marco Tomassini	University of Lausanne, Switzerland
Gregorio Toscano-Pulido	CINVESTAV-Tamaulipas, México
Pascal Van Hentenryck	Brown University, USA
Sebastien Verel	INRIA Lille-Nord Europe and University of Nice Sophia-Antipolis, France
Stefan Voß	University of Hamburg, Germany
Toby Walsh	NICTA and UNSW, Australia
David L. Woodruff	University of California, Davis, USA
Qingfu Zhang	University of Essex, UK

Additional Referees

Manuel Blanco Abello	Stefano Benedettini	Muneer Buckley
Samuel Rota Bulò	Ethan Burns	Marco Caserta
Camille Combier	Sabrina de Oliveira	Adam Ghandar
Stephane Gosselin	Jean-Philippe Hamiez	Franco Mascia
Eddy Parkinson	Nicola Rebagliati	Jordan Thayer

IMON Special Session Chairs

Dario Landa-Silva	University of Nottingham, UK
Qingfu Zhang	University of Essex, UK
David Wolfe Corne	Heriot-Watt University, UK
Hui Li	Xi'an Jiaotong University, China

LION-PP Special Session Chairs

Kate Smith-Miles	Monash University, Australia
Leo Lopes	Monash University, Australia

Self* EAs Special Session Chairs

Gabriela Ochoa	University of Nottingham, UK
Marc Schoenauer	INRIA Saclay - Ile-de-France and Microsoft/INRIA Joint Center, Saclay, France

LION-SWAP Special Session Chairs

Mauro Brunato	University of Trento, Italy
Youssef Hamadi	Microsoft Research, Cambridge, UK
Silvia Poles	EnginSoft, Italy
Andrea Schaerf	University of Udine, Italy

Web Chair

Franco Mascia	University of Trento, Italy

Steering Committee

Roberto Battiti	University of Trento, Italy
Holger Hoos	University of British Columbia, Canada
Mauro Brunato	University of Trento, Italy
Thomas Stützle	Université Libre de Bruxelles, Belgium
Christian Blum	Universitat Politècnica de Catalunya, Spain
Martin Charles Golumbic	CRI Haifa, Israel

Technical Co-sponsorship

Associazione Italiana per lIntelligenza Artificiale
http://www.aixia.it/

IEEE Computational Intelligence Society
http://www.ieee-cis.org/

Microsoft Research
http://research.microsoft.com/en-us/

Sapienza Università di Roma, Italy
http://www.uniroma1.it/

University of Trento, Italy
http://www.unitn.it/

Industrial Sponsorship

EnginSoft S.P.A.
http://www.enginsoft.com/

Local Organization Support

Reactive Search S.R.L.
http://www.reactive-search.com/

Table of Contents

Main Track (Regular Papers)

Multivariate Statistical Tests for Comparing Classification
Algorithms 1
    Olcay Taner Yıldız, Özlem Aslan, and Ethem Alpaydın

Using Hyperheuristics under a GP Framework for Financial
Forecasting 16
    Michael Kampouridis and Edward Tsang

On the Effect of Connectedness for Biobjective Multiple and Long Path
Problems 31
    Sébastien Verel, Arnaud Liefooghe, Jérémie Humeau,
    Laetitia Jourdan, and Clarisse Dhaenens

Improving Parallel Local Search for SAT 46
    Alejandro Arbelaez and Youssef Hamadi

Variable Neighborhood Search for the Time-Dependent Vehicle Routing
Problem with Soft Time Windows 61
    Stefanie Kritzinger, Fabien Tricoire, Karl F. Doerner, and
    Richard F. Hartl

Solving the Two-Dimensional Bin Packing Problem with a Probabilistic
Multi-start Heuristic 76
    Lukas Baumgartner, Verena Schmid, and Christian Blum

Genetic Diversity and Effective Crossover in Evolutionary
Many-objective Optimization 91
    Hiroyuki Sato, Hernán E. Aguirre, and Kiyoshi Tanaka

An Optimal Stopping Strategy for Online Calibration in Local
Search 106
    Gianluca Bontempi

Analyzing the Effect of Objective Correlation on the Efficient Set of
MNK-Landscapes 116
    Sébastien Verel, Arnaud Liefooghe, Laetitia Jourdan, and
    Clarisse Dhaenens

Instance-Based Parameter Tuning via Search Trajectory Similarity
Clustering 131
    Lindawati, Hoong Chuin Lau, and David Lo

Effective Probabilistic Stopping Rules for Randomized Metaheuristics:
GRASP Implementations 146
  Celso C. Ribeiro, Isabel Rosseti, and Reinaldo C. Souza

A Classifier-Assisted Framework for Expensive Optimization Problems:
A Knowledge-Mining Approach 161
  Yoel Tenne, Kazuhiro Izui, and Shinji Nishiwaki

Robust Gaussian Process-Based Global Optimization Using a Fully
Bayesian Expected Improvement Criterion 176
  Romain Benassi, Julien Bect, and Emmanuel Vazquez

Hierarchical Hidden Conditional Random Fields for Information
Extraction 191
  *Satoshi Kaneko, Akira Hayashi, Nobuo Suematsu, and
  Kazunori Iwata*

Solving Extremely Difficult MINLP Problems Using Adaptive
Resolution Micro-GA with Tabu Search 203
  *Asim Munawar, Mohamed Wahib, Masaharu Munetomo, and
  Kiyoshi Akama*

Adaptive Abnormality Detection on ECG Signal by Utilizing FLAC
Features 218
  Jiaxing Ye, Takumi Kobayashi, Tetsuya Higuchi, and Nobuyuki Otsu

Gravitational Interactions Optimization 226
  Juan J. Flores, Rodrigo López, and Julio Barrera

On the Neutrality of Flowshop Scheduling Fitness Landscapes 238
  *Marie-Eléonore Marmion, Clarisse Dhaenens, Laetitia Jourdan,
  Arnaud Liefooghe, and Sébastien Verel*

A Reinforcement Learning Approach for the Flexible Job Shop
Scheduling Problem 253
  Yailen Martínez, Ann Nowé, Juliett Suárez, and Rafael Bello

Supervised Learning Linear Priority Dispatch Rules for Job-Shop
Scheduling 263
  Helga Ingimundardottir and Thomas Philip Runarsson

Fine-Tuning Algorithm Parameters Using the Design of Experiments
Approach 278
  Aldy Gunawan, Hoong Chuin Lau, and Lindawati

MetaHybrid: Combining Metamodels and Gradient-Based Techniques
in a Hybrid Multi-Objective Genetic Algorithm 293
  Alessandro Turco

Designing Stream Cipher Systems Using Genetic Programming   308
    Wasan Shakr Awad

GPU-Based Multi-start Local Search Algorithms   321
    Thé Van Luong, Nouredine Melab, and El-Ghazali Talbi

Active Learning of Combinatorial Features for Interactive
Optimization   336
    Paolo Campigotto, Andrea Passerini, and Roberto Battiti

A Genetic Algorithm Hybridized with the Discrete Lagrangian Method
for Trap Escaping   351
    Madalina Raschip and Cornelius Croitoru

Greedy Local Improvement of SPEA2 Algorithm to Solve the
Multiobjective Capacitated Transshipment Problem   364
    Nabil Belgasmi, Lamjed Ben Said, and Khaled Ghedira

Hybrid Population-Based Incremental Learning Using Real Codes   379
    Sujin Bureerat

Pareto Autonomous Local Search   392
    Nadarajen Veerapen and Frédéric Saubion

Transforming Mathematical Models Using Declarative Reformulation
Rules   407
    Antonio Frangioni and Luis Perez Sanchez

Learning Heuristic Policies – A Reinforcement Learning Problem   423
    Thomas Philip Runarsson

Continuous Upper Confidence Trees   433
    *Adrien Couëtoux, Jean-Baptiste Hoock, Nataliya Sokolovska,
    Olivier Teytaud, and Nicolas Bonnard*

Main Track (Short Papers)

Towards an Intelligent Non-Stationary Performance Prediction of
Engineering Systems   446
    David J.J. Toal and Andy J. Keane

Local Search for Constrained Financial Portfolio Selection Problems
with Short Sellings   450
    Luca Di Gaspero, Giacomo di Tollo, Andrea Roli, and Andrea Schaerf

Clustering of Local Optima in Combinatorial Fitness Landscapes   454
    Gabriela Ochoa, Sébastien Verel, Fabio Daolio, and Marco Tomassini

Special Session: IMON

Multi-Objective Optimization with an Adaptive Resonance
Theory-Based Estimation of Distribution Algorithm: A Comparative
Study 458
    Luis Martí, Jesús García, Antonio Berlanga, and José M. Molina

Multi-Objective Differential Evolution with Adaptive Control of
Parameters and Operators 473
    Ke Li, Álvaro Fialho, and Sam Kwong

Distribution of Computational Effort in Parallel MOEA/D 488
    Juan J. Durillo, Qingfu Zhang, Antonio J. Nebro, and Enrique Alba

Multi Objective Genetic Programming for Feature Construction in
Classification Problems 503
    Mauro Castelli, Luca Manzoni, and Leonardo Vanneschi

Special Session: LION-PP

Sequential Model-Based Optimization for General Algorithm
Configuration 507
    Frank Hutter, Holger H. Hoos, and Kevin Leyton-Brown

Generalising Algorithm Performance in Instance Space: A Timetabling
Case Study... 524
    Kate Smith-Miles and Leo Lopes

Special Session: Self* EAs

A Hybrid Fish Swarm Optimisation Algorithm for Solving Examination
Timetabling Problems 539
    Hamza Turabieh and Salwani Abdullah

The Sandpile Mutation Operator for Genetic Algorithms 552
    *C.M. Fernandes, J.L.J. Laredo, A.M. Mora, A.C. Rosa, and
    J.J. Merelo*

Self-adaptation Techniques Applied to Multi-Objective Evolutionary
Algorithms... 567
    *Saúl Zapotecas Martínez, Edgar G. Yáñez Oropeza, and
    Carlos A. Coello Coello*

Analysing the Performance of Different Population Structures for an
Agent-based Evolutionary Algorithm 582
    *J.L.J. Laredo, J.J. Merelo, C.M. Fernandes, A.M. Mora, M.G.
    Arenas, P.A. Castillo, and P. Garcia-Sanchez*

Special Session: LION-SWAP

EDACC - An Advanced Platform for the Experiment Design,
Administration and Analysis of Empirical Algorithms 586
    Adrian Balint, Daniel Diepold, Daniel Gall, Simon Gerber,
    Gregor Kapler, and Robert Retz

HAL: A Framework for the Automated Analysis and Design of
High-Performance Algorithms 600
    Christopher Nell, Chris Fawcett, Holger H. Hoos, and
    Kevin Leyton-Brown

HYPERION – A Recursive Hyper-Heuristic Framework 616
    Jerry Swan, Ender Özcan, and Graham Kendall

The Cross-Domain Heuristic Search Challenge – An International
Research Competition 631
    Edmund K. Burke, Michel Gendreau, Matthew Hyde,
    Graham Kendall, Barry McCollum, Gabriela Ochoa,
    Andrew J. Parkes, and Sanja Petrovic

Author Index 635

Multivariate Statistical Tests for Comparing Classification Algorithms

Olcay Taner Yıldız[1], Özlem Aslan[2], and Ethem Alpaydın[2]

[1] Dept. of Computer Engineering, Işık University, TR-34980, Istanbul, Turkey
[2] Dept. of Computer Engineering, Boğaziçi University, TR-34342, Istanbul, Turkey

Abstract. The misclassification error which is usually used in tests to compare classification algorithms, does not make a distinction between the sources of error, namely, false positives and false negatives. Instead of summing these in a single number, we propose to collect multivariate statistics and use multivariate tests on them. Information retrieval uses the measures of precision and recall, and signal detection uses true positive rate (tpr) and false positive rate (fpr) and a multivariate test can also use such two values instead of combining them in a single value, such as error or average precision. For example, we can have bivariate tests for (precision, recall) or (tpr, fpr). We propose to use the pairwise test based on Hotelling's multivariate T^2 test to compare two algorithms or multivariate analysis of variance (MANOVA) to compare $L > 2$ algorithms. In our experiments, we show that the multivariate tests have higher power than the univariate error test, that is, they can detect differences that the error test cannot, and we also discuss how the decisions made by different multivariate tests differ, to be able to point out where to use which. We also show how multivariate or univariate pairwise tests can be used as post-hoc tests after MANOVA to find cliques of algorithms, or order them along separate dimensions.

1 Introduction

For a typical machine learning application, there are multiple candidate algorithms and we need to choose one among many. In supervised learning, this is typically done by comparing errors, and in classification with two classes, the misclassification error is the sum of false positives and false negatives (see Table 1(a)). However, misclassification error does not make a distinction between false positives and false negatives, and various other measures have been proposed depending on the type of error we focus on (see Table 1(b)). In information retrieval, the two measures used are precision and recall, and in signal detection, they are true positive rate (tpr) and false positive rate (fpr). People also use curves of these or areas under such curves. These different set of measures have different uses, as we will discuss later.

In comparing classification algorithms, we use statistical tests to make sure that the difference is *significant*, that is, big enough that it could not have happened by chance, or in other words, very unlikely to have been caused by

C.A. Coello Coello (Ed.): LION 5, LNCS 6683, pp. 1–15, 2011.
© Springer-Verlag Berlin Heidelberg 2011

Table 1. (a) 2×2 confusion matrix for two classes. (b) Different performance measures.

(a)

True class	Predicted class		
	Positive	Negative	Sum
Positive	tp	fn	p
Negative	fp	tn	n
Sum	p'	n'	

(b)

Name	Formula
error	$(fp+fn)/(p+n)$
accuracy	$(tp+tn)/(p+n)$
tpr	tp/p
fpr	fp/n
precision	tp/p'
recall	tp/p

chance – the so-called *p-value* of the test. To be able to measure the effect of chance (e.g., variance due to small changes in the training set), typically, one does training and validation a number of times, possibly by resampling using cross-validation. For example, with k training and validation dataset pairs, we train the classification algorithms on the k training sets and obtain the k confusion matrices on the validation sets. From these, we can for example calculate the k misclassification error values and to compare two algorithms, we can use a pairwise statistical test [1] to see whether the two algorithms lead to classifiers with equal expected error. When there are more than two to compare, one can use analysis of variance (ANOVA) to check if all have equal expected error. It is critical that such tests are *paired,* that is, we use the same training and validation data with all algorithms so that whatever difference we observe is due to the algorithm, and not due to any randomness in resampling the data.

We note the disadvantage of using error here; such tests cannot make a distinction between false positives and false negatives. Two classifiers may have the same error but one may have all its error due to false positives, the other all due to false negatives, and we will not be able to detect this difference if our comparison metric is simply the error; see Figure 1 for an example.

In this paper, we propose *multivariate tests* that can do comparison using multiple measures and not just a single one, i.e., error. That is, from the k confusion matrices, we will collect *multivariate statistics* such as a two-dimensional vector of (tpr, fpr) or (precision, recall), and do a bivariate test. We can also do a four-variate test using the whole 2×2 confusion matrix or any other vector of measurements. Statistical tests in the machine learning literature are all univariate; to the best of our knowledge, our use of multivariate tests in performance comparison of machine learning algorithms is the first.

The need to combine different measures have been noticed before. Average precision combines precision and recall, for example, Caruana et al. (2004) [2] compared different performance metrics such as accuracy, lift, F-Score, area under the ROC curve, average precision, precision/recall break-even point, squared error, cross entropy, and probability calibration; they showed that these metrics are correlated and proposed a new measure SAR as the average of Squared error, Accuracy and Roc area. Seliya et al. (2009) [3] calculated different measures too and going one step further proposed to combine them taking the correlation into

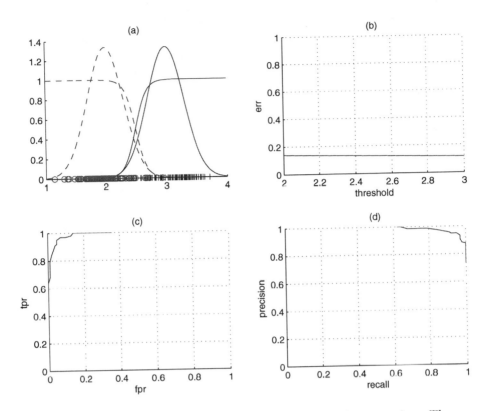

Fig. 1. Example showing that error is not the best measure in comparison. The negative and positive instances are normally distributed with their means at 2 and 3 respectively; both have standard deviation 0.3. In (a), we see the two densities, the posterior probabilities and 100 instances sampled from each. We have a classifier that chooses the positive class if the input is greater than a threshold (corresponding to a threshold on the posterior of the positive class) and what we then do, is move this threshold of decision gradually from 2 to 3 (corresponding to increasing the posterior threshold from 0 to 1). As we see in (b), the error does not change; the number of false positives decreases but the number of false negatives increase in equal amount. In (c) and (d), we see that if we use (tpr, fpr) and (precision, recall) as measures of performance, the values differ as the threshold is changed. As the threshold increases, the number of true positives decrease which decreases tpr and recall; but because false positives decrease, fpr decreases and precision increases. Note that in (c) and (d), as we increase the threshold, we move from the right to the left along the curves. (Tpr, fpr) and (precision, recall) can detect a difference due to different thresholds because they make a distinction between false positives and false negatives. For example, if we had two classifiers one with threshold at 2 and another with threshold at 3, a pairwise test on error would not be able to detect any difference between them, but tests on (tpr, fpr) or (precision, recall) would. The aim of this paper is the discussion of such tests.

account. Note however that these are for reporting performances only and they include no statistical methodology for testing or comparison, as we do here.

This paper is organized as follows: To compare two algorithms, we discuss the pairwise univariate test and the proposed multivariate test in Section 2. When there are $L > 2$ algorithms to compare, we can use univariate and multivariate ANOVA, as discussed in Section 3. We give our experimental results in Section 4 and conclude in Section 5.

2   Pairwise Comparison

Let us say we have two classification algorithms. We train and validate the two algorithms on k training/validation data folds and calculate the resulting k separate 2×2 confusion matrices $M_{ij}, i = 1, 2, j = 1, \ldots, k$, on the validation sets in the same format as shown in Table 1(a).

2.1   Univariate Case

If we want to compare in terms of error, for both algorithms and all k folds, we calculate $e_{ij} = fp_{ij} + fn_{ij}$ and then the paired difference between the errors

$$d_j = e_{1j} - e_{2j}$$

and we test if these differences come from a population with zero mean:

$$H_0 : \mu_d = 0 \text{ vs. } H_1 : \mu_d \neq 0$$

For the *univariate paired t test*, we calculate the average and the standard deviation:

$$\overline{d} = \sum_{j=1}^{k} d_j/k \; , \; s_d = \frac{\sum_j (d_j - \overline{d})^2}{k-1}$$

Under the null hypothesis that the two algorithms have the same expected error, we know that

$$t' = \sqrt{k}\frac{\overline{d}}{s_d} \tag{1}$$

is t distributed with $k - 1$ degrees of freedom. We reject H_0 if $|t'| > t_{\alpha/2, k-1}$ with $(1 - \alpha)100\,\%$ confidence.

2.2   Multivariate Case

If we do not want to reduce to a single statistic and want to use a set of values in comparison, we need a test that can use vectors instead of scalars. In such a case, we want to compare the means of two p-dimensional populations, that is, we want to test for the null hypothesis $H_0 : \boldsymbol{\mu}_1 - \boldsymbol{\mu}_2 = \mathbf{0}$. If we want to compare in terms of (tpr, fpr) or (precision, recall), then $p = 2$. Note that using the same setting, it is also possible to define a multivariate test on (sensitivity, specificity), or consider all four entries in the confusion matrix, in which case

$p = 4$. As before, we train and validate both algorithms with the same folds and use a paired test, except that now the test is multivariate.

Let us say $x_{ij} \in \Re^p$ is the performance vector containing p performance values. For the *multivariate paired Hotelling's test,* we calculate the paired difference vectors

$$d_j = x_{1j} - x_{2j}$$

and check if they come from a p-variate Gaussian with zero mean:

$$H_0 : \mu_d = 0 \text{ vs. } H_1 : \mu_d \neq 0$$

We calculate the average vector and the covariance matrix:

$$\overline{d} = \sum_{j=1}^{k} d_j / k , \mathbf{S}_d = \frac{1}{k-1} \sum_j (d_j - \overline{d})(d_j - \overline{d})^T$$

Under the null hypothesis that the two algorithms have the same expected behavior, we know that [4]

$$T'^2 = k\overline{d}^T \mathbf{S}_d^{-1} \overline{d} \tag{2}$$

is *Hotelling's* T^2 distributed with p and $k - 1$ degrees of freedom. We reject the null hypothesis if $T'^2 > T^2_{\alpha,p,k-1}$. Hotelling's $T^2(p, m)$ can be approximated using F distribution via the formula

$$\left(\frac{m - p + 1}{mp} \right) T^2_{p,m} \sim F_{m,m-p+1} \tag{3}$$

Note that we calculate our measures such as tpr, precision, and so on, from entries in the 2×2 confusion matrix; these are counts of indicator random variables (they are 0/1 Bernoulli random variables) caused by the same event (the trained classifier) and the total counts are then dependent binomial random variables. We know from the central limit theorem that the binomial converges to the Gaussian unless the sample (here, the validation set size) is very small and hence the assumption of joint multivariate normality makes sense. Remember that all parametric tests based on error also use the same assumption.

When $p = 1$, this multivariate test reduces to the univariate t test of Section 2.1. Just like $\overline{d}/s_d$ of (1) measuring the normalized distance in one dimension, $\overline{d}^T \mathbf{S}_d^{-1} \overline{d}$ of (2) measures the (squared) normalized distance in p dimensions.

If the multivariate test rejects, we can do p *post-hoc* univariate tests to check which one(s) of the variates cause(s) a rejection. For example, if a multivariate test on (precision, recall) rejects, we may want to check if the difference is due to a significant difference in precision, recall, or both. For testing difference in variate l, we use the univariate test in (1) and calculate

$$t'_l = \sqrt{k} \frac{\overline{d}_l}{\mathbf{S}_{d,ll}} \tag{4}$$

and reject $H_0 : \mu_{d,l} = 0$ if $|t'_l| > t_{\alpha/2,k-1}$.

Note that it may be the case that none of the univariate differences is significant whereas the multivariate one is, and the linear combination of variates that cause the maximum difference can be calculated as

$$w = S_d^{-1}\overline{d} \tag{5}$$

We can then see the effect of the different univariate dimensions by looking at the corresponding elements of w. The fact that this is the Fisher's LDA direction is not accidental—we are looking for the direction that maximizes the separation of two groups of data.

3 Analysis of Variance

If we have $L > 2$ algorithms to compare, we test whether they have the same expected performance. In the univariate case, we reduce the confusion matrices to error values and compare them; in the multivariate case, we compare vectors of performance values.

3.1 Univariate Case

Given L populations, we test for

$$H_0 : \mu_1 = \mu_2 = \cdots = \mu_L \text{ vs. } H_1 : \mu_r \neq \mu_s \text{ for one pair } r, s$$

Let us say that $e_{ij}, i = 1, \ldots, L, j = 1, \ldots, k$, denotes the error of algorithm i on validation fold j. $e_{i\cdot} = \sum_j e_{ij}/k$ denotes the average error of algorithm i, and $e_{\cdot\cdot} = \sum_i e_{i\cdot}/L$ denotes the overall average. The univariate ANOVA calculates

$$\begin{aligned} F' &= \frac{MSH}{MSE} = \frac{SSH/(L-1)}{SSE/L(k-1)} \\ &= \frac{(\sum_i e_{i\cdot}^2/k - e_{\cdot\cdot}/Lk)(L-1)}{(\sum_{i,j} e_{ij}^2 - \sum_i e_{i\cdot}^2/k)/L(k-1)} \end{aligned} \tag{6}$$

which, under the null hypothesis, is F distributed with $L-1$ and $L(k-1)$ degrees of freedom. We reject H_0 if $F' > F_{\alpha, L-1, L(k-1)}$.

If ANOVA rejects and we know that there is at least one pair that is significantly different, we can use the pairwise test of Section 2.1 as a post-hoc test on all pairs r, s to check which pair(s) lead(s) to the significant difference in error.

3.2 Multivariate Case

Given L populations, we test for

$$H_0 : \boldsymbol{\mu}_1 = \boldsymbol{\mu}_2 = \cdots = \boldsymbol{\mu}_L \text{ vs. } H_1 : \boldsymbol{\mu}_r \neq \boldsymbol{\mu}_s \text{ for one pair } r, s.$$

Let us say that $x_{ij}, i = 1, \ldots, L, j = 1, \ldots, k$ denotes the p-dimensional performance vector of algorithm i on validation fold j. The multivariate ANOVA (MANOVA) calculates the two matrices of between- and within-scatter:

$$\mathbf{H} = k \sum_{i=1}^{L} (\overline{x}_{i.} - \overline{x}_{..})(\overline{x}_{i.} - \overline{x}_{..})^T$$

$$\mathbf{E} = \sum_{i=1}^{L} \sum_{j=1}^{k} (x_{ij} - \overline{x}_{i.})(x_{ij} - \overline{x}_{i.})^T$$

Then

$$\Lambda' = \frac{|\mathbf{E}|}{|\mathbf{E} + \mathbf{H}|} \tag{7}$$

is *Wilks'* Λ distributed with $p, L-1, L(k-1)$ degrees of freedom [4]. We reject H_0 if $\Lambda' \leq \Lambda_{\alpha,p,L-1,L(k-1)}$. Note that rejection is for small values of Λ': If the sample mean vectors are equal, we expect $\mathbf{H}$ to be $\mathbf{0}$ and Λ' to approach 1; as the sample means become more spread, $\mathbf{H}$ becomes "larger" than $\mathbf{E}$ and Λ' approaches 0.

Wilks' Λ can be approximated using χ^2 distribution via the formula

$$\left(\frac{p - n + 1}{2} - m\right) \log \Lambda_{p,m,n} \sim \chi^2_{np} \tag{8}$$

If MANOVA rejects, we can do p separate univariate ANOVA on each of the individual variates as we discussed in Section 3.1, or the difference may be due to some linear combination of the variates: The mean vectors occupy a space whose dimensionality is given by $s = \min(p, L-1)$; its dimensions are the eigenvectors of $\mathbf{E}^{-1}\mathbf{H}$ and we have

$$\Lambda = \prod_{i=1}^{s} \frac{1}{1 + \lambda_i}$$

where λ_i are the corresponding sorted eigenvalues. The analysis of the eigenvalues and the corresponding variates of the eigenvectors allow us to pinpoint the causes if MANOVA rejects. For example, if $\lambda_1 / \sum_i \lambda_i > 0.9$, there is collinearity, i.e., the means lie on a single discriminant, $z = w^T x$, where w is the eigenvector with the largest eigenvalue λ_1.

We can also do a set of pairwise multivariate tests as we have discussed in Section 2.2 after MANOVA rejects, to see which pairs (or groups) of algorithms have comparable performance vectors.

4 Experiments

4.1 Setup

We use a total of 36 two-class datasets where 27 of them (*artificial, australian, breast, bupa, credit, cylinder, german, haberman, heart, hepatitis, horse, ironosphere, krvskp, magic, mammographic, monks, mushroom, parkinsons, pima, polyadenylation, promoters, satellite47, spambase, spect, tictactoe, transfusion, vote*) are from the UCI repository [5], three (*ringnorm, titanic, twonorm*) are from the Delve repository [6], and six (*acceptors, ads, dlbcl, donors, musk2, prostatetumor*) are Bioinformatics datasets [7]. We use 10-fold cross-validation and five algorithms: (1) *c45*: C4.5 decision tree. (2) *svm*: Support vector machine (SVM) with a linear kernel [8].(3) *lda*: Linear discriminant classifier. (4) *qda*: Quadratic discriminant classifier. (5) *knn*: k-nearest neighbor with $k = 20$.

4.2   Results

Univariate vs. Multivariate testing. In the first part of our experiments, we compare the univariate k-fold paired t test ($k = 10$) on error which we name UniErr, with our proposed multivariate pairwise test using (tpr, fpr), which we name MultiTF.

Figure 2 shows the example where the univariate test fails to reject and MultiTF rejects the null hypothesis that the two classifiers *lda* and *qda* have the same mean on the *breast* dataset. Figure 2(a) shows the (tpr, fpr) scatter plots of the ten runs each of the two methods and the isoprobability contours of the fitted bivariate Gaussians. We see that LDA has higher fpr whereas QDA has lower tpr, that is, higher false negative rate. We see in Figure 2(b) that the classifiers have comparable overall error histograms: LDA has more false positives, QDA has more false negatives, but overall they have comparable error. We see in Figure 2(c) that the contour plot of the covariance matrix of the paired differences has its mean far from (0,0) and that is why the multivariate test rejects the null hypothesis that the means are the same, whereas in Figure 2(d), we see that histogram of the differences of errors has its mean close to 0 and the univariate test fails to reject the null hypothesis that the means are equal.

MultiTF vs. MultiPR. In the second part of our experiments, we see the effect of different measures on the multivariate test and compare MultiTF with the multivariate test using (precision, recall) that we name MultiPR; this will help us identify which one to use in which context.

Figure 3 shows an example where MultiTF rejects and MultiPR fails to reject the null hypothesis that *c45* and *qda* have the same mean on the *pima* dataset. In Figures 3(a) and (b), the x axes are the same because tpr and recall are the same; the two differ in the y axes and that helps us understand why the two decisions are different. Although with respect to (tpr, fpr), the mean of *c45* and *qda* seem to be close to each other (Fig. 3(a)), their difference is significantly large compared to their standard deviations and this causes a rejection. They are close enough in the (precision, recall) space (Fig. 3(b)) and hence MultiPR does not reject. In calculating precision, we divide by p', and in calculating fpr, we divide by n; here, n is larger than p' and hence, the variance of fpr is smaller, which makes the difference significant.

We can also see this by comparing Figures 3(d) and (e): In (d), we see that (0,0) lies on the outermost contour indicating that the probability that we see a difference as large is small and hence we reject the null hypothesis; in (e), (0,0)

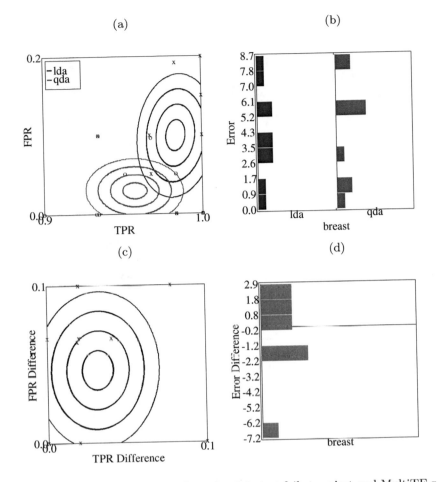

Fig. 2. The example case where the univariate test fails to reject and MultiTF rejects the null hypothesis that *lda* and *qda* have the same mean on the *breast* dataset. (a) shows the isoprobability contour plots of the Gaussians fitted to performance data from two algorithms and (c) shows the distribution of their paired difference; (b) and (d) show the corresponding error histograms and the histogram of paired error differences respectively. Roughly speaking, the multivariate test rejects if the mean of the differences is far from $(0,0)$, compared to the scale of the covariance matrix of differences; just as the univariate test rejects if the mean of the differences is far from 0, compared to the standard deviation of differences.

is close enough to the center of the contours and the probability that we see such a difference is not small and hence we do not reject.

If the univariate post-hoc tests are performed, we see that the algorithms are significantly different in terms of fpr with a p-value of 0.006. The corresponding elements of w (equation 5) are (tpr : 0.499, fpr : -25.898) and (precision : 2.014, recall : -4.374), which shows that fpr is the important one.

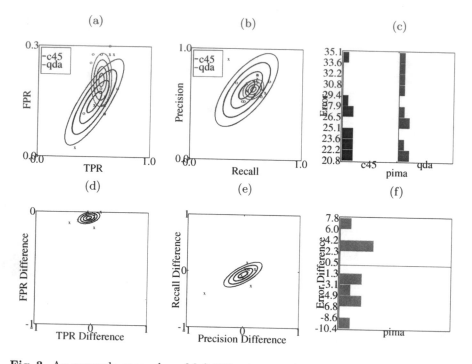

Fig. 3. An example case where MultiTF rejects and MultiPR fails to reject the null hypothesis that the two classifiers, *c45* and *qda*, have the same mean on the *pima* dataset. (a) and (d) show the isoprobability contour plots of the fitted Gaussians and of the difference with respect to (tpr, fpr); (b) and (e) show the same with respect to (precision, recall); (c) and (f) show the corresponding histogram of the error rates and the differences in the error rates.

The error distributions of the algorithms are also similar to each other and the univariate test also fails to reject the null hypothesis that the error rates of those algorithms are equal (see Figs. 3(c) and (f)).

(Precision, recall) and (tpr, fpr) metric pairs have different application areas. In (precision, recall), we are basically interested in how well we classify the positive examples, whereas in (tpr, fpr), in trying to minimize fpr, we also want to increase the true negatives. To show the difference between them, we did two experiments: In Figure 4(a), we simply add more and more true negatives to a classifier. In such a case, we see that this has no effect on precision and recall, but decreases fpr. When compared with the classifier without any additional true negatives, MultiPR does not reject but MultiTF starts rejecting after a point.

It is known that (precision, recall) is sensitive to class skewness [9], whereas (tpr, fpr) is not. In Figure 4(b)), we slowly change the ratio p/n, and we see that because precision uses values from both rows, it changes; however (tpr, fpr) do not change since they use values from only one row. Compared with the classifier with the original ratio, MultiTF does not reject (because the rates do not change), but MultiPR starts rejecting after a point.

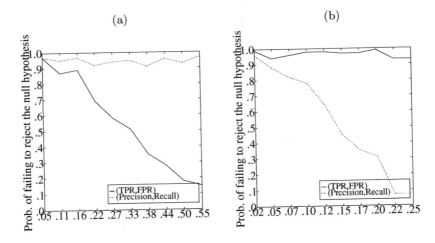

Fig. 4. In (a), when we add more and more true negatives ($tn \leftarrow tn\,(1+\lambda)$), precision and recall do not change, but fpr ($=fp\,/\,(n + \lambda\; tn)$) decreases and MultiTF test starts rejecting the null hypothesis. In (b), we change the ratio $\frac{p}{n} = \frac{(tp+fn)(1-\alpha)}{(fp+tn)(1+\alpha)}$ while keeping tpr and fpr the same ($tp \leftarrow tp\,(1-\alpha)$, $fn \leftarrow fn\,(1-\alpha)$, $fp \leftarrow fp\,(1+\alpha)$, $tn \leftarrow tn\,(1+\alpha)$), we see that precision changes and MultiPR starts rejecting. Plotted values are proportions of failures to reject in 100 independent runs.

If we are doing an information retrieval task with a query such as, "Find me all images of tigers," adding additional non-tiger images to the database does not have any effect on our measure of performance (as long as we have no difficulty in recognizing them as non-tigers and do not retrieve them), and hence we use precision and recall. If we want to differentiate between two types of targets, for example, cars and tanks, our accuracy on these different targets is important, and we use tpr and fpr.

Comparison of multiple algorithms. In the third part of our experiments, we use the univariate and multivariate tests to compare $L > 2$ classification algorithms. For the univariate case, if ANOVA rejects, we can do $L(L-1)/2$ pairwise univariate tests to find difference between pairs and also cliques, i.e., subsets of algorithms in which all pairwise tests fail to reject.

In the case of a univariate test, we can also write down an order by comparing the means. For this, we sort the algorithms in terms of average error in ascending order and then try to find groups where there is no statistically significant difference between the smallest and largest means in the group, which we check by applying a pairwise univariate test to these two at the ends. If this the case, we underline the group. We first try all five, if there is a difference between the first and the fifth, we try the two groups of four leaving out the two extremes, and so on.

If MANOVA rejects, similarly, we can do the pairwise multivariate tests and find cliques. We can also do univariate tests on the dimensions separately and

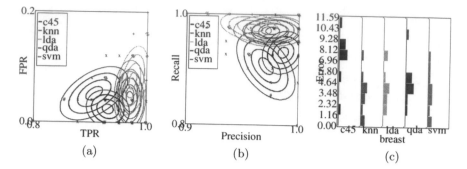

Fig. 5. Comparison of five algorithms on *breast*. (a) and (b) show the isoprobability contour plots of the fitted bivariate Gaussians with respect to (tpr, fpr) and (precision, recall) respectively; (c) shows the corresponding histogram of the error rates.

Table 2. Tabular representation of post-hoc univariate and MultiTF/MultiPR test results on *breast* dataset. 1 stands for a failure to reject the null hypothesis.

	c45	lda	qda	svm	knn		c45	lda	qda	svm	knn
c45		0	0	0	0	c45		0	0	0	0
lda	0		1	1	1	lda	0		0	1	1
qda	0	1		1	1	qda	0	0		0	0
svm	0	1	1		1	svm	0	1	0		1
knn	0	1	1	1		knn	0	1	0	1	

try to find orderings, as discussed above for error. For example, if MANOVA on (precision, recall) on five algorithms reject, we can try to find groups and orderings in terms of precision and recall separately.

Figure 5 shows the first example case on *breast* dataset. Both ANOVA and MANOVA reject the null hypothesis. According to post-hoc test results, the univariate test finds a single clique of four algorithms (*knn, lda, qda, svm*). On the other hand, both multivariate post-hoc tests (MultiTF and MultiPR) find a single clique of three algorithms (*knn, lda, svm*). Table 2 shows the results of all pairwise tests between five algorithms.

The univariate orderings found are as follows:

error	*knn svm lda qda c45*
tpr, recall	*lda knn svm qda c45*
fpr	*qda knn svm c45 lda*
precision	*qda knn svm c45 lda*

The clique found by multivariate tests (*knn, lda, svm*) appears as a single group with respect to tpr and recall. Although (*knn, svm*) appear together, *lda* is separate from that group when the criterion is fpr or precision. We see that these different measures are able to detect differences that error cannot, and that the differences vary depending on what performance measure we concentrate on.

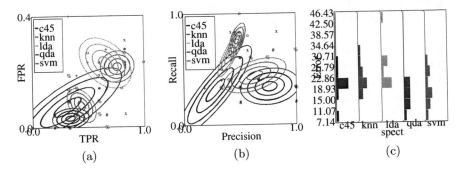

Fig. 6. Comparison of five algorithms on *spect*. (a) and (b) show the isoprobability contour plots of the fitted bivariate Gaussians with respect to (tpr, fpr) and (precision, recall) respectively; (c) shows the corresponding histogram of the error rates.

Table 3. Tabular representation of post-hoc MultiTF and MultiPR test results on *spect* dataset. **1** stands for failing to reject the null hypothesis.

	c45	*lda*	*qda*	*svm*	*knn*		*c45*	*lda*	*qda*	*svm*	*knn*
c45		0	1	1	0	*c45*		0	0	0	0
lda	0		0	0	1	*lda*	0		0	0	1
qda	1	0		1	0	*qda*	0	0		1	0
svm	1	0	1		0	*svm*	0	0	1		0
knn	0	1	0	0		*knn*	0	1	0	0	

Figure 6 shows the second example case where we compare all algorithms on *spect*. Again, both ANOVA and MANOVA reject the null hypothesis. According to the post-hoc tests, the univariate test finds five different cliques (one clique of three and four cliques of two algorithms): (*c45, qda, svm*), (*knn c45*), (*lda, c45*), (*lda, knn*), (*svm, knn*). On this dataset, the decisions of the two multivariate tests, MultiTF and MultiPR, are different from each other. MultiTF finds two cliques: (*c45, qda, svm*) and (*lda, knn*), whereas MultiPR finds the same cliques except *c45* is missing in one clique: (*qda, svm*) and (*lda, knn*). Table 3 shows the results of the multivariate pairwise tests between five algorithms.

The univariate ordering of the five classifiers are as follows:

error	*qda svm c45 knn lda*
tpr, recall	*knn lda svm qda c45*
fpr	*qda c45 svm knn lda*
precision	*qda svm knn lda c45*

The first clique found by MultiTF (*c45, qda, svm*) appears as a single group with respect to both tpr and fpr, whereas the second clique (*lda, knn*) form a group only with respect to fpr. Similarly, the first clique found by MultiPR (*qda, svm*) appears as a single group with respect to recall and precision, whereas the second clique (*lda, knn*) form a group only with respect to precision.

Using the full 2×2 confusion matrix. Instead of using (tpr, fpr) or (precision, recall), one can also use the full 2×2 confusion matrix using the same multivariate test in four dimensions. Note however that though the matrix contains four numbers, because p and n are fixed, the degree of freedom is two and that going to four dimensions is unnecessary. In our pairwise comparison experiments, we see that in 2582 cases out of 2740, the rank of the 2×2 confusion matrix S_d is indeed 2. Only in 98 cases, the rank is 1: This case occurs if the ratio tp / tn is the same for all folds, and in 60 cases, the rank is 4: This case occurs if the number of positive and/or negative instances is not exactly divisible by k, resulting in a difference between the positive and/or negative instances going from one fold to another.

It can be shown that when we use the 2×2 confusion matrix, the test statistic calculated will be the same as that of MultiTF. The other values fn and tn are fixed because we have $(p = tp + fn)$ and $(n = fp + tn)$ and they do not change going from one fold to another due to stratification, reducing the dimensionality to two, and it can be shown that MultiTF uses scaled versions of the counts used by Multi 2×2 but both return the same value. As explained above, there are cases when the stratification is not exact, but such cases are rare and do not affect the overall result.

5   Conclusions

In this paper, we propose to use multivariate tests to compare the performances of classification algorithms. Doing this, we can consider entries in the confusion matrix separately without needing to sum them up in a cumulative measure such as error or accuracy, which may hide certain differences in the behavior of the algorithms. Though multivariate pairwise tests and multivariate ANOVA have been known in the statistical literature, to the best of our knowledge, their use in performance comparison of machine learning algorithms is new.

There are a number of advantages to testing p variables multivariately rather than p separate univariate testing [4], as has also been shown in our experimental results above: (1) The use of p univariate tests inflates the type I error rate, unless we do some sort of correction (which in turn decreases power). (2) The univariate tests ignore correlations between variables, whereas the multivariate test uses the covariance information. (3) The multivariate test has higher power: Sometimes the p univariate tests may fail to detect a difference whereas the multivariate difference may be significant. (4) The multivariate test (pairwise test or MANOVA) constructs linear combinations of variables that reveals how the variables unite to reject the hypothesis.

The use of k-fold cross-validation to obtain k set of performance values comes with a caveat. Because all k training/validation sets are resampled from the same set, they overlap, and these k set of measurements are not really independent. This is true both for the univariate t test [1] and the multivariate test. Nadeu and Bengio (2003) [10] and Bouckaert and Frank (2004) [11] discuss a variance-correction term. We note that the resampling procedure used to generate the k

data folds is orthogonal to the test which uses these results and that our proposed multivariate test can be used with any improved resampling procedure.

We calculate (tpr, fpr) or (precision, recall) values for a specific threshold value. To have an overall comparison, for example, we can use s different thresholds (as done in a ROC curve) and calculate a pair for each value and get an overall $2s$ dimensional vector and again use the multivariate test. This is an interesting research direction. In this paper, we discuss how two or more algorithms can be compared on a single dataset. Demsar (2006) [12] discusses the comparison of algorithms over multiple datasets and an interesting future direction will be to extend our proposed multivariate test for this.

Acknowledgments

This work has been supported by TÜBİTAK 109E186.

References

1. Dietterich, T.G.: Approximate statistical tests for comparing supervised classification learning classifiers. Neural Computation 10, 1895–1923 (1998)
2. Caruana, R., Niculescu-Mizil, A., Crew, G., Ksikes, A.: Ensemble selection from libraries of models. In: Proceedings of the International Conference on Machine Learning, ICML 2004, pp. 137–144 (2004)
3. Seliya, N., Khoshgoftaar, T.M., Hulse, J.V.: Aggregating performance metrics for classifier evaluation. In: Proceedings of the 10th IEEE International Conference on Information Reuse and Integration (2009)
4. Rencher, A.C.: Methods of Multivariate Analysis. Wiley and Sons, New York (1995)
5. Blake, C., Merz, C.: UCI repository of machine learning databases (2000)
6. Hinton, G.H.: Delve project, data for evaluating learning in valid experiments (1996)
7. Statnikov, A., Aliferis, C., Tsamardinos, I., Hardin, D., Levy, S.: A comprehensive evaluation of multicategory classification methods for microarray gene expression cancer diagnosis. Bioinformatics 21, 631–643 (2005)
8. Chang, C.C., Lin, C.J.: LIBSVM: a library for support vector machines (2001)
9. Davis, J., Goadrich, M.: The relationship between precision-recall and roc curves. In: Proceedings of the 23rd International Conference on Machine Learning, vol. 148, pp. 233–240 (2006)
10. Nadeau, C., Bengio, Y.: Inference for the generalization error. Machine Learning 52, 239–281 (2003)
11. Bouckaert, R., Frank, E.: Evaluating the replicability of significance tests for comparing learning algorithms. In: Dai, H., Srikant, R., Zhang, C. (eds.) PAKDD 2004. LNCS (LNAI), vol. 3056, pp. 3–12. Springer, Heidelberg (2004)
12. Demsar, J.: Statistical comparisons of classifiers over multiple data sets. Journal of Machine Learning Research 7, 1–30 (2006)

Using Hyperheuristics under a GP Framework for Financial Forecasting

Michael Kampouridis[1] and Edward Tsang[2]

[1] School of Computer Science and Electronic Engineering, University of Essex,
Wivenhoe Park, CO4 3SQ, UK
mkampo@essex.ac.uk
http://kampouridis.net

[2] Centre for Computational Finance and Economic Agents, University of Essex,
Wivenhoe Park, CO4 3SQ, UK
edward@essex.ac.uk
http://www.bracil.net/edward/

Abstract. Hyperheuristics have successfully been used in the past for a number of search and optimization problems. To the best of our knowledge, they have not been used for financial forecasting. In this paper we use a simple hyperheuristics framework to investigate whether we can improve the performance of a financial forecasting tool called EDDIE 8. EDDIE 8 allows the GP (Genetic Programming) to search in the search space of indicators for solutions, instead of using pre-specified ones; as a result, its search area is quite big and sometimes solutions can be missed due to ineffective search. We thus use two different heuristics and two different mutators combined under a simple hyperheuristics framework. We run experiments under five datasets from FTSE 100 and discover that on average, the new version can return improved solutions. In addition, the rate of missing opportunities reaches it's minimum value, under all datasets tested in this paper. This is a very important finding, because it indicates that thanks to the hyperheuristics EDDIE 8 has the potential of missing less forecasting opportunities. Finally, results suggest that thanks to the introduction of hyperheuristics, the search has become more effective and more areas of the space have been explored.

Keywords: Hyperheuristics, Genetic Programming, Financial Forecasting.

1 Introduction

Financial forecasting is an important area in computational finance [29]. There are numerous works that attempt to forecast the future price movements of a stock; several examples can be found in [10,7]. A number of different methods have been used for forecasting. Such examples are for instance, Support Vector Machines [25], Fuzzy Logic [15] and Neural Networks [6]. Genetic Programming [18,24] (GP) is an evolutionary technique that has widely been used for financial

C.A. Coello Coello (Ed.): LION 5, LNCS 6683, pp. 16–30, 2011.

forecasting. Some recent examples are [26,6,1,12], where GP was used for time series forecasting.

In a previous paper [16], we presented EDDIE 8 (ED8), which was an extension of the financial forecasting tool EDDIE (Evolutionary Dynamic Data Investment Evaluator) [27,28]. EDDIE is a machine learning tool that uses Genetic Programming to make its predictions. The novelty of ED8 was in its extended grammar, which allowed the GP to search in the space of indicators to form its Genetic Decision Trees. In this way, ED8 was not constrained in using pre-specified indicators, but it was left up to the GP to choose the optimal ones. We then proceeded to compare ED8 with its predecessor, which used indicators that were pre-specified by the user. Results showed that thanks to the new grammar, ED8 could find new and improved solutions. However, those results also suggested that ED8's performance could have been compromised by the enlarged search space. With the old grammar, which was also discussed in [16], EDDIE used 6 indicators from technical analysis with two pre-specified period lengths. For instance, if one of the indicators was Moving Average, then the two period lengths used would be 12 and 50 days. On the contrary, ED8 could use any period within a given parameterized range, which for our experiments was set to 2-65 days. Thus, the GP could come up with any indicator within that range, and not just with 12 and 50 days. As we can see, the search space of ED8 was much bigger than the one of its predecessor. With the old grammar, the GP would have to combine only 12 indicators (6 indicators with 2 periods each) to form trees; on the other hand, ED8 would have to combine $6 \times (65 - 1) = 384$ indicators. The difficulty of ED8 in making the appropriate indicators combinations was obvious. In addition, the search space of ED8 was also much bigger. For instance, let us assume that the training data consisted of 1000 data points. Then, with the old grammar, the GP would have to search in a space of $12 \times 1000 = 12,000$ points. On the other hand, ED8 would have to search in the much larger space of 384,000 data points (384×1000). It was therefore obvious that we needed to find new ways that would make the search more effective in such large search areas.

In this paper, we want to investigate if a hyperheuristics framework can address this issue. Hyperheuristics is a well-known method that has been used in a variety of search and optimization problems [23], such as transportation [14], scheduling [11], and timetabling [8]. To the best of our knowledge, *hyperheuristics have not been used before for a financial forecasting problem.* We thus use a simple framework that utilizes simple hill climbing, simulated annealing, random mutation, and weighted random mutation. We are interested in investigating whether hyperheuristics can improve ED8's performance, and whether the search space can be better explored. The rest of this paper is organized as follows: Section 2 presents the ED8 algorithm, Sect. 3 presents the hyperheuristics framework, along with its heuristics and operators, Sect. 4 presents the experimental setup, Sect. 5 presents and discusses the results, and finally, Sect. 6 concludes this paper and also discusses future work.

2   Presentation of EDDIE 8

EDDIE is a forecasting tool, which learns and extracts knowledge from a set of data. The kind of question ED8 tries to answer is 'will the price increase within the n following days by r%'? The user first feeds the system with a set of past data; EDDIE then uses this data and through a GP process, it produces and evolves Genetic Decision Trees (GDTs), which make recommendations of buy (1) or not-to-buy (0).

The set of data used is composed of three parts: daily closing price of a stock, a number of attributes and signals. Stocks' daily closing prices can be obtained on-line in websites such as http : //finance.yahoo.com and also from financial statistics databases like *Datastream*. The attributes are indicators commonly used in technical analysis [13]; which indicators to use depends on the user and his belief of their relevance to the prediction. The technical indicators that we use in this work are: Moving Average (MA), Trade Break Out (TBR), Filter (FLR), Volatility (Vol), Momentum (Mom), and Momentum Moving Average (MomMA).[1]

The signals are calculated by looking ahead of the closing price for a time horizon of n days, trying to detect if there is an increase of the price by $r\%$ [27]. For this set of experiments, n was set to 20 and r to 4%. In other words, the GP is trying to use some of the above indicators to forecast whether the daily closing price iss going to increase by 4% within the following 20 days.

After we feed the data to the system, EDDIE creates and evolves a population of GDTs. Figure 1 presents the Backus Normal Form (BNF) [4] (grammar) of ED8. As we can see, the root of the tree is an If-Then-Else statement. The first branch is either a boolean (testing whether a technical indicator is greater than/less than/equal to a value), or a logic operator (and, or, not), which can hold multiple boolean conditions. The 'Then' and 'Else' branches can be a new GDT, or a decision, to buy or not-to-buy (denoted by 1 and 0).

As we can see from the grammar in Fig. 1, there is a function called *VarConstructor*, which takes two children. The first one is the indicator, and the second one is the Period. Period is an integer within the parameterized range [MinP, MaxP] that the user specifies. As a result, ED8 can return decision trees with indicators like 15 days Moving Average, 17 days Volatility, etc. The period is not an issue and it is up to ED8, and as a consequence up to the GP and the evolutionary process, to decide which lengths are more valuable for the prediction. A sample GDT is presented in Fig. 2. As we can observe, the periods 12 and 50 are now in a leaf node, and thus are subject to genetic operators, such as crossover and mutation.

Depending on the classification of the predictions, we can have four cases: True Positive (TP), False Positive (FP), True Negative (TN), and False Negative (FN). As a result, we can use the metrics presented in Equations (1), (2) and (3).

---

[1] We use these indicators because they have been proved to be quite useful in developing GDTs in previous works like [21], [2] and [3]. Of course, there is no reason why not use other information like fundamentals or limit order book. However, the aim of this work is not to find the ultimate indicators for financial forecasting.

<Tree> ::= If-then-else <Condition> <Tree> <Tree> | Decision
<Condition> ::= <Condition> "And" <Condition> |
         <Condition> "Or" <Condition> |
         "Not" <Condition> |
         VarConstructor <RelationOperation> Threshold
<VarConstructor> ::= MA period | TBR period | FLR period | Vol period |
         Mom period | MomMA period
<RelationOperation> ::= ">" | "<" | "="
Terminals:
         MA, TBR, FLR, Vol, Mom, MomMA are function symbols
         Period is an integer within a parameterized range, [MinP, MaxP]
         Decision is an integer, Positive or Negative implemented
         Threshold is a real number

Fig. 1. The Backus Normal Form of ED8

Rate of Correctness

$$RC = \frac{TP + TN}{TP + TN + FP + FN} \tag{1}$$

Rate of Missing Chances

$$RMC = \frac{FN}{FN + TP} \tag{2}$$

Rate of Failure

$$RF = \frac{FP}{FP + TP} \tag{3}$$

The above metrics combined give the following fitness function, presented in Equation (4):

$$ff = w_1 * RC - w_2 * RMC - w_3 * RF \tag{4}$$

where w_1, w_2 and w_3 are the weights for RC, RMC and RF respectively. These weights are given in order to reflect the preferences of investors. For instance, a conservative investor would want to avoid failure; thus a higher weight for RF should be used. For our experiments, we chose to include strategies that mainly focus on correctness and reduced failure. Thus these weights have been set to 0.6, 0.1 and 0.3 respectively.

The fitness function is a constrained one, which allows EDDIE to achieve lower RF. The effectiveness of this constrained fitness function has been discussed in [28,20]. The constraint is denoted by R, which consists of two elements represented by percentage, given by

$$R = [Cmin, Cmax],$$

where $C_{min} = \frac{P_{min}}{N_{tr}} \times 100\%$, $C_{max} = \frac{P_{max}}{N_{tr}} \times 100\%$, and $0 \leq C_{min} \leq C_{max} \leq 100\%$. N_{tr} is the total number of training data cases, P_{min} is the minimum number of positive position predictions required, and P_{max} is the maximum number of positive position predictions required.

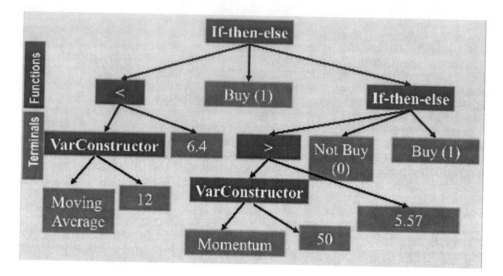

Fig. 2. Sample GDT generated by EDDIE 8

Therefore, a constrained of $R = [50, 65]$ would mean that the percentage of positive signals that a GDT predicts[2] should fall into this range. When this happens, then w_1 remains as it is (i.e. 0.6 in our experiments). Otherwise, w_1 takes the value of zero.

During the evolutionary procedure, we allow three operators: crossover, mutation and reproduction. After reaching the last generation, the best-so-far GDT, in terms of fitness, is applied to the testing data.

This concludes this short presentation of ED8. In the next section we briefly present the heuristics and operators used in our framework, and then present the hyperheuristics framework itself.

3 Hyperheuristics Framework

3.1 Heuristics and Operators

We use two heuristics, namely simple hill climbing (SHC) and simulated annealing (SA), and two GP operators, namely random mutation ($Rnd\ Mut$) and weighted random mutation ($W.\ Rnd\ Mut$). However, we do not argue that the above techniques are the optimal ones for the purposes of our experiments. Other heuristics and operators could also be chosen. Nevertheless, the purpose of this paper is not to look for the most effective heuristics or operators, but to investigate whether and how these heuristics combined under a hyperheuristics

---

[2] As we have mentioned, each GDT makes recommendations of buy (1) or not-to-buy (0). The former denotes a positive signal and the latter a negative. Thus, within the range of the training period, which is t days, a GDT will have returned a number of positive signals

framework can be used to improve the performance of ED8. We leave it to future research to investigate for even more appropriate heuristics.

Let us now start by explaining how the first heuristic is used (SHC). First of all, a leaf which contains a period is randomly selected. Let us assume that this period is $p = 12$ (days). Then the simple hill climber increases the period by 1. If there is an improvement in the fitness, the process is completed and the period returned to the leaf is $p + 1$, which in this example is 13. If, on the other hand, there is no improvement, then the period p is reduced by 1. Again, if this has resulted to an improvement, then the process ends and the new period is $p - 1$, which in this example is 11. If again there is no improvement, then the initial period $p = 12$ is returned and the process is terminated. The purpose of using this heuristic is quite obvious: we are interested in investigating how a marginal change in the period can affect the performance of a tree.

The motivation behind the use of the remaining three heuristics/operators is to expand the search to other areas of the search space, and not just the neighborhood of a selected period. The second heuristic is the classic simulated annealing. A leaf from a tree is again randomly chosen. Then the standard algorithm of simulated annealing is used [17,9]. We thus allow swaps among the different periods (2-65 days). The initial temperature t is set to a value such that around 65% of the inferior moves are accepted. The temperature is gradually reduced according to the following formula: $t = (1 - f) \times t$, where t denotes the temperature and f is the annealing temperature reduction factor, which is equal to 0.1. At each temperature a maximum of 10 iterations are executed. The acceptance criterion is the Metropolis criterion [22].

The two operators used in our framework are *Rnd Mut* and *W. Rnd Mut*. They both randomly select a leaf which contains a period. Random mutation then mutates this leaf to another period. This operator is the typical GP mutation operator and it allows us to randomly explore different areas of the search space. On the other hand, weighted random mutation offers a roulette-wheel-like selection, where the new period, which is going to replace a leaf, is probabilistically selected, based on its occurrence. However, it should be highlighted that *W. Rnd Mut* does not just focus on the period, but on the indicator itself. So when we select a period p for mutation, we also take into account the accompanying indicator (e.g. Moving Average). We then calculate the period occurrence under that specific indicator. The reason for doing this is that we aim to target indicators as a whole and not just periods. In this way, the new period p' that is going to replace of the old period p is a period that has a high occurrence under this specific indicator. *W. Rnd Mut* therefore introduces useful indicators to the population, rather than just introducing useful periods, like the other 3 heuristics/mutators do.

3.2   The Framework

In this simple framework, all low level heuristics are used simultaneously. The low level heuristics include both the heuristics and operators described in the previous section. Inspired by the Population Based Incremental Learning algorithm [5] and the alike Estimation of Distribution Algorithms [19,30], all four low level heuristics are initially given a weight w of being selected, where $w = 25\%$. Then depending on the result on the performance of a tree after the implication of a heuristic, the following cases can occur:

1. Increase in performance
   (a) By using a new period
   (b) By using a pre-existing period
2. No change in performance
   (a) By using a new period
   (b) By using a pre-existing period
3. Decrease in performance

As we can see, we are not only interested in improving the performance of a GDT. We are also interested in whether this improvement comes from a new or from a pre-existing period. The reason for this is obvious: one of the goals of our experiments is to have better exploration of the search space. We thus want to reward a heuristic that allowed a new period to be invoked in the population.

Let us denote the reward/punishment after the implication of a heuristic by r. Then the weight w for each one of the above cases (1-3) is updated as follows:

1. Increase in performance
   (a) $w = w + r$ (new period)
   (b) $w = w + r/2$ (pre-existing period)
2. No change in performance
   (a) $w = w + r/5$ (new period)
   (b) $w = w - r/5$ (pre-existing period)
3. Decrease in performance
   $w = w - r$

The highest reward is offered when there is an increase in the performance and a new period has been used (1-a). When an improvement is caused by the use of a period that is already being used by other GDTs, half of the reward is offered (1-b). In the case of no change in performance, we still offer a small reward (equal to $\frac{r}{5}$) if a new period has been used (2-a), since a new area of the search space has been explored. If no new period has been used, then a small punishment is invoked (equal to $-\frac{r}{5}$) (2-b). Finally, there is also a punishment in the case of decrease in the performance (3).

4   Experimental Setup

The data we feed to ED8 consist of daily closing prices. These closing prices are from 5 arbitrary stocks from FTSE100. These stocks are: British Petroleum (BP), Carnival, Hammerson, Imperial Tobacco, and Xstrara. The training period is 1000 days and the testing period 300.

The GP parameters are presented in Table 1. For statistical purposes, we run the GP for 50 times. Thus, the process that is followed is that we create a

Table 1. GP Parameters

GP Parameters	
Max Initial Depth	6
Max Depth	8
Generations	50
Population size	500
Tournament size	2
Reproduction probability	0.1
Crossover probability	0.9
Mutation probability	0.01
Period (EDDIE 8)	[2,65]

population of 500 GDTs, which are evolved for 50 generations, over a training period of 1000 days. At the last generation, the best performing GDT in terms of fitness is saved and applied to the testing period. As we have already said, this procedure is done for 50 individual runs.

In addition, we should emphasize that we require that the datasets have a satisfactory number of actual positive signals. By this we mean that we are neither interested in datasets with a very low number of signals, neither with an extremely high one. Such cases would be categorized as chance discovery, where people are interested in predicting rare events, such as a stock market crash. Clearly this is not the case in our current work, where we use EDDIE for investment opportunities forecasting. We are thus interested in datasets that have opportunities around 50-70% (i.e. 50-70% of actual positive signals). Therefore, we need to calibrate the values of r and n (see Sect. 2) accordingly, so that we can obtain the above percentage from our data. For our experiments, the value of n is set to 20 days. The value of r varies, depending on the dataset. This is because one dataset might reach a percentage of 50-70% with $r = 4\%$, whereas another one might need a higher or lower r value. Accordingly, we need to calibrate the value of the R constraint, so that EDDIE produces GDTs that forecast positive signals in a range which includes the percentage of the *actual* positive signals of the dataset we are experimenting with. R thus takes values in the range of $[-5\%, +5\%]$ of the number of positive signals that the dataset has. For instance, if under $r = 4\%$ and $n = 20$ days, a dataset has 60% of actual positive signals, then R would be set to [55,65].

Finally, Table 2 presents the parameters of the hyperheuristics framework. The probability of applying hyperheuristics is set for this work at 35%. Thus, 35% of the GDTs' periods can be updated through hyperheuristics at each generation. We did not want to set a higher probability, because this could increase the computational times. At the moment, the initial weight and reward are set to 0.25 and 0.005, respectively. The former is set to 0.25 because we want all heuristics to have equal chances of being selected when the process starts. The reward is set arbitrarily. We leave it to future research to investigate whether the test results can be affected by different parameter values.

Table 2. Hyperheuristics Parameters

Hyperheuristics Parameters	
Hyperheuristics probability	0.35
Initial Weight	0.25
Reward/Punishment	0.005

5   Results

We ran both the traditional version of EDDIE 8 (ED8) and the one that uses hyperheuristics (ED8-HH) for 50 individual times and present the results in this section. As we mentioned at the beginning of this paper, the goal of our experiments is twofold: (a) to investigate whether hyperheuristics can improve the performance of the trees that ED8 uses and (b) to investigate whether hyperheuristics can offer better exploration of the search space.

Let us begin with the performance of the GDTs. Table 3 presents the average and optimal results over the 50 runs for both ED8 and ED8-HH for the 5 stocks. The first row of each dataset presents the ED8 average results and the second row presents the ED8-HH ones. The third and fourth row of each dataset present the optimal values of the metrics for ED8 and ED8-HH, respectively. Optimal value can be either a maximum or a minimum, depending on the metric. Hence, because Fitness and RC are maximization problems, *optimal* refers to the *maximum* value of these metrics, over the 50 individual runs. On the other hand, because RMC and RF are minimization problems, *optimal* refers to the *minimum* value of these two metrics. In order to judge if the average results (rows 1 and 2) are significant, we also ran Kolmogorov-Smirnov tests at 5% significance level. The p-values of the tests are presented in Table 4. Thus, when there is a significantly better average value in Table 3, this is denoted by bold fonts. In addition, a higher value for the optimal results is underlined.

A first observation from Table 3 is that on average, there seems to be a 'tie' between ED8 and ED8-HH. ED8 is doing significantly better in the average Fitness and RF of BP, and in the average RMC of Hammerson. On the other hand, ED8-HH is doing significantly better in the average values of RMC of Carnival and Imperial Tobacco. The remaining metrics of the other stocks have insignificant differences at the 5% level. It is not very easy to draw safe conclusions by just looking at the average results.

However, the picture gets clearer when we look at the optimal values of the metrics. ED8-HH is doing better in 12 cases (BP: RMC; Carnival: RMC; Hammerson: Fitness, RC, RMC, RF; Imp. Tobacco: Fitness, RMC, RF; Xstrata: RC, RMC), whereas ED8 is doing better only in 7 cases. What is even more interesting though, is the large and consistent improvement in the minimum value of RMC, for all 5 stocks. In fact, as we can see from Table 3, the minimum RMC under ED8-HH is always 0. This is a very important finding, because it indicates that ED8-HH has the potential of never missing any forecasting opportunities.

Table 3. Average and optimal results over 50 runs for ED8 and ED8-HH for 5 FTSE 100 stocks. Each stock presents results in four rows: one for the average values of the metrics for ED8, one for the average values for ED8-HH, one for the optimal values of the metrics for ED8, and one for the optimal values for ED8-HH. A significantly better value (at 5% significance level) between ED8 and ED8-HH for the average results is presented in bold fonts, whereas a better value for the optimal results is underlined.

Stock		Fitness	RC	RMC	RF
BP	(ED8 Avg)	**0.2005**	0.5303	0.4756	**0.2338**
	(ED8-HH Avg)	0.1767	0.5299	0.4519	0.3203
	(ED8 Opt)	<u>0.3341</u>	<u>0.6900</u>	0.2523	0.1691
	(ED8-HH Opt)	0.2850	0.6500	0	<u>0.1176</u>
Carnival	(ED8 Avg)	0.1871	0.5607	0.3531	0.3801
	(ED8-HH Avg)	0.1900	0.5607	**0.2892**	0.3917
	(ED8 Opt)	<u>0.2511</u>	<u>0.6300</u>	0.1734	<u>0.1728</u>
	(ED8-HH Opt)	0.2470	0.6267	0	0.2414
Hammerson	(ED8 Avg)	0.2488	0.6071	**0.2331**	0.3073
	(ED8-HH Avg)	0.2164	0.5675	0.3507	0.2967
	(ED8 Opt)	0.3311	0.7033	0.0340	0.2472
	(ED8-HH Opt)	<u>0.3450</u>	<u>0.7200</u>	0	<u>0.1818</u>
Imp.Tobacco	(ED8 Avg)	0.1832	0.5245	0.6488	0.2222
	(ED8-HH Avg)	0.1946	0.5395	**0.5910**	0.2332
	(ED8 Opt)	0.2790	0.6533	0.2595	0.0270
	(ED8-HH Opt)	<u>0.2959</u>	0.6533	0	<u>0.0222</u>
Xstrata	(ED8 Avg)	0.2419	0.5807	0.3264	0.2462
	(ED8-HH Avg)	0.2359	0.5741	0.3330	0.2508
	(ED8 Opt)	<u>0.3571</u>	0.7267	0.0664	<u>0.0400</u>
	(ED8-HH Opt)	0.2510	<u>0.7600</u>	0	0.0714

Table 4. p-values of the Kolmogorov-Smirnov non-parametric tests. A p-value that is less than 0.05 denotes significantly different distributions.

Stock	Fitness	RC	RMC	RF
BP	0.0171	0.9541	0.5077	0.0001
Carnival	0.1546	0.5077	0.0317	0.1546
Hammerson	0.1546	0.0560	0.0089	0.2408
Imp.Tobacco	0.6779	0.0560	0.0317	0.3584
Xstrata	0.3584	0.2408	0.3584	0.5077

Moreover, Figure 3 presents the average over the 50 runs of the percentage of extinct indicators for each generation. To be more specific, the trees of each generation use a number of indicators from the available population. As we mentioned in Sect. 1, ED8 and ED8-HH use 384 indicators. An 'extinct indicator' is therefore defined as the indicator that is not currently used by any of the GDTs in the population. From Fig. 3 we can see that at generation 0, the percentage of extinct indicators is very low for both ED8 and ED8-HH (around 2%) for all 5 datasets. This means that very few indicators are not being used

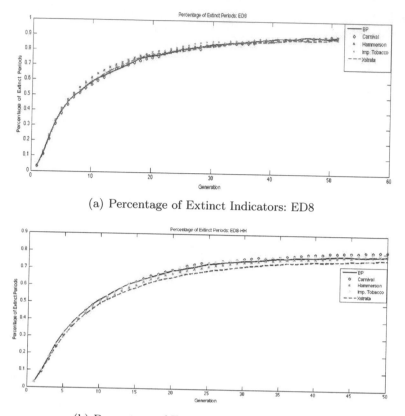

(a) Percentage of Extinct Indicators: ED8

(b) Percentage of Extinct Indicators: ED8-HH

Fig. 3. Percentage of Extinct Indicators for ED8 (a) and ED8-HH (b)

by the trees at generation 0. As evolution proceeds, we can observe that the percentage of extinct indicators increases, and thus less and less indicators are used by the GDTs. This is normal, because it means that the GDTs have found some useful indicators and are focusing on them. However, by the end of the evolutionary process, we can see that the percentage of extinct indicators has increased to around 90% for ED8. This means that the GDTs are using only 10% of the available 384 indicators, i.e. around 38 indicators, and are thus only taking advantage of a very small area of the search space. So all 500 GDTs are using only these indicators, which indicates very low diversity in the population. When we look at ED8-HH's statistics, we see that throughout the evolutionary process, the percentage of extinct indicators is constantly lower by around 5-10%. This indicates that ED8-HH constantly explores more areas of the search space than ED8. By generation 50, this percentage is around 80%. ED8-HH has thus managed to keep 'alive' an extra 10% (in total 20%) of the indicators. This better exploration has of course led to more diversity in the population and could be a reason of ED8-HH's improved performance that we saw earlier in Table 3.

Table 5. Performance of each heuristic. First row presents the percentage of improvement to the GDTs' performance introduced by the specific heuristic. The second row presents the percentage of improvement that was caused by an existing indicator and the third row the percentage of improvement caused by a new indicator. Results are on average of 50 runs.

Stock		Rnd Mut	SA	SHC	W. Rnd Mut
BP	(Improvement)	0.0572	0.0560	0.0522	0.0552
	(Existing Ind.)	0.8825	0.7760	0.7068	0.8565
	(New Ind.)	0.1175	0.2240	0.2932	0.1435
Carnival	(Improvement)	0.0602	0.0748	0.0768	0.0536
	(Existing Ind.)	0.8967	0.7988	0.7887	0.8655
	(New Ind.)	0.1033	0.2012	0.2113	0.1345
Hammerson	(Improvement)	0.0541	0.0711	0.0777	0.0412
	(Existing Ind.)	0.8333	0.8311	0.8500	0.9062
	(New Ind.)	0.1667	0.1689	0.1500	0.0938
Imp. Tobacco	(Improvement)	0.0339	0.0298	0.0512	0.0297
	(Existing Ind.)	0.8930	0.8450	0.8978	0.9522
	(New Ind.)	0.1070	0.1550	0.1022	0.0478
Xstrata	(Improvement)	0.0508	0.0705	0.0728	0.0669
	(Existing Ind.)	0.9635	0.8121	0.8004	0.8778
	(New Ind.)	0.0365	0.1879	0.1996	0.1222

Furthermore, Table 5 presents information about the individual heuristics. For each stock, we present information in three rows. The first one presents the percentage of improvement caused by the relevant heuristic, on the GDTs' performance. Results are on average of 50 runs. Thus, the first element of the table informs us that *Rnd Mut* has on average improved the BP's GDTs performance by 5.72%. Similarly, *SA*, *SHC* and *W. Rnd Mut* have improved the performance by around 5%. Same rates of improvement apply to the remaining 4 stocks, too, varying in around 3-8%. Overall, all four heuristics seem to have equivalent contribution to the GDTs' improvement in performance. In addition, *SHC* seems to be the most consistent one, always having an improvement above 5%.

The second and third row of each stock present the percentage of improvements that was caused either by an existing (second row) or by a new period (third row). Therefore, BP's 88.25% in *Rnd Mut* means that 88.25% of the improvements came from periods that were already being used by the population of GDTs. In other words, already-successful periods were located by the heuristics and re-used by the GDTs. The remaining 11.75% of the improvements came by new periods, which did not belong to the current GP population. As we explained earlier, we are especially interested in this figure, because it allows diversity in the population. From our experiments, *SA* and *SHC* seem to be more consistent in allowing new periods in the population, with percentages varying in the range of 10.22-29.32%. On the other hand, the two mutators seem to be better in re-using successful existing periods.

Finally, we should say that a single run of ED8 lasted approximately 3 minutes, whereas a single run of ED8-HH lasted approximately 13 minutes. It is obvious that the latter is more computationally intensive. However, we consider that the extra wait time is worthy, because of the improvements we saw in performance and search effectiveness. Future work could focus on improving the computational time of ED8-HH.

6   Conclusion

Hyperheuristics have been used in the past for several search and optimization problems, but not for financial forecasting. In this paper we used a simple hyperheuristics framework to investigate whether we could improve the performance of a financial forecasting tool called EDDIE 8 (ED8). ED8 allows the GP to search in the search space of indicators for solutions; as a result, its search space is quite big and sometimes solutions can be missed due to ineffective search. We thus used two different heuristics and two GP mutators combined under a simple hyperheuristics framework and discovered that hyperheuristics returned improved solutions. In addition, ED8-HH's minimum RMC reached it's minimum value of 0, under all 5 stocks tested in this paper. ED8-HH has thus the potential of never missing forecasting opportunities. Finally, results suggested that thanks to the introduction of hyperheuristics, more areas of the search space were explored, which led to higher diversity in the GP population.

Overall, we can characterize the results as encouraging. However, tests took place under a small sample (five stocks) and thus more experiments need to be done under more datasets. Furthermore, as our experiments took place under a simple hyperheuristics framework, we are interested in investigating the effects of more complex frameworks. At the same time, we want to examine the effects of different heuristics. Our goal is to show that under a more sophisticated framework, and with the use of more heuristics, the search can become even more effective, resulting to even higher performance of the GDTs.

Acknowledgments

The version has been revised in light of two anonymous referees' very helpful reviews, for which the authors are very grateful. The EPSRC grant with number EP/P563361/0 is also gratefully acknowledged.

References

1. Agapitos, A., O'Neill, M., Brabazon, A.: Evolutionary learning of technical trading rules without data-mining bias. In: Schaefer, R., Cotta, C., Kołodziej, J., Rudolph, G. (eds.) PPSN XI. LNCS, vol. 6238, pp. 294–303. Springer, Heidelberg (2010)
2. Allen, F., Karjalainen, R.: Using genetic algorithms to find technical trading rules. Journal of Financial Economics 51, 245–271 (1999)

3. Austin, M., Bates, G., Dempster, M., Leemans, V., Williams, S.: Adaptive systems for foreign exchange trading. Quantitative Finance 4(4), 37–45 (2004)
4. Backus, J.: The syntax and semantics of the proposed international algebraic language of Zurich. In: International Conference on Information Processing, pp. 125–132. UNESCO (1959)
5. Baluja, S.: Population-based incremental learning: a method for integrating genetic search based function optimisation and competitive learning, technical Report, Carnegie Mellon University (1994)
6. Bernal-Urbina, M., Flores-Méndez, A.: Time series forecasting through polynomial artificial neural networks and genetic programming. In: Proceedings of the IEEE Congress on Evolutionary Computation, Hong Kong, pp. 3324–3329 (June 2008)
7. Binner, J., Kendall, G., Chen, S.H. (eds.): Applications of Artificial Intelligence in Finance and Economics, Advances in Econometrics, vol. 19. Elsevier, Amsterdam (2004)
8. Burke, E., MacCloumn, B., Meisels, A., Petrovic, S., Qu, R.: A graph-based hyper heuristic for timetabling problems. European Journal of Operational Research 176, 177–192 (2006)
9. Cerny, V.: A thermodynamical approach to the travelling salesman problem: an efficient simulation algorithm. Journal of Optimization Theory and Applications 45, 41–51 (1985)
10. Chen, S.H.: Genetic Algorithms and Genetic Programming in Computational Finance. Springer, New York (2002)
11. Cowling, P., Chakhlevitch, K.: Hyperheuristics for managing a large collection of low level heuristics to schedule personnel, vol. 2, pp. 1214–1221 (December 2003)
12. Dempsey, I., O'Neill, M., Brabazon, A.: Live trading with grammatical evolution. In: Proceedings of the Grammatical Evolution Workshop (2004)
13. Edwards, R., Magee, J.: Technical analysis of stock trends. New York Institute of Finance (1992)
14. Hart, E., Ross, P., Nelson, J.: Solving a real-world problem using an evolving heuristically driven schedule builder. Evol. Comput. 6(1), 61–80 (1998)
15. Kablan, A.: Adaptive neuro fuzzy inference systems for high frequency financial trading and forecasting, pp. 105–110 (October 2009)
16. Kampouridis, M., Tsang, E.: EDDIE for investment opportunities forecasting: Extending the search space of the GP. In: Proceedings of the IEEE Conference on Evolutionary Computation, Barcelona, Spain, pp. 2019–2026 (2010)
17. Kirkpatrick, S., Gelatt Jr., C., Vecchi, M.: Optimization by simulated annealing. Science 220(4598), 671–680 (1983)
18. Koza, J.: Genetic Programming: On the programming of computers by means of natural selection. MIT Press, Cambridge (1992)
19. Larranaga, P., Lozano, J.: Estimation of Distribution Algorithms: A New Tool for Evolutionary Computation. Kluwer, Norwell (2001)
20. Li, J.: FGP: A Genetic Programming-ased Financial Forecasting Tool. Ph.D. thesis, Department of Computer Science, University of Essex (2001)
21. Martinez-Jaramillo, S.: Artificial Financial Markets: An agent-based Approach to Reproduce Stylized Facts and to study the Red Queen Effect. Ph.D. thesis, CFFEA, University of Essex (2007)
22. Metropolis, N., Rosenbluth, A.W., Rosenbluth, M.N., Teller, A.H., Teller, E.: Equation of calculations by fast computing machines. Journal of Chemical Physics 21, 1087–1092 (1953)
23. Özcan, E., Bilgin, B., Korkmaz, E.E.: A comprehensive analysis of hyper-heuristics. Intelligent Data Analysis 12(1), 3–23 (2008)

24. Poli, R., Langdon, W., McPhee, N.: A Field Guide to Genetic Programming. Lulu.com (2008)
25. Sapankevych, N., Sankar, R.: Time series prediction using support vector machines: A survey. IEEE Computational Intelligence Magazine 4(2), 24–38 (2009)
26. Sharma, V., Srinivasan, D.: Evolutionary computation and economic time series forecasting. In: Proceedings of the IEEE Conference on Evolutionary Computation, Singapore, September 25-28, pp. 188–195 (2007)
27. Tsang, E., Li, J., Markose, S., Er, H., Salhi, A., Iori, G.: EDDIE in financial decision making. Journal of Management and Economics 4(4) (2000)
28. Tsang, E., Markose, S., Er, H.: Chance discovery in stock index option and future arbitrage. New Mathematics and Natural Computation 1(3), 435–447 (2005)
29. Tsang, E., Martinez-Jaramillo, S.: Computational finance. IEEE Computational Intelligence Society Newsletter, 3–8 (2004)
30. Zhang, Q., Sun, J., Tsang, E.: Evolutionary algorithm with guided mutation for the maximum clique problem. IEEE Transactions on Evolutionary Computation 9(2), 192–200 (2005)

On the Effect of Connectedness for Biobjective Multiple and Long Path Problems

Sébastien Verel[1,2], Arnaud Liefooghe[1,3], Jérémie Humeau[1,4],
Laetitia Jourdan[1], and Clarisse Dhaenens[1,3]

[1] INRIA Lille-Nord Europe, France
[2] Université Nice Sophia Antipolis, I3S – CNRS, France
[3] Université Lille 1, LIFL – CNRS, France
[4] École des Mines de Douai, IA department, France
verel@i3s.unice.fr, arnaud.liefooghe@univ-lille1.fr,
jeremie.humeau@mines-douai.fr, laetitia.jourdan@inria.fr,
clarisse.dhaenens@lifl.fr

Abstract. Recently, the property of connectedness has been claimed to give a strong motivation on the design of local search techniques for multiobjective combinatorial optimization. Indeed, when connectedness holds, a basic Pareto local search, initialized with at least one non-dominated solution, allows to identify the efficient set exhaustively. However, this becomes quickly infeasible in practice as the number of efficient solutions typically grows exponentially with the instance size. As a consequence, we generally have to deal with a limited-size approximation, ideally a representative sample of efficient solutions. In this paper, we propose the biobjective long and multiple path problems. We show experimentally that, on the first problem, even if the efficient set is connected, a local search may be outperformed by a simple evolutionary algorithm in the sampling of the efficient set. At the opposite, on the second problem, a local search algorithm may successfully approximate a disconnected efficient set. Then, we argue that connectedness is not the single property to study for the design of multiobjective local search algorithms. This work opens new discussions on a proper definition of multiobjective fitness landscapes.

1 Introduction

The single-objective long path problem [1] has been introduced to show that a problem instance can be difficult to solve for a hillclimber-like heuristic even if the search space is unimodal, *i.e.* the single local optimum is the global optimum. For such a problem, a hillclimber guarantees to reach the global optimum, but the length of the path to get it is exponential in the dimension of the search space. As a consequence, a hillclimbing-based heuristic cannot expect to solve the problem in polynomial time. The 'path length' takes then place in the rank of problem difficulty, on the same level as multimodality, ruggedness, deceptivity, and so on. Rudolph [2] demonstrated that the long path problem can be solved in a polynomial expected amount of time for a $(1+1)$ evolutionary algorithm (EA)

C.A. Coello Coello (Ed.): LION 5, LNCS 6683, pp. 31–45, 2011.
© Springer-Verlag Berlin Heidelberg 2011

which is able to mutate more than one bit at a time. This $(1+1)$ EA is able to take some shortcuts on the outside of the path so that it makes the computation more efficient. However, it does not change the argument that, even for unimodal problems, the path length to the global optimum must be taken into account in the design of efficient local search algorithms.

Like in single-objective optimization, the structure of the search space can explain the difficulty for multiobjective local search methods. In multiobjective combinatorial optimization (MoCO), the efficient set is the set of solutions which are not dominated by any other feasible solution. It is often claimed that the structure of this efficient set plays a crucial role for the development of efficient local search methods [3]. Connectedness is related to the property that efficient solutions are connected (at distance 1) with respect to a neighborhood relation [4]. This property has later been extended to the notion of cluster, where distances can take higher values [5]. When connectedness holds, it becomes possible to find all the efficient solutions by means of the iterative exploration of the neighborhood of the current approximation set by starting by one (or more) solution(s) from the efficient set. This strategy coincides with the Pareto Local Search (PLS) algorithm [6], initialized with one efficient solution, and then acts like an exact approach. However, a common knowledge is that, for most MoCO problems, the number of non-dominated solutions is not polynomial in the size of the problem instance [7], so that a PLS algorithm can take an exponential time to identify the efficient set once the later contains an exponential number of solutions. Then, the goal of the optimization process is often to identify a representative sample set, containing a limited number of efficient solutions.

In this work, we argue that connectedness is not the only feature which explains the difficulty of MoCO for search algorithms. Analogously to the single-objective long path problems, where a hillclimbing algorithm is outperformed by a simple EA, even if the search space is unimodal, we here oppose straightforward extensions of those algorithms, a hillclimbing algorithm and a simple EA, in a multiobjective context. On one side, PLS extends a single-objective hillclimber in terms of Pareto dominance [6]. At the opposite, we use an adaptation of the Simple Evolutionary Multiobjective Optimization (SEMO) algorithm [8]. Both approaches are initialized with one solution from the efficient set, corresponding to an extreme point of the Pareto front. In this paper, we propose the definition of the biobjective long path problem (k-lp^2) and of the biobjective multiple path problem (k-mp^2). With k-lp^2, we show experimentally that, even if the efficient set is connected, the runtime required by PLS to find a reasonably good approximation (in terms of hypervolume [9]) is larger than for SEMO, and becomes computationally prohibitive for large-size instances. Furthermore, we construct k-mp^2 instances where the efficient set is completely disconnected, but some additional shortcuts are available to walk from one non-dominated solution to the others. In this case, we show experimentally that PLS can find a good approximation in a significantly less amount of time than SEMO. Indeed, both algorithms differ in the way they sample the efficient set. For k-lp^2, PLS can only follow the path defined by the connectedness property while SEMO is able to take some shortcuts outside of the path. For k-mp^2, PLS takes advantage of the multiple paths, defined outside the efficient set, which are temporally non-dominated and that lead to further non-dominated solutions.

The reminder of the paper is organized as follows. First, some notions related to MoCO, connectedness and long path problems are briefly presented in the next section. Section 3 introduces the class of biobjective long path problems, for which the efficient set is fully connected and exponential in the size of the problem instance. Next, the class of multiple path problems is presented in Section 4. It handles an exponential number of disconnected efficient solutions. Our experiments illustrate that PLS appears to be outperformed by SEMO for biobjective long path problems, while more surprisingly, the opposite occurs for multiple path problems. This work leads to further investigations on a proper definition of fitness landscapes for MoCO, not only with regards to the efficient set itself, but also to the way that leads to its approximation.

2 Background

2.1 Multiobjective Combinatorial Optimization

A multiobjective optimization problem can be defined by a set of $m \geq 2$ objective functions $(f_1, f_2, \ldots, f_m)$, and a set X of feasible solutions in the *decision space*. In the combinatorial case, X is a discrete set. Let $Z = f(X)$ denote the set of feasible outcome vectors in the *objective space*. To each solution $x \in X$ is assigned an objective vector on the basis of a vector function $f : X \to Z$ with $f(x) = (f_1(x), f_2(x), \ldots, f_m(x))$. Without loss of generality, we here assume that all m objective functions are to be maximized. A solution $x \in X$ is said to *dominate* a solution $x' \in X$, denoted by $x \succ x'$, iff $\forall i \in \{1, 2, \ldots, m\}$, $f_i(x) \geq f_i(x')$ and $\exists j \in \{1, 2, \ldots, m\}$ such as $f_j(x) > f_j(x')$. A solution $x \in X$ is said to be *efficient* (or *Pareto optimal, non-dominated*) if there does not exist any other solution $x' \in X$ such that x' dominates x. The set of all efficient solutions is called the *efficient set* and its mapping in the objective space is called the *Pareto front*. A possible approach in MoCO is to find a minimal set of efficient solutions, such that strictly one solution maps to each non-dominated vector. However, generating the entire efficient set of a MoCO problem is usually infeasible for two main reasons. First, the number of efficient solutions is typically exponential in the size of the problem instance [7]. In that sense, most MoCO problems are said to be intractable. Second, deciding if a feasible solution belongs to the efficient set is known to be NP-complete for numerous MoCO problems [10], even if none of its single-objective counterpart is NP-hard. Therefore, the overall goal is often to identify a good efficient set approximation, ideally a subpart of the efficient set. To this end, heuristic approaches have received a growing interest in the last decades.

2.2 Local Search and Connectedness

A *neighborhood structure* is a function $\mathcal{N} : X \to 2^X$ that assigns a set of solutions $\mathcal{N}(x) \subset X$ to any solution $x \in X$. $\mathcal{N}(x)$ is called the *neighborhood* of x, and a solution $x' \in \mathcal{N}(x)$ is called a *neighbor* of x. Local search algorithms for MoCO, like the Pareto Local Search (PLS) [6], generally combine the use of such a neighborhood structure with the management of an archive (or population) of mutually non-dominated solutions found so far. The basic idea is to iteratively improve this archive by exploring the neighborhood of its own content until no further improvement is possible, or until another stopping condition is fulfilled.

Recently, local search approaches have been successfully applied to MoCO problems. Some structural properties of the landscape seem to allow the search space to be explored in an effective way. Such a property, related to the efficient set, is *connectedness* [3,4]. As argued by the original authors, it could provide a theoretical justification for the design of multiobjective local search. Let us define a graph such that each node represents an efficient solution, and an edge connects a pair of nodes if the corresponding solutions are neighbors with respect to a given neighborhood relation [4]. The efficient set is said to be *connected* if there exists a path between every pair of nodes in the graph. Paquete and Stützle [5] extended this notion by introducing an arbitrary distance separating two efficient solutions (*i.e.* the minimal number of neighbors to visit to go from one solution to another). Unfortunately, in the general case, rather negative results have been reported in the literature for some classical MoCO problems [3,4]. However, in practice, many empirical results show that efficient solutions for some MoCO problems are strongly clustered with respect to more classical neighborhood structures from combinatorial optimization, see for instance [5]. Indeed, in the case of connectedness, by starting with one or more non-dominated solutions, it becomes possible to find all the efficient solutions through a basic iterative neighborhood exploration procedure, like PLS. However, we show in this paper that connectedness is not the only property to deal with when searching for an approximation of the efficient set.

2.3   The Single-Objective Long k-Path Problem

The long path problem has been introduced by Horn et al. [1] to design unimodal landscapes where the path length to reach the global optimum is exponential in the size of the problem instance. The long k-path is defined on bit strings of size l. Let $P_{l,k}$ be a long k-path of dimension l, and $P_{l,k}(i)$ the ith solution on this path. The long k-path of dimension 1 is only made of two solutions $P_{1,k} = (0, 1)$, and the path of dimension $l + k$ can be defined by recursion:

$$P_{l+k,k}(i) = \begin{cases} 0^k P_{l,k}(i) & \text{if } 0 \leq i < s_{l,k} \\ 0^{k-j}1^j P_{l,k}(s_{l,k} - 1) & \text{if } s_{l,k} \leq i < s_{l,k} + k - 1 \text{ with } j = i - s_{l,k} + 1 \\ 1^k P_{l,k}(s_{l+k,k} - 1 - i) & \text{if } s_{l,k} + k - 1 \leq i < s_{l+k,k} \end{cases}$$

where $s_{l,k} = |P_{l,k}| = 2s_{l-k,k} + (k - 1) = (k + 1)2^{(l-1)/k} - k + 1$ is the length of the k-path of dimension l. The fitness function of the long k-path problem (to be maximized) is defined as follows. For all $x \in \{0, 1\}^l$:

$$f(x) = \begin{cases} l + i \text{ if } x \in P_{l,k} \text{ and } x = P_{l,k}(i) \\ |x|_0 \text{ if } x \notin P_{l,k} \end{cases}$$

where $|x|_0$ is the number of '0' in the bit string x. In the long k-path, a shortcut can be found by flipping k consecutive bits. For a hillclimbing algorithm which chooses the best solution in the neighborhood defined by Hamming distance 1, the number of iterations to reach the global optimum matches the length of the path, $s_{l,k}$. The number of evaluations is then $(l \cdot s_{l,k})$ for a hillclimber. On the contrary, a $(1 + 1)$ EA which flips each bit with a probability $p = 1/l$ at each iteration is found the global optimum in polynomial expected running time $\mathcal{O}(l^{k+1}/k)$ [2][1].

3   The Biobjective Long k-Path Problem

In this section, we propose a biobjective problem where the efficient set is connected, but so huge that the full enumeration of it cannot be made in polynomial time. We define the *biobjective long k-path problem* to show that the required runtime to sample a connected efficient set can be very long for a simple local search algorithm.

3.1   Definition

The biobjective long k-path problem (k-lp^2) is defined on a bit string of length l, with an objective function vector of dimension 2. Each objective function corresponds to a 'single' long k-path problem, which is to be maximized. The k-lp^2 is built such that the efficient set matches the path $P_{l,k}$. The objective function vector of k-lp^2 is defined as follows. For all $x \in \{0,1\}^l$:

$$f(x) = (f_1(x), f_2(x)) = \begin{cases} h_{l,k}(i) & \text{if } x \in P_{l,k} \text{ and } x = P_{l,k}(i) \\ (|x|_0, |x|_0) & \text{if } x \notin P_{l,k} \end{cases}$$

where h is the function which associates each integer i to the point of coordinates $(l + i, l + s_{l,k} - 1 - i)$ in the objective space. So, the first objective is the fitness function of the single-objective long k-path problem.

The efficient set of k-lp^2 corresponds to the path $P_{l,k}$ (see Fig. 1). By construction, all solutions in $P_{l,k}$ are neighbors with respect to Hamming distance 1, so that the efficient set is connected. The size of $P_{l,k}$ is $s_{l,k} = (k+1)2^{(l-1)/k} - k + 1$, which cannot be enumerated in a polynomial number of evaluations in the general case. The efficient set of k-lp^2 is then (i) connected and (ii) intractable. Let us now experimentally examine the ability of search algorithms to identify a good approximation of it.

3.2   Experimental Analysis

Ingredients. For the single-objective long path problems, existing studies are based on the comparison of a hillclimber and of a $(1 + 1)$ EA [2]. Then, we will

---

[1] The lower bound of the expected runtime could be exponential when $k = \sqrt{l-1}$ [11].

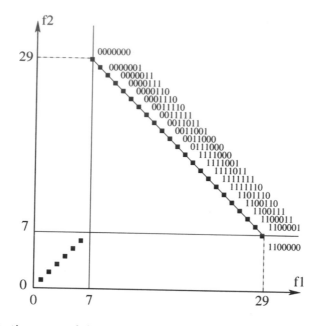

Fig. 1. Objective space of the biobjective long 2-path problem of dimension $l = 7$

here consider straightforward multiobjective extensions of these approaches, respectively a PLS- and a SEMO-like algorithm. They are both adapted to the path problems (k-lp^2 and k-mp^2) introduced in this paper, and they will be respectively denoted by PLS$_p$ and SEMO$_p$ to differentiate them from their original implementation. A pseudo-code is given in Algorithm 1 and Algorithm 2, respectively. At each PLS$_p$ iteration, one solution is chosen at random from the archive. All solutions located at Hamming distance 1 are evaluated and are checked for insertion in the archive. For the problem under study, note that at most two neighbors are located on the long path, with one of them being already found at a previous iteration. The current solution is then marked as *visited* in order to avoid a useless revaluation of its neighborhood. At each SEMO$_p$ step, one solution is randomly chosen from the archive. Each bit of this solution is independently flipped with a probability $p = 1/l$, and the obtained solution is checked for insertion in the archive. In PLS$_p$, the whole neighborhood is explored while in SEMO$_p$, all solutions are potentially reachable with respect to different probabilities[2]. In order to take advantage of the connectedness property, the archive of both algorithms is initialized with one solution from the efficient set: the bit string $(0, 0, \ldots, 0)$ of size l.

However, the efficient set of k-lp^2 is intractable. It becomes then impracticable to use an unbounded archive for large-size problem instances. As a consequence, contrary to the original approaches, we here maintain a *bounded archive* of size M in our implementation of the algorithms. Our attempt is not to compare different

---

[2] In SEMO, the neighborhood operator is generally supposed to be ergodic [8].

Algorithm 1. PLS_p

$A \leftarrow \{0^l\}$
repeat
 select $x \in A$ at random such that x is not *visited*
 set x to *visited*
 for all x' such that $|x - x'|_1 = 1$ **do**
  updateArchive (A, x')
 end for
until $I_H^\star - I_H(A) < \epsilon \cdot I_H^\star$

Algorithm 2. SEMO_p

$A \leftarrow \{0^l\}$
repeat
 select $x \in A$ at random
 create x' by flipping each bit of x with a probability $p = 1/l$
 updateArchive (A, x')
until $I_H^\star - I_H(A) < \epsilon \cdot I_H^\star$

---

bounded archiving techniques, but rather to limit the number of evaluations required for computing a reasonably good approximation of the efficient set. So, we define a nearly ideal archiving method to find such an approximation for the particular case of k-lp^2. If the Pareto front was linear, an 'optimal' approximation of size M contains uniformly distributed points over the segment $[(l, l + s_{l,k} - 1), (l + s_{l,k} - 1, l)]$ in the objective space. Note that, in our case, those points do not necessarily correspond to feasible solutions in the decision space. The distance between 2 solutions with respect to the first objective is then $\delta = (s_{l,k} - 1)/(M - 1)$. The bounded archiving technique under consideration is given in Algorithm 3. First, dominated solutions are always discarded. If the number of non-dominated solutions becomes too large, the solution with the lowest first objective value which is too close from the previous one (*i.e.* the difference with respect to the first objective is below δ) is removed from the archive. If this rule does not hold for any solution, the penultimate solution (with respect to the order defined by objective 1) is removed (not the last one). Of course, such an archiving technique is k-lp^2-specific, but it does not introduce any bias within heuristic rules generally defined by existing diversity-based archiving approaches.

Experimental Design. The algorithms are compared in terms of the required number of evaluations to attain a reasonable approximation of the efficient set. The cost related to archiving is then ignored, as we want to focus on the complexity of algorithms independently of the archiving strategy. The stopping criteria is based on a percentage of hypervolume I_H [9] covered by the solutions from the archive. For k-lp^2, an upper bound of the maximal hypervolume $(I_H^\star)$ for an approximation of size M can be computed by uniformly distributing M points over the Pareto front, that is $I_H^\star = \delta^2(M + 1)M/2$, (l, l) being the reference

---

Algorithm 3. Bounded archiving

---

updateArchive(A, x):
  for all $a \in A$ **do**
    if $x \succ a$ **then**
      $A \leftarrow A \setminus \{a\}$
    end if
  end for
  if not $\exists a \in A : a \succ x$ **then**
    $A \leftarrow A \cup \{x\}$
    if $|A| > M$ **then**
      reduceArchive(A)
    end if
  end if

reduceArchive(A):
  Sort A in the increasing order w.r.t f_1-values: $A = \{a_1, a_2, a_3, \ldots\}$
  $i \leftarrow 2$
  while $|A| > M$ **do**
    if $i = |A|$ **then**
      $A \leftarrow A \setminus \{a_{|A|-1}\}$
    else if $f_1(a_i) - f_1(a_{i-1}) < \delta$ **then**
      $A \leftarrow A \setminus \{a_i\}$
    else
      $i \leftarrow i + 1$
    end if
  end while

---

point. Once the hypervolume covered by the current archive $I_H(A)$ is below an ϵ-value from I_H^*, the algorithm stops.

The experimental study has been conducted with $k = 2$ and dimensions $l = \{19, 29, 39, 49, 59\}$. We use an archive of size $M = 100$, and the required approximation to be found is less than $\epsilon = 2\%$ of the maximal hypervolume. In other words, at least 98% of the best-possible approximation is covered in terms of hypervolume. The archive is initialized with a bit string where all bits are set to '0'. The number of evaluations is reported over 30 independent runs.

Results and Discussion. Fig. 2 shows the average and the standard deviation of the number of evaluations for each algorithm. The number of evaluations required by PLS_p seems to grow exponentially with the dimension l. It could be interpreted as follows. To approximate the efficient set, PLS_p follows the long path. When the archive reaches its maximum size, the archiving technique let one solution at an 'optimal' position in the objective space at every δ iteration. So, at a given iteration i, the current hypervolume is approximately $I_H(A) \approx \delta^2(2M + 1 - j) \cdot j/2$, where $j = \lceil i/\delta \rceil$. Then, the stopping criteria is reached at the end of the long path only, so that the number of evaluations is more than exponential in the dimension of the problem instance (l times larger). For SEMO_p, the number of evaluations increases from 20.10^3 evaluations for $l = 19$ to 250.10^3 for $l = 59$. The computational effort required by SEMO_p and by PLS_p is different of several orders of magnitude. For SEMO_p, it is difficult to pretend that the runtime is polynomial or not, nevertheless the number of evaluations remains huge. The increase is higher than quadratic and seems to fit a cubic curve.

To summarize, SEMO_p can sample the efficient set more easily than PLS_p by taking shortcuts out of the long path. From the SEMO_p point of view, the efficient set is k-connected [5]: one efficient solution can be reached by flipping k bits of another efficient solution. The computational difference between the two algorithms can be explained by different structures of the graph of efficient

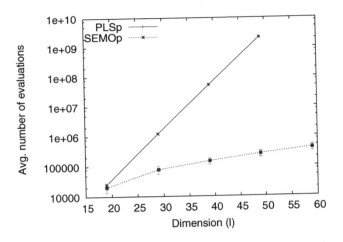

Fig. 2. Average value and standard deviation of the number of evaluations for PLS_p and $SEMO_p$ on biobjective long 2-path problems (log y-scale)

solutions. For PLS_p, it is linear, and for $SEMO_p$, the distance between 2 efficient solutions in the graph is much smaller than the distance in the objective space. This result suggests that the connectedness property is not fully satisfactorily to explain the degree of difficulty of the problem. The structure of the graph of efficient solutions induced by the neighborhood relation should also be taken into account. In the next section, we will show that the structure of this graph is still not enough to explain all the difficulties.

4   The Biobjective Multiple k-Path Problem

In the biobjective long k-path, the efficient set is connected, intractable and difficult to sample. In this section, we define the biobjective multiple k-path problem (k-mp^2) where the efficient set is still intractable but not connected anymore, while easier to sample for a PLS-like algorithm.

4.1   Definition

The idea is to modify k-lp^2 in order to make the efficient set disconnected (with respect to Hamming distance 1), and to add some shortcuts out of the path that guide the search towards efficient solutions. A k-mp^2 instance of dimension l is defined for bit strings of size l such that $(l-1)/k \in \mathbb{N}$, with k being an even integer value. First, let us define the additional paths, called *extra paths*. Let $D_{l,k}$ and $U_{l,k}$ be the extra paths of the k-path of dimension l. Let $u \in (0^k|1^k)^*$ be a concatenation of 1^k and 0^k. $D_{l,k}(u,j,i)$ (resp. $U_{l,k}(u,j,i)$) is the jth solution on the extra path from solution $P_{l,k}(i_0) = u0^k P_{l-|u|-k,k}(i)$ to solution $P_{l,k}(i_1) = u1^k P_{l-|u|-k,k}(i)$ of the long k-path (resp. from $P_{l,k}(i_1)$ to $P_{l,k}(i_0)$). D

and U are defined like the bridges in the single-objective long path problem [1]. $\forall p \in [0..\frac{l-1-k}{k}]$, $\forall u \in (0^k|1^k)^p$, $\forall i \in [0..s_{l-(p+1)k,k} - 1]$, $\forall j \in [1..k-1]$:

$$\begin{cases} D_{l,k}(u,j,i) = u0^{k-j}1^j P_{l-(p+1)k,k}(i) \\ U_{l,k}(u,j,i) = u1^{k-j}0^j P_{l-(p+1)k,k}(i) \end{cases}$$

The sequence of neighboring solutions $(D_{l,k}(u,1,i),\dots,D_{l,k}(u,k-1,i))$ is the extra path to go from solution $P_{l,k}(i_0)$ to solution $P_{l,k}(i_1)$. Respectively, the sequence $(U_{l,k}(u,1,i),\dots,U_{l,k}(u,k-1,i))$ allows to go from $P_{l,k}(i_1)$ to $P_{l,k}(i_0)$. For k an even number, i_0 and i_1 have the same parity: i_0 is even iff i_1 is even.

In k-mp^2, the efficient set corresponds to the set of solutions $P_{l,k}(i)$ in the long path where i is an even number. The efficient set is then fully disconnected with respect to Hamming distance 1. Solutions $P_{l,k}(2n+1)$ which are out of the efficient set are translated by a vector $(-0.5, -0.5)$ 'under' the solutions $P_{l,k}(2n+2)$, so that they become dominated. As a consequence, a solution $P_{l,k}(2n+1)$ leads to, but is dominated by, the efficient solution $P_{l,k}(2n+2)$. However, $P_{l,k}(2n+1)$ and $P_{l,k}(2n)$ are mutually non-dominated. In the same way, the extra paths to go from $P_{l,k}(i_0)$ to $P_{l,k}(i_1)$ are put on the first diagonal of the square enclosed by $(x_{i_1} - 1, y_{i_1} - 1)$ and (x_{i_1}, y_{i_1}). More formally, the fitness function of the k-mp^2 can be defined as follows. For all $x \in \{0,1\}^l$:

$$f(x) = \begin{cases} h_{l,k}(i) & \text{if } x \in P_{l,k} \text{ and } x = P_{l,k}(i) \text{ and } i \text{ even} \\ h_{l,k}(i+1) - (0.5, 0.5) & \text{if } x \in P_{l,k} \text{ and } x = P_{l,k}(i) \text{ and } i \text{ odd} \\ h_{l,k}(i_1) - (\frac{k-j}{k}, \frac{k-j}{k}) & \text{if } x \in D_{l,k} \text{ and } x \notin P_{l,k} \text{ and} \\ & x = D_{l,k}(u,j,i) \text{ with } P_{l,k}(i_1) = u1^k P_{l,k}(i) \\ h_{l,k}(i_0) - (\frac{k-j}{k}, \frac{k-j}{k}) & \text{if } x \in U_{l,k} \text{ and} \\ & x = U_{l,k}(u,j,i) \text{ with } P_{l,k}(i_0) = u0^k P_{l,k}(i) \\ (|x|_0, |x|_0) & \text{otherwise} \end{cases}$$

Fig. 3 illustrates the extra paths starting from one solution. Fig. 4 shows the objective space of a k-mp^2 instance. For $j < k-1$, solution $D_{l,k}(u,j,i)$ is a neighbor of solution $D_{l,k}(u,j+1,i)$ and is dominated by it. As well, solution $D(u,k-1,i)$ is a neighbor of the efficient solution $P_{l,k}(i_1)$ and is dominated by it. However, all $D_{l,k}(u,j,i)$ and $P_{l,k}(i_0)$ are mutually non-dominated. The extra paths D (Down) lead to a further solution in the long path, and the extra paths U (Up) are the backward paths of the extra paths D. With those extra paths, an algorithm based on one bit-flipping can reach an efficient solution easily, just by following the sequence defined by the set of mutually non-dominated solutions found so far.

4.2   Experimental Analysis

The experimental study is conducted with the same approaches and parameters defined for the biobjective long path problem on the previous section. Fig. 5 shows the average value and the standard deviation of the number of evaluations for each algorithm. Fig. 6 allows to compare the number of evaluations with the

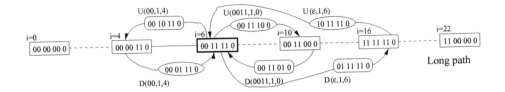

Fig. 3. Extra paths linking the solution $P_{7,2}(6)$ of k-mp^2 of dimension 7. Solutions in a rectangle are along the long path (*i.e.* the efficient set). Solutions in an ellipse are in the extra paths leading to solution $P_{7,2}(6)$ at the same position $(12.5, 22.5)$ in the objective space. The solutions in a rounded rectangle are in extra paths beginning at the solution $P_{7,2}(6)$ translated by $(-0.5, -0.5)$ in the objective space to their destination solution. The length of extra paths is 1. Each solution is labelled by D and U.

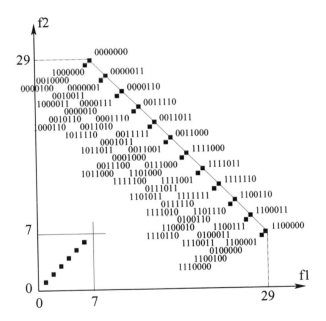

Fig. 4. Objective space of the biobjective multiple 2-path problem of dimension $l = 7$

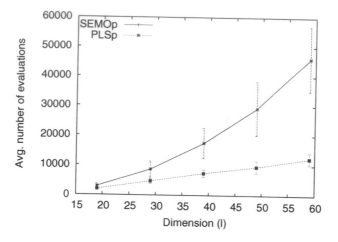

Fig. 5. Average value and standard deviation of the number of evaluations for PLS$_p$ and SEMO$_p$ on biobjective multiple 2-path problems

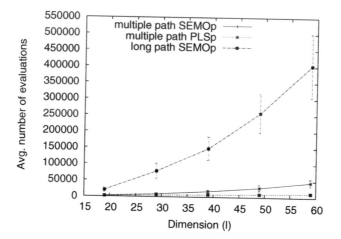

Fig. 6. Average value and standard deviation of the number of evaluations for PLS$_p$ and SEMO$_p$ on biobjective multiple 2-path problems compared to the SEMO$_p$ on biobjective long 2-path

previous problem. Contrary to the results obtained for the long 2-path problem, PLS$_p$ here clearly outperforms SEMO$_p$ which needs 3 times more evaluations for dimension $l = 49$. For PLS$_p$, the number of evaluations increases linearly with the dimension of the problem instance. PLS$_p$ can find easily the same shortcuts than SEMO$_p$, and the latter now loses computational resources to explore dominated solution and to evaluate the neighborhood of some solutions from the archive more than once. The curves on the right show that it is much easier to sample the efficient set of the multiple 2-path than for the long 2-path problem: for dimension 49, nearly 27 times more evaluations are required between $SEMO_p$ for k-lp^2 and PLS_p for k-mp^2.

This is the main results of this study. The extra paths guide the search process to efficient solutions distributed all over the Pareto front. The extra solutions are not in the efficient set and do not appear on the graph of efficient solutions, but they are the keys to explain the performances of local search approaches. Indeed, efficient solutions can now be reached very quickly by following the extra paths, this explains the good performances of the algorithms. Features from the efficient set (connectedness, etc.) are independent of the solutions from the extra paths. Hence, the features of the efficient set are not the only key issue to explain the success of local search for MoCO.

5   Conclusions and Future Works

In this paper, we proposed two new classes of biobjective combinatorial optimization problems, the long and the multiple path problems, in order to demonstrate empirically that connectedness is not the only key issue that characterizes the difficulty of a multiobjective combinatorial optimization problem. In other words, connectedness is not the 'Holy Grail' of search space features when the efficient set is intractable, and when the goal is to find a limited-size approximation. Indeed, on the long path problems, where the efficient set is intractable and connected, our experiments show that the running time to approximate it is exponential for a Pareto-based local search (PLS), and polynomial for a simple Pareto-based evolutionary algorithm (SEMO). On the multiple path problems, where the efficient set is still intractable but disconnected, PLS now outperforms SEMO, which seems rather unexpected at first sight. This suggests two new considerations to measure the difficulty of finding a good efficient set approximation:

- First, the structure of the graph of efficient solutions induced by the neighborhood relation defined by the algorithm should also be taken into account. In the long path problems, this graph is a huge line for PLS whereas it is highly connected for SEMO. Extending the notion of cluster on the efficient graph as defined by Paquete and Stützle [5], we should study a graph where an edge between efficient solutions is defined as the probability to reach one solution from the other.
- Second, the solutions outside the efficient set should also be considered. In the multiple path problems, some solutions outside of the efficient set are temporally non-dominated so that they are saved into the archive during the search process. They help to approximate the (disconnected) efficient set.

In some sense, the fitness landscape of biobjective multiple path problems is unimodal, with a number of short paths leading to good solutions. On the contrary, the biobjective long path problem can be characterized by a unimodal landscape where the path to good solutions is intractable.

Clearly, following the work of Horoba and Neumann [12], the next step will consist in leading a rigorous runtime analysis of PLS and SEMO for both the multiple and the long path problems. The actual bounded archiving method is probably too specific, and seems very difficult to study rigorously. Then, in order to do so, we certainly have to change this strategy with the concept of ϵ-dominance, for instance. It is also possible to extend the biobjective path problems proposed in this paper to a larger objective space dimension (more than 2 objective functions), or with a larger 'disconnectedness' (delete more than one solution over two). The next challenge will be to define a relevant definition of fitness landscape in order to better understand the difficulty of multiobjective combinatorial optimization problems. Given that the goal is here to find a set of solutions, we believe that another way to do so would be to analyze a fitness landscape where the search space consists of sets of solutions. A solution would then be a set of bit strings instead of a single bit string for the problems under study in this paper. Therefore, we plan to formally define fitness landscapes for the recent proposal of *set-based* multiobjective optimization [13].

Acknowledgments. The authors are grateful to Dr. Dirk Thierens for useful suggestions on the relation between intractable efficient sets and long path problems. They would also like to thank Dr. Luis Paquete for fruitful discussion on the subject of this work.

References

1. Horn, J., Goldberg, D., Deb, K.: Long path problems. In: Davidor, Y., Männer, R., Schwefel, H.-P. (eds.) PPSN 1994. LNCS, vol. 866, pp. 149–158. Springer, Heidelberg (1994)
2. Rudolph, G.: How mutation and selection solve long path problems in polynomial expected time. Evolutionary Computation 4(2), 195–205 (1996)
3. Gorski, J., Klamroth, K., Ruzika, S.: Connectedness of efficient solutions in multiple objective combinatorial optimization. Technical Report 102/2006, University of Kaiserslautern, Department of Mathematics (2006)
4. Ehrgott, M., Klamroth, K.: Connectedness of efficient solutions in multiple criteria combinatorial optimization. European Journal of Operational Research 97(1), 159–166 (1997)
5. Paquete, L., Stützle, T.: Clusters of non-dominated solutions in multiobjective combinatorial optimization: An experimental analysis. In: Multiobjective Programming and Goal Programming. LNEMS, vol. 618, pp. 69–77. Springer, Heidelberg (2009)
6. Paquete, L., Chiarandini, M., Stützle, T.: Pareto local optimum sets in the biobjective traveling salesman problem: An experimental study. In: Metaheuristics for Multiobjective Optimisation. LNEMS, vol. 535, pp. 177–199. Springer, Heidelberg (2004)
7. Ehrgott, M.: Multicriteria optimization, 2nd edn. Springer, Heidelberg (2005)
8. Laumanns, M., Thiele, L., Zitzler, E.: Running time analysis of evolutionary algorithms on a simplified multiobjective knapsack problem. Natural Computing: an International Journal 3(1), 37–51 (2004)
9. Zitzler, E., Thiele, L.: Multiobjective evolutionary algorithms: A comparative case study and the strength pareto approach. IEEE Transactions on Evolutionary Computation 3(4), 257–271 (1999)

10. Serafini, P.: Some considerations about computational complexity for multiobjective combinatorial problems. In: Recent Advances and Historical Development of Vector Optimization. LNEMS, vol. 294. Springer, Heidelberg (1986)
11. Droste, S., Jansen, T., Wegener, I.: On the optimization of unimodal functions with the $(1 + 1)$ evolutionary algorithm. In: Eiben, A.E., Bäck, T., Schoenauer, M., Schwefel, H.-P. (eds.) PPSN 1998. LNCS, vol. 1498, pp. 13–22. Springer, Heidelberg (1998)
12. Horoba, C., Neumann, F.: Additive approximations of pareto-optimal sets by evolutionary multi-objective algorithms. In: Tenth Workshop on Foundations of Genetic Algorithms (FOGA 2009), pp. 79–86. ACM, New York (2009)
13. Zitzler, E., Thiele, L., Bader, J.: On set-based multiobjective optimization. IEEE Transactions on Evolutionary Computation 14(1), 58–79 (2010)

Improving Parallel Local Search for SAT

Alejandro Arbelaez[1] and Youssef Hamadi[2,3]

[1] Microsoft-INRIA joint-lab, Orsay France
alejandro.arbelaez@inria.fr
[2] Microsoft Research, Cambridge United Kingdom
[3] LIX École Polytechnique, F91128 Palaiseau, France
youssefh@microsoft.com

Abstract. In this work, our objective is to study the impact of knowledge sharing on the performance of portfolio-based parallel local search algorithms. Our work is motivated by the demonstrated importance of clause-sharing in the performance of complete parallel SAT solvers. Unlike complete solvers, state-of-the-art local search algorithms for SAT are not able to generate redundant clauses during their execution. In our settings, each member of the portfolio shares its best configuration (i.e., one which minimizes conflicting clauses) in a common structure. At each restart point, instead of classically generating a random configuration to start with, each algorithm aggregates the shared knowledge to carefully craft a new starting point. We present several aggregation strategies and evaluate them on a large set of problems.

Keywords: local search, SAT solving, parallelism.

1 Introduction

Complete parallel solvers for the propositional satisfiability problem have received significant attention recently. These solvers can be divided into two main categories the classical divide-and-conquer model and the portfolio-based approach. The first one, typically divides the search space into several sub-spaces while the second one lets algorithms compete on the original formula [1]. Both take advantage of the modern SAT solving architecture [2], to exchange the conflict-clauses generated in the system and improve the overall performance.

This push towards parallelism in complete SAT solvers has been motivated by their practical applicability. Indeed, many domains, from software verification to computational biology and automated planning rely on their performance. On the contrary, since local search techniques only outperform complete ones on random SAT instances, their parallelizing has not received much attention so far. The main contribution on the parallelization of local search algorithms for SAT solving basically executes a portfolio of independent algorithms which compete without any communication between them. In our settings, each member of the portfolio shares its best configuration (i.e., one which minimizes the number of conflicting clauses) in a common structure. At each restart point, instead of

C.A. Coello Coello (Ed.): LION 5, LNCS 6683, pp. 46–60, 2011.
© Springer-Verlag Berlin Heidelberg 2011

classically generating a random configuration to start with, each algorithm aggregates the shared knowledge to carefully craft a new starting point. We present several aggregation strategies and evaluate them on a large set of instances.

This paper is organized as follows: background material is presented in section 2. Section 3 describes previous work on parallel SAT and cooperative algorithms. Section 4 presents our methodology and our aggregation strategies, section 5 evaluates them, and section 6 presents some concluding remarks and future directions of research.

2 Background

2.1 The Propositional Satisfiability Problem

The Propositional Satisfiability Problem (SAT) can be represented by a pair $\langle \mathcal{V}, \mathcal{C} \rangle$ where, $\mathcal{V}$ indicates a set of boolean variables and $\mathcal{C}$ a set of clauses representing a propositional *conjunctive-normal form* (CNF).

Solving a SAT problem involves finding a solution i.e., a truth assignment for each variable such that all clauses are satisfied, or demonstrating that no such assignment can be found. If a solution exist the problem is stated as satisfied and unsatisfied otherwise. Currently, there are two well established techniques for solving SAT problems, complete and incomplete techniques [3], the former is developed on top of the DPLL algorithm. It combines a tree-based search with constraint propagation, conflict-clause learning, and intelligent backtracking while the latter is based on local search algorithms to quickly find a truth assignment for a given satisfiable instance [4].

2.2 Local Search for SAT

Algorithm 1 describes a traditional local search algorithm for SAT solving, it starts with a random truth assignment for each variable in the formula F (*initial-configuration* line 2), and the key point of local search algorithms is depicted in lines (3-9) here the algorithm flips the most appropriate variable candidate until a solution is found or a given number of flips is reached (MaxFlips), after this process the algorithm restarts itself with a new (fresh) random configuration.

As one may expect, a critical part of the algorithm is the variable selection function (*select-variable*) which indicates the next variable to be flipped in the current iteration of the algorithm. Broadly speaking, there are two main categories of variable selection functions, the first one motivated by the GSAT algorithm [5] is based on the following score function:

$$score(x) = make(x) - break(x)$$

Intuitively $make(x)$ indicates the number of clauses that are currently satisfied but flipping x become unsatisfied, and $break(x)$ indicates the number of clauses that are unsatisfied but flipping x become satisfied. In this way, local search algorithms select the variable with minimal score value (preferably with negative value), because flipping this variable would most likely increase the chances of solving the instance.

The second category of variable selection functions is the Walksat-based one [6] which includes a diversification strategy in order to avoid local minimums, this extension selects, at random, an unsatisfied clause and then picks a variable from that clause. The variable that is generally picked will result in the fewest previously satisfied clauses becoming unsatisfied, with some probability of picking one of the variables at random.

Algorithm 1. Local Search For SAT (CNF formula F, Max-Flips, Max-Tries)

```
 1: for try := 1 to Max-Tries do
 2:     A := initial-configuration(F).
 3:     for flip := 1 to Max-Flips do
 4:         if A satisfies F then
 5:             return A
 6:         end if
 7:         x := select-variable(A)
 8:         A := A with x flipped
 9:     end for
10: end for
11: return 'No solution found'
```

2.3 Refinements

This section briefly reviews the main characteristics of state-of-the-art local search solvers for SAT solving. As pointed out above these algorithms are developed to deal with the variable selection function and are mainly devoted to avoid getting trapped in a local minima. This way, the following list describes several well-known mechanisms for selecting the most appropriate variable to flip at a given state of the search.

- *Novelty* [7] firstly selects an unsatisfied clause c and from c selects the best v_{best} and second best v_{2best} variable candidates, if v_{best} is not the latest flipped variable in c then Novelty flips this variable, otherwise v_{2best} is flipped with a given probability p and v_{best} with probability $1 - p$. Important extensions to this algorithm can be found in *Novelty+*, *Novelty++* and *Novelty+p*.
- G^2WSAT [8] (G2) uses a list of promising decreasing variables to determine the next variable to be flipped and if the list of decreasing variables is empty the algorithm uses *Novelty++* as a backup heuristic. $G^2WSAT+p$ (G2+p) uses a similar strategy that G^2WSAT however in this case the backup solver is *Novelty+p*.
- *Scaling and Probabilistic Smoothing* (SAPS) [9] implements a multiplicative increase rule to dynamically modify the penalty for unsatisfied clauses and with a given probability P_{smooth} this penalty value is adjusted according to a given smoothing factor ρ.

- *Pure Additive Weighting Scheme* (PAWS) [10] implements an additive increase rule to dynamically modify the penalty for unsatisfied clauses and if a given clause penalty has been changed a given number of times this penalty value is adjusted.
- *Reactive SAPS* (RSAPS) [9] extends SAPS by adding an automatic tuning mechanism to identify suitable values for the smoothing factor ρ.
- *Adaptive Novelty+* (AN+) [11] uses an adaptive mechanism to properly tune the noise parameter of Walksat-like algorithms (e.g, *Novelty+*)
- *Adaptive $G^2 WSAT$* (AG2) [12] aims to integrate an adaptive noise mechanism into the $G^2 WSAT$ algorithm. Similarly, *Adaptive $G^2 WSAT+p$* (AG2+p) also uses an adaptive noise mechanism into the $G^2 WSAT+p$ algorithm.

3   Previous Work

In this section, we review the most important contributions devoted to parallel SAT solving and cooperative algorithms.

3.1   Complete Methods for Parallel SAT

GrADSAT [13] is a parallel SAT solver based on the zChaff solver and equipped with a master-slave architecture in which the problem space is divided into subspaces, these sub-spaces are solved by independent zChaff clients and learnt clauses whose size (i.e., number of literals) is less or equal to a given limit are exchanged between clients. The technique organizes load-balancing through a work stealing technique which allows the master to push work to idle clients.

Unlike other parallel solvers for SAT which divide the initial problem space into sub-spaces, ManySAT [1] is a portfolio-based parallel solver where independent DPLL algorithms are launched in parallel to solve a given problem instance. Each algorithm in the portfolio implements a different and complementary restart strategy, polarity heuristic and learning scheme. In addition, the first version of the algorithm exchanges learnt clauses whose size is less or equal to a given limit. It is worth mentioning that ManySAT won the 2008 SAT Race, the 2009 SAT Competition and was placed second in the 2010 SAT Race (all these in the parallel track). Interestingly all the algorithms successfully qualified in the 2010 parallel track were based on a Portfolio architecture.

In [14] the authors proposed a hybrid algorithm which starts with a traditional DPLL algorithm to divide the problem space into sub-spaces. Each sub-space is then allocated to a given local search algorithm (Walksat).

3.2   Incomplete Methods for Parallel SAT

PGSAT [15] is a parallel version of the GSAT algorithm. The entire set of variables is randomly divided into τ subsets and allocated to different processors. In this way at each iteration, if no global solution has been obtained, the ith processor uses the GSAT score function (see section 2) to select and flip the best variable for the ith subset. Another contribution to this parallelization architecture is described in [16] where the authors aim to combine PGSAT and random walk, therefore at each iteration, with a given probability wp an unsatisfiable clause c is selected and a random variable from c is flipped and with probability $1-wp$. PGSAT is used to flip τ variables in parallel at a cost of reconciling partial configurations to test if a solution has been found.

gNovelty+ (v.2) [17], belongs to the portfolio approach, this algorithm executes n independent copies of the *gNovelty+ (v.2)* algorithm in parallel, until at least one of them finds a solution or a given timeout is reached. This algorithm was the only parallel local search solver presented in the *random* category of the 2009 SAT Competition[1].

In [18], Kroc et al., studied the application of a parallel hybrid algorithm to deal with the max-SAT problem. This algorithm combines a complete solver (minisat) and an incomplete one (Walksat). Broadly speaking both solvers are launched in parallel and minisat is used to guide Walksat to promising regions of the search space by means of suggesting values for the selected variables.

3.3   Cooperative Algorithms

In [19] a set of algorithms running in parallel exchange hints (i.e., partial valid solutions) to solve hard graph coloring instances. To this end, they share a blackboard where they can write a hint with a given probability q and read a hint with a given probability p.

In [20] the authors studied a sequential cooperative algorithm to deal with the office-space-allocation problem. In this paper cooperation takes place when a given algorithm is not able to improve its own best solution, at this point a cooperative mechanism is used to explore suitable partial solutions stored by individual heuristics. This algorithm is also equipped with a diversification strategy to explore different regions of the search space.

Although *Averaging in Previous Near Solutions* [21] is not a cooperative algorithm by itself, this method is used to determine the initial configuration for the ith restart in the GSAT algorithm. Broadly speaking, the initial configuration is computed by performing a bitwise average between variables of the best solution found during the previous restart ($restart_{i-1}$) and two restarts before ($restart_{i-2}$). That is, variables with same values in both configurations are reused, and the extra set of variables are initialized with random values. Since overtime, configurations with a few conflicting clauses tend to become similar, all the variables are randomly initialized after a given number of restarts.

4   Knowledge Sharing in Parallel Local Search for SAT

Our objective is to extend a parallel portfolio of state-of-the-art local search solvers for SAT with knowledge sharing or cooperation. Each algorithm is going to share with others the best configuration it has found so far with its respective cost (number of unsatisfied clauses) in a shared pair $\langle M, C \rangle$.

---

[1] http://www.satcompetition.org/2009/

$$M = \begin{pmatrix} X_{11} & X_{12} & \cdots & X_{1n} \\ X_{21} & X_{22} & \cdots & X_{2n} \\ \vdots & \vdots & \vdots & \vdots \\ X_{c1} & X_{c2} & \cdots & X_{cn} \end{pmatrix} \qquad C = [C_1, C_2, \ldots, C_c]$$

Where n indicates the total number of variables of the problem and c indicates the number of local search algorithms in the portfolio. In the following we are associating local search algorithms and processing cores. Each element X_{ji} in the matrix indicates the ith variable of the best configuration found so far by the jth core. Similarly, the jth element in C indicates the cost for the respective configuration in M.

These best configurations can be exploited by each local search to build a new initial configuration. In the following, we propose seven strategies to determine the initial configuration (cf. function *initial-configuration* in algorithm 1).

4.1 Using Best Known Configurations

In this section, we propose three methods to build the new initial configuration *init* by aggregating best known configurations. In this way, we define $init_i$ for all the variables $X_i, i \in [1..n]$ as follows:

1. *Agree*: if there exists a value v such that $v = X_{ji}$ for all $j \in [1..c]$ then $init_i = v$, otherwise a random value is used.
2. *Majority*: if there exists two values v and v' such that $|\{X_{ji} = v | j \in [1..c]\}| > |\{X_{ji} = v' | j \in [1..c]\}|$ then $init_i = v$, otherwise a random value is used.
3. *Prob*: $init_i = 1$ with probability $p_{ones} = \frac{ones}{c}$ and $init_i = 0$ with probability $1 - p_{ones}$, where $ones = |\{X_{ji} = 1 | j \in [1..c]\}|$.

4.2 Weighting Best Known Configurations

In contrast with our previous methods where all best known solutions are treated equally important, the methods proposed in this section use a weighting mechanism to consider the cost of best known configurations. The computation of the initial configuration *init* uses one of the following two weighting systems: *Ranking* and *Normalized Performance*, where values from better configurations are most likely to be used.

Ranking. This method sorts the configurations of the shared matrix from worst to best according to their cost. The worst ranked one gets weight of 1 (i.e., $RankW_1 = 1$), and the best ranked c (i.e., $RankW_c = c$).

Normalized Performance. This method assigns weights ($NormW$) considering a normalized value of the number of unsatisfied clauses of the configuration:

$$NormW_j = \frac{|\mathcal{C}| - C_j}{|\mathcal{C}|}$$

Using the previous two weighting mechanisms, we define the following four extra methods to determine initial configurations.

To this end, we define $\Phi(val, Weight) = \sum_{k \in \{j \mid X_{ji}=val\}} Weight_k$.

1. *Majority RankW*: if there exists two values v and v' such that $\Phi(v, RankW) > \Phi(v', RankW)$ then $init_i = v$, otherwise a random value is used.
2. *Majority NormalizedW*: if there exists two values v and v' such that $\Phi(v, NormW) > \Phi(v', NormW)$ then $init_i = v$, otherwise a random value is used.
3. *Prob RankW*: $init_i = 1$ with probability $P_{Rones} = \frac{Rones}{Rones+Rzeros}$ and $init_i = 0$ with probability $1 - P_{Rones}$, where $Rones = \Phi(1, RankW)$ and $Rzeros = \Phi(0, RankW)$.
4. *Prob NormalizedW*: $init_i = 1$ with probability $P_{Nones} = \frac{Nones}{Nones+Nzeros}$ and $init_i = 0$ with probability $1 - P_{Nones}$, where $Nones = \Phi(1, NormW)$ and $Nzeros = \Phi(0, NormW)$

4.3 Restart Policy

As mentioned earlier on, shared knowledge is exploited when a given algorithm is restarted. At this point the current working configuration of a given algorithm is re-initialized according to a given aggregation strategy. However, it is important to restrict cooperation since it adds overheads and more importantly tend to generate similar configurations. In this context, we propose a new restart policy to avoid re-initializing the working configuration again and again. This new policy re-initializes the working configuration for a given restart (i.e., every MaxFlips) if and only if, performance improvements in best known solutions have been observed during the latest restart window. This new restart policy is formally described in the following definition, where we assume that bc_{ki} is the cost of the best known configuration for a given algorithm i up to the $(k-1)^{th}$ restart.

Definition 1. *At a given restart k for a given algorithm i the working configuration is reinitialized iff there exists an algorithm q such that $bc_{kq} \neq bc_{(k-1)q}$ and $q \neq i$.*

5 Experiments

5.1 Experimental Settings

We conducted experiments using instances from the RANDOM category of the 2009 SAT competition. Since state-of-the-art local search solvers are unable to solve UNSAT instances, we filtered out these instances. We also removed instances whose status was reported as UNKNOWN in the competition. This way, we collected 359 satisfiable instances.

We decided to build our parallel portfolio on UBCSAT-1.1, a well known local search library which provides efficient implementation of the latest local search for SAT algorithms [22]. We did preliminary experiments to extract from this library the 8 algorithms which perform best on our set of problems. From that, we defined the following three baseline portfolio constructions where algorithms are independent searches without cooperation. The first one *pcores-PAWS* uses *p* copies of the best single algorithm (PAWS), the second portfolio *4cores-No sharing* uses the best subset of 4 algorithms (PAWS, G2+p, AG2, AG2+p) and the last one *8cores-No sharing* uses all the 8 algorithms (PAWS, G2+p, AG2, AG2+p, G2, SAPS, RSAPS, AN+). All the algorithms were used with their default parameters, and without any restart. Indeed these techniques are equipped with important diversification strategies and usually perform better when the restart flag is switched off (i.e., MaxFlips=∞).

On the other hand, the previous knowledge aggregation mechanisms were built on top of a portfolio with 4 algorithms (same algorithms as *4cores-No sharing*) and a portfolio with 8 algorithms (same algorithms as *8cores-No sharing*). There, we used the modified restart policy described in section 4.3 with *MaxFlips* set to 10^6.

All tests were conducted on a cluster of 8 Linux Mandriva machines with 8 GB of RAM and two quad-core (8 cores) 2.33 Ghz Intel Processors. In all the experiments, we used a timeout of 5 minutes (300 seconds) for each algorithm in the portfolio, so that for each experiment the total CPU time was set to $c \times 300$ seconds, where c indicates the number of algorithms in the portfolio.

We executed each instance 10 times (each time with a different random seed) and reported two metrics, the *Penalized Average Runtime* (PAR) [23] which computes the average runtime overall instances, but where unsolved instances are considered as $10\times$ the cutoff time, and the runtime for each instance which is calculated as the median across the 10 runs. Overall, our experiments for these 359 SAT instances took 187 days of CPU time.

5.2 Practical Performances with 4 Cores

Fig. 1 shows the results of each aggregation strategy using a portfolio with 4 cores, comparatively to the 4 cores baseline portfolios. The x-axis gives the number of problems solved and the y-axis presents the cumulated runtime.

As expected, the portfolio with the top 4 best algorithms (*4cores-No Sharing*) performs better (309) that the one with 4 copies of the best algorithms (*4cores-PAWS*) (275).

The performance of the portfolios with knowledge sharing is quite good. Overall, it seems that adding a weighting mechanism can often hurt the performance of the underlying aggregation strategy. Among the weighting options, it seems that the Normalized Performance performs better. The best portfolio implements the *Prob* strategy without any weighting (329). This corresponds to a gain of 20 problems against the corresponding *4cores-No Sharing* baseline.

A detailed examination of *4cores-Prob* and *4cores-No Sharing* is presented in Figs. 2 and 3. These Figures show, respectively, a runtime and a best configuration cost comparison. In both figures, points below (resp. above) the diagonal line indicate that *4cores-Prob* performs better (resp. worse) than *4cores-No Sharing*. In the runtime comparison, we observe that easy instances are correlated

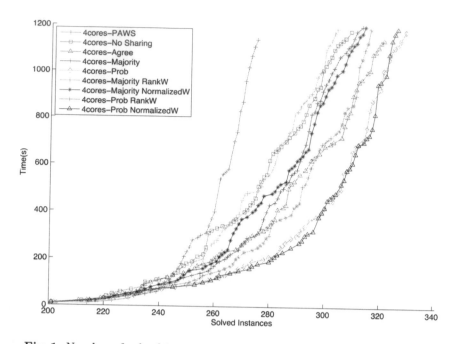

Fig. 1. Number of solved instances using 4 cores in a given amount of time

as they require few steps to be solved, and for the remaining set of instances *4cores-Prob* usually exhibits a better performance. On the other hand, the second figure shows that when the instances are not solved, the median cost of the best configuration (number of unsatisfied clauses) found by *4cores-Prob* is usually better than for *4cores-No Sharing*. Notice that some points are overlapped because the two strategies reported the same cost.

All the experiments using 4 cores are summarized in Table 1, reporting for each portfolio the number of solved instances (#solved), the median time across all instances (median time), the *Penalized Average Runtime* (PAR) and the total number of instances that timed out in all the 10 runs (never solved). These results confirm that sharing best known configurations outperforms independent searches, for instance *4cores-Prob* and *4cores-Prob NormalizedW* solved respectively 20 and 17 more instances than *4cores-No Sharing* and all the cooperative strategies (except *4cores-Majority RankW*) exhibit better PAR. Interestingly, 4cores-PAWS exhibited the best median runtime overall the experiments with 4 cores, this fact suggests that PAWS by itself is able to quickly solve an important number of instances. Moreover, only 2 instances timeout in all the 10 runs for *4cores-Agree* and *4cores-Prob NormalizedW* against 7 for *4cores-No Sharing*. Notice that this Table also includes *1core-PAWS*, the best sequential local search on this set of problems. The PAR score for *1core-PAWS* is lower than the other values of the table because this portfolio uses only 1 algorithm, therefore the timeout is only 300 seconds, while 4 cores portfolios use a timeout of 1200 seconds.

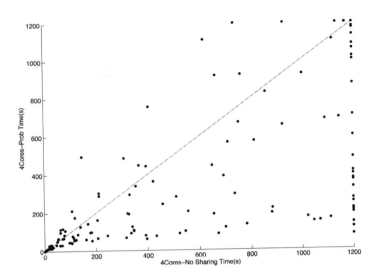

Fig. 2. Runtime comparison, each point indicates the runtime to solve a given instance using *4cores-Prob* (y-axis) and *4cores-No Sharing* (x-axis)

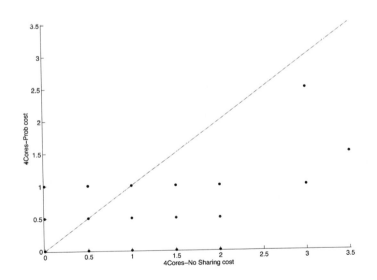

Fig. 3. Best configuration cost comparison on unsolved instances. Each point indicates the best configuration (median) cost of a given instance using *4cores-Prob* (y-axis) and *4cores-No Sharing* (x-axis)

Table 1. Overall evaluation using 4 cores

Strategy	#solved	median time	PAR	never solved
1core-PAWS	249	1.76	911.17	71
4cores-PAWS	275	**1.63**	2915.19	61
4cores-No Sharing	309	2.19	1901.00	7
4cores-Agree	321	2.54	1431.33	**2**
4cores-Majority	313	2.53	1724.94	11
4cores-Prob	**329**	2.51	**1257.93**	4
4cores-Majority RankW	304	2.47	1930.61	11
4cores-Majority NormalizedW	314	2.48	1807.42	9
4cores-Prob RankW	316	2.53	1621.33	7
4cores-Prob NormalizedW	326	2.50	1261.82	2

5.3   Practical Performances with 8 Cores

We now move on to portfolios with 8 cores. The results of these experiments are depicted in Fig. 4 indicating the total number of solved instances within a given amount of time. As in previous experiments, we report the results of baseline portfolios *8cores-No Sharing* and *8cores-PAWS*, and in this case we focus the experiments on *Prob* and *Prob NormalizedW* (the best two strategies using 4 cores). We can observe that the cooperative portfolios largely outperform the non-cooperative ones.

Table 2 summarizes these results, and once again it includes the best individual algorithm running in a single core. We can remark that *8cores-Prob* and *8cores-Prob NormalizedW* solve respectively 24 and 16 more instances than *8cores-No Sharing*. Furthermore, it shows that knowledge sharing portfolios are faster than individual searches, with a PAR of 3743.63 seconds for *8cores-No Sharing* against respectively 2247.97 for *8cores-Prob* and 2295.99 for *8cores-Prob NormalizedW* . Finally, it is also important to note that only 1 instance timed out in all the 10 runs for *8cores-Prob NormalizedW* against 8 for *8cores-No Sharing*.

Extensive experimental results presented in this paper show that *Prob* (4 and 8 cores) exhibited the overall best performance. We attribute this to the fact that the probability component of this method balances the exploitation of best solutions found so far with the exploration of other values for the variables, helping in this way, to diversify the new starting configuration.

5.4   Hardware Impact

In this section, we wanted to assess the inherent slowdown caused by increased cache, and bus contingency when more processing cores are used at the same time. To this end we decided to run our PAWS baseline portfolio where each independent algorithm uses the same random seed on respectively 1, 4 and 8 cores. Since all the algorithms are executing the same search, this experiment

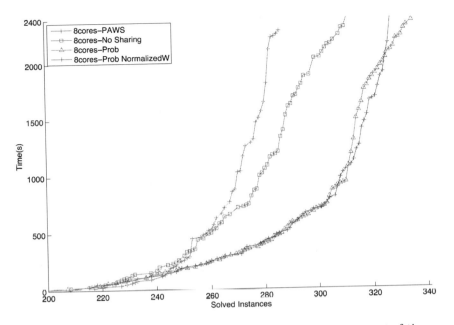

Fig. 4. Number of solved instances using 8 cores in a given amount of time

Table 2. Overall evaluation using 8 cores

Strategy	#solved	median time	PAR	never solved
1core-PAWS	249	1.76	911.17	71
8cores-PAWS	286	**2.00**	5213.84	56
8cores-No Sharing	311	2.33	3743.63	8
8cores-Prob	**335**	2.45	**2247.97**	2
8cores-Prob NormalizedW	327	2.47	2295.99	1

measures the slowdown caused by hardware limitations. The results are presented in Fig. 5.

The first case executes a single copy of PAWS with a timeout of 300 seconds, the second case executes 4 parallel copies of PAWS with a timeout of 1200 seconds (4 × 300) and the third case executes 8 parallel copies of PAWS with a timeout of 2400 seconds (8 × 300).

Finally, we estimate the runtime of each instance as the median across 10 runs (each time with the same seed) divided by the number of cores. In this figure, it can be observed that the performance overhead is almost not distinguishable between 1 and 4 cores (red points). However, the overhead between 1 and 8 cores is important for difficult instances (black points).

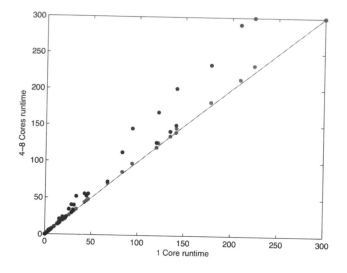

Fig. 5. Runtime comparison using parallel local search portfolios made of respectively 1, 4, and 8 identical copies of PAWS (same random seed). Red points indicate the performance of 4 cores vs 1 core. Black points indicate the performance of 8 cores vs 1 core, points above the blue line indicate that 1 core is faster.

6   Conclusions and Future Work

In this work, our objective was to integrate knowledge sharing strategies in parallel local search for SAT. We were motivated by the recent developments in parallel DPLL solvers. We decided to restrict the information shared to the best configuration found so far by the algorithms in a portfolio. From that we defined several simple knowledge aggregation strategies along a specific lazy restart policy which creates a new initial configuration when a fix cutoff is meet and when the quality of the shared information has been improved.

Extensive experiments were done on a large number of instances coming from the latest SAT competition. They showed that adding the proposed sharing policies improves the performance of a parallel portfolio, this improvement is exhibited in both number of solved instances and the *Penalized Average Runtime* (PAR). It is also reflected in the best configuration cost of problems which could not be solved within the time limit.

We believe that our work represents a very first step in the incorporation of knowledge sharing strategies in parallel local search for SAT. Further work will investigate the use of additional information to exchange, for instance: tabu-list, the age and score of a variable, information on local minima, etc. It should also investigate the best way to integrate this extra knowledge in the course of a given algorithm. As said earlier, state-of-the-art local search perform better when they do not restart. Incorporating extra information without forcing the algorithm to restart is likely to be important.

Acknowledgements

We would like to thank Said Jabbour and Ibrahim Abdoulahi for helpful discussions about parallel SAT solving and the anonymous reviewers for their comments which helped to improve this paper.

References

1. Hamadi, Y., Jabbour, S., Sais, L.: ManySAT: A Parallel SAT Solver. Journal on Satisfiability, Boolean Modeling and Computation, JSAT 6, 245–262 (2009)
2. Moskewicz, M.W., Madigan, C.F., Zhao, Y., Zhang, L., Malik, S.: Chaff: Engineering an Efficient SAT Solver. In: Proceedings of the 38th Design Automation Conference (DAC 2001), pp. 530–535 (2001)
3. Bordeaux, L., Hamadi, Y., Zhang, L.: Propositional Satisfiability and Constraint Programming: A Comparative Survey. ACM Comput. Surv. 38(4) (2006)
4. Hoos, H.H., Stützle, T.: Local Search Algorithms for SAT: An Empirical Evaluation. J. Autom. Reasoning 24(4), 421–481 (2000)
5. Selman, B., Levesque, H.J., Mitchell, D.G.: A New Method for Solving Hard Satisfiability Problems. In: AAAI (ed.), pp. 440–446 (1992)
6. Selman, B., Kautz, H.A., Cohen, B.: Noise Strategies for Improving Local Search. In: AAAI, pp. 337–343 (1994)
7. McAllester, D.A., Selman, B., Kautz, H.A.: Evidence for Invariants in Local Search. In: AAAI/IAAI, pp. 321–326 (1997)
8. Li, C.M., Huang, W.Q.: Diversification and Determinism in Local Search for Satisfiability. In: Bacchus, F., Walsh, T. (eds.) SAT 2005. LNCS, vol. 3569, pp. 158–172. Springer, Heidelberg (2005)
9. Hutter, F., Tompkins, D.A.D., Hoos, H.H.: Scaling and Probabilistic Smoothing: Efficient Dynamic Local Search for SAT. In: Van Hentenryck, P. (ed.) CP 2002. LNCS, vol. 2470, pp. 233–248. Springer, Heidelberg (2002)
10. Thornton, J., Pham, D.N., Bain, S., Ferreira Jr, V.: Additive versus Multiplicative Clause Weighting for SAT. In: McGuinness, D.L., Ferguson, G. (eds.) AAAI, pp. 191–196. AAAI Press/The MIT Press, San Jose, California, USA (2004)
11. Hoos, H.H.: An Adaptive Noise Mechanism for WalkSAT. In: AAAI/IAAI, pp. 655–660 (2002)
12. Li, C.M., Wei, W., Zhang, H.: Combining Adaptive Noise and Look-Ahead in Local Search. In: Marques-Silva, J., Sakallah, K.A. (eds.) SAT 2007. LNCS, vol. 4501, pp. 121–133. Springer, Heidelberg (2007)
13. Chrabakh, W., Wolski, R.: GridSAT: A System for Solving Satisfiability Problems Using a Computational Grid. Parallel Computing 32(9), 660–687 (2006)
14. Zhang, W., Huang, Z., Zhang, J.: Parallel Execution of Stochastic Search Procedures on Reduced SAT Instances. In: Ishizuka, M., Sattar, A. (eds.) PRICAI 2002. LNCS (LNAI), vol. 2417, pp. 108–117. Springer, Heidelberg (2002)
15. Roli, A.: Criticality and Parallelism in Structured SAT Instances. In: Van Hentenryck, P. (ed.) CP 2002. LNCS, vol. 2470, pp. 714–719. Springer, Heidelberg (2002)
16. Roli, A., Blesa, M.J., Blum, C.: Random Walk and Parallelism in Local Search. In: Metaheuristic International Conference (MIC 2005), Vienna, Austria (2005)
17. Pham, D.N., Gretton, C.: gNovelty+ (v.2). In: Solver Description, SAT Competition 2009 (2009)

18. Kroc, L., Sabharwal, A., Gomes, C.P., Selman, B.: Integrating Systematic and Local Search Paradigms: A New Strategy for MaxSAT. In: Boutilier, C. (ed.) IJCAI, Pasadena, California, pp. 544–551 (July 2009)
19. Hogg, T., Williams, C.P.: Solving the Really Hard Problems with Cooperative Search. In: AAAI, pp. 231–236 (1993)
20. Silva, D.L., Burke, E.K.: Asynchronous Cooperative Local Search for the Office-Space-Allocation Problem. INFORMS Journal on Computing 19(4), 575–587 (2007)
21. Selman, B., Kautz, H.A.: Domain-Independent Extensions to GSAT: Solving Large Structured Satisfiability Problems. In: IJCAI, pp. 290–295 (1993)
22. Tompkins, D.A.D., Hoos, H.H.: UBCSAT: An Implementation and Experimentation Environment for SLS Algorithms for SAT and MAX-SAT. In: Hoos, H.H., Mitchell, D.G. (eds.) SAT 2004. LNCS, vol. 3542, pp. 306–320. Springer, Heidelberg (2005)
23. Hutter, F., Hoos, H.H., Leyton-Brown, K.: Tradeoffs in the Empirical Evaluation of Competing Algorithm Designs. Annals of Mathematics and Artificial Intelligence (AMAI), Special Issue on Learning and Intelligent Optimization (2010)

Variable Neighborhood Search for the Time-Dependent Vehicle Routing Problem with Soft Time Windows

Stefanie Kritzinger[1], Fabien Tricoire[1], Karl F. Doerner[1,2],
and Richard F. Hartl[1]

[1] Department of Business Adminstration, University of Vienna,
Bruenner Strasse 72, 1210 Vienna, Austria
{stefanie.kritzinger,fabien.tricoire,karl.doerner,
richard.hartl}@univie.ac.at
http://www.univie.ac.at/bwl/prod
[2] Department of Production and Logistics, Johannes Kepler University Linz,
Altenberger Strasse 69, 4040 Linz, Austria

Abstract. In this paper we present a variable neighborhood search for time-dependent vehicle routing problems with time windows. Unlike the well-studied routing problems with constant travel times, in the time-dependent case the travel time depends on the time of the day. This assumption approaches reality, in particular for urban areas where travel times typically vary during the day, e.g., because of traffic congestion due to rush hours. An experimental evaluation for the vehicle routing problem with soft time windows with and without time dependent travel times is performed and it is shown that taking time-dependent travel times into account provides substantial improvements of the considered objective function.

Keywords: Vehicle routing problem, time-dependent travel times, time windows, variable neighborhood search.

1 Introduction

Most of the vehicle routing problems (VRPs) reported in the literature assume constant travel times although the travel time between two locations does not only depend on the traveled distance. In fact, it depends on many other factors including the time when the travel starts. For simulations of situations close to real-world conditions, different factors, e.g., traffic congestion due to rush hours, are not negligible because of their influence on travel speeds and travel times. Routing problems considering time-dependent travel times are called time-dependent vehicle routing problems (TD-VRPs).

The TD-VRP basically consists of finding a set of routes of minimized travel time made by a fleet of vehicles starting from a specified depot, visiting a set of geographically distributed customers and finishing the route at the depot in consideration of capacity and tour length restrictions. The problem considered here

C.A. Coello Coello (Ed.): LION 5, LNCS 6683, pp. 61–75, 2011.

is motivated by a logistic service in an urban area, where goods are distributed to different customer locations. Each route must start and end within the time window assigned to the central depot, as opposed to the customers to which soft time windows are associated. A soft time window is characterized as: if the vehicle arrives too early, it has to wait to start its service; if the vehicle arrives too late, the difference between the arrival time and the latest service time, the so called tardiness, is penalized. The cost to be minimized is a weighted sum of the total travel time over all routes, plus the total tardiness over all customers. This VRP is denoted as time-dependent vehicle routing problem with soft time windows (TD-VRPSTW).

As an illustrative example, let us consider Vienna, the capital of Austria with about 1.7 million inhabitants and an area of approximately 415 km^2. As in each other city, the traffic characteristics vary during the day. Let us compare Fig. 1 and Fig. 2. Both show a speed map of Vienna but at different times of the day. On the left hand side a speed map at nine o'clock in the evening is displayed. Normally it is a time where the traffic flows with no obstruction. The light grey area in the middle shows the city center of Vienna, where vehicles are not allowed to go faster than 30 km/h. In the dark grey area, the outer districts of Vienna, the maximum speed is 60 km/h and in the black squares (250 meter × 250 meter), mostly the city highways, a speed higher than 60 km/h is allowed. On

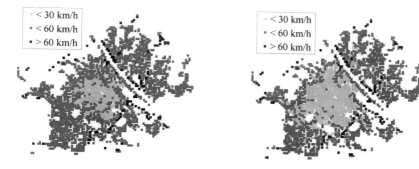

Fig. 1. Speed map of Vienna at 9pm   **Fig. 2.** Speed map of Vienna at 8am

the right hand side a speed map at 8 o'clock in the morning during the rush hour is presented. It can be seen that the light grey area expands: heavier traffic in the city center implies slower speed. This fact leads us to a longer travel time for the same distance. Therefore it makes sense to involve travel times dependent on the time of the day in the delivery plan creation to imitate conditions close to real-world.

The first approach to consider varying travel times for the VRP as well as the traveling sales man problem was performed by Malandraki and Daskin [10]. They describe each arc with a step function distribution of the travel time and propose a mixed integer programming approach and a nearest neighbor heuristic for optimization.

As a feature of the real-world problem, Ichoua et al. [9] introduce the First-In-First-Out (FIFO) property. That is, if two vehicles leave from the same location for the same destination traveling on the same path, the one that leaves first will always arrive earlier, no matter how speed changes on the arcs during the travel. In order to enforce FIFO, they use step functions for the travel speed within a tabu search heuristic for the time-dependent vehicle routing problem with time windows (TD-VRPTW). This model is also formulated in a dynamic setting, where not all service requests are known before the start of the optimization.

As Fleischmann et al. [4] criticize, the disadvantage of models with varying speeds and constant distances is that a potential change of the shortest path is usually not considered. With time varying speeds, it might happen that taking other links requires less travel time.

Besides the minimization of the total travel time, Donati et al. [3] also consider the minimization of the number of routes. Here, a multi ant colony system is modeled to solve the classic VRP and its extension for the time-dependency. They show that when dealing with, e.g., hard delivery time windows for customers, the known solutions for the classic case become infeasible. If there are no hard time constraints, the classic solutions become suboptimal.

An iterated local search algorithm is developed by Hashimoto et al. [8] for a vehicle routing problem with time windows (VRPTW) with time-dependent traveling times and costs. As local search in the neighborhood, they consider slight modifications of the standard neighborhood called 2-opt*, cross exchange and Or-opt. Additionally, they apply a filtering method that restricts the size of the neighborhood to avoid many solutions having no expectation of improvement. Hashimoto et al. [8] show that their algorithm is highly efficient for the given data as well as for artificially generated instances.

Soler et al. [14] transform theoretically the TD-VRPTW into an asymmetric capacitated vehicle routing problem. The time and the cost of traversing an arc depend on the period of time at which the traversing starts.

This paper is organized as follows: in Section 2 we proceed with the problem description of the VRP with soft time windows (VRPSTW) in detail and its extension to the time-dependent case. Section 3 explains precisely the variable neighborhood search (VNS) we use for solving the VRPSTW and the TD-VRPSTW. Section 4 is dedicated to computational results and application to a simulated real-world setting and Section 5 concludes.

2   Problem Description

The VRPSTW is a well-known generalization of the VRP and can be stated as follows. Let us assume that an undirected complete graph $G = (V, E)$ is given where $V = \{0, 1, \ldots, n\}$ is the set of $n+1$ vertices and E is the set of edges. Vertex 0 is the depot and the vertex set $V' = V \setminus \{0\}$ represents the n customers. Each customer i has a demand d_i. The nonnegative travel cost for each edge $(i, j) \in E$ is denoted by d_{ij}. Travel costs satisfy the triangular inequality

$$d_{ij} + d_{jk} \geq d_{ik} \tag{1}$$

for all nodes $\{i, j, k\} \in V$. A set of m identical vehicles of capacity Q are available at the depot to supply the demand of all customers. A vehicle route starts from the depot, visits a number of customers such that the total demand of the visited customers does not exceed the vehicle capacity, and returns to the depot within the time window assigned to the depot. Each customer $i \in V \setminus \{0\}$ should be visited in a given time window $[e_i, l_i]$, where e_i is the earliest start of service and l_i is the latest start of service. The customers have soft time windows, i.e., a vehicle can arrive before the earliest service time e_i and after the latest service time l_i. If the vehicle arrives too early, it has to wait to start its service; if the vehicle arrives too late, the tardiness is penalized in the objective function. The objective value of solution x is:

$$obj(x) = c(x) + \gamma c_\gamma(x), \tag{2}$$

where $c(x)$ is the total travel time over all routes and $c_\gamma(x)$ is the total tardiness over all customers $i \in V \setminus \{0\}$ multiplied by the tardiness parameter γ. The tardiness $c_\gamma(x_i)$ of each customer i with arrival time t_i is calculated as $(t_i - l_i)^+ = \max\{0, t_i - l_i\}$.

The TD-VRPSTW is a generalization of the VRPSTW with time-dependency of travel times on the time of the day. The cost of traversing an arc depends on the period of time at which we start to traverse it. It allows stronger approximations of the real-world conditions where travel times are subject to more variations over time, i.e., traffic congestions. For the realization of time-dependent travel times, the time horizon of the considered routing problem is divided into p time intervals $T_1, \ldots, T_p$ with different travel speeds.

Besides thinking of the time when the traveling starts it makes sense that the speed has to be adapted when another time period is entered. An efficient and simple way of calculating the travel times is introduced by Ichoua et al. [9]. Instead of the assumption of a constant travel speed over the entire length of an arc, the speed changes when a vehicle crosses the boundary between two consecutive time periods. In case of time-dependency, the FIFO property has to be fulfilled. That is, leaving a node earlier guarantees that one will arrive earlier

1. set t to t_0
   set d to d_{ij}
   set t' to $t + d/v_k$

2. while $(t' > \overline{t_k})$ do
   (a) set d to $d - v_k(\overline{t_k} - t)$
   (b) set t to $\overline{t_k}$
   (c) set t' to $t + d/v_{k+1}$
   (d) set k to $k + 1$

3. return $(t' - t_0)$

Algorithm 1. Calculation of time-dependent travel times generalized in [9]

at destination. Hence, there is no useless waiting time. The FIFO property is assured by using a step function for the speed distribution, from which the travel times are then calculated, instead of a step function for the travel time distribution. The calculation of time-dependent travel times in Algorithm 1 is a simplified version of the calculation of time-dependent travel times done in [9].

Suppose that the vehicle leaves node i at time t_0 in time period $T_k =]\underline{t_k}, \overline{t_k}]$. In Algorithm 1, it is assumed that d_{ij} is the distance between i and j, and v_k is the travel speed associated with time period T_k. Also, t denotes the current time and t' denotes the arrival time. After the initialization of the current time t and the distance d, the arrival time t' at customer j is set to $t + d/v_k$, where the quotient d/v_k denotes the travel time from i to j within the time period T_k. If the boundary to the next period $\overline{t_k}$ is crossed, a recalculation of the travel time between $\overline{t_k}$ and j has to be made. The total travel time from i to j is the difference of t_0 and t'.

The example in Fig. 3 shows that the travel speed changes when a boundary between two consecutive time periods is crossed. The kilometers go on the x-axis, the time goes on the y-axis. Each time interval is a time period. If the travel speed is not adjusted when another time period is entered, it is possible that a vehicle can wait if its speed will increase in the next time period, although the vehicle could have used the current speed to get closer to its destination until the time of speed changes. In Fig. 4 vehicle v_2 is allowed to wait at node i before starting the traveling. If the speed is not adopted when another time period is entered, vehicle v_2 arrives earlier at node j than vehicle v_1, which starts traveling immediately.

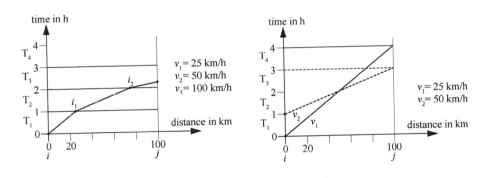

Fig. 3. Changing travel speed when a boundary between two time periods is crossed

Fig. 4. Waiting is allowed to increase the speed afterwards

3   Solution Method

In this chapter we describe the VNS algorithm we use for the solution process. VNS was first proposed in 1997 by Mladenović and Hansen [11]. In the last years this metaheuristic has gained popularity for solving combinatorial and global optimization problems and it has a widely spread field of applications, e.g., routing and scheduling problems, industrial applications or design problems in communication.

The basic scheme of the VNS is *initialization, shaking, local search* and *acceptance decision*. For initialization a finite set of pre-selected neighborhood structures N_κ, $\kappa = 1, \ldots, \kappa_{max}$, is defined, where $N_{\kappa+1}$ is typically larger than N_κ. Further an initial solution is obtained randomly or heuristically. As stopping condition, e.g., a limit on the CPU time, a limit on the number of iterations or a limit on the number of iterations between two improvements is configured. The initialization is followed by a so-called shaking step, that randomly selects a solution from the first neighborhood of the incumbent solution. A local search procedure starting from this created solution is performed to obtain a local optimum. Within the acceptance decision step it has to be decided to move or not to the new local optimal solution. In this simplified case, if the current solution is better than the incumbent solution, the incumbent solution is replaced by the current solution and the search continues with the shaking step within the first neighborhood, otherwise the search proceeds with the next neighborhood. A more complicated acceptance decision is used in Section 3.4. For a more precise description on VNS see Mladenović and Hansen [11] and Hansen and Mladenović [5,6,7].

In the following subsections we describe the different components of the VNS implemented for the TD-VRPSTW.

3.1   Initial Solution

The initial solution is obtained by a simple construction algorithm. After the customers have been ordered with respect to the center of their time window $\frac{1}{2}(e_i + l_i)$, the routes are constructed by sequentially adding the customers to the route with minimal costs. As we start with a fixed fleet size at the beginning, the obtained initial solution allows infeasibility in capacity and route length.

3.2   Shaking

The building blocks in the VNS is the construction of the set of neighborhoods used for shaking. The neighborhood operator is characterized by the ability of perturbing the incumbent solution while important parts of the incumbent solution are kept unchanged.

A popular and effective neighborhood for VRPs is based on the cross-exchange operator introduced by Taillard et al. [16] (see Fig. 5). The guiding idea of this exchange is to take two segments of different routes and exchange them. For sequence inversion, the icross-exchange operator proposed in Bräysy [1] is used (see Fig.6).

We use maximum sequence length as a parameter, in order to produce different nested neighborhoods based on the same structure. Let C_k denote the number of customers assigned to route k, then the maximum sequence length for each neighborhood κ is $\min(\kappa, C_k)$. Note that the maximum sequence length cannot exceed C_k for any given route k.

For the shaking step, we randomly choose with the same probability between the four possible variants of reinserting the segments directly or inverted as illustrated in Fig. 5, 6 and 7.

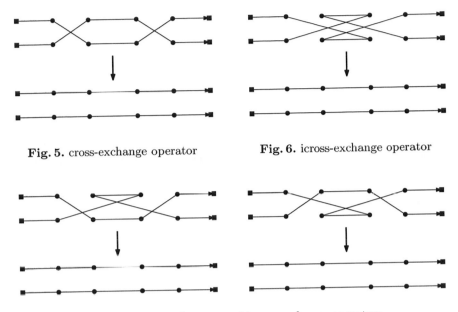

Fig. 5. cross-exchange operator   **Fig. 6.** icross-exchange operator

Fig. 7. Mixture of cross- and icross-exchange operators

3.3   Local Search

The solution obtained through shaking has to undergo a local search procedure to come up with a local optimum afterwards. While the shaking steps focus on exchanging customers between routes, the local search only searches for improvements within the routes that were modified in the shaking step. We apply one of the three different local search methods, 2-opt, 3-opt* and Or-opt. For detailed description see the survey of Bräysy and Gendreau [2].

In general a 2-opt heuristic consistently inverts sequences within one route. The 3-opt* heuristic moves subsequences without inversion to other positions within one route. An Or-opt local search systematically moves subsequences up to a sequence length of three with and without inversion to other positions within one route.

For each iteration we use one of the mentioned local search procedures as it is explained in Table 1. Each local search restarts immediately after an improving move was found.

Table 1. Application of local search procedures

	local search
(iteration mod 3) = 0	2-opt
(iteration mod 3) = 1	3-opt*
(iteration mod 3) = 2	Or-opt

3.4   Acceptance Decision

After the shaking and the local search procedures have been performed, the obtained solution has to be compared to the incumbent solution to decide to accept it or not. The acceptance criterion in the basic VNS is to accept only improvements. However, that way the search can easily get stuck in a local optimum. Thus, in many cases it has been shown to be essential to also have a strategy of accepting non-improving solutions under certain conditions. We implement an approach to accept non-improving solutions based on threshold accepting (TA) used by Polacek et al. in [12]. A solution yielding an improvement is always accepted. Moreover ascending moves are accepted after a certain number of iterations counted from the last accepted move, but only if the cost increase is below a certain threshold. This threshold is given by θ percent of the incumbent solution.

An important characteristic of our VNS is the ability to deal with infeasible solutions. Infeasibility occurs if the total capacity or the tour duration exceed a specific limit or if the time windows of the customers are violated. Therefore the solution has to be evaluated for acceptance. The evaluation function is specified in the following way:

$$f(x) = obj(x) + \alpha c_\alpha(x) + \beta c_\beta(x). \tag{3}$$

The evaluation function $f(x)$ for the solution x sums up the objective value $obj(x)$ (see (2)) and the penalty terms which consist of the violation value of the capacity $c_\alpha(x)$ and violation value of the route length $c_\beta(x)$ multiplied by the corresponding penalty parameters α and β. In order to consider hard time windows, one must give a large value to γ.

4   Computational Results

For the computational experiments we use the Solomon's 100-customer Euclidean problems [15]. In these problems, customers are generated within a $[0, 100]^2$ square. The customer locations of the instance class C1 and C2 are clustered in groups, the customer locations of the instance class R1 and R2 are randomly generated and the customer locations of the instance class RC1 and RC2 are a mix of clustered in groups and randomly generated customer locations. In the problem sets C1, R1 and RC1, only a few customers can be serviced on each route due to a small time window at the depot, contrary to the problem sets C2, R2 and RC2, where many customers can be serviced by the same vehicle due to a long scheduling horizon.

For evaluating the solution we set our penalty parameters $\alpha = \beta = 100$ following Polacek et al. [12]. For dealing with hard time windows at the customer locations we set the tardiness parameter γ also to 100. We show that in our case a feasible final solution is guaranteed. We stop the solution process either after 10^6 iterations or after one hour. For TA we allow a degradation of the evaluation function of 5% after 8000 steps without improvement. Our construction heuristic requires a fixed fleet size, so we start in these experiments with 20 percent more vehicles than the number of vehicles in the optimal solution. During the solution process in the shaking step there is always one empty route available for no restriction on the fleet size.

In Tables 2 and 3 our best solution and the average solution over 10 runs are presented and compared with the optimal solution, if available.

Table 2. Comparison to the optimal solution on problems of instance class 1

Problem	Our best solution	Our avg. solution over 10 runs	Optimal solution	Best gap	Avg. gap
C101	10/828.94	10/828.94	10/827.3	0.20%	0.20%
C102	10/828.94	10/828.94	10/827.3	0.20%	0.20%
C103	10/828.07	10/828.07	10/826.3	0.21%	0.21%
C104	10/824.78	10/824.93	10/822.9	0.23%	0.25%
C105	10/828.94	10/828.94	10/827.3	0.20%	0.20%
C106	10/828.94	10/828.94	10/827.3	0.20%	0.20%
C107	10/828.94	10/828.94	10/827.3	0.20%	0.20%
C108	10/828.94	10/828.94	10/827.3	0.20%	0.20%
C109	10/828.94	10/828.94	10/827.3	0.20%	0.20%
Average	10/828.38	10/828.39	10/826.7	0.20%	0.21%
R101	20/1642.88	20/1647.46	20/1637.70	0.32%	0.60%
R102	18/1472.81	18/1474.53	18/1466.60	0.42%	0.54%
R103	14/1213.62	14.1/1217.08	14/1208.70	0.41%	0.69%
R104	11/976.61	11/983.43	11/971.50	0.53%	1.23%
R105	15/1360.78	15.4/1367.36	15/1355.30	0.40%	0.89%
R106	13/1241.35	13/1246.17	13/1234.60	0.55%	0.94%
R107	11/1076.23	11/1079.85	11/1064.60	1.09%	1.43%
R108	10/943.24	10.1/952.11	10/932.10	1.19%	2.15%
R109	13/1151.84	12.8/1153.90	13/1146.90	0.43%	0.61%
R110	12/1080.20	12/1085.01	12/1068.00	1.14%	1.59%
R111	12/1053.50	12/1058.89	12/1048.70	0.46%	0.97%
R112	10/959.08	10/962.50	10/948.60	1.10%	1.47%
Average	13.25/1181.01	13.28/1185.69	13.25/1173.61	0.63%	1.03%
RC101	16/1639.39	16.2/1650.02	15/1619.80	1.21%	1.87%
RC102	14/1461.49	14.3/1478.54	14/1457.40	0.28%	1.45%
RC103	11/1272.49	11.9/1280.48	11/1258.00	1.15%	1.79%
RC104	10/1136.67	10.2/1143.76	10/1132.3	0.39%	1.01%
RC105	15/1523.19	15.3/1537.06	15/1513.70	0.63%	1.54%
RC106	13/1379.99	12.9/1387.97	13/1372.7	0.53%	1.11%
RC107	12/1212.83	12/1216.49	12/1207.80	0.42%	0.72%
RC108	11/1119.84	11/1135.65	11/1114.20	0.51%	1.93%
Average	12.75/1342.86	12.98/1353.75	12.63/1334.49	0.63%	1.44%

The entry of the solution is of the form *number of vehicles/objective value*.

Table 3. Comparison to the optimal solution on problems of instance class 2

Problem	Our best solution	Our avg. solution over 10 runs	Optimal solution	Best gap	Avg. gap
C201	3/591.56	3/591.56	3/589.1	0.42%	0.42%
C202	3/591.56	3/595.33	3/589.1	0.42%	1.06%
C203	3/591.17	3/591.17	3/588.7	0.42%	0.42%
C204	3/590.60	3/596.08	3/588.1	0.42%	1.36%
C205	3/588.88	3/588.88	3/586.4	0.42%	0.42%
C206	3/588.49	3/588.49	3/586.0	0.43%	0.43%
C207	3/588.29	3/588.29	3/585.8	0.42%	0.42%
C208	3/588.32	3/588.32	3/585.8	0.43%	0.43%
Average	3/589.86	3/591.02	3/587.38	0.42%	0.62%
R201	8/1157.54	7.7/1158.25	8/1143.20	1.25%	1.32%
R202	7/1039.09	6.2/1043.71	8/1029.6	0.92%	1.37%
R203	6/874.87	5.7/883.07	6/870.8	0.47%	1.41%
R204	5/738.61	4.6/742.16	5/731.3	1.00%	1.48%
R205	5/959.19	5.4/967.09	5/949.8	0.99%	1.82%
R206	5/884.79	5.3/893.62	5/875.9	1.02%	2.01%
R207	5/818.98	4.2/824.77	3/794.0	3.15%	3.88%
R208	4/710.83	3.3/722.27	3/701.2*	1.37%	3.00%
R209	5/860.11	5/869.05	5/854.8	0.62%	1.67%
R210	5/914.62	5.5/919.74	6/900.5	1.57%	2.14%
R211	4/756.51	4.4/762.51	4/746.7	1.31%	2.12%
Average	5.36/883.19	5.73/889.66	5.27/872.53	1.22%	1.96%
RC201	7/1280.71	7.6/1288.28	9/1261.80	1.50%	2.10%
RC202	8/1099.54	7.5/1106.96	8/1092.30	0.66%	1.34%
RC203	6/937.68	5.4/943.91	5/923.7	1.51%	2.19%
RC204	4/790.68	4/798.86	4/783.5	0.92%	1.96%
RC205	7/1158.67	7/1162.55	7/1154.0	0.40%	0.74%
RC206	6/1057.83	6.2/1073.58	7/1051.1	0.64%	2.14%
RC207	6/968.00	5.8/976.50	6/962.9	0.53%	1.41%
RC208	5/784.03	4.6/786.58	4/776.5	0.97%	1.30%
Average	6.13/1009.64	6.01/1017.15	6.25/1000.73	0.89%	1.64%

* No proven optimal solution.
The entry of the solution is of the form *number of vehicles/objective value*.

Although we often start with an infeasible initial solution, we always end up with a feasible one. Mostly, the infeasibility is due to time window violations at the customer locations. The gaps between the our best solutions and the optimal one is smaller than one percent in 73% of the instance problems. For the instance class C, we gain good approximation of the optimal solution with an average gap of 0.31%. Apart from one problem of instance class 1, problem RC101, we end up at least once with the number of routes of the optimal solution without using any vehicle minimizing operation. For problems R202, R210, RC201 and R206, we receive solutions with less vehicles than used in the optimal solution.

As impulse to create a real-world urban setting we make use of a statistic of the average speed of the city highways in Vienna as shown in Fig. 8. For a more detailed data analysis see [13]. We delimit our observation to the time horizon of a standard working day from 7am to 7pm. One can see that between the morning and evening rush hours the speed decreases up to 8km/h compared to the speed in the middle of the day of about 90km/h. Of course, it influences the solution quality enormously if the difference of the speeds between rush hours

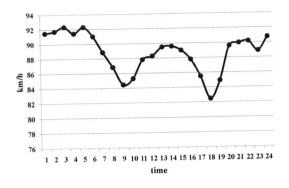

Fig. 8. Average speed of Vienna's highways

and the middle of the day is huge. Therefore we construct different scenarios of time-dependency to show the consequences in the solution quality. Note that the behavior of the speed in the inner-city varies to a greater extent than on the city highways.

Hence, we transfer the Solomon test instances to a 12 hour working day with 3 periods, the morning rush hour from 7am to 10am, the middle of the day from 10am to 4pm and the evening rush hour from 4pm to 7pm. Further, we create four different types of scenarios as proposed in Table 4 by considering an average speed of 1 in each row.

Table 4. Travel speed in Scenarios 1 - 4

Scenario	7am - 10am	10am - 4pm	4pm - 7pm
1	1	1	1
2	0.8	1.2	0.8
3	0.6	1.4	0.6
4	0.4	1.6	0.4

Scenario 1 is the benchmark and is assumed to have constant travel speed over the considered time horizon - it is related to the already discussed VRPTW. Scenario 2 has already a weak time-dependency: during the morning and evening period the vehicles travel at a slightly slower speed than during the middle of the day. Scenarios 3 and 4 have an increasing level of time-dependency.

With the defined scenarios we perform two different numerical tests. First, we show that ignoring time-dependent travel times leads to poor performance. To do this, we run VNS with Scenario 1, that means a standard VRPSTW with constant travel times. The whole parameter setting is taken as it stands for the evaluation of the VRPTW, except for the tardiness parameter γ which is set to 1 because we deal with soft time windows. Then, we evaluate the solutions with our time-dependent Scenarios 2, 3 and 4. Clearly, the averages of 10 runs in Table 5, where the instance class 1 is considered, and Table 6, where the instance class 2 is considered, show that the solution quality is worse. Already in Scenario 2, the scenario with the weakest time-dependent travel times, a high rate of infeasible solutions is evaluated. In all cases the high percentage of infeasible solutions is due to route length violations. One can see, the stronger the time-dependent aspect becomes, the more infeasible the solutions are and the higher the total travel time is. This observation is due to the fact that the route length increases,

Table 5. Problems of instance class 1: Solution of Scenario 1 evaluated with time-dependent aspect of Scenarios 2 - 4

Problem	Scenario 1	Scenario 2	Scenario 3	Scenario 4	
	847.13	896.72	1055.10	1515.63	total travel time
	0.07	10.58	52.14	506.76	tardiness
C1	847.20	907.30	1107.24	2022.39	objective value (2)
	0	6.29	21.34	59.36	route length violation
	0 %	50 %	61.11 %	83.33 %	infeasible solutions
	10.03				fleet size
	1175.35	1206.53	1318.29	1576.66	total travel time
	17.56	27.54	110.23	392.74	tardiness
R1	1192.91	1234.07	1428.52	1969.40	objective value (2)
	0	5.48	41.20	181.28	route length violation
	0 %	67.67 %	96.83 %	100 %	infeasible solutions
	13.36				fleet size
	1341.06	1403.82	1558.12	1867.48	total travel time
	19.37	54.02	196.39	536.56	tardiness
RC1	1360.43	1457.84	1754.51	2404.04	objective value (2)
	0	18.02	92.37	305.54	route length violation
	0 %	93.75 %	100 %	100 %	infeasible solutions
	13.04				fleet size

Table 6. Problems of instance class 2: Solution of Scenario 1 evaluated with time-dependent aspect of Scenarios 2 - 4

Problem	Scenario 1	Scenario 2	Scenario 3	Scenario 4	
	734.44	767.81	877.85	1144.87	total travel time
	147.43	158.33	326.90	1126.08	tardiness
C2	881.87	926.14	1204.75	2270.95	objective value (2)
	0	4.46	31.64	172.01	route length violation
	0 %	18.75 %	43.75 %	93.75 %	infeasible solutions
	3.10				fleet size
	918.16	922.34	999.05	1194.96	total travel time
	2.48	3.82	19.65	80.60	tardiness
R2	920.64	926.16	1018.70	1275.56	objective value (2)
	0	0.24	1.69	7.76	route length violation
	0 %	0 %	0 %	9.09 %	infeasible solutions
	5.11				fleet size
	1053.98	1056.36	1143.29	1374.75	total travel time
	3.66	10.70	52.60	231.23	tardiness
RC2	1057.64	1067.06	1195.89	1605.98	objective value (2)
	0	2.70	13.18	57.12	route length violation
	0 %	0 %	37.50 %	62.50 %	infeasible solutions
	5.81				fleet size

if the speed decreases in the morning and evening rush hour. This characteristic does not influence route length so much, because if a vehicle arrives too early at the customer, it has to wait to start its service until the earliest service time.

Second, we show that taking time-dependency into account improves the evaluation function significantly. Hence, we rerun VNS with the time-dependent aspect for each scenario again 10 times. Compared to the solutions before we observe substantial improvements of our solutions. In Tables 7, 8 and 9, the results obtained in the first numerical analysis without time-dependent consideration are compared to the results obtained in the second numerical analysis with time-dependent consideration. In most of the instance classes, considering time-dependency improves all values well. For instance class C, the tardiness completely disappears for Scenario 2 or decreases enormously for Scenarios 3 and 4. For the instance classes R and RC, the tardiness is improved at least by 70% for all scenarios. If there are infeasible solutions, the infeasibility, again in the route length, is small. For the instance class C1, there is no more infeasibility in route length for all scenarios. Route length violation can still be found in R1 and RC1. This can be avoided by increasing the route length penalty β.

A result of the numerical analysis without time-dependent consideration is that the route length violation for the problems of instance class 1 and instance class C2 is very high. To avoid this, more vehicles are provided. Therefore, in the numerical analysis with time-dependent considerations, the number of vehicles increases for the problems of the mentioned instance classes. Conversely, instance classes R2 and RC2 show a decreasing fleet size for analysis with time-dependent considerations. This observation is the outcome of the partial randomly distributed customers with wide spread time windows.

Knowing the travel time in advance, it makes an approximation of the real world conditions more realistic.

Table 7. Comparison of solutions of Scenario 2

Problem	without TD	with TD	Problem	without TD	with TD	
C1	896.72	916.46	C2	767.81	679.05	total travel time
	10.58	0		158.33	0	tardiness
	907.30	916.46		926.14	679.05	objective value (2)
	6.29	0		4.46	0	route length violation
	50 %	0 %		18.75 %	0 %	infeasible solutions
	10.03	10.79		3.10	3.70	fleet size
R1	1206.53	1199.98	R2	922.34	900.27	total travel time
	27.54	17.70		3.82	2.10	tardiness
	1234.07	1217.68		926.16	902.37	objective value (2)
	5.48	0		0.24	0	route length violation
	67.67 %	0 %		0 %	0 %	infeasible solutions
	13.36	13.18		5.11	5.04	fleet size
RC1	1403.82	1411.01	RC2	1056.36	1029.68	total travel time
	54.02	13.93		10.70	3.37	tardiness
	1457.84	1424.94		1067.06	1033.05	objective value (2)
	18.02	0.12		2.70	0	route length violation
	93.75 %	12.50 %		0 %	0 %	infeasible solutions
	13.04	13.34		5.81	5.63	fleet size

Table 8. Comparison of solutions of Scenario 3

Problem	without TD	with TD	Problem	without TD	with TD	
	1055.10	1073.74		877.85	771.56	total travel time
	52.14	11.14		326.90	0	tardiness
C1	1107.24	1084.78	C2	1204.75	771.56	objective value (2)
	21.34	0		31.64	0	route length violation
	61.11 %	0 %		43.75 %	0 %	infeasible solutions
	10.03	11.28		3.10	4.06	fleet size
	1318.29	1266.99		999.05	910.36	total travel time
	110.23	41.62		19.65	3.97	tardiness
R1	1428.52	1308.61	R2	1018.70	914.33	objective value (2)
	41.20	4.76		1.69	0	route length violation
	96.83 %	33.33 %		0 %	0 %	infeasible solutions
	13.36	14.21		5.11	4.79	fleet size
	1558.12	1511.30		1143.29	1074.20	total travel time
	196.39	21.77		52.60	5.64	tardiness
RC1	1754.51	1533.07	RC2	1195.89	1079.85	objective value (2)
	92.37	5.77		13.18	0	route length violation
	100 %	30 %		37.50 %	0 %	infeasible solutions
	13.04	14.75		5.81	5.48	fleet size

Table 9. Comparison of solutions of Scenario 4

Problem	without TD	with TD	Problem	without TD	with TD	
	1515.63	1481.20		1144.87	993.39	total travel time
	506.76	51.98		1126.08	0.01	tardiness
C1	2022.39	1533.18	C2	2270.95	993.40	objective value (2)
	59.36	0		172.01	0	route length violation
	83.33 %	0 %		93.75 %	0 %	infeasible solutions
	10.03	12.53		3.10	4.25	fleet size
	1576.66	1432.97		1194.96	998.60	total travel time
	392.74	117.14		80.60	15.80	tardiness
R1	1969.40	1550.11	R2	1275.56	1014.40	objective value (2)
	181.28	39.09		7.76	0	route length violation
	100 %	66.67 %		9.09 %	0 %	infeasible solutions
	13.36	14.54		5.11	4.79	fleet size
	1867.48	1659.90		1374.75	1269.63	total travel time
	536.56	42.85		231.23	17.47	tardiness
RC1	2404.04	1702.75	RC2	1605.98	1287.10	objective value (2)
	305.54	17.76		57.12	1.24	route length violation
	100 %	50 %		62.50 %	12.50 %	infeasible solutions
	13.04	15.33		5.81	4.95	fleet size

5   Conclusion

This paper presents a VNS for TD-VRPSTW fit to a distribution problem in cities. An experimental evaluation was performed with and without time-dependent travel times. The results show that taking time-dependent travel times into account while optimizing provides substantial improvements in the total travel time. We have shown that in most cases the tardiness can be completely avoided and that the violation of route length is extensively minimized.

Acknowledgments

This work is supported by the Austrian Science Fund (FWF) under grants L510-N13 and L628-N15 (Translational Research Programs). Special thanks go to Verena Schmid, who provided the illustrative example [13].

References

1. Bräysy, O.: A Reactive Variable Neighborhood Search for the Vehicle-Routing Problem with Time Windows. INFORMS Journal on Computing 15(4), 347–368 (2003)
2. Bräysy, O., Gendreau, M.: Vehicle Routing Problem with Time Windows, Part I: Route Construction and Local Search Algorithms. Transportation Science 39, 104–118 (2005)
3. Donati, A.V., Montemanni, R., Casagrande, N., Rizzoli, A.E., Gambardella, L.M.: Time dependent vehicle routing problem with a multi ant colony system. European Journal of Operational Research 185, 1174–1191 (2008)
4. Fleischmann, B., Gietz, M., Gnutzmann, S.: Time-Varying Travel Times in Vehicle Routing. Transportation Science 38, 160–173 (2004)
5. Hansen, P., Mladenović, N.: Variable Neighborhood Search. In: Pardalos, P.M., Resende, M.G.C. (eds.) Handbook of Applied Optimization, pp. 221–234. Oxford University Press, New York (2000)
6. Hansen, P., Mladenović, N.: Variable Neighborhood Search: Principles and applications. European Journal of Operational Research 130, 449–467 (2001)
7. Hansen, P., Mladenović, N., Moreno Pérez, J.A.: Variable neighborhood search: methods and applications. Annals of Operations Research 175, 367–407 (2010)
8. Hashimoto, H., Yagiura, M., Ibaraki, T.: An iterated local search algorithm for the time-dependent vehicle routing problem with time windows. Discrete Optimization 5, 434–456 (2008)
9. Ichoua, S., Gendreau, M., Potvin, J.-Y.: Vehicle dispatching with time-dependent travel times. European Journal of Operational Research 144, 379–396 (2003)
10. Malandraki, C., Daskin, M.S.: Time Dependent Vehicle Routing Problems: Formulations, Properties and Heuristic Algorithms. Transportation Science 26, 185–200 (1992)
11. Mladenović, N., Hansen, P.: Variable Neighborhood Search. Computers & Operations Research 24, 1097–1100 (1997)
12. Polacek, M., Doerner, K.F., Hartl, R.F., Reimann, M.: A Variable Neighborhood Search for the Multi Depot Vehicle Routing Problem with Time Windows. Journal of Heuristics 10, 613–627 (2004)
13. Schmid, V., Doerner, K.F.: Ambulance location and relocation problems with time-dependent travel times. European Journal of Operational Research 207, 1293–1303 (2010)
14. Soler, D., Albiach, J., Martínez, E.: A way to optimally solve a time-dependent Vehicle Routing Problem with Time Windows. Operations Research Letters 37, 37–42 (2009)
15. Solomon, M.M.: Algorithms for the vehicle routing and scheduling problems with time constraints. Operations Research 35(2), 254–265 (1987)
16. Taillard, E.D., Badeau, P., Gendreau, M., Potvin, J.Y.: A Tabu Search Heuristic for the Vehicle Routing Problem with Soft Time Windows. Transportation Science 31, 170–186 (1997)

Solving the Two-Dimensional Bin Packing Problem with a Probabilistic Multi-start Heuristic*

Lukas Baumgartner[1], Verena Schmid[1], and Christian Blum[2]

[1] Department of Business Administration, Universität Wien, Vienna, Austria
{lukas.baumgartner,verena.schmid}@univie.ac.at
[2] ALBCOM Research Group, Universitat Politècnica de Catalunya, Barcelona, Spain
cblum@lsi.upc.edu

Abstract. The two-dimensional bin packing problem (2BP) consists in packing a set of rectangular items into rectangular, equally-sized bins. The problem is NP-hard and has a multitude of real world applications. We consider the case where the items are oriented and guillotine cutting is free. In this paper we first present a review of well-know heuristics for the 2BP and then propose a new ILP model for the problem. Moreover, we develop a multi-start algorithm based on a probabilistic version of the LGFi heuristic from the literature. Results are compared to other well-known heuristics, using data sets provided in the literature. The obtained experimental results show that the proposed algorithm returns excellent solutions. With an average percentage deviation of 1.8% from the best know lower bounds it outperformes the other algorithms by $1.1\% - 5.7\%$. Also for 3 of the 500 instances we tested a new upper bound was found.

Keywords: two-dimensional bin packing, integer linear programming, heuristics.

1 Introduction

The two-dimensional bin packing problem (2BP) consists in packing a set of n rectangular items $j \in Q = \{1, \ldots, n\}$ into bins of height H and width W. The total number of bins is unlimited. Each item j is characterized by its height h_j and its width w_j. Items have to be packed so that they do not overlap. The goal is to minimize the number of used bins. Many real world applications exist for the 2BP such as, for example, cutting glass, wood or metal and packing in the context of transportation or warehousing (see [1] [2]).

According to Lodi et. al [3] there are four different cases of the 2BP. The differences between these four cases are derived from two aspects: (1) the 90°

* This work was supported by the binational grant *Acciones Integradas* ES16-2009 (Austria) and MEC HA2008-0005 (Spain), and by grant TIN2007-66523 (FORMAL-ISM) of the Spanish government. In addition, Christian Blum acknowledges support from the *Ramón y Cajal* program of the Spanish Government of which he is a research fellow.

C.A. Coello Coello (Ed.): LION 5, LNCS 6683, pp. 76–90, 2011.

rotation of items may be allowed, or not, and (2) guillotine cutting may be required or free. The four problem cases can be characterized as follows:

- 2BP|O|G: The items are oriented and guillotine cutting is required.
- 2BP|O|F: The items are oriented and guillotine cuttings is free.
- 2BP|R|G: The items can be rotated by 90° and guillotine cutting is required.
- 2BP|R|F: The items can be rotated by 90° and guillotine cutting is free.

In this paper we exclusively focus on the 2BP|O|F case, that is, in the remainder of the paper the abbreviation 2BP will refer to this problem version. Concerning the complexity of the 2BP, Garey and Johnson classified the problem as NP-hard [4]. For further reading by Lodi et al. [5] [6], Lodi [7] and Dowsland & Dowsland [8] provide a good overview over the 2BP by presenting different models, heuristics, exact algorithms, metaheuristics, lower and upper bounds.

1.1 Organization of the Paper

The main purpose of this paper is to present a multi-start algorithm based on a probabilistic extension of the LGFi heuristic from the literature. However, in Section 2 we present related heuristics for solving the 2BP. In Section 3 we first present a new ILP model for the tackled problem. Our algoritm proposal is then presented in Section 4. Finally, an experimental evaluation is provided in Section 5, while conclusions and an outlook to the future are given in Section 6.

2 Related Work

Concerning heuristic solution methods we mainly distinguish between one-phase and two-phase approaches. One-phase algorithms pack the items directly into the bins, whereas two-phase algorithms first pack the items into levels of one infinitely high strip with width W and then stack these levels into the bins.

Level-packing algorithms place items next to each other in each level. Hereby, the bottom of the first level is the bottom of the bin. For the next level the bottom is a horizontal line coinciding with the tallest item of the level below. Therefore, items can only be placed besides each other in each level, in contrast to packing items on top of each other.

Well known level-packing algorithms are NEXT-FIT DECREASING HEIGHT (NFDH), FIRST-FIT DECREASING HEIGHT (FFDH) and BEST-FIT DECREASING HEIGHT (BFDH) [9]. These strategies were originally developed as algorithms for the one-dimensional bin packing problem, but have also been adapted to strip packing problems and as components of heuristics for the two-dimensional bin packing problem, which we will present in the following. For all three heuristics the items must first be sorted by non-increasing height. Then they are packed in this order.

NFDH packs the current item in the leftmost position of the current level, unless it does not fit. In this case, it creates a new level, which becomes the new current level, where the item will be packed in the leftmost position. In contrast,

FFDH packs the current item as follows. Starting from the first level (among the currently available levels), FFDH tries to accomodate the current item, which is finally packed into the first level in which it still fits. As in the case of NFDH, the current item is always placed in the leftmost possible position. If no level can accomodate the current item a new level is created. Finally, BFDH works as follows. For the current item, BFDH chooses among the available levels the one where the distance from the right side of the item to the right side of the bin is the smallest. If the current item does not fit in any available level, a new level is created. In general, NFDH is the fastest among these three heuristics, but it produces the worst solutions. The opposite is the case for BFDH, while FFDH is a compromise between these two.

Next we shortly describe two-phase level-packing algorithms which are based on the three heuristics described above. HYBRID NEXT-FIT (HNF) [10] is based on NFDH, HYBRID FIRST-FIT (HFF) [11] on FFDH and FINITE BEST-STRIP (FBS) [12], which is also sometimes referred to as HYBRID BEST-FIT, is based on BFDH. In the first phase of all three algorithms the levels are created by the algorithm on which they are based. Then the levels are packed into bins. This is done using the same strategy as was used for the packing of the items into the levels.

Further two-phase level-packing algorithms are FLOOR CEILING (FC) [3] and KNAPSACK PACKING (KP) [3]. In the first phase of KP the levels are packed by solving a knapsack problem. In the second phase these levels are packed into bins. For the first phase the tallest unpacked item, say j, initializes the level. In terms of the knapsack problem the remaining horizontal distance up to the right bin border, $W - w_j$, is the capacity. Moreover, the width w_i of an unpacked item i is regarded as its weight, while the items' area $w_i \cdot h_i$ is regarded as its value (or profit). This results in a knapsack problem which is then solved. This procedure is repeated until all items are packed into levels. In the second phase the remaining one-dimensional bin packing problem is solved by using a heuristic such as BEST-FIT DECREASING or an exact algorithm.

The FC algorithm can be seen as an improvement over the FBS algorithm. Again items are packed into levels in the first phase, and these levels are packed into bins in the second phase. First, the items are sorted by non-increasing height. The tallest unpacked item initializes the level and a horizontal line coinciding with the top edge of this item is the ceiling of that level. Remaining items are packed from left to right on the floor and from right to left on the ceiling. The first item on the ceiling must not fit on the floor of that level. FC tries to pack the current item first on a ceiling (if allowed) following a best-fit strategy. If not possible it tries to pack it on a floor and if that is not possible it initializes a new level. The second phase is the same as in *KP*.

One-phase non-level-packing algorithms are ALTERNATE DIRECTION (AD) [3], BOTTOM-LEFT FILL (BLF) [13], IMPROVED LOWEST GAP FILL (LGFi) [14] and TOUCHING PERIMETER (TP) [3]. In the following we describe these techniques shortly. AD sorts the items by non-increasing heights and initializes L bins, where L is a lower bound of the two-dimensional bin packing problem. It then fills the bottom border of the bins from left to right using a best-fit decreasing strategy. Then one bin after another is being filled. In this context items are packed in bands from left to right and from right to left until no items can be packed into the current bin anymore.

BLF initializes bins by placing the first item at the bottom left corner. The top left and bottom right corners of already placed items are positions where new items could be inserted. BLF tries to place the items starting from the lowest to the highest available position. When postions with an equal height are encountered, the position closer to the left is tried first.

LGFi has a preprocessing and a packing stage. In the preprocessing stage, items are sorted by non-increasing area as a first criterion, and in a case of tie by non increasing absolute difference between height and width of the item. In the packing stage a bin is initialized with the first unpacked item, which is placed at the bottom left corner. Now items are packed on the bottom leftmost position. If possible, an item is chosen such that either the horizontal gap, or the vertical gap to the top, is filled. If this is not possible, the largest fitting item is placed at this position. This is repeated until all items are packed.

TP first sorts the items by non-increasing area and initializes L bins, where L is a computed lower bound for the related two-dimensional bin packing problem. Furthermore, depending on a certain position, a score is associated to each item: the percentage of the edges of the item touching either an edge of another item or the border of the bin. Each item is now tried on different positions in the bin and for each position the corresponding score is calculated. The item is then placed at the position at which the score is highest.

TABU SEARCH (TS) [15] [5] is a meta-heuristic and therefor cannot be classified as a one- or two-phase algorithm.

Tabu Search uses lists containing moves which are considered forbidden to be used again for a certain amount of iterations. First a starting solution is created using a heuristic such as FBS, KP, AD...etc. and a lower bound for the problem instance is calculated. TS then selects a target bin b, which it tries to empty. Therefore it defines a subset S containing an item i from bin b and k other bins. Using a heuristic, such as the ones mentioned before, it now repacks the subset S and if it can be packed in k or less bins the move is executed and added to the tabu list. This is repeated with all combinations of i and k, where k can be increased up to a fixed number, until either the lower bound is reached or the algorithms is considered stuck and has to be restarted by randomly moving packed items into empty bins.

EXTREME POINT-BASED HEURISTICS FOR THREE-DIMENSIONAL BIN PACKING (C-EPBFD) [16] is a heuristic originally designed for the three-dimensional bin packing problem.

This heuristic uses extreme points to determine all points in the bin where items can be placed. Extreme points can either be corners of the already placed items or points generated by the extended edges of the placed items. These points are updated every time an item is placed into the bin. For placing the items a modified version of BFDH is used.

3   A New ILP Model

Inspired by the models proposed in Pisinger and Sigurd [17] and Puchinger and Raidl [18] we present in the following a new ILP model for the 2BP. For this purpose, we denote by $Q = \{1, \ldots, n\}$ the set of all items and the set of all bins. W and H refer to the bin-width and the bin-height, while w_i and h_i refer to the width and the height of item $i \in Q$. W, H, w_i and h_i are all integer.

The binary decision variable α_{ik} evaluates to 1 if item i is packed into bin k, and 0 otherwise. Only variables α_{ik} where $i \geq k$ are created so that only $\frac{n^2+n}{2}$ instead of n^2 have to be initialized. Furthermore items α_{ik} indicate if bins are opened or not. A bin is considered open if the item with the same index as the bin is placed in that bin. For example item 1 cannot be placed in bin 3 but only in bin 1. Item 3 can be placed in bin 3, in bin 2 in case item 2 is placed in bin 2, or in bin 1, which is always open as item 1 can only be placed in bin 1. It is easy to see that, even with this restricted variable set, all combinations of items packed into one bin are still possible. The integer variables x_i and y_i decide the x- and y-coordinates of each item within a bin. For the overlapping constraints, which we will introduce in the next paragraph, we need the binary variables ul_{ij}, ua_{ij}, ur_{ij} and uu_{ij}. Each one of these four variables decides if item i has to be to the left (ul_{ij}), above (ua_{ij}), to the right (ur_{ij}) or underneath (uu_{ij}) item j. Only variables for $i < j$ are created so that only $\frac{n^2-n}{2}$ instead of n^2 have to be initialized for each variable. This can be done because if item i has to be to the left of item j, item j automacitally has to be to the right of item i which makes it unnecassary to initialize the corresponding variable of item j.

$$Z = \sum_{i=0}^{n} \alpha_{ii} \rightarrow min \tag{1}$$

$$\sum_{k=0}^{n} \alpha_{ik} = 1 \qquad\qquad i, k \in Q; i \geq k \tag{2}$$

$$\alpha_{ik} \leq \alpha_{kk} \qquad\qquad i, k \in Q; i \geq k \tag{3}$$

$$x_i + w_i \leq W \qquad\qquad i \in Q \tag{4}$$

$$y_i + h_i \leq H \qquad\qquad i \in Q \tag{5}$$

$$ul_{ij} + ua_{ij} + ur_{ij} + uu_{ij} = 1 \qquad\qquad i, j \in Q; i < j \tag{6}$$

$$x_i + w_i \leq x_j + W \cdot (3 - ul_{ij} - \alpha_{ik} - \alpha_{jk}) \qquad i, j, k \in Q; k \leq i < j \tag{7}$$

$$y_i + H \cdot (3 - ua_{ij} - \alpha_{ik} - \alpha_{jk}) \geq y_j + h_j \qquad i, j, k \in Q; k \leq i < j \tag{8}$$

$$x_i + W \cdot (3 - ur_{ij} - \alpha_{ik} - \alpha_{jk}) \geq x_j + w_j \qquad i, j, k \in Q; k \leq i < j \tag{9}$$

$$y_i + h_i \leq y_j + H \cdot (3 - uu_{ij} - \alpha_{ik} - \alpha_{jk}) \qquad i, j, k \in Q; k \leq i < j \tag{10}$$

The objective function (1) minimizes the number of bins used. The constraint (2) ensures that each item is assigned to one bin. That an item i can only be assigned to an open/initialized bin is ensured by (3). That each item is placed within the bin is ensured by inequations (4) and (5). Equation (6) states that item i has to be placed either to the left, above, to the right or underneath item j. The last four equations (7)-(10) ensure that two items do not overlap if assigned to the same bin.

4   The Proposed Algorithm

The algorithm that we present in this paper is a multi-start heuristic based on a probabilistic version of LGFi, which was developed by Wong and Lee in [14]. LGFi itself is an improved version of the LGF heuristic presented by Lee in [19]. Note that LGFi is a two-stage heuristic. In the preprocessing stage items are sorted into a list, while in the packing stage the items are packed from that list into bins. The main difference between LGFi and our probabilistic version of LGFi (P-LGFi) is that the items are not chosen in a deterministic way but rather in a probalistic manner. It cannot be seen as a two-phase heuristic as it has two stages (preprocessing and packing) but only one phase in which items are packed (packing stage).

4.1   Probabilistic LGFi

In the following we first outline the probabilistic way of using LGFi that we developed. This concerns in particular the preprocessing stage. Remember that the preprocessing stage is supposed to generate an ordered list of all items. However, instead of doing that deterministically, as in LGFi, P-LGFi does that in a probabilistic manner.

The Preprocessing Stage: First, for each item i the area (a_i) and the absolute difference between height and width (d_i) must be calculated:

$$a_i = w_i \cdot h_i \tag{11}$$
$$d_i = |w_i - h_i| \tag{12}$$

Then, on the basis of a_i and d_i, a value v_i is computed for each item i:

$$v_i = (\lambda \cdot a_i - d_i)^{\kappa} \tag{13}$$

Hereby, λ and κ are parameters. Larger values of λ result in the fact that items with larger areas receive higher v-values, that is, with increasing λ the importance of the area grows in comparison to the absolute difference between width and height. Note that for the computational experiments presented in the following section we used $\lambda = 100$. Concerning κ, larger values of κ increase the difference between the v-values of different items. In other words, when $\kappa = 0.1$ the v-values of all items will be very similar to each other, while when $\kappa = 10$, for example, the v-values are characterized by large differences.

Based on the v-values, an ordered list of all items is then generated in a probabilistic way from left to right. At each step, let $\mathcal{I} \subseteq \mathcal{Q}$ be the set of items that are not yet assigned to the list. An item $i \in \mathcal{I}$ is chosen according to probabilities p_i (for all $i \in \mathcal{I}$) by roulette-wheel-selection. The probabilities p_i are calculated proportional to the v-values:

$$p_i = \frac{v_i}{\sum_{i \in \mathcal{I}} v_i} \tag{14}$$

The result of this process is a list of all items, which is used for the packing stage.

The Packing Stage: The first bin is initialized by placing the first item from the list obtained in the preprocessing stage at the bottom left corner of the bin (Figure 1(a)).

Now the bottom leftmost point in the bin, on which no item is placed, is choosen as the current point. From this point there are two gaps, one horizontal and one vertical. The horizontal gap is the distance between the point and the right border of the bin or the left edge of the first item between the point and the right border of the bin. The distance between the point and the upper border of the bin defines the value of the vertical gap. Which ever one of those two is smaller is the current gap (Figure 1(b)).

The current gap is compared to either the widths of the items, if the horizontal gap is the current gap, or to the heights of the items, if the vertical gap is the current gap. The heuristic compares the current gap against the width and height of all unpacked items (Figure 1(c)).

If any item fills the gap completely it is packed with its bottom left corner on the current point and the next bottom leftmost point is determined. If no item is able to fill the gap completely the heuristic goes through the list one more time and picks the first item whose height is less or equal than the vertical gap and whose width is less or equal than the horizontal gap (Figure 1(d)).

If still no item fits on this position a certain area has to be declared as wastage, which works as follows. A wastage area with the width of the horizontal gap is created. The height of it is chosen so that the area continuously touches either an edge of an item or the border of the bin on both sides (Figure 1(e)).

The heuristic now searches for a new point and tries to place an unpacked item from the list again in the same way as described above. This is done until the bin is completely filled with either items or areas declared as wastage. A new bin is initialized with the first unpacked item placed on the bottom left corner of the new bin. This is repeated until all items are packed into bins.

4.2   Multi-start Algorithm

As mentioned above, the P-LGFi heuristic that we developed can be used in a simple multi-start fashion. In particular, at each iteration the multi-start algorithm executes the probabilistic LGFi heuristic once. The best solution obtained in this way is stored and provided as output of the algorithm when the stopping criterion has been reached. In this work we used a fixed number of iterations as stopping criterion. This algorithm is denoted in the following as MULTI-START PROBABILISTIC IMPROVED LOWEST GAP FILL (MP-LGFi).

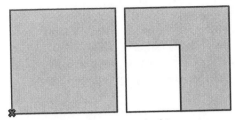

(a) Initializing the bin

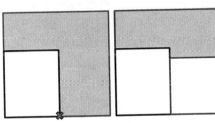

(b) Identifying the current gap

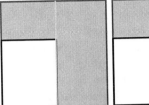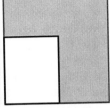

(c) Placing a perfectly fitting item

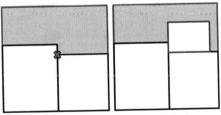

(d) Placing the first item which fits

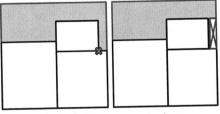

(e) Declaring an area wastage

Fig. 1. Differnt stages of packing

5   Experimental Evaluation

MP-LGFi was implemented using Microsoft Visual C++ 2008. All experiments were performed on an Intel® Xeon® X5500 @ 2.67 GHz with 3 GB of RAM. The proposed algorithm was tested on instances provided in the literature. After an initial study of the algorithm behaviour, a detailed experimental evaluation is presented.

5.1   Problem Instances

Ten classes of problem instances for the 2BP are provided in the literature. A first instance set, containing six classes (I-VI), was proposed by Berkey and Wang in [12]. For each of these classes, the widths and heights of the items were chosen uniformly at random from the intervals presented in Table 1. Moreover, the classes differ in the width (W) and the height (H) of the bins. Instance sizes, in terms of the number of items, are taken from $\{20, 40, 60, 80, 100\}$. Berkey and Wang provided 10 instances for each combination of a class with an instance size. This results in a total of 300 problem instances.

Table 1. Specification of instance classes I-VI (as provided by [12])

Class	w_j	h_j	W	H
I	[1,10]	[1,10]	10	10
II	[1,10]	[1,10]	30	30
III	[1,35]	[1,35]	40	40
IV	[1,35]	[1,35]	100	100
V	[1,100]	[1,100]	100	100
VI	[1,100]	[1,100]	300	300

The second instance set, consisting of classes VII-X, was introduced by Martello and Vigo in [20]. In general, they considered four different types of items, as presented in Table 2. The four item types differ in the limits for the width w_i and the height h_i of an item. Then, based on these four item types, Martello and Vigo introduced four classes of instances which differ in the percentage of items they contain from each type. As an example, let us consider an instance of class VII. 70% of the items of such an instance are of type 1, 10% of the items are of type 2, further 10% of the items are of type 3, and the remaining 10% of the items are of type 4. These percentages are given per class in Table 3. As in the case of the first instance set, instance sizes are taken from $\{20, 40, 60, 80, 100\}$. The instance set by Martello and Vigo consists of 10 instances for each combination of a class with an instance size. This results in a total of 200 problem instances.

These alltogether 500 instances can be downloaded from `http://www.or.deis.unibo.it/research.html`.

Table 2. Item types for classes VII-X (as introduced in [20])

Item type	w_j	h_j	W	H
1	$[\frac{2}{3} \cdot W, W]$	$[1, \frac{1}{2} \cdot H]$	100	100
2	$[1, \frac{1}{2} \cdot W]$	$[\frac{2}{3} \cdot H, H]$	100	100
3	$[\frac{1}{2} \cdot W, W]$	$[\frac{1}{2} \cdot H, H]$	100	100
4	$[1, \frac{1}{2} \cdot W]$	$[1, \frac{1}{2} \cdot H]$	100	100

Table 3. Specification of instance classes VII-X (as provided by [20])

Class	Type 1	Type 2	Type 3	Type 4
VII	70%	10%	10%	10%
VIII	10%	70%	10%	10%
IX	10%	10%	70%	10%
X	10%	10%	10%	70%

5.2 Parameter Setting

Before conducting a full experimental evaluation of MP-LGFi, we first wanted to understand certain aspects of the behaviour of the algorithm. More specifically, we were interested in the influence of the value of parameter κ as well as in the run-time behaviour of the algorithm. Concerning κ, remember that rather high values result in random sequences of all the items that are very similar to the deterministic sequence generated by LFGi. This means that the higher the value of κ, the less probabilistic is our version of LGFi. Intuitively, we expected that values close to zero do not work very well, because the degree of stochasticity is too high. We also expected that values that are too high do not work very well, because the resulting sequences are too similar to the deterministic sequence of LGFi. In order to confirm this intuition, we applied MP-LGFi with a limit of 100 iterations three times to each of the 500 instances. This was done for $\kappa \in \{0.1, 0.5, 1, 2, 3, 4, 5, 6, 7, 8, 9, 10\}$. For each κ we calculated the average percentage deviation of the corresponding results with respect to the optimal (respectively, best known) solutions. The obtained results are graphically shown in Figure 2. They show indeed that our initial intuition appears to be true: MP-LGFi seems to work best for intermediate values of κ, that is, for values in $\{4, 5, 6\}$. Therefore, we chose the setting of $\kappa = 5$ for all the remaining experiments.

As mentioned above, we also studied the run-time behaviour of the algorithm. For this purpose we applied MP-LGFi (with $\kappa = 5$) thrice to each of the 500 problem instances, using an iteration limit of 20000 iterations. The aggregated results are shown graphically in Figure 3. The results show that most improvements are obtained during the first 100 iterations. Further significant improvements are achieved until around 5000 iterations. After that the results almost do not improve. Given this behaviour, we chose an iteration limit of 10000 iterations for the final set of experiments.

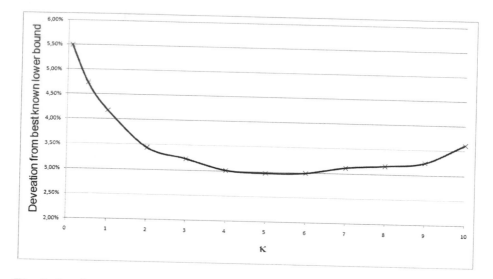

Fig. 2. Results averaged over all 500 instances for different values of κ (x-axis). The y-axis provides the average percent deviation of the corresponding results with respect to the best known lower bounds.

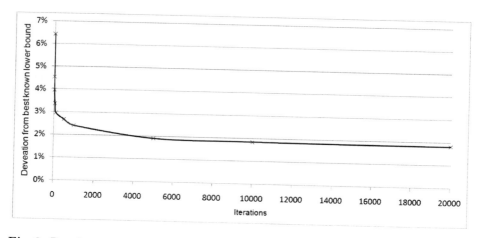

Fig. 3. Results averaged over all 500 instances for different iteration limits (x-axis). The y-axis provides the average percent deviation of the corresponding results with respect to the best known lower bounds.

5.3   Computational Results

Table 4 provides numerical results of MP-LGFi in comparison to IMPROVED LOWEST GAP FILL (LGFi) [14] and TABU SEARCH (TS) [15] [5]. For each of the three algorithms the results are shown averaged over the 10 instances for each combination of class and instance size. The values in the columns with heading (q) are the ratio between the obtained solution and the lower bound of the respective two-dimensional bin packing problem. Therefore, the lower a value in the columns with heading (q) the better. Also note that in a case in which for all 10 instances a solution was obtained whose value matches the one of the lower bound, the corresponding q-value is 1.000. In other words, 1.000 is the best possible q-value. In the case of MP-LGFi, where the average results of three runs are shown, we also provide information about the corresponding standard deviations (columns with heading σ), the average time when the best solution was found (columns with heading t_b), and the average total runtime in seconds (columns with heading t). Moreover, the last line for each class gives the average of each algorithm for all instance sizes. The best result(s) for each combination of instance size and class are shown in bold.

Table 4. Numerical results for all 500 instances. The results are shown as averages over the 10 instances for each combination of instance size and class. In addition to the results of MP-LGFi, the table also presents the results of FC, AD, LGFi and TS.

	FC	AD	LGFi	TS	MP-LGFi					FC	AD	LGFi	TS	MP-LGFi			
	q	q	q	q	q	sigma	tb	t		q	q	q	q	q	sigma	tb	t
Class I									**Class VI**								
20	1.120	1.120	1.110	1.060	**1.000**	0.000	0.0	0.1	20	**1.000**	**1.000**	**1.000**	**1.000**	**1.000**	0.000	0.0	0.0
40	1.080	1.090	1.060	1.060	**1.000**	0.000	0.0	0.3	40	1.400	1.400	1.400	1.400	**1.200**	0.000	10.5	124.4
60	1.070	1.070	1.050	1.040	**1.017**	0.000	0.0	0.5	60	1.100	1.050	1.100	1.050	**1.000**	0.000	9.0	4.5
80	1.060	1.060	1.040	1.050	**1.004**	0.000	0.0	0.4	80	**1.000**	**1.000**	**1.000**	**1.000**	**1.000**	0.000	0.1	0.1
100	1.060	1.050	1.030	1.040	**1.000**	0.000	0.0	0.8	100	1.100	1.070	1.100	1.070	**1.067**	0.000	0.3	234.8
Average	1.078	1.078	1.059	1.050	**1.004**	0.000	0.0	0.4	Average	1.120	1.104	1.120	1.104	**1.053**	0.000	4.0	72.8
Class II									**Class VII**								
20	1.100	**1.000**	**1.000**	**1.000**	**1.000**	0.000	0.0	0	20	1.080	1.100	1.100	1.040	**1.000**	0.000	0.0	9.0
40	1.100	1.100	1.100	1.100	**1.000**	0.000	0.0	0	40	1.090	1.100	1.070	1.060	**1.020**	0.000	4.3	17.0
60	1.100	1.100	1.100	1.100	**1.000**	0.000	0.0	0	60	1.070	1.070	1.040	1.050	**1.019**	0.000	0.7	38.0
80	1.070	1.070	1.030	1.070	**1.000**	0.000	0.0	0	80	1.060	1.060	1.060	1.040	**1.037**	0.000	0.0	128.4
100	1.030	1.030	1.030	1.030	**1.000**	0.000	0.0	0	100	1.040	1.040	1.030	1.030	**1.009**	0.002	43.8	63.8
Average	1.080	1.060	1.053	1.060	**1.000**	0.000	0.0	0	Average	1.068	1.074	1.059	1.044	**1.017**	0.000	9.8	51.3
Class III									**Class VIII**								
20	1.180	1.200	1.230	1.200	**1.022**	0.019	0.0	1.3	20	1.160	1.130	1.120	1.060	**1.000**	0.000	0.4	13.8
40	1.140	1.150	1.170	1.110	**1.033**	0.007	0.3	1.2	40	1.070	1.080	1.080	1.030	**1.009**	0.000	0.1	7.2
60	1.110	1.130	1.100	1.050	**1.032**	0.000	0.0	4	60	1.060	1.060	1.060	1.020	**1.013**	0.000	5.8	25.0
80	1.100	1.100	1.070	1.080	**1.030**	0.003	1.4	6.4	80	1.060	1.060	1.040	1.020	**1.005**	0.000	3.0	12.0
100	1.090	1.090	1.090	1.090	**1.026**	0.003	1.4	9.1	100	1.060	1.060	1.050	1.040	**1.015**	0.000	1.0	58.7
Average	1.124	1.134	1.131	1.106	**1.029**	0.006	0.6	4.4	Average	1.082	1.078	1.068	1.034	**1.008**	0.000	2.1	23.3
Class IV									**Class IX**								
20	**1.000**	**1.000**	**1.000**	**1.000**	**1.000**	0.000	0.0	0	20	1.010	1.010	1.010	**1.000**	**1.000**	0.000	0.0	0.0
40	**1.000**	**1.000**	**1.000**	**1.000**	**1.000**	0.000	0.0	0	40	1.020	1.020	1.010	1.010	**1.000**	0.000	0.0	52.4
60	**1.100**	1.150	**1.100**	1.150	**1.100**	0.000	0.0	5.7	60	1.020	1.020	1.010	1.010	**1.000**	0.000	0.0	53.2
80	1.100	1.100	1.100	1.100	**1.033**	0.000	0.1	3.6	80	1.020	1.020	1.010	1.010	**1.000**	0.000	0.1	110.0
100	1.100	1.030	1.070	1.030	**1.000**	0.000	2.3	1.1	100	1.010	1.010	1.010	1.010	**1.000**	0.000	0.1	87.9
Average	1.060	1.056	1.053	1.056	**1.027**	0.000	0.5	2.1	Average	1.016	1.016	1.012	1.008	**1.000**	0.000	0.0	60.7
Class V									**Class X**								
20	1.140	1.140	1.110	1.110	**1.000**	0.000	0.1	22.7	20	1.140	1.100	1.130	1.100	**1.000**	0.000	0.0	7.8
40	1.110	1.110	1.100	1.040	**1.000**	0.000	3.6	25.8	40	1.090	1.090	1.090	1.060	**1.000**	0.000	0.0	9.7
60	1.100	1.100	1.090	1.060	**1.009**	0.004	13.7	49.6	60	1.080	1.110	1.110	1.070	**1.053**	0.000	0.0	57.9
80	1.090	1.090	1.080	1.060	**1.026**	0.000	10.2	103.4	80	1.110	1.100	1.090	1.060	**1.056**	0.000	0.0	74.7
100	1.090	1.090	1.090	1.080	**1.035**	0.000	1.2	146.4	100	1.090	1.100	1.080	1.080	**1.054**	0.007	0.2	94.7
Average	1.106	1.106	1.092	1.070	**1.014**	0.001	5.8	69.6	Average	1.102	1.100	1.100	1.074	**1.033**	0.001	0.1	49.0
									Total Average	1.084	1.081	1.075	1.061	**1.018**	0.001	2.3	33.4

Table 5. Numerical results for all 500 instances. The results are shown as averages over the 10 instances for each combination of instance size and class. In addition to the results of MP-LGFi, the table also presents the results of C-EPBFD.

	C-EPBFD	MP-LGFi		C-EPBFD	MP-LGFi
	q	q		q	q
Class I	1.019	**1.004**	Class VI	1.093	**1.037**
Class II	1.040	**1.000**	Class VII	1.030	**1.020**
Class III	1.047	**1.023**	Class VIII	1.022	**1.010**
Class IV	1.101	**1.025**	Class IX	**1.000**	**1.000**
Class V	1.031	**1.019**	Class X	1.059	**1.041**
			Total Average	1.023	**1.012**

Table 5 compares the results of MP-LGFi and EXTREME POINT-BASED HEURISTICS FOR THREE-DIMENSIONAL BIN PACKING (C-EPBFD) [16]. But unlike Table 4, where the values represent the mean of the gap to the lower bound, the values in this Table are the gap to the lower bound of the respective mean for each combination of class and items. Therefor only the average of the classes are compared as the results of C-EPBFD cannot be compared to the other heuristics.

The comparison between the different algorithms shows that MP-LGFi nearly always outperforms the competitors. Only in a few cases concerning classes II, IV, VI, VII, IX and X, other heuristics are able to match the results of MP-LGFi. When averaging over the gaps for the whole instance set LGFi achieves a value of 7.5%, TS a value of 6.1% and MP-LGFi a value of 1.8%. When calculating the gap over the average number of bins used for the whole isntance set C-EPBFD generates a value of 2.3% and MP-LGFi a value of 1.2%. Therefore, MP-LGFi is clearly a new state-of-the-art algorithm for the considered instance sets. In our opinion, MP-LGFi can be seen as a prime example for the fact that sometimes a simple heuristic can outperform more sophisticated techniques, such as—in this case—a tabu search metaheuristic or an extreme point-based heuristic.

Further MP-LGFi managed to find a new best upper bound for three of the 500 instances, reducing the number of instances where the upper bound does not match the lower bound from 68 to 65. The Upper bound was lowered for instance 398 (Class 8, Instance 8, 100 Items) from 29 to 28, 197 (Class 4, Instance 7, 100 Items) from 4 to 3 and 187 (Class 4, Instance 7, Items 80) from 4 to 3.

6   Conclusions

In this paper we have dealt with the two-dimensional bin packing problem with oriented items and free guillotine cutting (2BP|O|F). A first contribution of this work has been the presentation of a new ILP model for this problem. Moreover, we developed a simple multi-start algorithm based on a probabilistic version of an existing heuristic from the literature. With an average percentage deviation of

1.8% it shows that the proposed algorithm is currently a state-of-the-art method for the 2BP|O|F, as it outperforms other algorithms by 1.1% − 5.7% and found 3 new upper bounds for the 500 instances tested.

In the future we envisage several possible improvements of the proposed algorithm. Most notably we plan to add a learning component to the algorithm in order to take profit from the search history.

References

1. Hopper, E., Turton, B.: A genetic algorithm for a 2d industrial packing problem. Computers and Industrial Engineering 37(1-2), 375–378 (1999)
2. Sweeney, P.E., Paternoster, E.R.: Cutting and packing problems: A categorized, application-orientated research bibliography. The Journal of the Operational Research Society 43(7), 691–706 (1992)
3. Lodi, A., Martello, S., Vigo, D.: Heuristic and metaheuristic approaches for a class of two-dimensional bin packing problems. INFORMS Journal on Computing 11(4), 345–357 (1999)
4. Garey, M.R., Johnson, D.S.: Computers and Intractability: A Guide to the Theory of NP-Completeness. W.H. Freeman, New York (1979)
5. Lodi, A., Martello, S., Vigo, D.: Recent advances on two-dimensional bin packing problems. Discrete Applied Mathematics 123(1-3), 379–396 (2002)
6. Lodi, A., Martello, S., Vigo, D.: Two-dimensional packing problems: A survey. European Journal of Operational Research 141(2), 241–252 (2002)
7. Lodi, A.: Algorithms for Two-Dimensional Bin Packing and Assignment Problems. PhD thesis, Università degli Studio di Bologna (1996-1999)
8. Dowsland, K.A., Dowsland, W.B.: Packing problems. European Journal of Operational Research 56(1), 2–14 (1992)
9. Coffman Jr., E.G., Garey, M.R., Johnson, D.S., Tarjan, R.E.: Performance bounds for level-oriented two-dimensional packing algorithms. SIAM Journal on Computing 9(4), 808–826 (1980)
10. Frenk, J.B.G., Galambos, G.: Hybrid next-fit algorithm for the two-dimensional rectangle bin-packing problem. Computing 39(3), 201–217 (1987)
11. Chung, F.R.K., Garey, M.R., Johnson, D.S.: On packing two-dimensional bins. SIAM Journal on Algebraic and Discrete Methods 3(1), 66–76 (1982)
12. Berkey, J.O., Wang, P.Y.: Two dimensional finite bin packing algorithms. Journal of the Operational Research Society 38(5), 423–429 (1987)
13. Baker, B.S., Coffman Jr., E.G., Rivest, R.L.: Orthogonal packings in two dimensions. SIAM Journal on Computing 9(4), 846–855 (1980)
14. Wong, L., Lee, L.S.: Heuristic placement routines for two-dimensional bin packing problem. Journal of Mathematics and Statistics 5(4), 334–341 (2009)
15. Lodi, A., Martello, S., Vigo, D.: Approximation algorithms for the oriented two-dimensional bin packing problem. European Journal of Operational Research 112(1), 158–166 (1999)
16. Crainic, T.G., Perboli, G., Tadei, R.: Extreme point-based heuristics for three-dimensional bin packing. Informs Journal on Computing 20(3), 368–384 (2008)
17. Pisinger, D., Sigurd, M.: Using decomposition techniques and constraint programming for solving the two-dimensional bin-packing problem. INFORMS Journal on Computing 19(1), 36–51 (2007)

18. Puchinger, J., Raidl, G.: Models and algorithms for three-stage two-dimensional bin packing. European Journal of Operational Research 183(3), 1304–1327 (2007)
19. Lee, L.S.: A genetic algorithm for two-dimensional bin packing problem. MathDigest 2(1), 34–39 (2008)
20. Martello, S., Vigo, D.: Exact solution of the two-dimensional finite bin packing problem. Management Science 44(3), 388–399 (1998)

Genetic Diversity and Effective Crossover in Evolutionary Many-objective Optimization

Hiroyuki Sato[1], Hernán E. Aguirre[2,3], and Kiyoshi Tanaka[3]

[1] Faculty of Informatics and Engineering, The University of Electro-Communications
1-5-1 Chofugaoka, Chofu, Tokyo 182-8585 Japan
[2] International Young Researcher Empowerment Center, Shinshu University
4-17-1 Wakasato, Nagano, 380-8553 Japan
[3] Faculty of Engineering, Shinshu University
4-17-1 Wakasato, Nagano, 380-8553 Japan

Abstract. In this work, we analyze genetic diversity of Pareto optimal solutions (POS) and study effective crossover operators in evolutionary many-objective optimization. First we examine the diversity of genes in the true POS on many-objective 0/1 knapsack problems with up to 20 items (bits), showing that genes in POS become noticeably diverse as we increase the number of objectives. We also verify the effectiveness of conventional two-point crossover, Local Recombination that selects mating parents based on proximity in objective space, and two-point and uniform crossover operators Controlling the maximum number of Crossed Genes (CCG). We use NSGA-II, SPEA2, IBEA$_{\epsilon+}$ and MSOPS, which adopt different selection methods, and many-objective 0/1 knapsack problems with $n = \{100, 250, 500, 750, 1000\}$ items (bits) and $m = \{2, 4, 6, 8, 10\}$ objectives to verify the search performance of each crossover operator. Simulation results reveal that Local Recombination and CCG operators significantly improve search performance especially for NSGA-II and MSOPS, which have high diversity of genes in the population. Also, results show that CCG operators achieve higher search performance than Local Recombination for $m \geq 4$ objectives and that their effectiveness becomes larger as the number of objectives m increases.

1   Introduction

The research interest of the multi-objective evolutionary algorithm (MOEA) [1] community has rapidly shifted to develop effective algorithms for many-objective optimization problems (MaOPs) because more objective functions should be considered and optimized in recent complex applications. However, in general, MOEAs noticeably deteriorate their search performance as we increase the number of objectives to more than 4 [2,3], especially Pareto dominance-based MOEAs such as NSGA-II [4] and SPEA2 [5]. This is because these MOEAs meet difficulty to rank solutions in the population, i.e., most of the solutions become non-dominated and the same rank is assigned to them, which seriously spoils proper selection pressure required in the evolution process. To overcome this

C.A. Coello Coello (Ed.): LION 5, LNCS 6683, pp. 91–105, 2011.

problem, several studies have been made on methods to determine the superiority of non-dominated solutions in a more effective manner in order to strengthen parent selection pressure [6,7].

Contrary to these studies, in this work we focus on genetic diversity in Pareto optimal solutions (POS) in MaOPs. It is well known that in MaOPs the number of non-dominated solutions increases considerably with the number of objectives. However, not much is known about the distribution of those solutions in decision space, how selection shapes that distribution, and how both the distribution in variable space and selection in objective space influence the effectiveness of genetic operators of MOEAs in MaOPs. This work is an important step towards understating these important issues in many-objective optimization.

In this work, first we analyze genetic diversity in the true POS obtained by exhaustive search on many-objective 0/1 knapsack problem with $n = \{10, 15, 20\}$ bits (items), showing that genes in POS become noticeably diverse in the same way as the ratio of POS in feasible solution space increases with the number of objectives. In MOEAs, if genes of solutions in the population become noticeably diverse, conventional recombination might become too disruptive and decrease its effectiveness. In this work, we verify the effectiveness of conventional two-point crossover, Local Recombination that selects mating parents based on proximity in objective space, and two-point and uniform crossover operators Controlling the maximum number of Crossed Genes (CCG). To verify the search performance of each crossover operator, we use NSGA-II [4], SPEA2 [5], IBEA$_{\epsilon+}$ [9] and MSOPS [2], well known MOEAs that adopt different selection methods, and many-objective 0/1 knapsack problems with $n = \{100, 250, 500, 750, 1000\}$ items (bits) and $m = \{2, 4, 6, 8, 10\}$ objectives.

2   Analysis of Pareto Optimal Solutions in Many-objective 0/1 Knapsack Problem

First, we analyze many-objective 0/1 knapsack problems [10] by observing the number of Pareto optimal solutions |POS| and their genetic diversity in discrete solution space. Here, we generate problems with $n = \{10, 15, 20\}$ items (bits) and $m = 2 \sim 20$ objectives, setting the feasibility ratio $\phi = 0.5$. We generate 90 problems for each combination of parameters m and n, find all true POS by exhaustive search in solution space S, and analyze average results.

Fig.1 shows the ratio $|POS|/|\mathcal{F}|$ of true POS in feasible solution space $\mathcal{F}$ ($\subseteq S$). From these results, we can see that the ratio of POS in $\mathcal{F}$ increases significantly with the number of objectives m. Also, the ratio of POS in $\mathcal{F}$ decreases as the solution space expands with n (2^n). Next, to observe the genetic diversity of POS, Fig.2 shows the average hamming distance of POS. Here, we also plot the average hamming distance of all solutions in the solution space S as a horizontal line. From these results, note that the average hamming distance of POS noticeably increases with the number of objectives m. In case of a small number of objectives m, the ratio of POS in $\mathcal{F}$ is relatively low and the average hamming distance of POS is short compared to the average hamming distance of

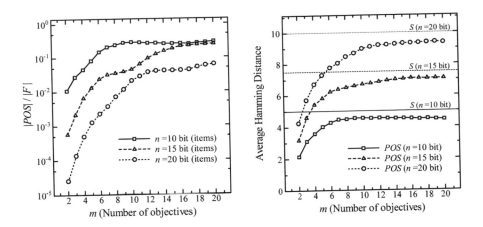

Fig. 1. Ratio of true Pareto optimal solutions POS in feasible solution space $\mathcal{F}$ **Fig. 2.** Average hamming distance of true POS

all solutions $\mathcal{S}$, suggesting that POS are distributed in a relatively narrow region in the solution space $\mathcal{S}$. On the other hand, the ratio of POS in $\mathcal{F}$ increases with the number of objectives m and the average hamming distance also increases, approaching the average hamming distance of all solution in $\mathcal{S}$. For example, on $n = 20$ bits and $m = 20$ objectives, around 6% of feasible solutions become POS and the average hamming distance of POS is 9.36 bits, which is very close to the 10 bits average hamming distance of all solutions in $\mathcal{S}$. This tendency is also observed in problems with $n = \{10, 15\}$ bits, where the average hamming distances of POS $\{4.45, 7.06\}$ are close to the average hamming distance of $\mathcal{S}$ $\{5.0, 7.5\}$, respectively. These results suggest that POS come to be distributed nearly uniformly in solution space by increasing m. That is, we can expect that genes become noticeably diverse in the population during evolutionary many-objective optimization. Also, the exploitation effectiveness of the conventional recombination might decrease if difference of genes between two parents becomes too large.

3   Mating Based on Proximity in Objective Space

3.1   Related Works

To realize effective recombination of solutions in MOEAs, several studies that apply crossover for two parents located near each other in the objective function space have been made. NCGA (Neighborhood Cultivation GA) introduce neighborhood crossover in the objective function space [11]. In NCGA, after sorting solutions in the population according to one objective function value, two neighboring solutions become a pair for crossover. The improvement of convergence and diversity of obtained POS by NCGA on continuous and combinatorial 0/1 knapsack problems with two objectives functions has been reported in [11]. In another study, Local Recombination [8] selects pairs of parents by considering

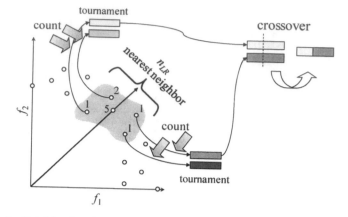

Fig. 3. Neighborhood creation and mating for Local Recombination [8]

nearness of the search direction of solutions, using a locality parameter n_{LR}. When we use a small n_{LR}, parents are selected with high locality in the objective space. Increasing n_{LR}, the neighborhood expands and in the extreme it comes to be the entire parent population. That is, in the extreme we have conventional recombination, because mates are selected without restriction from the entire parent population. The effectiveness of Local Recombination has been verified on $m = 2 \sim 4$ objective problems [8]D In [12], a mating scheme was proposed to select one pair of solutions for crossover by first selecting various candidates performing multiple binary tournaments, and then picking two of them based on their distance in the objective space. This method controls the balance between convergence and diversity of obtained solutions by the mating scheme. Additionally, MOEA/D [13] utilizes multiple scalarization functions to find POS, selecting pair of solutions for recombination from solutions that maximize neighbor scalarization functions.

In terms of avoiding the inefficient recombination of solutions having very different objective function values, it is thought that these methods bring similar effects for the search performance of MOEAs. In problems where there is some correlation between objective and variable space, it is expected that these methods can effectively apply crossover to solutions that have relatively similar gene structure even in MaOP. In this paper we focus on Local Recombination [8] which controls locality for recombination with parameter n_{LR} and verify its effectiveness on MaOPs.

3.2   Local Recombination

To create n_{LR} neighborhoods which have similar search direction, original Local Recombination utilizes angle information in polar coordinate vector transformed from objective function values [8]. In this work, we use search direction $\boldsymbol{d}(\boldsymbol{x}) = (d_1(\boldsymbol{x}), d_2(\boldsymbol{x}), \cdots, d_m(\boldsymbol{x}))$ calculated by the fraction of each objective function value, namely

$$d_i(\boldsymbol{x}) = f_i(\boldsymbol{x}) / \sum_{j=1}^{m} f_j(\boldsymbol{x}) \qquad (i = 1, 2, \cdots, m). \tag{1}$$

Then, we calculate the Euclidean distance between $\boldsymbol{d}(\boldsymbol{x})$ and the search direction of other solutions, and create a sub-population S_{LR} of n_{LR} neighboring solutions, as shown in Fig.3. Note that n_{LR} is the locality parameter for recombination. Similar to [8], mating is performed within the neighborhood S_{LR} and then recombination followed by mutation are carried out. We enforce equal participation in the tournaments. To accomplish that we keep for each individual in the parent population $\mathcal{P}_t$ a counter showing the number of times it has participated in a tournament and select the individuals that will undergo a binary tournament randomly from among those with smallest value in its counter. Note that the individual's counters are not re-initialized until all offspring $\mathcal{Q}_t$ have been created. Varying the number of elements in the neighborhood $n_{LR} \leq |\mathcal{Q}_t|$ we can control the degree of locality for recombination. In the extreme, $n_{LR} = |\mathcal{Q}_t|$, we have conventional recombination. Note that we refer to conventional recombination as global recombination, because the neighborhood for mating considers the entire parent population.

4   Controlling Crossed Genes for Crossover

4.1   Problem of Local Recombination in MaOPs

Since Local Recombination selects mates having similar search direction, the probability that some selected pairs of solutions have similar genes structure increases. However, as mentioned in section 2, the diversity of genes in solutions noticeably increase in MaOPs. In this case, even if we select neighborhood solutions in objective space for recombination, it is expected that they have a large difference in genes and recombination might be inefficient. To solve this problem, in this work we consider methods to restrict the length of crossed genes when we apply crossover in MaOPs.

4.2   CCG for Two-Point Crossover (CCG$_{TX}$)

When we apply the conventional one- or two-point crossover for individuals with n genes, the length of crossed genes vary in the range $[0, n]$ by randomly chosen the crossover point(s). To restrict the variation of genes in crossover for parents having large difference in gene structure, in this work we propose controlling crossed genes (CCG) for crossover. In this section we explain CCG for two-point crossover (CCG$_{TX}$). CCG$_{TX}$ controls the length of crossed genes by using a user-defined parameter α_t. Fig.4 shows the conceptual diagram of CCG$_{TX}$. First we randomly select parents $\boldsymbol{A}$ and $\boldsymbol{B}$ from the parent population $\mathcal{P}_t$, and randomly choose the 1st crossover point p_1. Then, we randomly determine the length of the crossed genes l in the range $[0, \alpha_t \cdot n]$. In case of $p_1 + l \leq n$, the second crossover point is set to $p_2 = p_1 + l$. In case of $p_1 + l > n$, the second crossover

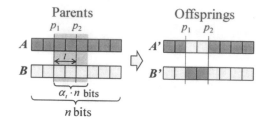

Fig. 4. Controlling crossed genes for two-point crossover (CCG$_{\text{TX}}$)

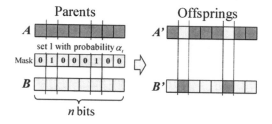

Fig. 5. Controlling crossed genes for uniform crossover (CCG$_{\text{UX}}$)

point is set to $p_2 = p_1 + l - n$. Here, the possible range of the parameter α_t is $[0.0, 1.0]$. In this method, when we utilize a small α_t, the maximum length of crossed segment becomes short. On the other hand, when we utilize a large α_t, the maximum length of crossed segment becomes long. In the case of $\alpha_t = 0.0$, since the length of the crossed segment becomes $\alpha_t \cdot n = 0$, the solutions search is equivalent to only mutation without crossover. Also, in the case of $\alpha_t = 1.0$, the maximum length of the crossed segment become $\alpha_t \cdot n = n$. This case is equivalent to the conventional two-point crossover. In this work we verify the effects of CCG$_{\text{TX}}$ in MOEA as we vary α_t in the range $\alpha_t \in [0.0, 1.0]$.

4.3 CCG for Uniform Crossover (CCG$_{\text{UX}}$)

Next, we explain a method for CCG in uniform crossover (CCG$_{\text{UX}}$). As shown in Fig.5, for uniform crossover we randomly select two parents from the parent population and generate a n bit mask [14,15]. For offspring $\boldsymbol{A'}$, if mask bit is 0, the gene is copied from parent $\boldsymbol{A}$. If mask bit is 1, the gene is copied from parent $\boldsymbol{B}$. Similarly, for offspring $\boldsymbol{B'}$, if mask bit is 0, the gene is copied from parent $\boldsymbol{B}$. If mask bit is 1, the gene is copied from parent $\boldsymbol{A}$. To control the number of crossed genes, in this work we control the probability of 1 in the mask by using the parameter α_u. The possible range of α_u is $[0, 1]$, and $\alpha_u = 0.5$ indicates typical uniform crossover [14]. In this method, when we utilize a small α_u, the number of crossed genes becomes small. On the other hand, when we utilize a large α_u, the number of crossed genes becomes large. $\alpha_u = 0.0$ is equivalent to only mutation without crossover. Also, $\alpha_u = 1.0$ is equivalent to $\alpha_u = 0.0$ because all gene are exchanged in this crossover. In this work we verify the effects of CCG$_{\text{UX}}$ in MOEA as we vary α_u in the range $\alpha_u \in [0.0, 0.5]$

5   Preparation

5.1   Algorithms and Selection Methods

To verify the effectiveness of Local Recombination [8], CCG$_{TX}$ and CCG$_{UX}$, in this work we implement them in NSGA-II [4], SPEA2 [5], IBEA$_{\epsilon+}$ [9] and MSOPS [2], which use different selection methods. NSGA-II and SPEA2 are dominance based MOEAs that use Pareto dominance to determine the superiority of solutions in parent selection. IBEA$_{\epsilon+}$ (Indicator-based Evolutionary Algorithm) introduces fine grained ranking of solutions by calculating fitness value based on the indicator $I_{\epsilon+}$ which measure the degree of superiority for each solution in the population [9]. MSOPS (Multiple single objective Pareto sampling) aggregates fitness vector with multiple weight vectors, and reflects the ranking of solutions calculated for each weight vector in parent selection [2].

According to a previous performance comparison [16], in NSGA-II the convergence of obtained POS gradually deteriorates increasing the number of objectives m, but the diversity of POS significantly increases. On the other hand, POS obtained by IBEA$_{\epsilon+}$ achieves extremely high convergence but scarce diversity. In contrast, MSOPS realizes a well-balanced search between convergence and diversity in MaOPs.

5.2   Problems, Parameters and Metrics

In this paper we use many-objective 0/1 knapsack problems [10] as benchmark problem. We generate problems with $m = \{2, 4, 6, 8, 10\}$ objectives, $n = \{100, 250, 500, 750, 1000\}$ items, and feasibility ratio $\phi = 0.5$. For all algorithms to be compared, we adopt crossover with a crossover rate $P_c = 1.0$, and apply bit-flipping mutation with a mutation rate $P_m = 1/n$. In the following experiments, we show the average performance with 30 runs, each of which spent $T = 2,000$ generations. Population size is set to $N = 200$ ($|P_t| = |Q_t| = 100$). In IBEA$_{\epsilon+}$, scaling parameter κ is set to 0.05 similar to [9]. Also, in MSOPS, we use $W = 100$ uniformly distributed weight vectors [7], which maximizes $Hypervolume$ (HV) [17] in the experiments.

In this work, to evaluate the search performance of MOEAs we use HV, which measures the m-dimensional volume of the region enclosed by the obtained non-dominated solutions and a dominated reference point in objective space. Here we use $r = (0, 0, \cdots, 0)$ as the reference point. Obtained POS showing a higher value of hypervolume can be considered as a better set of solutions from both convergence and diversity viewpoints. To calculate the hypervolume, we use the improved dimension-sweep algorithm proposed by Fonseca et al. [18], which significantly reduces computational time especially for large m. To provide additional information separately on convergence and diversity of the obtained POS, in this work we also use $Norm$ [19] and $Maximum$ $Spread$ (MS) [17], respectively. Higher value of $Norm$ generally means higher convergence to true POS. Although $Norm$ cannot precisely reflect local features of the distribution of the obtained POS, we can observe the general convergence tendency of POS from their values. On the other hand, higher MS indicates better diversity in POS, i.e. a widely spread Pareto front.

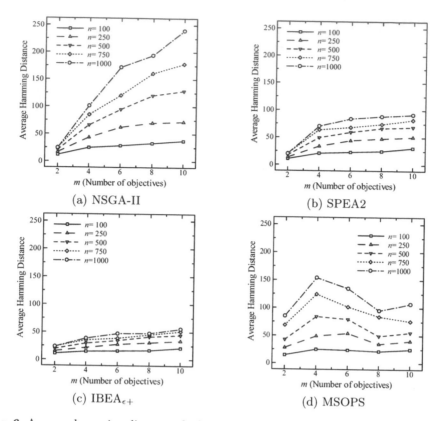

Fig. 6. Average hamming distance of solutions in the population at the final generation

6   Experimental Results and Discussion

6.1   Diversity of Genes in the Population Obtained by Conventional Crossover

First, we observe the diversity of genes in the population at the final generation when conventional two-point crossover is used. Fig.6 shows the average hamming distance of solutions in the population obtained by NSGA-II, SPEA2, IBEA$_{\epsilon+}$ and MSOPS on many-objective 0/1 knapsack problems with $m = \{2, 4, 6, 8, 10\}$ objectives and $n = \{100, 250, 500, 750, 1000\}$ items (bits).

For NSGA-II, SPEA2 and IBEA$_{\epsilon+}$, we can see that the average hamming distance increases as we increase the number of objectives m. This tendency is similar to the aforementioned results obtained by exhaustive search on $n = \{10, 15, 20\}$ bits problems, as shown in Fig.2. On the other hand, although MSOPS shows higher average hamming distance than other MOEAs in $m = \{2, 4\}$ objectives, the average hamming distance tendency is to become short in $m \geq 6$. Also, we can see that the average hamming distance obtained by IBEA$_{\epsilon+}$ is the shortest in all MOEA compared in Fig.6. That is, the population obtained by IBEA$_{\epsilon+}$ is distributed in narrow region of solution space. On the other hand, the population obtained by NSGA-II shows the highest average hamming distance. That is, the population obtained by NSGA-II is widely distributed in solution space. In the case of $m = 10$ objectives and $n = 1,000$ bits, note that the average hamming distance obtained by NSGA-II becomes around 250 bits at the final generation. In this case, if we randomly select two solutions from the population as parents, they will be different in 250 bits out of 1,000 bits. Thus, since diversity of genes in the population obtained by NSGA-II is significantly high in MaOPs, the likelihood that the conventional recombination becomes too disruptive is also high, making it an inefficient genetic operator for solutions search.

6.2   Effects of Local Recombination in MaOPs

Next, we observe the effects of Local Recombination [8] in NSGA-II, SPEA2, IBEA$_{\epsilon+}$ and MSOPS on problems with $n = 1,000$ items (bits) and $m = \{2, 4, 6, 8, 10\}$ objectives. Figs.7~10 show results on HV as a combined metric of convergence and diversity, $Norm$ as a measure of convergence, and MS as a measure of diversity, varying the locality of recombination n_{LR}. In the case of $n_{LR} = 4$, tournament selection for recombination is performed in highest locality. Increasing n_{LR} the locality of recombination decrease, and the conventional recombination is applied when we utilize $n_{LR} = 100$. After we select a pair of parents, the conventional two-point crossover is applied. In these figures, all the plots are normalized by the results of NSGA-II using the conventional two-point crossover.

First, from results of HV in Figs.7~10 (a), we can see that IBEA$_{\epsilon+}$ and MSOPS with conventional recombination ($n_{LR} = 100$) achieve higher HV than dominance based NSGA-II and SPEA2 with conventional recombination as we increase the number of objectives m. When we decrease n_{LR} and enhance the locality of selected pair of parents in objective space, we see improvements on HV by NSGA-II and MSOPS, but not by SPEA2 and IBEA$_{\epsilon+}$. Improvement of HV by NSGA-II becomes significant as we increase the number of objectives m. In the case of $m = 8$ objectives, although HV obtained by NSGA-II with conventional recombination ($n_{LR} = 100$) is lower than IBEA$_{\epsilon+}$ and MSOPS, NSGA-II using Local Recombination with $n_{LR} = 10$ achieves higher HV than MSOPS and comparative with IBEA$_{\epsilon+}$.

Next, from results of $Norm$ in Figs.7~10 (b), SPEA2, IBEA$_{\epsilon+}$ and MSOPS achieve higher $Norm$ than NSGA-II especially for large number of objectives m. We can see that small improvement in $Norm$ is obtained by NSGA-II as we decrease n_{LR}, but SPEA2, IBEA$_{\epsilon+}$ and MSOPS do not improve $Norm$ by varying n_{LR}.

Next, from results of MS in Figs.7~10 (c), all MOEA improve MS by decreasing n_{LR}. MS obtained by SPEA2 and IBEA$_{\epsilon+}$ are relatively lower than NSGA-II and MSOPS. Consequently, average hamming distance of the population by SPEA2 and IBEA$_{\epsilon+}$ becomes short in Fig.6, and these populations are distributed in a relatively narrow region in objective/solution space.

Summarizing, it is difficult to obtain effectiveness of Local Recombination in SPEA2 and IBEA$_{\epsilon+}$ which evolve less diverse populations. On the other hand, although NSGA-II obtain well-spread solutions, convergence towards Pareto

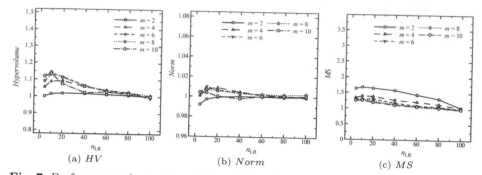

Fig. 7. Performance obtained by NSGA-II [4] with Local Recombination ($n = 1,000$)

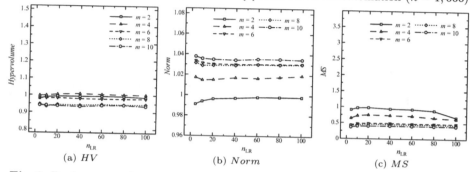

Fig. 8. Performance obtained by SPEA2 [5] with Local Recombination ($n = 1,000$)

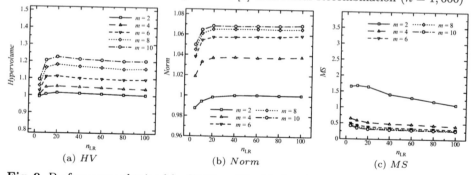

Fig. 9. Performance obtained by IBEA$_{\epsilon+}$ [9] with Local Recombination ($n = 1,000$)

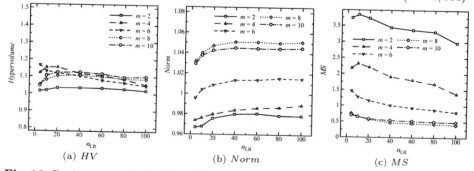

Fig. 10. Performance obtained by MSOPS [2] with Local Recombination ($n = 1,000$)

optimal front is not enough. In contrast, MSOPS achieves well-balanced search between convergence and diversity of obtained solutions. Since NSGA-II and MSOPS achieve relatively high diversity of solutions in objective/solution space, average hamming distance becomes large in Fig.6. Thus, the effectiveness of Local Recombination becomes clear in NSGA-II and MSOPS which evolve well-spread populations. Also, the effectiveness becomes significant increasing the number of objectives m.

6.3 Effects of CCG$_{TX}$ in MaOPs

Next, we observe the effects of CCG$_{TX}$ in NSGA-II, SPEA2, IBEA$_{\epsilon+}$ and MSOPS. Figs.11~14 shows results on HV, $Norm$ and MS varying the parameter α_t. Similar to the previous section, all plots are normalized by the results of NSGA-II using conventional two-point crossover.

First, from results of HV in Figs.11~14 (a), we can see that there is no improvement when we use CCG$_{TX}$ varying α_t in SPEA2 and IBEA$_{\epsilon+}$, which evolve less diverse solutions in the population. On the other hand, NSGA-II and MSOPS having well-spread population significantly improve HV when we set small α_t. Compared with HV achieved by Local Recombination shown in Figs.7~10 (a), the maximum HV obtained by NSGA-II with CCG$_{TX}$ is higher than the maximum HV obtained by NSGA-II with Local Recombination. The same is true for MSOPS with CCG$_{TX}$ compared to MSOPS with Local Recombination. Next, from results of $Norm$ in Figs.11~14 (b), it can be seen that NSGA-II, SPEA2 and MSOPS improve the convergence of obtained POS by using smaller α_t. Also, from results of MS in Figs.11~14 (c), as general tendency, we can see that MS improves by decreasing α_t. It is interesting to note that, although MSOPS achieves the highest HV in $m = 6$ objectives, NSGA-II achieves highest HV in $m = \{8, 10\}$ objectives problems. This is because deterioration of MS in MSOPS becomes significant for large number of objectives. From these results, we conclude that the effectiveness of CCG$_{TX}$ becomes significant especially for NSGA-II and MSOPS because these MOEAs evolve well-spread solutions in the population. Also, HV obtained by CCG$_{TX}$ is higher than HV obtained by Local Recombination especially for large number of objectives. This is because crossover under Local Recombination still could be too disruptive, especially for large m, whereas CCG$_{TX}$ can control better the number of genes being crossed.

6.4 Effects of CCG$_{UX}$ in MaOPs

Next, we observe the effects of CCG$_{UX}$ in NSGA-II, SPEA2, IBEA$_{\epsilon+}$ and MSOPS. Figs.15~18 shows results on HV, $Norm$, and MS varying the parameter α_u. Similar to previous sections, all plots are normalized by the results of NSGA-II using conventional two-point crossover.

Results obtained by CCG$_{UX}$ have similar tendency to results obtained by CCG$_{TX}$ shown in Figs.11~14. However, values of $Norm$ obtained by CCG$_{UX}$ become higher than CCG$_{TX}$. Consequently, CCG$_{UX}$ achieves higher HV than CCG$_{TX}$ due to the improvement of convergence. Overall, values of MS obtained by CCG$_{UX}$ are similar to MS obtained by CCG$_{TX}$.

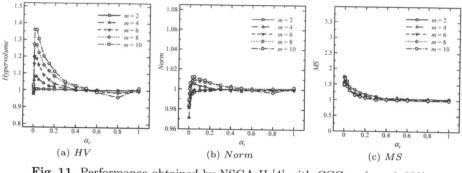

Fig. 11. Performance obtained by NSGA-II [4] with CCG$_{TX}$ ($n = 1,000$)

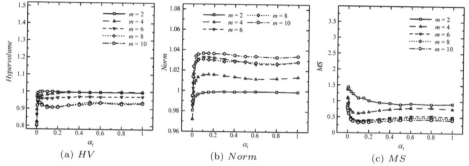

Fig. 12. Performance obtained by SPEA2 [5] with CCG$_{TX}$ ($n = 1,000$)

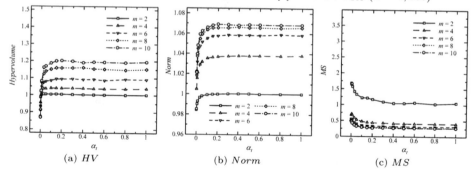

Fig. 13. Performance obtained by IBEA$_{\epsilon+}$ [9] with CCG$_{TX}$ ($n = 1,000$)

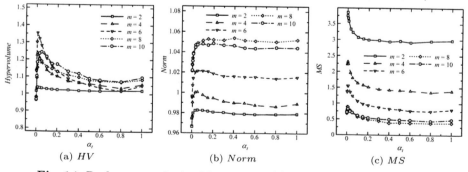

Fig. 14. Performance obtained by MSOPS [2] with CCG$_{TX}$ ($n = 1,000$)

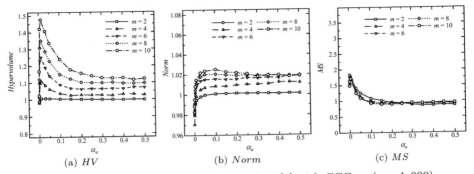

Fig. 15. Performance obtained by NSGA-II [4] with CCG$_{\mathrm{UX}}$ ($n = 1,000$)

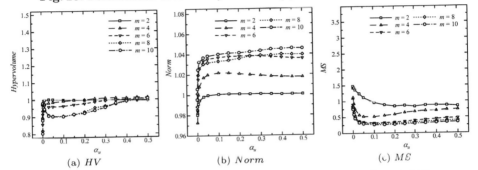

Fig. 16. Performance obtained by SPEA2 [5] with CCG$_{\mathrm{UX}}$ ($n = 1,000$)

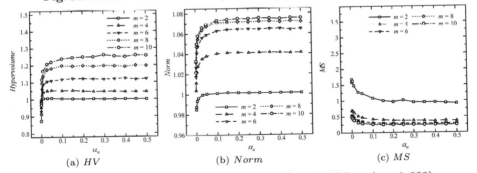

Fig. 17. Performance obtained by IBEA$_{\epsilon+}$ [9] with CCG$_{\mathrm{UX}}$ ($n = 1,000$)

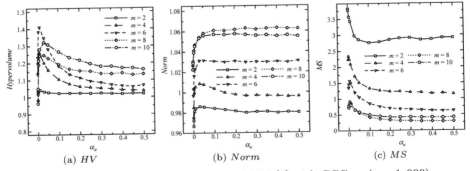

Fig. 18. Performance obtained by MSOPS [2] with CCG$_{\mathrm{UX}}$ ($n = 1,000$)

These results reveal that Local Recombination, CCG_{TX}, and CCG_{UX} improve the search performance significantly. Also, the effectiveness is emphasized when we apply these crossover operators to NSGA-II and MSOPS, which evolve well-spread solutions in objective/solution space. Furthermore, the effectiveness of CCG operators is higher than Local Recombination especially for large number of objectives. Additionally, CCG_{UX} achieves higher HV than CCG_{TX} by enhancing the convergence of obtained POS toward Pareto optimal front. This is because CCG_{UX} can control more precisely the number of genes being crossed than CCG_{TX}.

7   Conclusions

In this work, we have analyzed genetic diversity of Pareto optimal solutions in many-objective optimization problems and studied the effectiveness of crossover for many-objective optimization. First, we analyzed the true Pareto optimal solutions obtained by exhaustive search on many-objective 0/1 knapsack problem with $n = \{10, 15, 20\}$ bits, verifying that the ratio of Pareto optimal solutions in feasible solution space increases with the number of objectives. Also, we observed that genes of Pareto optimal solutions become noticeably diverse, and Pareto optimal solutions come to be distributed nearly uniformly in solution space by increasing the number of objectives m. Then, we used NSGA-II, SPEA2, $IBEA_{\epsilon+}$, and MSPOS, well known multi-objective evolutionary algorithms that adopt different selection methods, to analyze the search performance of conventional recombination, Local Recombination that selects mating parents based on proximity in objective space, and crossover operators Controlling the maximum number of Crossed Genes (CCG). Simulation results on many-objective 0/1 knapsack problems with $m = \{2, 4, 6, 8, 10\}$ objectives reveal that Local Recombination and CCG operators significantly improve search performance, especially for NSGA-II which have high diversity of genes in the population. CCG operators achieve higher search performance than Local Recombination for $m \geq 4$ objectives problems. Also, the effectiveness becomes more significant as the number of objectives m increase.

As future works, we should further analyze Pareto optimal solutions on various many-optimization problems to understand better the relationship between variable and objective space and how to reflect that in the genetic operators the algorithm use. Also, we want to study the effect of the proposed CCG operations in other MOEAs.

References

1. Deb, K.: Multi-Objective Optimization using Evolutionary Algorithms. John Wiley & Sons, Chichester (2001)
2. Hughes, E.J.: Evolutionary Many-Objective Optimisation: Many Once or One Many? In: Proc. IEEE Congress on Evolutionary Computation (CEC 2005), pp. 222–227 (September 2005)

3. Aguirre, H., Tanaka, K.: Working Principles, Behavior, and Performance of MOEAs on MNK-Landscapes. European Journal of Operational Research 181(3), 1670–1690 (2007)

4. Deb, K., Agrawal, S., Pratap, A., Meyarivan, T.: A Fast Elitist Non-Dominated Sorting Genetic Algorithm for Multi-Objective Optimization: NSGA-II, KanGAL report 200001 (2000)

5. Zitzler, E., Laumanns, M., Thiele, L.: SPEA2: Improving the Strength Pareto Evolutionary Algorithm. TIK-Report (103) (2001)

6. Ishibuchi, H., Tsukamoto, N., Nojima, Y.: Evolutionary many-objective optimization: A short review. In: Proc. of 2008 IEEE Congress on Evolutionary Computation (CEC 2008), pp. 2424–2431 (2008)

7. Wagner, T., Beume, N., Naujoks, B.: Pareto-, Aggregation-, and Indicator-Based Methods in Many-Objective Optimization. In: Obayashi, S., Deb, K., Poloni, C., Hiroyasu, T., Murata, T. (eds.) EMO 2007. LNCS, vol. 4403, pp. 742–756. Springer, Heidelberg (2007)

8. Sato, H., Aguirre, H., Tanaka, K.: Local Dominance and Local Recombination in MOEAs on 0/1 Multiobjective Knapsack Problems. European Jour. on Operational Research 181(3), 1670–1690 (2007)

9. Zitzler, E., Künzli, S.: Indicator-Based Selection in Multiobjective Search. In: Yao, X., Burke, E.K., Lozano, J.A., Smith, J., Merelo-Guervós, J.J., Bullinaria, J.A., Rowe, J.E., Tiño, P., Kabán, A., Schwefel, H.-P. (eds.) PPSN 2004. LNCS, vol. 3242, pp. 832–842. Springer, Heidelberg (2004)

10. Zitzler, E., Thiele, L.: Multiobjective optimization using evolutionary algorithms - A comparative case study. In: Eiben, A.E., Bäck, T., Schoenauer, M., Schwefel, H.-P. (eds.) PPSN 1998. LNCS, vol. 1498, pp. 292–304. Springer, Heidelberg (1998)

11. Watanabe, S., Hiroyasu, T., Miki, M.: Neighborhood Cultivation Genetic Algorithm for Multi-Objective Optimization Problems. In: Proc. Genetic and Evolutionary Computation Conference (GECCO 2002), pp. 458–465 (2002)

12. Ishibuchi, H., Shibata, Y.: Mating Scheme for Controlling the Diversity-Convergence Balance for Multiobjective Optimization. In: Deb, K., et al. (eds.) GECCO 2004. LNCS, vol. 3102, pp. 1259–1271. Springer, Heidelberg (2004)

13. Zhang, Q., Li, H.: MOEA/D: A Multi-objective Evolutionary Algorithm Based on Decomposition. IEEE Trans. on Evolutionary Computation 11(6), 712–731 (2007)

14. Syswerda, G.: Uniform Crossover in Genetic Algorithms. In: Proc. of the Third International Conference on Genetic Algorithms (ICGA 1989), pp. 2–9 (1989)

15. Spears, W., De Jong, K.A.: An analysis of multi-point crossover. In: Proc. Foundations of Genetic Algorithms (1990)

16. Sato, H., Aguirre, H., Tanaka, K.: Pareto Partial Dominance MOEA and Hybrid Archiving Strategy Included CDAS in Many-Objective Optimization. In: Proc. IEEE Congress on Evolutionary Computation (CEC 2010), pp. 3720–3727 (2010)

17. Zitzler, E.: Evolutionary Algorithms for Multiobjective Optimization: Methods and Applications, PhD thesis, Swiss Federal Institute of Technology, Zurich (1999)

18. Fonseca, C., Paquete, L., López-Ibáñez, M.: An Improved Dimension-sweep Algorithm for the Hypervolume Indicator. In: Proc. 2006 IEEE Congress on Evolutionary Computation, pp. 1157–1163 (2006)

19. Sato, M., Aguirre, H., Tanaka, K.: Effects of δ-Similar Elimination and Controlled Elitism in the NSGA-II Multiobjective Evolutionary Algorithm. In: Proc. IEEE Congress on Evolutionary Computation (CEC 2006), pp. 3980–3398 (2006)

An Optimal Stopping Strategy for Online Calibration in Local Search

Gianluca Bontempi

Machine Learning Group, Département d'Informatique
Faculté des Sciences, ULB, Université Libre de Bruxelles
1050 Bruxelles - Belgium
gbonte@ulb.ac.be

Abstract. This paper formalizes the problem of choosing online the number of explorations in a local search algorithm as a last-success problem. In this family of stochastic problems the events of interest belong to two categories (success or failure) and the objective consists in predicting when the last success will take place. The application to a local search setting is immediate if we identify the success with the detection of a new local optimum. Being able to predict when the last optimum will be found allows a computational gain by reducing the amount of iterations carried out in the neighborhood of the current solution. The paper proposes a new algorithm for online calibration of the number of iterations during exploration and assesses it with a set of continuous optimisation tasks.

1   Introduction

A stochastic local search algorithm [5] starts from some given solution and tries to find a better solution by performing a number of function evaluations in an appropriately defined neighborhood of the current solution. In case a better solution is found, it replaces the current solution and the local search is continued from there. The stopping criterion for the local exploration is often defined in terms of a maximal number of local function evaluations. This number is a central parameter in a local search algorithm and its setting is not an easy task, since a too low value could prevent the algorithm from finding betters solutions while a too big value would waste precious computational ressources. In an ideal case we would like to stop exploring as soon as the best solution in the given neighborhood is found.

The issue of when to stop in a stochastic setting appears in several applied problems. These problems are formalized by a stochastic setting where the event of interest may take value in one of the two categories: success (1) or failure (0). The objective is to determine when the last success will take place. A well-known instance is the classical *secretary problem* [4] where the employer wants to stop making interviews as soon as the best candidate has been met.

The exploration in local search is an analogous problem since it is useless to continue explore when the best value in the neighborhood of the current state

C.A. Coello Coello (Ed.): LION 5, LNCS 6683, pp. 106–115, 2011.
© Springer-Verlag Berlin Heidelberg 2011

has already been attained. In this context the success event corresponds to the discovery of a better neighbouring solution.

A brilliant and computationally efficient solution to the optimal stopping problem has been proposed by the Odds algorithm of Bruss [3]. This algorithm applies to sequences of n independent events $\mathbf{I}_k, k = 1, \ldots, n$ and consists in a compact formula to derive from the n probabilities of success $p_k = P(\mathbf{I}_k = 1)$ the moment at which the probability of the last success is maximal. In other terms the Bruss algorithm returns an iteration value s such that if we stop at the first success we meet after s steps, the probability that this success will be the last is maximal.

In order to be applied, the Odds formulation requires the probability of success of each event $\mathbf{I}_k$ where $\mathbf{I}_k = 1$ stands for the fact that a new optimum was found by performing k additional explorations. This quantity is not immediately available in a generic optimization task by local search. Our approach consists then in estimating this quantity from data (i.e. a set of N observed value functions) by using a nonparametric approach to compute $P(\mathbf{I}_k = 1)$.

The use of observed function evaluations in order to improve the performance of a local search algorithm is not new. Boyan and Moore [2] proposed a learning approach to improve local search by estimating the best objective function value that can be attained in a neighbourhood of a certain solution. In their approach the goal is to associate to each intermediate solution exploration a measure of quality: this measure can be used to replace the original objective function and make the search smoother. Automated parameter tuning techniques (for a detailed review see [6]) aim also to calibrate the parameters by learning offline the dependency between the value of parameters and the algorithm performance for a class of problems. Examples are the racing algorithm for configuring metaheuristics proposed by [7], the ILS algorithm proposed by [6] or the CALIBRA system proposed by [1].

What is original in our approach with respect to the state-of-the-art is that, by taking advantage of the optimal stopping theory and restricting to consider a specific parameter (i.e. the duration of the exploration phase), we can define an online calibration procedure which is not limited to a specific class of problems nor requires the fitting of a model linking parameter value and performance. The only element which is taken into consideration is the distribution of the function values during an exploration phase which makes possible to define how the probability of finding a better minimum evolves with time. Optimal stopping algorithms can then be used to decide online when it is optimal to stop exploration.

It is important however to remark that, like any online calibration activity, our algorithm demands a computational overhead in order to calculate the optimal number of local function evaluations. Overall, the calibration will be beneficial if the calibration time is compensated by the gain deriving by the smaller number of evaluations.

The rest of the paper is structured as follows. In the next section, we will introduce the basics of optimal stopping and the formula proposed by Bruss.

Section 3 will discuss how the optimal stopping algorithm can be instantiated to address a local search problem. A toy example to visualise the approach is presented in Section 4. Experiments on a set of continuous optimisation tasks to assess the added value of the approach in terms of exploration strategy are presented in Section 5.

2   The Bruss Algorithm

Let us consider a sequence of n independent events and the related indicator functions $\mathbf{I}_k$ such that $\mathbf{I}_k = 1$ means that the kth event is a success. Let us denote $p_k = \text{Prob}\{\mathbf{I}_k = 1\}$, $q_k = 1 - p_k$ and $r_k = p_k/q_k$. An optimal stopping rule is a rule which returns a value K such that the probability $\text{Prob}\{\mathbf{I}_K = 1, \mathbf{I}_{K+1} = 0, \dots, \mathbf{I}_n = 0\}$ is maximised.

Bruss [3] demonstrates that the optimal rule for stopping consists in stopping at the first index (if any) K such that $I_K = 1$ and $K \geq s$ where

$$s = \sup\left\{1, \sup\left\{1 \leq k \leq n : \sum_{j=k}^{n} r_j \geq 1\right\}\right\} \tag{1}$$

In other terms the stopping algorithm lists in reversed order the terms r_k and computes $R_k = r_n + r_{n-1} + \cdots + r_k$. The algorithms returns the value s as the first value k when R_k equals or exceeds 1.

3   The Estimation of the Probability of Success in Local Exploration

Let us consider a continuous optimization problem

$$x^* = \arg\min f(x), \quad x \in \mathbb{R}^d$$

where the function f is available. Suppose we adopt a stochastic local search strategy. Once initialized with the solution $x^{(0)}$, the SLS algorithm iterates these two phases: (i) a random search (exploration phase) in the neighborhood $\mathcal{N}(x^{(i)})$ of $x^{(i)}$ up to a stopping criterion is met (typically a maximum number n of iterations is reached) and (ii) an update of the solution (exploitation phase) to $x^{(i+1)}$ where $x^{(i+1)} = \arg\min_{x_k, k=1,\dots,n} f(x_k)$ is the best solution in the neighborhood.

Let $\mathbf{x}_k$ be the random variable denoting the kth solution assessed in the neighborhood of $x^{(i)}$. If we intend to apply the optimal stopping terminology to the context of local search exploration, a success event corresponds to the discovery at the step k of the exploration phase of a solution x_k whose function value is smaller than all the $f(x_j), j < k$ assessed so far.

The adoption of the Bruss algorithm to address the problem of stopping in local search exploration requires then the availability of the values $p_k, k = 1, \dots, n$ where

$$p_k = \text{Prob}\left\{f(\mathbf{x}_k) < \min_{1 \leq j < k} f(\mathbf{x}_j)\right\} \quad \text{and} \quad \mathbf{x}_k, \mathbf{x}_j \in \mathcal{N}(x^{(i)}). \tag{2}$$

An analytical computation of the terms p_k is typically not feasible for complex nonlinear functions f. For that reason we propose in this paper a data based resampling approach to compute the values p_k on the basis of a small number of explorations. The rationale of our approach consists in collecting first a training set of values $D_N = \{f(x_k)\}$ with $k = 1, \ldots, N$ with N sufficiently large to allow a reliable estimation of the distribution of $f(\mathbf{x}_k)$. The parametric bootstrap estimation of p_k for $k > N$ is obtained by generating first B samples distributed according to the empirical distribution of $f(\mathbf{x}_k)$ and then counting the frequency of the event characterized by the kth value smaller than the minimum of the previous ones. Once the estimates $\hat{p}_k$ and $\hat{r}_k$ are computed we proceed with a plug-in estimation of the optimal number of exploration steps by using the equation (1).

The resulting exploration algorithm implementing the optimal stopping criterion when the current best solution is $x^{(i)}$ is resumed in Algorithm 1. Note that the nonparametric estimation of the probability of success is performed in the for loop at lines 8-15 and that the repeat loop in 20-23 has the role of waiting for the first success event after the sth iteration.

Algorithm 1: SLS exploration with optimal stopping

1: **Input:** n: max number of explorations
2: **for** $j = 1$ to N **do**
3:     Sample $x_j \in \mathcal{N}(x^{(i)})$ and compute $f(x_j)$
4:     $D_N = D_N \cup f(x_j)$
5: **end for**
6: **for** $k = 2$ to n **do**
7:     $\hat{p}_k = 0$
8:     **for** $b = 1$ to B **do**
9:         Generate an iid vector V_b of size k according to
          the empirical distribution of D_N
10:        **if** $V_b[k] < \min(V_b[1 : k - 1])$ **then**
11:            $\hat{p}_k = \hat{p}_k + 1$
12:        **end if**
13:    **end for**
14:    $\hat{p}_k = \frac{\hat{p}_k}{B}, \hat{r}_k = \frac{\hat{p}_k}{1 - \hat{p}_k}$
15: **end for**
16: Compute s by (1)
17: **for** $j = N + 1$ to s **do**
18:    Sample $x_j \in \mathcal{N}(x^{(i)})$ and compute $f(x_j)$
19: **end for**
20: **repeat**
21:    Sample $x_j \in \mathcal{N}(x^{(i)})$ and compute $f(x_j)$
22:    $j = j + 1$
23: **until** $(f(x_j) < \min_{h=1,\ldots,j-1} f(x_h))$ OR $(j \geq n)$

---

4   Illustration of the Approach

In order to ilustrate how the Odds algorithm works in a specific local search, let us consider a toy example where two univariate ($d = 1$) functions f_1 and f_2 have to be minimised in the neighbourhood of the current state $x^{(i)} = 0.5$. Suppose that the landscapes of the two functions in $\mathcal{N}(x^{(i)})$ are different. The first landscape, illustrated in Figure 1a, is a plateau where most of the values of the function f_1 are close to the local minimum. The second one, illustrated in Figure 1b, is a *reverse* plateau where the function f_2 is essentially constant apart from two peaks at boundaries which correspond to two local minima. Intuitively, the first landscape is the least interesting to explore. A small number of repetitions is sufficient to realise that the neighborhood is not worthy to be explored further. The second landscape instead is more interesting from a minimisation point of view and would deserve a larger number of exploration steps. Let us see how this intuitive notion can be put in the language of optimal stopping. In order to visualise the notions introduced in the previous section we will have recourse to the illustration of the densities of $f_l(\mathbf{x}_k)$ and $\min_{1 \leq j < k} f_l(\mathbf{x}_j)$ ($l = 1, 2$) when $\mathbf{x}_k$ and $\mathbf{x}_j$ are sampled uniformly in the neighborhood of $x^{(i)}$. Note that all the densities are obtained by Monte Carlo simulation.

The density functions $f_1(\mathbf{x}_k)$ and $f_2(\mathbf{x}_k)$, for $\mathbf{x}_k$ sampled uniformly in the neighborhood of $x^{(i)}$, are illustrated in Figures 2a and 2b, respectively. The density functions of the random variables $\min_{1 \leq j < k} f_1(\mathbf{x}_j)$ and $\min_{1 \leq j < k} f_2(\mathbf{x}_j)$ for $k = 5$ and $k = 10$ are illustrated in Figures 3 and 4, respectively. It is interesting to remark that in the case of the plateau landscape the distribution of $f_1(\mathbf{x})$ and $\min(f_1(\mathbf{x})$ tend to overlap while this is not the case for the reverse plateau. In qualitative terms this means that in the plateau configuration it is less probable that a new exploration ($\mathbf{x}_k$) returns an objective function value ($f_1(\mathbf{x}_k)$) which is lower than the ones found so far ($\min_{j < k}(f_1(\mathbf{x}_j))$.

Quantitatively this can be shown by estimating by Monte Carlo the values of p_k (Equation 2) for $k = 5$ and $k = 10$. We obtain in the case of the plateau $p_5 = \text{Prob}\{f_1(\mathbf{x}_5) < \min_{1 \leq j < 5} f_1(\mathbf{x}_j)\} \approx 0.185$ and $p_{10} \approx 0.081$. In the case of reverse plateau the probabilities are bigger: $p_5 = \text{Prob}\{f_2(\mathbf{x}_5) < \min_{1 \leq j < 5} f_2(\mathbf{x}_j)\} \approx 0.197$ and $p_{10} \approx 0.1$. This implies that the Odds algorithm will return an higher value of s for the function f_2 than for the function f_1, then proposing a number of local iterations larger in the case of the reverse plateau and consequently confirming the initial intuition.

5   Experiments

The experimental session aims at assessing the optimisation performance of the local search strategy for a fixed budget H of function evaluations in two alternative configurations: in the first one a fixed number of iterations n of the exploration phase is set a priori while in the second one the number of iterations is adaptively determined by the optimal stopping algorithm described in the previous section. In order to assess the effectiveness of the proposed procedure we

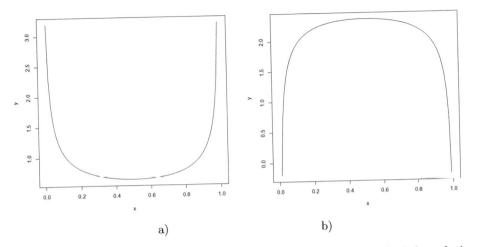

a)   b)

Fig. 1. Left: plateau landscape of function f_1 in the neighborhood of the solution $x^{(i)} = 0.5$. Right: reverse plateau landscape of function f_2 in the neighborhood of the solution $x^{(i)} = 0.5$.

consider a set of 10 multidimensional test functions commonly used in continuous optimisation [8]. The set of functions, to be minimised, is detailed in Table 1. For each of these functions we consider all the values d of the dimensionality between 2 and 40.

We performed a set of local searches where the total number of function evaluations $H = 500d$ is set as a function of the dimensionality, n takes values in the set $\{50, 100, 200, 300, 500\}$ and the exploration step consists in a random sampling according to a Normal distribution centered in the best solution so far and with a standard deviation σ. In order to consider different exploration settings σ is taken equal to $\frac{3u}{\sqrt{d}}$ where u is uniformly sampled in $[0, 1]$. Note also that the number N of evaluations required before estimating the probability of success in Algorithm 1 is set to 50. This implies that the number of iterations proposed by the Odds algorithm is always greater or equal than 50. Since the number returned by the Odds strategy is contained in the range of fixed values n, this allows a fair evaluation of the Odds strategy with respect to local search strategies relying on a fixed number of evaluations. For each value of d we performed 25 paired repetitions such that the initial conditions and the sequence of exploration steps performed by the different local searches are identical. The only allowed difference is the amount of local functions evaluations allocated to each exploration phases.

The results are organized in two tables: Table 2 shows the attained minima of the test functions, averaged over multiple runs and over different values of d. Table 3 presents the attained minima for the different values d of the dimensionality and averaged over the test functions. The bold notation is used when the minimum attained (on average) by the fixed strategy is significantly different (paired permutation test, pv< 0.05) from the one attained by the Odds strategy.

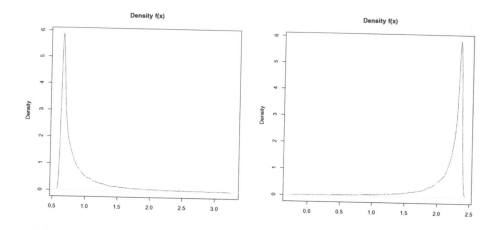

Fig. 2. Density of $f_1(\mathbf{x}_k)$ (left) and $f_2(\mathbf{x}_k)$ (right) for $\mathbf{x}_k$ sampled uniformly in the neighborhood of $x^{(i)}$

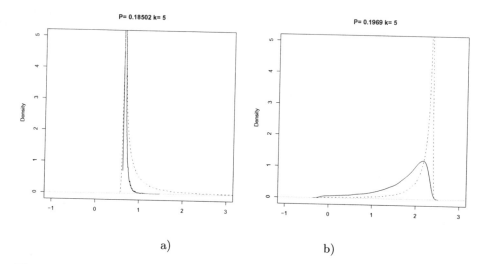

Fig. 3. Density (black line) of $\min_{1 \leq j < k} f_1(\mathbf{x}_j)$ (left) and $\min_{1 \leq j < k} f_2(\mathbf{x}_j)$ (right) for $k = 5$ and $\mathbf{x}_j$ sampled uniformly in the neighborhood of $x^{(i)}$. For the sake of comparison the density of $f_l(\mathbf{x}_k), l = 1, 2$ is reported in dotted line.

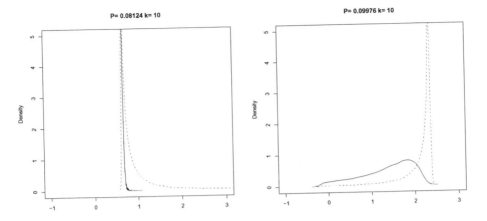

Fig. 4. Density (black line) of $\min_{1 \leq j < k} f_1(\mathbf{x}_j)$ (left) and $\min_{1 \leq j < k} f_2(\mathbf{x}_j)$ (right) for $k = 10$ and $\mathbf{x}_j$ sampled uniformly in the neighborhood of $x^{(i)}$. For the sake of comparison the density of $f_l(\mathbf{x}_k), l = 1, 2$ is reported in dotted line.

Table 1. Benchmark nonlinear test functions

Name	Function
Rosenbrock	$f(x) = \sum_{i=1}^{d-1} \left[100(x_{i+1} - x_i^2)^2 + (1 - x_i)^2 \right]$
Michalewicz	$f(x) = -\sum_{i=1}^{d} \sin(x_i) \left[\sin\left(\frac{i x_i^2}{\pi}\right) \right]^{20}$
Langermann	$f(x) = \sum_{i=1}^{5} c_i \exp\left[-\frac{1}{\pi} \sum_{j=1}^{d} (x_j \quad a_{ij})^2 \right] \cos\left[\pi \sum_{j=1}^{d} (x_j - a_{ij})^2 \right]$
De Jong	$f(x) = \sum_{i=1}^{d} x_i^2$
axis-parallel	$f(x) = \sum_{i=1}^{d} (i x_i^2)$
rotated hyper-ell	$f(x) = \sum_{i=1}^{d} \sum_{j=1}^{i} x_j^2$
Rastrigin	$f(x) = 10d + \sum_{i=1}^{d} \left[x_i^2 - 10 \cos(2\pi x_i) \right]$
Schwefel	$f(x) = \sum_{i=1}^{d} \left[-x_i \sin\left(\sqrt{|x_i|} \right) \right]$
Griewangk	$f(x) = \frac{1}{4000} \sum_{i=1}^{d} x_i^2 - \prod_{i=1}^{d} \cos\left(\frac{x_i}{\sqrt{i}} \right) + 1$
Ackley	$f(x) = -20 \exp\left(-0.2 \sqrt{\frac{1}{d} \sum_{i=1}^{d} x_i^2} \right) - \exp\left(\frac{1}{d} \sum_{i=1}^{d} \cos(2\pi x_i) \right) + 20 + \exp(1)$

The lines W-L contains the number of times that the Odds strategy is significantly better (wins) or worse (loss) than the strategy with a fixed number of exploration iterations.

The experimental results show that:

- if we consider exploration phases with a fixed number of iterations, we observe that i) the optimal strategy depends on the considered data set and, 2) on average the best strategy is the one with $n = 500$.
- the Odds strategy is competitive with the $n = 500$ strategy,
- the Odds strategy is significantly better than the other exploration strategies, when the results are stratified both per function and per dimensionality.

Since the Odds strategy is competitive with the best fixed approach, this means that the Odds strategy is able to adapt in an efficient manner the number of

Table 2. Attained minima averaged over different values of d and different runs. The last line accounts for the number of times that the Odds strategy is better-worse than the local search with a fixed number n of iterations per exploration.

Name	ODD	$n = 500$	$n = 300$	$n = 200$	$n = 100$	$n = 50$
Rosenbrock	256.73	258.91	**264.3**	**269.53**	**285.06**	**316.1**
Michalewicz	-8.3	-8.32	-8.31	-8.26	**-8.15**	**-7.95**
Langermann	-1.32	**-1.15**	**-1.21**	**-1.27**	-1.32	-1.29
De Jong	1.55	1.56	**1.58**	**1.63**	**1.68**	**1.8**
axis-parallel	21.46	**20.95**	21.54	**22.34**	**23.49**	**25.3**
rotated hyper-ell	21.09	20.99	21.06	**21.7**	**22.95**	**24.79**
Rastrigin	90.76	**89.16**	91.01	**92.66**	**94.86**	**99.44**
Schwefel	9152.33	**9180.05**	**9174.5**	**9167.23**	**9157.84**	**9146.79**
Griewangk	4.15e-4	4.18e-4	4.19e-4	**4.31e-4**	**4.46e-4**	**4.78e-4**
Ackley	1.82	**1.8**	1.82	**1.84**	**1.86**	**1.92**
W-L		2-3	4-0	9-0	9-0	8-1

Table 3. Attained minima averaged over different test functions and different runs. The last line accounts for the number of times that the Odd strategy is better-worse than the local search with a fixed number n of iterations per exploration.

d	ODD	$n = 500$	$n = 300$	$n = 200$	$n = 100$	$n = 50$
2	124.29	124.38	**124.39**	**124.36**	124.31	124.32
10	293.5	**294.35**	**294.34**	**294.18**	**294.09**	**294.55**
20	469.92	470.28	470.19	470.1	470.98	**472.69**
30	733.63	735.74	734.92	734.8	735.2	**739.64**
40	910.92	912.39	912.03	912.22	914.03	**919.98**
W-L		1-0	2-0	3-0	1-0	4-0

local iterations to different functions, neighbourhood sizes, dimensionality and local landscapes. Note that such result is obtained in an online fashion without using any apriori knowledge of the problem or previous offline analysis of the algorithm performance.

6   Conclusion and Future Work

An important issue in optimisation is the automatic calibration of the hyperparameters of the algorithms. A stochastic search algorithm produces during its execution a bunch of data about the evolution of the objective function and the attained minimum. It is then intuitive to take advantage of the information hidden within this data to know more about the algorithm performance and try to better calibrate its parameters. So far, most of the calibration techniques require an off-line multi-instances procedure. This paper showed that optimal stopping theory can play an important role if we want to calibrate online the duration of the exploration phase. This work shows some promising, yet preliminary results, in the case of continuous optimisation. Several issues are still open and are worthy to be investigated further in the future: among them, we mention the comparison with state-of-the-art offline methods, the implementation of alternative estimation procedures of the probability of success and the extension to combinatorial optimisation tasks.

Acknowledgments

The author wish to thank Souhaib Ben Taieb as well as the three anonymous reviewers for useful comments and remarks.

References

1. Adenso-Daz, B., Laguna, M.: Fine-tuning of algorithms using fractional experimental design and local search. Operations Research 54 (2006)
2. Boyan, J.A., Moore, A.W.: Learning evaluation functions to improve optimization by local search. Journal of Machine Learning Research 1, 77–112 (2001)
3. Bruss, F.T.: Sum the odds to one and stop. Annals of Probability 28, 1384–1391 (2000)
4. Freeman, P.R.: The secretary problem and its extensions: a review. International Statistical Review 51, 189–206 (1983)
5. Hoos, H.H., Stuetzle, T.: Stochastic Local Search. Foundations and Applications. Morgan Kaufmann, San Francisco (2004)
6. Hutter, F., Hamadi, Y.: Parameter adjustment based on performance prediction: Towards an instance-aware problem solver. Technical report, Department of Computer Science University of British Columbia (2005)
7. Paquete, L., Birattari, M., Stuetzle, T., Varrentrapp, K.: A racing algorithm for configuring metaheuristics. In: Proceedings of the Genetic and Evolutionary Computation Conference (GECCO 2002), pp. 11–18. Morgan Kaufmann Publishers, San Francisco (2002)
8. Molga, M., Smutnicki, C.: Test functions for optimization needs. Technical report (2005), http://www.zsd.ict.pwr.wroc.pl/files/docs/functions.pdf

Analyzing the Effect of Objective Correlation on the Efficient Set of MNK-Landscapes

Sébastien Verel[1,3], Arnaud Liefooghe[2,3],
Laetitia Jourdan[3], and Clarisse Dhaenens[2,3]

[1] University of Nice Sophia Antipolis – CNRS, France
[2] Université Lille 1, LIFL – CNRS, France
[3] INRIA Lille-Nord Europe, France
verel@i3s.unice.fr, arnaud.liefooghe@univ-lille1.fr,
laetitia.jourdan@inria.fr, clarisse.dhaenens@lifl.fr

Abstract. In multiobjective combinatorial optimization, there exists two main classes of metaheuristics, based either on multiple aggregations, or on a dominance relation. As in the single-objective case, the structure of the search space can explain the difficulty for multiobjective metaheuristics, and guide the design of such methods. In this work we analyze the properties of multiobjective combinatorial search spaces. In particular, we focus on the features related the efficient set, and we pay a particular attention to the correlation between objectives. Few benchmark takes such objective correlation into account. Here, we define a general method to design multiobjective problems with correlation. As an example, we extend the well-known multiobjective NK-landscapes. By measuring different properties of the search space, we show the importance of considering the objective correlation on the design of metaheuristics.

1   Introduction

Multiobjective combinatorial optimization (MoCO) problems, where several criteria have to be optimized simultaneously, receive more and more interest in the field of search algorithms. One of the main issues in multiobjective optimization is the Pareto dominance relation, which gives a partial order between feasible solutions. Roughly speaking, a given solution dominates another solution if it is better according to all objective functions. A possible approach in solving a multiobjective problem consists in finding the whole set of non-dominated solutions, called the *efficient set*, or a subset that is close to it. This efficient set plays a central role in the structure of the search space.

The design of metaheuristics for multiobjective combinatorial optimization is a real challenge, as it is problem-dependent. Like in single-objective optimization, the structure of the search space can explain the ability of multiobjective metaheuristics. Two main classes of multiobjective metaheuristics can be distinguished. The first ones, known as scalar approaches, are based on multiple scalarized aggregations of the objective functions. However, they are only able

C.A. Coello Coello (Ed.): LION 5, LNCS 6683, pp. 116–130, 2011.
© Springer-Verlag Berlin Heidelberg 2011

to find a subset of efficient solutions, called supported efficient solutions. The second ones, known as Pareto-based approaches, directly or indirectly focus the search on the Pareto dominance relation. Moreover, when the size of the efficient set is too large, a metaheuristic should manipulate a limited-size solution set during the search, and this limit is related to the size of the efficient set. In addition, connectedness is related to the property that efficient solutions are connected with respect to a neighborhood relation [1]. When connectedness holds, it becomes possible to find the whole efficient set by iteratively exploring the neighborhood of the current approximation, initialized with at least one efficient solution. This strategy is often used explicitly, or implicitly by Pareto-based approaches. For the design of metaheuristics for MoCO, three main questions, related to the efficient set properties, are of our interest in this paper:

(*i*) What is the cardinality of the efficient set? Can we pretend to identify or approximate the whole set of efficient solutions, or should we consider a mechanism to bound the size of the approximation set?

(*ii*) How many efficient solutions are supported? Is a scalar approach able to identify or approximate enough efficient solutions?

(*iii*) Are efficient solutions connected with respect to a neighborhood operator? Is it possible to identify or approximate additional efficient solutions by a simple local search initialized with a subpart of the efficient set?

In particular we want to study such properties according to the objective correlation, as it seems to largely affect the solutions of MoCO problems [2] and the behavior of metaheuristics [3]. Few benchmark takes the correlation between objectives into account. To the best of our knowledge, the multiobjective quadratic assignment problem [4] should be the single one. In this problem, a parameter can tune the correlation between different pairs of objectives. Another well-known benchmark, the multiobjective NK-landscapes [5] facilitate the study of problem structure in multiobjective optimization. In this class, the epistatic degree, which is the degree of non-linearity of the problem, can be tuned very precisely. In this work, in order to study the problem structure, and in particular the structure of the efficient set, we define a general method to tune the correlation between all pairs of objectives very precisely. As an example, we define the multiobjective ρMNK-landscapes, an extension of multiobjective NK-landscapes with objective correlation. With such a benchmark, we can study the problem structure according to the objective space dimension, the epistasis and especially the objective correlation, and then highlight some guidelines for the design of efficient multiobjective metaheuristics.

In summary, the contributions of this work can be stated as follows. First, we propose a method to precisely tune the correlation between objective functions. It is applied to the design of MNK-landscapes, but it can easily be generalized to other problems. Second, we show the influence of the objective correlation on some properties of the efficient set (and its image in the objective space): its size, the proportion of supported solutions, and the connectedness of efficient solutions. Third, we bring those properties with the design of local search metaheuristics in order to help the practitioner to make proper choices between several classes of methodologies. The reminder of the paper is organized as follows. Section 2 is dedicated to multiobjective combinatorial optimization, multi-objective metaheuristics, as well as single- and multi-objective NK-landscapes. Section 3 presents the design of ρMNK-landscapes. We conduct a theoretical analysis and an experimental study to show the sharpness of the objective correlation. Section 4 deeply analyzes the efficient set structure on this new class of problems according to the objective space dimension, the non-linearity and especially the objective correlation. The consequence on the design of multiobjective metaheuristics are discussed in the last section.

2   Background

2.1   Multiobjective Combinatorial Optimization

A large number of real-world optimization problems are multiobjective by nature, because several criteria have to be considered simultaneously. A MoCO problem can be defined by a set of $M \geq 2$ objective functions $(f_1, f_2, \ldots, f_M)$, and a discrete set X of feasible solutions in the *decision space*. Let $Z = f(X) \subseteq \mathbb{R}^M$ be the set of feasible outcome vectors in the *objective space*. In a maximization context, a solution $x \in X$ dominates a solution $x' \in X$, denoted by $x \succ x'$, iff $\forall i \in \{1, 2, \ldots, M\}$, $f_i(x) \geq f_i(x')$ and $\exists j \in \{1, 2, \ldots, M\}$ such as $f_j(x) > f_j(x')$. A solution $x \in X$ is said to be *efficient* (or *non-dominated, Pareto optimal*), if there does not exist any other solution $x' \in X$ such that x' dominates x. The set of all efficient solutions is called the *efficient set* (or *Pareto optimal set*), denoted by X_E, and its mapping in the objective space is called the *Pareto front*. A possible approach in MoCO is to identify a minimal complete efficient set, *i.e.* one efficient solution mapping to each point of the Pareto front.

   However, generating the entire efficient set of a MoCO problem is often infeasible for two main reasons [6]. First, for most MoCO problems, the number of efficient solutions is known to be exponential in the size of the problem instance. In that sense, most MoCO problems are said to be *intractable*. Second, deciding if a feasible solution belongs to the efficient set is NP-complete for numerous MoCO problems, even if none of its single-objective counterpart is NP-hard. Therefore, the overall goal is often to identify a good efficient set approximation. To this end, metaheuristics in general, and evolutionary algorithms in particular, have received a growing interest since the late eighties, and multiobjective metaheuristics still constitute an active research area.

2.2   Metaheuristics for Multiobjective Combinatorial Optimization

Two main classes of metaheuristics for MoCO can be distinguished, see for instance [7]. The first ones, known as scalar approaches, are based on multiple scalarized aggregations of the objective functions. The second ones, known as Pareto-based approaches, directly or indirectly focus the search on the Pareto dominance relation (or a slight modification of it). These two kinds of approaches can also be hybridized in a two-phase way.

Initial approaches dealing with MoCO are based on successive transformations of the original multiobjective problem into single-objective ones by means of a scalarization strategy. Most of the time, *scalar approaches* are based on a weighted-sum aggregation of the objective functions, that can be defined as follows. $\forall x \in X$: $f_\lambda(x) = \sum_{i=1}^{M} \lambda_i\, f_i(x)$ where $\lambda_i > 0$ for all $i \in \{1, \ldots, M\}$. The problem is now to identify a (single) solution that maximizes f_λ. For any given weighting coefficient vector λ, if $x^\star = arg\max_{x \in X} f_\lambda(x)$, then $x^\star$ is an efficient solution. Multiple weighting coefficient vectors can be iteratively defined so that several non-dominated solutions are identified (or approximated). For each scalarization, the corresponding solution is incorporated into an approximation set, whose dominated solutions are then discarded. However, in the combinatorial case, a number of efficient solutions are not optimal for any definition of f_λ. They are known as *non-supported (efficient) solutions*. On the contrary, there exists *supported (efficient) solutions* whose corresponding objective vectors are located on the convex hull of the Pareto front. The set of all supported efficient solutions will be denoted by X_{SE}. As a consequence, the proportion of non-supported solutions over the efficient set has a direct implication on the ability of scalar approaches to find a proper non-dominated set approximation.

Over the years, other types of approaches were proposed. They are based on the explicit or implicit use of the Pareto dominance relation, that allows to define a partial order between feasible solutions. The basic idea is to maintain a set solutions (typically a population or an archive of mutually non-dominated solutions). The content of this set is then iteratively updated with new solutions built by means of variation or neighborhood operators. The update of this set is based on a specific decision on which solutions to accept or to choose for further manipulation. This process is iterated until no further improvement is possible or another stopping condition is fulfilled. In the end, this set corresponds to the approximation outputted by the algorithm. The implicit goal is to identify an approximation whose image in the objective space is (*i*) close to and (*ii*) well-spread along the Pareto front. However, as the number of efficient solutions is often intractable, we generally have to design specific strategies to limit the size of the approximation set [8]. As a consequence, the *cardinality* of the efficient set also plays a major role on the design of multiobjective metaheuristics.

More recently, the neighborhood structure of the efficient set has been claimed to play a crucial role for the development of efficient metaheuristics. One of these properties is known as *connectedness* [1,9]. Let us define a graph such that each node represents an efficient solution, and an edge connects a pair of nodes if the corresponding solutions are neighbors with respect to a given neighborhood operator [1]. This graph is called the *efficient graph*. A neighborhood operator is a function $\mathcal{N} : X \to 2^X$ that assigns a set of solutions $\mathcal{N}(x) \subset X$ to any solution $x \in X$. $\mathcal{N}(x)$ is called the *neighborhood* of x, and a solution $x' \in \mathcal{N}(x)$ is called a *neighbor* of x. The efficient set is said to be *connected* if there exists a path between every pair of nodes in the graph. In other words, each efficient solution is located in the neighborhood of at least one other solution from the efficient set. This property has later been extended to the notion of cluster by introducing an arbitrary distance separating two efficient solutions [10]. When connectedness holds, it becomes possible to find all the efficient solutions by means of the iterative exploration of the neighborhood of the current approximation by starting with one (or more) solution(s) from the efficient set. This gives rise to a *two-phase* approach: (*i*) identify a number of (typically supported) non-dominated solutions (*ii*) improve the set of non-dominated solutions by exploring their neighborhood.

2.3   *NK*- and *MNK*-Landscapes

The family of NK-landscapes [11] is a problem-independent model used for constructing multimodal landscapes. N refers to the number of (binary) genes in the genotype (*i.e.* the string length) and K to the number of genes that influence a particular gene from the string (the epistatic interactions). By increasing the value of K from 0 to $(N-1)$, NK-landscapes can be gradually tuned from smooth to rugged. The fitness function (to be maximized) of a NK-landscape $f_{NK} : \{0,1\}^N \rightarrow [0,1)$ is defined on binary strings with N bits. An 'atom' with fixed epistasis level is represented by a fitness component $f_i : \{0,1\}^{K+1} \rightarrow [0,1)$ associated to each bit $i \in N$. Its value depends on the allele at bit i and also on the alleles at K other epistatic positions (K must fall between 0 and $N-1$). The fitness $f_{NK}(x)$ of a solution $x \in \{0,1\}^N$ corresponds to the mean value of its N fitness components f_i:

$$f_{NK}(x) = \frac{1}{N} \sum_{i=1}^{N} f_i(x_i, x_{i_1}, \ldots, x_{i_K})$$

where $\{i_1, \ldots, i_K\} \subset \{1, \ldots, i-1, i+1, \ldots, N\}$. Several ways have been proposed to set the K bits from the bit string of size N. Two possibilities are mainly used: adjacent and random neighborhoods. With an adjacent neighborhood, the K bits nearest to the bit $i \in N$ are chosen (the genotype is taken to have periodic boundaries). With a random neighborhood, the K bits are chosen randomly on the bit string. Each fitness component f_i is specified by extension, *i.e.* a number $y^i_{x_i, x_{i_1}, \ldots, x_{i_K}}$ from $[0,1)$ is associated with each element $(x_i, x_{i_1}, \ldots, x_{i_K})$ from $\{0,1\}^{K+1}$. Those numbers are uniformly distributed in the range $[0,1)$.

More recently, a multiobjective variant of NK-landscapes (namely MNK-landscapes) [5] have been defined with a set of M fitness functions:

$$\forall m \in [1, M], \ f_{NK_m}(x) = \frac{1}{N} \sum_{i=1}^{N} f_{m,i}(x_i, x_{i_{m,1}}, \ldots, x_{i_{m,K_m}})$$

The numbers of epistasis links K_m can theoretically be different for each fitness function. But in practice, the same epistasis degree $K_m = K$ for all $m \in [1, M]$ is used. Each fitness component $f_{m,i}$ is specified by extension with the numbers $y^{m,i}_{x_i, x_{i_{m,1}}, \ldots, x_{i_{m,K_m}}}$. In the original MNK-landscapes [5], these numbers are randomly and independently drawn from $[0,1)$. As a consequence, it is very unlikely that two different solutions map to the same point in the objective space.

3 ρMNK-Landscapes: Multiobjective NK-Landscapes with Correlation

In this section, we define the $CMNK$- and the ρMNK-landscapes, which are based on the MNK-landscapes [5]. In this multiobjective model, the correlation between objective functions can be precisely tuned by a correlation matrix. It allows to study the simultaneous influence of objective space dimension, non-linearity and objective correlation on the main properties of multiobjective fitness landscapes. The construction of landscapes is defined and the analytic proof of the correlation between objectives, completed with an experimental study, are given. Note that the proposed approach to tune the objective correlation can be applied to other MoCO problems where the objective functions are summing objectives, share the same definition, but are computed with different cost or profit matrices. This is the case, for instance, of the multiobjective knapsack, traveling salesman and quadratic assignment problems [4,6].

3.1 Definition

In the proposed $CMNK$-landscapes, the epistasis structure is identical for all the objective functions: $\forall m \in [1, M]$, $K_m = K$ and $\forall m \in [1, M]$, $\forall j \in [1, K_m]$, $i_{m,j} = i_j$. The fitness components are not defined independently. The numbers $(y^{1,i}_{x_i, x_{i_1}, \ldots, x_{i_K}}, \ldots, y^{M,i}_{x_i, x_{i_1}, \ldots, x_{i_K}})$ follow a multivariate uniform law of dimension M, defined by a correlation matrix C. Thus, the y's follow a multidimensional law with uniform marginals and the correlations between $y^{m,i}_{\ldots}$'s are defined by the matrix C. So, the four parameters of the family of $CMNK$-landscapes are (i) the number of objective functions M, (ii) the length of the bit string N, (iii) the number of epistatic links K, and (iv) the correlation matrix C.

The matrix C is a symmetric positive-definite matrix where $\frac{M(M-1)}{2}$ numbers can be defined. In order to limit the number of free numbers in matrix C, we define the matrix $C_\rho = (c_{np})$ which has the same correlation between all the objectives: $c_{nn} = 1$ for all n, and $c_{np} = \rho$ for all $n \neq p$. In this case, we denote $CMNK$-landscapes by ρMNK-landscapes, and the original MNK-landscapes are equivalent to ρMNK-landscapes with $\rho = 0$. However, it is not possible to have the matrix C_ρ for all ρ between $[-1, 1]$. C_ρ must be positive-definite: $\forall u \in \mathbb{R}^M$, $u^t C_\rho u \geq 0$. So, ρ must be greater than $\frac{-1}{M-1}$. For two-objective problems, all the correlations between $[-1, 1]$ are possible. However, for three-objective problems, the correlation ρ must fall in $[-0.5, 1]$. Of course, if one wants to study very negative correlations between some pairs of objectives, it is possible to design a matrix C that keeps the condition that C is positive-definite.

To generate random variables with uniform marginals and a specified correlation matrix C, we follow the work of Hotelling [12]. We first generate $(Z_1, \ldots, Z_M)$ a multinormal laws of means 0 and correlation matrix $R = 2 \sin(\frac{\pi}{6} C)$. Then, the values $z_i = \Phi(Z_i)$ are uniformly distributed with a correlation matrix C, where Φ is the univariate normal cumulative density function. Note that this is not the only way to generate a multivariate uniform law.

3.2   Correlation between Objective Functions

The construction of $CMNK$-landscapes defines correlation between the y's but not directly between the objectives. In this section, we prove by algebra that the correlation between objectives is tuned by the matrix C. This proof is followed by an experimental analysis.

Theoretical analysis. Let $F_m = (f_{mNK}(x))$ be the fitness vector values of the 2^N solutions with respect to objective m. The correlation between objective n and p is: $cor(F_n, F_p) = \frac{cov(F_n, F_p)}{\sigma_n \sigma_p}$ where σ_n and σ_p are the standard deviations of fitness values over the landscape of the nth and p^{th} NK fitness functions. F_n (resp. F_p) corresponds to the average value of the N vectors F_{ni} (resp. F_{pj}) of fitness component values:

$$cov(F_n, F_p) = \frac{1}{N^2} \sum_{i,j=1}^{N} cov(F_{ni}, F_{pj})$$

By definition, when $i \neq j$, $cov(F_{ni}, F_{pj}) = 0$ and $cov(F_{ni}, F_{pi}) = c_{np} \cdot \sigma_{ni} \cdot \sigma_{pi}$, where c_{np} is the correlation defined in the matrix C, and σ_{ni} (resp. σ_{pi}) is the standard deviation of fitness component i. The correlation between objectives n and p becomes:

$$cor(F_n, F_p) = c_{np} \frac{\sum_{i=1}^{N} \sigma_{ni} \sigma_{pi}}{N^2 \sigma_n \sigma_p}$$

By construction of the fitness functions, the following relation between standard deviations stands $\sigma_n^2 = \frac{1}{N} \sum_{i=1}^{N} \sigma_{ni}^2$ (resp. for σ_p^2). On average, the σ_{ni} are equal to the standard deviation of the uniform law on $[0, 1)$.

$$E(cor(F_n, F_p)) = c_{np} \qquad (1)$$

Then, the average of the correlations between objective functions are given by the matrix C. In the ρMNK-landscapes, the parameter ρ allows to tune very precisely the correlation between all pairs of objectives.

Experimental study. In order to enumerate the search space exhaustively, we conduct an empirical study for $N = 18$. In order to minimize the influence of the random creation of landscapes, we considered 30 different and independent landscapes for each parameter combinations: ρ, M, N and K. The measures reported are the average over these 30 landscapes. The remaining set of parameters are given in Table 1. Figure 1 shows the average[1] of the Spearman correlation coefficient according to the parameters ρ, M and K. This confirms the result of equation (1), the correlation coefficients are very close to the expected value ρ.

Then, in the ρMNK-landscapes, the parameter ρ tunes very precisely the correlation, and, in addition to the correlated multiobjective quadratic assignment

---

[1] For $M > 2$, there are several correlation coefficients. We report here the average correlation coefficients over all the objectives (these values are all very close).

Table 1. Parameters used in the paper for the experimental analysis

Parameter	Values
N	18
M	$\{2, 3, 5\}$
K	$\{2, 4, 6, 8, 10\}$
ρ	$\{-0.9, -0.7, -0.4, -0.2, 0.0, 0.2, 0.4, 0.7, 0.9\}$ such that $\rho \geq \frac{-1}{M-1}$

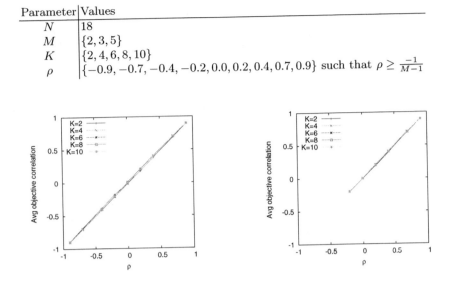

Fig. 1. Average values of the correlation between objectives according to the parameter ρ. The number of objectives is $M - 2$ (left) and $M - 5$ (right).

problem [4], it is possible to tune this correlation between all pairs of objectives. In the following, we study the influence of epistasis, number of objective and objective correlation on the properties of the efficient set for the ρMNK-landscapes model.

4   Analysis of the Efficient Set Properties

In this section, we conduct experiments on the ρMNK-landscapes in order to study different properties of the efficient set: its cardinality, the number of supported solutions and connectedness-related features. The instances under study are defined by the parameter setting given in Table 1.

4.1   Cardinality of the Efficient Set

Figure 2 shows the proportion of efficient solutions in the search space according to parameters K, ρ and M of ρMNK-landscapes. First of all, the epistatic parameter K does not seem to have a major influence on the results. At the opposite, the objective correlation ρ modifies the number of efficient solutions to several orders of magnitude. Indeed, the proportion decreases from 10^{-4} to 10^{-5} ($\rho \in [-1, 1]$) for two-objective problems, and from 10^{-1} to 10^{-5} ($\rho \in [-0.2, 1]$) for $M = 5$. With respect to the number of objective functions ($M = 2, 3$, and 5), the size increases of several decades according to M. For a negative objective correlation ($\rho = -0.2$), the proportion goes from 10^{-4} to 10^{-1} whereas it goes from 10^{-5} to 10^{-4} for a positive correlation ($\rho = 0.9$).

The influence of objective correlation on the efficient size becomes as important as the number of dimension of objective space. A lot of solutions becomes efficient when the anti-correlation is high. Now, let us suppose that we want to set or to bound the size of the approximation set by 100. Such a parameter setting is often used while handling a population or an archive of non-dominated solutions in a multiobjective metaheuristic. For the ρMNK-landscapes, the proportion of non-dominated solutions over the search space should be roughly around $4 \cdot 10^{-4}$ (this goes up to $8 \cdot 10^{-4}$ for 200 solutions). Whatever the correlation value ρ, a 100$-$solution approximation set always allows to store all the efficient set for two-objective problems. However, this is not the case for a higher dimension of the objective space. For instance, for $M = 5$, 100 solutions suffice to store the whole efficient set for a high objective correlation only ($\rho > 0.5$). In other words, for $\rho < 0.5$, we cannot pretend to identify the whole efficient set exhaustively by handling a 100$-$solution approximation set.

To summarize, when the number of objective increases, and even more when the objectives are in conflict, the size of the efficient set becomes very large, and then tend to be intractable. In this case, it is not reasonable to pretend to identify the whole efficient set, and a limited-size approximation should be considered. This first result shows the importance to design a benchmark where the objective correlation can be tuned precisely, even when $M > 2$. Such a property should be taken into consideration for the development of metaheuristics, when the number of objective becomes too large, and when there is a high anti-correlation between objective functions. A special attention should be paid with regards to the size of the approximation set handled by the search approach.

4.2   Number of Supported Efficient Solutions

Figure 3 shows the proportion of supported solutions in the search space according to parameters K, ρ and M of ρMNK-landscapes. Mainly, this number follows the size of the efficient set: the epistatic parameters K has low influence on the size. When the objective space dimension increases or the objective correlation decreases, the number of supported solutions gets higher. The difference with the size of the efficient set becomes more clear in Figure 4. It gives the proportion of supported solutions over the efficient set. This proportion is nearly independent of the epistasis degree of the problem (K). However, when the objective correlation increases, this proportion increases. For a high objective correlation ($\rho = 0.9$), nearly all solutions become supported (this is even the case for some instances). The same observation can be made with the number of objectives. The number of supported solution increases with the cardinality of the efficient set, but the former increases faster than the latter.

While putting this property in relation with the design of a metaheuristic, we can conclude that scalar approaches should become more appropriate when the number of objective is low, and when the objective correlation is high.

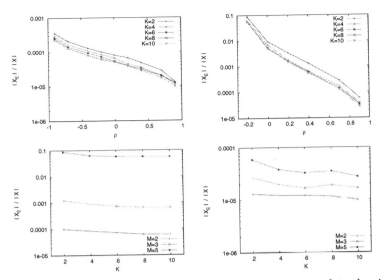

Fig. 2. Average ratio of the number of efficient solutions compared to the size of the search space (2^N) according to parameter ρ (top left $M = 2$, right $M = 5$), and according to parameter K for different number of objectives (bottom left $\rho = -0.2$, right $\rho = 0.9$). Notice the log y-scale.

4.3   Connectedness of the Efficient Set

In this section, the efficient graph (see Section 2.2), *i.e.* the graph of efficient solutions where edges are induced by a given neighborhood operator, is analyzed.

Firstly, the efficient graph can be composed of several connected components. In this case, all the efficient solutions are not connected with respect to the neighborhood relation. Figure 5 shows the average ratio of the larger connected component size induced by Hamming distance 1. Nearby all solutions of the efficient graph are in the same component when the objective space dimension is high ($M = 5$) and when the objective correlation is negative ($\rho = -0.2$). At first sight, such a result seems to be explained by the very large size of the efficient set obtained for those parameters (see Section 4.1). However, we compared this result to the size of the larger component of a graph of same size, but where the nodes are now random solutions. We found out that this size is much smaller than the one of the efficient graph, in particular when the epistatic degree is low (170 times larger for $M = 5$, $\rho = -0.2$, and $K = 4$). Consequently, the ratio size of the larger component is not the consequence of the number of efficient solutions only.

Contrary to the size of the efficient set, the size of the largest connected component seems to depend on the epistatic degree K. Indeed, this size decreases when K increases. As an example, for $M = 2$ and $\rho = -0.4$, the ratio size is 0.42 for $K = 2$ and lower than 0.1 for $K = 10$. When the epistatic degree is low, the objective values of neighboring solutions are correlated, and this correlation decreases with the epistatic degree [13]. This could explain our experimental result: If a solution is efficient, the probability that one of its neighbors is also efficient gets higher when the epistatic degree gets lower.

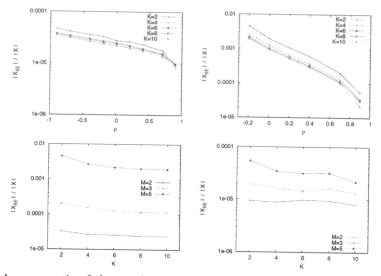

Fig. 3. Average ratio of the number of supported efficient solutions compared to the size of the search space (2^N) according to parameter ρ (top left $M = 2$, right $M = 5$), and according to parameter K for different number of objectives (bottom left $\rho = -0.2$, right $\rho = 0.9$). Notice the log y-scale.

The objective correlation and the number of objective functions also affect the size of the largest connected component. But the variation is different with respect to the number of objective functions. For $M = 2$, the ratio of the larger component size increases when the objective correlation increases (apart from $K = 2$). For $M = 5$, the ratio decreases when the objective correlation increases. As a consequence, excepting when the efficient set is intractable (that is, when there is a high objective space dimension and a high anti-correlation degree), we cannot expect to reach all the efficient solutions by iteratively exploring the neighborhood of an approximation set initialized with one non-dominated solution. However, when there are several connected components for the efficient graph based on Hamming distance 1 (see the definition of cluster in Section 2.2), the distance between those components could be small.

When efficient solutions are connected with respect to a neighborhood structure related to Hamming distance k and not $k - 1$, the efficient set is then said to be k-connected [10]. When the minimal distance k is around 9, which is the average distance between random solutions, we can say that the distance between efficient solutions is large. Figure 6 shows the average minimal distance k to connect all the efficient solutions. This minimal distance k increases when the epistatic degree increases. As an example, for $\rho = -0.2$, the average distance is equals to 4.3 and 2 for dimension 2 and 5, respectively, when $K = 2$, whereas it is equal to 7.1 and 2.8, respectively, when $K = 10$. These results meet the previous ones on the largest component size: At the same time, the size of the larger component decreases, and the distance between efficient solutions increases.

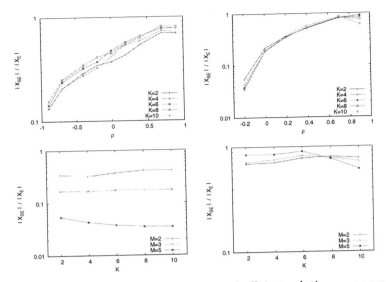

Fig. 4. Average ratio of the number of supported efficient solutions compared to the size of the efficient set according to parameter ρ (top left $M = 2$, right $M = 5$), and according to parameter K for different number of objectives (bottom left $\rho = -0.2$, right $\rho = 0.9$). Notice the log y-scale.

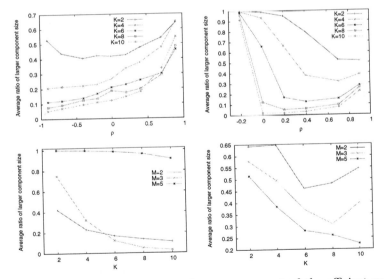

Fig. 5. Average ratio of the size of the larger component of the efficient graph and Hamming distance of 1 to the size of the efficient set according to parameter ρ (top left $M = 2$, right $M = 5$), and according to parameter K for different number of objectives (bottom left $\rho = -0.2$, right $\rho = 0.9$).

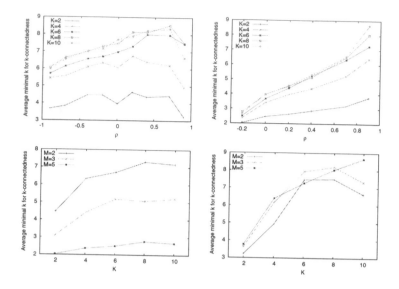

Fig. 6. Average of the minimal Hamming distance to connect all the efficient solutions according to parameter ρ (top left $M = 2$, right $M = 5$), and according to parameter K for different number of objectives (bottom left $\rho = -0.2$, right $\rho = 0.9$).

The average k-connectedness increases also when the objective correlation increases. For an objective space dimension 5 and a negative objective correlation $\rho = -0.2$, it could be possible to reach all non-dominated solutions from another one, as the average minimal distance is lower than 3. At the opposite, when the objective correlation is positive, it should be easier to find a new non-dominated solution by restarting the search from a random solution, rather than exploring the neighborhood of a given non-dominated solution such as the distance is around the third of the bit string length. When objectives are correlated, less solutions are to be found, but knowing some of them will not help to find more. Then, the design of an efficient metaheuristic has to be different according to the objective correlation. In a *two-phase* approach, the number of starting solutions and the size of the neighborhood can be tuned according to correlation between objectives following this study.

5   Discussion

In this paper, we analyzed the consequence of the objective space dimension, the non-linearity, and the objective correlation on the structure of multiobjective combinatorial search spaces for the design of metaheuristics. We proposed a new method to design a multiobjective combinatorial benchmark where the correlation between all pairs of objectives can be tuned very precisely. As an example, we defined the ρMNK-landscapes which extend the multiobjective NK-landscapes.

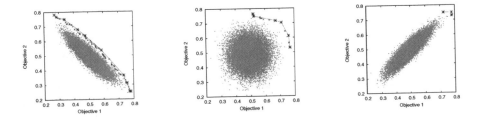

Fig. 7. The objective space (maximization problem) for three landscapes. The number of objective is $M = 2$, the length of bit string remains $N = 18$, the epistasis parameters is $K = 4$. From left to right, the correlation increases from negative correlation to positive correlation ($\rho = -0.9, 0.0$ and 0.9). The green points are random solutions of the search space (10% of the size), the red points are the solutions of the efficient set, and blue are the supported solutions of the efficient set.

Figure 7 shows three examples of ρMNK-landscapes in the objective space. The number of objective is 2, the parameter K is 4, and length of the bit string is 18. This gives a summary of our results in a more intuitive way. When the objective correlation is negative, the objectives are in conflict (feasible solutions are in green). The efficient set size (in red) is large, and the problem could become intractable. In this case, a metaheuristic has to find a limited-size approximation of the efficient set only. When the objective correlation is null, as in [5], the image of the search space in the objective space can be represented as a multidimensional 'bowl'. The objectives are independent. When the objective correlation is positive, there exists few solutions in the efficient set. Nearly all solutions become supported. Indeed, when the number of objectives is low, and when the objective correlation is high, efficient solutions are supported. We can conclude that scalar approaches should become more appropriate in such a case. The connectedness property is not represented in the last figure. The size of larger connected component and the minimal distance to connect all the efficient solutions depend on the objective space dimension, the epistatic degree, and also on the objective correlation. A two-phase strategy, starting from some efficient (supported) solutions, and exploring their neighborhood at a given distance, can be tuned according to the results of this work.

Bringing those properties with the design of local search metaheuristics help to make proper choices between several classes of methodologies. This analysis shows the importance of the objective correlation on the design of benchmark problems, in particular when the number of objectives is higher than 2. In future works, we will use some sample technics to study the ρMNK-landscapes of larger size. We will also compare our results on the properties of search space with the performance of different metaheuristics. However, the efficient set does not cover all the search space properties, so next works will focus on the properties related to the Pareto local optima, and to the Pareto local optimum sets.

References

1. Ehrgott, M., Klamroth, K.: Connectedness of efficient solutions in multiple criteria combinatorial optimization. European Journal of Operational Research 97(1), 159–166 (1997)
2. Mote, J., Olson, I.M.D.L.: A parametric approach to solving bicriterion shortest path problems. European Journal of Operational Research 53(1), 81–92 (1991)
3. Paquete, L., Stützle, T.: A study of stochastic local search algorithms for the biobjective QAP with correlated flow matrices. European Journal of Operational Research 169(3), 943–959 (2006)
4. Knowles, J., Corne, D.: Instance generators and test suites for the multiobjective quadratic assignment problem. In: Fonseca, C.M., Fleming, P.J., Zitzler, E., Deb, K., Thiele, L. (eds.) EMO 2003. LNCS, vol. 2632, pp. 295–310. Springer, Heidelberg (2003)
5. Aguirre, H.E., Tanaka, K.: Working principles, behavior, and performance of MOEAs on MNK-landscapes. European Journal of Operational Research 181(3), 1670–1690 (2007)
6. Ehrgott, M.: Multicriteria optimization, 2nd edn. Springer, Heidelberg (2005)
7. Paquete, L., Stützle, T.: Stochastic local search algorithms for multiobjective combinatorial optimization: A review. In: Handbook of Approximation Algorithms and Metaheuristics. Computer & Information Science Series, vol. 13, Chapman & Hall / CRC (2007)
8. Knowles, J., Corne, D.: Bounded Pareto archiving: Theory and practice. In: Metaheuristics for Multiobjective Optimisation. LNEMS, vol. 535, pp. 39–64. Springer, Heidelberg (2004)
9. Gorski, J., Klamroth, K., Ruzika, S.: Connectedness of efficient solutions in multiple objective combinatorial optimization. Technical Report 102/2006, University of Kaiserslautern, Department of Mathematics (2006)
10. Paquete, L., Stützle, T.: Clusters of non-dominated solutions in multiobjective combinatorial optimization: An experimental analysis. In: Multiobjective Programming and Goal Programming. LNEMS, vol. 618, pp. 69–77. Springer, Heidelberg (2009)
11. Kauffman, S.A.: The Origins of Order. Oxford University Press, New York (1993)
12. Hotelling, H., Pabst, M.R.: Rank correlation and tests of significance involving no assumptions of normality. Ann. Math. Stat. 7, 29–43 (1936)
13. Weinberger, E.D.: Correlated and uncorrelatated fitness landscapes and how to tell the difference. Biological Cybernetics 63, 325–336 (1990)

Instance-Based Parameter Tuning via Search Trajectory Similarity Clustering

Lindawati, Hoong Chuin Lau, and David Lo

School of Information Systems, Singapore Management University, Singapore
lindawati.2008@phdis.smu.edu.sg, {hclau,davidlo}@smu.edu.sg

Abstract. This paper is concerned with automated tuning of parameters in local-search based meta-heuristics. Several generic approaches have been introduced in the literature that returns a "one-size-fits-all" parameter configuration for all instances. This is unsatisfactory since different instances may require the algorithm to use very different parameter configurations in order to find good solutions. There have been approaches that perform instance-based automated tuning, but they are usually problem-specific. In this paper, we propose CluPaTra, a generic (problem-independent) approach to perform parameter tuning, based on CLUstering instances with similar PAtterns according to their search TRAjectories. We propose representing a search trajectory as a directed sequence and apply a well-studied sequence alignment technique to cluster instances based on the similarity of their respective search trajectories. We verify our work on the Traveling Salesman Problem (TSP) and Quadratic Assignment Problem (QAP). Experimental results show that CluPaTra offers significant improvement compared to ParamILS (a one-size-fits-all approach). CluPaTra is statistically significantly better compared with clustering using simple problem-specific features; and in comparison with the tuning of QAP instances based on a well-known distance and flow metric classification, we show that they are statistically comparable.

Keywords: instance-based automated tuning parameter, search trajectory, sequence alignment, instance clustering.

1 Introduction

In the last decade there has been a dramatic rise in the design and application of meta-heuristics such as tabu search and simulated annealing to solve combinatorial optimization problems (COP) in many practical applications. The effectiveness of a meta-heuristic algorithm hinges on its parameter configurations. For example, a tabu search will perform differently with different tabu lengths. Previous studies revealed that only 10% of the time is spent on algorithm design and test; while the rest of the development time is spent on fine-tuning the parameter settings [1]. The latter process is either a laborious manual exercise by the algorithm designer, or an automated procedure. The key challenge in

C.A. Coello Coello (Ed.): LION 5, LNCS 6683, pp. 131–145, 2011.
© Springer-Verlag Berlin Heidelberg 2011

automated tuning is the large parameter configuration space on even a handful of parameters.

Given an algorithm (which we call the target algorithm) to solve a given COP, it has been observed that different problem instances require different parameter configurations in order for the algorithm to find good solutions (e.g. [6,19,24]). An interesting research question is whether there are patterns or rules governing the choice of parameter configurations, and whether such patterns can be learnt.

Several approaches have been proposed to automate the tuning problem, such as the Racing Algorithm by Birratari et al. [3], Decision Tree Classification Approach by Srivastava and Mediratta [22], CALIBRA by Andenso-Daz and Laguna [1], ParamILS by Hutter et al. [12,13] and Randomized Convex Search (RCS) by Lau and Xiao [14]. These are generic approaches which can be used for various COP problems. One common shortcoming of such approaches is that they produce a **one-size-fits-all** configuration for all instances, which may not perform well on large and diverse instances. On the other hand, approaches by Patterson and Kautz [19], Hutter and Hamadi [11], Gagliolo and Schmidhuber [6] and Xu et al. [24] attempted to deal with **instance-based** automatic tuning. However, those approaches are less general in the sense that each of them can only solve a particular problem by making use of problem-specific features. For example, SATzilla constructs per-instance algorithm portfolios for SAT [24]. SATzilla07 uses 48 features, most of which are SAT-specific features. The caveat is that feature selection is itself a very complex problem in general which cannot be done automatically but rather must rely on the knowledge of a domain expert.

Rather than ambitiously attempting instance-based tuning which we believe to be a computationally prohibitive and unachievable task in the near future because of the large parameter configuration space and large number of instances, we turn towards a cluster-based treatment. Our goal extends a preliminary work on features-based tuning proposed in [14] where instances are clustered according to some problem-specific features, but unlike [14], we do not rely on problem-specific features; rather, we propose a generic approach where we make use of the search trajectory patterns as a feature. A search trajectory pattern is defined as the path that the target algorithm follows as it searches from an initial solution to its neighbor iteratively [10]. We then apply a standard clustering algorithm to segment the training set of instances into clusters based on their search trajectory patterns similarity.

Motivated by earlier works on the tight correlation between fitness landscape and search trajectories [7,8], and the tight correlation between the fitness landscape and algorithm performance [20], our bold conjecture in this paper is that trajectory patterns themselves are correlated with parameter configurations; in other words, we believe that if a parameter configuration works well for a particular instance, then it will also work well for instances with similar fitness landscapes (which can be inferred from the trajectory patterns). Consequently, we train our automated tuning algorithm by first performing clustering on problem instances based on their search trajectories similarity, and then apply existing one-size-fits-all algorithms (such as CALIBRA, ParamILS or RCS) to derive the best parameter configurations for the respective clusters. Subsequently, given an arbitrary instance, we first map its search trajectory to the closest cluster. The tuned parameter configuration for that cluster is then returned as the parameter configuration for this instance. The result is a fine-grained tuning algorithm that does not produce a one-size-fits-all parameter configuration, but rather instance (or rather cluster)-based parameter configurations. Even though strictly speaking, our approach is cluster-specific rather than instance-specific, it is a big leap from one-size-fits-all schemes. Arguably, our approach, taken to the extreme, can potentially produce instance-based tuning; although we do not know how to scale it well at the moment. The major contributions in this paper are summarized as follows:

- We propose *CluPaTra*, a novel instance-based problem-independent automated parameter tuning approach based on clustering of patterns of instances by their search trajectories.
- A search trajectory can be derived readily from a local-search based algorithm without incurring extra computation (other than the task of storing these solutions as the local search discovers them). Hence our approach can be applied to tune any local search-based target algorithm to solve a given problem.
- We tap into the rich depository of machine learning and data mining, utilizing a clustering method based on two well-studied techniques, sequence alignment and hierarchical clustering. We apply sequence alignment to calculate a similarity score between a pair of instance search trajectories, and hierarchical clustering to form the clusters.

CluPaTra is verified with experiments on two classical COPs - Traveling Salesman Problem (TSP) and Quadratic Assignment Problem (QAP). For TSP, our target algorithm is the classical Iterated Local Search (ILS) algorithm (implemented by [8]), whereas for QAP we use a relatively new hybrid metaheuristic algorithm proposed in [18]. These choices are made on the dual intent to benchmark our approach against best published results (showing that it is capable of producing results compatible to the best-found results), as well as to demonstrate how our approach can yield significant improvement when applied to tune a newly designed algorithm.

2   Preliminaries

In this section, we formally define the Automated Parameter Configuration problem, followed by the concepts of the one-size-fits-all and instance-based configurators.

2.1   Automated Parameter Configuration Problem

Let $\mathcal{A}$ be the target algorithm with n number of parameters to be tuned based on a given set of training instances I. Each parameter x_i can assume a value taken from a (either continuous or discrete) interval $[a_i, b_i]$ in parameter configuration space Θ. Let the vector $\mathbf{x} = [x_1, x_2, ..., x_n]$ represent a parameter configuration and $\mathcal{H}$ be a performance metric function that maps $\mathbf{x}$ to a numeric score computed over a set of instances (see details below). The automated parameter configuration problem is thus an optimization problem seeking to find $\mathbf{x} \in \Theta$ that minimizes $\mathcal{H}(\mathbf{x})$.

Notice that unlike standard optimization problems, the function $\mathcal{H}$ is a meta-function on $\mathbf{x}$ is typically highly non-linear and very expensive to compute. Furthermore, as the parameter space may be extremely large (even for discrete values, the size is equal to $(b_1 - a_1)(b_2 - a_2) \cdots (b_n - a_n)$), it is generally impractical to execute a tuning algorithm based on full factorial exploration of good parameter values. As in [13], to avoid confusion between a algorithm whose performance is being optimized and an algorithm used to tune it, we refer to the former as the *target algorithm* and the latter as the *configurator*.

2.2   One-Size-Fits-All Configurator

Since a one-size-fits-all configurator (such as ParamILS) only produces a single parameter configuration for a set of instances I, it calculates the function $\mathcal{H}$ by using a specific statistic (such as mean or standard deviation) measured over the entire set (or distribution) of problem instances. We define the one-size-fits-all configurator as follows.

Definition 1 (One-Size-Fits-All Configurator). *Given a target algorithm $\mathcal{A}$, a set of training instances I, a set of testing instances I_t, a parameter configuration space Θ and a meta-function $\mathcal{H}$ to measure algorithm $\mathcal{A}$ performance, a one-size-fits-all configurator finds a parameter configuration $\mathbf{x} \in \Theta$ such that $\mathcal{H}$ is minimized over the entire set (or distribution) of I. Subsequently, given a testing instance in I_t, that parameter configuration $\mathbf{x}$ will be used to execute $\mathcal{A}$.*

2.3   Instance-Based Configurator

In this paper, we are concerned with clustering of problem instances. Hence, using the same notation as the one-size-fits-all configurator, we define the instance-based configurator as follows.

Definition 2 (Instance-Based Configurator). *Given a target algorithm $\mathcal{A}$, a set of training instances I, a set of testing instances I_t, a parameter configuration space Θ and a meta-function $\mathcal{H}$ to measure algorithm $\mathcal{A}$ performance, an instance-based configurator creates a set of clusters C from I and finds a parameter configuration $\mathbf{x}_c$ for each cluster $c \in C$ that minimizes $\mathcal{H}$ for the set of instances in the respective cluster. For a given testing instance in I_t, it will find the most similar cluster $c \in C$ and return the parameter configuration $\mathbf{x}_c$ which will be used to execute $\mathcal{A}$.*

2.4   Performance Metric

We now define the performance metric function $\mathcal{H}$, for both the training and testing instances. For training, this value is measured over all training instances, while for testing, ditto test instances.

Definition 3 (Performance Metric). *Let i be a problem instance, and $\mathcal{A}_x(i)$ be the objective value of the corresponding solution obtained by $\mathcal{A}$ when executed under the configuration x. Let $OPT(i)$ denote either (a) the known global optimal value of i, or (b) where the global optimal value is unknown, the best known value. $\mathcal{H}(x)$ is defined as the mean percentage deviation of $\mathcal{A}_x(i)$ from $OPT(i)$, for all problem instance i in question (training/testing). Obviously, the lower the deviation value the better it is.*

3   Solution Approach

In this section, we present our solution approach *CluPaTra* by first defining the search trajectory similarity and describing *CluPaTra* major components, namely: search trajectory representation, similarity calculation, and the clustering method, followed by the overall steps for the training and testing phases.

3.1   Search Trajectory Similarity

A search trajectory is defined as a path of solutions that the target algorithm $\mathcal{A}$ finds as it searches through the neighborhood search space. Two or more search trajectories are similar if some fragments (several number of consecutive moves) of the path have identical solution's attributes. An example of solution's attributes is the deviation of its objective value from global optimum (or best known) value (see section 3.2). The longer the fragments the more similar it is.

As an example, Fig. 1 shows a search trajectory obtained by 10 consecutive moves of the ILS algorithm for three TSP instances, namely: kroa100, bier127 and eil51 with two very different parameter configurations, namely: configuration I and configuration II. Observe that for the same configuration, kroa100 and bier127 have similar search trajectories, while eil51 has a very different trajectory. Observe also that even when different configurations result in different search trajectories for a given instance, the similarity between kroa100 and bier127's trajectories are preserved. This similarity property is what we need that allows us to perform clustering of instances using an arbitrary parameter configuration.

Since there is a tight correlation between fitness landscape (or commonly known as search space) and search trajectories [7,8], and the tight correlation between the fitness landscape and algorithm performance [20], we assume that instances with similar search trajectories will need the same parameter configuration.

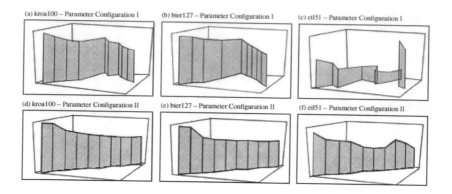

Fig. 1. Search Trajectories of 3 TSP instances kroa100, bier127 and eli51 using two very different parameter configuration (with z-axis as the objective value and x and y axis as the search space)

3.2 Search Trajectory Representation

We present the search trajectory as a directed sequence of symbols, each representing a solution along the trajectory. A symbol encodes a combination of two solution attributes, namely: the position type and its percentage deviation of quality from OPT (as defined in section 2.4).

The position type represents in a sense the local property of a solution with respect to its search neighborhood, and is defined based on the topology of the local neighborhood [10]. There are 7 position types, determined by evaluating the solution objective value with all its local direct neighbors' objective values - whether it is better, worse or the same. The 7 positions types are given in Table 1.

The deviation of solution quality measures in a sense the global property of the solution (since it is compared with the global value OPT). If the global optimum value is unknown, we use the best known value. Granted however that

Table 1. Position Types of Solution

Position Type Label	Symbol	<	=	>
SLMIN (strict local min)	S	+	-	-
LMIN (local min)	M	+	+	-
IPLat (interior plateau)	I	-	+	-
SLOPE	P	+	-	+
LEDGE	L	+	+	+
LMAX (local max)	X	-	+	+
SLMAX (strict local max)	A	-	-	+

'+' = present, '-' = absent; referring to the presence of neighbors with larger ('<'), equal ('=') and smaller ('>') objective values

the best known value is not the same as the global optimal value, it provides a reasonably good upper bound (for a minimization problem); and since our aim is to find similar patterns of the transition from one solution to the next, and not to measure the actual absolute performance of the algorithm, the best known value suffices in providing a good proxy to the global optimal value for our purpose of representing the trajectory.

These two attributes are combined into a symbol with the first two digits being the deviation of the solution quality and the last digit being the position type. Note that the attributes are generic in the sense that they can be easily retrieved/computed from any local-search-based algorithm albeit different problems. Being mindful that some target algorithms may have cycles and (random) restarts, we intentionally add two additional symbols: 'CYCLE' and 'JUMP'; 'CYCLE' is used when the target algorithm returns to a position that has been found previously, while 'JUMP' is used when the local search is restarted.

In order to obtain the search trajectory for a given problem instance, we naturally need to execute the target algorithm with a certain parameter configuration and record all the solutions visited. We refer to this configuration as the *initial sequence configuration*.

An example of the sequence representing the eil51 search trajectory in Fig. 1 is *15L-11L-09L-07L-07P-06P-04S-05L-J-21L-19L*. Notice that after position 8, the target algorithm performs a random restart, hence we add 'JUMP' symbol after position 8.

3.3   Similarity Calculation

Having represented trajectories by linear sequences, it is natural to use pairwise sequence alignment to determine the similarity between a pair of trajectories. In pairwise sequence alignment [9], the symbols of one sequence will be matched with those of the other sequence while respecting the sequential order in the two sequences. It can also allow gaps to occur if symbols do not match. There are two kinds of alignment strategies: local and global. In local alignment, only portions of the sequences are aligned, whereas global alignment aligns over the entire length of the sequences. Because search trajectory sequences have varying lengths, we find local alignment best fits our need.

To measure the similarity score between two search trajectory sequences, a metric based on the best alignment is used. The matched symbol contributes a positive score (+1), while a gap contributes a negative score (-1). The sum of the scores is taken as the maximal similarity score of the two sequences. We may find situations as follows: (a) a search trajectory sequence is a subsequence of another one thus having a very high similarity score or (b) longer sequences get higher similarity score. To avoid these situations, the final similarity score will be divided by $\frac{1}{2} \times (|Sequence_1| + |Sequence_2|)$.

Our sequence alignment is implemented using standard dynamic programming [9], with a complexity of $O(n^2)$. As an example, the sequence alignment for the kroa100 and bier127 search trajectories from Fig. 1 is illustrated in Table 2.

Table 2. Example of Sequence Alignment from 2 TSP instances search trajectory, kroa100 and bier127

kroa100	19L	19P	18P	17P	16P	15P	14P	13P	11P	10P		
		|	|	|	|	|	|	|	|	|		
bier127		19P	18P	17P		15P		13P	11P	10P	09P	08P
score		+1	+1	+1	-1	+1	-1	+1	+1	+1		

To cluster instances (see the subsection below), we need to compute similarity scores for all possible pairs of training instances. Hence, the total time complexity for sequence alignment is $O(m^2 \times n^2)$, where n is the maximum sequence length of the sequences and m is the number of instances in the training set.

3.4 Clustering Method

Here, our goal is to group similar instances according to their search trajectory similarity. A typical clustering algorithm requires a distance measure between data points. For the distance measure we use $\frac{1}{similarity\ score}$. After such a measure is known, a standard clustering algorithm could be employed. For our purpose, we adopt the well-known hierarchical clustering approach AGNES (AGglomerative NESting) to cluster the instances [9]. AGNES works by placing each instance initially in a cluster of its own. It then iteratively merges two closest clusters (i.e., a pair of clusters with the smallest distance) resulting in lesser number of clusters of larger sizes. The process is repeated until all nodes belong to the same cluster unless a termination condition applies. Examples of termination conditions are minimal number of cluster is reached or the maximal inter-cluster distance goes below a certain value. The complexity of AGNES is $O(n^2)$ with n being the number of instances.

Since the learning is unsupervised, we need to determine the number of clusters to be used. For this purpose, we apply the L method from [21] which makes use of an evaluation graph where the x-axis is the number of clusters and the y-axis is the value of the evaluation function at x clusters. The evaluation function can be any evaluation metric based on distance, similarity, error or quality. In this paper, we use the average distance among all instances in two different clusters. It determine the number of clusters by finding the point that has minimum root mean square error for both the left and right side. It is calculated using the following formula:

$$c^* = min \left[\frac{RMSE(L)}{n_L} + \frac{RMSE(R)}{n_R} \right] \tag{1}$$

where:

Notation	Definition
$RMSE(L)$	root mean squared error of points in the left side of c
n_L	number of points in the left side of c
$RMSE(R)$	root mean squared error of points in the right side of c
n_R	number of points in the right side of c

This method only requires AGNES algorithm to be run once, since all the clusters created by AGNES can be recorded in one run. And since we want to produce a compact set of clusters, we limit the number of clusters to be less then 10. Thus, for the x-axis, we only use the number of clusters from 1 to 10.

3.5   Training and Testing Phases

The steps involved in the training and testing phases are shown in Fig. 2 (which are quite self-explanatory, and details are skipped in the interest of space).

Procedure TrainingPhase
Inputs: A: Target Algorithm;
   I: Training instances;
   Θ: Parameter Configuration Space;
   $\mathbf{x}_i$: Initial Sequence Configuration;
Outputs: C: A set of clusters;
   $\mathbf{X}$: Parameter configurations for each cluster in C;
Method:
1: Let $TRAJ = $ A search trajectory from A for I using $\mathbf{x}_i$;
2: Let $SEQ = $ A transformation from $TRAJ$ to sequence;
3: Let $Score = $ A mapping from I x I to scores;
4: For each (i,j) in I X I
5:      Let $s_1 = $ SEQ(i);
6:      Let $s_2 = $ SEQ(j);
7:      Score[s_1,s_2] = similarity(s_1,s_2);
8: Let $C = $ Run AGNES using Score;
9: Let $\mathbf{X} = $ A mapping from clusters to configuration;
10: For each cluster c in $Clusters$;
11:    $\mathbf{X}[c] = $ Run One-size-fits-all configurator on instances in c with respect to Θ;
12: Output C, $\mathbf{X}$;

Procedure TestingPhase
Inputs: i: An Arbitrary Testing instance;
   C: Set of clusters;
   $\mathbf{X}$: Parameter configurations for each cluster in C;
Outputs: $BestConfig$: A recommended configuration;
Method:
1: Let $Score = $ A mapping from C to scores;
2: For each cluster c in C
3:      $Score[c] = $ Average similarity from i to each instance in c;
4: Let $BestClust = c$, where for all c' not equals to c in C, $Score[c] >= Score[c']$;
5: Let $BestConfig = \mathbf{X}[BestClust]$;
6: Output $BestConfig$;

Fig. 2. Training and Testing Phase

4   Experimental Design

In this section, we provide information on the experiments presented in the following section. First, we present our experiment settings. Second, we present our validity and statistical significant measurement. And finally, we describe the low-level details of our experimental setup.

4.1   Experiment Settings

Here we briefly explain the target algorithm, one-size-fits-all configurator, benchmark instances and initial sequence configuration.

Target Algorithm. We used two different target algorithms respectively for solving two different problems. The first algorithm is a variant of a well-known Iterated Local Search (ILS) algorithm [15] for solving the classical TSP, as implemented in [8]. It has 5 discrete-value parameters to be tuned. The second algorithm is a new hybrid Simulated Annealing and Tabu Search (SA-TS) algorithm for solving QAP (presented in [18]). It has 4 parameters; some are discrete while the others are continuous.

One-Size-Fits-All Configurator. In order to derive meaningful experimental comparison, we deliberately chose to use ParamILS [13] as our configurator. ParamILS is itself an iterated local search algorithm used for tuning discrete parameters. Since ParamILS works only with discrete parameters, we first discretize the values of the parameters if the target algorithm has parameters that assume continuous values.

Benchmark Instances. For TSP, we applied our target algorithm to 70 benchmark instances extracted from TSPLib. Fifty six random instances were used as training instances and the remaining 14 instances as testing instances. The problem size (the number of cities) varies from 51 to 3038. For QAP, we used 50 benchmark instances from QAPLib, and randomly picked 40 instances for training and 10 for testing. The problem size (number of facilities) varied from 20 to 150.

Initial Sequence Configuration. The initial sequence configuration is a random configuration from the configuration space Θ.

4.2   Validity and Statistical Significant Measurement

To ensure unbiased evaluation, we used a 5-fold cross-validation [9]. The overall result is recorded to be the average performance over all iterations. We also performed a statistical test to compare the significance of our result. We performed a t-test [17]; we consider p-value below 0.05 to be statistically significant (confidence level 5%).

4.3   Experimental Setup

All experiments were performed on a 1.7GHz Pentium-4 machine running Windows XP. We measured runtime as the CPU time on this machine. As an input to the one-size-fits-all configuratior, we fairly set cutoff times of 10 seconds per run for TSP target algorithm and 100 seconds for QAP target algorithm and allowed each configuration process to execute the target algorithm for a maximum of two CPU hours and to call the target algorithm for a maximum of 10 x n times, where n is the number of instances in the cluster.

5   Empirical Evaluation

In this section, we present our experiment results on the effectiveness of *CluPaTra*. First, we compare *CluPaTra* against one-size-fits-all configurator. Then, to analyze the effectiveness of our generic feature, we compare it with simple specific feature. In addition to that, we also compare *CluPaTra* against an existing classification of QAP instances based on distance and flow metrics [23]. Next we analyze the effect of different initial sequence configurations to our result. We also present the computational time of *CluPaTra*. Finally, a brief discussion regarding the experiment is presented. For the entire experiment, we measure the performance by using the performance metric described in Definition 3.

5.1   Performance Comparison

We evaluated the effectiveness of *CluPaTra* against the vanilla one-size-fits-all configurator (ParamILS). In Fig. 3a, we show the performance achieved by the two approaches for two target algorithms. This result is an average from each of the 5-fold results. The average improvement using *CluPaTra* is 7.78% for TSP training instances, 12.31% for TSP testing instances, 14% for QAP training instances and 21.78% for QAP testing instances. *CluPaTra* performed better and the difference was statistically significant.

5.2   Comparison on Feature Selection

To evaluate the effectiveness of the generic feature (i.e. search trajectory) used by *CluPaTra*, we compared *CluPaTra* with a simple problem-specific feature clustering for TSP and QAP, and a known instance classification for QAP.

First, we compared *CluPaTra* with simple specific feature clustering (SpecFeat). For the specific feature cluster, we used the number of cities (for TSP) [14] and the number of facilities (for QAP). Besides using different features, steps in training and testing phase for both approaches are the same. In Fig. 3b, we present the average performance achieved using 5-fold cross-validation by the two approaches for two target algorithm. *CluPaTra* always perform better and the differences are statistically significant.

Next, we compared *CluPaTra* against an existing well-studied classification of QAP instances based on the distance and flow metrics, due to [23]. We refer

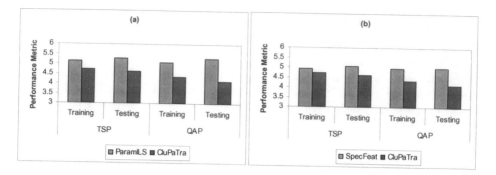

Fig. 3. Performance Comparison (a) *CluPaTra* and ParamILS, (b) *CluPaTra* and Specific Feature

to this as the Natural Cluster (Natural). (We conducted this comparison only for QAP since the classification of QAP benchmark instances is already well-studied and well characterized). Under this classification, QAP instances are divided into 5 groups: (1) random and uniform distances and flows, (2) random flows on grids, (3) real-life problems, (4) characteristics of real-life problems and (5) non-uniform, random problems. Due to the target algorithm limitation (it does not solve groups (4) and (5) problems), we can only provide results on groups (1), (2) and (3). The average performance of 5 folds is shown in Table 3. It shows that *CluPaTra* performs slightly better but the results are not statistically significant. Nonetheless, we can claim that the performance of *CluPaTra* is not inferior to tuning based on the natural classification.

Table 3. Comparison between *CluPaTra* and Natural Clustering

	CluPaTra	Natural	Difference(%)	p-value
Training	4.36	4.54	3.96	0.84108
Testing	4.13	4.14	0.24	0.61976

5.3 Sensitivity Analysis on Different Initial Sequence Configurations

Being mindful that our results may be biased depending on the initial sequence configuration used (to find the trajectories), we consider here two different parameter configurations - a configuration derived from running ParamILS versus a random configuration. Table 4 shows no statistically significant difference between these two initial sequence configurations.

5.4 Computational Results

The two most time-consuming processes in the training phase are those of calculating the similarity of trajectories and running the one-size-fits-all configurator

Table 4. Comparison between Different Initial Sequence Configurations

		Random	ParamILS	Different(%)	p-value
TSP	Training	4.74	4.66	1.69	0.94486
	Testing	4.63	4.72	(1.95)	0.40181
QAP	Training	4.36	3.97	8.94	0.1499
	Testing	4.13	4.5	(8.96)	0.54516

(ParamILS) for each cluster. For the 56 TSP instances with a maximal sequence length of 1560, the time taken for similarity computation is approximately 3 minutes; and for 40 QAP instances with a maximal sequence length of 520, the time taken is approximately 1 minute. For all clusters in one fold, the time needed to run ParamILS was approximately 44 minutes for TSP and 1 hour and 45 minutes for QAP. The total time needed to run the training phase for each fold is hence approximately 48 minutes on TSP and 1 hour and 47 minutes on QAP. While the time needed to run ParamILS alone for each fold is approximately 45 minutes on TSP and 1 hour and 40 minutes on QAP.

For the testing phase, we need to find the best cluster to fit the testing instances. For TSP instances, it took approximately 1.5 minutes in total; while for QAP instances, it took approximately 42 seconds.

5.5 Discussion

As shown from the results, compared to the vanilla one-size-fits-all configuration ParamILS, *CluPaTra* gives a significant improvement in performance (with respect to the performance metric we defined) with a small additional computation time. The additional computation time is needed to cluster the instances (approximately 6.66% for TSP and 7% for QAP from ParamILS run time). Based on this observation, we claim that dividing the instances into cluster using *CluPaTra* before running one-size-fits-all configurator provides a better parameter configuration for each instance and significantly improves the performance with minor additional computational time.

The effectiveness of using the search trajectory as the generic feature is evaluated by comparing it with problem-specific features. For the simple specific feature tried (number of cities for TSP and number of facilities for QAP), our approach is significantly better. Furthermore for QAP, we benchmarked against the natural clustering proposed in [23]. *CluPaTra* is statistically equivalent with the existing natural clustering approach. This shows that search trajectory can be used as a generic feature to cluster the instances without deep prior knowledge of the problem structure. We also evaluated the effect of different initial sequence configurations. Even though different initial sequence configurations may create different search trajectories, the effect of different initial sequence configurations is not significant.

6   Conclusion and Future Works

In this paper, we presented *CluPaTra*, a computationally efficient approach for generic instance-based configurator via clustering of patterns according to the instance search trajectories. We verified our approach on TSP and QAP and observed a significant improvement compared to a vanilla one-size-fits-all approach.

We see two limitations of our proposed approach. First, in terms of scope, our approach can only be applied to target algorithms which are local-search-based, since our approach uses search trajectory as feature. Second, there is an inherent computational bottleneck introduced by the method used for sequence alignment whose worst-case time complexity is $O(m^2 \times n^2)$ (where m is the number of instances in the training set and n is the maximum length of the sequences). As future work, one may investigate the effects of exploiting a less computationally intensive sequence alignment algorithm such as [5] or limit the length of the sequences.

There are also a number of challenges that remain to be explored. On the feature selection method, we proposed a single generic feature, search trajectory. It will be interesting to see if the accuracy can be improved if we combine several fitness landscape features, such as fitness distance correlation, run time distribution and density of local optima. On the metric and clustering method, we use only one metric (sequence alignment) and one clustering method (agglomerative clustering). It may also be interesting to learn how different possible metrics and how different clustering methods can influence the performance. And on a separate front, our approach is to learn to set parameter values based on training instances. This contrasts and complements the volume of works which seek to adaptively adjust the parameter configuration dynamically during search (such as the works of reactive search by Battiti (e.g. [2]) and many others). In adaptive scenario, the parameter values are modified in response to the search algorithm's behavior during its execution. It will be interesting to see if synergies can be exploited to create better instance-based configurators.

References

1. Adenso-Diaz, B., Laguna, M.: Fine-Tuning of Algorithms Using Fractional Experimental Design and Local Search. Operations Research 54(1), 99–114 (2006)
2. Battiti, R., Brunato, M., Campigotto, P.: Learning While Optimizing an Unknown Fitness Surface. In: Maniezzo, V., Battiti, R., Watson, J.-P. (eds.) LION 2007 II. LNCS, vol. 5313, pp. 25–40. Springer, Heidelberg (2008)
3. Birattari, M., Stuzle, T., Paquete, L., Varrentrapp, K.: A Racing Algorithm for Configuring Metaheuristics. In: Genetic and Evolutionary Computation Conference, pp. 11–18. Morgan Kaufmann, San Francisco (2002)
4. Coy, S.P., Golden, B.L., Runger, G.C., Wasil, E.A.: Using Experimental Design to Find Effective Parameter Setting for Heuristics. Journal of Heuristic 7(1), 77–97 (2001)
5. Edgar, R.C.: MUSCLE: multiple sequence alignment with high accuracy and high throughput. Nucleic Acids Research 35(5), 1792–1797 (2004)

6. Gagliolo, M., Schmidhuber, J.: Dynamic Algorithm Portfolio. In: Amato, C., Bernstein, D., Zilberstein, S. (eds.) Ninth International Symposium on Artificial Intelligence and Mathematics (2006)
7. Halim, S., Yap, R., Lau, H.C.: Viz: A Visual Analysis Suite for Explaining Local Search Behavior. In: 19th Annual ACM Symposium on User Interface Software and Technology, pp. 57–66. ACM, New York (2006)
8. Halim, S., Yap, R., Lau, H.C.: An Integrated White+Black Box Approach for Designing and Tuning Stochastic Local Search. In: Bessière, C. (ed.) CP 2007. LNCS, vol. 4741, pp. 332–347. Springer, Heidelberg (2007)
9. Han, J., Kamber, M.: Data Mining: Concept and Techniques, 2nd edn. Morgan Kaufman, San Francisco (2006)
10. Hoos, H.H., Stutzle, T.: Stochastic Local Search: Foundation and Application, 1st edn. Morgan Kaufman, San Francisco (2004)
11. Hutter, F., Hamadi, Y.: Parameter Adjustment Based on Performance Prediction: Towards an Instance-Aware Problem Solver. Technical Report, Microsoft Research (2005)
12. Hutter, F., Hoos, H.H., Stutzle, T.: Automatic Algorithm Configuration based on Local Search. In: 22nd National Conference on Artifical Intelligence, pp. 1152–1157. AAAI Press, Menlo Park (2007)
13. Hutter, F., Hoos, H.H., Leyton-Brown, K., Stutzle, T.: ParamILS: An Automatic Algorithm Configuration Framework. Journal of Artificial Intelligence Research 36, 267–306 (2009)
14. Lau, H.C., Xiao, F.: Enhancing the Speed and Accuracy of Automated Parameter Tuning in Heuristic Design. In: 8th Metaheuristics International Conference (2009)
15. Lourenco, H.R., Martin, O.C., Stutzle, T.: Iterated Local Search. In: Glover, F., Kochenberger, G.A. (eds.) Handbook of Metaheuristics. International Series in Operations Research & Management Science, vol. 57, pp. 320–353. Springer, Heidelberg (2003)
16. Merz, P., Freisleben, B.: Fitness Landscape Analysis and Memetic Algorithms for the Quadratic Assignment Problem. IEEE Transactions on Evolutionary Computation 4, 337–351 (2000)
17. Montgomery, D.C., Runger, G.C.: Applied Statistics and Probability for Engineers, 2nd edn. John Wiley & Son, Chichester (1999)
18. Ng, K.M., Gunawan, A., Poh, K.L.: A hybrid algorithm for the quadratic assignment problem. In: International Conf. on Scientific Computing (2008)
19. Patterson, D.J., Lautz, H.: Auto-WalkSAT: A Self-Tuning Implementation of WalkSAT. Electronic Notes in Discrete Mathematics 9, 360–368 (2001)
20. Reeves, C.R.: Landscapes, operators and heuristic search. Annals of Operations Research 86(1), 473–490 (1999)
21. Salvador, S., Chan, P.: Determining the Number of Clusters/Segments in Hierarchical Clustering/Segmentation Algorithms. In: 16th IEEE International Conference on Tools with Artificial Intelligence, pp. 576–584 (2004)
22. Srivastava, B., Mediratta, A.: Domain-dependent parameter selection of search-based algorithms compatible with user performance criteria. In: 20th National Conference on Artificial Intelligence, pp. 1386–1391. AAAI Press, Pennsylvania (2005)
23. Taillard, E.D.: Comparison of Iterative Searches for The Quadratic Assignment Problem. Location Science 3(2), 87–105 (1995)
24. Xu, L., Hutter, F., Hoos, H.H., Leyton-Brown, K.: SATzilla: Portfolio-based Algorithm Selection for SAT. Journal of Artificial Intelligence Research 32, 565–606 (2008)

Effective Probabilistic Stopping Rules for Randomized Metaheuristics: GRASP Implementations

Celso C. Ribeiro[1], Isabel Rosseti[1], and Reinaldo C. Souza[2]

[1] Department of Computer Science, Universidade Federal Fluminense,
Rua Passo da Pátria 156, Niterói, RJ 24210-240, Brazil
[2] Department of Electrical Engineering, Pontifícia Universidade Católica do Rio de
Janeiro, Rio de Janeiro, RJ 22453-900, Brazil
{celso,rosseti}@ic.uff.br, reinaldo@ele.puc-rio.br

Abstract. The main drawback of most metaheuristics is the absence of effective stopping criteria. Most implementations stop after performing a given maximum number of iterations or a given maximum number of consecutive iterations without improvement in the best known solution value, or after the stabilization of the set of elite solutions found along the search. We propose probabilistic stopping rules for randomized metaheuristics such as GRASP and VNS. We first show experimentally that the solution values obtained by GRASP fit a Normal distribution. Next, we use this approximation to obtain an online estimation of the number of solutions that might be at least as good as the best known at the time of the current iteration. This estimation is used to implement effective stopping rules based on the trade off between solution quality and the time needed to find a solution that might improve the best found to date. This strategy is illustrated and validated by a computational study reporting results obtained with some GRASP heuristics.

1 Introduction and Motivation

Metaheuristics are general high-level procedures that coordinate simple heuristics and rules to find good approximate solutions to computationally difficult combinatorial optimization problems. Among them, we find simulated annealing, tabu search, GRASP, VNS, and others. They are based on distinct paradigms and offer different mechanisms to escape from locally optimal solutions, contrarily to greedy algorithms or local search methods. Metaheuristics are among the most effective solution strategies for solving combinatorial optimization problems in practice and they have been applied to a very large variety of areas and situations. The customization (or instantiation) of some metaheuristic to a given problem yields a heuristic to the latter.

A number of principles and building blocks blended into different and often innovative strategies are common to different metaheuristics. Randomization plays a very important role in algorithm design. Metaheuristics such as simulated annealing, GRASP, VNS, and genetic algorithms rely on randomization to

C.A. Coello Coello (Ed.): LION 5, LNCS 6683, pp. 146–160, 2011.

sample the search space. Randomization can also be used to break ties, so as that different trajectories can be followed from the same initial solution in multistart methods or to sample fractions of large neighborhoods. One particularly important use of randomization appears in the context of greedy randomized algorithms, which are based on the same principle of pure greedy algorithms, but make use of randomization to build different solutions at different runs.

Greedy randomized algorithms are used in the construction phase of GRASP heuristics or to create initial solutions to population metaheuristics such as genetic algorithms or scatter search. Randomization is also a major component of metaheuristics such as simulated annealing and VNS, in which a solution in the neighborhood of the current one is randomly generated at each iteration.

The main drawback of most metaheuristics is often the absence of effective stopping criteria. Most of their implementations stop after performing a given maximum number of iterations or a given maximum number of consecutive iterations without improvement in the best known solution value, or after the stabilization of the set of elite solutions found along the search. In some cases the algorithm may perform an exaggerated and non-necessary number of iterations, when the optimal solution is quickly found (as it often happens in GRASP implementations). In other situations, the algorithm may stop just before the iteration that could find an optimal solution. Dual bounds may be used to implement quality-based stopping rules, but they are often hard to compute or very far from the optimal values, which make them unusable in both situations.

Bayesian stopping rules proposed in the past were not followed by enough computational results to sufficiently validate their effectiveness or to give evidence of their efficiency. Bartkutė et al. [1,2] made use of order statistics, keeping the value of the k-th best solution found. A probabilistic criterion is used to infer with some confidence that this value will not change further. The method proposed for estimating the optimal value with an associated confidence interval is implemented for optimality testing and stopping in continuous optimization and in a simulated annealing algorithm for the bin-packing problem. The authors observed that the confidence interval for the minimum value can be estimated with admissible accuracy when the number of iterations is increased.

Boender and Rinnooy Kan [3] observed that the most efficient methods for global optimization are based on starting a local optimization routine from an appropriate subset of uniformly distributed starting points. As the number of local optima is frequently unknown in advance, it is a crucial problem when to stop the sequence of sampling and searching. By viewing a set of observed minima as a sample from a generalized multinomial distribution whose cells correspond to the local optima of the objective function, they obtain the posterior distribution of the number of local optima and of the relative size of their regions of attraction. This information is used to construct sequential Bayesian stopping rules which find the optimal trade off between reliability and computational effort.

In Dorea [5] a stochastic algorithm for estimating the global minimum of a function is described and two types of stopping rules are derived. The first is based on the estimation of the region of attraction of the global minimum, while the second is based on the existence of an asymptotic distribution of properly normalized estimators. Hart [12] described sequential stopping rules for several stochastic algorithms that estimate the global minimum of a function. Stopping rules are described for pure random search and stratified random search. These stopping rules use an estimate of the probability measure of the ϵ-close points to terminate these algorithms when a specified confidence has been achieved. Numerical results indicate that these stopping rules require fewer samples and are more reliable than the previous stopping rules for these algorithms. They can also be applied to multistart local search and stratified multistart local search. Numerical results on a standard test set show that these stopping rules can perform as well as Bayesian stopping rules for multistart local search. The authors claimed an improvement on the results in [5].

Orsenigo and Vercellis [15] developed a Bayesian framework for stopping rules aimed at controlling the number of iterations in a GRASP heuristic. Two different prior distributions are proposed and stopping conditions are explicitly derived in analytical form. The authors claimed that the stopping rules lead to an optimal trade off between accuracy and computational effort, saving from unnecessary iterations and still achieving good approximations.

In another context, stopping rules have also been discussed in [6,28]. The statistical estimation of optimal values for combinatorial optimization problems as a way to evaluate the performance of heuristics was also addressed in [16,25].

We propose effective probabilistic stopping rules for randomized metaheuristics. In the next section, we give a template for a GRASP heuristic and we describe the optimization problems and test instances that have been used in our computational experiments. In Section 3, we assume that the solution values obtained by a GRASP procedure fit a Normal distribution. This hypothesis is validated experimentally for all problems and test instances described in the previous section. In Section 4, we first show how this Normal approximation can be used to give an online estimation of the number of solutions that might be at least as good as the currently best known solution. This estimation is used to implement effective stopping rules based on the time needed to find a solution that might improve the incumbent. The robustness of this strategy is illustrated and validated by a computational study reporting results obtained with some GRASP implementations. Concluding remarks are made in the last section.

2   GRASP and Experimental Environment

We consider in what follows a general combinatorial optimization problem of minimizing $f(x)$ over all solutions $x \in F$, which is defined by a ground set $E = \{e_1, \ldots, e_n\}$, a set of feasible solutions $F \subseteq 2^E$, and an objective function $f : 2^E \to \mathbb{R}$. The ground set E, the objective function f, and the constraints defining the set of feasible solutions F are defined and specific for each problem. We seek an optimal solution $x^* \in F$ such that $f(x^*) \leq f(x)$, $\forall x \in F$.

GRASP (which stands for *greedy randomized adaptive search procedures*) [8], is a multi-start metaheuristic, in which each iteration consists of two phases: construction and local search. The construction phase builds a feasible solution.

The local search phase investigates its neighborhood until a local minimum is found. The best overall solution is kept as the result; see [18,21,19,20].

The pseudo-code in Figure 1 gives a template illustrating the main blocks of a GRASP procedure for minimization, in which MaxIterations iterations are performed and Seed is used as the initial seed for the pseudo-random number generator.

```
procedure GRASP(MaxIterations, Seed)
1.    Set f* ← ∞;
2.    for k = 1, ..., MaxIterations do
3.         x ← GreedyRandomizedAlgorithm(Seed);
4.         x ← LocalSearch(x);
5.         if f(x) < f* then begin; x* ← x; f* ← f(x); end;
6.         fₖ ← f(x);
7.    end;
8.    return x*;
end.
```

Fig. 1. Template of a GRASP heuristic for minimization

An especially appealing characteristic of GRASP is the ease with which it can be implemented. Few parameters need to be set and tuned, and therefore development can focus on implementing efficient data structures to assure quick iterations. Basic implementations of GRASP rely exclusively on two parameters: the stopping criterion (usually set as a predefined number of iterations) and the parameter used to limit the size of the restricted candidate list within the greedy randomized algorithm used by the construction phase. In spite of its simplicity and ease of implementation, GRASP is a very effective metaheuristic and produces the best known solutions for many problems, see [9,10,11].

Two combinatorial optimization problems have been used in the experiments reported in this paper: the 2-path network design problem and the p-median problem. They are both described below.

Given a connected undirected graph $G = (V, E)$ with non-negative weights associated with its edges, together with a set of formed by K pairs of origin-destination nodes, the 2-path network design problem consists of finding a minimum weighted subset of edges containing a path formed by at most two edges between every origin-destination pair. Applications can be found in the design of communication networks, in which paths with few edges are sought to enforce high reliability and small delays. Its decision version was proved to be NP-complete by Dahl and Johannessen [4]. The GRASP heuristic that has been used in the computational experiments was firstly presented in [23,24]. Data of the four instances involved in the experiments are summarized in Table 1.

Given a set F of m potential facilities, a set U of n customers, a distance function $d : U \times F \to \mathbb{R}$, and a constant $p \leq m$, the p-median problem consists of determining which p facilities to open so as to minimize the sum of the distances from each costumer to its closest open facility. It is a well-known NP-hard

Table 1. Test instances for the 2-path network design problem

Instance	$|V|$	$|E|$	K
2pndp50	50	1,225	500
2pndp70	70	2,415	700
2pndp90	90	4,005	900
2pndp200	200	19,900	2000

problem [14], with numerous applications in location [26] and clustering [17,27]. The GRASP heuristic that has been used in the computational experiments with the p median problem was firstly presented in [22]. Data of the four instances involved in the experiments are summarized in Table 2.

Table 2. Test instances for the p-median problem

Instance	m	n	p
pmed10	200	800	67
pmed15	300	1800	100
pmed25	500	5000	167
pmed30	600	7200	200

3   Normal Approximation for GRASP Iterations

We assume that the solution values obtained by a GRASP procedure fit a Normal distribution. This hypothesis is validated experimentally for all problems and test instances described in the previous section. Let $f_1, \ldots, f_N$ be a sample formed by all solution values obtained along N GRASP iterations. We assume that the null (H_0) and alternative (H_1) hypotheses are:

H_0: the sample $f_1, \ldots, f_N$ follows a Normal distribution; and
H_1: the sample $f_1, \ldots, f_N$ does not follow a Normal distribution.

The chi-square test is the most commonly used to determine if a given set of observations fits a specified distribution. It is very general and can be used to fit both discrete or continuous distributions [13]. First, a histogram of the sample data is estimated. Next, the observed frequencies are compared with those obtained from the specified density function. If the histogram is formed by k cells, let o_i and e_i be the observed and expected frequencies for the i-th cell, with $i = 1, \ldots, k$. The test consists of computing

$$D = \sum_{i=1}^{k} \frac{(o_i - e_i)^2}{e_i}.$$

(1)

It can be shown that, under the null hypothesis, D follows a chi-square distribution with $k-1$ degrees of freedom. Since the mean and the standard deviation are unknown, they should be estimated from the sample. As a consequence, two degrees of freedom are lost to compensate for that. The null hypothesis that the observations come from the specified distribution cannot be rejected at a level of significance α if D is less than $\chi^2_{[1-\alpha;k-3]}$.

Let m and S be, respectively, the average and the standard deviation of the sample $f_1,\ldots,f_N$. A normalized sample $f'_i = (f_i-m)/S$ is obtained by subtracting the average m from each value f_i and dividing the result by the standard deviation S, for $i = 1,\ldots,N$. Then, the null hypothesis that the original sample fits a Normal distribution with mean m and standard deviation S is equivalent to compare the normalized sample with the $N(0,1)$ distribution.

We show below that the solution values obtained along N GRASP iterations fit a Normal distribution, for all problems and test instances presented in Section 2. In all experiments, we used $\alpha = 0.1$ and $k = 14$, corresponding to a histogram with the intervals $(-\infty, -3)$, $[-3.0, -2.5)$, $[-2.5, -2.0)$, $[-2.0, -1.5)$, $[-1.5, -1.0)$, $[-1, -0.5)$, $[-0.5, 0.0)$, $[0.0, 0.5)$, $[0.5, 1.0)$, $[1.0, 1.5)$, $[1.5, 2.0)$, $[2.0, 2.5)$, $[2.5, 3.0)$, and $[3.0, \infty)$. For each instance, we illustrate the Normal fittings after $N = 50$, 100, 500, 1000, 5000, and 10000 iterations.

Table 3 reports on the application of the chi-square test to the four instances of the 2-path network design problem after $N = 50$ iterations. We observe that already after as few as 50 iterations the solution values obtained by the heuristic fit very close a Normal distribution.

To further illustrate that this close fitting is maintained when the number of iterations increase, we present in Table 4 the main statistics for each instance and for increasing values of the number $N = 50$, 100, 500, 1000, 5000, and 10000 of iterations: mean, standard deviation, skewness (η_3), and kurtosis (η_4). The skewness and the kurtosis are computed as follows [7]:

$$\eta_3 = \frac{\sqrt{N} \cdot \sum_{i=1}^{N}(f_i - m)^3}{[\sum_{i=1}^{N}(f_i - m)^2]^{3/2}} \quad \text{and} \quad \eta_4 = \frac{N \cdot \sum_{i=1}^{N}(f_i - m)^4}{[\sum_{i=1}^{N}(f_i - m)^2]^2}.$$

The skewness measures the symmetry of the original data, while the kurtosis measures the shape of the fitted distribution. Ideally, they should be equal to 0 and 3, respectively, in the case of a perfect Normal fitting. We first notice that the mean value consistently converges very quickly to a steady-state value when the number of iterations increases. Furthermore, the mean after 50 iterations is already very close to that of the Normal fitting after 10000 iterations. The skewness values are consistently very close to 0, while the measured kurtosis of the sample is always close to 3.

Table 3. Chi-square test for 90% confidence level: 2-path network design problem

Instance	Iterations	D	$\chi^2_{[1-\alpha;k-3]}$
2pndp50	50	0.398049	17.275000
2pndp70	50	0.119183	17.275000
2pndp90	50	0.174208	17.275000
2pndp200	50	0.414327	17.275000

Table 4. Statistics for Normal fittings: 2-path network design problem

Instance	Iterations	Mean	Std. dev.	Skewness	Kurtosis
	50	372.920000	7.583772	0.060352	3.065799
	100	373.550000	7.235157	-0.082404	2.897830
2pndp50	500	373.802000	7.318661	-0.002923	2.942312
	1000	373.854000	7.192127	0.044952	3.007478
	5000	374.031400	7.442044	0.019068	3.065486
	10000	374.063500	7.487167	-0.010021	3.068129
	50	540.080000	9.180065	0.411839	2.775086
	100	538.990000	8.584282	0.314778	2.821599
2pndp70	500	538.334000	8.789451	0.184305	3.146800
	1000	537.967000	8.637703	0.099512	3.007691
	5000	538.576600	8.638989	0.076935	3.016206
	10000	538.675600	8.713436	0.062057	2.969389
	50	698.100000	9.353609	-0.020075	2.932646
	100	700.790000	9.891709	-0.197567	2.612179
2pndp90	500	701.766000	9.248310	-0.035663	2.883188
	1000	702.023000	9.293141	-0.120806	2.753207
	5000	702.281000	9.149319	0.059303	2.896096
	10000	702.332600	9.196813	0.022076	2.938744
	50	1599.240000	13.019309	0.690802	3.311439
	100	1600.060000	14.179436	0.393329	2.685849
2pndp200	500	1597.626000	13.052744	0.157841	3.008731
	1000	1597.727000	12.828035	0.083604	3.009355
	5000	1598.313200	13.017984	0.057133	3.002759
	10000	1598.366100	13.066900	0.008450	3.019011

Figure 3 displays the Normal distributions fitted for the three first instances for each number of iterations. Together with the above statistics, these plots illustrate the robustness of the Normal fittings to the solution values obtained along the iterations of the GRASP heuristic for the 2-path network design problem.

Table 5 reports the application of the chi-square test to the four instances of the p-median problem after $N = 50$ iterations. As before, we observe that already after as few as 50 iterations the solution values obtained by the heuristic for this problem also fit very close a Normal distribution.

Table 6 gives the same statistics for each instance of the p-median problem and for increasing values of the number $N = 50, 100, 500, 1000, 5000,$ and 10000 of iterations. As for the previous problem, we notice that the mean value consistently converges very quickly to a steady-state value when the number of iterations increases. Furthermore, the mean after 50 iterations is already very close to that of the Normal fitting after 10000 iterations. Once again, the skewness values are consistently very close to 0, while the measured kurtosis of the sample is always close to 3. Figure 4 displays the Normal distributions fitted for the three first instances for each number of iterations. Once again, these results illustrate the robustness of the Normal fittings to the solution values obtained along the iterations of the GRASP heuristic for the p-median problem.

Table 5. Chi-square test for 90% confidence level: p-median problem

Instance	Iterations	D	$\chi^2_{[1-\alpha;k-3]}$
pmed10	50	0.196116	17.275000
pmed15	50	0.167526	17.275000
pmed25	50	0.249443	17.275000
pmed30	50	0.160131	17.275000

Table 6. Statistics for Normal fittings: p-median problem

Instance	Iterations	Mean	Std. dev.	Skewness	Kurtosis
pmed10 $p = 67$	50	1622.020000	57.844097	-0.179163	3.255009
	100	1620.890000	59.932611	-0.364414	3.304588
	500	1620.332000	63.484721	0.111186	3.142248
	1000	1619.075000	64.402076	0.074091	2.964164
	5000	1617.875200	63.499795	0.043152	2.951273
	10000	1618.415400	63.415181	0.087909	2.955408
pmed15 $p = 100$	50	2170.500000	58.880642	-0.041262	1.949923
	100	2168.450000	65.313609	0.270892	2.693553
	500	2173.060000	65.881958	0.202400	2.828056
	1000	2173.484000	65.590272	0.129234	2.784433
	5000	2174.860000	64.639604	0.086450	2.940204
	10000	2175.651600	65.101495	0.096328	2.954639
pmed25 $p = 167$	50	2277.780000	54.782220	0.330959	3.028905
	100	2279.610000	58.034799	0.360133	3.466265
	500	2271.546000	56.029848	0.219415	3.311486
	1000	2274.182000	56.915366	0.081878	3.068963
	5000	2276.305200	56.985195	-0.041096	3.108109
	10000	2277.151600	57.583524	-0.041570	3.073374
pmed30 $p = 200$	50	2434.660000	57.809899	-0.130383	2.961249
	100	2446.560000	57.292464	-0.259531	2.667470
	500	2444.638000	56.109134	-0.189935	2.691882
	1000	2441.465000	57.265005	-0.053183	2.858399
	5000	2441.340400	54.941836	-0.013377	3.054188
	10000	2441.277700	54.978827	0.006407	3.066879

Similar experiments have been performed for other problems and test instances, such as the quadratic assignment and the set k-covering problems, with results of the same caliber. We conclude this section by observing that the null hypothesis cannot be rejected with 90% of confidence. Therefore, we may approximate the solution values obtained by a GRASP heuristic by a Normal distribution that can be progressively fitted and improved as more iterations are performed. This approximation will be used in the next section to establish and validate a probabilistic stopping rule for GRASP heuristics.

4   Probabilistic Stopping Rule

We show in this section that the Normal distribution fitted to the solution values obtained along the GRASP iterations can be used to give an online estimation of

Table 7. Stopping criterion vs. estimated and counted number of solutions at least as good as the incumbent after $N = 1,000,000$ additional iterations

Problem	Instance	Threshold β	Probability $F_X^k(\overline{UB})$	Estimation $\hat{N}^{\leq}$	Count $N^{\leq}$
2-path	2pndp50	10^{-3}	0.000701657	701	738
		10^{-4}	0.000001326	1	0
		10^{-5}	0.000001326	1	0
	2pndp70	10^{-3}	0.000655383	655	465
		10^{-4}	0.000036147	36	26
		10^{-5}	0.000005363	5	4
	2pndp90	10^{-3}	0.000322033	322	190
		10^{-4}	0.000014878	14	7
		10^{-5}	0.000001265	1	0
	2pndp200	10^{-3}	0.000525545	525	503
		10^{-4}	0.000098792	98	95
		10^{-5}	0.000000853	0	1
p-median	pmed10	10^{-3}	0.000181323	181	47
		10^{-4}	0.000088594	88	16
		10^{-5}	0.000007667	7	0
	pmed15	10^{-3}	0.000331692	331	123
		10^{-4}	0.000028636	28	7
		10^{-5}	0.000005236	5	0
	pmed25	10^{-3}	0.000293215	293	211
		10^{-4}	0.000053319	53	31
		10^{-5}	0.000008891	8	3
	pmed30	10^{-3}	0.000569064	569	310
		10^{-4}	0.000028080	28	8
		10^{-5}	0.000000790	0	0

the number of solutions that might be at least as good as the best known solution at the time of the current iteration. This estimation is used to implement an effective stopping rule based on the time needed to find a solution that might improve the incumbent. The robustness of the proposed strategy is illustrated and validated by a computational study reporting the results obtained.

We denote by X the random variable representing the value of the local minimum obtained at each iteration. We recall that $f_1, \ldots, f_k$ is a sample formed by the solution values obtained along the k first iterations. Let m^k and S^k be, respectively, the estimated mean and standard deviation of $f_1, \ldots, f_k$. As already established, we assume that X fits a Normal distribution $N(m^k, S^k)$ with average m^k and standard deviation S^k, whose probability density function and cumulative probability distribution are, respectively, $f_X^k(.)$ and $F_X^k(.)$.

Let UB^k be the value of the best solution found along the k first iterations. Therefore, the probability of finding a solution value smaller than or equal to UB^k in the next iteration can be estimated by $F_X^k(UB^k) = \int_{-\infty}^{UB^k} f_X^k(\tau)d\tau$. This estimation is periodically updated or whenever the best solution value improves.

```
procedure GRASP(β, Seed)
1.    Set f* ← ∞;
2.    Set k ← 0;
3.    repeat
4.         x ← GreedyRandomizedAlgorithm(Seed);
5.         x ← LocalSearch(x);
6.         if f(x) < f* then begin; x* ← x; f* ← f(x); end;
7.         k ← k + 1;
8.         f_k ← f(x);
9.         UB^k ← f*;
10.        Update the average m^k and the standard deviation S^k of f_1, ..., f_k;
11.        Compute the estimate F_X^k(UB^k) = F_X^k(f*) = ∫_{-∞}^{f*} f_X^k(τ)dτ;
12.   until F_X^k(f*) < β;
13.   return x*;
end.
```

Fig. 2. Template of a GRASP heuristic for minimization with the probabilistic stopping criterion

We propose the following stopping rule: for any given threshold β, stop the GRASP iterations whenever $F_X^h(UB^h) \leq \beta$. In other words, the iterations will be interrupted whenever the probability of finding a solution at least as good as the current best becomes less than or equal to β.

To assess the effectiveness of this stopping rule, we have devised and performed the following experiment for each problem and test instance considered in Section 3. For each value of the threshold $\beta = 10^{-3}$, 10^{-4}, and 10^{-5}, we run the GRASP heuristic until $F_X^k(UB^k)$ becomes less than or equal to β. Let us denote by $\bar{k}$ the iteration counter when this condition is met and by $\overline{UB}$ the best known solution at this time. At this point, we may estimate by $\hat{N}^{\leq} = \lfloor N \cdot F_X^{\bar{k}}(\overline{UB}) \rfloor$ the number of solutions whose value will be at least as good as $\overline{UB}$ if N additional iterations are performed. We empirically set $N = 1,000,000$. Next, we perform N additional iterations and we count the number $N^{\leq}$ of solutions whose value is less than or equal to $F_X^{\bar{k}}(\overline{UB})$.

The computational results displayed in Table 7 show that $\hat{N}^{\leq} = \lfloor N \cdot F_X^{\bar{k}}(\overline{UB}) \rfloor$ is a good estimation for the number $N^{\leq}$ of solutions that might be found after N additional iterations whose value is less than or equal to the best value at the time the algorithm would stop for each threshold value β. The probability $F_X^k(UB^k)$ may be used to estimate the number of iterations that must be performed by the algorithm to find a new solution at least as good as the currently best one. Since the user is able to account for the average time taken by each GRASP iteration, the threshold defining the stopping criterion can either be fixed or determined online so as to bound the computation time when the probability of finding improving solutions becomes very small.

The pseudo-code in Figure 2 extends the previous template of a GRASP procedure for minimization, implementing the termination rule based on stopping

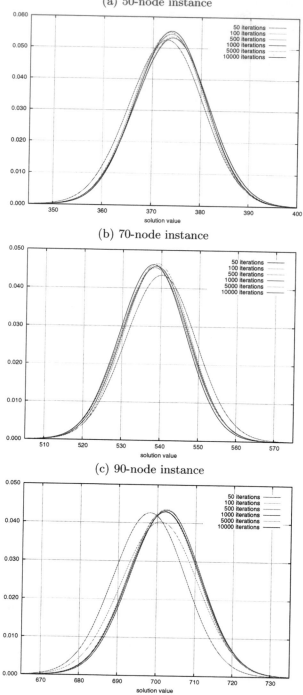

Fig. 3. Fitted probability density functions for the 2-path network design problem

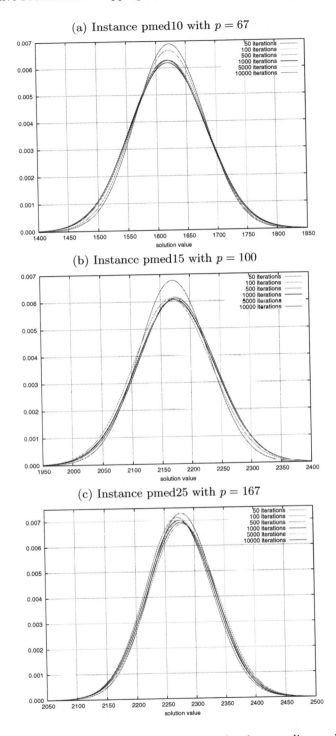

Fig. 4. Fitted probability density functions for the p-median problem

the GRASP iterations whenever the probability $F_X^k(UB^k)$ of improving the best known solution value gets smaller than or equal to β. Lines 8 and 9 update the sample $f_1, \ldots, f_k$ and the best known solution value $UB^k = f^*$ at each iteration k. The mean m^k and the standard deviation s^k of the fitted Normal distribution in iteration k are estimated in line 10. The probability of finding a solution whose value is better than the currently best known solution value is computed in line 11 and used in the stopping criterion implemented in line 12.

The threshold β used to implement the stopping criterion may either be a fixed parameter or iteratively computed. In the last case, it will be computed considering the probability of finding an improving solution (or, alternatively, the estimated number of iterations to find an improving solution) and the average computation time per iteration.

We also notice that since the average time consumed by each GRASP iteration is known, another promising avenue of research consists in investigating stopping rules based on estimating the amount of time needed to probabilistically improve the best solution found by each percent point.

5   Concluding Remarks

The main drawback of most metaheuristics is often the absence of effective stopping criteria. Most of their implementations stop after performing a given maximum number of iterations or a given maximum number of consecutive iterations without improvement in the best solution value, or after the stabilization of a population of solutions or of a set of elite solutions found along the search. In some cases, the algorithm may perform an exaggerated and non-necessary number of iterations. In other situations, the algorithm may stop just before the iteration that could find a better, or even optimal, solution.

Bayesian stopping rules proposed in the past were not followed by enough computational results to sufficiently validate their effectiveness or to give evidence of their efficiency. In this paper, we proposed effective probabilistic stopping rules for randomized metaheuristics.

We first showed experimentally that the solution values obtained by a GRASP heuristic fit a Normal distribution. Next, we used the above Normal approximation to estimate the probability of finding a solution at least as good as the currently best known solution at any iteration. With this probability, we have been able to estimate the number of iterations that must be performed by the algorithm to find a new solution at least as good as the currently best one.

We proposed a stopping rule based on the trade off between this estimation and the time needed to find a solution that might improve the current best one. GRASP iterations will be interrupted whenever the probability of finding a solution at least as good as the current best becomes smaller than or equal a certain threshold.

The robustness of this strategy was illustrated and validated by a computational study reporting results obtained with GRASP implementations for two combinatorial optimization problems. Similar results already obtained for other problems, such as the quadratic assignment and the set k-covering problems, will be reported elsewhere in an extended version of this work.

Since the average time consumed by each GRASP iteration is known, another promising avenue of research consists in investigating stopping rules based on estimating the amount of time needed to probabilistically improve the best solution found by each percent point. We notice that the approach proposed in this paper can be extended and applied not only to GRASP, but also to other metaheuristics that rely on randomization to sample the search space.

Acknowledgments. The authors are grateful to M.G.C. Resende and R. Werneck for making available their GRASP code for solving the p-median problem.

References

1. Bartkutė, V., Felinskas, G., Sakalauskas, L.: Optimality testing in stochastic and heuristic algorithms. Technical report, Vilnius Gediminas Technical University, pp. 4–10 (2006)
2. Bartkutė, V., Sakalauskas, L.: Statistical inferences for termination of markov type random search algorithms. Journal of Optimization Theory and Applications 141, 475–493 (2009)
3. Boender, C.G.E., Rinnooy Kan, A.H.G.: Bayesian stopping rules for multistart global optimization methods. Mathematical Programming 37, 59–80 (1987)
4. Dahl, G., Johannessen, B.: The 2-path network problem. Networks 43, 190–199 (2004)
5. Dorea, C.: Stopping rules for a random optimization method. SIAM Journal on Control and Optimization 28, 841–850 (1990)
6. Duin, C., Voss, S.: The Pilot method: A strategy for heuristic repetition with application to the Steiner problem in graphs. Networks 34, 181–191 (1999)
7. Evans, M., Hastings, N., Peacock, B.: Statistical Distributions, 3rd edn. Wiley, New York (2000)
8. Feo, T.A., Resende, M.G.C.: Greedy randomized adaptive search procedures. Journal of Global Optimization 6, 109–133 (1995)
9. Festa, P., Resende, M.G.C.: GRASP: An annotated bibliography. In: Ribeiro, C.C., Hansen, P. (eds.) Essays and Surveys in Metaheuristics, pp. 325–367. Kluwer Academic Publishers, Dordrecht (2002)
10. Festa, P., Resende, M.G.C.: An annotated bibliography of GRASP, Part I: Algorithms. International Transactions in Operational Research 16, 1–24 (2009)
11. Festa, P., Resende, M.G.C.: An annotated bibliography of GRASP, Part II: Applications. International Transactions in Operational Research 16, 131–172 (2009)
12. Hart, W.E.: Sequential stopping rules for random optimization methods with applications to multistart local search. SIAM Journal on Optimization 9, 270–290 (1998)
13. Jain, R.: The Art of Computer Systems Performance Analysis: Techniques for Experimental Design, Measurement, Simulation, and Modeling. Wiley, New York (1991)
14. Kariv, O., Hakimi, L.: An algorithmic approach to nework location problems, Part II: The p-medians. SIAM Journal of Applied Mathematics 37, 539–560 (1979)
15. Orsenigo, C., Vercellis, C.: Bayesian stopping rules for greedy randomized procedures. Journal of Global Optimization 36, 365–377 (2006)

16. Rardin, R.L., Uzsoy, R.: Experimental evaluation of heuristic optimization algorithms: A tutorial. Journal of Heuristics 7, 261–304 (2001)
17. Rao, M.R.: Cluster analysis and mathematical programming. Journal of the American Statistical Association 66, 622–626 (1971)
18. Resende, M.G.C., Ribeiro, C.C.: GRASP. In: Burke, E.K., Kendall, G. (eds.) Search Methodologies, 2nd edn. Springer, Heidelberg (to appear)
19. Resende, M.G.C., Ribeiro, C.C.: A GRASP with path-relinking for private virtual circuit routing. Networks 41, 104–114 (2003)
20. Resende, M.G.C., Ribeiro, C.C.: GRASP with path-relinking: Recent advances and applications. In: Ibaraki, T., Nonobe, K., Yagiura, M. (eds.) Metaheuristics: Progress as Real Problem Solvers, pp. 29–63. Springer, Heidelberg (2005)
21. Resende, M.G.C., Ribeiro, C.C.: Greedy randomized adaptive search procedures: Advances, hybridizations, and applications. In: Gendreau, M., Potvin, J.-Y. (eds.) Handbook of Metaheuristics, 2nd edn., pp. 283–319. Springer, Heidelberg (2010)
22. Resende, M.G.C., Werneck, R.F.: A hybrid heuristc for the p-median problem. Journal of Heuristics 10, 59–88 (2004)
23. Ribeiro, C.C., Rosseti, I.: A parallel GRASP heuristic for the 2-path network design problem. In: Monien, B., Feldmann, R.L. (eds.) Euro-Par 2002. LNCS, vol. 2400, pp. 922–926. Springer, Heidelberg (2002)
24. Ribeiro, C.C., Rosseti, I.: Efficient parallel cooperative implementations of GRASP heuristics. Parallel Computing 33, 21–35 (2007)
25. Serifoglu, F.S., Ulusoy, G.: Multiprocessor task scheduling in multistage hybrid flow-shops: A genetic algorithm approach. Journal of the Operational Research Society 55, 504–512 (2004)
26. Tansel, B.C., Francis, R.L., Lowe, T.J.: Location on networks: A survey. Management Science 29, 482–511 (1983)
27. Vinod, H.D.: Integer programming and the theory of groups. Journal of the American Statistical Association 64, 506–519 (1969)
28. Voss, S., Fink, A., Duin, C.: Looking ahead with the Pilot method. Annals of Operations Research 136, 285–302 (2005)

A Classifier-Assisted Framework for Expensive Optimization Problems: A Knowledge-Mining Approach

Yoel Tenne, Kazuhiro Izui, and Shinji Nishiwaki

Kyoto University, Kyoto, Japan
yoel.tenne@ky3.ecs.kyoto-u.ac.jp,
{izui,shinji}@prec.kyoto-u.ac.jp

Abstract. Real-world engineering design optimization problems often rely on computationally-expensive simulations to replace laboratory experiments. A common optimization approach is to approximate the expensive simulation with a computationally cheaper model resulting in a model-assisted optimization algorithm. A prevalent issue in such optimization problems is that the simulation may crash for some input vectors, a scenario which increases the optimization difficulty and results in wasted computer resources. While a common approach to handle such vectors is to assign them a penalized fitness and incorporate them in the model training set this can result in severe model deformation and degrade the optimization efficacy. As an alternative we propose a classifier-assisted framework where a classifier is incorporated into the optimization search and biases the optimizer away from vectors predicted to crash to simulator and with no model deformation. Performance analysis shows the proposed framework improves performance with respect to the penalty approach and that it may be possible to 'knowledge-mine' the classifier as a post-optimization stage to gain new insights into the problem being solved.

1 Introduction

Nowadays researchers replace real-world laboratory experiments with computer simulations to reduce the time and cost of the engineering design process. In this setup the design process is effectively an optimization problem having two distinct features:

a) Objective values are obtained from the simulation which is often a legacy code or a commercial software available only as an executable. As such the simulation is treated as a *'black-box'* function (no analytic expression for the function or its derivatives).

b) Each simulation run is *expensive*, that is, it requires large computational resources (anywhere from minutes to weeks of CPU time) and so only a small number of evaluations can be made.

Accordingly, these scenarios are often referred to as *expensive optimization problems* [26].

Besides the two issues mentioned above such problems introduce another challenge: the simulation may 'crash' and fail to return an objective value (fitness) for some vectors

C.A. Coello Coello (Ed.): LION 5, LNCS 6683, pp. 161–175, 2011.

(candidate designs). We refer to such vectors as *simulator-infeasible* (SI) while vectors for which the simulation completes successfully are *simulator-feasible* (SF). Encountering SI vectors during an optimization search has two main implications: a) the objective function is now discontinuous which is problematic for optimizers requiring continuous functions (such as SQP) and b) such vectors can consume a large portion of the optimization budget without improving the fitness landscape and so the optimization search may stagnate.

To effectively handle such SI vectors we propose a framework which uses both a model *and* a classifier during the optimization search. The classifier is continuously trained using all evaluated vectors (SI and SF) and its role is to predict if a new candidate solution is SI or not. The framework then leverages on the classifier's prediction to bias the search to vectors predicted to be SF. Analysis also shows that besides improving the search the classifier can also provide new insights into the problem being solved.

The remainder of this paper is as follows: Sect. 2 reviews expensive optimization problems and relevant computational intelligence approaches, Sect. 3 describes the proposed framework, Sect. 4 gives a detailed performance analysis and lastly Sect. 5 summarizes the paper.

2 Background

2.1 Expensive Optimization Problems

Expensive optimization problems, that is, where objective values are obtained from a computer simulation with a lengthy run time, arise in diverse domains across engineering and science. The high computational cost of each simulator run implies that only a small number of such function evaluations can be made during the entire search. This is particularly challenging for a computational intelligence (CI) optimizer (such as an evolutionary algorithm (EA), particle swarm optimizer (PSO), simulated annealing (SA) and alike) which often requires many thousands of function evaluations to obtain a good solution.

A common approach to combat expensive evaluations is *modelling*, that is, where a computationally cheaper approximation of the objective function is trained using previously evaluated solutions and is used during the search instead of calling the true (expensive) function. Examples of models include quadratics [18], radial basis functions (RBFs) [3], artificial neural networks (ANNs) [1] and Kriging [7]. CI algorithms which use models are commonly termed *model-assisted* or *surrogate-assisted* and the literature is rich with variants [26].

While models alleviate the bottleneck of a high computational cost they introduce a challenge of inaccurate objective values: since function evaluations are expensive the training sample is small which leads to an inaccurate model [5]. Model inaccuracy implies that the optimizer is searching on a deformed landscape with a possibly *false optimum* (an optimum of the model which is not an optimum of the true expensive function) [12]. As such model-assisted algorithms must manage this inherent inaccuracy in order to be effective. One approach to handle model inaccuracy is with the *trust-region* (TR) framework which has a long standing history in nonlinear programming (and unrelated to expensive black-box optimization) [5]. The TR is a sequential approach where starting from an initial guess $\mathbf{x}^{(0)}$ then at each iteration $i = 0, 1, \ldots$ a model is trained and the framework performs a *trial step* where it seeks an optimum of the model $m(\mathbf{x})$ constrained to the TR $\mathscr{T}$ where

$$\mathscr{T} = \{\mathbf{x} : \|\mathbf{x} - \mathbf{x}^{(i)}\|_2 \leqslant \Delta\}, \tag{1}$$

where Δ is the TR radius. This defines the constrained optimization problem

$$\begin{aligned} \min \quad & m(\mathbf{x}) \\ \text{s.t.} \quad & \mathbf{x} \in \mathscr{T} \end{aligned} \tag{2}$$

which gives a minimizer $\mathbf{x}_m$. The success of the trial step is gauged by the merit value

$$\rho = \frac{f(\mathbf{x}^{(i)}) - f(\mathbf{x}_m)}{m(\mathbf{x}^{(i)}) - m(\mathbf{x}_m)}, \tag{3}$$

where $\rho > 0$ indicates the trial was successful. The TR is then updated based on ρ, for example the TR is expanded if $\rho > 0$ but is contracted otherwise [19, 25].

2.2 Simulator Infeasible Vectors

As mentioned in Sect. 1 this study focuses on expensive optimization problems with vectors which 'crash' the simulation. Such vectors pose the risk of consuming a significant portion of the optimization budget without providing new objective values, a scenario which can lead to search stagnation. Several studies have acknowledged the difficulties such vectors induce, for example [14] mentioned 'inputs combinations which are likely to crash the simulator', [20] studied a multidisciplinary optimization problem with 'unevaluable points' which 'cause the simulator to crash', [6] mentioned 'virtual constraints' where 'function evaluations fail at certain points' and additional examples include [2, 10].

With respect to handling such vectors [20] proposed using a classifier to screen vectors before evaluating them. Those classified as SI were assigned a 'death penalty', that is, a fictitious and highly penalized fitness which resulted in them being quickly eliminated from the population. The study did not consider using models but the EA called the expensive function directly. A related approach was used in [10] in the context of airfoil shape optimization where SI vectors were severely penalized and incorporated into the model in order to bias the search away from them. In [2] such vectors were simply excluded from the training sample of the model. While these approaches offer a workable solution they suffer from two main drawbacks: a) eliminating SI vectors discards valuable (and expensive to obtain) information regarding the fitness landscape while b) incorporating highly-penalized vectors into the model may severely deform it and introduce false optima. As an example Fig. 1 compares Kriging models of the Rosenbrock function with and without SI vectors (penalty taken as the worst fitness of the SF ones). It follows that incorporating the SI vectors with a penalized fitness had severely deformed the model.

The demerits of such approaches have motivated studying various alternatives. As mentioned above, in [20] the authors proposed using a classifier to screen candidate

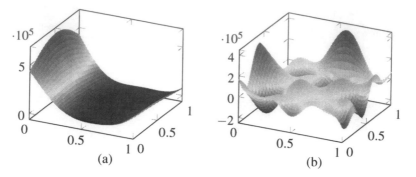

Fig. 1. Comparison of models with and without penalized SI vectors. The objective function is Rosenbrock and the model is Kriging: (a) baseline sample of 30 vectors all SF (b) 20 SI vectors were added, assigned the worst function from the baseline 30 value and incorporated into model.

vectors prior to evaluation if they are expected to be SI or not and those predicted to be SI were assigned a death penalty. Also along the classification concept [24] proposed a dual model approach: one for the objective function and one which interpolates a penalty between SI vectors and where vectors predicted to be SI received a high penalty and vice-versa. Other studies have explored the use of classifiers for constrained non-linear programming (but did not focus on handling SI vectors) [11]. Further exploring the use of classifiers [27] studied a preliminary classifier-assisted framework for handling SI vectors (termed there 'undefined') and applied it to an airfoil shape optimization problem.

3   Proposed Framework

Following the above discussion we propose a classifier-assisted framework for expensive optimization problems in the presence of SI vectors. It leverages on the TR framework as a rigorous approach to manage both the model and classifier and to ensure convergence to an optimum of true (expensive) objective function. We briefly describe the model and classifiers used and then the mechanics of the framework.

3.1   The Model

The proposed framework can accommodate any model and in this study we have used Kriging which is a statistical approach to interpolation [14]. This particular model was chosen since it is widely used in real-world applications [7, 21]. Given a set of evaluated vectors $\mathbf{x}_i$, $i = 1 \ldots k$, the model is trained such that is exactly interpolates the observed values, that is, $m(\mathbf{x}_i) = f(\mathbf{x}_i)$. The model combines a global 'drift' function with a local correction based on correlation between neighbouring sites. Using a constant drift function gives the Kriging model

$$m(\mathbf{x}) = \beta + c(\mathbf{x}),\tag{4}$$

with the drift function β and point-wise local correction c given by a stationary Gaussian process with mean zero and covariance

$$Cov[C(\mathbf{x})C(\mathbf{y})] = \sigma^2 \mathbf{R} \tag{5}$$

where $\mathbf{R}$ is the symmetric $k \times k$ correlation matrix between all sample vectors, that is, $R_{i,j}$ is the correlation between vectors i and j and following the discussion in [14] we have used the Gaussian correlation function

$$\mathscr{R}(\theta, \mathbf{x}, \mathbf{y}) = \prod_{i=1}^{d} \exp\left(-\theta (x_i - y_i)^2\right). \tag{6}$$

The model prediction is then

$$m(\mathbf{x}) = \hat{\beta} + \mathbf{r}^{\mathrm{T}} \mathbf{R}(\mathbf{f} - \mathbf{1})\hat{\beta} \tag{7}$$

where $\hat{\beta}$ is the estimated drift coefficient, $\mathbf{f}$ is the vector of objective values and $\mathbf{1}$ is a vector with all elements equal 1. The estimated drift coefficient $\hat{\beta}$ and variance $\hat{\sigma}^2$ are obtained from

$$\hat{\beta} = \left(\mathbf{1R}^{-1}\mathbf{1}\right)^{-1} \mathbf{1}^{\mathrm{T}} \mathbf{R}^{-1} \mathbf{f} \tag{8}$$

$$\hat{\sigma}^2 = \frac{1}{n}\left[(\mathbf{f} - \mathbf{1}\hat{\beta})\mathbf{R}^{-1}(\mathbf{f} - \mathbf{1}\hat{\beta})\right]. \tag{9}$$

Fully defining the model requires the correlation parameter θ which is commonly taken as the maximizer of the model likelihood

$$\theta^\star : \min - \left(n\log(\sigma^2) + \log(|\mathbf{R}|)\right). \tag{10}$$

3.2 The Classifier

As mentioned in Sect. 1, the proposed framework also uses a classifier to predict if candidate vectors are SI. Briefly, a classifier maps inputs vectors into one of several 'groups' based on some similarly measure [9].

We consider two representative classifiers. The first is the *nearest neighbour* (NN) classifier [16] which assigns the new vector the same class as its closest training vector (measured by a distance $d(\mathbf{x}, \mathbf{y})$ such as the l_2 norm), namely:

$$c(\mathbf{x}_{new}) = F(\mathbf{x}_{\mathrm{NN}}) : d(\mathbf{x}_{new}, \mathbf{x}_{\mathrm{NN}}) = \min d\left(\mathbf{x}_{new}, \mathbf{x}_i\right), i = 1 \ldots k. \tag{11}$$

where $c(\mathbf{x})$ is the class assigned by the classifier and $F(\mathbf{x}_{\mathrm{NN}})$ is the class of the NN vector. An extension of the algorithm is to observe the most common class among the k nearest neighbours (k-NN) of the new vector and assign it that class. A merit of k-NN classifiers is that they do not require any training and have no parameters to calibrate (besides the user-prescribed parameter k).

The second classifier is the *support vector machine* (SVM) which projects the data into a high-dimensional space where it can be more easily separated [28]. In a two-class problem an SVM tries to find the best classification function for the training data. For a linearly separable training set a linear classification function is the separating hyper-plane passing through the middle of the two classes. Once the classifier (hyperplane) is fixed then new vectors are classified based on the sign of the classifier output (± 1). There are many such hyperplanes so an SVM adds the condition that the function (hyperplane) maximizes the margin between the two classes (geometrically the distance between the hyperplane and the nearest vectors to it from each class) by maximizing the following Lagrangian:

$$L_P = \frac{1}{2}\|\mathbf{w}\| - \sum_{i=1}^{K} \alpha_i y_i (\mathbf{w} \cdot \mathbf{x}_i + b) + \sum_{i=1}^{K} \alpha_i \qquad (12)$$

where $y_i = \pm 1$ is the class of each training vector, $\alpha_i \geqslant 0$ and the derivatives of L_p with respect to α_i are zero. The vector $\mathbf{w}$ and scalar b define the hyperplane.

3.3   The Framework

The proposed framework begins by sampling an initial set of points using a Latin hyper-cube design (LHD) [17] which ensures the points are space-filling and hence improve the model accuracy.

The main optimization loop then begins where the framework first trains a Kriging model (Sect. 3.1) using only the SF vectors in the cache and then trains a classifier using all cached vectors (both SF and SI) (Sect. 3.2). At this stage the framework performs a TR trial step where it uses a real-coded EA to search for an optimum of the model. However the EA does not receive the fitness value directly from the model but instead from the objective function $\hat{m}(\mathbf{x})$ where

$$\hat{m}(\mathbf{x}) = \begin{cases} m(\mathbf{x}) & \text{if } c(\mathbf{x}) \text{ is SF} \\ \tau & \text{if } c(\mathbf{x}) \text{ is SI} \end{cases} \qquad (13)$$

where $m(\mathbf{x})$ is the model-predicted objective value, τ is a penalized fitness taken to be the worst function value from the initial LHD sample and $c(\mathbf{x})$ is the classifier prediction. In this setup the EA receives the model prediction if the classifier predicts a vector is SF but receives the penalized fitness otherwise. A merit of this setup is that the knowledge about the SI vectors is preserved in the classifier but they are *not* incorporated into the model (with a penalized fitness) and hence do not deform the model (Sect. 2.1).

The proposed framework can accommodate any CI optimizer and we use the real-coded EA from [4] as it is representative of many other real-coded EA variants. Since evaluating the model is computationally cheap (a fraction of a second) the EA uses a population size of 100 for a lengthy 100 generations to improve the search efficacy. For its operators the EA used stochastic universal selection (SUS) with $p = 0.7$, intermediate recombination with $p = 0.7$ and the Breeder Genetic Algorithm (BGA) mutation with $p = 0.1$ and 10% elitism.

The EA is invoked and yields $\mathbf{x}^\star$ an optimum of the model which is then evaluated with the true (expensive) function (at a cost of one function evaluation), obtaining $f(\mathbf{x}^\star)$. Next, the framework updates the TR based on the success of the trial step with the following steps:

- if $f(\mathbf{x}^\star) < f(\mathbf{x}_c)$: the search was successful since the EA found a better solution. As such the TR is centred at the new vector and the TR radius is enlarged to search in a wider region since the model appears to be accurate.
- if $f(\mathbf{x}^\star) \geqslant f(\mathbf{x}_c)$ *and* there are sufficient SF points inside the TR: the search was unsuccessful but since there is a sufficient number of points in the TR the model is considered accurate enough to justify contracting the TR.
- if $f(\mathbf{x}^\star) \geqslant f(\mathbf{x}_c)$ *and* there are insufficient SF points inside the TR: the search was unsuccessful but this may be due to poor model accuracy in the TR. As such the framework adds a new point ($\mathbf{x}_n$) inside the TR to improve the local model accuracy. The procedure for adding the point is explained below.

The above tests differ from the classical TR framework by accounting for the number of points in the TR since contracting the TR even when the model accuracy is poor (small number of points in the TR) may lead to premature convergence [5]. As such, checking the number of points in the TR is an additional measure to account for uncertainty due to the model approximation error. Based on experimentation the threshold number of points was taken as $\max(5, 0.1d)$ where d is the problem dimension.

As explained above the framework may add a new point ($\mathbf{x}_n$) to improve the model in the TR. To achieve this the new interior point should be placed in a region sparse with points so it adds information about the model in a relatively unexplored region. To find such a point the framework generates a LHD sample of points in the TR and selects the point having the largest minimum distance to all interior points (a max-min criterion [13]), that is

$$\mathbf{x}_n : \max_{\mathbf{x} \in \mathcal{T}} \min_{\mathbf{x}_i \in \mathcal{T}} \{ \|\mathbf{x} - \mathbf{x}_i\|_2 \} \tag{14}$$

where $\mathbf{x}_i$, $i = 1 \ldots l$ are cached points (vectors) which are in the TR.

Besides ensuring convergence to a true optimum of the objective function the TR framework offers another merit: it has an intrinsic mechanism to measure the model accuracy, at least with respect to its ability to predict an optimum, as formulated by the TR trial steps. The use of such an accuracy measure precludes the need to assess the model accuracy (at least with respect to its ability to predict an optimum) by other means such as cross-validation [15]. Of course, the TR trial step does not indicate the model accuracy over the entire TR (or the search space).

We have also considered handling the case where the classifier hampers the optimization by 'masking' an optimum predicted by the model (the classifier may mask an optimum and prevent the optimizer from reaching it). To monitor the effect of the classifier the proposed framework adds an additional step: if no progress has been made for u consecutive optimization iterations (termed *unsuccessful iterations*) then a pseudo-search is made where the classifier is 'disabled' and the EA searches using the model only (based on experiments we have used $u = 5$). The obtained optimum is then compared to the one in the original search (with the classifier) but the optimum is not evaluated with the expensive function. If they differ then the classifier is affecting the search and this motivates improving the classifier accuracy to reduce the chances it 'masks' a better solution and prevents the optimizer from reaching it. To improve the classifier accuracy the framework generates a LHD sample of vectors in the box defined by the extremal coordinates of the optimum predicted with the classifier and the optimum found with the classifier disabled. Similarly to improving the model locally, the framework selects a vector from this sample based on the max-min distance criterion to existing vectors in the box. Lastly, if the TR has been contracted for v consecutive times this can indicate convergence to a local optimum and so the framework samples a point in the entire search space to improve the global accuracy of the model and classifier and to assist in discovering possibly new optima. As above, the new point is selected using the max-min criterion but with respect to all cached vectors. Based on experimentation we have used $v = 2$. To complete the description Algorithm 1 gives the algorithm for the proposed framework.

Algorithm 1: Proposed Framework

---

generate an initial LHD sample;
evaluate and cache vectors;
repeat

    TR centre: $\mathbf{x}_c \leftarrow$ best vector;
    train a model using SF vectors in cache;
    train a classifier using all vectors in cache;
    search for the model optimum with an EA (fitness modified by classifier);
    evaluate the predicted optimum ($\mathbf{x}^*$);
    /* manage the TR, model and classifier   */
    if *new optimum is better than TR centre* **then**
        └ increase the TR radius

    else if *new optimum is not better than TR centre* and *insufficient points in TR* **then**
        └ add a new point in the TR to improve the model;

    else if *new optimum not better than TR centre* and *sufficient points in TR* **then**
        └ decrease the TR radius;

    /* check the effect of the classifier   */
    if *u consecutive unsuccessful iterations* **then**
        search for the model optimum but with the classifier disabled;
        if different from $\mathbf{x}^*$ then add a point to improve the classifier;

    /* check search stagnation   */
    if *v consecutive TR contractions* **then**
        └ add a point globally to improve the model and classifier;

until *optimization budget exhausted*;

---

4   Performance Analysis

4.1   Test Problem and Benchmarks

We test the efficacy of the proposed framework on a problem of airfoil shape optimization as it is representative of real-world engineering problems *and* contains SI vectors as explained below. In this problem the goal is to find an airfoil shape which maximizes the lift coefficient (c_l) and minimizes the aerodynamic drag coefficient (c_d) at some prescribed flight conditions (flight altitude, speed and angle of attack (AOA) which indicates the angle between the airfoil chord and the aircraft velocity). Also, between 0.2

to 0.8 of the chord length the airfoil's minimum thickness (t) must be equal to or larger than a critical value $t^* = 0.1$ to ensure structural integrity. The objective function is

$$f = -\frac{c_l}{c_d} + p_t \tag{15a}$$

where p_t is a penalty for airfoils which violate the thickness constraint and defined as

$$p_t = \begin{cases} \dfrac{t^*}{t} \cdot \left| \dfrac{c_l}{c_d} \right| & \text{if } t < t^* \\ 0 & \text{otherwise} \end{cases} \tag{15b}$$

Airfoils were represented with the Parametric Sections (PARSEC) parameterization [23] which defines 11 design variables representing geometrical features (Figure 2). To ensure a closed airfoil shape we have set $dz_{TE} = 0$ (the PARSEC variable) while bounds on the other variables were set based on [27]. To obtain the lift and drag of candidate airfoils we used XFoil–a computational fluid dynamics simulation for analysis of subsonic isolated airfoils [8]. Each airfoil evaluation required up to 30 seconds on a desktop computer.

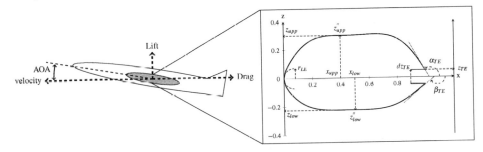

Fig. 2. Physical quantities (left) and the PARSEC design variables (right) in the airfoil optimization problem

To change the prevalence of SI vectors we have used four AOA settings ($2°,5°,10°$ and $15°$) since due to the mechanics of the simulation and the underlying physics higher AOA values result in more frequent simulation crashes. For the proposed framework we have used two variants:

- P-SVM: proposed framework with an SVM classifier (Gaussian kernels) and
- P-KNN: proposed framework with a k-NN classifier ($k = 3$).

We have also benchmarked these two variants against penalty-based variants (termed the *reference algorithms*) which use the same optimization steps *except* that they do not use a classifier but instead assign the SI vectors a penalized fitness and incorporate them into the model training sample. The two variants used were:

- R-1: reference algorithm with the penalized fitness as the worst objective value from the initial LHD sample
- R-10: as above but uses 10 times the worst objective value.

Table 1. Statistics for Best Objective Value

AOA		P-SVM	P-k-NN	R-1	R-10
2	mean	**-2.851e+02**	-9.981e+01	-9.637e+01	-2.550e+02
	SD	5.408e+02	3.179e+01	2.660e+01	8.370e+02
	median	-9.806e+01	-9.869e+01	-9.768e+01	**-1.126e+02**
	min	-2.375e+03	-1.835e+02	-1.670e+02	-4.604e+03
	max	-3.181e+01	-1.084e+01	-3.425e+01	-3.070e+01
5	mean	**-1.600e+03**	-1.496e+03	-2.820e+02	-3.104e+02
	SD	7.392e+03	4.761e+03	3.973e+02	5.424e+02
	median	**-8.810e+01**	-1.016e+02	-9.320e+01	-8.405e+01
	min	-3.999e+04	-2.488e+04	-1.515e+03	-2.210e+03
	max	-3.595e+01	-1.879e+01	-3.635e+01	-1.518e+01
10	mean	**-2.416e+01**	-2.293e+01	-1.697e+01	-2.103e+01
	SD	1.969e+01	1.198e+01	9.226e+00	1.102e+01
	median	-2.017e+01	**-2.090e+01**	-1.478e+01	-1.636e+01
	min	-1.216e+02	-7.533e+01	-4.130e+01	-4.233e+01
	max	-1.237e+01	-8.926e+00	-6.545e+00	-4.784e+00
15	mean	**-6.079e+00**	-5.172e+00	-5.690e+00	-6.043e+00
	SD	2.038e+00	1.497e+00	2.111e+00	3.200e+00
	median	**-5.322e+00**	-5.321e+00	-5.009e+00	-4.598e+00
	min	-9.923e+00	-8.834e+00	-1.119e+01	-1.701e+01
	max	-3.441e+00	-2.720e+00	-3.272e+00	-3.183e+00

P-SVM: proposed approach, SVM classifier.
P-k-NN: proposed approach, k-NN classifier.
R-1: reference approach, penalty = worst objective.
R-10: reference approach, penalty = 10× worst objective.

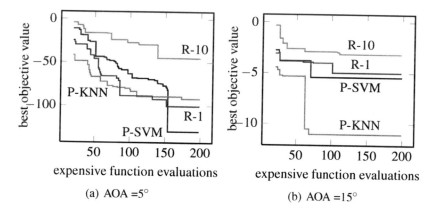

(a) AOA =5°   (b) AOA =15°

Fig. 3. Convergence trends the algorithms

We have chosen this benchmarking setup since it highlights the effect of the proposed classifier-assisted approach (the reference algorithms simply disabled the classifier effect and used penalized vectors in the training set). We specifically did not try to select the best model type, best optimizer or to calibrate any algorithm parameters to the specific test problem but instead we study the contribution of adding the proposed framework to a typical optimization setup (a real-coded EA with a Kriging model). In all tests the limit was 200 function evaluations (simulation runs) and for valid statistical analysis we have repeated each algorithm–AOA combination for 30 times.

Table 1 shows the statistics for the best objective function with the best mean and median emphasized at each AOA. It follows the proposed framework with an SVM classifier (P-SVM) had the best mean in all cases and best median in two cases. The k-NN variant had the best median for AOA=10°. Overall results show the proposed framework outperformed the reference (penalty-based) algorithms and also indicate the demerits of the penalty approach, namely: a) incorporating penalized vectors into the model can deform the landscape and hinder performance and b) performance can be sensitive to the penalty value but an optimal penalty value is unknown a-priori and may be difficult to obtain given the tight optimization budget.

Statistical significance analysis (at the $\alpha = 0.05$ level) shows that performance gains of the P-SVM variant were not statistically-significant at AOA=2° since at a low AOA settings the SI vectors are less frequent and so the proposed framework operates during most of the optimum very similar or identically to the reference algorithms. At higher AOAs where SI vectors are more prevalent the situation changes: at AOA=5° gains were significant over the R-10 variant and at AOA=10° over the R-1 and R-10 variants. At AOA=15° gains were borderline significant over the R-1 and significant over the R-10 variant.

To visualize the convergence trends for each algorithm Fig. 3 compares the representative tests for AOA=5° and 15°, showing the variants of the proposed framework performed well.

We have also studied the number of SI vectors encountered during the search by each algorithm. Table 2 gives the test statistics for the number of SI vectors from which it follows the proposed classifier-assisted approach consistently obtained a competitive number of SI vectors which is similar or better to that of the penalty approach variants. This indicates the proposed approach both obtained a better final solution and also reduced the number of failed evaluations (and hence the amount of wasted computer resources).

4.2 Knowledge-Mining the Classifier

The classifier adds a machine-learning component to the optimization algorithm and so we explore the option of 'knowledge-mining' the classifier as a post-processing stage to the optimization in order to gain new insights into the problem. Specifically, we are interested in understating: a) how SI vectors are distributed in the search space and b) why some vectors crash the simulation.

To understand how SI vectors are distributed we applied the following procedure. After an optimization search was completed we applied the classifier to a new LHD sample of vectors (without evaluating them with the expensive function). Since the vectors are 11D we visualize their distribution by projecting them to a 2D scatter plot using the Sammon mapping (a dimensionality-reduction procedure) [22]. The mapping preserves the proximity relations between vectors (data points) such that adjacent high-dimensional points will be mapped into adjacent low-dimensional ones and vice versa, resulting in a topologically consistent projection. Starting from a (possibly random) distribution of low-dimensional points, the mapping algorithm iteratively updates these points to minimize the *Sammon stress function*

$$C = \frac{1}{\sum_{i=1}^{k} \delta(\mathbf{x}_i, \mathbf{x}_j)} \sum_{i=1}^{k} \sum_{j<i}^{k} \frac{\left(\delta(\mathbf{x}_i, \mathbf{x}_j) - \delta(\hat{\mathbf{x}}_i, \hat{\mathbf{x}}_j)\right)^2}{\delta(\mathbf{x}_i, \mathbf{x}_j)} \tag{16}$$

Table 2. Statistics for the Number of SI Vectors

AOA		P-SVM	P-k-NN	R-1	R-10
2	mean	3.410e+01	3.090e+01	**2.552e+01**	3.043e+01
	SD	1.832e+01	1.429e+01	1.123e+01	1.582e+01
	median	2.800e+01	2.550e+01	**2.300e+01**	2.600e+01
	min	1.700e+01	1.300e+01	1.300e+01	1.400e+01
	max	9.500e+01	7.900e+01	5.900e+01	7.900e+01
5	mean	**4.624e+01**	5.837e+01	4.833e+01	7.050e+01
	SD	1.274e+01	2.055e+01	2.002e+01	2.533e+01
	median	4.900e+01	5.750e+01	**4.650e+01**	6.900e+01
	min	1.800e+01	2.300e+01	2.200e+01	3.600e+01
	max	6.800e+01	1.070e+02	9.700e+01	1.440e+02
10	mean	**9.647e+01**	1.066e+02	1.242e+02	1.214e+02
	SD	1.578e+01	1.819e+01	2.578e+01	2.568e+01
	median	**1.000e+02**	1.070e+02	1.280e+02	1.185e+02
	min	5.500e+01	6.500e+01	6.300e+01	6.400e+01
	max	1.210e+02	1.490e+02	1.770e+02	1.730e+02
15	mean	1.159e+02	**1.109e+02**	1.369e+02	1.249e+02
	SD	1.836e+01	2.757e+01	2.768e+01	3.501e+01
	median	**1.180e+02**	1.135e+02	1.440e+02	1.300e+02
	min	6.200e+01	4.000e+01	8.200e+01	3.600e+01
	max	1.430e+02	1.570e+02	1.800e+02	1.770e+02

P-SVM: proposed approach, SVM classifier.
P-k-NN: proposed approach, k-NN classifier.
R-1: reference approach, penalty = worst objective.
R-10: reference approach, penalty = 10× worst objective.

where δ is a distance measure (typically l_2), $\mathbf{x}$ is a high-dimensional vector (original data) and $\hat{\mathbf{x}}$ is a low-dimensional (projected) vector. Figure 4 shows an example for AOA=5°. The background scatter plot shows the projection of the vectors evaluated during the search (and which trained the classifier) while the foreground plot shows the prediction of an SVM classifier on a new LHD sample of 100 vectors which were *not* evaluated with the simulation. The three inset plots show the airfoils corresponding to vectors classified as SI. This analysis indicate that the classifier 'learns' the limitations of the simulation code and can predict which airfoil geometries will likely crash it. The Sammon mapping shows the SI are expected to be scattered over the entire search space.

We have also explored the option of using histograms of the design variables to identify for each variable if there is a 'critical' range of values which is likely to result in a SI vector. For example, analyzing the variables histograms for the vectors in the above sample shows that vectors classified as SI often had the variable x_{upp} in the range $0.2 \ldots 0.28$. Plotting a few of the vectors with x_{upp} in this critical range showed that they correspond to irregularly shaped airfoils which highlights the effect of the variable. With this insight it is now possible to refine the variable bounds and to reduce the number of failed evaluations in future optimization runs.

Overall, these experiments show that exploratory data analysis procedures (which do not require any additional expensive evaluations) were able to 'knowledge-mine' the classifier and to yield new insights into the problem being solved.

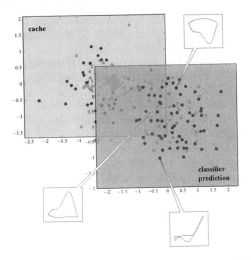

Fig. 4. Classifier prediction for SF and SI vectors at AOA=5

5 Summary

Real world engineering design optimization problems often rely on computationally-expensive simulations to replace laboratory experiments. A common optimization approach is to approximate the expensive simulation with a computationally cheaper model resulting in a model-assisted optimization algorithm. A prevalent issue in such optimization problems is that the simulation may crash for some input vectors, a scenario which increases the optimization difficulty and results in wasted computer resources. While a common approach to handle such vectors is to assign them a penalized fitness and incorporate them in the model training set this can result in severe model deformation and degrade the optimization efficacy. As an alternative we have proposed a classifier-assisted framework where a classifier is incorporated into the optimization search and biases the optimizer away from vectors predicted to crash to simulator and with no model deformation. Performance analysis showed the proposed framework improved performance with respect to the penalty approach and that it was possible to 'knowledge-mine' the classifier as a post-optimization stage to gain new insights into the problem being solved.

Acknowledgement

The first author thanks the Japan Society for Promotion of Science for its fellowship support.

References

1. Bishop, C.M.: Neural Networks for Pattern Recognition. Oxford University Press, New York (1995)

2. Büche, D., Schraudolph, N.N., Koumoutsakos, P.: Accelerating evolutionary algorithms with Gaussian process fitness function models. IEEE Transactions on Systems, Man, and Cybernetics–Part C 35(2), 183–194 (2005)
3. Buhmann, M.D.: Radial Basis Functions Theory and Implementations. Cambridge Monographs on Applied and Computational Mathematics, vol. (12). Cambridge University Press, Cambridge (2003)
4. Chipperfield, A., Fleming, P., Pohlheim, H., Fonseca, C.: Genetic Algorithm TOOLBOX For Use with MATLAB, Version 1.2. Department of Automatic Control and Systems Engineering, University of Sheffield, Sheffield (1994)
5. Conn, A.R., Gould, N.I.M., Toint, P.L.: Trust Region Methods. SIAM, Philadelphia (2000)
6. Conn, A.R., Scheinberg, K., Toint, P.L.: A derivative free optimization algorithm in practice. In: Proceedings of the Seventh AIAA/USAF/NASA/ISSMO Symposium on Multidisciplinary Analysis and Optimization. American Institute of Aeronautics and Astronautics, Reston (1998); AIAA Paper AIAA-1998-4718
7. Cressie, N.A.C.: Statistics for Spatial Data. Wiley, New York (1993)
8. Drela, M., Youngren, H.: XFOIL 6.9 User Primer. Department of Aeronautics and Astronautics, Massachusetts Institute of Technology, Cambridge, MA (2001)
9. Duda, R.O., Hart, P.E., Stork, D.G.: Pattern Classification, 2nd edn. Wiley, second edn (2001)
10. Emmerich, M.T.M., Giotis, A., Özdemir, M., Bäck, T., Giannakoglou, K.: Metamodel-assisted evolution strategies. In: Guervós, J.J.M., Adamidis, P.A., Beyer, H.-G., Fernández-Villacañas, J.-L., Schwefel, H.-P. (eds.) PPSN 2002. LNCS, vol. 2439, pp. 361–370. Springer, Heidelberg (2002)
11. Handoko, S., Kwoh, C.K., Ong, Y.S.: Feasibility structure modeling: An effective chaperon for constrained memetic algorithms. IEEE Transactions on Evolutionary Computation 14(5), 740–758 (2010)
12. Jin, Y., Olhofer, M., Sendhoff, B.: A framework for evolutionary optimization with approximate fitness functions. IEEE Transactions on Evolutionary Computation 6(5), 481–494 (2002)
13. Johnson, M.E., Moore, L.M., Ylvisaker, D.: Minimax and maximin distance designs. Journal of Statistical Planning and Inference 26(2), 131–148 (1990)
14. Koehler, J.R., Owen, A.B.: Computer experiments. In: Ghosh, S., Rao, C.R., Krishnaiah, P.R. (eds.) Handbook of Statistics, pp. 261–308. Elsevier, Amsterdam (1996)
15. Linhart, H., Zucchini, W.: Model Selection. Wiley Series in Probability and Mathematical Statistics. Wiley-Interscience Publication, New York (1986)
16. MacQueen, J.B.: Some methods for classification and analysis of multivariate observations. In: Proceedings of 5th Berkeley Symposium on Mathematical Statistics and Probability, pp. 281–297. University of California Press, Berkeley (1967)
17. McKay, M.D., Beckman, R.J., Conover, W.J.: A comparison of three methods for selecting values of input variables in the analysis of output from a computer code. Technometrics 21(2), 239–245 (1979)
18. Myers, R.H., Montgomery, D.C.: Response Surface Methodology: Process and Product Optimization Using Designed Experiments. John Wiley and Sons, New York (1995)
19. Ong, Y.S., Nair, P.B., Keane, A.J.: Evolutionary optimization of computationally expensive problems via surrogate modeling. AIAA Journal 41(4), 687–696 (2003)
20. Rasheed, K., Hirsh, H., Gelsey, A.: A genetic algorithm for continuous design space search. Artificial Intelligence in Engineering 11, 295–305 (1997)
21. Sacks, J., Welch, W.J., Mitchell, T.J., Wynn, H.P.: Design and analysis of computer experiments. Statistical Science 4(4), 409–435 (1989)
22. Sammon, J. J.W.: A nonlinear mapping for data structure analysis. IEEE Transactions on Computers C-18(5), 401–409 (1969)

23. Sobieszczansk-Sobieski, J., Haftka, R.: Multidisciplinary aerospace design optimization: Survey of recent developments. Structural Optimization 14(1), 1–23 (1997)
24. Tenne, Y., Armfield, S.W.: A versatile surrogate-assisted memetic algorithm for optimization of computationally expensive functions and its engineering applications. In: Yang, A., Shan, Y., Thu Bui, L. (eds.) Success in Evolutionary Computation. SCI, vol. 92, pp. 43–72. Springer, Heidelberg (2008)
25. Tenne, Y., Armfield, S.W.: A framework for memetic optimization using variable global and local surrogate models. Journal of Soft Computing 13(8) (2009)
26. Tenne, Y., Goh, C.K. (eds.): Computational Intelligence in Expensive Optimization Problems, Evolutionary Learning and Optimization, vol. 2. Springer, Heidelberg (2010), http://www.springerlink.com/content/v81864
27. Tenne, Y., Izui, K., Nishiwaki, S.: Handling undefined vectors in expensive optimization problems. In: Di Chio, C., Cagnoni, S., Cotta, C., Ebner, M., Ekárt, A., Esparcia-Alcazar, A.I., Goh, C.-K., Merelo, J.J., Neri, F., Preuß, M., Togelius, J., Yannakakis, G.N. (eds.) EvoApplicatons 2010. LNCS, vol. 6024, pp. 582–591. Springer, Heidelberg (2010)
28. Vapnik, V.N.: Statistical Learning Theory. Wiley-Interscience Publication, Hoboken (1998)

Robust Gaussian Process-Based Global Optimization Using a Fully Bayesian Expected Improvement Criterion

Romain Benassi, Julien Bect, and Emmanuel Vazquez

SUPELEC
Gif-sur-Yvette, France

Abstract. We consider the problem of optimizing a real-valued continuous function f, which is supposed to be expensive to evaluate and, consequently, can only be evaluated a limited number of times. This article focuses on the Bayesian approach to this problem, which consists in combining evaluation results and prior information about f in order to efficiently select new evaluation points, as long as the budget for evaluations is not exhausted.

The algorithm called efficient global optimization (EGO), proposed by Jones, Schonlau and Welch (*J. Global Optim.*, 13(4):455–492, 1998), is one of the most popular Bayesian optimization algorithms. It is based on a sampling criterion called the expected improvement (EI), which assumes a Gaussian process prior about f. In the EGO algorithm, the parameters of the covariance of the Gaussian process are estimated from the evaluation results by maximum likelihood, and these parameters are then plugged in the EI sampling criterion. However, it is well-known that this plug-in strategy can lead to very disappointing results when the evaluation results do not carry enough information about f to estimate the parameters in a satisfactory manner.

We advocate a fully Bayesian approach to this problem, and derive an analytical expression for the EI criterion in the case of Student predictive distributions. Numerical experiments show that the fully Bayesian approach makes EI-based optimization more robust while maintaining an average loss similar to that of the EGO algorithm.

1 Introduction

Let f be a continuous real-valued function defined on some compact space $\mathbb{X} \subset \mathbb{R}^d$. We consider the problem of finding the maximum of f, when f is supposed to be expensive to evaluate because one evaluation takes a long time or a large amount of resources. In this case, the optimization of f must be carried out using a limited number of evaluations. More precisely, given a budget of N evaluations of f, our objective is to choose sequentially N evaluation points $X_1, \ldots, X_N \in \mathbb{X}$ so that $\varepsilon(\underline{X}_N, f) = M - M_N$ is small, where $\underline{X}_N$ stands for $(X_1, \ldots, X_N)$, $M = \max_{x \in \mathbb{X}} f(x)$ and $M_N = f(X_1) \vee \cdots \vee f(X_N)$.

In this article, we adopt a Bayesian approach to this sequential decision problem: the unknown function f is considered as a sample path of a real-valued

C.A. Coello Coello (Ed.): LION 5, LNCS 6683, pp. 176–190, 2011.

random process ξ defined on some probability space $(\Omega, \mathcal{B}, \mathsf{P}_0)$ with parameter $x \in \mathbb{X}$, and a good strategy is a strategy that achieves, or gets close to, the Bayes risk $r_{\mathrm{B}} := \inf_{\underline{X}_N} \mathsf{E}_0 \left(\varepsilon(\underline{X}_N, \xi) \right)$, where E_0 denotes the expectation with respect to P_0 and the infimum is taken over the set of all sequential strategies. The reader is referred to the books [1,2,3,4,5] for a broader view on the field of global optimization.

It is well-known [6,7,8,9,10,11,12] that an optimal Bayesian optimization strategy, i.e. a strategy $\underline{X}_N^\star$ such that $\mathsf{E}_0 \left(\varepsilon(\underline{X}_N^\star, \xi) \right) = r_{\mathrm{B}}$, can be formally obtained by dynamic programming. Let E_n, $n = 1, 2, \ldots$, denote the conditional expectation with respect to the σ-algebra $\mathcal{F}_n$ generated by the random variables $X_1, \xi(X_1), \ldots, X_n, \xi(X_n)$. Denote by $R_N = \mathsf{E}_N \left(\varepsilon(\underline{X}_N, \xi) \right)$ the terminal risk and define by backward induction

$$R_n = \min_{x \in \mathbb{X}} \mathsf{E}_n \left(R_{n+1} \mid X_{n+1} = x \right), \quad n = N - 1, \ldots, 0. \tag{1}$$

Then, we have $R_0 = r_{\mathrm{B}}$, and the strategy $\underline{X}_N^\star$ defined by

$$X_{n+1}^\star = \operatorname*{argmin}_{x \in \mathbb{X}} \mathsf{E}_n \left(R_{n+1} \mid X_{n+1} = x \right), \quad n = 1, \ldots, N - 1, \tag{2}$$

is optimal. Unfortunately, solving (1)–(2) over an horizon N of more than a few steps is not numerically tractable, for both the space of possible actions and the space of possible outcomes at each step are continuous.

A natural way of dealing with this problem is to consider a suboptimal one-step lookahead strategy; see, e.g., [13, chapter 6]. This leads to choosing each new evaluation point according to

$$\begin{aligned} X_{n+1} &= \operatorname*{argmin}_{x \in \mathbb{X}} \mathsf{E}_n \left(M - M_{n+1} \mid X_{n+1} = x \right) \\ &= \operatorname*{argmax}_{x \in \mathbb{X}} \mathsf{E}_n \left(M_{n+1} \mid X_{n+1} = x \right) \\ &= \operatorname*{argmax}_{x \subset \mathbb{X}} \rho_n(x) := \mathsf{E}_n \left((\xi(X_{n+1}) - M_n)_+ \mid X_{n+1} = x \right), \end{aligned} \tag{3}$$

where $(z)_+ = 0 \vee z$. The sampling criterion ρ_n, introduced by J. Mockus [6] and popularized through the EGO algorithm [14], is known as the *expected improvement* (EI).

When ξ is a Gaussian process, or in other words, when a Gaussian process prior is chosen for f, it is well-known that the EI can be written in closed form, with the consequence that the maximization of ρ_n can be carried out with a moderate computational effort. However, a Gaussian process prior carries a high amount of information about f and it is often difficult to elicit such a prior before any evaluation is made. As a result, the covariance function of ξ is usually assumed to belong to some parametric class of positive definite functions, the value of the parameters assumed to be unknown. In the EGO algorithm, the parameters are estimated from the evaluation results by maximum likelihood, and then plugged in the EI sampling criterion (computed for a Gaussian process with known covariance function). It has been reported [15] that this plug-in strategy can lead to very disappointing results when the evaluation results do not carry enough information about f to estimate the parameters satisfactorily. We advocate a fully Bayesian approach to this problem, following the steps of Locatelli [9, 16] and, more recently, Osborne and co-authors [17, 18, 19].

The paper is organized as follows. Section 2 recalls the expression of the EI criterion in the case of a Gaussian process prior with known covariance function, and describes the plug-in approach used in the EGO algorithm to handle the parameters of the covariance function when it is only assumed to belong to some parametric class. Section 3 explains how a fully Bayesian approach can be adopted in this problem, in order to take into account the uncertainty on the parameters of the covariance function. Section 4 presents a new closed-form expression of the EI criterion for Student predictive densities, which arises naturally when a conjugate inverse-gamma prior is used for the variance parameter of the Gaussian process prior. Section 5 illustrates with numerical results the benefits of the fully Bayesian approach, focusing more particularly on the tail of the error distribution, i.e., on the occurrence of large errors.

Nota bene. *The analytical expression of the expected improvement for Student predictive distributions, presented in Section 4, has in fact already been obtained by Williams, Santner and Notz [20] in the special case of an improper Jeffrey prior on the variance. We warmly thank Frank Hutter for pointing out this paper to us during the LION5 conference.*

2   Efficient Global Optimization

2.1   The Expected Improvement Sampling Criterion for a Gaussian Process

Recall that the distribution of a Gaussian process ξ is uniquely determined by its mean function $m(x) := \mathsf{E}_0(\xi(x))$, $x \in \mathbb{X}$, and its covariance function $k(x, y) := \mathsf{E}_0 \left((\xi(x) - m(x))(\xi(y) - m(y)) \right)$, $x, y \in \mathbb{X}$. Hereafter, we assume that the mean function is constant on $\mathbb{X}$ and write $\xi \sim \mathrm{GP}\,(m,\,k)$ to denote that ξ is a Gaussian process with mean function $m(x) = m \in \mathbb{R}$ and covariance function k.

Proposition 1. *Let k be a stationary covariance function written as $k(x, y) = \sigma^2 r(x - y)$, $x, y \in \mathbb{X}$, where $\sigma^2 > 0$ and $r(0) = 1$ (hence, r is a correlation function). Assume that $\xi \mid m \sim \mathrm{GP}\,(m, k)$ and $m \sim \mathcal{U}(\mathbb{R})$, where $\mathcal{U}(\mathbb{R})$ denotes the (improper) uniform distribution over $\mathbb{R}$. Then, for all $x \in \mathbb{X}$,*

$$\xi(x) \mid \mathcal{F}_n \sim \mathcal{N}\left(\widehat{\xi}_n(x),\, s_n^2(x) \right),$$

where

$$\widehat{\xi}_n(x) = \widehat{m}_n + r_n(x)^{\mathsf{T}} R_n^{-1}(\underline{\xi}_n - \widehat{m}_n \mathbb{1}_n), \tag{4}$$

with

$$
\begin{cases}
\underline{\xi}_n = (\xi(X_1), \ldots, \xi(X_n))^{\mathsf{T}}, \\[2mm]
\mathbb{1}_n = (1, \ldots, 1)^{\mathsf{T}} \in \mathbb{R}^n, \\[2mm]
R_n \text{ the correlation matrix of } \underline{\xi}_n, \\[2mm]
r_n(x) \text{ the correlation vector between } \xi(x) \text{ and } \underline{\xi}_n, \\[2mm]
\widehat{m}_n = \dfrac{\mathbb{1}_n^{\mathsf{T}} R_n^{-1} \underline{\xi}_n}{\mathbb{1}_n^{\mathsf{T}} R_n^{-1} \mathbb{1}_n}, \text{ the weighted least squares estimate of } m,
\end{cases}
$$

and

$$
s_n^2(x) = \sigma^2 \kappa_n^2(x), \tag{5}
$$

with

$$
\kappa_n^2(x) = 1 - r_n(x)^{\mathsf{T}} R_n^{-1} r_n(x) + \frac{(1 - r_n(x)^{\mathsf{T}} R_n^{-1} \mathbb{1}_n)^2}{\mathbb{1}_n^{\mathsf{T}} R_n^{-1} \mathbb{1}_n}. \tag{6}
$$

Proposition 2. *Under the assumptions of Proposition 1, the expected improvement can be written as*

$$
\rho_n(x) = \begin{cases}
s_n(x) \, \Phi'\left(\frac{\widehat{\xi}_n(x) - M_n}{s_n(x)}\right) + (\widehat{\xi}_n(x) - M_n)\, \Phi\left(\frac{\widehat{\xi}_n(x) - M_n}{s_n(x)}\right) & \text{if } s_n(x) > 0, \\[3mm]
\left(\widehat{\xi}_n(x) - M_n\right)_+ & \text{if } s_n(x) = 0.
\end{cases} \tag{7}
$$

where Φ denotes the Gaussian cumulative distribution function.

Propositions 1 and 2 show that, given a set of evaluation points and a Gaussian prior, the EI sampling criterion can be computed with a moderate amount of resources (computing (4) at q different points in $\mathbb{X}$ involves $O(qn^2)$ operations).

However, it is rare that a user has enough information about f in order to choose an adequate covariance function k before any evaluation is made. The approach generally taken consists in choosing k in a parametrized class of covariance functions and estimating the parameters of k from the evaluation results.

2.2   Classical Parametrized Covariance Functions

There are chiefly three classes of parametrized covariance functions in the literature of Gaussian processes for modeling computer experiments. These are the class of the so-called Gaussian covariances, the class of the exponential covariances, and that of the Matérn covariances. Using Matérn covariances makes it possible to tune the mean square differentiability of ξ, which is not the case with the exponential and Gaussian covariances.

Define $\upsilon_\nu : \mathbb{R}^+ \to \mathbb{R}^+$ such that, $\forall h \geq 0$,

$$
\upsilon_\nu(h) = \frac{1}{2^{\nu-1}\Gamma(\nu)} \left(2\nu^{1/2}h\right)^{\nu} \mathcal{K}_\nu\left(2\nu^{1/2}h\right), \tag{8}
$$

where Γ is the Gamma function and $\mathcal{K}_\nu$ is the modified Bessel function of the second kind of order ν. The parameter $\nu > 0$ controls regularity at the origin of υ_ν.

The anisotropic form of the Matérn covariance on $\mathbb{R}^d$ may be written as $k_\theta(x,y) = \sigma^2 r_\theta(x,y)$, with

$$r_\theta(x,y) = \upsilon_\nu \left(\sqrt{\sum_{i=1}^{d} \frac{(x_{[i]} - y_{[i]})^2}{\beta_i^2}} \right), \quad x, y \in \mathbb{R}^d, \tag{9}$$

where the positive scalar σ^2 is a variance parameter (we have $k_\theta(x,x) = \sigma^2$), $x_{[i]}, y_{[i]}$ denote the ith coordinate of x and y, the positive scalars β_i represent scale or *range* parameters of the covariance, or in other words, characteristic correlation lengths, and finally $\theta = (\nu, \beta_1, \ldots, \beta_d) \in \mathbb{R}_+^{d+1}$ denotes the parameter vector of the Matérn covariance. Note that an isotropic form of the Matérn covariance is obtained by setting $\beta_1 = \ldots = \beta_d = \beta$. Then, the parameter vector of the Matérn covariance is $\theta = (\nu, \beta) \in \mathbb{R}_+^2$.

2.3   The EGO Algorithm

The approach taken in the EGO (efficient global optimization) algorithm [21,23, 22,14] consists in estimating the unknown parameters of the covariance function by maximum likelihood, after each new evaluation. Then, the EI sampling criterion is computed using the current value of the parameters of the covariance. EGO can therefore be viewed as a plug-in approach.

Remark 1 (about maximum likelihood estimation of the parameters of a covariance function of a Gaussian process). Recall that, for $\xi \sim \text{GP}(m, k_\theta)$ with $k_\theta(x,y) = \sigma^2 r_\theta(x,y)$, the likelihood of the evaluation results can be written as

$$\ell_n(\underline{\xi}_n; m, \sigma^2, \theta) = \frac{1}{(2\pi\sigma^2)^{n/2}|R_n(\theta)|^{1/2}} e^{-\frac{1}{\sigma^2}(\underline{\xi}_n - m\mathbb{1}_n)^\top R_n(\theta)^{-1}(\underline{\xi}_n - m\mathbb{1}_n)}, \tag{10}$$

where $R_n(\theta)$ stands for the correlation matrix of $\underline{\xi}_n$, parametrized by θ. Note that setting to zero the partial derivatives of ℓ_n with respect to m and σ^2 yields the following maximum likelihood estimates for m and σ^2:

$$\widehat{m}(\theta) = \frac{\mathbb{1}_n^\top R_n(\theta)^{-1} \underline{\xi}_n}{\mathbb{1}_n^\top R_n(\theta)^{-1} \mathbb{1}_n}, \tag{11}$$

$$\widehat{\sigma}^2(\theta) = \frac{1}{n} \left(\underline{\xi}_n - \widehat{m}\mathbb{1}_n \right)^\top R_n(\theta)^{-1} \left(\underline{\xi}_n - \widehat{m}\mathbb{1}_n \right). \tag{12}$$

Thus the maximum likelihood estimate of θ can be obtained by maximizing the profile likelihood $\theta \mapsto \ell_n(\underline{\xi}_n; \widehat{m}(\theta), \widehat{\sigma}^2(\theta), \theta)$.

2.4   The Case of Deceptive Functions

Deceptive functions is a term coined by D. Jones (see [15,25]) to describe functions that appear to be "flat" based on evaluation results. In fact, any function can potentially appear to be flat depending on how it is sampled.

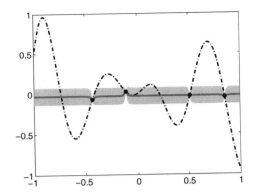

Fig. 1. Example of a deceptive sampling of a function (dashdot line). Evaluation points (black dots) are chosen such that the value of the function is around zero at these points. After having estimated the parameters of the covariance function by maximum likelihood, the prediction is very flat (solid line) and confidence intervals derived from the standard deviation of the error of prediction (gray area) are severely underestimated.

When the available evaluation results do not bring enough information on the objective function f to estimate the parameters of the covariance function with a reasonable precision, the variance of the error of prediction can be severely under-estimated as depicted in Figure 1. As will be shown in Section 5.1, this can lead to very unsatisfactory behaviors of the EGO algorithm, which tends to waste lots of evalutions in local search around the current maxima (exploitation), very early in the optimization procedure, to the detriment of global search (exploration).

3   Fully Bayesian One-Step Lookahead Optimization

It has been emphasized in Section 1 that the rationale behind the EI criterion is of a Bayesian decision-theoretic nature. Indeed, maximizing the EI criterion at iteration n is equivalent to minimizing the expected loss $\mathsf{E}_n\left(\max(\xi) - M_{n+1}\right)$, where the expectation is taken with respect to the value of the next evaluation, which is unknown and therefore modeled as a random variable.

In a *fully Bayesian* setting, *all* the unknown parameters of the model have to be given prior distributions. This has already been done for the unknown mean m in Proposition 1. Let π_0 denote the prior distribution of the vector of covariance parameters $\theta' = (\sigma^2, \theta)$, and let π_n, $n = 1, \ldots, N$, denote the corresponding posterior distributions. According to Bayes' rule, the posterior distribution of $\xi(x)$ is a mixture of Gaussian distributions $\mathcal{N}\left(\widehat{\xi}_n(x; \theta'), s_n^2(x; \theta')\right)$ weighted by $\pi_n(\mathrm{d}\theta')$. The expected improvement criterion for this model can thus be written, using the tower property of conditional expectations, as

$$\mathsf{E}_n\left((\xi(x) - M_n)_+\right) = \mathsf{E}_n\left(\mathsf{E}_n\left((\xi(x) - M_n)_+ \mid \theta'\right)\right)$$

$$= \int \rho_n(x; \theta')\, \pi_n\,(\mathrm{d}\theta') . \tag{13}$$

Note that the plug-in EI criterion of Section 2.3 can be seen as an approximation of the fully Bayesian criterion (13):

$$\int \rho_n(x; \theta')\, \pi_n\,(\mathrm{d}\theta') \approx \rho_n(x; \widehat{\theta}'_n),$$

which is justified only if the posterior distribution is concentrated enough around the MLE estimate $\widehat{\theta}'_n$. In the general case, we claim that it is safer to use the fully Bayesian criterion (13), since the corresponding expected loss integrates the uncertainty related to the fact that θ' is not exactly known. This claim will be supported by the numerical results of Section 5.

When π_0 is a finitely supported discrete distribution, the posterior distribution π_n—and therefore the integral (13)—can be computed exactly using Bayes' rule. For more general prior distribution, the integral can be approximated by stochastic techniques like MCMC sampling or SMC sampling (see [27, 26, 28] and the references therein). An alternative approach using Bayesian quadrature rules [29] has been proposed in [18, 17, 19]. In all cases, the EI criterion is approximated by an expression of the form $\sum_i w_i \rho_n(x; \theta'_i)$, which amounts to saying that π_n is approximated by the discrete distribution $\sum_i w_i \delta_{\theta'_i}$.

Remark 2. Although fully Bayesian approaches for Gaussian process models have been proposed in the literature for more than two decades (see [30, 31] and the references therein), surprisingly little has been written from this perspective in the context of Bayesian global optimization. An early attempt in this direction can be found in [16, 9], where the variance parameter of a Brownian motion is given an inverse-gamma prior and then integrated out as in (13). More recently, the fully Bayesian approach has been developed in a more general way by [18, 17, 19], but the important connection of (13) with the usual (Gaussian) EI criterion was not clearly established.

Remark 3. Discrete mixtures of Gaussian distributions and the corresponding EI criterion have also been introduced in [32] to allow for the use of several parametric classes of covariance functions, in order to provide increased robustness with respect to the choice of a particular class. The approach is not Bayesian, however, since the weights in the mixture are not posterior probabilities.

4   Student EI

Let us consider the case of a Gaussian process ξ with unknown mean m and covariance function of the form $k(x, y) = \sigma^2 r(x, y)$. We assume that m and σ^2 are independent, with m uniformly distributed on $\mathbb{R}$ (as in Proposition 1) and σ^2 following an inverse-gamma distribution with shape parameter a_0 and scale parameter b_0, hereafter denoted by $\mathrm{IG}\,(a_0, b_0)$. We shall prove that, in this setting, the EI criterion still has an explicit analytical expression, which is a generalization of the usual EI criterion given in Proposition 2.

First, recall that the prior chosen for σ^2 is conjugate [33]:

Proposition 3. *The conditional distribution of σ^2 given $\mathcal{F}_n$ is* IG (a_n, b_n), *with*

$$a_n = a_0 + \frac{n-1}{2},$$

$$b_n = b_0 + \frac{1}{2}\left(\underline{\xi}_n - \widehat{m}_n \mathbb{1}_n\right)^{\mathsf{T}} R_n^{-1}\left(\underline{\xi}_n - \widehat{m}_n \mathbb{1}_n\right).$$

Using this result and the fact that $\xi(x) \mid \sigma^2, \underline{\xi}_n \sim \mathcal{N}\left(0, \sigma^2 \kappa_n^2(x)\right)$, it is easy to show that the predictive distribution of $\xi(x)$ is a Student distribution. More precisely:

Proposition 4. *Let t_η denote the Student distribution with $\eta > 0$ degrees of freedom. Then, for all $x \in \mathbb{X}$,*

$$\frac{\xi(x) - \widehat{\xi}_n(x)}{\gamma_n(x)} \mid \mathcal{F}_n \sim t_{\eta_n},$$

with $\eta_n = 2a_n$, and $\gamma_n^2(x) = b_n/a_n \, \kappa_n^2(x)$.

In other words, the predictive distribution at x is a location-scale Student distribution with η_n degrees of freedom, location parameter $\widehat{\xi}_n(x)$ and scale parameter $\gamma_n(x)$. The following result is the key to our EI criterion for Student predictive distributions:

Lemma 1. *Let $T \sim t_\eta$ with $\eta > 0$. Then*

$$\mathsf{E}\left((T + u)_+\right) = \begin{cases} +\infty & \text{if } \eta \leq 1, \\ \frac{\eta + u^2}{\eta - 1} F_\eta'(u) + u \, F_\eta(u) & \text{otherwise,} \end{cases}$$

where F_η is the cumulative distribution function of t_η.

Combining Lemma 1 and Proposition 4 finally yields an explicit expression of the EI criterion:

Theorem 1. *Under the assumptions of this section, for all $x \in \mathbb{X}$,*

$$\mathsf{E}_n\left((\xi(x) - M_n)_+\right) = \gamma_n(x)\left(\frac{\eta_n + u^2}{\eta_n - 1} F_{\eta_n}'(u) + u \, F_{\eta_n}(u)\right), \qquad (14)$$

with $u = (\widehat{\xi}_n(x) - M_n)/\gamma_n(x)$.

It has been assumed, up to this point, that the only unknown parameter in the covariance function is the variance σ^2. More generally, assume that $k(x, y) = \sigma^2 r(x, y; \theta)$: in this case we proceed by conditioning as in Section 3. Indeed, assume that θ is independent from (m, σ^2) with a prior distribution π_0. Let us denote by $\tilde{\rho}_n(x; \theta) = \mathsf{E}_n\left(\left(\xi(x) - M_n\right)_+ \mid \theta\right)$ the value of the EI criterion at x provided by Theorem 1 when the value of the unknown parameter is θ. Then

$$\mathsf{E}_n\left(\left(\xi(x) - M_n\right)_+\right) = \mathsf{E}_n\left(\tilde{\rho}_n\left(x; \theta\right)\right) = \int \tilde{\rho}_n(x; \theta)\, \pi_n(\mathrm{d}\theta), \qquad (15)$$

where π_n denotes the posterior distribution of θ after n evaluations. As explained in Section 3, the integral (15) boils down to a finite sum that can be computed exactly (using Bayes' rule) when the prior π_0 has a finite support; in the general case, approximation techniques have to be used.

5   Numerical Experiments

5.1   Optimization of a Deceptive Function

Experiment. Consider the objective function $f : \mathbb{X} = [-1, 1] \to \mathbb{R}$ defined by

$$f(x) = x\left(\sin(10x + 1) + 0.1\sin(15x)\right), \quad \forall x \in \mathbb{X}.$$

We choose an initial set of four evaluation points with abscissas -0.43, -0.11, 0.515 and 0.85, as shown in Figure 1. Our objective is to compare the evaluation points chosen by the plug-in approach (i.e., the EGO algorithm) and those chosen by the fully Bayesian algorithm (FBA) proposed in Section 4.

In both approaches, we consider a Matérn covariance function with a known regularity parameter $\nu = 2$ (see Section 2.2). In the approach of Section 4, we choose an inverse gamma distribution $IG(0.2, 12)$ for σ^2. Since $\mathbb{X}$ has dimension one, there is only one range parameter β. To simplify the implementation of the approach proposed, we shall assume that β has a finite support distribution. More precisely, define a $\beta_{\min}$ and a $\beta_{\max}$, such that $\beta_{\min} < \beta_{\max}$, and set, for all $i = 0, \dots, I$, $\beta_i = \beta_{\min}\left(\frac{\beta_{\max}}{\beta_{\min}}\right)^{i/I}$. We assume a uniform prior distribution over the β_is, with $\beta_{\min} = 2 \times 10^{-3}$, $\beta_{\max} = 2$ and $I = 100$.

The optimization of the two sampling criteria is performed by a Monte Carlo approach. More precisely, we generate once and for all a set of $q = 600$ candidate points uniformly distributed over $\mathbb{X}$ and the search for the maximum of each sampling criterion is carried out at each iteration by determining the value of the sampling criterion over this finite set (the same set of points is used for both criteria).

Table 1. Parameters used for building the testbeds of Gaussian-process sample-paths

Parameter \ Testbed	$\mathcal{T}_1$	$\mathcal{T}_2$
Dimension d	1	4
Number of sample paths L	20000	20000
Variance σ^2	1.0	1.0
Regularity ν	2.5	2.5
Scale $\beta = (\beta_1, \dots, \beta_d)$	0.1	(0.7, 0.7, 0.7, 0.7)

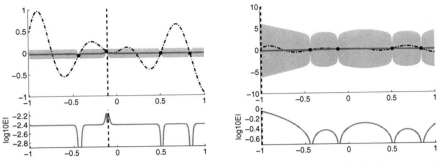

(a) parameters estimated by MLE   (b) Bayesian approach for the parameters

Fig. 2. A comparison of a) EGO and b) FBA at iteration 1. Top: objective function (dashdot line), prediction (solid line), 95% confidence intervals derived from the standard deviation (gray area), sampling points (dots) and position of the next evaluation (vertical dashed line). Bottom: EI criterion.

Results. Figures 2, 3 and 4 show that the standard deviation of the error of prediction is severely underestimated when using the EGO algorithm, as a result of the maximum likelihood estimation of the parameters of the covariance from a deceptive set of evaluation points. If the uncertainty about the covariance parameters is taken into account, as explained above, the standard deviation of the error is more satisfactory. Figures 3 and 4 show that the maximum is approximated satisfactorily after only four iterations with FBA, whereas EGO needs nine more iterations before making an evaluation in the neighborhood of the maximizer. Indeed, we observe that EGO stays in the neighborhood of a local optimum for a long time, while $\mathbb{X}$ remains unexplored. This behavior is not desirable in a context of expensive-to-evaluate functions.

5.2 Comparison on Sample Paths of a Gaussian Process

Experiment. In order to assess the performances of EGO and FBA from a statistical point of view, we study the convergence to the maximum using both algorithms on a set of sample paths of a Gaussian process.

We have built several testbeds $\mathcal{T}_k$, $k = 1, 2, \ldots$, of functions $f_{k,l}$, $l = 1, \ldots, L$, corresponding to sample paths of a Gaussian process, with zero-mean and a Matérn covariance function, simulated on a set of $q = 600$ points in $[0, 1]^d$ generated using a Latin hypercube sampling (LHS), with different values for d and for the parameters of the covariance. Here, due to the lack of room, we present only the results obtained for two testbeds in dimension 1 and 4 (the actual parameters are provided in Table 1).

We shall compare the performance of EGO and FBA based on the approximation error $\varepsilon(\underline{X}_n, f_{k,l})$, $l = 1, \ldots, L$. For reference, we also provide the results obtained with two other strategies. The first strategy corresponds to using an EI criterion with the same values for the parameters of the covariance func-

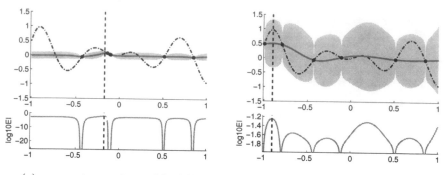

(a) parameters estimated by MLE   (b) Bayesian approach for the parameters

Fig. 3. Iteration 3 (see Figure 2 for details)

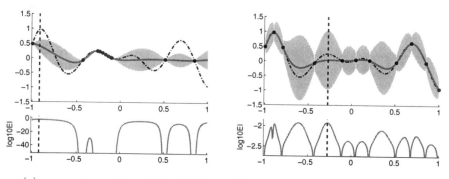

(a) parameters estimated by MLE   (b) Bayesian approach for the parameters

Fig. 4. Iteration 8 (see Figure 2 for details)

tion of ξ than those used to generate the sample paths in the testbeds. In principle this strategy ought to perform very well. The second strategy corresponds to space-filling sampling, which is not necessarily a good optimization strategy.

For FBA, we choose the same priors as those described in Section 5.1. More precisely, whatever be the dimension d, we choose an isotropic covariance function (with only one scale parameter) and we set $\beta_{min} = 1/400$ and $\beta_{max} = 2\sqrt{d}$.

Results. Figures 5(a) and 6(a) show that EGO and FBA have very similar average performances. In fact, both of them perform almost as well, in this experiment, as the reference strategy where the true parameters are assumed to be known. Comparing the tails of complementary cumulative distribution function of the error $\max f - M_n$ makes it clear, however, that using a fully Bayesian approach brings a significant reduction of the occurrence of large errors with respect to the EGO algorithm. In other words, the fully Bayesian approach appears to be statistically more robust than the plug-in approach, while retaining the same average performance.

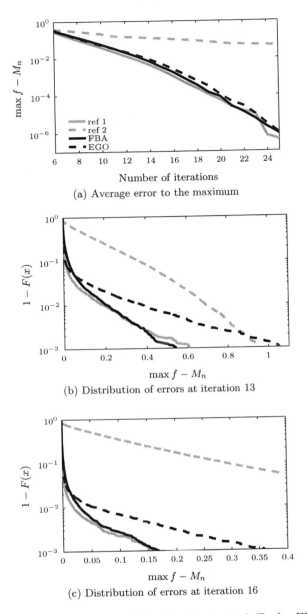

(a) Average error to the maximum

(b) Distribution of errors at iteration 13

(c) Distribution of errors at iteration 16

Fig. 5. Average results and error distributions for testbed $\mathcal{T}_1$, for FBA (solid black line), EGO (dashed black line), the EI with the parameters used to generate sample paths (solid gray line), the space-filling strategy (dashed gray line). More precisely, (a) represents the average approximation error as a function of the number of evaluation points. In (b) and (c), $F(x)$ stands for the cumulative distribution function of the approximation error. We plot $1 - F(x)$ in logarithmic scale in order to analyze the behavior of the tail of the distribution (big errors with small probabilities of occurrence). Small values for $1 - F(x)$ mean better results.

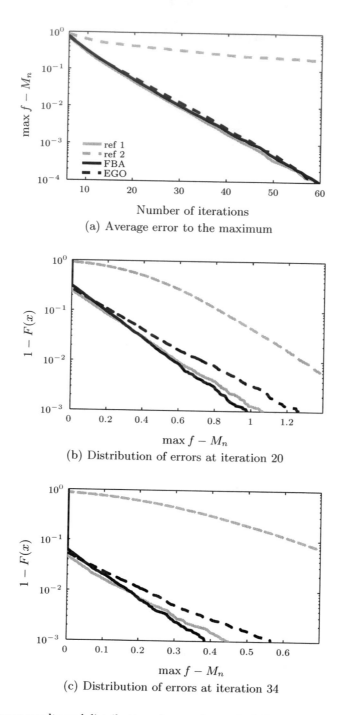

(a) Average error to the maximum

(b) Distribution of errors at iteration 20

(c) Distribution of errors at iteration 34

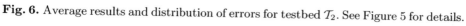

Fig. 6. Average results and distribution of errors for testbed $\mathcal{T}_2$. See Figure 5 for details.

References

1. Törn, A., Zilinskas, A. : Global Optimization. Springer, Berlin (1989)
2. Pintér, J.D. : Global optimization. Continuous and Lipschitz optimization : algorithms, implementations and applications. Springer, Heidelberg (1996)
3. Zhigljavsky, A., Zilinskas, A. : Stochastic global optimization. Springer, Heidelberg (2007)
4. Conn, A.R., Scheinberg, K., Vicente, L.N. : Introduction to derivative-free optimization. SIAM, Philadelphia (2009)
5. Tenne, Y., Goh, C.K. : Computational intelligence in optimization : applications and implementations. Springer, Heidelberg (2010)
6. Mockus, J., Tiesis, V., Zilinskas, A. : The application of Bayesian methods for seeking the extremum. In : Dixon, L., Szego, G. (eds.) Towards Global Optimization, vol. 2, pp. 117–129. Elsevier, Amsterdam (1978)
7. Mockus, J. : Bayesian approach to Global Optimization : Theory and Applications. Kluwer Acad. Publ., Dordrecht (1989)
8. Betrò, B. : Bayesian methods in global optimization. Journal of Global Optimization 1, 1–14 (1991)
9. Locatelli, M., Schoen, F. : An adaptive stochastic global optimization algorithm for one-dimensional functions. Annals of Operations Research 58(4), 261–278 (1995)
10. Auger, A., Teytaud, O. : Continuous lunches are free plus the design of optimal optimization algorithms. Algorithmica 57(1), 121–146 (2008)
11. Ginsbourger, D., Le Riche, R. : Towards Gaussian process-based optimization with finite time horizon. In : mODa 9 Advances in Model-Oriented Design and Analysis. Contribution to Statistics, pp. 89–96. Springer, Heidelberg (2010)
12. Grünewälder, S., Audibert, J.-Y., Opper, M., Shawe-Taylor, J. : Regret bounds for Gaussian process bandit problems. In : Proceedings of the 13th International Conference on Artificial Intelligence and Statistics (AISTATS 2010). JMLR W&CP, vol. 9, pp. 273–280 (2010)
13. Bertsekas, D.P. : Dynamic programming and optimal control. Athena Scientific, Belmont (1995)
14. Jones, D.R., Schonlau, M., Welch, W.J. : Efficient global optimization of expensive black-box functions. Journal of Global Optimization 13(4), 455–492 (1998)
15. Forrester, A.I.J., Jones, D.R. : Global optimization of deceptive functions with sparse sampling. In : 12th AIAA/ISSMO Multidisciplinary Analysis and Optimization Conference, September 10-12 (2008)
16. Locatelli, M. : Bayesian algorithms for one-dimensional global optimization. Journal of Global Optimization 10(1), 57–76 (1997)
17. Osborne, M.A. : Bayesian Gaussian Processes for Sequential Prediction Optimisation and Quadrature. PhD thesis, University of Oxford (2010)
18. Osborne, M.A., Garnett, R., Roberts, S.J. : Gaussian processes for global optimization. In : 3rd International Conference on Learning and Intelligent Optimization (LION3), Online Proceedings, Trento, Italy (2009)
19. Osborne, M.A., Roberts, S.J., Rogers, A., Ramchurn, S.D., Jennings, N.R. : Towards real-time information processing of sensor network data using computationally efficient multi-output Gaussian processes. In : Proceedings of the 7th International Conference on Information Processing in Sensor Networks, pp. 109–120. IEEE Computer Society, Los Alamitos (2008)
20. Williams, B., Santner, T., Notz, W. : Sequential Design of Computer Experiments to Minimize Integrated Response Functions. Statistica Sinica 10(4), 1133–1152 (2000)

21. Schonlau, M. : Computer experiments and global optimization. PhD thesis, University of Waterloo, Waterloo, Ontario, Canada (1997)
22. Schonlau, M., Welch, W.J. : Global optimization with nonparametric function fitting. In : Proceedings of the ASA, Section on Physical and Engineering Sciences, pp. 183–186. Amer. Statist. Assoc. (1996)
23. Schonlau, M., Welch, W.J., Jones, D.R. : A data analytic approach to Bayesian global optimization. In : Proceedings of the ASA, Section on Physical and Engineering Sciences, pp. 186–191. Amer. Statist. Assoc. (1997)
24. Forrester, A.I.J., Keane, A.J. : Recent advances in surrogate-based optimization. Progress in Aerospace Sciences 45(1-3), 50–79 (2009)
25. Jones, D.R. : A taxonomy of global optimization methods based on response surfaces. Journal of Global Optimization 21(4), 345–383 (2001)
26. Robert, C.P., Casella, G. : Monte Carlo statistical methods. Springer, Heidelberg (2004)
27. Del Moral, P., Doucet, A., Jasra, A. : Sequential Monte Carlo samplers. Journal of the Royal Statistical Society : Series B (Statistical Methodology) 68(3), 411–436 (2006)
28. Liu, J.S. : Monte Carlo strategies in scientific computing. Springer, Heidelberg (2008)
29. O'Hagan, A. : Bayes-Hermite quadrature. Journal of Statistical Planning and Inference 29(3), 245–260 (1991)
30. O'Hagan, A. : Curve Fitting and Optimal Design for Prediction. Journal of the Royal Statistical Society : Series B (Statistical Methodology) 40(1), 1–42 (1978)
31. Handcock, M.S., Stein, M.L. : A Bayesian analysis of Kriging. Technometrics 35(4), 403–410 (1993)
32. Ginsbourger, D., Helbert, C., Carraro, L. : Discrete mixtures of kernels for kriging-based optimization. Quality and Reliability Engineering International 24, 681–691 (2008)
33. O'Hagan, A. : Some Bayesian numerical analysis. In : Bayesian Statistics 4 : Proceedings of the Fourth Valencia International Meeting, April 15-20, 1991. Oxford University Press, Oxford (1992)

Hierarchical Hidden Conditional Random Fields for Information Extraction

Satoshi Kaneko, Akira Hayashi, Nobuo Suematsu, and Kazunori Iwata

Graduate School of Information Sciences, Hiroshima City University,
3-4-1 Ozuka-higashi, Asaminami-ku, Hiroshima 731-3194, Japan
kaneko@prl.info.hiroshima-cu.ac.jp

Abstract. Hidden Markov Models (HMMs) are very popular generative models for time series data. Recent work, however, has shown that for many tasks Conditional Random Fields (CRFs), a type of discriminative model, perform better than HMMs. Information extraction is the task of automatically extracting instances of specified classes or relations from text. A method for information extraction using Hierarchical Hidden Markov Models (HHMMs) has already been proposed. HHMMs, a generalization of HMMs, are generative models with a hierarchical state structure. In previous research, we developed the Hierarchical Hidden Conditional Random Field (HHCRF), a discriminative model corresponding to HHMMs. In this paper, we propose information extraction using HHCRFs, and then compare the performance of HHMMs and HHCRFs through an experiment.

1 Introduction

1.1 Hierarchical Hidden Conditional Random Fields

Hidden Markov Models (HMMs) are very popular generative models for sequence data. Recent work, however, has shown that Conditional Random Fields (CRFs), a type of discriminative model, perform better than HMMs on many tasks [1]. There are several differences between CRFs and HMMs. (1) HMMs are generative models and thus model the joint probability of input (i.e., observations) and output data (i.e., states), whereas CRFs are discriminative models that model the conditional probability of output data given the input data. (2) HMMs make independent assumptions on observations given states, whereas CRFs do not. (3) For model parameter estimation, HMMs do not need the states, whereas CRFs do.

Hierarchical HMMs (HHMMs) are a generalization of HMMs with a hierarchical structure [2]. Murphy [3] showed that an HHMM is a special kind of Dynamic Bayesian Network (DBN) and derived an efficient inference algorithm. In previous research, we developed Hierarchical Hidden Conditional Random Fields (HHCRFs), a discriminative model corresponding to HHMMs. In addition, it has been shown that HHCRFs achieve better performance than HHMMs on certain tasks [4] [5].

C.A. Coello Coello (Ed.): LION 5, LNCS 6683, pp. 191–202, 2011.

1.2   Information Extraction

Information extraction is the task of automatically extracting instances of specified classes or relations from text. Systems for information extraction are usually built using machine learning techniques. Single layer models such as HMMs have been used for information extraction in previous works. Most of the work in learning HMMs for information extraction has focused on tasks with semi-structured text sources in which English grammar does not play a key role.

Skounakis et al. [6] considered the task of extracting information from abstracts of biology articles. In this domain, it is important that the learned models are able to represent regularities in the grammatical structure of a sentence. They proposed an approach using HHMMs. Hierarchical models have multiple levels of states, which describe input sequences at different levels of granularity. In the model they used, the upper level represents sentences at the level of phrases, while the lower level represents sentences at the level of individual words.

In this paper, we first propose information extraction using HHCRFs, and then compare their performance with that of HHMMs in extracting instances of three binary relations from abstracts of scientific articles. An example of the binary relations we use in our experiments is the *subcellular − localization* relation, which represents the location of a particular protein within a cell. We refer to the domains of this relation as PROTEIN and LOCATION and to an instance of a relation as a tuple. Given the sentence, "This enzyme, UBC6, localizes to the endoplasmic reticulum, with the catalytic domain facing the cytosol", for example, PROTEIN "UBC6" and LOCATION "endoplasmic reticulum" should be extracted. The sentence asserts that protein UBC6 is found in the subcellular compartment called the endoplasmic reticulum.

1.3   Paper Organization

In Section 2, we review the HHMM, and represent it as a DBN. Then, in Section 3, we define HHCRFs and explain their training algorithm. In Section 4, we discuss the sentence representation for information extraction, while in Section 5, we explain the architecture and the learning and inference methods for both HHMMs and HHCRFs. Experimental results are given in Section 6, and we conclude in Section 7.

2   HHMMs

Hierarchical HMMs (HHMMs) are a generalization of HMMs with a hierarchical structure [2]. HHMMs have three kinds of states: internal, production, and end states. They also have three kinds of transitions: vertical, horizontal, and forced transitions. Murphy [3] showed that an HHMM is a special kind of DBN, and derived an efficient inference algorithm. In what follows, we show how to represent an HHMM as a DBN.

(a)   (b)

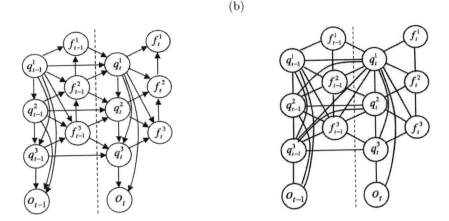

Fig. 1. (a) An HHMM represented as a DBN. (b) An HHCRF represented as an undirected graph. Both (a) and (b) describe only the part of the model between $t-1$ and t.

2.1   Representing an HHMM as a DBN

We can represent an HHMM as a DBN as shown in Fig. 1(a). (We assume for simplicity that all production states are at the bottom of the hierarchy.) The state of the HHMM is denoted by q_t^d ($d \in \{1, \ldots, D\}$), where d is the hierarchy index, with the top level having $d = 1$, and the bottom level $d = D$.

f_t^d is the indicator variable, which is equal to 1 if q_t^d has transited to its end state; otherwise it is 0. Note that if $f_t^d = 1$, then $f_t^{d'} = 1$ for all $d' > d$; hence, the number of indicator variables equal to 0 denotes the level of the hierarchy we are currently on. The indicator variables play an important role in representing the HHMM as a DBN.

Defined below are the transition and output probability distributions. These complete the definition of the model. When q_t^d has transited to its end state, $f_t^d = 1$. This is the signal that the states at the upper levels can be changed. Furthermore, it is a signal that the next value of q_{t+1}^d should be determined by a vertical transition, instead of a horizontal transition. Formally, we denote these as follows:

$$p(q_t^d = j' | q_{t-1}^d = j, f_{t-1}^{d+1} = b, f_{t-1}^d = f, q_t^{1:d-1} = i) = \begin{cases} \delta(j, j') & \text{if } b = 0 \\ A_i^d(j, j') & \text{if } b = 1 \text{ and } f = 0 \\ \pi_i^d(j') & \text{if } b = 1 \text{ and } f = 1 \end{cases}$$

$$p(f_t^d = 1 | q_t^d = j, q_t^{1:d-1} = i, f_t^{d+1} = b) = \begin{cases} 0 & \text{if } b = 0 \\ Ae^d(i, j) & \text{if } b = 1 \end{cases} \quad (1)$$

$$p(o_t = s | q_t^D = i) = B[s|i]$$

where the state vector $q_t^{1:d} = \{q_t^1, \ldots, q_t^d\}_{d \in \{1, \ldots, D\}}$ is represented by an integer i (i.e., i is the index for the "mega state"). In Eq. (1), we assume the dummy state $q_t^0 = 0$ (i.e., the root state) for notational convenience. We also assume dummy indicator variables $f_0^{2:D} = 1$ and $f_t^{D+1} = 1$ for the first slice and bottom level, respectively.

$\delta(j, j')$ is Kroneckers delta. $A_i^d(j, j')$ is the horizontal transition probability into the j'-th state (except into an end state) from the j-th state at level d. $\pi_i^d(j')$ is the vertical transition probability into the j'-th state from the i-th state at level d. $Ae^d(i, j)$ is the horizontal transition probability into an end state from the j-th state at level d.

$B[s|i]$ is the output probability of observation s at the bottom level of the i-th state.

3   HHCRFs

3.1   Model

HHCRFs are undirected graphical models (as shown in Fig. 1(b)) that encode the conditional probability distribution:

$$p(Q^{1:D}, F^{1:D}|O; \Lambda) = \frac{1}{Z(O; \Lambda)} \exp\left(\sum_{k=1}^{K} \lambda_k \Phi_k(Q^{1:D}, F^{1:D}, O)\right) \qquad (2)$$

where we represent the state sequence $Q^{1:D} = \{Q^1, \ldots, Q^D\}$ and the indicator variable sequence $F^{1:D} = \{F^1, \ldots, F^D\}$. $O = \{o_1, \ldots, o_T\}$ is the sequence data (observations) and $\Lambda = \{\lambda_1, \ldots, \lambda_K\}$ is the model parameter. $Z(O; \Lambda)$ is the partition function that ensures that $p(Q^{1:D}, F^{1:D}|O; \Lambda)$ is properly normalized.

$$Z(O; \Lambda) = \sum_{Q^{1:D}} \sum_{F^{1:D}} \exp\left(\sum_{k=1}^{K} \lambda_k \Phi_k(Q^{1:D}, F^{1:D}, O)\right) \qquad (3)$$

$\Phi_k(Q^{1:D}, F^{1:D}, O)$ is a feature function that can be arbitrarily selected.

To compare the performance of HHCRFs with that of HHMMs, which have a Markov structure in the state sequence, we restrict the feature function as $\Phi_k(Q^{1:D}, F^{1:D}, O) = \sum_{t=1}^{T} \phi_k(q_{t-1}^{1:D}, q_t^{1:D}, f_{t-1}^{1:D}, f_t^{1:D}, o_t)$ to make the model structure equivalent to that of HHMMs. The different feature functions $\phi_k(q_{t-1}^{1:D}, q_t^{1:D}, f_{t-1}^{1:D}, f_t^{1:D}, o_t)$ are as follows.

$$\phi_{j,j',i,d}^{(Hor)}(q_{t-1}^{1:D}, q_t^{1:D}, f_{t-1}^{1:D}, f_t^{1:D}, o_t) = \left(\delta(q_{t-1}^d = j) \cdot \delta(q_t^d = j') \cdot \delta(q_t^{1:d-1} = i)\right.$$

$$\left. \cdot \, \delta(f_{t-1}^{d+1} = 1) \cdot \delta(f_{t-1}^d = 0)\right) \quad \forall j, \forall j', \forall i, \forall d$$

$$\phi_{i,j',d}^{(Ver)}(q_{t-1}^{1:D}, q_t^{1:D}, f_{t-1}^{1:D}, f_t^{1:D}, o_t) = \left(\delta(q_t^{d-1} = i) \cdot \delta(q_t^d = j')\right.$$

$$\left. \cdot \, \delta(f_{t-1}^{d+1} = 1) \cdot \delta(f_{t-1}^d = 1)\right) \quad \forall i, \forall j', \forall d$$

$$\phi_{i,j,d}^{(End)}(q_{t-1}^{1:D}, q_t^{1:D}, f_{t-1}^{1:D}, f_t^{1:D}, o_t) = \Big(\delta(q_t^{1:d-1} = i) \cdot \delta(q_t^d = j)$$

$$\cdot \delta(f_t^{d+1} = 1) \cdot \delta(f_t^d = 1)\Big) \qquad \forall_i, \forall_j, \forall_d$$

$$\phi_{i,s}^{(Obs)}(q_{t-1}^{1:D}, q_t^{1:D}, f_{t-1}^{1:D}, f_t^{1:D}, o_t) = \delta(q_t^{1:D} = i)\delta(o_t = s) \qquad \forall_i$$

$$(4)$$

where $\delta(q = q')$ is equal to 1 when $q = q'$ and 0 otherwise. The first three feature functions are transition features. $\phi_{j,j',i,d}^{(Hor)}$ counts the horizontal transitions into the j'-th state (except into an end state) from the j-th state at level d. $\phi_{i,j',d}^{(Ver)}$ counts the vertical transitions into the j'-th state from the i-th state at level d. $\phi_{i,j,d}^{(End)}$ counts the horizontal transitions into an end state from the j-th state at level d. $\phi_i^{(Obs)}$ counts the output at the i-th state.

It can be shown that setting parameter Λ (i.e., the weight of the feature functions) as follows gives the conditional probability distribution induced by HHMMs with the transition probability distributions and the output probability distributions defined in Eq. (1):

$$\lambda_{j,j',i,d}^{(Hor)} = \log A_i^d(j, j')$$
$$\lambda_{i,j',d}^{(Ver)} = \log \pi_i^d(j')$$
$$\lambda_{i,j,d}^{(End)} = \log Ae^d(i, j)$$
$$\lambda_{i,s}^{Obs} = \log B[s|i]$$

$$(5)$$

3.2   Parameter Estimation

Exactly as in HHMMs, parameter estimation for HHCRFs is based on the maximum likelihood principle given a training set $\mathcal{D} = \{O^{(n)}, Q^{1:D(n)}, F^{1:D(n)}\}_{n=1}^N$. The difference is that we maximize the conditional probability distribution $p(Q^{1:D}, F^{1:D}|O; \Lambda)$ for HHCRFs, whereas we maximize the joint probability distribution $p(Q^{1:D}, F^{1:D}, O; \Lambda_1)$ for HHMMs. Here, Λ_1 is the parameter for HHMMs. The conditional log-likelihood for HHCRFs is given below.

$$\mathcal{L}(\Lambda) = \sum_{n=1}^N \log p(Q^{1:D(n)}, F^{1:D(n)}|O^{(n)}; \Lambda)$$

$$= \sum_{n=1}^N \left(\sum_{k=1}^K \lambda_k \Phi_k(Q^{1:D(n)}, F^{1:D(n)}, O^{(n)})\right)$$

$$- \sum_{n=1}^N \log Z(O^{(n)}; \Lambda) \qquad (6)$$

The gradient of Eq. (6), which is needed for estimating parameter $\hat{\Lambda}$, is given by

$$\frac{\partial \mathcal{L}}{\partial \lambda_k} = \sum_{n=1}^{N} \Phi_k(Q^{1:D^{(n)}}, F^{1:D^{(n)}}, O^{(n)})$$

$$- \sum_{n=1}^{N} \sum_{Q^{1:D}} \sum_{F^{1:D}} \Phi_k(Q^{1:D}, F^{1:D}, O^{(n)}) p(Q^{1:D}, F^{1:D}|O^{(n)}; \Lambda) \qquad (7)$$

The right hand side of Eq. (7) is the difference between the expectation of feature values under the actual distribution and that under the model distribution $p(Q^{1:D}, F^{1:D}|O^{(n)}; \Lambda)$. The first expectation, the first term of the equation, can be computed using the junction tree algorithm [7], or by converting the hierarchical model to a flat model with mega states and applying the backward-forward-backward algorithm [8].

The sufficient statistics to compute the second expectation are the transition probabilities $\{p(q_{t-1}^{1:D}, q_t^{1:D}, f_{t-1}^{1:D}, f_t^{1:D}|O^{(n)}; \Lambda)|1 \leq t \leq T\}$ and the occupancy probabilities $\{p(q_t^{1:D}, f_t^{1:D}|O^{(n)}; \Lambda)|1 \leq t \leq T\}$, which can be computed using the junction tree algorithm, or by converting the hierarchical model to a flat model with mega states and applying the forward-backward algorithm. (Once again, we use the latter method in our experiment.)

4   Sentence Representation

In previous works on single level time series models (HMMs) for natural language tasks, the passages of text to be processed were represented as a sequence of tokens. Skounakis et al. [6] showed that representing the sentence structure in the learned hierarchical models (HHMMS) provides better extraction. Their approach is based on using syntactic parses of all sentences to be processed.

We follow their approach for sentence representation, with our representation providing a two-level description. The upper level represents each sentence as a sequence of phrase segments, while the lower level represents individual tokens, together with their part-of-speech (POS) tags. In positive training examples, if a segment contains a word or words that belong to a domain in a target tuple, the segment and the words of interest are annotated with the corresponding domain. We refer to these annotations as labels. Test instances do not contain labels; these labels need to be predicted by the learned model.

Fig. 2 depicts a sentence containing an instance of a $subcellular - localization$ relation and its annotated segments. The sentence is segmented into typed phrases, while each phrase is segmented into words typed with part-of-speech tags. The labels are shown in red and green next to the typed phrases and POS tags.

Phrase number	Phrase	POS	Word
1	NP_SEGMENT	DET	this
		UNK	enzyme
2	NP_SEGMENT:PROTEIN	UNK:PROTEIN	ubc6
3	VP_SEGMENT	V	localizes
4	PP_SEGMENT	PREP	to
5	NP_SEGMENT:LOCATION	ART	the
		N:LOCATION	endoplasmic
		N:LOCATION	reticulum
6	PP_SEGMENT	PREP	with
7	NP_SEGMENT	ART	the
		N	catalytic
		UNK	domain
8	VP_SEGMENT	V	facing
9	NP_SEGMENT	ART	the
		N	cytosol

Fig. 2. Input representation for a sentence containing a tuple

5 Hierarchical Models for Information Extraction

5.1 Upper and Lower Levels

The hierarchical models, HHMMs and HHCRFs, have two levels. At the "coarse" level, our hierarchical models represent a sentence as a sequence of phrases. Thus, we can think of the upper level as a single level model whose states emit phrases. We refer to this single level model as the phrase model and its states as phrase states. At the "fine" level, each phrase is represented as a sequence of words. This is achieved by embedding another single level model within each phrase state. We refer to this embedded single level model as a word model and its states as word states. Fig. 3 shows a transition graph between phrase states, while Fig. 4 shows a transition graph between word states. The phrase states in Fig. 3 are depicted as circles, while the word states in Fig. 4 are depicted as trapeziums. To explain a sentence, the phrase model first follows a transition from the START state to some phrase state q_i, uses the word model of q_i to emit the first phrase of the sentence, then transitions to another phrase state q_j, emits another phrase using the word model of q_j, and so on, until it moves to the END state of the phrase model. Note that only word states have direct emissions.

5.2 Model Learning

Training sentences can be classified into two types, positive and negative sentences. Positive sentences include a tuple (an instance of a particular relation) in

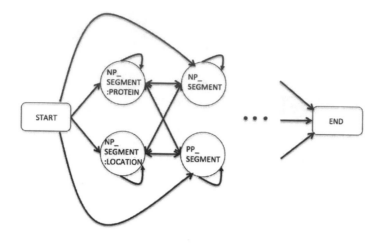

Fig. 3. Phrase model for the *subcellular − localization* relation

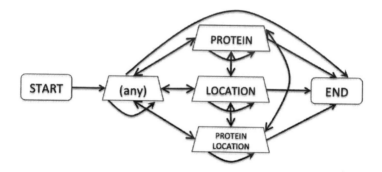

Fig. 4. Word model for the *subcellular − localization* relation

the sentences, while negative sentences do not. We use both positive and negative sentences for learning the models.

In the case of HHMMs, a pair of HHMMs, one positive and one negative, are learned. The positive HHMM learns parameters using only positive sentences, whereas the negative HHMM learns parameters using only negative sentences.

In the case of HHCRFs, being a discriminative model, the HHCRF can learn both positive and negative sentences with only one model. Our HHCRFs have an indicator variable PN in addition to regular parameters to indicate whether a given sentence includes a tuple. The conditional probability that a tuple is included in the sentence is described as follows.

$$P(PN = positive, Q^{1:D}, F^{1:D}|O, \Lambda) \tag{8}$$

5.3   Inference

Once the model has been trained, we can predict whether a sentence includes a tuple using the forward algorithm (the forward part of the forward-backward algorithm), and then we can predict the position of the tuple using Viterbi algorithm.

In the case of HHMMs, we can compare the likelihood for positive and negative HHMMs for a given test sentence. If the likelihood of the positive HHMM is greater than that of the negative HHMM, we infer that the test sentence includes a tuple. The forward algorithm is used to calculate the likelihoods. For positive test sentences, we then apply Viterbi algorithm to the positive HHMM. We extract a tuple from the given sentence if the Viterbi path goes through states with labels for all the domains of the relation.

In the case of HHCRFs, we can predict whether the sentence includes a tuple by comparing the conditional probabilities $P(PN = positive/negative|O, \Lambda)$. The forward algorithm is used to calculate the conditional probabilities and then, we use Viterbi algorithm. Viterbi algorithm is used with parameter $PN = positive$ for those test sentences predicted as being positive.

6   Experiments

6.1   Data

To compare the performance of HHCRFs with that of HHMMs in information extraction, we evaluated the HHMMs and HHCRFs using three data sets assembled by Skounakis [6] from the biomedical literature. The first set contains instances of the $subcellular - localization$ relation, which represents the location of a particular protein within a cell. The second set, which we refer to as the $disorder - association$ data set, characterizes a binary relation between genes and disorders. The third set, which we refer to as the $protein - interaction$ data set, characterizes physical interactions between pairs of proteins. We selected 300 positive and 300 negative sentences from each of the three sets.

We use five-fold cross-validation to measure the accuracy of each approach. We map all numbers to a special NUMBER token and all words that occur only once in a training set to an OUT-OF-VOCAB token. Also, all punctuation is discarded. The same preprocessing is done on test sentences, with the exception that words not encountered in the training set are mapped to the OUT-OF-VOCAB token. The vocabulary is the same for all emitting states in the models, and all parameters are smoothed using $m - estimates$ [9].

6.2   Retrieved Results

We use the lower levels most likely Viterbi path to predict the positions of the tuple words for the relation. The most likely Viterbi path is returned as a retrieved result, if the following two conditions hold.

- The forward algorithm predicts that the given test sentence is positive and includes a tuple.
- The confidence measure for the most likely path is above the threshold.

For the most likely path for a sentence snt, we calculate the confidence measure as follows.

$$c(snt) = \frac{\delta_n(|snt|)}{\alpha_n(|snt|)} \tag{9}$$

Let n be the length of the most likely path. $\delta_n(|snt|)$ is the probability of the most likely path up to the n-th state, given by the Viterbi algorithm, and $\alpha_n(|snt|)$ is the total probability of the sequence, calculated by the forward algorithm.

We consider the retrieved result to be correct if the following hold.

- The given test sentence is positive, i.e., the sentence includes a tuple (an instance of the relation).
- The positions of the labels predicted by the most likely path correspond to the actual positions of the tuple words in the sentence.

6.3   Performance Evaluation

To evaluate our models, we construct *precision − recall* graphs. *Precision* is defined as the fraction of the number of the sentences for which the positions of tuple words were predicted correctly over the sentences predicted to include a tuple by the model. *Recall* is defined as the fraction of the number of the sentences for which the positions of tuple words were predicted correctly over the total number of positive sentences. We constructed precision-recall curves by varying the threshold for the confidence measure defined in (9).

6.4   Results

Fig. 5(a), (b), and (c) show the precision-recall graphs for the three data sets. Each figure shows graphs for both the HHMM and HHCRF. The shapes of these precision-recall curves differ from the more common precision-recall curves with respect to the following. (1) We predict not only whether a sentence includes a tuple, but also the position of the tuple. Hence, we cannot assume that all of the most likely paths are retrieval results and thus, the recall cannot be 1.0. (2) The recall takes its maximum value when the threshold for the confidence measure is 0. (3) The precision does not necessarily increase with the threshold for the confidence measure, because the confidence measure is only an indication of accuracy. Because of (1) and (2),the curves end before the precision reaches 0. Because of (3), it is possible that both precision and recall increase together on the graph. It is also possible that they both decrease together. Precision and recall do not necessarily have a trade-off relation.

For the first two data sets, HHCRFs achieve a higher precision than HHMMs given the same recall, as can be seen in Fig. 5(a) and (b). For the last data set,

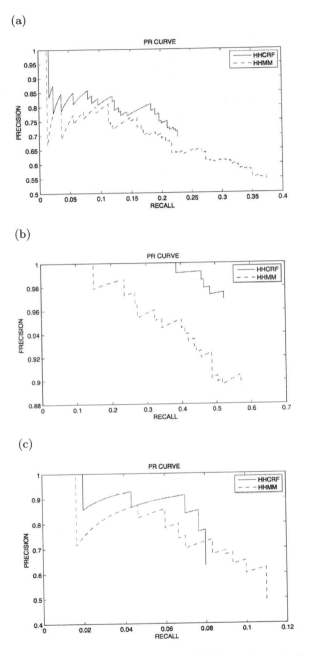

Fig. 5. Precision-recall curves: (a) HHMM and HHCRF on the subcellular-localization data set, (b) HHMM and HHCRF on the disorder-association data set, (c) HHMM and HHCRF on the protein-interaction data set

according to Fig. 5(c), HHCRFs show higher precision except for one location. Furthermore, as shown in the figures, the maximum recall value for HHCRFs is smaller than that for HHMMs. We summarize the results as follows.

- As long as the recall value is below the maximum recall value for HHCRFs, HHCRFs make fewer errors than HHMMs.
- If the recall value lies between the maximum recall value for HHCRFs and that for HHMMs, HHCRFs are not able to extract more tuple words correctly even if the threshold is removed.

In these experiments, we expect extraction precision to improve when using HHCRFs instead of HHMMs. We assume that HHCRFs have higher precision, since they are discriminative models with an indicator variable PN.

7   Conclusion

Information extraction is the task of automatically extracting instances of specified classes or relations from text. Skounakis et al. [6] proposed a method for information extraction using HHMMs. In this paper, we have proposed information extraction using an HHCRF, which is a discriminative model corresponding to the HHMM.

In the experiment, the maximum recall value for HHCRFs is smaller than that for HHMMs. However, HHCRFs achieve higher precision than HHMMs with the same recall. We presume that this is because the HHCRF is a model that is discriminatively trained to predict whether or not a sentence is positive.

References

1. Lafferty, J.D., McCallum, A., Pereira, F.C.N.: Conditional random fields: Probabilistic models for segmenting and labeling sequence data. In: Proc. 18th Int. Conf. Machine Learning (2001)
2. Fine, S., Singer, Y., Tishby, N.: The hierarchical hidden Markov model: Analysis and applications. Machine Learning 32(1) (1998)
3. Murphy, K., Paskin, M.: Linear time inference in hierarchical HMMs. In: Advances in Neural Information Processing Systems, vol. 14 (2001)
4. Sugiura, T., Goto, N., Hayashi, A.: A discriminative model corresponding to hierarchical hMMs. In: Yin, H., Tino, P., Corchado, E., Byrne, W., Yao, X. (eds.) IDEAL 2007. LNCS, vol. 4881, pp. 375–384. Springer, Heidelberg (2007)
5. Tamada, H., Hayashi, A.: Sports video segmentation using a hierarchical hidden CRF. In: Köppen, M., Kasabov, N., Coghill, G. (eds.) ICONIP 2008. LNCS, vol. 5506, pp. 715–722. Springer, Heidelberg (2009)
6. Skounakis, M., Craven, M., Ray, S.: Hierarchical hidden Markov models for information extraction. In: Proc. 18th Int. Joint Conf. Artificial Intelligence (2003)
7. Huang, C., Darwiche, A.: Inference in belief networks: A procedural guide. Int. J. of Approximate Reasoning 15(3) (1996)
8. Scheffer, T., Decomain, C., Wrobel, S.: Active hidden markov models for information extraction. In: Hoffmann, F., Adams, N., Fisher, D., Guimarães, G., Hand, D.J. (eds.) IDA 2001. LNCS, vol. 2189, p. 309. Springer, Heidelberg (2001)
9. Cestnik, B.: Estimating probabilities. In: Proc. 9th European Conf. Artificial Intelligence (1990)

Solving Extremely Difficult MINLP Problems Using Adaptive Resolution Micro-GA with Tabu Search

Asim Munawar[1], Mohamed Wahib[1],
Masaharu Munetomo[2], and Kiyoshi Akama[2]

[1] Graduate School of Information Science and Technology,
Hokkaido University, Sapporo, Japan
{asim,wahib}@ist.hokudai.ac.jp
[2] Information Initiative Institute, Hokkaido University, Sapporo, Japan
{munetomo,akama}@iic.hokudai.ac.jp

Abstract. Non convex mixed integer non-linear programming problems (MINLPs) are the most general form of global optimization problems. Such problems involve both discrete and continuous variables with several active non-linear equality and inequality constraints. In this paper, a new approach for solving MINLPs is presented using adaptive resolution based micro genetic algorithms with local search. Niching is incorporated in the algorithm by using a technique inspired from the tabu search algorithm. The proposed algorithm adaptively controls the intensity of the genetic search in a given sub-solution space, i.e. promising regions are searched more intensely as compared to other regions. The algorithm reduces the chances of convergence to a local minimum by maintaining a list of already visited minima and penalizing their neighborhoods. This technique is inspired from the tabu list strategy used in the tabu search algorithm. The proposed technique was able to find the best-known solutions to extremely difficult MINLP/NLP problems in a competitive amount of time. The results section discusses the performance of the algorithm and the effect of different operators by using a variety of MINLP/NLPs from different problem domains.

Keywords: Mixed Integer Non-Linear Programming (MINLP), micro Genetic Algorithms (mGA), Tabu Search (TS), niching.

1 Introduction

Mixed integer non-linear programming problems (MINLP) are the most generalized form of single-objective global optimization problems. They contain both continuous and integer decision variables, and involve non-linear objective function and constraints setting no limit to the complexity of the problems. MINLPs are difficult to solve [6]:

1. They involve both discrete (integer) and continuous (floating point) variables.
2. Objective function & constraints are non-linear, generating potential non-convexities.
3. They involve active equality and inequality constraints.

Many real world constrained optimization problems are modeled as MINLPs e.g. heat and mass exchange networks, batch plant design and scheduling, design of interplanetary spacecraft trajectories etc. In a mathematical form an MINLP problem can be given as:

C.A. Coello Coello (Ed.): LION 5, LNCS 6683, pp. 203–217, 2011.
© Springer-Verlag Berlin Heidelberg 2011

Minimize $f(x, y)$ $x \in \mathbb{N}^{n_{disc}}, y \in \mathbb{R}^{n_{cont}}, n_{disc} \in \mathbb{N}, n_{cont} \in \mathbb{N}$

Subject to: $g_i(x, y) = 0,$ $i = 1, ..., m_{eq} \in \mathbb{N}$

$\quad\quad\quad\quad g_i(x, y) \geq 0,$ $i = m_{eq} + 1, ..., m \in \mathbb{N}$

$\quad\quad\quad\quad x_l \leq x \leq x_u,$ $x_l, x_u \in \mathbb{N}$

$\quad\quad\quad\quad y_l \leq y \leq y_u,$ $y_l, y_u \in \mathbb{R}$

Where, $f(x, y)$ is the objective function, x is a vector of n_{disc} discrete variables, y is a vector of n_{cont} continuous variables, m_{eq} & m are the number of equality and total constraints respectively, x_l, x_u, y_l, y_u are the lower and upper bounds for the discrete and continuous variables respectively.

Genetic Algorithms (GAs) are population based search and optimization methods that mimic the process of natural evolution. They fall in the category of stochastic global optimization algorithms. Over the recent years GAs have been successfully applied to solve different MINLPs [6,10,27]. GAs are easy to implement and are black box optimizers (BBOs) as they do not require any auxiliary information like continuity or differentiability of functions. They are robust and usually do not get trapped in a local optima. However, like other stochastic methods GAs may need a large number of fitness evaluations because of the combinatorial nature of sampling multidimensional space. Nonetheless, GAs have proven effective for the solution of MINLPs [6,27].

The existing genetic algorithms to solve MINLPs concentrate on a set of problems from a particular domain and carry no promise to perform well on a problem from an entirely new domain. The main motivation behind this paper is to develop a method that is generalized enough to solve difficult MINLP/NLPs taken from different problem domains in a black-box fashion without user intervention. We address the problems that are far more difficult than the MINLPs solved in the literature of solving MINLPs using GAs. Our technique is based on a recursive adaptive resolution micro GA (arGA) with local search (LS). The basic idea is to locate the regions of interest and intensify the genetic search in those areas without revisiting the same areas redundantly. We use the entropy measure of each continuous variable to determine the size of the critical area around a promising individual. The entropy measure is also used to perform an adaptive resolution based local search. This local search tries to find a better solution in the neighborhoods of an individual using multiple resolutions. In order to avoid revisits to previously visited local optima we use a technique inspired from the tabu search algorithm. We maintain a finite list of visited local optima and penalize their neighborhoods for a specific number of iterations. This generates a niching effect and encourages the algorithm to search in unexplored areas. Without this niching the algorithm does not work for difficult problems. We have used oracle penalty method [25] for constraint handling. Oracle penalty method is an advanced penalty method that depends on a single easily controllable input parameter called Ω. We have verified the efficiency of the algorithm by solving a variety of difficult MINLPs.

The rest of the paper is organized as follows: In the next section we will discuss some of the existing GAs for solving of MINLPs. In Sect. 3 we explain the proposed arGA and the arLS operators. Section 4 gives some results to show the advantage of using the proposed algorithm. We conclude the paper in Sect. 5 with some guidelines for possible improvements in the algorithm.

2   Related Work

The solvers for MINLPs can be categorized as deterministic and stochastic methods. Deterministic techniques have been extensively used to solve MINLPs. Branch & bound, outer approximation, and extended cutting plan methods are some of the famous deterministic techniques. Grossman (2002) [13] gives a detailed review of the deterministic techniques for solving MINLPs. Deterministic methods usually guarantee the global optimality at the expense of long execution times depending on the problem complexity. Deterministic methods are usually not BBOs as they often require the problem to be reduced in a particular form e.g. removal of non-convexities, initialization of optimizers etc. This often requires the knowledge about the problem structure.

Stochastic methods based on metaheuristics search techniques are true BBOs as they do not require any information about the mathematical model of the optimization problem. Although such algorithms carries no guarantee for reaching global optimality, due to their robustness and ease of implementation they are widely used to solve difficult optimization problems, yet their applications in MINLPs remain small. A recent approach on MINLPs by ant colony optimization is done by MIDACO [24].

Two main concepts of evolution, natural selection and genetic dynamics, inspired the development of GAs. Basic principles of GAs were laid down by J.H. Holland [17] and his colleagues in 1975 and were elaborated in detail by D.E. Goldberg (1989) [11]. GAs are flexible and can easily be used with other algorithms in a hybrid fashion [7,9].

2.1   GAs for Solving MINLP Problems

Simple GA is not able to solve even the easiest MINLPs. There are two approaches that can be used to enable a GA to solve MINLPs: First approach is the use of advanced genetic operators to ensure the desired convergence of the algorithm, Second approach is hybridization of GAs with deterministic or LS methods [7].

A. Ponsich et al. (2008) [23] gives some guidelines for GA implementation in batch plant design problems. Batch plant design problem is a real world problem that is modeled as a non-convex MINLP. Two main issues discussed in this study are the specific encoding methods and efficient constraint handling. The research uses similar encoding for both integer and continuous part and claims the mixed real-discrete encoding method to be the best option. For constraint handling the paper suggests elimination for small problems but appropriate penalization for the complex problems. However, finding appropriate penalization factor is not always easy for the case of MINLPs where the constraints are non-convex and numerous. Another research on the use of GAs for similar problem is given in M. Danish et al. (2006) [6]. The paper uses tournament selection, SBX crossover, polynomial mutation and variable elitism operator along with distance based dynamic penalty with anti-distortion. The authors claim to solve six difficult MINLP problems by using the proposed method. T. Young et al. (2007) [27] suggests an information guided GA (IGA) approach. It implements the information theory to the mutation stage of the GAs to refresh the premature population. Local search is also performed to increase the efficiency. The paper uses an adaptive penalty scheme to handle constraints [21] and solves 5 popular benchmark problems using the suggested scheme. V.B. Gantovnik et al. (2005) [10] uses GAs to solve a problem to design of fiber reinforced composite shell. The suggested approach tries to reduce the number of fitness and constraint function evaluations by using tree based data structures for efficient search in the memory to avoid redundant fitness calculations. It suggests the use of multivariate approximation for continuous variables to avoid unnecessary exact analyses for points close to previous values.

Apart from the research work done in the area there is at least one commercial product that uses GAs to solve MINLP problems. The product is known as GENO or General Evolutionary Numerical Optimizer [1].

As opposed to the existing techniques, the proposed approach uses unbiased genetic operators that are applicable to problems from a wide variety of problem domains. Moreover, we try to solve some of the extremely difficult problems that to the best of author's knowledge have never attempted before using genetic algorithms.

3   The Proposed Algorithm

The proposed approach adopts three guiding principles: (a) areas around better solutions have a greater chance of having an even better solution, (b) constraints must be handled using an advanced penalty method that do not get trapped in a single feasible region, and (c) revisiting local optima results in waist of time and therefore must be avoided. Using these principles as guidelines, the algorithm uses a hierarchical approach for vigorously searching the promising sub-solution spaces by adaptive resolution GA combined with an adaptive resolution LS operator. In order to eliminate redundancy in revisiting already visited local optima we have used an operator inspired from the working of tabu search algorithm. The oracle penalty method is used to handle constraints.

3.1   Variables Encoding and Genetic Operators

Encoding is one of the most important design factors for GAs, as it limits the kind of genetic operators that can be used by the algorithm. In our approach, we use different encoding for the real and the integer part. In the proposed approach the continuous part of the problem is encoded as real numbers with double precision while the discrete part is encoded as binary numbers of fixed user defined length. This kind of encoding allows to perform real genetic operators on the continuous variables while binary genetic operators are applied to the discrete part.

We employ simple one-point crossover for the discrete part while SBX crossover [2] is applied to the real part of the chromosome. Binary mutations are achieved by simple bit flipping operation while polynomial mutation is used for the continuous part of the individuals. Tournament selection of size T is used as a selection operator along with sharing operator that acts as a niching technique to avoid early convergence. Elitist replacement is used for the insertion of new individuals in the population.

3.2   Constraint Handling

Penalty methods are used to handle the problem constraints. Such methods transform a constrained problem to an unconstrained problem by adding the weighted sum of constraint violations to the original fitness function. Death or static penalty methods are the most commonly used penalty methods. Although easy to use these methods are not able to achieve good performance for tightly constrained problems. We have used oracle penalty method [25] for constraint handling. Oracle method depends only on one parameter, named Ω, which is selected as the best equivalent or just slightly greater than the optimal (feasible) objective function value for a given problem. As for most real-world problems this value is unknown a priori, we start with a value of $\Omega = 1e^6$. We keep on improving the value of Ω by assigning the best known feasible fitness value of the previous run. Mathematically the oracle penalty function can be represented as:

$$p(x) = \begin{cases} \alpha \cdot |f(x) - \Omega| + (1 - \alpha) \cdot res(x) \,, & \text{if } f(x) > \Omega \text{ or } res(x) > 0 \\ \\ -|f(x) - \Omega| & , \text{ if } f(x) \leq \Omega \text{ and } res(x) = 0 \end{cases}$$

where α is given by:

$$\alpha = \begin{cases} \dfrac{|f(x)-\Omega| \cdot \frac{6\sqrt{3}-2}{6\sqrt{3}} - res(x)}{|f(x)-\Omega| - res(x)} & , \text{ if } f(x) > \Omega \text{ and } res(x) < \frac{|f(x)-\Omega|}{3} \\ \\ 1 - \dfrac{1}{2\sqrt{\frac{|f(x)-\Omega|}{res(x)}}} & , \text{ if } f(x) > \Omega \text{ and } \frac{|f(x)-\Omega|}{3} \leq res(x) \leq |f(x) - \Omega| \\ \\ \dfrac{1}{2}\sqrt{\dfrac{|f(x)-\Omega|}{res(x)}} & , \text{ if } f(x) > \Omega \text{ and } res(x) > |f(x) - \Omega| \\ \\ 0 & , \text{ if } f(x) \leq \Omega \end{cases}$$

Shape of the oracle penalty function is shown in Fig. 1. The oracle penalty function is good at dealing with the non-convexities in both equality and inequality constraints.

3.3 Micro GA

The algorithm relies on micro GAs [20] as opposed to the conventional GAs. Micro GAs maintain a very small population ($\lesssim 20$) that is re-initialized after every few generations (between 10 and 100). We call this re-initialization of the population a restart. A restart can inherit some information from the previous run in order to improve the performance. This re-initialization of the population every few generations can be considered as a mutation operator. The re-initialization of the population may or may not be completely random. We define a proximity parameter that determines the proximity of the newly initialized individuals to the best known individual of the previous restart. Conventional GAs maintain a large population with sizes of approximately $(1/k) \cdot 2^k$ for binary encoding, where k is the average size of the schema of interest (effectively the average number of bits per parameter, i.e. approximately equal to $nchrome/nparam$, rounded to the nearest integer) [12]. The large population size ensures with high probability that the required genetic material is present in the initial population. In our observation, for extremely complex MINLP/NLPs a large population slows down the overall process as a convergence to local optima may lead the whole population to converge to this point. While in the case of micro GAs a small population may converge to a local optima but the next restart will have a good chance of jumping to another area. Moreover, as the population size is very small this process happens very quickly. Therefore,

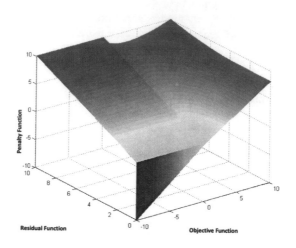

Fig. 1. The oracle penalty function [25]

even though the micro GAs require many restarts before they succeed to achieve all the required genetic information to reach the best or near best solution the process is much faster than their conventional counterparts. Individuals carried on from the previous run allow the building blocks of two different restarts to mix with each other, a step required by the schema theorem of GAs. In the results section we discuss the effects of population sizing on the solution quality.

3.4 Adaptive Resolution Approach

Adaptive resolution is a recursive approach to divide the solution space in search for a better solution. The recursion terminates when no better solution is found. The size of the sub-solution space is calculated by using the entropy value of each variable. The recursive nature of arGA is shown in Fig. 2. The probability of adaptive resolution is proportional to the fitness of the individuals i.e. an individual with a good overall fitness has a greater chance of getting selected for the adaptive resolution search.

Using Entropy to control Resolution. In order to control the size of the sub-solution space a vector of real numbers γ is defined, γ is the same size as number of continuous variables. γ value for each variable is calculated using the information entropy. According to Shannon's definition of information entropy [26], for a variable V which can randomly take a value v from a set $\mathbb{V}$, the information entropy of the set $\mathbb{V}$ is:

$$E(V) = -\sum_{v \in \mathbb{V}} p(v) ln p(v)$$

If V can only take a narrow range of values, $p(v)$ for these values is ≈ 1. For other values of V, $p(v)$ is close to zero. Therefore, $E(V)$ will be close to zero. In contrast, if V can take many different values each time with a small $p(v)$, $E(V)$ will be close to 1.

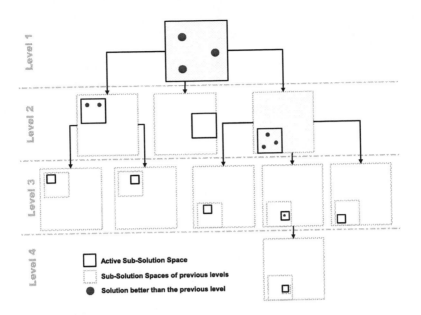

Fig. 2. Recursive behavior of adaptive resolution technique

Measuring entropy using the above equation is simple for discrete variables, but the entropy must be redefined for real numbers by discritizing the range of each variable. If we have $i = 0, \ldots, I$ real variables with lower and upper bound (L_i, U_i) such that $L_i \leq V_i \leq U_i$. For each variable V_i, we divide the solution space into R sections of equal size. Let $S = \{s_{r,i} | i = 1, \ldots, I, r = 1, \ldots, R\}$ and $s_{r,i} = [L_i^r, U_i^r]$, where:

$$L_i^r = L_i + \frac{r-1}{R}(U_i - L_i) \qquad\qquad U_i^r = U_i - \frac{R-r}{R}(U_i - L_i)$$

for $i = 1, \ldots, I$ and $r = 1, \ldots, R$. Probability that the variable V_i takes the value in subspace $s_{r,i} = [L_i^r, U_i^r]$ is given by $P_{r,i} = P(V_i = v_i | v_i \in s_{r,i})$. The total entropy of the set V_i is:

$$E(V_i) = -\sum_{r=1}^{R} P_{r,i} log(P_{r,i})$$

So, for a variable V_i we define $\gamma_i = E(V_i)$. Furthermore, in each iteration of arGA the value of γ for each variable is halved, hence reducing the size of sub-solution space to half. This technique is similar to the well known bisection based reduction technique.

3.5 Local Search

We have used an asynchronous local search to improve the efficiency of the algorithm. LS is applied only to a specified percentage of the individuals. This probability of local search is kept low (0.01 to 0.1). The proposed local search is adaptive resolution version of the widely used hill-climbing algorithm. The algorithm for the local search used in

---

Algorithm 1. Adaptive resolution local search algorithm

---

inputs

$X = [X_{bin}, X_{real}]$ {X_{bin} is a binary vector of size N_{bin}, and X_{real} is a binary vector of size N_{real}}
γ {A real number vector of size N_{real}. contains the resolution for each real variable}
N_{LS}, N_{rLS} {number of overall local search iterations, and number iterations for each real variable}

---

for $i = 1$ to N_{LS} **do**
  nextEval = INFINITY;
  ———————— *binary local search* ————————
  for $j = 1$ to N_{bin} **do**
    ¬ $X_{bin}[j]$;
    if nextEval > objFunc(X) **then**   nextNode ← X,   nextEval ← objFunc(X);
    ¬ $X_{bin}[j]$;
  end for
  ———————— *real local search* ————————
  for $j = 1$ to N_{real} **do**
    for $k = 1$ to N_{rLS} **do**
      $X_{real}[j]$ —= $k \cdot \gamma[j]$;
      if nextEval > objFunc(X) **then**   nextNode ← X,   nextEval ← objFunc(X);
      $X_{real}[j]$ += $2 \cdot k \cdot \gamma[j]$;
      if nextEval > objFunc(X) **then**   nextNode ← X,   nextEval ← objFunc(X);
      $X_{real}[j]$ —= $k \cdot \gamma[j]$;
    end for
  end for
  if objFunc(X) > nextEval **then**   X ← nextNode;   **else**   break;
end for

---

arGA is shown in Algorithm 1. The effect of this asynchronous operator is discussed in the results section.

3.6   Avoiding Redundancy

Avoiding redundant search near already visited local optima is a key to better and faster search. We have used a simple technique inspired from tabu search algorithm to avoid this redundancy. The algorithm used for avoiding redundancy is shown in Algorithm 2. It maintains a finite list of visited local optima. The neighborhoods of these individuals are penalized in every fitness evaluation. As the list is of finite size the individual can be removed from the list and get a second chance of getting searched by the algorithm.

4   Results

4.1   Environment and Parameters

All the experiments were performed over an Intel Core 2 Duo 3.3GHz CPU with 4GB of RAM. The implementation is serial but uses the auto parallelization performed by

---

Algorithm 2. Algorithm for avoiding redundancy

Initialize counter
Loop until the maximum number of restarts is achieved
  Run GA
  If no better solution is found in the current restart
    Increment the counter
    If counter exceeds the maximum number of similar runs (M_r)
      Apply arGA to the neighborhood just to make sure that this is the local optima
      If a better solution is found by the arGA
        Reset counter and continue the loop
      Else
        Insert the solution into a finite size Queue
End Loop

---

the compiler. The algorithm is controlled and configured by various input parameters. All the parameters are preconfigured to an appropriate value. However, an advanced user can modify the parameters by accessing the parameter's input file. Nomenclature of the parameters is given below:

G, P, R	Maximum number of allowed generations, population and restarts
nC, nD, L_c	Number of continuous, discrete variables & Chromosome length (nC + nD)
m_{eq}, m	Number of equality constraints & total number of active constraints
L_i, U_i	Lower & Upper bound for the ith continuous variable respectively
l_j, u_j, b_j	Lower & upper bound and bits required for the jth discrete variable
P, P_c	Penalty function and penalty configuration
P_c, P_{mb}, P_{mr}	Crossover probability, binary mutation & real mutation probability
P_{ls}, N_{ls}	Local search probability and number of LS iterations
T	Tournament size for the selection operator
Pr	Proximity parameter for sampling of population in consecutive runs
η_c, η_m	Crossover & mutation probability distribution index
Ω	Initial value of the oracle for oracle penalty method

Table 1. Results obtained by applying the proposed algorithm on different optimization problems

Problem	Restarts$_{mean}$	Eval$_{mean}$	Feasible	Optimal	f_{best}	f_{worst}	f_{mean}	Time$_{mean}$
1	4	3097	30	30	99.3	99.3	99.3	0m0.039s
2	19	16122	30	30	3.56	3.56	3.56	0m0.206s
3	56	36160	30	30	-30665.54	-30665.54	-30665.54	0m0.612s
4	101	185368	30	30	67.9	67.9	67.9	0m4.424s
5	12932	16039145	30	21	4.93	5.87	5.01	57m7.8s
6	16231	20832415	30	13	1,580,428	1,508,958	1,561,445	1h48m
7	23029	26028521	30	5	6.04	13.7	11.3	2h37m8.3s
8	19823	21459135	30	17	8.6	16.93	14.14	1h58m53s
9	17982	17328828	30	16	8.3	26.12	13.22	1hr39m43s
10	18929	17981459	30	16	1.3	22.35	15.82	1hr45m2s
11	11781	14982393	29	19	18.9	26.52	20.01	48m45.2s

I_a	Number of levels for performing adaptive resolution genetic algorithm
I_r	Number of iterations of adaptive resolution local search for continuous variables
St	Stopping criteria
M_f, M_t	Maximum allowed fitness evaluations & execution time
M_G, M_R	Maximum allowed generations & runs without any improvement in fitness
Q	Size of the queue for the tabu list
N	Size of the neighborhood that needs to be penalized
R_e	Number of partitions to discretize the continuous range for entropy calculations
L_i^r, U_i^r	Lower & upper range of the rth partition of the ith variable
E	Allowed error between the calculated and the best known result
E_p	Residual accuracy

We consider MINLP/NLPs from a wide variety of problem domains. The problems vary in difficulty levels starting from simple to extremely difficult problems. The benchmarking problems used in the results section are shown in the appendix.

4.2   Results and Discussion

Results obtained by applying the proposed algorithm on the MINLP/NLP benchmark problems can be found in Table 1. Note that all the results are an average of 30 independent runs under identical circumstances. Table below explains the abbreviations used in Table 1.

Abbreviation	Explanation
Problem	problem name used in the literature
Restarts$_{mean}$	average number of restarts
Feasible	number of feasible solutions found out of 30 test runs
Optimal	number of optimal solutions found out of 30 test runs
f_{best}	best (feasible) objective function value found out of 30 test runs
f_{worst}	worst (feasible) objective function value found out of 30 test runs
f_{mean}	average objective function value over all runs with a feasible solution
Time$_{mean}$	average CPU-time over all runs with a feasible solution
Eval$_{mean}$	average number of evaluations over all runs with a feasible solution

With so many parameters for the algorithm calibration it is vital to carefully study the effect of each and every important parameter on the total execution time and solution quality. The preset values of the parameters used for the above experimentation are as follows: $G = 10$, $P = 10$, $R = 30000$, $P_c = 0.5$, $P_{mb} = 0.2$, $P_{mr} = 0.3$, $P_{ls} = 0.01$, $\Omega = 1e6$, $\eta_c = 2$, $\eta_m = 100$, $P_r = 1e2$, $M_G = 6$, $M_R = 30$, $I_a = I_r = 8$, $E_p = 0.01$, $E = 1\%$, $R_e = 8$, $Q = 20$, $P = $ "oracle penalty", $St = $ "best solution found or no improvement", $N_n = 1$, $T = 3$. The values of these parameters are selected empirically by studying the effect of the different parameters on the output.

Figure 3 shows the relationship between the average number of evaluations and average number of restarts vs. the generation/population (G, P) pair. Even though the number of average restarts decreased with the increase of G and P, the total number of fitness evaluations increased by a significant amount. As the total number of fitness evaluations is directly related to the total execution time, larger values of G and P results in longer execution time. The results provide a solid ground for the use of micro GAs instead of conventional GAs.

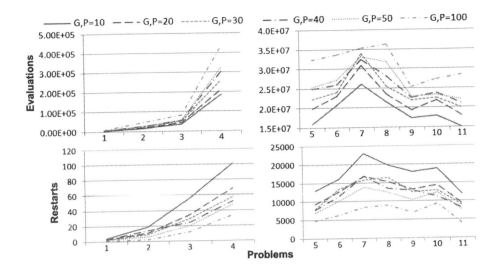

Fig. 3. Effect of generation-population (G,P) pair on the total number of evaluations and average number of restarts

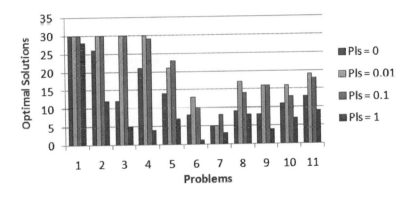

Fig. 4. Effect of local search probability (P_{ls}) on the number of optimal solutions found

Figure 4 shows the importance of optimizing the probability of LS P_{ls}. It is clear that $P_{ls} = 0.01$ is the optimal value as it results in the maximum number of optimal solutions found. Keeping the value of $P_{ls} \geq 0.1$ forces the algorithm towards local optima.

Figure 5 shows the effect of arGA iterations on the total number of optimal solutions found in 30 runs. arGA is a kind of non-deterministic local search algorithm. The figure depicts the importance of the arGA step.

Figure 6 shows the efficiency of the queue based approach to avoid redundancy in searching. The figure shows that for $Q = 0$ the algorithm is not able to find any optimal result for some problems, while for other difficult optimization problems the performance remains low. Hence, the quality of the results is tightly related with the value of Q.

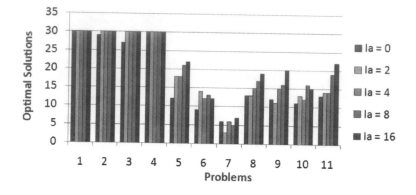

Fig. 5. Effect of arGA iterations (I_a) on the number of optimal solutions found

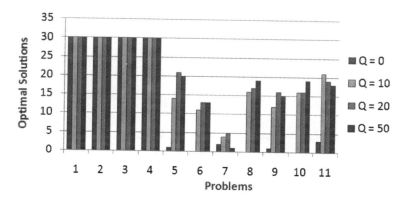

Fig. 6. Effect of queue size (Q) on the number of optimal solutions found.

5 Conclusions and Future Work

In this paper we suggested a new technique to solve extremely difficult MINLP problems. We were able to find the best known solutions to the problems from different domains in a reasonable amount of time. Adaptive resolution technique combined with microGAs is able to find good local optima in a very rough multidimensional terrain. This technique combined with an operator inspired by the tabu list in tabu search does the trick. Local optima are stored in a list for a specific number of generations in order to avoid redundant searching in already searched areas. Even though the algorithm depends on many different operators and input parameters one of the most important operators is the restarting of the algorithm whenever stuck. This might sound simple but microGA with many restarts produces much better results than a normal GA with larger population and greater number of generations.

As a future work it would be interesting to calibrate (optimize) the input parameters using the same algorithm. The parameter tuning in this case will become a very complicated MINLP. The fitness function in this case would be cumulative performance of the algorithm over a set of benchmarking MINLP/NLP problems.

Acknowledgments. We would like to thank Martin Schlüter for his help regarding the MINLPs. We are also thankful to Kanpur Genetic Algorithms Laboratory for the free codes for C code of single-objective GA; we have built our code using this code as the foundation. This work is supported by Grant-in-Aid for Scientific Research (C) by MEXT, Japan.

References

1. Geno: General evolutionary numerical optimizer, http://tomopt.com/tomlab/products/geno/
2. Agrawal, R.B., Deb, K., Deb, K., Agrawal, R.B.: Simulated binary crossover for continuous search space (1995)
3. Babu, B., Angira, A.: A differential evolution approach for global optimisation of minlp problems. In: Proceedings of the Fourth Asia Pacific Conference on Simulated Evolution and Learning (SEAL 2002), Singapore, pp. 880–884 (2002)
4. Coello, C.C.: Constraint-handling using an evolutionary multi-objective optimisation technique. Civil Engineering and Environmental Systems 17, 319–346 (2000)
5. Colville, A.: A comparative study of non-linear programming codes. Tech. Rep. Report 320-2949, IBM Scientific Centre, New York
6. Danish, M., Kumar, S., Qamareen, A., Kumar, S.: Optimal solution of minlp problems using modified genetic algorithm. Chemical Product and Process Modeling 1(1) (2006)
7. El-mihoub, T.A., Hopgood, A.A., Nolle, L., Battersby, A.: Hybrid genetic algorithms: A review. Engineering Letters 13(12), 124–137 (2006)
8. Floudas, C., Aggarwal, A., Ciric, A.: Global optimum search for nonconvex nlp and minlp problems. Computers & Chemical Engineering 13(10), 1117–1132 (1989)
9. French, A.P., Robinson, A.C., Wilson, J.M.: Using a hybrid genetic-algorithm/branch and bound approach to solve feasibility and optimization integer programming problems. Journal of Heuristics 7(6), 551–564 (2001)
10. Gantovnik, V.B., Gurdal, Z., Watson, L.T., Anderson-Cook, C.M.: A genetic algorithm for mixed integer nonlinear programming problems using separate constraint approximations. Departmental Technical Report TR-03-22, Computer Science, Virginia Polytechnic Institute and State University (2005)
11. Goldberg, D.: Genetic Algorithms in Search, Optimization, and Machine Learning. Addison-Wesley Professional, Reading (1989)
12. Goldberg, D., Deb, K., Clark, J.: Genetic algorithms, noise, and the sizing of populations. Complex Systems 6, 333–362 (1991)
13. Grossmann, I.E.: Review of nonlinear mixed-integer and disjunctive programming techniques. Optimization and Engineering 3, 227–252 (2002)
14. (GTOP), E.S.A.E.G.O.T.P., Solutions, http://www.esa.int/gsp/ACT/inf/op/globopt.htm
15. Himmelblau, D.: Applied Nonlinear Programming. McGraw-Hill, New York (1972)
16. Hock, W., Schittkowski, K.: Test examples for non-linear programming codes. LNEMS, vol. 187. Springer, Berlin (1981)
17. Holland, J.: Adaptation in natural and artificial systems. University of Michigan Press, Ann Arbor (1975)

18. Homaifar, A., Lai, S.: Constrained optimisation via genetic algorithms. Simulation 62, 242–254 (1994)
19. Kocis, G., Grossmann, I.: A modelling and decomposition strategy for the minlp optimisation of process flow sheets. Computers and Chemical Engineering 13, 797–819 (1989)
20. Krishnakumar, K.: Micro-genetic algorithms for stationary and non-stationary function optimization. In: SPIE: Intelligent Control and Adaptive Systems, vol. 1196, pp. 289–296 (1989)
21. Lemonge, A.C., Barbosa, H.J.: An adaptive penalty scheme for genetic algorithms in structural optimization. International Journal for Numerical Methods in Engineering 59(5), 703–736 (2004)
22. Michalewicz, Z., Fogel, D.: How to solve it. In: Modern Heuristics. Springer, Berlin (2000)
23. Ponsich, A., Azzaro-Pantel, C., Domenech, S., Pibouleau, L.: Some guidelines for genetic algorithm implementation in minlp batch plant design problems. In: Advances in Metaheuristics for Hard Optimization. Natural Computing Series, pp. 293–316. Springer, Heidelberg (2008) ISSN 1619-7127
24. Schlueter, M.: Midaco: Global optimization software for mixed integer nonlinear programming (2009), http://www.midaco-solver.com
25. Schlueter, M., Gerdts, M.: The oracle penalty method. Journal of Global Optimization 47(2), 293–325 (2010)
26. Shannon, C.: A mathematical theory of communication. Bell System Technical Journal 27 (1948)
27. Young, C., Zheng, Y., Yeh, C., Jang, S.: Information-guided genetic algorithm approach to the solution of minlp problems. Industrial & Engineering Chemistry Research 46(5), 1527–1537 (2007)

Appendix: Problems

Problem 1. Originally proposed by Kocis et al. (1989) [19]. It is a process synthesis model simulation. The latest effort to solve the problem appears to be that by Angira and Babu (2002) [3] who used a differential evolution algorithm. The best known fitness value is 99.245209. The problem definition is as follows:

$$\text{Minimize}_x: J(x) = 7.5x_3 + 5.5(1 - x_3) + 7x_1 + 6x_2 + 50\frac{(1-x_3)}{0.8[1-\exp(-0.4x_2)]} + 50\frac{x_3}{0.9[1-\exp(-0.5x_1)]}$$

Subject to:
$$x_1 \leq 10x_3 \qquad 0.9[1 - \exp(-0.5x_1)] - 2x_3 \leq 0$$
$$x_2 \leq 10(1 - x_3) \quad 0.8[1 - \exp(-0.4x_2)] - 2(1 - x_3) \leq 0$$
$$x_1 \in [0, \infty); x_2 \in [0, \infty); x_3 \in \{0, 1\}$$

Problem 2. This problem was originally proposed by Floudas et al. (1989) [8]. The latest effort appears to be that by Angira and Babu (2002) [3] who used a differential evolution algorithm.

$$\text{Minimize}_{x,y}:$$
$$J(x, y) = (y_1 - 1)^2 + (y_2 - 1)^2 + (y_3 - 1)^2 - \ln(y_4 + 1) + (x_1 - 1)^2 + (x_2 - 2)^2 + (x_3 - 3)^2$$

Subject to:
$$y_1 + y_2 + y_3 + x_1 + x_2 + x_3 - 5 \leq 0 \quad y_3^2 + x_1^2 + x_2^2 + x_3^2 - 5.5 \leq 0$$
$$x_1 + y_1 - 1.2 \leq 0 \qquad\qquad\qquad x_2 + y_2 - 1.8 \leq 0$$
$$x_3 + y_3 - 2.5 \leq 0 \qquad\qquad\qquad x_1 + y_4 - 1.2 \leq 0$$
$$x_1^2 + y_2^2 - 1.64 \leq 0 \qquad\qquad\quad x_3^2 + y_3^2 - 4.25 \leq 0$$
$$x_3^2 + y_2^2 - 4.64 \leq 0$$
$$x \in [0, \infty) \qquad\qquad\qquad\qquad\quad y \in \{0, 1\}$$

Problem 3. The original source of this problem is reputed to be the Proctor and Gamble Corporation, and the earliest reference appears to be Colville (1968) [5]. It has featured in many empirical studies on numerical optimization including Himmelblau (1972) [15], Hock and Schittkowski (1981) [16], Homaifar et al. (1994) [18], Michalewicz and

Fogel (2000) [22], and Coello Coello (2000) [4]. The best known solution still remains as that reported by Hock and Schittkowski (1981) [16]

Minimize$_{x,y}$: $J(x) = 5.3578547x_3^2 + 0.8356891x_1x_3 + 37.293239x_1 - 40792.141$
Subject to: $0 \leq 85.334407 + 0.0056858x_2x_3 + 0.0006262x_1x_4 - 0.0022053x_3x_5 \leq 92$
                $90 \leq 80.51249 + 0.0071317x_2x_5 + 0.0029955x_1x_2 + 0.0021813x_3^2 \leq 110$
                $20 \leq 9.300961 + 0.0047026x_3x_5 + 0.0012547x_1x_3 - 0.0019085x_3x_4 \leq 25$
                $x_1 \in [78, 102]; x_2 \in [33, 45]; x_3 \in [27, 45]; x_4 \in [27, 45]; x_5 \in [27, 45]$

Problem 4. Is a problem with 23 constraints in total out of which 2 are equality constraints. Number of discrete variables is 8 while the number of continuous variables is 9. Best known fitness so far is 67.998977252444.

Problem 5 - 11. These are NLP space mission trajectory design problems that are taken from the ESA (European space agency) GTOP (Global Optimization Trajectory Problems) Database [14]. Table 3 gives the details of these problems and the best known results so far. The table also shows the solvers that are attributed for finding the best known result. The variable bounds for some of the problems were reduced to make them more suitable for the algorithm and avoid some technical glitches in the simulator.

Table 3. Details of problems 5 to 11

Problem	m_{eq}	m	nC	nD	B_f	Best known solution attributed to
5	0	4	6	0	4.9307	Manfred Stickel (Max-Planck-Institut fuer Astronomie) by using a modified version of Particle Swarm Optimization (April 2006)
6	0	8	6	0	1,581,950	M. Schlueter, M. Gerdts (University of Birmingham, UK) by using MIDACO solver.
7	0	0	26	0	4.254	F. Biscani and D. Izzo (ESTEC Advanced Concepts Team) by using PaGMO (December 2009)
8	0	0	18	0	8.630	F. Biscani, M. Rucinski and D.Izzo (ESTEC Advanced Concepts Team) by using PaGMO, a new version of DiGMO based on the asynchronous island model (Feb 2009)
9	0	0	22	0	8.383	M. Schlueter, J. Fiala, M. Gerdts (University of Birmingham, UK) by using MIDACO solver (May 2009)
10	0	0	22	0	1.343	M. Vasile, E. Minisc (University of Glasgow) (Sep 2008)
11	0	2	12	0	18.19	T., Vinko, D., Izzo using DiGMO (March 2008)

Adaptive Abnormality Detection on ECG Signal by Utilizing FLAC Features

Jiaxing Ye[1], Takumi Kobayashi[2], Tetsuya Higuchi[1,2], and Nobuyuki Otsu[2]

[1] Department of Computer Science, University of Tsukuba, Japan
[2] National Institute of Advanced Industrial Science and Technology (AIST), Japan
{jiaxing.you,takumi.kobayashi,t-higuchi,otsu.n}@aist.go.jp

Abstract. In this paper we propose a self-adaptive algorithm for noise robust abnormality detection on ECG data. For extracting features from ECG signals, we propose a feature extraction method by characterizing the magnitude, frequency and phase information of ECG signal as well as the temporal dynamics in time and frequency domains. At abnormality detection stage, we employ the subspace method for adaptively modeling the principal pattern subspace of ECG signal in unsupervised manner. Then, we measure the dissimilarity between the test signal and the trained major pattern subspace. The atypical periods can be effectively discerned based on such dissimilarity degree. The experimental results validate the effectiveness of the proposed approach for mining abnormalities of ECG signal including promising performance, high efficiency and robust to noise.

Keywords: ECG signal processing, time-frequency analysis, local auto-correlation, self-adaptive algorithm, subspace method.

1 Introduction

The electrocardiographic (ECG) is the chart interpretation of electrical activity of the heart over time and is externally captured by skin electrodes. ECG signals may be recorded over a long timescale (i.e., few days). It cost pretty expensive labor for investigating the ECG data manually as well as huge storage space to keep it. In recent years, the signal processing techniques on ECG data have been extensively studied and contribute significantly for diagnosing cardiac diseases.

Mostly, ECG signal processing system comprises several components such as preprocessing, detection, or compression stage, several signal processing approaches have involved for specific goal accordingly. For example, low pass filter is employed for removing the baseline wander of ECG signal and band stop filter is utilized to get rid of powerline interference of Direct Current (DC) [1, 2]. Further developed analytical schemes are adopted for detecting the typical deflections of ECG signal which is called QRS [3]. For reducing the storage space of ECG signal, lots of signal analysis tools have been studied for addressing the data compression problem such as principle component analysis and wavelet transform [4]. Recently, more machine learning approaches have been introduced to enhance the performance of ECG

C.A. Coello Coello (Ed.): LION 5, LNCS 6683, pp. 218–225, 2011.
© Springer-Verlag Berlin Heidelberg 2011

processing system such as Support Vector Machines (SVM) [5], Artificial Neural Networks (ANN) and Genetic Algorithms (GA) [6].

In this work, we propose a self-adaptive framework for detecting abnormalities of ECG signal. The goal is to discern the aberrant periods from the whole ECG data. Note that the abnormalities refer to all kinds of disordered variations corresponding to prominent regular recurrent ECG signal. For extracting the dynamic properties of ECG signal, we modify our previous work of Fourier Local Auto-Correlation (FLAC) methodology [7]. The FLAC is initially proposed for representing audio signal and proved to work effective for modeling unstructured sound with wide variations. In this work we generalize FLAC features for time-series signal processing with some modifications characterizing the features of ECG signal. Unlike most time-frequency analysis schemes characterize the magnitude spectrum only; The FLAC feature extraction method is based on the complex spectrogram of ECG signal and takes advantage both of the magnitude and the phase components without losing any information. In addition, the FLAC extract the temporal dynamics on the time-frequency plane by calculating correlations of respective frequency components at adjacent positions on the ECG spectrum plane, which are favorable for describing non-stationary signals with unpredictable variations such as ECG signals.

The detection problem can be interpreted as two-set classification between normal and aberrant ECG waveforms. The subspace method was proved effective for describing and solving the abnormal detection problem [8]. To deal with complex FLAC feature vectors, we utilize the reformed type of subspace method, called complex subspace method. We utilize complex subspace method to extract the prominent characteristics of ECG signals which assumed to be normal heart beating. In addition, the noise reduction procedure is crucial for the performance of the ECG signal processing system. The subspace method can adapt to such noise variations by constructing subspace statistically in unsupervised manner. The trained "normal" subspace is sensitive to the occurrences of irregular variations in ECG data, which are reflected in deviation distances between the ECG signal feature vectors and the trained subspace. These distance values provide effective evidence for detecting aberrant periods. Benefited from the self-adaptation of subspace method, the proposed framework can achieve promising detection performance on original ECG record. The experimental results clearly validate the effectiveness of proposed methodology.

2   Architecture of the Proposed Framework

Fig.1 shows the brief flow chart of the proposed system. Firstly, the ECG signal is transformed to complex spectrum by short-time Fourier transform (STFT), and then a series of frame sequences of short-time spectra are obtained. Before extracting FLAC features, a spectral preprocessing procedure is applied based on prior-knowledge of ECG signal. Then we compute FLAC features considering neighborhoods along both time and Mel-frequency coordinate to extract the temporal dynamic information. At detection stage, all FLAC feature vectors extracted from ECG signal are utilized to train the "main pattern" subspace. The regular recurrent ECG signals are assumed to produce pretty low deviation distance to the trained subspace. Conversely, the aberrant variations of ECG signal would exhibit distinct deviation. In other word, such deviation distance values manifest the dissimilarity degree between input signal and the trained major pattern subspace. A threshold on the deviation distance values can be determined to efficiently detect the atypical sections of ECG signal.

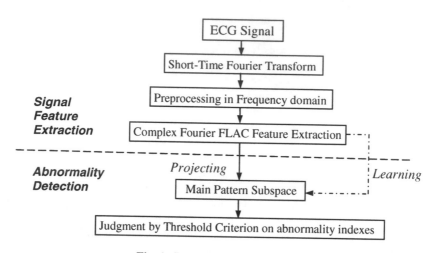

Fig. 1. Chart flow of proposed system

2.1 Preprocessing in Frequency Domain

In this part, the prior-knowledge of ECG signal is employed to enhance the efficiency and the performance of detection. The maximum frequency of ECG signal is generally fixed to 70Hz. We remove the frequency components from complex spectrogram which exceed that maximum. Owing to reducing the computation time, we compress the frequency component by adding up several neighbor frequency entries' values as the compressed frequency feature. The length of summation is optimized and fixed to 8 according to experimental results. With such compression procedure, we can accelerate the algorithm as well as achieve high performance according to experimental results.

2.2 Local Auto-correlation on Complex Fourier Values (FLAC) for ECG

In [7] we proposed FLAC, a methodology for extracting acoustic features by characterizing joint temporal dynamic features in time and frequency domains as well as taking advantage both of magnitude and phase information. We introduce the mechanism of FLAC features for time-series data explicitly in this section.

Let time and frequency be denoted by t and v, respectively. Note that the frequency information corresponds to the already finished preprocessing procedure. $f(r)$ denotes complex spectrogram at position $r=(t,v)$ on such two-dimensional plane. We employ the local auto-correlation function to extract the features based on complex spectrogram:

$$x_{t,v}(a) = f^*(r)f(r+a), \qquad (1)$$

where a is a displacement vector indicating local neighborhoods and f^* denotes the complex conjugate. We limit a within 2x2 region on time-frequency plane as the local neighborhoods are assumed to be highly correlated. The combination patterns of r and $r+a$ are shown in Fig. 3. FLAC can extract phase information as follows. These patterns enable us to extract plenty of in-domain and cross-domain temporal dynamic features. The complex values $f(r)$ and $f(r+a)$ are represented by $Ae^{-j\theta}$ and $Be^{-j\varphi}$ where A and B are magnitudes and θ and φ are phases. Then the complex FLAC feature is described by:

$$x_{t,v} = Ae^{-i\theta}Be^{i\varphi} = ABe^{i(\varphi-\theta)}. \qquad (2)$$

This is based on multiplication of magnitudes and difference of phases. Such correlation of complex values provides joint features of magnitudes as well as those of phases which are robust to phase shift by considering the difference. Note that No.1 pattern produces ordinal magnitude-based feature like power spectrum, while, No.2~4 provide dynamic features in time and frequency domains. In the end we concatenate all pattern feature vectors to the (long) feature vector at each frame (t) as feature vector which effective for representing ECG signals. Comparing to the acoustic FLAC features, the FLAC feature in this work removed the Mel filter bank and maintain the whole spectral information.

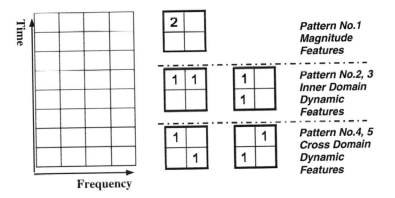

Fig. 3. Time-frequency local auto-correlation feature patterns in FLAC features

2.3 Complex Subspace Method

There are two goals for adopting subspace method [8] in our framework. First is to "learn" the "mainstream" pattern of ECG signal in unsupervised manner and model those features by a subspace statistically. Secondly, we investigate the ECG signal feature vectors by measuring the deviation distances (dissimilarity degrees) to the trained subspace and extract only "non-mainstream" periods which present distinct deviation distances. Such distance values provide effective evidences to discern atypical periods. This section provides a thorough explanation to the procedure. Note that for noise signals, subspace method can incorporate the characteristics of noise adaptively into the trained subspace.

Let $x_i(i=1,\cdots,n) \in C^M$ denote M-dimensional complex feature vectors. We calculate eigenvalues $\Lambda = diag(\lambda_1,\cdots,\lambda_M)$ and eigenvectors $U = [u_1,\cdots,u_M], u \in C^M$ by:

$$R_{Cov_x} U = U \Lambda, \qquad R_{Cov_x} \Box \overset{n}{\underset{i=1}{E}}\{x_i x_i^*\}, \tag{3}$$

where $x_i^*, i \in (1,\cdots,n)$ are conjugate transpose of x_i and R_{Cov} refers to the covariance matrix of feature vectors. A phase shift of x_i is denoted by $x_i e^{j\theta}$ and the robustness of complex subspace method to phase shift can be simply proved as:

$$\overset{n}{\underset{i=1}{E}}\{(x_i e^{j\theta})\cdot(x_i e^{j\theta})^*\} = \overset{n}{\underset{i=1}{E}}\{x_i x_i^* e^{j\theta} e^{-j\theta}\} = \overset{n}{\underset{i=1}{E}}\{x_i x_i^*\} \tag{4}$$

We sort eigenvectors by eigenvalues in decreasing order. These eigenvalues denote the significance of corresponding eigenvectors for expressing the time-series data. The contribution rate of η_K is defined as:

$$\eta_K \Box \sum_{i=1}^{K} \lambda_i / \sum_{i=1}^{M} \lambda_i. \tag{5}$$

We keep first K eigenvectors $U_K = [u_1,\cdots,u_K]$ with contribution rate of $\eta_K > 0.99$ to express the main patterns of ECG data. We represent main pattern subspace of ECG signal as S and its projection operator as $P = U_K U_K^*$. The project operator onto the ortho-complement subspace of S is denoted by $P_\perp = I_M - P$. The deviation distance between signal feature vector and S can be measured as:

$$d^2 = \|P_\perp x\|^2 = x^*(I_M - U_K U_K^*)x = x^* x - x^* U_K U_K^* x. \tag{6}$$

The robustness to phase shift is also achieved in this distance:

$$d^2 = \|P_\perp(x\cdot e^{j\theta})\|^2 = x^* e^{-j\theta} e^{j\theta} x - x^* U_K e^{-j\theta} e^{j\theta} U_K^* x$$
$$= x^* x - x^* U_K U_K^* x. \tag{7}$$

The following mathematical forms manifest that only the abnormal ECG feature vectors lead to distinct deviation distances while projecting them onto the trained subspace.

$$\text{Suppose: } x = x_1^{(N)} + \cdots + x_n^{(N)} + x^{(A)}.$$
$$(\textit{N: Normal ECG, A: Abnormal ECG}) \tag{9}$$

$$\text{Then: } d^2 = \|P_\perp x\|^2 = \|P_\perp(x_1^{(N)} + \cdots + x_n^{(N)}) + P_\perp x^{(S)}\|^2 \tag{10}$$
$$= \|0 + P_\perp x^{(S)}\|^2.$$

Subsequently, we define the sequence of d as abnormality indexes for representing dissimilarity degrees between input feature and regular ECG signal. According to experimental tests, the threshold is determined based on d as:

$$Threshold = mean(d) + std(d). \tag{11}$$

3 Experiments

To validate the proposed framework, we conducted abnormality detection experiments on ECG signal. The ECG recordings were extracted from MIT-BIH arrhythmia ECG database [9]. The MIT-BIH arrhythmia ECG database consists of 48 half hour excerpts ambulatory ECG recordings, which were obtained from 47 subjects, including men and women of various ages. The ECG waveforms were digitized at 360 samples per second. Meanwhile, the reference annotations were utilized for evaluating the detection performance. Note that the annotations presented all detailed information of abnormality types such as premature ventricular contraction (PVC), premature atrial contraction (PAC), etc. In our case, we involve the rough classification between normal and abnormal only, so we reformed the labels two to normal and abnormal sections.

The length of analysis window in short-time Fourier transform was set to 512 points of ECG data with 128 points overlapped. The contribution rate was fixed to 0.99 for constructing main pattern subspace of ECG signal. For compressing the frequency components, we chose the summation of 8 neighbor entries as new feature in frequency domain.

For a more intelligible understanding of the proposed scheme, we provide the illustration of abnormality detection on 200-second ECG data. Note that the test ECG signal was the original data without any separated preprocessing steps, such as baseline wander removal or powerline interference suppression. The detection result by the proposed methodology was depicted in Fig.3. Meanwhile, the method achieved pretty high efficiency that it cost only 0.0017 second for detecting one second's data points (360) of ECG signal by utilizing Core™2 Quad Processor Q6600 2.4GHz.

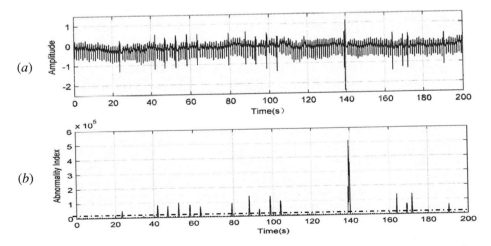

Fig. 3. Abnormality Detection on clean ECG signal. (*a*). ECG data waveform. (*b*). Abnormality indexes detected by proposed framework with threshold (dashed line)

In the case of real-world ECG signal processing, noise always existed. For validating self-adaptation characteristic of the proposed scheme to the noise in ECG data, we conducted abnormality detection experiments on noisy ECG signal. The source data of ECG was the same as in the previous experiment and distorted manually by additive Gaussian noise with 20dB SNR. Fig.4 (*a*) plotted the waveform of ECG signal. According to the detection result presented in Fig.4 (*b*), the noise can be effectively encoded by subspace method and the detection performance was maintained stably.

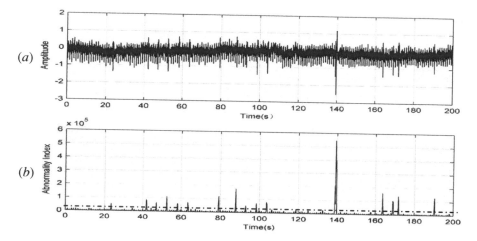

Fig. 4. Abnormality Detection on Noisy ECG signal. (*a*). Noisy ECG data waveform. (*b*). Abnormality indexes detected by proposed framework with threshold (dashed line).

Subsequently, we conducted extensive experiments on ECG signal extracted from the MIT-BIH arrhythmia ECG database. We selected six sequences of ECG data and each contains 10 minutes' ECG data. The experiments were conducted twice as on the clean (original) ECG signal and the noisy ECG at 20dB SNR. The criterions of Recall rate (RC) and False Alarm rate (FA) were adopted to evaluate the performance of the proposed detection approach. The detailed results of experiments are presented in Table.1.

According to all the experimental results listed above, the effectiveness of the proposed detection algorithm was validated as well as the self-adaptation characteristic to noise interference. In addition, our approach doesn't need any preprocessing and present pretty high efficiency. It could be utilized as an automatic filter to target the candidates for further examination.

Tabel 1. Experimental result of abnormalities detection by proposed scheme (%)

	No. 213		No. 214		No. 215		No. 217		No. 221		No. 223	
	RC	*FA*	*RC*	*FA*	*RC*	*FA*	*RC*	*FA*	*RC*	*FA*	*RC*	*FA*
Clean ECG	95.14	4.41	88.16	3.00	88.76	13.84	100	15.53	99.05	4.63	90.48	8.47
Noisy ECG	93.75	5.47	86.18	3.03	87.16	18.40	100	18.12	99.05	4.66	90.37	11.48

4 Conclusions

In this paper, we proposed a novel adaptive abnormality detection scheme for ECG signal processing. For extracting the dynamic properties from ECG signal, we employed the method of computing local auto-correlation on complex Fourier values (FLAC) to capture the temporal dynamics in time and frequency domain as well as to take advantages both of the magnitude and the phase information. To detect aberrant sections as well as to cope with the FLAC complex feature vectors, we employed the complex subspace method in unsupervised manner. The experimental results demonstrate that the proposed methodology can effectively adapt to noise variations with high efficiency and promising detection performance. The proposed scheme can also be generalized for other applications of abnormality mining in addition to ECG signal.

References

1. Rahman, M.Z.U., Shaik, R.A., Rama Koti Reddy, D.V.: Noise Cancellation in ECG Signals using Computationally Simplified Adaptive Filtering Techniques: Application to Biotelemetry. Int. J. Signal Processing 3(5), 120–131 (2009)
2. Blanco-Velasco, M., Weng, B., Barner, K.E.: ECG signal denoising and baseline wander correction based on the empirical mode decomposition. Computers in Biology and Medicine 38(1), 1–13 (2008)
3. Chouhan, V.S., Mehta, S.S.: Threshold-based Detection of P and T-wave in ECG using New Feature Signal. Int. J. Computer Science and Network Security 8(2), 144–153 (2008)
4. Mohammadpour, T.I., Mollaei, M.R.K.: ECG Compression with Thresholding of 2-D Wavelet Transform Coefficients and Run Length Coding. Euro. J. Scientific Research 27(2), 248–257 (2009)
5. Mehta, S.S., Lingayat, N.S.: Support Vector Machine for Cardiac Beat Detection in Single Lead Electrocardiogram. IAENG Int. J. of Applied Mathematics 36(2), 4–11 (2007)
6. Karpagachelvi, S., Arthanari, M., Sivakumar, M.: ECG Feature Extraction Techniques - A Survey Approach. Int. J. Computer Science and Information Security 8(1), 76–80 (2010)
7. Jiaxing.Ye, T.: Kobayashi and T. Higuchi: Audio-based Sports Highlight Detection by Fourier Local Auto-Correlations. In: 10th INTERSPEECH 2010, International Speech Communication Association, pp. 2198–2201 (2010)
8. Nanri, T., Otsu, N.: Unsupervised abnormality detection in video surveillance. In: IAPR Conference on Machine Vision Applications, pp. 574–577 (2005)
9. MIT-BIH arrhythmia ECG database,
   http://www.physionet.org/physiobank/database/mitdb

Gravitational Interactions Optimization

Juan J. Flores[1], Rodrigo López[1], and Julio Barrera[2]

[1] Universidad Michoacana de San Nicolás de Hidalgo
División de Estudios de Posgrado, Facultad de Ingeniería Eléctrica
[2] CINVESTAV-IPN
Departamento de Computación
Evolutionary Computation Group
Av. IPN No. 2508, Col. San Pedro Zacatenco
México, D.F. 07360, Mexico
juanf@umich.mx, rlopez@faraday.fie.umich.mx, julio.barrera@gmail.com

Abstract. Evolutionary computation is inspired by nature in order to formulate metaheuristics capable to optimize several kinds of problems. A family of algorithms has emerged based on this idea; e.g. genetic algorithms, evolutionary strategies, particle swarm optimization (PSO), ant colony optimization (ACO), etc. In this paper we show a population-based metaheuristic inspired on the gravitational forces produced by the interaction of the masses of a set of bodies. We explored the physics knowledge in order to find useful analogies to design an optimization metaheuristic. The proposed algorithm is capable to find the optima of unimodal and multimodal functions commonly used to benchmark evolutionary algorithms. We show that the proposed algorithm (Gravitational Interactions Optimization - GIO) works and outperforms PSO with niches in both cases. Our algorithm does not depend on a radius parameter and does not need to use niches to solve multimodal problems. We compare GIO with other metaheuristics with respect to the mean number of evaluations needed to find the optima.

Keywords: Optimization, gravitational interactions, evolutionary computation, metaheuristic.

1 Introduction

Multimodal optimization problems deal with objective functions that commonly contain more than one global optima and several local optima. In order to find all the global optima in multimodal problems with classical methods, one typically runs a given method several times with different starting points, expecting to find all the global optima. However, these techniques do not guarantee the location of all optima. Therefore, this kind of techniques are not the best way to explore multimodal functions with complex and large search spaces. In the evolutionary computation literature exists a variety of metaheuristics challenging the typical problems of classical optimization. E.g. in particle swarm optimization with niches the best particle makes a niche with all particles within a radius

C.A. Coello Coello (Ed.): LION 5, LNCS 6683, pp. 226–237, 2011.

r, until the niche is full; it then selects the next best no niched particle and its closest particles to form the second niche; the process repeats until all particles are assigned to a niche. Objective function stretching, introduced by Parsopolous [9], [10] is another algorithm whose strategy is to modify the fitness landscape in order to remove local optima and avoid the premature convergence in PSO. In a minimization problem, a possible local minimum is stretched to overcome a local maximum allowing to explore other sections of the search space identifying new solutions. GSA introduced by Rashedi [11], is a gravitational memory-less (does not include a cognitive component in the model) metaheuristic capable to find only one global optima in unimodal and multimodal problems with more than one global optima, where a heavier mass means a better solution and the gravitational constant G is used to adjust the accuracy search.

The rest of the paper is organized as follows: Section 2 compares GIO with GSA and CSS, two other metaheuristics very similar to our proposal. Section 3 reviews Newton's universal gravitation law. Section 4 proposes Gravitational Interactions Optimization, the main contribution of this paper. Section 5 presents the experimental framework and the obtained results. Finally, Section 6 presents our conclusions.

2   Review GSA GIO and CSS

GIO has similarities with other two nature-inspired algorithms: GSA (Gravitational Search Algorithm), inspired on gravitational interactions and CSS Charge Search System (CSS) inspired on electrostatic dynamics laws. Our work is very similar to theirs, but since we have publications from around the same time [1], [3], and have not had any personal communication with their respective authors, we can state that this work is independent from the works of CSS and GSA.

First GSA and CSS are very similar to GIO, assigning masses and charges respectively to bodies, according to the fitness function in the place the body is located, in order to determine the evolution.

The gravitational constant G in GSA decreases exponentially with time, using a decay constant α. We think that is not a good idea because this function is not autoadaptative, it makes the method dependent on one more parameter. In an attempt to increase its exploration capabilities, when GSA is determining the total force exerted to a body, it weights each component (the force exerted by each other body) by a random number. This situation, in the worst case, destroys the underlying metaphor, i.e. the gravitational interaction.

Another point is that GSA uses Kbest agents in order to minimize computing time, although the complexity of the algorithm is not reduced. On the other hand, we allow all masses (agents) to interact with each other.

The CSS algorithm tries to imitate the electrostatic dynamics in order to optimize unimodal and multimodal functions, assigning charge to the particles in a similar way as GSA and GIO assign masses to the bodies. We think that the high level of detail of the charges of the particles in CSS is unnecessary, and the parameter a is very large for some functions; the estimation of this parameter could be arbitrary. We think that CSS lost the sense of electromagnetism dynamics when they assign binary flags to determine the direction of the attraction of the bodies. All charges are positive, and still attract each other, departing from the electrostatic metaphor and making it look more like a gravitational one.

Another important difference is that both CSS and GSA aim to locate the global optimum for multimodal functions, while GIO's main interest is to determine all local and global optima for multimodal functions.

In our work we explore the properties of gravitational interactions in order to make a useful metaheuristic to find optima in unimodal and multimodal problems.

3   Newton's Law of Universal Gravitation

The attraction force of two particles is proportional to their masses and inversely proportional to their distance. The Law of Universal Gravitation was proposed by Isaac Newton [8]. This law is stated in Definition 1.

DEFINITION 1 *The force between any two particles having masses m_1 and m_2, separated by a distance r, is an attraction acting along the line joining the particles and has G magnitude shown in Equation (1).*

$$F = G\frac{m_1 m_2}{r^2} \tag{1}$$

where G is a universal gravitational constant.

The forces between two particles with mass are an action-reaction pair. Two particles with masses m_1 and m_2 exert attracting forces F_{12} and F_{21} towards each other whose magnitudes are equal but their directions are opposed.

The gravitational constant G is an empirical physical constant involved in the computation of the gravitational attraction between particles with masses, which when determined by the maximum deflection method [12] yields.

$$G = 6.673 \times 10^{-11} N(m/kg)^2 \tag{2}$$

The gravitational force is extremely weak compared to other fundamental forces; e.g. the electromagnetic force is 39 orders of magnitude greater than the gravity force.

Newton's law of universal gravitation can be written in vectorial notation, which considers both: The force of the masses and the direction of each force. The vectorial notation is shown in Equation (3).

$$F_{12} = -G\frac{m_1 m_2}{|r_{12}|^2}\hat{r}_{12} \tag{3}$$

where F_{12} is the force exerted by m_1 on m_2, G is the gravitational constant, m_1 and m_2 are the masses of the particles, $|r_{12}|$ is the euclidean distance between particles m_1 and m_2, $\hat{r}_{12}$ is the unit vector, defined as $\frac{r_2-r_1}{|r_2-r_1|}$, and r_1 and r_2 are the locations of particles m_1 and m_2 (see Figure 1).

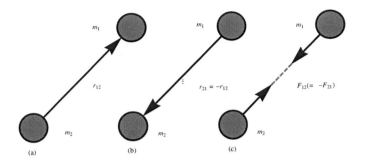

Fig. 1. (a) The force exerted on m_2 (by m_1), F_{21}, is directed opposite to the displacement, r_{12}, of m_2 from m_1. (b) The force exerted on m_1 (by m_2), F_{12}, is directed opposite to the displacement, r_{21}, of m_1 from m_2. (c) $F_{21} = -F_{12}$, the forces being an action-reaction pair.

4 Gravitational Interactions Optimization

In order to find one or more optima there exists a large variety of evolutionary algorithms, e.g. genetic algorithms (GA) [4], evolutionary strategies (ES) [5], ant colony optimization (ACO) [2], particle swarm optimization (PSO) [6], electrostatic PSO (EPSO) [1], etc. There exist works related to the design of metaheuristics that take into account distances in order to determine the cluster membership of the particles, computing and maximizing a ratio for all particles in the swarm with respect to the particle to be updated, e.g. FER-PSO [7]. We propose a Gravitational Interaction Optimization metaheuristic (GIO) capable of solving optimization problems. The motivation of the design of this metaheuristic is to find useful properties and anolgies that can relate optimization problems with Newton's gravitational theory. In the approach presented in this paper, we abduct the interactions exhibited by a set of bodies and use them to guide the search for the global optimum in an optimization problem.

4.1 Gravitational Interactions for Unimodal Optimization

GIO is a population-based metaheuristic where a set of bodies are initially dispersed along the search space with a uniform random distribution. The fitness of bodies located on the search space are mapped as masses in a Gravitational field where the solutions are evolved. Each body stores its current position B and possibly its best position so far B^b, according to the fitness function. Bodies are allowed to interact in a synchronous discrete manner for a number of epochs. The body interactions follow Newton's gravitational law and move each body to a new location in such way that the whole population tends to reach the global optimum (or multiple local optima for multi-modal problems).

The fitness function is a mapping that transforms a vector $X = (x_1, x_2, \ldots, x_n)$ to a scalar $f(X)$. This mapping associates the fitness value $f(X)$ to each location

$X = (x_1 \cdots x_n)$ of the search space. We assign a body B to every location X in the search space where an individual of the population is found. Body B is assigned a mass, whose magnitude is a function of the fitness of its location.

Newton's law of universal gravitation describes the attraction forces that exist between two punctual bodies with masses (described in vectorial form in 3). Substituting we obtain Equation (4).

$$F_{ij} = \frac{M\left(f(B_i)\right) \cdot M\left(f(B_j)\right)}{|B_i - B_j|^2} \hat{B}_{ij} \tag{4}$$

where B_i is the position of the ith body and B_j is the jth body that contributes exerting a force on the mass B_i; $|B_i - B_j|$ is the euclidean distance and B_{ij} is the unit vector between bodies B_i and B_j; $f(B_i)$ is the fitness of body B_i, M is the mapping function that associates the fitness value f of domain $\{x : x \in \Re\}$ to a mass of codomain $\{y : y \in (0, 1]\}$ for each position of the body B_i. This mapping is computed using Equation (5).

$$M(f(B_i)) = \left(\frac{f(B_i) - \mathbf{min}f(B)}{\mathbf{max}f(B) - \mathbf{min}f(B)}(1 - mapMin) + mapMin\right)^2 \tag{5}$$

where $\mathbf{min}f(B)$ is the minimum fitness value of the positions of the bodies so far, $\mathbf{max}f(B)$ is the maximum fitness value of the positions so far. $mapMin$ is a constant with a small positive value near zero, such that $(1 - mapMin)$ reescales the fitness value $f(B_i)$ to a mass in the interval $[mapMin, 1)$. The result is squared to emphasize the best and worst fitnesses.

One characteristic of the proposed method is the full interaction; i.e each body B_i interacts with every other body B_j through their masses. Interactions contribute to their displacement, according to the resultant force. Equation (6) computes the resultant force exerted on body B_i by the bodies B_j.

$$F_i = \sum_{j=1}^{n} \frac{M\left(f(B_i)\right) \cdot M\left(f(B_j^b)\right)}{|B_i - B_j^b|^2} \hat{B_i B_j^b} \tag{6}$$

where F_i is the resultant force of the sum of all vector forces between $M(B_i)$ and $M(B_j^b)$, $|B_i - B_j^b|$ is the Euclidean distance between the current positions of body B_i and the best position so far of the body B_j. In order to avoid numerical errors we compute the force between masses $M(B_i)$ and $M(B_j^b)$ only if $|B_i - B_j| \geq \times 10^{-5}$ (if the distance is smaller than that, we suppose both bodies collided already and are located in the same place; we are assuming punctual masses), $\hat{B_i B_j^b}$ is the unit vector that directs the force. In order to estimate a displacement that could enhance the solution of particle B_i, it is neccesary to solve Equation (4) for B_j. Assuming that we want to find a location of the body B_k with $M\left(f(B_k)\right) = 1$, B_k is computed using Equation (7).

$$B_k = \sqrt{\frac{M(f(B_i))}{|F_i|}} \hat{F}_i \tag{7}$$

To update the position of the bodies we use Equations (8) and (9).

$$V_{t+1} = \chi \left(V + R \cdot C \cdot B_k \right) \tag{8}$$

$$B_{t+1} = B + V_{t+1} \tag{9}$$

where V is the current velocity of B_i, R is a random real number generated in the range of $[0, 1)$ and is multiplied by the gravitational interaction coefficient C, in order to expect random exploration distances with mean $\mu \approx 1$, we set $C = 2.01$, this displacement is constrained multiplying by a constant with a value of 0.86, in order to ensure convergence. B_k is the main displacement computed by Equation (7).

Using Newton's law $F = ma$, we can compute velocity and from there displacement, as in GSA [11]. Using this scheme though, when bodies are far apart, forces are small, and the resulting displacement is also small; when the method is converging, bodies are closer to each other, producing larger forces, therefore larger displacements. This leads to a divergent or at least non-convergent behavior. GSA solves this problem by assuming G as a linearly decreasing function of time. As in classical mechanics, we consider G a constant and use a heuristic solution to this problem: where should a body of unitary mass be located to produce the same resulting force in B_i? We use that location as B_i's new location.

Using this heuristic, when bodies are far apart from each other, forces are small, as they would be produced by a unitary mass located far away. When the method is converging and masses are close together, resulting forces are larger, as produced by a close unitary mass, resulting in small displacements. This heuristic leads to very a convenient convergence scheme, where exploration takes place at the beginning and exploitation at end of the process.

The complete GIO algorithm is described the Algorithms 1, 2, and 3. Algorithm 1 computes the the total force exerted by the masses $M(f(B_j))$ and $M(f(B_i))$; in order to prevent premature convergence and division by 0, we compute only those pairs of bodies with a distance greater than ϵ. Algorithm 2 computes the velocities of the bodies, receives the bodies and computes the resultant force that attracts the mass assigned to B_i. In order to prevent a division by 0 we compute the distance only if $|Ftotal| > 0$, the new velocity is computed by Equation (8), and finally we update the velocity associated to B_i. Algorithm 3 computes the new positions B of each iteration t; this algorithm takes as parameters the search range, the number of bodies $nBodies$, and the maximum number of iterations $maxIter$. The algorithm computes the velocities with $computeVelocities(bodies)$ using Algorithm 2, and updates the their positions with $updatePosition()$, which implements Equation (9), $limitPositions()$ limits the positions of the bodies to the search space defined by the search range; $updateFitness()$ updates the fitness according to the new positions of the bodies; finally, we update the best position so far with $updateB^b()$.

This scheme develops good results for unimodal problems. The performace results of the algorithm presented in this section are presented in Section 5.

Algorithm 1. computeFtotal(index)

1: $i \leftarrow index$
2: $Ftotal \leftarrow 0$
3: **for** $j \leftarrow 1$ to $nBodies$ **do**
4:     **if** $distance(B_i, B_j^b) > \epsilon$ **then**
5:         $Ftotal \leftarrow Ftotal + \hat{B_{i,j}^b} M(f(B_i)) M(f(B_j^b))/distance(B_i, B_j^b)^2$
6:     **end if**
7: **end for**
8: **return** $Ftotal$

Algorithm 2. computeVelocities(bodies)

1: **for** $i \leftarrow 1$ to $nBodies$ **do**
2:     $Ftotal \leftarrow computeFtotal(i)$
3:     **if** $|Ftotal| > 0$ **then**
4:         $distance \leftarrow \sqrt{M(f(B_i))}/|Ftotal|$
5:     **else**
6:         $distance \leftarrow 0$
7:     **end if**
8:     $V_{new} \leftarrow \chi(V + R \cdot C \cdot distance \cdot \hat{Ftotal})$
9:     $updateVelocity(B_i, V_{new})$
10: **end for**
11: **return** $Ftotal$

Algorithm 3. GIO(ranges, nBodies,maxIter)

1: $bodies \leftarrow initializeParticles(nBodies, ranges)$
2: **for** $t \leftarrow 0$ to $maxIter$ **do**
3:     $computeVelocities(bodies)$
4:     $limitVelocity()$
5:     $updatePosition()$
6:     $limitPosition()$
7:     $updateFitness()$
8:     $updateB^b()$
9: **end for**

4.2   Gravitational Interactions for Multimodal Optimization

In the previous Subsection we showed the basic steps of the gravitational interactions metaheuristic. This scheme works well for unimodal problems. For multimodal problems it is necessary to add a cognitive component analogous to the one used in PSO [6]; the cognitive component is a constant that gives a weight to each body's memory. The new positions of the bodies are computed in order to find more than one optima with Equations (10) and (11).

Adding the cognitive component to Equation (8) and using the constriction factor χ (Equation (12)) [6], makes the new Equation (10) capable to find more than one optimum in multimodal problems. The effect of this component is to make the local search more robust, restricting the bodies to local search, unless the gravitational forces of a cluster of masses overcome the force exerted by its cognitive component.

$$V_{new} = \chi \left(V + C_1 \cdot R_1 \cdot (B^b - B) + C_2 \cdot R_2 \cdot B_k \right) \tag{10}$$

$$B_{new} = B + V_{new} \tag{11}$$

where, analogous to PSO, C_1 and C_2 are the cognitive and the gravitational interaction constants, R_1 and R_2 are real random numbers variables in the $[0, 1)$ range and χ is the inertia constraint (Proposed by Clerk [6]). The inertia constraint is used to avoid the bodies to explore out of the search space computed by Equation (12).

$$\chi = \frac{2\kappa}{|2 - \phi - \sqrt{\phi^2 - 4\phi}|} \tag{12}$$

where $\phi = C_1 + C_2 > 4$, κ is an arbitrary value in the range of $(0, 1]$ [6]. In our algorithm we set $C_1 = C_2 = 2.01$. The constriction factor in our algorithm contributes to convergence through the iterations.

To make multimodal Gravitational Interactions Algorithm (Algorithms 1, 2, and 3, described in the previous subsection) work for multimodal optimization problems, we replace line 8 of Algorithm 2 by Equation 10.

5   Experiments

In order to test the performance of the Gravitational Interactions Optimization algorithm for unimodal and multimodal functions, we tested both versions with some functions commonly used to measure the performance of different kinds of metaheuristics.

5.1   Test Functions

We show the performance of unimodal and multimodal Gravitational Interactions Optimization with 3 unimodal and 4 multimodal functions. The test functions used are shown in the Table 1.

For unimodal optimization we used the functions in Figure 2; $U1$ is the Goldstein and Price function (Figure 2(a)), $U2$ is the Booth function (Figure 2(b)), and $U3$ is the 4 variable Colville Function. For multimodal optimization we used the functions in Figure 3; $M1$ is the Branin's RCOS Function with 3 global optima (Figure 3(a)), $M2$ is the 6 global maximum univariable Deb's function (Figure 3(b)), $M3$ is Himmelblau's function with 4 global optima (Figure 3(c)), $M4$ is the Six-Hump cammelback function with 2 global optima and 4 local optima (Figure 3(d)).

Table 1. Test functions used for our experiments

Unimodal Test Functions		
U1	$U1 = [1 + (1 + (x + y + 1)^2)(19 - 14x + 3y^2 + 6xy + 3y^2)] \cdot$ $[(30 + (2x - 3y)^2)(18 - 32x + 12x^2 + 48y - 36xy + 27y^2)]$	$-2 \leq x, y \leq 2$
U2	$U2 = (x + 2y - 7)^2 + (2x + y - 5)^2$	$-10 \leq x, y \leq 10$
U3	$U3 = -1100 \cdot (w^2 - x)^2 + (w - 1)^2 + (y - 1)^2 + 90 \cdot (y^2 - z)^2 + \cdot$ $10.1 \cdot ((x - 1)^2 + (z - 1)^2) + 19.8 \cdot (x^{-1}) \cdot (z - 1)$	$-10 \leq w, x, y, z \leq 10$
Multimodal Test Functions		
M1	$M1 = -\left((y - \frac{5.1x^2}{4\pi^2} + \frac{5x}{\pi} - 6)^2 + 10(1 - \frac{1}{8\pi})Cos(x) + 10\right)$	$-5 \leq x \leq 10$ $0 \leq y \leq 15$
M2	$M2 = Sin(5\pi x)^6$	$-0 \leq x \leq 1$
M3	$M3 = -(x^2 + y - 11)^2 - (x + y^2 - 7)^2$	$-6 \leq x, y \leq 6$
M4	$M4 = -4\left((4 - 2.1x^2 + \frac{x^4}{3})x^2 + xy + (-4 + 4y^2)y^2\right)$	$-1.9 \leq x \leq 1.9$ $-1.1 \leq x \leq 1.1$

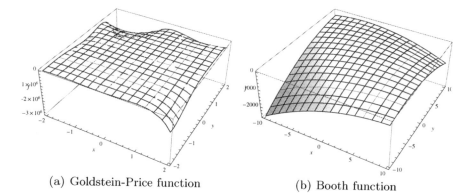

(a) Goldstein-Price function   (b) Booth function

Fig. 2. Fitness landscape of two test functions with one optima used to measure the performance of Unimodal Gravitational Interactions

5.2 Results

In our experiments we consider $\epsilon = 1 \times 10^{-3}$ to be an acceptable error to determine if the solution obtained had reached the optimum. We used 100 bodies for a maximum of 1000 iterations, we used as stop condition the inability of all the bodies to enhance their fitness memory solutions by 1×10^{-4}, or when the algorithm found all the optima. Each experiment was repeated 30 times.

PSO with niches requires two extra parameters: the radius r, and the maximum number of particles per niche $nMax$. To solve $M1$ we set $r = 0.5$ and $nMax = 50$, to solve $M2$ we set $r = 0.1$ and $nMax = 15$, to solve $M3$ we set $r = 0.5$ and $nMax = 30$, and $M4$ with $r = 0.5$ and $nMax = 25$.

The performance of Gravitational Interaction Optimization (GIO) is compared with Particle Swarm Optimization with niches (NPSO) in Table 2; this table includes the mean and the standard deviation of evaluations required to

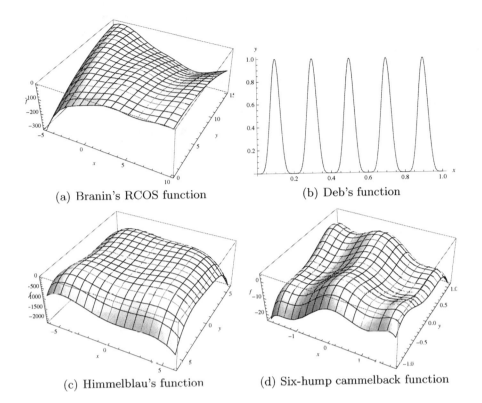

(a) Branin's RCOS function

(b) Deb's function

(c) Himmelblau's function

(d) Six-hump cammelback function

Fig. 3. Fitness landscape of multimodal test functions used in our experiments

find all the global optima (column **Evaluations**) and the percentage of successes (column **Success**) to finding all the optima.

Table 2. Results of our experiments

Functions	PSO			GIO Unimodal		
	Evaluations		Success	**Evaluations**		Success
	μ	σ		μ	σ	
$U1$	1,394.44	399.22	20%	5,653.33	711.838	100%
$U2$	1,130.77	330.11	60%	6,057.55	3,984.54	70%
$U3$	764.00	777.75	83%	530.00	208.69	100%
	NPSO			**GIO Multimodal**		
	Evaluations			**Evaluations**		
	μ	σ		μ	σ	
$M1$	2,529.17	764.13	80%	2,803.33	972.90	100%
$M2$	276.66	81.72	100%	390.00	88.44	100%
$M3$	3,400.00	0.00	00.3%	2,323.33	288.496	100%
$M4$	1,136.67	303.41	100%	1,600.00	501.721	100%

The obtained results show that Unimodal and Multimodal Gravitational Interactions have a higher probability to converge to global optima than PSO and PSO with niches with a similar number of evaluations. GIO gets to the correct results avoiding premature convergence present in PSO. We proved the GIO algorithm replacing the Equation 7 by the acceleration Equation proposed in [11] respect the Newton's law gravity multiplied by the gravity constant showed in Equation 2. The results were not better than GIO but we think that Equation (7) could give us a more accurate behavior of the gravitational constant G through the iterations.

6 Conclusions

We presented a new heuristic, GIO, which has proven to be more reliable than PSO. GIO needs no additional parameters like the radius and the maximum number of particles in a niche used in PSO with niches. To solve problems with high dimensions using PSO the radius is determined by trial and error, because we can not plot the objective function and make a visual analysis.

The same algorithm, GIO, is used for unimodal and multimodal cases. When used in its general form. (i.e. including the cognitive component), GIO solves both cases without the need of any a-priori information. Adding the cognitive component allows us to solve both, unimodal and multimodal optimization problems, while GSA can only solve unimodal problems.

Furthermore, GIO has proven to find all optima in a multimodal problem, while GSA can only determine one of them.

References

1. Barrera, J., Coello Coello, C.A.: A particle swarm optimization method for multimodal optimization based on electrostatic interaction. In: Aguirre, A.H., Borja, R.M., Garciá, C.A.R. (eds.) MICAI 2009. LNCS, vol. 5845, pp. 622–632. Springer, Heidelberg (2009)
2. Colorni, A., Dorigo, M., Maniezzo, V.: Distributed optimization by ant colonies. In: Proceedings of the Parallel Problem Solving from Nature Conference. Elsevier Publishing, Amsterdam (1992)
3. Flores, J.J., Farías, R.L., Barrera, J.: Particle swarm optimization with gravitational interactions for multimodal and unimodal problems. In: Sidorov, G., Hernández Aguirre, A., Reyes García, C.A. (eds.) MICAI 2010, Part II. LNCS, vol. 6438, pp. 361–370. Springer, Heidelberg (2010)
4. Goldberg, D.E.: Genetic Algorithms in Search, Optimization, and Machine Learning, 1st edn. Addison-Wesley Professional, Reading (1989)
5. Ingo, R.: Evolutionsstrategie 1994. PhD thesis, Technische Universität Berlin (1994)
6. Kennedy, J., Eberhart, R.: Swarm Intelligence. In: Evolutionary Computation. Morgan Kaufmann Publisher, San Francisco (2001)
7. Li, X.: A multimodal particle swarm optimizer based on fitness euclidean-distance ratio. In: Proceedings of the 9th Annual Conference on Genetic and Evolutionary Computation (GECCO 2007), pp. 78–85. ACM, New York (2007)

8. Newton, I.: Newtons Principia Mathematica. Física. Ediciones Altaya, S.A., 21 edition (1968)
9. Parsopoulos, K.E., Magoulas, G.D., Uxbridge, U.P., Vrahatis, M.N., Plagianakos, V.P.: Stretching technique for obtaining global minimizers through particle swarm optimization. In: Proceedings of the Particle Swarm Optimization Workshop, pp. 22–29 (2001)
10. Parsopoulos, K.E., Plagianakos, V.P., Magoulas, G.D., Vrahatis, M.N.: Improving the particle swarm optimizer by function "stretching". Nonconvex Optimization and its Applications 54, 445–458 (2001)
11. Rashedi, E., Nezamabadi-pour, H., Saryazdi, S.: Gsa: A gravitational search algorithm. Information Sciences 179(13), 2232–2248 (2009)
12. Robert, H., David, R.: Physics Part I. Physics (1966)

On the Neutrality of
Flowshop Scheduling Fitness Landscapes

Marie-Eléonore Marmion[1,2], Clarisse Dhaenens[1,2], Laetitia Jourdan[1],
Arnaud Liefooghe[1,2], and Sébastien Verel[1,3]

[1] INRIA Lille-Nord Europe, France
[2] Université Lille 1, LIFL – CNRS, France
[3] University of Nice Sophia Antipolis – CNRS, France
marie-eleonore.marmion@inria.fr, clarisse.dhaenens@lifl.fr,
laetitia.jourdan@inria.fr, arnaud.liefooghe@univ-lille1.fr,
verel@i3s.unice.fr

Abstract. Solving efficiently complex problems using metaheuristics, and in particular local search algorithms, requires incorporating knowledge about the problem to solve. In this paper, the permutation flowshop problem is studied. It is well known that in such problems, several solutions may have the same fitness value. As this neutrality property is an important issue, it should be taken into account during the design of search methods. Then, in the context of the permutation flowshop, a deep landscape analysis focused on the neutrality property is driven and propositions on the way to use this neutrality in order to guide the search efficiently are given.

1 Motivations

Scheduling problems form one of the most important class of combinatorial optimization problems. They arise in situations where a set of operations (tasks) have to be performed on a set of resources (machines), optimizing a given quality criterion. Flowshop problems constitute a special case of scheduling problems in which an operation must pass through all the set of resources before being completed. Such scheduling problems are often difficult to solve, because of the large search space they induce, and then represent a great challenge for combinatorial optimization. Therefore many optimization methods have been proposed so far and experimented on a set of widely-used benchmark instances. Regarding, the minimization of makespan in flowshop problems, iterated local search (ILS) approaches seem to achieve very good performance. In particular, Stützle's ILS [1] stays one of the references of the literature. It has been listed as one of the best performing metaheuristics on a review of heuristic approaches for the flowshop problem investigated in the paper [2]. More recently, Ruiz and Stützle [3] have proposed an iterated greedy algorithm to solve the flowshop problem, based on similar mechanisms, and they have shown that is outperforms the classical metaheuristics for this problem.

C.A. Coello Coello (Ed.): LION 5, LNCS 6683, pp. 238–252, 2011.

The aim of the paper is to analyze characteristics of the flowshop problems in order to understand and to explain why Stützle's method achieves such good performance. A quick analysis shows that the neutrality is high in those problems and we want to explain how this neutrality influences the behavior of heuristic methods. It will then become possible to propose mechanisms that are able to exploit this neutrality.

The method proposed by Stützle consists of an Iterated Local Search (ILS) approach based on the *insertion* neighborhood operator. This operator is argued to be the best one by the original author, as it produces better results than the *transpose* operator, for example, while allowing a faster evaluation compared to the *exchange* operator. The method starts from a solution constructed using a greedy heuristic (the NEH heuristic), initially proposed by Nawaz et al. [4]. Next, the local search algorithm, based on a *first improvement* exploration of the neighborhood, is iterated until a local minimum is reached. Then, between each local search, a small perturbation is applied on the current solution using random applications of the *transpose* and *exchange* neighborhood operators. An important characteristic of this approach is the acceptance criterion of the ILS algorithm, which is based on the Metropolis condition (as in simulated annealing). Indeed, such a condition allows to accept a solution with a same or worse fitness value than the current one.

Hence, the contributions of this work are the following ones. On the one hand, the specific problem of flowshop scheduling is deeply studied in terms of landscape analysis and neutrality. On the other hand, some propositions are drawn in order to exploit neutrality in the design of a local search algorithm. Of course, these considerations are still valid for other combinatorial optimization problems with a neutrality.

The paper is organized as follows. Section 2 is dedicated to the presentation of the flowshop scheduling problem investigated in this paper, and of the required notions about neutrality analysis in fitness landscapes. Section 3 presents the neutral networks analysis for the permutation flowshop problem under study, whereas Section 4 gives some hints on how to exploit the neutrality property in order to solve such problems efficiently by means of local search algorithms. Finally, the last section is devoted to discussion and future works.

2   Background

2.1   Definition of the Permutation Flowshop Scheduling Problem

The Flowshop Scheduling Problem (FSP) is one of the most investigated scheduling problem from the literature. The problem consists in scheduling N jobs $\{J_1, J_2, \ldots, J_N\}$ on M machines $\{M_1, M_2, \ldots, M_M\}$. Machines are critical resources, *i.e.* two jobs cannot be assigned to the same machine at the same time. A job J_i is composed of M tasks $\{t_{i1}, t_{i2}, \ldots, t_{iM}\}$, where t_{ij} is the jth task of J_i, requiring machine M_j. A processing time p_{ij} is associated with each task t_{ij}. We here focus on a permutation FSP, where the operating sequences of the jobs are

Table 1. Notations used in the paper

Notation	Description
S	Set of feasible solutions in the search space
s	A feasible solution $s \in S$
C_{max}	Makespan
N	Number of jobs
M	Number of machines
$\{J_1, J_2, \ldots, J_N\}$	Set of Jobs
$\{M_1, M_2, \ldots, M_M\}$	Set of Machines
$\{t_{i1}, t_{i2}, \ldots, t_{iM}\}$	Tasks
$\{p_{i1}, p_{i2}, \ldots, p_{iM}\}$	Processing times
$\{C_{i1}, C_{i2}, \ldots, C_{iM}\}$	Completion dates

identical and unidirectional for every machine. As consequence, a feasible solution can be represented by a permutation π_N of size N (the ordered sequence of scheduled jobs), and the size of the search space is then $|S| = N!$.

In this study, we will consider that the makespan, *i.e.* the total completion time, is the objective function to be minimized. Let C_{ij} be the completion date of task t_{ij}, the makespan (C_{max}) can be computed as follows:

$$C_{max} = \max_{i \in \{1, \ldots, N\}} \{C_{iM}\}$$

According to Graham et al. [5], the problem under study can be denoted by $F/perm/C_{max}$. The FSP can be solved in polynomial time by the Johnson's algorithm for two machines [6]. However, in the general case, minimizing the makespan has been proven to be NP-hard for three machines and more [7]. As a consequence, large-size problem instances can generally not be solved to optimality, and then metaheuristics may appear to be good candidates to obtain well-performing solutions.

Benchmark Instances. Experiments will be driven using a set of benchmark instances originally proposed by Taillard [8] and widely used in the literature [1,2]. We investigate different values of the number of jobs $N \in \{20, 50, 100, 200\}$ and of the number of machines $M \in \{5, 10, 20\}$. The processing time t_{ij} of job $i \in N$ and machine $j \in M$ is generated randomly, according to a uniform distribution $\mathcal{U}([0; 99])$. For each problem size ($N \times M$), ten instances are available. Note that, as mentioned on the Taillard's website[1], very few instances with 20 machines have been solved to optimality. For 5- and 10-machine instances, optimal solutions have been found, requiring for some of them a very long computational time. Hence, the number of machines seems to be very determinant in the problem difficulty. That is the reason why the results of the paper will be exposed separately for each number of machines.

---

[1] http://mistic.heig-vd.ch/taillard/problemes.dir/ordonnancement.dir/ordonnancement.html

2.2 Neighborhood and Local Search

The design of local search metaheuristics requires a proper definition of a neighborhood structure for the problem under consideration. A *neighborhood structure* is a mapping function $\mathcal{N} : S \rightarrow 2^S$ that assigns a set of solutions $\mathcal{N}(s) \subset S$ to any feasible solution $s \in S$. $\mathcal{N}(s)$ is called the *neighborhood* of s, and a solution $s' \in \mathcal{N}(s)$ is called a *neighbor* of s. A neighbor results of the application of a *move operator* performing a small perturbation to solution s. This neighborhood operator is a key issue for the local search efficiency.

For the FSP, we will consider the *insertion operator*. This operator is known to be one of the best neighborhood structure for the FSP [1,2]. It can be defined as follows. A job located at position i is inserted at position $j \neq i$. The jobs located between positions i and j are shifted, as illustrated in Figure 1. The number of neighbors per solution is $(N - 1)^2$, where N stands for the size of the permutation (and corresponds to the number of jobs).

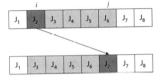

Fig. 1. Illustration of the *insertion neighborhood* operator for the FSP. The job located at position i is inserted at position j, all the jobs located between i and j are shifted to the left.

2.3 Fitness Landscape

Fitness landscape with neutrality. In order to study the typology of problems, the fitness landscape notion has been introduced [9]. A landscape is a triplet $(S, \mathcal{N}, f)$ where S is a set of admissible solutions (*i.e.* a search space), $\mathcal{N} : S \longrightarrow 2^{|S|}$, a neighborhood operator, is a function that assigns to every $s \in S$ a set of neighbors $\mathcal{N}(s)$, and $f : S \longrightarrow \mathbb{R}$ is a fitness function that can be pictured as the *height* of the corresponding solutions. In our study, the search space is composed of permutations of size N so that its size is $N!$.

Neutral neighbor. A neutral neighbor of s is a neighbor solution s' with the same fitness value $f(s)$. Given a solution $s \in S$, its set of neutral neighbors is defined by:

$$\mathcal{N}_n(s) = \{s' \in \mathcal{N}(s) \mid f(s') = f(s)\}$$

The neutral degree of a solution is the number of its neutral neighbors. A fitness landscape is said to be neutral if there are many solutions with a high neutral degree $|V_n(s)|$. The landscape is then composed of several sub-graphs of solutions with the same fitness value. Sometimes, another definition of neutral neighbor is used in which the fitness values are allowed to differ by a small amount. Here we stick to the strict definition given above as the fitness of flowshop (makespan) is discretized (it is an integer value).

Neutral network. A neutral network, denoted as NN, is a connected sub-graph whose vertices are solutions with the same fitness value. Two vertices in a NN are connected if they are neutral neighbors. With the insertion operator, for all solutions x and y, if $x \in \mathcal{N}(y)$ then $y \in \mathcal{N}(x)$. So in this case, the neutral networks are the equivalent classes of the relation $R(x, y)$ iff $(x \in \mathcal{N}(y)$ and $f(x) = f(y))$. We denote the neutral network of a solution s by $NN(s)$. A *portal* in a NN is a solution which has at least one neighbor with a better fitness, *i.e.* a lower fitness value in a minimization context.

Local optimum. A solution s^* is a local optimum iff no neighbor has a better fitness value: $\forall s \in \mathcal{N}(s^*)$, $f(s^*) \leq f(s)$. When all solutions on a neutral network are local optima, the NN is a local optima neutral network.

Measures of neutral fitness. The average or the distribution of neutral degrees over the landscape is used to test the level of neutrality of the problem. This measure plays an important role in the dynamics of metaheuristics [10,11,12]. When the fitness landscape is neutral, the main features of the landscape can be described by its neutral networks. Due to the number and the size of neutral networks, they are sampled by *neutral walks*. A neutral walk $W_{neut} = (s_0, s_1, \ldots, s_m)$ from s to s' is a sequence of solutions belonging to S where $s_0 = s$ and $s_m = s'$ and for all $i \in [0, m-1]$, s_{i+1} is a neighbor of s_i and $f(s_{i+1}) = f(s_i)$.

A way to describe neutral networks NN is given by the *autocorrelation of neutral degree* along a neutral random walk [13]. From neutral degrees collected along this neutral walk, we computed its autocorrelation function $\rho(k)$ [14], that is the correlation coefficient of the neutral degree between the solutions s_i and s_{i+k} for all possible i. The autocorrelation measures the correlation structure of a NN. If the first correlation coefficient $\rho(1)$ is close to 1, the variation of neutral degree is low ; and so, there are some areas in NN of solutions which have close neutral degrees, which shows that NN are not random graphs.

Another interesting information to determine if a local search could find a better solution on a neutral network, is the position of portals. The number of steps before finding a portal during a neutral random walk is a good indicator of the probability to find better solution(s) according to the computational cost to find it, *i.e.* the number of evaluations.

Moreover, to design a local search which explores the neutral networks in an efficient way, we need to find some information around the NN where, *a priori*, there is a lack of information. *Evolvability* is defined by Altenberg [15] as "the ability of random variations to sometimes produce improvement". The concept of evolvability could be difficult to define in combinatorial optimization. For example, the evolvability could be the minimum fitness which can be reached in the neighborhood. In this work, we choose to define the evolvability of a solution as the average fitness in its neighborhood. It gives the expectation of fitness reachable after a random move. The *autocorrelation of evolvability* [16] allows to measure the information around neutral networks. This autocorrelation is the autocorrelation function of a evolvability measure collected during a neutral random walk. When this correlation is large, the solutions which are close from each other on a neutral network have evolvabilities which are close too. So, the evolvability could guide the search on neutral networks such as the fitness guides the search in the landscape where the autocorrelation of fitness values is large [14].

3 Neutral Networks Analysis for the Permutation Flowshop Scheduling Problem

3.1 Experimental Design

To analyze neutral networks, for each instance of Taillard's benchmarks, 30 different neutral walks were performed. The neutral walks all start from a local optimum. It has been obtained by a steepest descent algorithm initialized with a random solution. The length of each neutral walk depends on the length of the descents which lead to local optima. We consider 10 times the maximal length found on the 30 descents. In the following, the results are presented according to the number of jobs (N) and the number of machines (M). For each problem size, an average value and the corresponding standard deviation are represented. By the term size, we mean both the number of jobs (N) and the number of machines (M). This average value is computed from the means obtained from the 10 instances of the same size, themselves calculated from the values given by the 30 neutral walks.

3.2 Neutral Degree

In this section, we first measure the neutral degree of the FSP. Then, we describe the structure of the neutral networks (NN).

Figure 2 shows the average neutral degree to the size of the neighborhood $(N-1)^2$, collected along the 30 neutral walks. Whatever the number of machines, the neutral degree ratio increases when the number of jobs increases. This ratio is higher for small number of machines. For 5-machine, and for 100- or 200-job and 10-machine instances, the neutral degree is huge, higher than 20%. For 100- or 200-job and 20-machine instances, the ratio seems to be very low (3.9%), but the number of neighbors with same fitness value is significant (about 382 and 1544 neutral neighbors for 100 and 200 jobs, respectively). There is no local optimum without a neighbor with the same fitness value, which means that each local optimum belongs to a local optima neutral network. The neutral degree is high enough to describe the fitness landscape with neutral networks.

A neutral walk corresponds to a sequence of neighbor solutions on a NN of the fitness landscape, where all solutions share the same fitness value. During those neutral walks, we compute the autocorrelation of the neutral degree (see Section 2.3). Figure 3 shows the first autocorrelation coefficient for 5, 10 and 20 machines with respect to the number of jobs. In order to prove that those correlations are significative, we compare them to a null model. It consists of shuffling the same values of neutral degrees collected during the neutral walks.

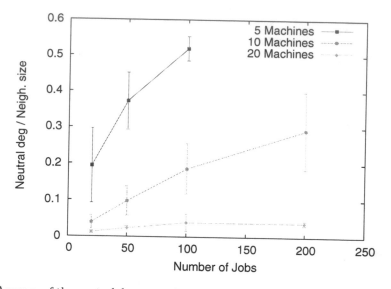

Fig. 2. Average of the neutral degree to the neighborhood size according to the number of jobs

Then, the autocorrelation of this model is compared to the original one. For all sizes, the first autocorrelation coefficient of the null model is below 0.01. Therefore, we can conclude that the autocorrelation is a consequence of the succession of solutions encountered during the walk.

Obviously, for 50, 100 and 200 jobs, the neutral degree is highly correlated (higher than 0.7). Moreover, the standard deviations are very low, which indicates that the average values reflect properly this property on instances of same size. For 20-job and 5- or 10-machine instances, the standard deviation gets higher. This can possibly be explained by a higher correlation.

Nevertheless, these values allow us to conclude that the neutral degree of a solution is partially linked to the one of its neighbor solutions. Let us remark that the correlation for 20-job 20-machine instances is very low, due to the small average value of the neutral degree for this size.

The first conclusions of this analysis is that (*i*) there exists a high neutrality over the fitness landscape, particularly for large-size instances (*ii*) the neutral networks, defined as the graphs of neighbor solutions with the same fitness value, are not random. As a consequence, we should not expect to explore the neutral networks efficiently with a random walk. Hence, heuristic methods should exploit the information available in the neighborhood of the solutions.

3.3   Typology of Neutral Networks

A metaheuristic such as ILS visits several local optima. In the previous section, we have seen that the local optima often belong to a NN. A natural question

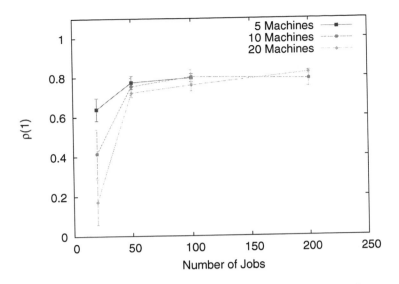

Fig. 3. First autocorrelation coefficient $\rho(1)$ computed between s_i and s_{i+1} of the neutral degree according to the number of jobs

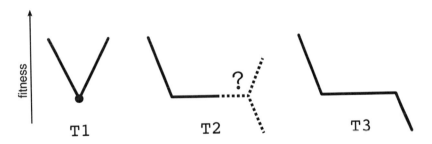

Fig. 4. Typology of neutral networks (minimization problem)

arises when the metaheuristic reaches a NN: Is it possible to escape from this NN? In this section, we classify the local optima NN in three different types, and we analyze their size.

Three types of NN typologies may exist (see Figure 4):

1. The local optimum is the single solution on the NN (type T1), *i.e.* it has no neighbor with the same fitness value, we call it a degenerated NN.
2. The neutral walk from the local optimum did not show any neighbors with a better fitness values for all the solutions encountered along the neutral walk (type T2). Of course, as the whole NN has not been enumerated, we can not decide if it is possible to escape from them.
3. At least one solution having a neighbor with better fitness value than the local optimum fitness is found along the neutral walk (type T3).

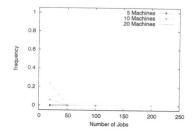

Fig. 5. Average frequency of the number of degenerated neutral networks with a single solution (type T1) according to the number of jobs

Fig. 6. Average frequency of the number of neutral networks where no portal was found (type T2) according to the number of jobs

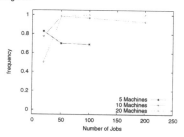

Fig. 7. Average frequency of the number of neutral networks where at least one portal was found (type T3) according to the number of jobs

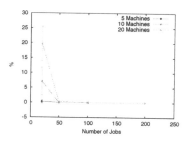

Fig. 8. Average percentage of solutions visited at least twice along the neutral walk according to the number of jobs

Figures 5, 6 and 7 show the proportion of NN of each type (T1, T2 or T3) counted along the neutral walks. For 50-, 100- and 200-job instances, the neutral walks show only NN of types T2 and T3. No local optimum solution is alone on the NN. For 20-job instances, the number of type (T1) is also small, except for 20 machines (25% of type T1). Hence, the neutrality is important to keep in mind while solving such instances. The number of NN without any escaping solutions found (T2) is significative only for 5-machine instances (higher than 18%) and stays very low for 10- an 20-machine instances (lower than 6%). The 20-machine instances, which are known to be the hardest to solve optimally, are the ones where the probability to escape from local optimum by neutral exploration of the NN is close to one.

When the neutral networks size is very small, the number of visited solutions is very small. Indeed, a NN of type T2 or T3 could contain very few solutions and, the neutral walk could loop on some solutions. These situations have to be considered with attention. Figure 8 shows the average percentage of solutions visited more than once during the neutral walk. For the 50-, 100- and 200-job

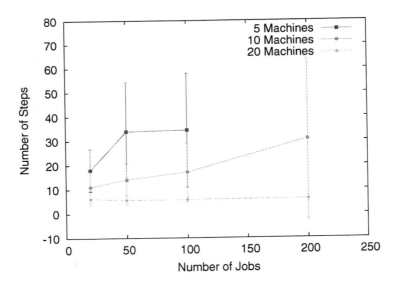

Fig. 9. Number of steps along the neutral random walk to reach the first portal according to number of jobs

instances, there is no re-visited solutions during the neutral walks. For 20-job and 20-machine instances, the number of re-visited solutions is approximatively 20% during neutral walks on NN of type T2 or T3. This result points out two remarks. First, the NN of local optima seems to be large for most instances. Second, the number of re-visited solutions is low, which means that the probability to escape the NN of type T2 is below the inverse of the size of the neutral walk.

In conclusion, for most instances, a metaheuristic could escape the local optimum by exploring the NN. The next section will show some hints on how to guide a metaheuristic on neutral networks.

4   Exploiting Neutrality to Solve the FSP

In the previous section, we proposed to use neutral exploration to escape from local optimum, as there exists solutions having neighbor(s) with a better fitness value around neutral networks. We called those solutions, portals. An efficient metaheuristic has to find such portal with a minimum number of evaluations. First, we study the number of steps to reach a portal, and then we propose an insight to get information to find them quickly.

4.1   Reaching Portals

As shown on Figure 7 at least 70% of neutral random walks for FSP with 50, 100 and 200 jobs can reach a portal (more than 90% for 10 and 20 machines). The performance of a metaheuristic which explores neutral networks highly depends on the probability to find a portal. Indeed, it could become more time consuming to consider a neutral walk than applying a smart restart.

Figure 9 gives the average number of steps to reach the first portal during the 30 neutral walks. The larger the number of machines, the less the number of steps is required by the neutral walk to reach a portal. For 20-machine instances, the neutral random walks need around 7 steps to reach a portal, which is very small compared to the length of the descents (19, 40, 64, 101 respectively for 20, 50, 100, and 200 jobs). For 5-machine instances, the length of the neutral walks is around the length of the descents. Hence, it is probably more advantageous to perform a neutral random exploration than a random restart. Moreover, the fitness value obtained after the neutral walk is better than after the descent. Consequently, if an *a priori* study highlights that a portal is supposed to be encountered quickly, a metaheuristic that takes into account information on the neutral walk should move on the NN, and then finally find an improving solution.

4.2   How to Guide the Search?

In the previous section, the role of neutrality was demonstrated by the correlation of the neutral degree between the neutral walk neighbors and the high frequency of neutral networks. Neutral networks lead, with very few steps, to a portal. The neutrality could give interesting information about the landscape in order to guide the search. However, since the neutral network is large, the search has to be guided to find quickly a portal and not to stagnate on the NN. Thus, proper information has to be collected and interpreted along the neutral walk to help the metaheuristic to take good decision: Is it more interesting to continue the neutral walk until a portal is reached or to restart? As suggested in Section 2.3, we compute the evolvability of a solution as the average fitness values of its neighbors for all visited solutions. We analyze the evolvability of solutions on neutral networks and we give some results about the correlation of evolvability and portals on a neutral network. This allows us to propose new ideas for the design of a metaheuristic.

During those neutral walks, we compute the evolvability of each solution along the neutral walk, and then its autocorrelation (see section 2.3). Figure 10 shows the first autocorrelation coefficient $\rho(1)$ for 5, 10 and 20 machines with respect to the number of jobs. In order to show that those correlations are significative, as in section 3.2, we compare them to a null model. For all sizes, the first autocorrelation coefficient of the null model is below 0.01. Therefore, we can conclude that the autocorrelation is a consequence of the succession of solutions encountered during the walk. The average fitness values of the neighbors are not distributed randomly: they can then be exploited by a metaheuristic.

The neutral networks present evolvability and portals. So, we can wonder if the evolvability would be able to guide a metaheuristic quickly to a portal. To test this hypothesis, along the neutral walks, we compute the correlation between the average fitness values in the neighborhood and the number of steps required to reach the closer portal of the walk. This is presented in Figure 11. The larger the number of machines, the higher (in absolute value) the negative correlation.

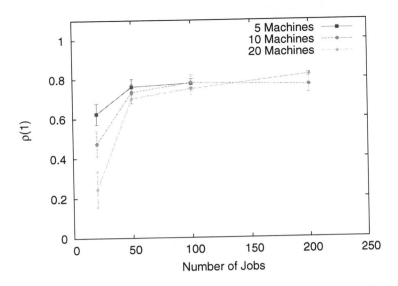

Fig. 10. First autocorrelation coefficient $\rho(1)$ of the average fitness values of neighbors solutions between s_i and s_{i+1} according to the number of jobs

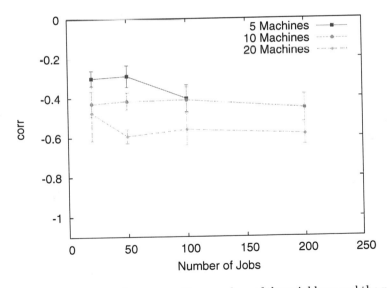

Fig. 11. Correlation between the average fitness values of the neighbors and the number of steps required to reach the closer portal according to the number of jobs

For 10- or 20-machine instances, this correlation belongs to $[-0.6; -0.4]$, so that it is significant for a metaheuristic to use such an information. The lower the average fitness values in the neighborhood, the closer a portal is. Consequently, we propose to design a metaheuristic that takes into account the neutrality by allowing the exploration of solutions along the neutral walk. Starting from a local optimum, it would choose the next neutral solution with the lower average fitness values of its neighbors. This would increase the probability to find a portal quickly, and then to continue the search process.

5   Discussion

In this work, we studied the neutrality of the FSP on a set of benchmark instances originally proposed by Taillard. Most of the instances have a high neutral degree: for a solution, the number of its neighbors with the same fitness value is significant in comparison to the neighborhood size. Starting from local optima, neutral walks have been performed. Each walk moves from a solution to another with the same fitness value and defines a neutral network that is shown to be structured. Indeed, the graph of neighbor solutions is not random and so a solution shares information with its neighbors. We show that a neutral walk leads easily to portals, solutions of the neutral network having a neighbor with a better fitness value. Furthermore, the evolvability, defined in this study as the average fitness values of the neighbors, is highly autocorrelated. It proves that this information is not random between the neighbor solutions and so it could be helpful to take it into account. Besides, improving the evolvability during the neutral walk often leads to a portal. This work completes the knowledge of FSP fitness landscape, and in particular, about its neutrality. Here, the neutrality has been shown for the FSP Taillard instances where the durations of jobs are integer values from $[0; 99]$. This is a specific choice which could have an impact on the difficulty of instances. Future works will consider other instance generators, and study the neutrality according to the instance parameters.

This work also helps to understand some experimental results on the efficiency of metaheuristics. In a study of iterated local search to solve the FSP [1], Stützle designs several efficient ILS, called ILS-S-PFSP and compares them to local search algorithms. He writes: "Experimentally, we found that rather small modifications [of the solution] are sufficient to yield very good performance". In section 4.1, we show that improving solutions can be reached very quickly applying *insertion* operator on a neutral network. So, Stützle's remark can be explained by the neutrality and the high probability on the neutral networks to move on a solution with an improving neighbor. Moreover, this works supports the experimentations on ILS design for 20-machine instances. The study of neutral walks highlights features that explain the efficient design of the ILS-S-PFSP. Indeed, remember that the ILS-S-PFSP, initialized with a random solution, applies a local search based on *insertion*-neighborhood mapping to get a local optimum, and then applies iteratively the steps (i) perturbation, (ii) local search, and (iii) acceptance criterion, until a termination condition is met. All acceptance criteria tested in ILS-S-PFSP are based on the Metropolis condition: they always accept a solution with equal fitness value. So the neutral moves are always accepted. Besides, Stützle work shows that the perturbation based on the application of several *swap* operators (also called *transpose* operators) is efficient. And, the *swap* neighborhood is included in the *insertion* neighborhood as the job i can be inserted at the positions $(i-1)$ or $(i+1)$. So, applying the *swap* operator several times could correspond to a walk on a neutral network defined by *insertion*-neighborhood relation. Thus, steps (i) and (iii) allow the ILS-S-PFSP to move on the neutral network that could be frequent for those FSP instances. Moreover, we show that the distance is small between a local optimum and a portal. So, such an ILS-S-PFSP is able to quickly improve the current best solution, which could explain its performances.

Furthermore, our work proposes to consider the neutrality to guide a meta-heuristic on the search space. The FSP instances shows neutrality, it is easy to encounter portals along a neutral walk and the evolvability leads quickly to them. With such information, a metaheuristic is proposed: first a local search is performed from a random solution, and then iteratively (i) the evolvability on the neutral network is optimized until a portal is found and (ii) the local search is applied to move to an other local optimum. The metaheuristic finishes when the termination criterion is met. Similar ideas have been ever tested on other problems with neutrality such as Max-SAT and NK-landscapes with neutrality [17]. A first attempt for developing such a strategy leads to the proposition of NILS [18] that has been successfully tested on flowshop problems.

References

1. Stützle, T.: Applying iterated local search to the permutation flow shop problem. Technical Report AIDA-98-04, FG Intellektik, TU Darmstadt (1998)
2. Ruiz, R., Maroto, C.: A comprehensive review and evaluation of permutation flow-shop heuristics. European Journal of Operational Research 165(2), 479–494 (2005)
3. Ruiz, R., Stützle, T.: A simple and effective iterated greedy algorithm for the permutation flowshop scheduling problem. European Journal of Operational Research 177(3), 2033–2049 (2007)
4. Nawaz, M., Enscore, E., Ham, I.: A heuristic algorithm for the m-machine, n-job flow-shop sequencing problem. Omega 11(1), 91–95 (1983)
5. Graham, R.L., Lawler, E.L., Lenstra, J.K., Rinnooy Kan, A.H.G.: Optimization and approximation in deterministic sequencing and scheduling: A survey. Annals of Discrete Mathematics 5, 287–326 (1979)
6. Johnson, S.M.: Optimal two- and three-stage production schedules with setup times included. Naval Research Logistics Quarterly 1, 61–68 (1954)
7. Lenstra, J.K., Rinnooy Kan, A.H.G., Brucker, P.: Complexity of machine scheduling problems. Annals of Discrete Mathematics 1, 343–362 (1977)
8. Taillard, E.: Benchmarks for basic scheduling problems. European Journal of Operational Research 64, 278–285 (1993)
9. Wright, S.: The roles of mutation, inbreeding, crossbreeding and selection in evolution. In: Jones, D. (ed.) Proceedings of the Sixth International Congress on Genetics, vol. 1 (1932)
10. Van Nimwegen, E., Crutchfield, J., Huynen, M.: Neutral evolution of mutational robustness. Proc. Nat. Acad. Sci. USA 96, 9716–9720 (1999)
11. Wilke, C.O.: Adaptative evolution on neutral networks. Bull. Math. Biol. 63, 715–730 (2001)

12. Vérel, S., Collard, P., Tomassini, M., Vanneschi, L.: Fitness landscape of the cellular automata majority problem: view from the "Olympus". Theor. Comp. Sci. 378, 54–77 (2007)
13. Bastolla, U., Porto, M., Roman, H.E., Vendruscolo, M.: Statiscal properties of neutral evolution. Journal Molecular Evolution 57(S), 103–119 (2003)
14. Weinberger, E.D.: Correlated and uncorrelatated fitness landscapes and how to tell the difference. Biological Cybernetics 63, 325–336 (1990)
15. Altenberg, L.: The evolution of evolvability in genetic programming. In: Kinnear Jr., K.E. (ed.) Advances in Genetic Programming, pp. 47–74. MIT Press, Cambridge (1994)
16. Verel, S., Collard, P., Clergue, M.: Measuring the Evolvability Landscape to study Neutrality. In: Keijzer, M., et al. (eds.) Poster at Genetic and Evolutionary Computation – GECCO 2006, pp. 613–614. ACM Press, Seattle (2006)
17. Verel, S., Collard, P., Clergue, M.: Scuba Search: when selection meets innovation. In: Evolutionary Computation, CEC 2004, pp. 924–931. IEEE Press, Portland (2004)
18. Marmion, M.E., Dhaenens, C., Jourdan, L., Liefooghe, A., Verel, S.: NILS: a neutrality-based iterated local search and its application to flowshop scheduling. In: 11th European Conference on Evolutionary Computation in Combinatorial Optimisation (EvoCOP11). LNCS, Springer, Heidelberg (2011)

A Reinforcement Learning Approach for the Flexible Job Shop Scheduling Problem

Yailen Martínez[1,2], Ann Nowé[1], Juliett Suárez[2], and Rafael Bello[2]

[1] CoMo Lab, Department of Computer Science, Vrije Universiteit Brussel, Belgium
{ymartine,ann.nowe}@vub.ac.be
[2] Department of Computer Science, Central University of Las Villas, Cuba
{yailenm,jsf,rbellop}@uclv.edu.cu

Abstract. In this work we present a Reinforcement Learning approach for the Flexible Job Shop Scheduling problem. The proposed approach follows the ideas of the hierarchical approaches and combines learning and optimization in order to achieve better results. Several problem instances were used to test the algorithm and to compare the results with those reported by previous approaches.

1 Introduction

Scheduling is a scientific domain concerning the allocation of tasks to a limited set of resources over time. The goal of scheduling is to maximize (or minimize) different optimization criteria such as the makespan or the tardiness. The scientific community usually classifies the problems according to different characteristics, for example, the number of machines (one machine, parallel machines), the shop type (Job Shop, Flow Shop or Open Shop) and so on. These kind of problems have captured the interest of many researchers from a number of different research communities for decades. To find a good schedule (or the best schedule) can be a very difficult task depending on the constraints of the problem and the environment. The Job Shop Scheduling Problem (JSSP) is one of the most popular scheduling models existing in practice, and it is also among the hardest combinatorial optimization problems [1]. The Flexible Job Shop Scheduling Problem (FJSSP) is a generalization of the classical JSSP, where operations are not processed by a fixed machine, but there is a choice between a set of available machines that can execute it. Therefore, the FJSSP has an extra decision step besides the sequencing, the job routing. To determine the job route means to choose, for each operation, which machine will execute it from the set of available ones.

Literature on flexible job shop scheduling is not rare, but approaches using learning based methods are. In the literature we find different (meta-)heuristic approaches for this problem, for example, Ant Colony Optimization [2] and Genetic Algorithms [3] [4].

In [5] Thomas Gabel and Martin Riedmiller suggested and analyzed the application of reinforcement learning techniques to solve the task of job shop

C.A. Coello Coello (Ed.): LION 5, LNCS 6683, pp. 253–262, 2011.

scheduling problems. They demonstrated that interpreting and solving this kind of problems as a multi-agent learning problem is beneficial for obtaining near-optimal solutions and can very well compete with alternative solution approaches.

Reinforcement Learning is the problem faced by an agent that must learn behavior through trial-and-error interactions with a dynamic environment. Each time the agent performs an action in its environment, a trainer may provide a reward or penalty to indicate the desirability of the resulting state. For example, when training an agent to play a game, the trainer might provide a positive reward when the game is won, negative when it is lost and zero in all other states. The task of the agent is to learn from this indirect, delayed reward, to choose sequences of actions that produce the greatest cumulative reward [6].

In this paper we present a Reinforcement Learning approach for the FJSSP. More specifically, we adopt the assign-then-sequence rule proposed by the hierarchical approaches and combine a two step learning algorithm with a mode optimization procedure in order to achieve better results.

The remainder of this paper is organized as follows. Section 2 introduces the problem formulation and a literature review on the subject is also given. Section 3 gives and overview on reinforcement learning and in Section 4 the algorithm is presented detailing what is done in each step. In Section 5 we present a computational study using some classical instances, comparing our results with some previous approaches. Some final conclusions and ideas for future work are given in Section 6.

2   Flexible Job Shop Scheduling Problem

2.1   Problem Formulation

The Flexible Job Shop Scheduling Problem consists of performing a set of n jobs $J = \{J_1, J_2, \ldots, J_n\}$ on a set of m machines $M = \{M_1, M_2, \ldots, M_m\}$. Each job J_i has an ordered set of o_i operations $O_i = \{O_{i,1}, O_{i,2}, \ldots, O_{i,o_i}\}$. Each operation $O_{i,j}$ can be performed on any among a subset of available machines $(M_{i,j} \subseteq M)$. Executing operation $O_{i,j}$ on machine M_k takes $p_{i,j,k}$ processing time. Operations of the same job have to respect the precedence constraints given by the operation sequence. A machine can only execute one operation at a time. An operation can only be executed on one machine and can not leave it before the treatment is finished. There are no precedence constraints among the operations of different jobs. The problem is to assign each operation to an appropriate machine (routing problem), and then to sequence the operations in the selected machines (sequencing problem) in order to minimize the makespan, i.e., the time needed to complete all the jobs, which is defined as $C_{max} = max\{C_i | 1 \leq i \leq n\}$: where C_i is the completion time of job J_i.

2.2   Previous Approaches

Different heuristic procedures have been developed in the last years for the FJSSP, for example, tabu search, dispatching rules, simulated annealing and genetic algorithms. According to the literature review, all these methods can be classified into two main categories: hierarchical approaches and integrated approaches, meaning that we have two different ways to deal with the problem.

The hierarchical approaches are based on the idea of decomposing the original problem in order to reduce its complexity. A typical decomposition is *"assign then sequence"*, meaning that the assignment of operations to machines and the sequencing of the operations on the resources are treated separately. Once the assignment is done (each operation has a machine assigned to execute it), the resulting sequencing problem is a classical JSSP. This approach is followed by Brandimarte [7], who was the first to use decomposition for the FJSSP, Kacem [4] and Pezzella [3] also followed this idea in the implementation of Genetic Algorithms.

Integrated approaches consider assignment and sequencing at the same time. The methods following this type of approach usually give better results but they are also more difficult to implement.

2.3 Dispatching Rules

As mentioned above, the complexity of the FJSSP gives raise to the search of heuristic algorithms able to provide good solutions. Dispatching rules are among the more frequently applied heuristics due to their ease of implementation and low time complexity.

A dispatching rule is a sequencing strategy by which a priority is assigned to each job waiting to be executed on a specific machine. Whenever a machine is available, a priority-based dispatching rule inspects the waiting jobs and the one with the highest priority is selected to be processed next [8]. Some of the most used dispatching rules are:

- Shortest Processing Time (SPT): The highest priority is given to the waiting operation with the shortest processing time.
- First In First Out (FIFO): The operation that arrived to the queue first receives the highest priority.
- Most Work Remaining (MWKR): Highest priority is given to the operation belonging to the job with the most total processing time remaining to be done.
- Earliest Due Date (EDD): The job due out first is processed first.

There are also some *composite* dispatching rules (CDR), which combine single dispatching rules and results have shown that a careful combination can perform better in terms of quality.

3   Reinforcement Learning

Reinforcement Learning (RL) is a technique that allows an agent to learn how to maximize a numerical reward signal. The learner is not told which actions to take, as in most forms of machine learning, but instead must discover which actions yield the most reward by trial-and-error. In the most interesting and challenging cases, actions may affect not only the immediate reward but also the next situation and, through that, all subsequent rewards. These two characteristics, trial-and-error search and delayed reward, are the two most important distinguishing features of RL [9].

In the standard RL paradigm, an agent is connected to its environment via perception and action, as depicted in Figure 1. In each step of interaction, the agent senses the current state s of its environment, and then selects an action a which may change this state. The action generates a reinforcement signal r, which is received by the agent. The task of the agent is to learn a policy for choosing actions in each state so that the maximal long-run cumulative reward is received.

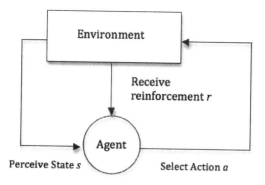

Fig. 1. The Reinforcement Learning Paradigm

One of the challenges that arise in RL is the trade-off between exploration and exploitation. To obtain a high reward, a RL agent must prefer actions that it has tried in the past and found to be effective in producing reward. But to discover such actions, it has to try actions that it has not selected before. The agent has to exploit what it already knows in order to obtain reward, but it also has to explore in order to make better action selections in the future. The dilemma is that neither exploration nor exploitation can be pursued exclusively without failing at the task. Therefore the agent must sample the available actions sufficiently and progressively favor those that appear to be best.

Some previous works showed the effectiveness of the Q-Learning algorithm in the solution of scheduling problems, more specifically the Job Shop Scheduling Problem [10] and the Parallel Machines Job Shop Scheduling Problem [11], that is why the Q-Learning was chosen among the different existing algorithms to solve the Flexible Job Shop Scheduling Problem.

3.1   Q-Learning

A well-known reinforcement learning algorithm is Q-Learning [12], which works by learning an action-value function that expresses the expected utility (i.e. cumulative reward) of taking a given action in a given state.

The core of the algorithm is a simple value iteration update, each state-action pair (s, a) has a Q-value associated. When action a is selected by the agent located in state s, the Q-value for that state-action pair is updated based on the reward received when selecting that action and the best Q-value for the subsequent state s'. The update rule for the state action pair (s, a) is the following:

$$Q(s,a) = Q(s,a) + \alpha[r + \gamma max_{a'}(Q(s',a')) - Q(s,a)] \tag{1}$$

In this expression, $\alpha \in \{0, 1\}$ is the learning rate and r the reward or penalty resulting from taking action a in state s. The learning rate α determines the degree by which the old value is updated. For example, if the learning rate is 0, then nothing is updated at all. If, on the other hand, $\alpha = 1$, then the old value is replaced by the new estimate. Usually a small value is chosen for the learning rate, for example, $\alpha = 0.1$. The discount factor (parameter γ) has a range value of 0 to 1 ($\gamma \in \{0, 1\}$). If γ is closer to zero, the agent will tend to consider only immediate reward. If γ is closer to one, the agent will consider future reward with greater weight.

4   The Proposed Approach: Learning / Optimization

The Learning/Optimization method is an offline scheduling approach divided in two steps. First, a two-stage learning method is applied to obtain feasible schedules, which are then used as initial data for the mode optimization procedure [13] developed during the second step.

The learning method implemented decomposes the problem following the assign-then-sequence approach. Therefore, we have two learning phases, during the first phase operations learn which is the most suitable machine and during the second phase machines learn in which order to execute the operations in order to minimize the makespan. For this, each phase has a Q-Learning algorithm associated and different dispatching rules are taken into account when giving rewards to the agents. As the process is being divided in two, we take into account the goal of each phase in order to decide where to place the agents and which are the possible actions. In the first phase, where the learning takes care of the routing, we have an agent per operation being responsible for choosing a proper machine to execute the corresponding operation, this machine is selected from the given set of available ones, and the selection is based on the processing time of the operation on the machine and also on the workload of the machine so far.

It could also be possible to have an agent per job, which would be responsible of selecting a proper machine for each of its operations. This is not the case for the second phase, where the learning algorithm takes care of the sequencing and each operation already knows where it has to be executed so, the main idea is to decide the order in which they will be processed on the machines, that is why in this phase we placed the agents on the different resources, and for these agents an action will be to choose an operation from the queue of operations waiting at the corresponding resource.

To start the algorithm every job releases its first operation at time 0, all these operations go to the machine they have assigned and start to be processed, if two or more operations go to the same machine then only one of them is selected and the rest remain in the queue until the machine is available again.

To choose the next action the agent takes into account the Q-Values associated to the possible operations to execute at that time step in the corresponding machine. According to the epsilon greedy policy, in order to balance exploration and exploitation, the agent has a small probability of selecting an action at random, and a higher probability of selecting the best action, in this case the operation with the highest Q-Value associated, in this step the dispatching rule taken into account to give reward to the agent is the Shortest Processing Time (SPT).

Once a feasible schedule is obtained, the mode optimization procedure is executed, which we refer to as the second step. This is a forward-backward procedure which tries to shift the schedule to the left in order to minimize the makespan, it has the following steps:

- Order the operations according to their end times (the time when they were ended in the schedule received as input).
- Taking into account the previous ordering, for each operation, choose the machine that will finish it first (shortest end time, not shortest processing time). The result is a backward schedule.
- Repeat steps 1 and 2 to obtain a forward schedule.

Once the mode optimization is executed, the quality of the solution is taken into account to give feedback to the agents of the learning phases.

4.1   Pseudo-code of the Algorithm

```
Step 1 - Learning
  Phase 1 - Routing
    For each operation - Choose a machine
  Phase 2 - Sequencing
    While there are operations to execute
      For each machine with operations in the queue
        Choose operation to execute
        Update Queues of the System
Step 2 - Execute the Mode Optimization Procedure
```

4.2   Example

Assuming that we have a small instance with 2 jobs and 3 machines, where Job_1 has 2 operations and Job_2 has 3 operations, and these operations can be executed by the following sets of machines, where each pair represents a possible machine and the corresponding processing time.

$$J_0O_0 \begin{cases} M_0, 10 \\ M_1, 15 \end{cases} \quad J_0O_1 \begin{cases} M_1, 12 \\ M_2, 18 \end{cases}$$

$$J_1O_0 \begin{cases} M_0, 20 \\ M_2, 25 \end{cases} \quad J_1O_1 \begin{cases} M_0, 25 \\ M_1, 18 \end{cases} \quad J_1O_2 \begin{cases} M_1, 15 \\ M_2, 25 \end{cases}$$

As mentioned in the description of the algorithm, the first learning phase takes care of the routing, meaning that the first step is to choose an appropriate machine for each operation. Let's say that after executing the first phase the resulting assignment is the following: $J_0O_0 - M_0$, $J_0O_1 - M_1$, $J_1O_0 - M_0$, $J_1O_1 - M_0$ and $J_1O_2 - M_2$. A possible schedule for this operation-machine assignment is shown in Figure 2. Applying the Mode Optimization Procedure to this schedule, the first step is to order the operations according to their end time, that will give us the following ordering: J_1O_2, J_1O_1, J_1O_0, J_0O_1 and J_0O_0. Taking into account this ordering, the operations will choose a machine to execute it basing the decision in the possible end time.

For example, J_1O_2 can choose between going to M_1 for 15 time steps or to M_2 for 25 time steps, obviously the best choice is M_1, meaning that M_1 will be busy between time 0 and 15.

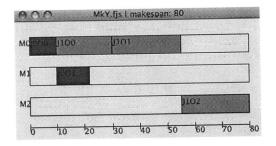

Fig. 2. Schedule

Then the next operation on the ordered list makes a choice, in this case J_1O_1 can choose between M_0 for 25 time steps and M_1 for 18, as this is the second operation of J_1 it can not start until the previous one is finished so, the starting time will be 15, the possible end times are 40 and 33, being the best choice M_1, which will be occupied from 15 to 33. When an operation from another job has to choose a machine has to respect this busy times but can search for an available slot of the size of the time it requires.

5   Experimental Results

5.1   Instances

The approach proposed in this paper was tested on a set of instances from literature [7]. The results shown below are those obtained for the set of Brandimarte

Table 1. Brandimarte Instances

Instance	Jobs	Machines	Lower Bound
Mk01	10	6	36
Mk02	10	6	24
Mk03	15	8	204
Mk04	15	8	48
Mk05	15	4	168
Mk06	10	15	33
Mk07	20	5	133
Mk08	20	10	523
Mk09	20	10	299
Mk10	20	15	165

instances, this dataset consists of 10 problems (Mk01-Mk10) with number of jobs ranging from 10 to 20 and a number of machines ranging from 4 to 15 (Table 1).

5.2 Parameters

Different parameter settings were studied before deciding which combination to use for the final experiments. The parameters involved on this study were the discount factor (λ) and epsilon (ϵ). The different combinations involved the following sets of values: $\lambda = \{0.8, 0.85, 0.9\}$ and $\epsilon = \{0.01, 0.1, 0.15, 0.2\}$.

After analyzing all the possibilities the best setting was picked, which resulted to be $\lambda = 0.8$ and $\epsilon = 0.1$, together with a discount factor $\alpha = 0.1$. The algorithm was executed for 1000 iterations.

5.3 Comparative Study

Table 2 shows a comparative study between the proposed approach and some results already reported. LB is the Lower Bound for each instance, taken from the original Brandimarte data. The algorithms used to compare our method (QL) are:

- GA: Genetic Algorithm [3], algorithm integrating different strategies for generating the initial population, selecting the individuals for reproduction and reproducing new individuals.
- ACO: Ant Colony Optimization [2], it provides an effective integration between the Ant Colony Optimization model and knowledge model.
- GEN: Abbreviation of GENACE, an architecture proposed in [14] where an effective integration between evolution and learning within a random search process is proposed.
- Brand: Tabu Search [7], a hierarchical algorithm for the flexible job shop scheduling based on the tabu search metaheuristic.

Table 3 shows the mean relative errors in % (MRE) of the different approaches used to compare our method with respect to the best-known lower bound. The relative error (RE) is defined as RE =[(MK -LB)/LB 100]%, where MK is the

Table 2. Experimental Results

Inst.	LB	GA	ACO	GEN	Brand	QL
Mk01	36	40	39	40	42	40
Mk02	24	26	29	29	32	26
Mk03	204	204	204	204	211	204
Mk04	48	60	65	67	81	66
Mk05	168	173	173	176	186	173
Mk06	33	63	67	67	86	62
Mk07	133	139	144	147	157	146
Mk08	523	523	523	523	523	523
Mk09	299	311	311	320	369	308
Mk10	165	212	229	229	296	225

best makespan obtained by the reported algorithm and LB is the best-known lower bound. The MRE takes into account the average of the results for the whole group of instances.

Table 3. MRE: Mean relative errors

	GA	ACO	GEN	Brand	QL
MRE	17,53	22,16	23,56	41,43	19,69

From the tables we can notice that the method proposed is able to find the best reported value for several instances(Mk01-Mk03, Mk05, Mk08). For the instances Mk04, Mk07 and Mk10 the genetic algorithm is better. For the instances Mk06 and Mk09 the algorithm is able to yield better results.

The cases where our algorithm did not report very good solutions where mainly instances for which a proper machine assignment was not found (multiple machines for the same operation with similar processing times). It is important to mention that the use of the mode optimization procedure helps when the operation-machine assignment developed during the first learning phase was adequate, this is a key step which influences the quality of the solution.

6 Conclusions and Future Work

In this paper we introduced a Reinforcement Learning Approach for the Flexible Job Shop Scheduling Problem. The learning process was combined with an optimization procedure in order to obtain better results. Different instances from literature were used in order to compare our method with some other existing approaches, results show that the method proposed is able to yield better results than some of the previous reported, except for the Genetic Algorithm in some of the instances. It will be interesting to combine different dispatching rules (*composite* dispatching rules) in order to get better results using only the learning algorithm.

References

1. Garey, M.R., Johnson, D.S., Sethi, R.: The Complexity of Flowshop and Jobshop Scheduling. Mathematics of Operations Research 1, 117–129 (1976)
2. Lining, X., Chen, Y.: A Knowledge-Based Ant Colony Optimization for Flexible Job Shop Scheduling Problems. Applied Soft Computing (2009)
3. Pezzella, F., Morganti, G., Ciaschetti, G.: A genetic algorithm for the flexible job-shop scheduling problem. Computers & Operations Research 35, 3202–3212 (2008)
4. Kacem, I., Hammadi, S., Borne, P.: Approach by localization and multiobjective evolutionary optimization for flexible job-shop scheduling problems. IEEE Transactions on Systems, Man, and Cybernetics, Part C 32, 1–13 (2002)
5. Gabel, T., Riedmiller, M.: On a successful application of multi-agent reinforcement learning to operations research benchmarks. IEEE Transactions on Systems, Man, and Cybernetics (2009)
6. Mitchell, T.: Machine Learning. McGraw-Hill Science/Engineering/Math (1997)
7. Brandimarte, P.: Routing and scheduling in a flexible job shop by tabu search. Annals of Operations Research 41, 157–183 (1993)
8. Nhu Binh, H., Joc Cing, T.: Evolving Dispatching Rules for solving the Flexible Job-Shop Problem. In: IEEE Congress on Evolutionary Computation (CEC 2005), vol. 3, pp. 2848–2855 (2005)
9. Sutton, R.S., Barto, A.G.: Reinforcement Learning: An Introduction. The MIT Press, Cambridge (1998)
10. Martínez, Y.: A Multi-Agent Learning Approach for the Job Shop Scheduling Problem. Master thesis, Vrije Universiteit Brussel (2008)
11. Martínez, Y., Wauters, T., De Causmaecker, P., Nowe, A., Verbeeck, K., Bello, R., Suarez, J.: Reinforcement Learning Approaches for the Parallel Machines Job Shop Scheduling Problem. In: Proceedings of the Cuba-Flanders Workshop on Machine Learning and Knowledge Discovery, Santa Clara, Cuba (2010)
12. Watkins, C., Dayan, P.: Technical note: Q-learning. Machine Learning 8, 279–292 (1992)
13. Peteghem, V., Vanhoucke, M.: A genetic algorithm for the multi-mode resource-constrained project scheduling problem. In: Working Papers of Faculty of Economics and Business Administration. Ghent University, Belgium (2008)
14. Ho, N.B., Tay, J.C., Lai, E.M.: An effective architecture for learning and evolving flexible job-shop schedules. European Journal of Operational Research 179, 316–333 (2007)

Supervised Learning Linear Priority Dispatch Rules for Job-Shop Scheduling

Helga Ingimundardottir and Thomas Philip Runarsson

School of Engineering and Natural Sciences, University of Iceland
{hei2,tpr}@hi.is

Abstract. This paper introduces a framework in which dispatching rules for job-shop scheduling problems are discovered by analysing the characteristics of optimal solutions. Training data is created via randomly generated job-shop problem instances and their corresponding optimal solution. Linear classification is applied in order to identify good choices from worse ones, at each dispatching time step, in a supervised learning fashion. The method is purely data-driven, thus less problem specific insights are needed from the human heuristic algorithm designer. Experimental studies show that the learned linear priority dispatching rules outperforms common single priority dispatching rules, with respect to minimum makespan.

1 Introduction

Hand crafting heuristics for NP-hard problems is a time-consuming trial and error process, requiring inductive reasoning or problem specific insights from their human designers. Furthermore, within a problems class, such as job-shop scheduling, it is possible to construct problem instances where one heuristic would outperform another. Given the ad-hoc nature of the heuristic design process there is clearly room for improving the process. Recently a number of attempt have been made to automate the heuristic design process. Here we focus on the job-shop problem. Various learning approaches have been applied to this task such as, reinforcement learning [1], evolutionary learning [2], and supervised learning [3,4]. The approach taken here is a supervised learning classifier approach.

In order to find an optimal (or near optimal) solution for job-shop scheduling problem (JSSP) one could either use exact methods or heuristics methods. Exact methods guarantee an optimal solution, however, JSSP is NP-hard [5]. Any exact algorithm generally suffers from the curse of dimensionality, which impedes the application in finding the global optimum in a reasonable amount of time. Heuristics are generally more time efficient but do not necessarily attain the global optimum. A common way of finding a good feasible solution for the JSSP is by applying heuristic dispatching rules, e.g., choosing a task corresponding to longest/shortest operation time; most/least successors; or ranked positional weight, i.e., sum of operation times of its predecessors. Ties are broken in an arbitrary fashion or by another heuristic rule. Recently it has been shown that

C.A. Coello Coello (Ed.): LION 5, LNCS 6683, pp. 263–277, 2011.
© Springer-Verlag Berlin Heidelberg 2011

combining dispatching rules is promising [2], however, there is large number of rules to choose from and so combinations requires expert knowledge or extensive trial-and-error. A summary of over 100 classical dispatching rules can be found in [6].

The alternative to hand-crafting heuristics for the JSSP, is to implement an automatic way of learning heuristics using a data driven approach. Data can be generated using a known heuristic, such an approach is taken in [3], where a LPT-heuristic is applied. Then a decision tree is used to create a dispatching rule with similar logic. However, this method cannot outperform the original LPT-heuristic used to guide the search. For instruction scheduling this drawback is confronted in [4,7] by using an optimal scheduler, computed off-line. The optimal solutions are used as training data and a decision tree learning algorithm applied as before. Preferring simple to complex models, the resulting dispatching rules gave significantly more optimal schedules than using popular heuristics in that field, and a lower worst-case factor from optimality. A similar approach is taken for timetable scheduling in [8] using case based reasoning. Training data is guided by the two best heuristics for timetable scheduling. The authors point out that in order for their framework to be successful, problem features need to be sufficiently explanatory and training data need to be selected carefully so they can suggest the appropriate solution for a specific range of new cases.

In this work we investigate an approach based on supervised learning on optimal schedules and illustrate its effectiveness by improving upon well known dispatch rules for job-shop scheduling. The approach differs from previous studies, as it uses a simple linear combination of features found using a linear classifier. The method of generating training data is also shown to be critical for the success of the method. In section 2 priority dispatch rules for the JSSP problem are discussed, followed by a description of the linear classifier in section 3. An experimental study is then presented in section 4. The paper concludes with a summary of main findings.

2   Priority Dispatch Rules for Job-Shop Scheduling

The job-shop scheduling task considered here is where n jobs are scheduled on a set of m machines, subject to the constraint that each job must follow a predefined machine order and that a machine can handle at most one job at a time. The objective is to schedule the jobs so as to minimize the maximum completion times, also known as the makespan.

Each job j has an indivisible operation time on machine a, $p(j, a)$, which is assumed to be integral, where $j \in \{1, .., n\}$ and $a \in \{1, .., m\}$. Starting time of job j on machine a is denoted $x_s(a, j)$ and its completion time is denoted x_f and

$$x_f(a, j) = x_s(a, j) + p(j, a) \tag{1}$$

Each job has a specified processing order through the machines, it is a permutation vector, σ, of $\{1, .., m\}$. Representing a job j can be processed on $\sigma(j, a)$ only after it has been completely processed on $\sigma(j, a - 1)$, i.e.,

$$x_s(\sigma(j,a),j) \geq x_f(\sigma(j,a-1),j) \quad j \in \{1,..,n\}, \ a \in \{2,..,m\} \tag{2}$$

The disjunctive condition that each machine can handle at most one job at a time is the following:

$$x_s(a,i) \geq x_f(a,j) \quad \text{or} \quad x_s(a,j) \geq x_f(a,i) \tag{3}$$

for all $i,j \in \{1,..,n\}$ and $a \in \{1,..,m\}$. The time in which machine a is idle between jobs j and $j-1$ is called slack time,

$$s(a,j) = x_s(a,j) - x_f(a,j-1). \tag{4}$$

The makespan is the maximum completion time

$$z = \max\{x_f(j,m) \mid j = 1,..,n\}. \tag{5}$$

Dispatching rules are of a construction heuristics, where one starts with an empty schedule and adds on one job at a time. When a machine is free the dispatching rule inspects the waiting jobs and selects the job with the highest priority. The priority may depend on which job has the most work remaining (MWKR); least work remaining (LWKR); shortest immediate processing time (SPT); and longest immediate processing time (LPT). These are the most effective dispatching rules. However there are many more available, e.g. randomly selecting an operation with equal possibility (RND); minimum slack time (MST); smallest slack per operation (S/OP); and using the aforementioned dispatching rules with predetermined weights. A survey of more than 100 of such rules was given in 1977 by [6]. It has recently been shown that a careful combination of basic dispatching rules can perform significantly better [9].

In order to apply a dispatching rule a number of features of the schedule being built must be computed. The features of particular interest were obtained from inspecting the aforementioned single priority-based dispatching rules. Some features are directly observed from the partial schedule. The temporal scheduling features applied in this paper for a job j to be dispatched on machine a are: 1) processing time for job j on its next machine u; 2) work remaining for job j; 3) start-time of job j; 4) end-time of j; 5) when machine a is next free; 6) current makespan for all jobs; 7) slack time for machine a; 8) slack time for all machines; and 9) slack time weighted w.r.t number of number of jobs already dispatched. Fig. 1 shows an example of a temporal partial schedule for a six job and six machine job-shop problem. The numbers in the boxes represent the job identification j. The width of the box illustrates the processing times for a given job for a particular machine M_i (on the vertical axis). The dashed boxes represent the resulting partial schedule for when a particular job is scheduled next. As one can see, there are 17 jobs already scheduled, and 6 potential jobs to be dispatched next. If the job with the shortest processing time were to be scheduled next then job 4 would be dispatched. A dispatch rule may need to perform a one-step look-ahead and observes features of the partial schedule to make a decision, for example by observing the resulting temporal makespan.

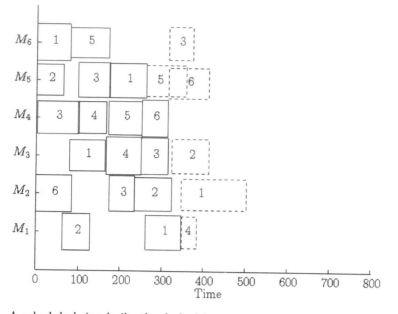

Fig. 1. A schedule being built, the dashed boxes represent six different possible jobs that could be scheduled next using a dispatch rule

These resulting observed features are sometimes referred to as an *after-state* or *post-decision state*. Other dispatch rules use features not directly observable from the current partial schedule, for example by assigning jobs with most total processing time remaining.

Problem instances are generated stochastically by fixing the number of jobs and machines and sampling a discrete processing time from the uniform distribution $U(R, 100)$. The machine order is a random permutation. Two different processing times were explored, namely $U(50, 100)$ and $U(1, 100)$ for all machines. For each processing time distribution 500 instances were generated for a six job and six machine job-shop problem. Their optimal solution were then found using the GNU linear programming kit [10]. The optimal solutions are used to determine which job should be dispatched in order to create an optimal schedule and which ones are not. When a job is dispatched the features of the partial schedule change. The aim of the linear learning algorithm, discussed in the following section, is to determine which features are better than others. That is, features created when a job is scheduled in order to build the known optimal solution as opposed to features generated by dispatching jobs that will result in a sub-optimal schedule.

3 Logistic Regression

The preference learning task of linear classification presented here is based on the work presented in [11,12]. The modification relates to how the point pairs are selected and the fact that a $L2$-regularized logistic regression is used.

Let $\phi^{(o)} \in \mathbb{R}^d$ denote the post-decision state when the job dispatched corresponds to an optimal schedule being built. All post-decisions states corresponding to suboptimal dispatches are denoted by $\phi^{(s)} \in \mathbb{R}^d$. One could label which feature sets were considered optimal, $\mathbf{z}_o = \phi^{(o)} - \phi^{(s)}$, and suboptimal, $\mathbf{z}_s = \phi^{(s)} - \phi^{(o)}$ by $y_o = +1$ and $y_s = -1$ respectively. Note, a negative example is only created as long as the job dispatched actually changed the resulting makespan, since there can exist situations in which more than one choice can be considered optimal.

The preference learning problem is specified by a set of preference pairs:

$$S = \left\{ \left\{ \phi^{(o)} - \phi_j^{(s)}, +1 \right\}_{k=1}^{\ell}, \left\{ \phi_j^{(s)} - \phi^{(o)}, -1 \right\}_{k=1}^{\ell} \mid \forall j \in J^{(k)} \right\} \subset \Phi \times Y \quad (6)$$

where $\Phi \subset \mathbb{R}^d$ is the training set of d features, $Y = \{-1, +1\}$ is the outcome space, $\ell = n \times m$ is the total number of dispatches and $j \in J^{(k)}$ are the possible suboptimal dispatches at dispatch (k). In this study, there are $d = 9$ features, and the training set is created from known optimal sequences of dispatch.

Now consider the model space $h \in \mathcal{H}$ of mappings from points to preferences. Each such function h induces an ordering $\succ$ on the points by the following rule:

$$\phi^{(o)} \succ \phi^{(s)} \quad \Leftrightarrow \quad h(\phi^{(o)}) > h(\phi^{(s)}) \quad (7)$$

where the symbol $\succ$ denotes "is preferrred to". The function used to induce the preference is defined by a linear function in the feature space:

$$h(\phi) = \sum_{i=1}^{d} w_i \phi_i. \quad (8)$$

Let $\mathbf{z}$ denote either $\phi^{(o)} - \phi^{(s)}$ with $y = +1$ or $\phi^{(s)} - \phi^{(o)}$ with $y = -1$ (positive or negative example respectively). Logistic regression learns the optimal parameters $\mathbf{w} \in \mathbb{R}^d$ determined by solving the following task:

$$\min_{\mathbf{w}} \quad \tfrac{1}{2}\langle \mathbf{w} \cdot \mathbf{w} \rangle + C \sum_{i=1}^{l} \log \left(1 + e^{-y_i \langle \mathbf{w} \cdot \mathbf{z}_i \rangle} \right) \quad (9)$$

where $C > 0$ is a penalty parameter, and the negative log-likelihood is due to the fact the given data points $\mathbf{z}$ and weights $\mathbf{w}$ are assumed to follow the probability model:

$$P(y = \pm 1 | \mathbf{z}, \mathbf{w}) = \frac{1}{1 + e^{-y \langle \mathbf{w} \cdot \mathbf{z} \rangle}}. \quad (10)$$

The logistic regression defined in (9) is solved iteratively, in particular using Trust Region Newton method [12], which generates a sequence $\{\mathbf{w}^{(k)}\}_{k=1}^{\infty}$ converging to the optimal solution $\mathbf{w}^*$ of (9).

The regulation parameter C in (9), controls the balance between model complexity and training errors, and must be chosen appropriately. It is also important to scale the features ϕ first. A standard method of doing so is by scaling the training set such that all points are in some range, typically $[-1, 1]$. That is, scaled $\tilde{\phi}$ is

$$\tilde{\phi}_i = 2(\phi_i - \underline{\phi}_i)/(\overline{\phi}_i - \underline{\phi}_i) - 1 \quad i = 1, \ldots, d \tag{11}$$

where $\underline{\phi}_i$, $\overline{\phi}_i$ are the maximum and minimum i-th component of all the feature variables in set Φ. Scaling makes the features less sensitive to process times.

Logistic regression makes optimal decisions regarding optimal dispatches and at the same time efficiently estimates a posteriori probabilities. The optimal $\mathbf{w}^*$ obtained from the training set, can be used on any new data point, ϕ, and their inner product is proportional to probability estimate (10). Hence, for each feasible job j that may be dispatched, ϕ_j denotes the corresponding post-decision state. The job chosen to be dispatched, j^*, is the one corresponding to the highest preference estimate, i.e

$$j^* = \underset{j}{\operatorname{argmax}}\, h(\phi_j) \tag{12}$$

where $h(\cdot)$ is the linear classification model (lin) obtained by the training data.

4   Experimental Study

In the experimental study we investigate the performance of the linear dispatching rules trained on problem instance generated using production times according to distributions $U(1, 100)$ and $U(50, 100)$. The resulting linear models is referred to as $lin_{U(1,100)}$ and $lin_{U(50,100)}$, respectively. These rules are compared with the single priority dispatching rules mentioned previously. The goal is to minimize the makespan, here the optimum makespan is denoted μ_{opt}, and the makespan obtained from a dispatching rule by μ_{DR}. Since the optimal makespan varies between problem instances the following performance measure is used:

$$\rho = \frac{\mu_{\text{DR}}}{\mu_{\text{opt}}} \tag{13}$$

which is always greater or equal to 1.

There were 500 problem instances generated using six machines and six jobs, for both $U(1, 100)$ and $U(50, 100)$ processing times distributions. Throughout the experimental study, a Kolmogorov-Smirnov goodness-of-fit hypothesis test with a significance level 0.05 is used to check if there is a statistical difference between the models in question.

4.1   Data Generation

An optimal sequence of job dispatches is known for each problem instance. The sequence indicates in which order the jobs should be dispatched. A job is placed at the earliest available time slot for its next machine, whilst still fulfilling constraints (2) and (3). Unfinished jobs are dispatched one at a time according to the optimal sequence. After each dispatch the schedule's current features are updated based on the half-finished schedule. This sequence of job assignments is by no means unique. Take for instance Fig. 1, let's say job #1 would be dispatched next, and in the next iteration job #2. Now this sequence would yield the same schedule as if job #2 would have been dispatched first and then job #1 in the next iteration. In this particular instance one could not infer that choosing job #1 is optimal and #2 is suboptimal (or vice versa) since they can both yield the same optimal solution, however the state of the schedule has changed and thus its features. Care must be taken in this case that neither resulting features are labeled as undesirable. Only the resulting features from a dispatch resulting in a suboptimal solution should be labeled undesirable. This is the approach taken here. Nevertheless, there may still be a chance that having dispatched a job resulting in a different makespan would have resulted in the same makespan if another optimal scheduling path were to have been chosen. That is, there are multiple optimal solutions to the same problem instance. We will ignore this for the current study, but note that our data may be slightly corrupted for this reason. In conclusion, at each time step a number of feature pair are created, they consist of the features resulting from optimal dispatch versus features resulting from suboptimal dispatches.

When building a complete schedule $n \times m$ dispatches must be made sequentially. At each dispatch iteration a number of data pairs are created which can then be multiplied by the number of problem instance created. We deliberately create a separate data set for each dispatch iterations, as our initial feeling is that dispatch rules used in the beginning of the schedule building process may not necessarily be the same as in the middle or end of the schedule. As a result we will have $n \times m$ linear scheduling rules for solving a $n \times m$ JSSP.

4.2   Training Size and Accuracy

Of the 500 schedule instances, 20% were devoted solely to validation, in order to optimize the parameters of the learning algorithm. Fig. 2 shows the ratio from optimum makespan, ρ in (13), of the validation set as a function of training size for both processing time distributions considered. As one might expect, a larger training set yields a better result. However, a training size of only 200 is deemed sufficient for both distributions, and will be used here on after, yielding the remaining unused 200 instances as its test set. The training accuracy reported by the lin-model during training with respect to choosing the optimal job at each time step is depicted in Fig. 3 for both data distribution considered. The models obtained from using the training set corresponding to $U(1, 100)$ and $U(50, 100)$ data distributions are referred to as $lin_{U(1,100)}$ and $lin_{U(50,100)}$, respectively. The training accuracy, that is the ability to dispatch jobs according to an optimal solution, increases as more jobs are dispatched. This seems reasonable since the features initially have little meaning and hence are contradictory. It becomes easier to predict good dispatches towards the end of the schedule. This illustrates the care needed in selecting training data for learning scheduling rules.

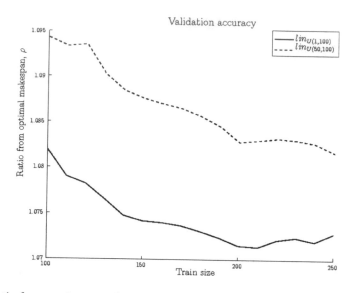

Fig. 2. Ratio from optimum makespan, ρ, for the validation set as a function of size of training set. Solid line represents model $lin_{U(1,100)}$ and dashed line represents model $lin_{U(50,100)}$

Fig. 3. Training accuracy as a function of sequence of dispatching decisions. Solid line represents model $lin_{U(1,100)}$ and dashed line represents data distributions $lin_{U(50,100)}$

Table 1. Mean value, standard deviation, median value, minimum and maximum values of the ratio from optimum makespan, ρ, using the test sets $U(1, 100)$ (top) and $U(50, 100)$ (bottom)

$U(1, 100)$	mean	std	med	min	max
$lin_{U(1,100)}$	1.0842	0.0536	1.0785	1.0000	1.2722
SPT	1.6707	0.2160	1.6365	1.1654	2.2500
$MWRM$	1.2595	0.1307	1.2350	1.0000	1.7288
$LWRM$	1.8589	0.2292	1.8368	1.2907	2.6906

$U(50, 100)$	mean	std	med	min	max
$lin_{U(50,100)}$	1.0724	0.0446	1.0713	1.0000	1.2159
SPT	1.7689	0.2514	1.7526	1.2047	2.5367
$MWRM$	1.1835	0.0994	1.1699	1.0217	1.5561
$LWRM$	1.9422	0.2465	1.9210	1.3916	2.6642

4.3   Comparison with Single Priority Dispatching Rules

The performance of the two learned linear priority dispatch rules, ($lin_{U(1,100)}$, $lin_{U(50,100)}$), are now compared with the three most common single priority-based dispatching rules from the literature, which dispatch according to: operation with shortest processing time (SPT), most work remaining ($MWRM$), and least work remaining ($LWRM$). Their ratio from optimum, (13), is depicted in Fig. 4, and corresponding statistical findings are presented in Table 1. Clearly model $lin_{U(R,100)}$ outperforms all conventional single priority-based dispatching rules, but of them $MWRM$ is the most successful. It is interesting to note that for both data distributions, the worst-case scenario (right tail of the distributions) for model $lin_{U(R,100)}$ is noticeably better than the mean obtained using dispatching rules SPT and $LWRM$, so the choice of an appropriate single dispatching rule is of paramount importance.

4.4   Robustness towards Data Distributions

All features are scaled according to (11), which may enable the dispatch rules to be less sensitive to the different processing time distributions. To examine this the dispatch rules $lin_{U(1,100)}$ and $lin_{U(50,100)}$ are tested on both $U(1, 100)$ and $U(50, 100)$ test sets. The statistics for ρ are presented in Table 2. There is no statistical difference between series #1 and #4, implying that when the dispatch rules are tested on their corresponding test set, they perform equally well. It is also noted that there is no statistical difference between series #2 and #4, implying that rule $lin_{U(50,100)}$ performed equally well on both test sets in question. However, when observing at the test sets, then in both cases there is a statistical difference between applying model $lin_{U(1,100)}$ or $lin_{U(50,100)}$, where the latter yielded a better results. This implies that the rules are actually not robust towards different data distributions in some cases. This is as one may have expected.

Table 2. Mean value, standard deviation, median value, minimum and maximum values of the ratio from optimum makespan, ρ, for the test sets $U(1, 100)$ and $U(50, 100)$, on both models $lin_{U(1,100)}$ and $lin_{U(50,100)}$

	model	test set	mean	std	med	min	max
#1	$lin_{U(1,100)}$	$U(1, 100)$	1.0844	0.0535	1.0786	1.0000	1.2722
#2	$lin_{U(50,100)}$	$U(1, 100)$	1.0709	0.0497	1.0626	1.0000	1.2503
#3	$lin_{U(1,100)}$	$U(50, 100)$	1.1429	0.1115	1.1158	1.0000	1.5963
#4	$lin_{U(50,100)}$	$U(50, 100)$	1.0724	0.0446	1.0713	1.0000	1.2159

Table 3. Feature description and mean weights for models $lin_{U(1,100)}$ and $lin_{U(50,100)}$

Weight	$lin_{U(1,100)}$	$lin_{U(50,100)}$	Feature description
$\bar{w}(1)$	-0.6712	-0.2220	processing time for job on machine
$\bar{w}(2)$	-0.9785	-0.9195	work remaining
$\bar{w}(3)$	-1.0549	-0.9059	start-time
$\bar{w}(4)$	-0.7128	-0.6274	end-time
$\bar{w}(5)$	-0.3268	0.0103	when machine is next free
$\bar{w}(6)$	1.8678	1.3710	current makespan
$\bar{w}(7)$	-1.5607	-1.6290	slack time for this particular machine
$\bar{w}(8)$	-0.7511	-0.7607	slack time for all machines
$\bar{w}(9)$	-0.2664	-0.3639	slack time weighted w.r.t. number of operations already assigned

Table 4. Mean value, standard deviation, median value, minimum and maximum values of the ratio from optimum makespan, ρ, on models $lin_{U(1,100)}$, $lin_{U(50,100)}$, $lin_{U(1,100),\text{fixed } w}$ and $lin_{U(50,100),\text{fixed } w}$ for corresponding test sets

	model	test set	mean	std	med	min	max
#1	$lin_{U(1,100)}$	$U(1, 100)$	1.0844	0.0535	1.0786	1.0000	1.2722
#2	$lin_{U(1,100),\text{fixed } w}$	$U(1, 100)$	1.0862	0.0580	1.0785	1.0000	1.2722
#3	$lin_{U(50,100)}$	$U(50, 100)$	1.0724	0.0446	1.0713	1.0000	1.2159
#4	$lin_{U(50,100),\text{fixed } w}$	$U(50, 100)$	1.0695	0.0459	1.0658	1.0000	1.2201

4.5   Fixed Weights

Here we are interested in examining the sensitivity of the weights found for our linear dispatching rules. The weights found for each feature at each sequential dispatching step for models $lin_{U(1,100)}$ and $lin_{U(50,100)}$ are depicted in Fig. 5. These weights are averaged and listed along side their corresponding features in Table 3. The sign and size of these weights are similar for both distributions, but with the exception of features 5 and 1. The average weights are now used throughout the sequence of dispatches, these models are called $lin_{U(1,100),\text{fixed } w}$ or $lin_{U(50,100),\text{fixed } w}$, respectively.

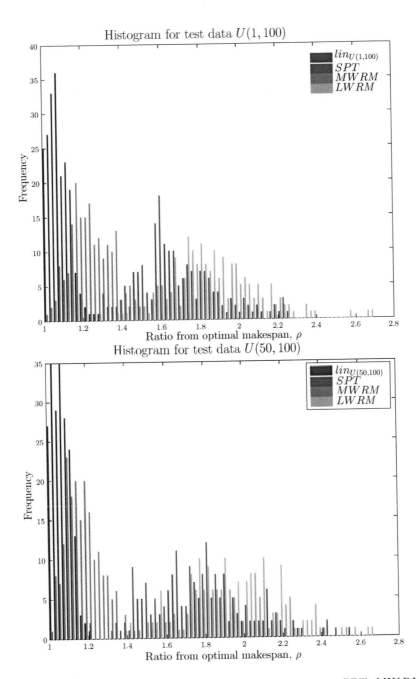

Fig. 4. Histogram of ratio ρ for the dispatching rules $lin_{U(R,100)}$, SPT, $MWRM$ and $LWRM$ for models $lin_{U(1,100)}$ (top) and $lin_{U(50,100)}$ (bottom)

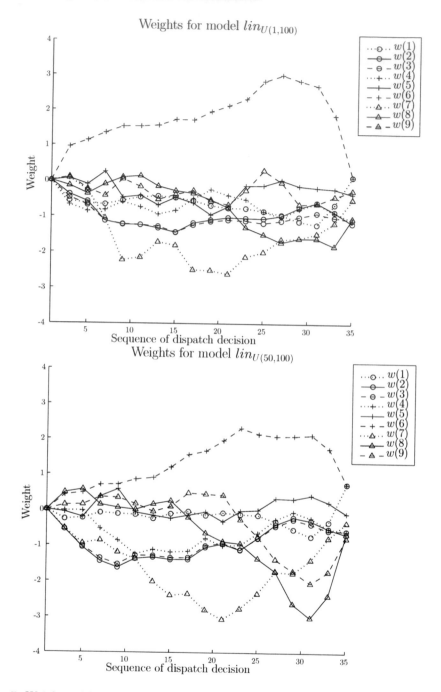

Fig. 5. Weights of features as a function of sequence of dispatching decisions, for test data $U(1, 100)$ (top) and $U(50, 100)$ (bottom)

Table 5. Mean value, standard deviation, median value, minimum and maximum values of the ratio from optimum makespan, ρ, for the test sets $U(1, 100)$ and $U(50, 100)$, on both fixed weight models $lin_{U(1,100),\text{fixed } w}$ and $lin_{U(50,100),\text{fixed } w}$

	model	test set	mean	std	med	min	max
#1	$lin_{U(1,100),\text{fixed } w}$	$U(1, 100)$	1.0862	0.0580	1.0785	1.0000	1.2722
#2	$lin_{U(50,100),\text{fixed } w}$	$U(1, 100)$	1.0706	0.0493	1.0597	1.0000	1.2204
#3	$lin_{U(1,100),\text{fixed } w}$	$U(50, 100)$	1.1356	0.0791	1.1296	1.0000	1.5284
#4	$lin_{U(50,100),\text{fixed } w}$	$U(50, 100)$	1.0695	0.0459	1.0658	1.0000	1.2201

Experimental results in Table 4 indicate that the weights could be held constant since there is no statistical difference between series #1 and #2 and series #3 and #4, i.e. no statistical difference between using varied or fixed weights for both data distributions. Hence, a simpler model using fixed weights should be preferred to the one of varied weights. The experiment described in section 4.4 is also repeated for fixed weights, and its results are listed in Table 5. As for varied weights (cf., Table 2), there is no statistical difference between models #2 and #4. However, unlike using varied weights, there exists a statistical difference between series #1 and #4. Again, looking at the test sets, in both cases there is statistical difference between applying model $lin_{U(1,100),\text{fixed } w}$ or $lin_{U(50,100),\text{fixed } w}$, where the latter yielded again the better result.

5 Summary and Conclusion

In this paper, a supervised learning linear priority dispatch rules (lin) is investigated to find optimal schedules for JSSP w.r.t. minimum makespan. The lin-model uses a heuristic strategy such that jobs are dispatched corresponding to the feature set that yielded the highest proportional probability output (12). The linear priority dispatch rules showed clear superiority towards single priority-based dispatch rules. The method of generating training data is critical for the framework's robustness.

The framework is not as robust with respect to different data distribution in some cases, and thus cannot be used interchangeably for training and testing and still maintain satisfactory results. Most features were of similar weight between the two data distributions (cf., Table 3), however, there are some slight discrepancies between the two distributions, e.g. $\bar{w}(5)$, which could explain the difference in performance between $lin_{U(1,00)}$ and $lin_{U(50,100)}$.

There is no statistical difference between using the linear model with varied or fixed weights when using a corresponding test set, so it is sufficient to apply only the mean varied weight, no optimization of the weight parameters is needed. It is noted that some of the robustness between data distribution is lost by using fixed weights. Hence, when dealing with a test set of known data distributions, it is sufficient to use the simpler fixed model $lin_{U(R,100),\text{fixed } w}$, however when the data distribution is not known beforehand, it is best to use the slightly more complex varied weights model, and inferring from the experimental data rather use $lin_{U(50,100)}$ to $lin_{U(1,100)}$.

It is possible for a JSSP problem to have more than one optimal solution. However for the purpose of this study, only one optimal solution used for generating training data is sufficient. But clearly the training data set is still corrupted because of multiple ways of representing the same or different (yet equally optimal w.r.t minimum makespan) optimal schedule. One way of overcoming this obstacle is applying mixed integer programming for each possible suboptimal choice, with the current schedule as its initial value to make it absolutely certain that the choice is indeed suboptimal or not.

The proposed approach of discovering learned linear priority dispatching rules introduced in this study, are only compared with three common single priority-based dispatching rules from the literature. Although they provide evidence of improved accuracy, other comparisons of learning approaches, e.g. genetic programming, regression trees and reinforcement learning, need to be looked further into.

Another possible direction of future research is to extend the obtained results to different types of scheduling problems, along with relevant features. The efficiency of this problem solver will ultimately depend on the skills of plausible reasoning and how effectively the features extrapolate patterns yielding rules concerning optimal solutions, if they exist.

The main drawback of this approach is in order for the framework to be applicable one needs to know optimal schedules and their corresponding features in order to learn the preference, which may be difficult if not impossible to compute beforehand for some instances of JSSP using exact methods.

References

1. Zhang, W., Dietterich, T.G.: A Reinforcement Learning Approach to Job-shop Scheduling. In: Proceedings of the Fourteenth International Joint Conference on Artificial Intelligence, pp. 1114–1120. Morgan Kaufmann, San Francisco (1995)
2. Tay, J., Ho, N.: Evolving dispatching rules using genetic programming for solving multi-objective flexible job-shop problems. Computers & Industrial Engineering 54(3), 453–473 (2008)
3. Li, X., Olafsson, S.: Discovering Dispatching Rules Using Data Mining. Journal of Scheduling 8(6), 515–527 (2005)
4. Malik, A.M., Russell, T., Chase, M., Beek, P.: Learning heuristics for basic block instruction scheduling. Journal of Heuristics 14(6), 549–569 (2007)
5. Garey, M., Johnson, D., Sethi, R.: The complexity of flowshop and jobshop scheduling. Mathematics of Operations Research 1(2), 117–129 (1976)
6. Panwalkar, S., Iskander, W.: A Survey of Scheduling Rules. Operations Research 25(1), 45–61 (1977)
7. Russell, T., Malik, A.M., Chase, M., van Beek, P.: Learning Heuristics for the Superblock Instruction Scheduling Problem. IEEE Transactions on Knowledge and Data Engineering 21(10), 1489–1502 (2009)

8. Burke, E., Petrovic, S., Qu, R.: Case-based heuristic selection for timetabling problems. Journal of Scheduling 9(2), 115–132 (2006)
9. Jayamohan, M.: Development and analysis of cost-based dispatching rules for job shop scheduling. European Journal of Operational Research 157(2), 307–321 (2004)
10. Makhorin, A.: GNU linear programming kit. Moscow Aviation Institute, Moscow, Russia, 38 (May 2009), Software available at
    `http://www.gnu.org/software/glpk/glpk.html`
11. Fan, R.e., Wang, X.r., Lin, C.j.: LIBLINEAR: A Library for Large Linear Classification. Corpus 9, 1871–1874 (2008), Software available at
    `http://www.csie.ntu.edu.tw/~cjlin/liblinear`
12. Lin, C.j., Weng, R.C.: Trust Region Newton Method for Large-Scale Logistic Regression. Journal of Machine Learning Research 9, 627–650 (2008)

Fine-Tuning Algorithm Parameters Using the Design of Experiments Approach

Aldy Gunawan, Hoong Chuin Lau, and Lindawati

School of Information Systems, Singapore Management University,
80 Stamford Road, S(178902), Singapore
{aldygunawan,hclau,lindawati.2008}@smu.edu.sg

Abstract. Optimizing parameter settings is an important task in algorithm design. Several automated parameter tuning procedures/configurators have been proposed in the literature, most of which work effectively when given a good initial range for the parameter values. In the Design of Experiments (DOE), a good initial range is known to lead to an optimum parameter setting. In this paper, we present a framework based on DOE to find a good initial range of parameter values for automated tuning. We use a factorial experiment design to first screen and rank all the parameters thereby allowing us to then focus on the parameter search space of the important parameters. A model based on the Response Surface methodology is then proposed to define the promising initial range for the important parameter values. We show how our approach can be embedded with existing automated parameter tuning configurators, namely ParamILS and RCS (Randomized Convex Search), to tune target algorithms and demonstrate that our proposed methodology leads to improvements in terms of the quality of the solutions.

Keywords: parameter tuning algorithm, design of experiments, response surface methodology.

1 Introduction

It is well-known that good parameter settings have a significant effect on the performance of an algorithm (Eiben et al., 1999; Hutter et al., 2010). For example, a simulated annealing algorithm is sensitive to the cooling factor, while a tabu search algorithm relies on a good choice of the tabu tenure. Many of the works we witness to date propose algorithms where the underlying parameters are set either arbitrarily without explanation, or conveniently choose parameter values that have been reported in previous studies.

In response to the need for a principled approach to find good parameter settings, several automated approaches have been proposed in recent years. For model-based approaches, Díaz and Laguna (2006) developed CALIBRA which employs a Taguchi fractional experimental design followed by a local search procedure. The former focuses on providing the starting point of the experiment, while the latter continues to search for the best parameter configuration. This procedure can only handle up to five parameters and focuses on the main effects of parameters without exploiting the

C.A. Coello Coello (Ed.): LION 5, LNCS 6683, pp. 278–292, 2011.

interaction effects between parameters. SPO+ (Hutter et al., 2010) is an improved model-based technique extended from the Sequential Parameter Optimization framework that constructs predictive performance models to focus attention on promising regions of a design space, aimed at tuning target algorithms with continuous parameters and a single problem instance at a time. F-Race (Birattari et al., 2002) is the specialization of the generic class of racing algorithms for configuration of metaheuristics.

For model-free approaches, Hutter et al. (2009) presented a local search approach, ParamILS, for algorithm configuration which is suited for discrete parameters. Again, ParamILS only considers changing one single parameter value at a time. Much potential in the use of statistical testing methods as well as RSM in algorithm configuration problems were also discussed. Randomized Convex Search (RCS) was recently proposed to handle both discrete and continuous parameter values (Lau and Xiao, 2009). The underlying assumption of RCS is that the points lie inside the convex hull of a certain number of the best points (parameter configurations).

The Design of Experiments (DOE) is a well-established statistical approach that involves experiment designs for the empirical modeling of processes (see for example Montgomery, 2005). Some typical applications of DOE include 1) evaluation and comparison of basic design configurations, 2) evaluation of different materials, and 3) selection of design parameters. The proposal for exploiting DOE for algorithm parameter tuning is in fact not new. Barr et al. (1995) discussed the design of computational experiments to test heuristic methods and provided guidelines for such experimentation. The performance of algorithm in computation experiments was affected by algorithm factors which include initial solution construction procedures and any parameters employed by the heuristic. The authors suggested the use of DOE in the process of planning an experiment.

Parsons and Johnson (1997) used statistical techniques, a central composite design embedded a fractional factorial design, to build a response surface for four parameters. This approach was applied to a genetic algorithm with applications to DNA sequence assembly. More recently, Ridge and Kudenko (2007) used the DOE approach to build a predictive model of the performance of a combinatorial optimization heuristic over a range of heuristic tuning parameter settings. However, the approach was only applicable to tuning Ant Colony System for the Travelling Salesman problem. There was no further comparison with other automated tuning approaches.

The Response Surface methodology (RSM) is a model-based approach within DOE that can be used to quantify the importance of each parameter, support interpolation of performance between parameter settings as well as extrapolation to previously-unseen regions of the parameter space (Hutter et al., 2010). Recently, Caserta and Voss (2009) adapted the RSM to fine-tune their Corridor Method for solving a block relocation problem in container terminal logistics. The values of parameters were restricted to discrete intervals due to the problem characteristics. Caserta and Voss (2010) presented a simple mechanism aimed at automatically fine tuning only a single parameter, the corridor width, of the corridor method for solving the DNA sequencing problem.

This paper describes a sequential experimental approach for screening and tuning algorithm parameters. Our approach is grounded on the DOE methodology as follows.

Consider an algorithm (called the target algorithm) to solve a particular problem that requires a number of parameters to be set prior to the execution of the algorithm. A factorial experiment design is applied to first screen and rank the parameters. Parameters which are determined to be *unimportant* (in that the solution quality is insensitive to the values of these parameters) are set to some constant values so that the resulting parameter space that needs to be explored is reduced. A first-order polynomial model based on RSM is then built to define the promising initial range for the important parameter values. We apply our proposed approach to two different automated tuning configurators, ParamILS (Hutter et al., 2009) and RCS (Lau and Xiao, 2009). Each configurator is applied to a target algorithm for solving the Traveling Salesman Problem (TSP) and Quadratic Assignment Problem (QAP), respectively.

In summary, the major contributions/highlights of this paper are as follows:

1. We propose the use of a factorial experiment design that enables to screen and rank the algorithm parameters. The screening process helps us to identify those unimportant parameters so they can be set into constant values. By focusing on important parameters, we reduce the parameter search space and target our search on the promising regions of the important parameter search space.
2. We propose the use of RSM to define the promising initial range for important parameter values that can be embedded to automated tuning procedures for improving the quality of solutions.

The remainder of this paper is organized as follows. Section 2 describes our proposed automated tuning framework. Section 3 provides a computational analysis of our proposed approach applied to two problems. Finally, we provide some concluding perspectives and future research plans in Section 4.

2   Automated Tuning Framework

The Automated Tuning problem is defined as follows:

Definition: Given a target algorithm TA parameterized by a set of parameters X with their respective intervals, a set of training instances I_{tr}, and a meta-function H(x) that measures the algorithm performance on a fixed parameter setting x over a set of problem instances, the goal is to determine a configuration x* such that H(x*) is minimized over I_{tr}.

In this paper, we assume all parameters to lie within numeric intervals. An example of the function value $H(x)$ is the average percentage deviation of the solution values obtained by *TA* using x as the parameter setting from the optimal values over the given set of instances. In our paper, the goal is to optimize x over the given set of training instances I_{tr} and subsequently verify the quality of this parameter setting on a set of testing instances.

A high-level view of our proposed automated tuning framework is given in Figure 1. The framework consists of three phases, (1) screening, (2) exploration, and (3) exploitation phases. In the following, we discuss the details of each phase.

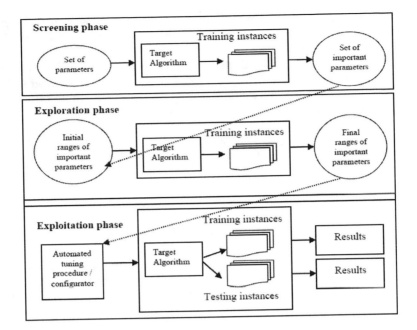

Fig. 1. Automated tuning framework

2.1 Screening Phase

Let k denote the number of parameters of the target algorithm to be tuned, and each parameter p_i (discrete or continuous) lies within a numeric interval $[l_i, u_i]$. In this phase, we perform screening to determine which parameters are significantly important thereby reducing the number of parameters under consideration. For this purpose, we apply a 2^k factorial design which consists of k parameters, where each parameter p_i only has two levels (l_i and u_i). A complete design requires $(2 \times 2 \times \ldots \times 2) \times n = n \times 2^k$ observations where n represents the number of replicates.

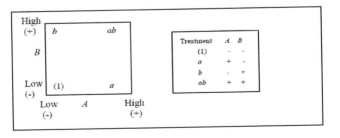

Fig. 2. The 2^2 factorial design

As an example, consider there are two parameters, A and B. Figure 2 shows the 2^2 design with treatment combinations are represented as the corners of the squares.

The signs + and − denote the values of l_i and u_i of each parameter p_i, respectively. In general, a treatment combination is represented by a series of lowercase letters (Montgomery, 2005). For example, treatment combination a indicates that parameters A and B are set to u_A and l_B, respectively. To estimate this treatment combination, we average n replications obtained. By using equations (1)–(3) and some other statistical testing (Montgomery, 2005), we can further examine the main effects of parameters A, B and the two-factor interaction AB as well.

$$A = \frac{1}{2n}[a + ab - b - (1)] \tag{1}$$

$$B = \frac{1}{2n}[b + ab - a - (1)] \tag{2}$$

$$AB = \frac{1}{2n}[ab + (1) - a - b] \tag{3}$$

The **importance** of a particular parameter is defined by conducting the test of significance on the main effect of the parameter. We choose a significance level ($\alpha = 5\%$) for our purpose. To further determine the **ranking** of the important parameters, we look at the absolute values of the main effects of those important parameters. By doing so, we can determine which parameters should be carefully controlled including the direction of adjustment for these parameters (see Figure 3 for illustration). The result in Figure 3 is obtained with the MINITAB statistical software.

Fractional Factorial Fit: Obj versus A, B, C

Estimated Effects and Coefficients for Obj (coded units)

Term	Effect	Coef	SE Coef	T	P
Constant		8.625	0.3062	28.17	0.000
A	1.750	0.875	0.3062	2.86	0.021
B	-3.000	-1.500	0.3062	-4.90	0.001
C	1.000	0.500	0.3062	1.63	0.141
A*B	-0.500	-0.250	0.3062	-0.82	0.438
A*C	-1.000	-0.500	0.3062	-1.63	0.141
B*C	-0.750	-0.375	0.3062	-1.22	0.256
A*B*C	0.250	0.125	0.3062	0.41	0.694

Analysis of Variance for Obj (coded units)

Source	DF	Seq SS	Adj SS	Adj MS	F	P
Main Effects	3	52.2500	52.2500	17.4167	11.61	0.003
2-Way Interactions	3	7.2500	7.2500	2.4167	1.61	0.262
3-Way Interactions	1	0.2500	0.2500	0.2500	0.17	0.694
Residual Error	8	12.0000	12.0000	1.5000		
Pure Error	8	12.0000	12.0000	1.5000		
Total	15	71.7500				

Fig. 3. Statistical results of the screening phase

From Figure 3, we observe that the main effects of A and B are significant since the p-values of both effects are less than 5%. In terms of ranking, B is the most dominant parameter, followed by A. Assuming that our objective function is a minimizing function, we modify the range of each significant parameter by the main effect value of the parameter. For instance, parameter A should be set to a *low* value since its coefficient is positive; hence the range of parameter A is modified to $[l'_A, u'_A] = [l_A, l_A + 2\Delta]$, where Δ is a constant. (This notation will become clear in the next section.)

For each unimportant parameter, we simply set to a constant value by the main effect value of the parameter; if the value is positive, we set the parameter to a low value (in our case, it is set to its lower bound l_c). The analysis of variance confirms our interpretation of the effect estimates. Both parameters A and B exhibit significant main effects.

2.2 Exploration Phase

Let m be the total number of important parameters ($m \leq k$) determined in the screening phase where each parameter p_i has a modified interval $[l'_i, u'_i]$ (as defined in Section 2.1) as well as its centre point value $(l'_i + u'_i)/2$. The Exploration phase is summarized in the following figure.

Procedure ExplorationPhase
Input: *TA*: Target Algorithm with m parameters,
$\quad\quad\Theta$: Parameter Configuration Space, defined by each parameter p_i having initial range
$\quad\quad\quad [l'_i, u'_i]$;
$\quad\quad I$: Set of Training Instances;
Output: Modified configuration space, each parameter with modified interval.

Procedure:
1: Run *TA* with respect to configuration space Θ on I;
2: Implement 2^{m+1} factorial design on m parameters ;
3: Conduct the interaction and curvature tests. If at least one of the tests is statistically significant, stop. Otherwise, go to Step 4;
4: Build a planar model of significant parameters;
5: Apply steepest descent to define a new centre point for each important parameter p_i;
6: Update the range of each important parameter p_i and generate a new $[l'_i, u'_i]$;
7: If at least one parameter p_i with either $l'_i < l_i$ or $u'_i > u_i$, stop. Otherwise, go to Step 1;

Fig. 4. Exploration phase

In essence, we begin with a small region and aim to find a "promising" range for important parameters using steepest descent on the response surface. The target algorithm is run with respect to the parameter configuration space Θ which contains 2^m+1 possible parameter settings (each parameter has two possible values, with an additional parameter setting defined by the centre point value of each parameter).

We apply a factorial experiment design in order to build a first-order (planar) model. The underlying assumption is that the region can be approximated by a planar model, which is a reasonable assumption when the region is sufficiently small and far from the optimum. The planar model is given by the following approximating function:

$$Y = \beta_0 + \beta_1 x_1 + ... + \beta_m x_m + \varepsilon \tag{4}$$

In order to test the significance of this model, we conduct two additional statistical tests:

— *Interaction test.* This test is mainly on testing whether any interaction between parameters. This can be done by looking at the significance of the estimated coefficient between two parameters (for instance, β_{ij}).
— *Curvature test.* This test is mainly on testing whether the planar model is adequate to represent the local response function.

As long as each test is not significant, we can always assume that the planar model is adequate to represent the true surface of parameters. We then continue the process by applying steepest descent that allows us to move rapidly to the vicinity of the optimum. More precisely, we move sequentially along the path of steepest descent in the direction of the maximum decrease in the response Y (Box and Wilson, 1951). The path is proportional to the signs and magnitudes of the equation (4). For example, if β_A (coefficient of parameter A) is the largest absolute coefficient value compared against other coefficient values, the step size of another parameter i is calculated by β_i/β_A. Several points along this path of steepest descent would be generated. A point with the minimum objective function value is then selected as the new centre point. A new set of l_i and u_i values for each parameter p_i as well as a new parameter configuration space Θ are then determined.

We illustrate the steepest descent step as follows. Assuming two parameters, A and B, where A has the larger absolute coefficient value (ties broken randomly). We first generate n possible values of x_A and x_B as follows: the values of x_A are set to arbitrary values (e.g., 0.1, 0.2, ..., 0.9), whereas the corresponding values of x_B are calculated by $(\beta_B/\beta_A) \times x_A$. Finally, the n possible parameter values for A and B are calculated as follows:

$$V_A^n = x_A \times \frac{(u_A - l_A)}{2} + \frac{(l_A + u_A)}{2} \tag{5}$$

$$V_B^n = x_B \times \frac{(u_B - l_B)}{2} + \frac{(l_B + u_B)}{2} \tag{6}$$

We then run the target algorithm with these n parameter values of A and B. The parameter setting with the minimum objective function value, denoted by V_A^{best} and V_B^{best}, is selected as a new centre point. The range of is parameter then modified as $[V_i^{best} - \Delta, V_i^{best} + \Delta]$ where Δ is a constant.

From statistical point of view, the region of planar local optimality is indicated by the existence of either interaction or curvature. Hence, we conduct the experiments until either interaction test or curvature test is statistically significant and proceed to the exploitation phase.

2.3 Exploitation Phase

In this phase, we drop the planarity assumption and devote our attention to finding the optimal point in the region output from the exploration phase. This is achieved by applying an automated tuning procedure, such as ParamILS (Hutter et al., 2009) or RCS (Lau and Xiao, 2009).

In this study, ParamILS is applied to tune the Iterated Local Search algorithm (Lourenco et al., 2003) for the Traveling Salesman Problem, while RCS is applied to the hybrid algorithm combining Simulated Annealing and Tabu Search (Ng et al., 2008) for the Quadratic Assignment Problem (QAP).

3   Experimental Results

In this section, we report a suite of computational results and analysis obtained from our proposed approach. All the experiments are run on a Intel (R) Core (TM)^2 Duo CPU 2.33 GHz with 1.96GB RAM that runs Microsoft Windows XP.

To evaluate the performance of our proposed automated tuning framework, we conduct two different experiments: 1) test ParamILS on Traveling Salesman Problem (TSP), and 2) test RCS on Quadratic Assignment Problem (QAP). For each experiment, two different scenarios, configurator+DOE (1st scenario) and configurator (2nd scenario), would be analyzed and compared. In this case, the amount of resources allocated (i.e. the number of iterations) are fixed. For instance, suppose the number of iterations of ParamILS and DOE are x and y respectively, the number of iterations of the 1st scenario is $x+y$, while the number of iterations of the 2nd scenario is set to z, with $z = x+y$.

The main purpose is to show that our approach can lead to improvements in terms of the gap (i.e. percentage deviation) between the average objective values of the solutions obtained by our approach against the best known solutions. We show that our proposed approach could provide better solutions for both discrete and continuous parameter values.

3.1   Traveling Salesman Problem (TSP)

The target algorithm to solve TSP is the Iterated Local Search (ILS). In this paper, we used the implementation from Halim et al. (2007). Four parameters that need to be tuned are as follows (Table 1):

— *Maximum_number_of_iterations* that limits the number of iterations for running the algorithm.
— *Perturbation_strength* that limits the number of times required for running the perturbation.
— *Non_improving_moves_tolerance* that limits the number of non-improving moves to be accepted.
— *Perturbation_choice* that selects the perturbation strategy.

Table 1. Parameter space for ILS on TSP

Parameters (p_i)	Range
Maximum_number_of_iterations (max_iter)	[100, 900]
Perturbation_strength (perturb)	[1, 10]
Non_improving_moves_tolerance (non_imprv)	[1, 10]
Perturbation_choice (opt_cho)	[3, 4]

In this screening process, the parameter space for *max_iter*, *perturb*, *non_improv* are reduced to [100, 500], [1, 5] and [1, 5] respectively. We started by selecting 47 instances from the 70 instances (TSPLIB) as training instances while the rest (23 instances) are treated as testing instances. For a particular parameter setting, we take the average of 10 runs on the training instances. The details of the experiment would be explained below.

3.1.1 Screening Phase

As described in Section 2.1, we focus on determining which parameters are significantly important. Figures 5 and 6 present the results of a 2^4 factorial design with $n = 10$ replicates using the factors mentioned in Table 1. The numerical estimates of the effects indicate that the effect of *max_iter*, *perturb*, and *non_imprv* are significant (with p-value < 5%), while the effect of *opt_cho* appears small. Based on the coefficient value of parameter *opt_cho* obtained, we decide to set the value of this parameter to its lower bound value (l_{opt_cho}). As we can see from Figure 6, only three parameters (*max_iter*, *perturb*, and *non_imprv*) have significant effects. The dotted line represents the cut-off limit associated with that significance level.

Fractional Factorial Fit: Dev versus Max_iter, Perturb, ...

Estimated Effects and Coefficients for Dev (coded units)

Term	Effect	Coef	SE Coef	T	P
Constant		3.9972	0.05187	77.06	0.000
Max_iter	-0.2430	-0.1215	0.05187	-2.34	0.021
Perturb	0.9924	0.4962	0.05187	9.57	0.000
Non_impr	0.2277	0.1139	0.05187	2.20	0.030
Opt_cho	0.1966	0.0983	0.05187	1.90	0.060
Max_iter*Perturb	-0.0676	-0.0338	0.05187	-0.65	0.515
Max_iter*Non_impr	-0.0225	-0.0112	0.05187	-0.22	0.829

Analysis of Variance for Dev (coded units)

Source	DF	Seq SS	Adj SS	Adj MS	F	P
Main Effects	4	45.376	45.3765	11.3441	26.35	0.000
2-Way Interactions	6	2.053	2.0528	0.3421	0.79	0.576
3-Way Interactions	4	0.992	0.9918	0.2479	0.58	0.681
4-Way Interactions	1	0.003	0.0025	0.0025	0.01	0.939
Residual Error	144	61.996	61.9964	0.4305		
Pure Error	144	61.996	61.9964	0.4305		
Total	159	110.420				

Fig. 5. Statistical results of the screening phase

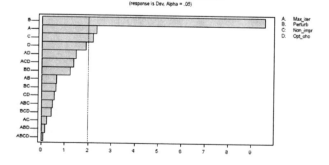

Fig. 6. Screening phase of ILS Algorithm

3.1.2 Exploration Phase

In this phase, we focus on three important parameters obtained from screening phase. We apply a factorial experiment design in order to build the first-order model. In order to test the significance of the first-order model, we conduct two additional statistical testing: interaction and curvature tests. As described earlier, as long as theses two additional tests are not significant, we can always assume that the first – order model is adequate to represent the true surface of parameters. Table 2 summarizes the parameter space of parameters along the path of the steepest descent.

Table 2. Parameter space for ILS Algorithm

Parameters	Range Exploration_1
max_iter	[400, 600]
Perturb	[1 ,3]
non_imprv	[4, 6]
opt_cho	3
Objective function value (%)	3.811

3.1.3 Exploitation Phase

In this phase, we use ParamILS to further explore neighbor parameters, given the information about the parameter values from exploration phase. Here, we would like to show that by using the DOE approach, we can provide a very good initial range for the parameter values.

Table 3. Parameter space for ILS on TSP

Parameters	Type	Range	
		ParamILS	ParamILS + DOE
Maximum_number_of_iteration	Discrete	[100, 900]	[400, 600]
Perturbation_strength	Discrete	[1, 10]	[1, 3]
Non_improving_moves_tolerance	Discrete	[1, 10]	[4, 6]
Perturbation_choice	Discrete	[3, 4]	3

Table 4. Parameter tuning for ILS on TSP

Algorithms	Mean
ParamILS (training instances)	2.653
ParamILS + DOE (training instances)	2.513
ParamILS (testing instances)	4.103
ParamILS + DOE (testing instances)	4.066

For comparison purpose, we also run ParamILS with the initial range for the parameter values (Table 3). The default parameter setting is based on the lower bound value of each parameter. The details tuning results for both ParamILS and ParamILS + DOE are given in Table 4. We observe that the results obtained by ParamILS + DOE are better than those of ParamILS. We can conclude that DOE approach could lead to improvements in terms of the solution quality. The percentage deviations between the average objective function value of the solutions obtained and the best known/optimal solutions are only 1.117 % and 1.710% for training and testing instances, respectively.

3.2 Quadratic Assignment Problem (QAP)

In this experiment, the target algorithm to solve QAP is the hybrid algorithm (Ng et al. 2008). The hybrid algorithm involves using the Greedy Randomized Adaptive Search Procedure (GRASP) to obtain an initial solution, and then using a combined Simulated Annealing (SA) and Tabu Search (TS) algorithm to improve the solution. There are four parameters to be tuned, which are listed as follows:

- Initial temperature of SA algorithm (*temp*)
- Cooling factor (*alpha*)
- Length of tabu list (*length*)
- Percentage of number of non-improvement iterations prior to intensification strategy (*pct*).

In order to evaluate the performance of our proposed approach, we decided to solve some benchmark problems from a library for research on the QAP (QAPLIB) which have been studied and solved by other researchers (Burkard et al., 1997). According to Taillard (1995), the instances of QAPLIB can be classified into four classes: unstructured (randomly generated) instances, grid-based distance matrix and real-life instances and real-life-like instances. Due to the limitation of the target algorithm that can only solve symmetric instances with zero diagonal values, we only focus on some instances from three classes: unstructured (randomly generated) instances, grid-based distance matrix and real-life instances.

3.2.1 Screening Phase

We selected a certain number of instances for training and testing instances for each class (Table 5). Table 6 summarizes the initial range for each parameter value. Only parameter *length* is a discrete parameter while the rest are continuous ones.

Table 5. Training and testing instances for each class

Class	Training instances	Testing instances
Unstructured (randomly generated) instances	11 instances	5 instances
Grid-based distance matrix	24 instances	11 instances
Real-life instances	14 instances	7 instances

Table 6. Parameter space for hybrid algorithm on QAP

Parameters	Type	Range
Temp	Continuous	[100, 7000]
Alpha	Continuous	[0.5, 0.95]
Length	Discrete	[5, 10]
Pct	Continuous	[0.01. 0.10]

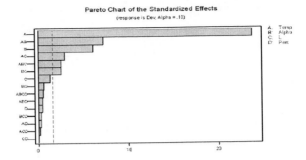

Fig. 7. Screening phase of the hybrid algorithm (unstructured instances)

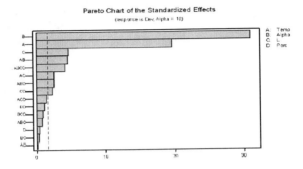

Fig. 8. Screening phase of the hybrid algorithm (grid-based distance matrix)

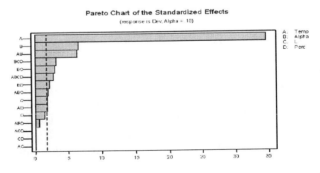

Fig. 9. Screening phase of the hybrid algorithm (real-life instances)

3.2.2 Exploration Phase

In this phase, we again focus on important parameters obtained from screening phase. By applying the same approach discussed in Section 3.1, we conduct the experiment until the first-order model is not appropriate for each class.

The parameter spaces of parameters along the path of the steepest descent are summarized in Tables 7, 8 and 9. We observe that the objective function value would decrease subsequently when we reach the promising region of the parameter values. The last column for each table represents the final range for each parameter that would be used as an input in exploitation phase.

Table 7. Parameter space for hybrid algorithm on QAP (unstructured instances)

Parameters	Range	
	Exploration_1	Exploration_2
Temp	[4000, 6000]	[4378, 6348]
Alpha	[0.85, 0.95]	[0.935, 0.945]
Length	5	5
Pct	0.01	0.01
Objective function value (%)	2.517	2.108

Table 8. Parameter space for hybrid algorithm on QAP (grid-based distance matrix)

Parameters	Range	
	Exploration_1	Exploration_2
Temp	[4000, 6000]	[4238, 6238]
Alpha	[0.85, 0.95]	[0.935, 0.945]
Length	[4, 6]	6
Pct	0.1	0.1
Objective function value (%)	0.591	0.425

Table 9. Parameter space for hybrid algorithm on QAP (real-life instances)

Parameters	Range
	Exploration_1
Temp	[4000, 6000]
Alpha	[0.85, 0.95]
Length	[4, 6]
Pct	0. 1
Objective function value (%)	9.255

3.2.3 Exploitation Phase

In this phase, the final range for parameter values obtained from exploration phase would be compared with the default configuration of RCS (Table 10). The results obtained by testing two different scenarios, RCS and RCS + DOE, are given in Table 11. We can conclude that RCS + DOE outperforms RCS in all groups of instances. We obtained improvements of results over RCS for both training and testing instances.

Table 10. Parameter space for hybrid algorithm on QAP

Parameters	Range			
	RCS	RCS + DOE (unstructured instances)	RCS + DOE (grid-based distance matrix)	RCS + DOE (real-life instances)
Temp	[100, 7000]	[4378, 6348]	[4238, 6238]	[4000, 6000]
Alpha	[0.5, 0.95]	[0.935, 0.945]	[0.935, 0.945]	[0.85, 0.95]
Length	[5, 10]	5	6	[4, 6]
Pct	[0.01. 0.10]	0.01	0.10	0.1

Table 11. Parameter Tuning for Hybrid Algorithm on QAP

Algorithms	Mean		
	(unstructured instances)	(grid-based distance matrix)	(real-life instances)
RCS (training instances)	1.100	0.630	3.264
RCS + DOE (training instances)	0.938	0.190	2.822
RCS (testing instances)	1.595	1.158	6.770
RCS + DOE (testing instances)	1.518	0.754	5.985

4 Conclusion

This paper proposes an automated tuning framework based on the Design of Experiments (DOE) approach. We demonstrate that our approach can be adapted to address the parameter tuning problem for target algorithms that find approximate solutions to two combinatorial optimization problems, TSP and QAP. We show that the proposed approach performs very well for both discrete and continuous parameter value settings.

One limitation of a factorial experiment design is that the number of experiments increases exponentially with the number of parameters. Fractional factorial designs offer a manageable alternative, which uses only some subset of a full factorial design's run.

In ParamILS and RCS, the neighborhoods of the current parameter setting are usually randomly selected. For future extensions to this work, we can consider using a second-order response surface model which is usually required when the experimenter is relatively close to the optimum.

References

1. Adenso-Diaz, B., Laguna, M.: Fine-Tuning of Algorithms Using Fractional Experimental Design and Local Search. Operations Research 54(1), 99–114 (2006)
2. Barr, R.S., Golden, B.L., Kelly, J.P., Resende, M.G.C., Stewart, W.R.: Designing and Reporting on Computational Experiments with Heuristic Methods. Journal of Heuristics 1, 9–32 (1995)
3. Birattari, M., Stützle, T., Paquete, L., Varrentrapp, K.: A Racing Algorithm for Configuring Metaheuristics. In: Proc. Of the Genetic and Evolutionary Computation Conference, pp. 11–18. Morgan Kaufmann, San Francisco (2002)
4. Box, G., Wilson, K.: On the Experimental Attainment of Optimum Conditions. Journal of the Royal Statistical Society Series b 13, 1–45 (1951)
5. Burkard, R.E., Karisch, S.E., Rendl, F.: QAPLIB – A Quadratic Assignment Problem Library. Journal of Global Optimization 10, 391–403 (1997)
6. Caserta, M., Voß, S.: A Math-Heuristic Algorithm for the DNA Sequencing Problem. In: Blum, C., Battiti, R. (eds.) LION 4. LNCS, vol. 6073, pp. 25–36. Springer, Heidelberg (2010)
7. Caserta, M., Voß, S.: Corridor Selection and Fine Tuning for the Corridor Method. In: Stützle, T. (ed.) LION 3. LNCS, vol. 5851, pp. 163–175. Springer, Heidelberg (2009)

8. Halim, S., Yap, R., Lau, H.C.: An Integrated White+Black Box Approach for Designing and Tuning Stochastic Local Search. In: Bessière, C. (ed.) CP 2007. LNCS, vol. 4741, pp. 332–347. Springer, Heidelberg (2007)
9. Hutter, F., Hoos, H.H., Leyton-Brown, K., Murphy, K.: Time-Bounded Sequential Parameter Optimization. In: Blum, C., Battiti, R. (eds.) LION 4. LNCS, vol. 6073, pp. 281–298. Springer, Heidelberg (2010)
10. Hutter, F., Hoos, H.H., Leyton-Brown, K., Stützle, T.: ParamILS: An Automatic Algorithm Configuration Framework. Journal of Artificial Intelligence Research 36, 267–306 (2009)
11. Lau, H.C., Xiao, F.: A Framework for Automated Parameter Tuning in Heuristic Design. In: 8th Metaheuristics International Conference, Hamburg, Germany (2009)
12. Lourenco, H.R., Martin, O.C., Stutzle, T.: Iterated Local Search. In: Glover, F., Kochenberger, G.A. (eds.) Handbook of Metaheuristics. International Series in Operations Research & Management Sci., vol. 57, pp. 320–353. Springer, Heidelberg (2003)
13. Montgomery, D.C.: Design and analysis of Experiments, 6th edn. John Wiley and Sons Inc., Chichester (2005)
14. Ng, K.M., Gunawan, A., Poh, K.L.: A hybrid Algorithm for the Quadratic Assignment Problem. In: Proc. International Conference on Conference on Scientific Computing, Nevada, USA, pp. 14–17 (2008)
15. Parsons, R., Johnson, M.: A Case Study in Experimental Design Applied to Genetic Algorithms with Application to DNA Sequence Assembly. Journal of Mathematical and Management Sciences 17(3), 369–396 (1997)
16. Ridge, E., Kudenko, D.: Tuning the Performance of the MMAS Heuristic. In: Stützle, T., Birattari, M., Hoos, H.H. (eds.) SLS 2007. LNCS, vol. 4638, pp. 46–60. Springer, Heidelberg (2007)
17. Taillard, E.D.: Comparison of Iterative Searches for the Quadratic Assignment Problem. Location Science 3(2), 87–105 (1995)

MetaHybrid: Combining Metamodels and Gradient-Based Techniques in a Hybrid Multi-Objective Genetic Algorithm

Alessandro Turco

ESTECO srl
alessandro.turco@esteco.com

Abstract. We propose a metamodel approach to the approximation of functions gradients within a hybrid genetic algorithm. The underlying structure is implemented in order to support parallel execution of the code: a genetic and a SQP algorithm run in different threads and can ask designs evaluations independently, but keeping all the available resources always working. A common archive collects the results and generates the population for the GA and the starting points for the SQP runs. A particular attention is dedicated to elitism and to constraints. The hybridization is performed through a modified $\epsilon-$constrained method. The general philosophy of the algorithm is to concentrate on not wasting information: metamodels, archiving and elitism, steady-state parallel evolution are key elements for this scope and they will be discussed in details. A preliminary but explanatory row of tests concludes the paper highlighting the benefits of this new approach.

Introduction

Hybridization is a common practice in several fields of optimization [2]. An hybrid algorithm is the combination of two (or more) different strategies for the solution of a single task. In principle, each single strategy could be applied independently to the problem, but it would focus only on some features. On the contrary, if the coupling is effective, a deeper comprehension is achievable and therefore a better or faster solution can be found.

Our proposal is a further development of this strategy. We start from a standard hybridization of a genetic algorithm (GA) with a gradient-based one and we enrich the implementation introducing several other techniques with the aim to exploit as much as possible the available resources. We consider as resources both the data collected and the computational efforts we can afford and they all do not have to be wasted.

Stored data can be very useful for each of the two main parts of the algorithm. The genetic algorithm needs a parent population fed by an efficient elitism operator: the information gained in the previous iterations must be distilled and made available for the following ones. We follow a previously detailed study on elitism [21] and we update some algorithmic choices. The hypothesized field is a

C.A. Coello Coello (Ed.): LION 5, LNCS 6683, pp. 293–307, 2011.

multi-objective framework and the target is to produce an approximate Pareto front accurate, uniform and well extended. However the same strategy can be successfully applied to single-objective problems, when robustness is the target and the optimization problem is multimodal. Also the gradient-based part of the algorithm can benefit from an appropriate managing of already collected information. We use a filter-based Sequential Quadratic Programming (SQP) algorithm [20]. The filter uses previously evaluated points to judge whether a new iterate can be accepted or not. The acceptance criterion involves at the same time the objective function and the constraints attainment.

Moreover, we propose a metamodel approach to gradients evaluation: we train a response surface (Radial Basis Function [3] or Polynomial SVD [16]) for each output variable using a suitable selection of points coming from previous iterates and we analytically extract gradients from the obtained metamodels. A finite differences approximation is most of the times more accurate, but it requires a large number of evaluations. The proposed benchmark tests highlight the benefits of our choice under different environmental conditions.

The efficient use of computing resources from the algorithmic and the implementation point of view is also taken in consideration. The use of a steady-state evolution helps in leaving idle as less resources as possible, but implies modification of some standard structures in the GA. Moreover, the use of a multi-threading framework has direct consequences on the hybridization mechanism: the two algorithms can run in parallel, while reading and writing data on a common (synchronized) archive.

Some concluding remarks ends the paper focusing on the novelties proposed. The algorithm presented is promising and we are planning to further investigate its potentials. But the most important message we hope to convey regards the intense use of all the available resources. Real-world problems cannot be solved using thousands of iterations as we are used to do dealing with mathematical benchmarks, since a single design evaluation can be very time (and not only) consuming. But this opens the door to heavy refinements of the algorithms which do not have anymore constraints on their own computational costs, under reasonable thresholds of course.

The paper is organized as follows: section §1 contains the details of the genetic algorithm which is hybridized with the SQP optimizer described in section §2. Metamodels are described in section §3. These elements are linked together in section §4, where the parallelization is explained and a global overview of the algorithm is presented. Section §5 concerns the preparation of the benchmark tests, while section §6 contains the results discussion and some final remarks.

1   GA Elements: Focus on Elitism

The backbone of the proposed MetaHybrid algorithm is a steady-state genetic algorithm. We assume as known the basic structure of a genetic algorithm, we simply cite the chosen operators: we work with variable-wise encoding, SBX crossover and probability based mutation; we use constraint-domination ranking and crowding distance [7].

We are interested in discussing elitism and selection. Given the ranking strategy, there remains two crucial choices: which are the individuals who will enter in the selection process? How does the selection process use the ranking? The MetaHybrid algorithm gathers the parent and the children population before ranking and then selects a new parent population using a refined idea of elitism The original mechanism is described in [7] where the population size is fixed and a sort of Darwinian law is strictly respected: only the best individuals survive.

Two different objections have been addressed to that original implementation of elitism: the controlled elitism approach [8] gives emphasis also to dominated but well-spread points in order to obtain a more uniform front. The variable population size [1] lets the population grow with the number of first-front points. The declared aim is to enhance the convergence speed and the extension of the obtained front. In a more recent paper [21] we propose to combine these two techniques switching from one to the other depending on the dimension of the actual non-dominated front.

The MetaHybrid algorithm implements a different combination which inherits the experience gained with the previous work. The population size is free to grow (up to an upper bound which is two times the initial size) but not to shrink. Whenever the number of points in the first front is bigger than the previous population size, all the non-dominated points become parents. If, in a successive iteration, the first-front size is less than the population size then all the non-dominated points become parents and the empty slots are filled by rarer points selected using the controlled elitism approach.

2 SQP Elements: Focus on Constraints

In this section we describe the chosen single-objective SQP algorithm and the modified $\epsilon-$constrained technique used to adapt it to a multi-objective environment. A detailed description of the basics of Sequential Quadratic Programming is out of the purposes of this paper, but we can sketch the main features and cite some useful readings.

The idea behind this class of algorithms is to use gradients information to build a quadratic approximation of the Lagrangian function associated to the objective function and the involved constraints around the current point [14]. A local optimization problem is defined and solved using this approximated function as objective function and a linearized version of the original constraints. The solution of this local problem will be the center of the following iteration.

This procedure is proved to be very efficient (quadratic rate of convergence) only when starting already nearby the solution of the problem, like the classic Newton method. There are however different choices for achieving global convergence [11]. We focus on the Filter technique introduced by Fletcher [10]. In order to avoid bad iterates due to inefficiencies of the local model or to flat gradients, this technique prescribes to keep memory of all the already accepted points and to judge the quality of a new one using the following criterion: the new point has to be non-dominated in a multi-objective framework where the first objective is the original one and a certain number of objectives are related to the constraints violations. The original algorithm of Fletcher uses only one "constraints-to-objective" built with the sum of all constraints violations. We use Adaptive Filter SQP [20] which is self adaptive and is able to recognize the order of magnitude of the constraints building an appropriate number of filter entries.

Several techniques for transforming a single-objective optimization strategy into a multi-objective one exist [13]. We use a mixing between the ϵ−constrained and the weighted sum methods with a particular attention to constraints attainment. The ϵ−constrained method consists in choosing one of the objectives and minimizing (or maximizing) it with one additional constraint for each of the remaining ones: the constraint is not to worsen the initial value for that objective more than ϵ. This method has been already successfully employed hybridizing GA and SQP algorithms [12,18]

We recall the original idea. Starting from a problem of the type:

$$\begin{cases} \min f_i(\mathbf{x}) & \text{for } i = 1 \ldots n, \\ g_j(\mathbf{x}) \leq 0 & \text{for } j = 1 \ldots m_i, \\ h_k(\mathbf{x}) = 0 & \text{for } k = 1 \ldots m_e, \end{cases} \tag{1}$$

the ϵ−constrained method will ask to solve the modified problem:

$$\begin{cases} \min f_\alpha(\mathbf{x}) & \alpha \in [1, n] \\ f_i(\mathbf{x}) < f_i(\mathbf{x}_0) + \epsilon & \alpha \neq i \in [1, n] \\ g_j(\mathbf{x}) \leq 0 & \text{for } j = 1 \ldots m_i, \\ h_k(\mathbf{x}) = 0 & \text{for } k = 1 \ldots m_e, \end{cases} \tag{2}$$

where $\mathbf{x}_0$ is the starting point for this problem and ϵ is a small parameter. This formulation can be applied only when $\mathbf{x}_0$ is a feasible point and if, for example, we know only one feasible point, then we will be able to reach only one Pareto point for each choice of α (assuming a perfectly deterministic SQP algorithm).

We propose a different implementation of this idea. First of all, we allow both feasible and unfeasible starting points. If $\mathbf{x}_0$ is feasible, then we fix n random numbers ϕ_i (summing up to 1) and solve:

$$\begin{cases} \min \sum_{i \in [1,n]} \phi_i f_i(\mathbf{x}) \\ f_i(\mathbf{x}) < f_i(\mathbf{x}_0) + \epsilon & i \in [1, n] \\ g_j(\mathbf{x}) \leq 0 & \text{for } j = 1 \ldots m_i, \\ h_k(\mathbf{x}) = 0 & \text{for } k = 1 \ldots m_e. \end{cases} \tag{3}$$

If otherwise the starting point is unfeasible, we remove the constraints on the objective functions obtaining a sort of feasibility recovering subproblem.

The multi-objective SQP and the GA can be combined, hybridized following several paths. A possible choice is the one implemented in [12,18]: once the population is updated, the new individuals pass through a run of the SQP algorithm before becoming parents for the subsequent generation. We follow a different strategy: we add the SQP run to the operators list. A parent individual will be the initial guess for the SQP run and the best iterate will become the child. The maximum number of generations is not fixed: if the new iterate point is not dominated (considering the original optimization problem) among all previous ones, the search continues. If otherwise, for a fixed number of iterations (we use 5), the newly generated points are dominated by the older ones, the SQP run is stopped.

The above ideas are implemented taking care of resource management under different points of view. The first and the most important element regards how we compute functions gradients and it will be the topic of the following section. All other features regarding the parallelization will be discussed in Sect. §4.

3 Metamodels Derivatives

The correspondence between input and output values in real-world problem usually is not accessible and must be considered as a black-box. Only in very favorable situations, for example, the employed finite elements solver is able to evaluate a design and to return also gradients relative to output variables. Therefore, when working with a gradient-based algorithm, derivatives must be approximated.

The most common and perhaps one of the more precise approximation technique is finite differences [19]. The idea is to rely on the basic definition of a derivative as the limit of a shrinking incremental ratio: evaluating some design around the configuration for which gradients are required, following an appropriate stencil and knowing the distances from the samples, it is possible to compute a very accurate approximate derivative for each coordinate direction and each output variable. The drawback of this technique is that even the less consuming stencil (forward or backward differences) requires the evaluation of one extra design for each input variable in order to build the complete gradients for a single configuration. Moreover, since these extra points must be very near to their reference one and must lie in strictly defined positions, the possibility of reusing already available information is extremely low and unpredictable.

Metamodels, or Response Surface Modeling (RSM), can help in performing this task without requiring extra evaluations since they can be trained over the database built during previous iterations. Metamodels are surrogates of the black-box functions which transform input variables to output variables. Among all the metamodel techniques there some which provides explicit formulas that can be analytical derived. In this work we use Polynomial Singular Value Decomposition [16] and Radial Basis Function [3] regression schemes: in both cases we end up with a set of weights assigned to a set of kernel explicit functions. The chain rule for derivation allow us to use this information to compute the required gradients.

Polynomial SVD technique selects the best approximating polynomial of the input variables over the training dataset. The maximum degree of the polynomial is fixed by the size of the dataset. Indeed, a linear system is obtained imposing the perfect fitting on the training points and the coefficients of the monomial are computed through its singular value decomposition. Therefore the metamodel obtained is nothing but the least squares solution of the above system. This technique is very effective when dealing with smooth functions, which unfortunately is not always the case. However the training time is very small compared to other RSMs and hence this technique is often used for a quick first glance on the problem. An hybrid algorithm can exploit this peculiarity profitably whenever the GA is stuck in some local minimum and the SQP must provide useful more than precise search directions.

Radial Basis Functions, on the contrary, are a powerful tool for multivariate scattered data interpolation and they can model efficiently non-smooth functions. If we consider an unknown function $f(\mathbf{x}) : \mathbb{R}^n \to \mathbb{R}$ and a training set of I couples $(\mathbf{x}_i, f(\mathbf{x}_i) := f_i)$, then the RBF interpolant approximating function will have the form:

$$\hat{f}(\mathbf{x}) = \sum_{j=1}^{I} c_j \phi(\|\mathbf{x} - \mathbf{x}_j\|/\delta),$$

where δ is a suitable scaling parameter, $\|\cdot\|$ is the standard Euclidean distance and $\phi(\cdot)$ is the radial or kernel function. Literature reports several choices for this function, in this work we use the Hardy's MultiQuadrics:

$$\phi(r) = \sqrt{r^2 + c^2},$$

where c is a scalar parameter that will be fixed during the training (together with δ) in order to maximize the precision of the metamodel.

Both techniques require a training dataset of truly evaluated points (we keep the size of the training database equal to the population size), however we can choose them among the already computed designs. The choice criterion must consider the metamodels structure and the information we want to extract from them. We use response surfaces in order to compute gradients and we train them from scratch each time a new gradient is required, therefore we are interested in a very restricted area of the design space. On the contrary, a training set composed by too near points would produce high errors due to side effects, extrapolation and overfitting. Moreover, a very refined search implies high computational costs either if we scan the whole database at each metamodel creation, or if we keep track of the closeness relations among points.

Our proposal is an incremental random search. We compute the hyper-volume of the input variable space and we start picking randomly points in the database: if the distance between the selected point and the location where the gradient is needed is less than 1% of the hyper-volume, we accept the point. If, after a fixed number of trials (usually, one tenth of the database size), the required number of training points has not been reached, then we relax the threshold accepting points up to 2% of the hyper-volume and so on. Moreover, the points explored by the SQP algorithm are added to the database in order to improve the precision of the metamodels in the region of interest.

4   Hybridization in a Parallel Environment

We designed the interaction (hybridization) between the genetic and the SQP algorithms supposing to work in a parallel computing environment. The resulting code can be run also using only one computing resource (exploiting the multi-threading capabilities of Java programming), but it is able to interface with an arbitrary large number of concurrent design evaluations without leaving resources idle and making new obtained information available immediately.

As already mentioned, the backbone of the MetaHybrid algorithm is a steady-state genetic evolution: a new child individual is produced as soon as an idle computing resource is found or a previously occupied one is freed. The parents are chosen randomly among the actual population. This set of individual can be updated without stopping the whole process and therefore avoiding the usual bottleneck due to sorting and selecting routines. We follow two schemata: an on-the-fly insertion (very quick, it does not create any delay) and a periodic update (which is more demanding, but the required effort can be considered negligible compared to the one required by design evaluation).

The on-the-fly update is performed each time a new child is evaluated. We check only if it is dominated by one of the parents: if this is not the case, this child become immediately part of the parent population. The periodic population update is performed each time the GA evaluates a number of points equal to the population size: it follows the elitism operator described in Sect. §1 including in the set of individuals to be sorted also the points coming from concurrently SQP steps.

The SQP algorithm is inserted in the GA evolution as an extra operator which acts only on some randomly selected non-dominated parents. Each run operates in a new thread executed in parallel with the GA by the master machine and the required designs evaluations will be performed by a dedicated resource (the SQP algorithm is by definition sequential and therefore it requires only one evaluation at the time). The points explored by the SQP algorithm runs are collected in a dedicated archive which will be scanned during the periodic population update.

The whole process can be logically decomposed as follows:

1. Creation of the parent population from an initial DOE (Design Of Experiments) or performing a tournament selection among the actual population.
2. Recombination: mutation, crossover and SQP operators transform parents into children.
3. Archiving: all the intermediate points generated by SQP are stored in a dedicated archive.
4. On-the-fly update: if a non-dominated point is created, it becomes immediately part of the parent population. Otherwise it is stored in the children archive.
5. Periodic update: the archives are gathered and elitism is applied. A new population is built and the loop is restarted.

However, as discussed above, this is not a true loop. Steps 2, 3 and 4 form an inner loop which evolve without waiting the completion of the outer one. This is possible thanks to a master-slave architecture. The (relative) long sorting operations are executed on the master node while some points are being evaluated by the slaves.

5   Tests

We validate the proposed algorithm testing it on six different benchmarks. We chose three unconstrained problems (ZDT4 [23], CEC '09 UP2 [22] and Sym-Part [17]), two constrained ones (a modified version of OSY [15], where we rotate the problem and we enlarge the input variables range in order to avoid side-effects and CTP2 [9]) and a single-objective multi-modal problem (a ten-atom Lennard-Jones cluster [4]). All these problems are quite known and widely used for benchmarking optimization algorithm, therefore we will not report here their formulation which can be found in the cited papers.

We measure the quality of the achieved non-dominated fronts comparing them to the true Pareto fronts using IGD performance metric [24]. We run each algorithm 10 times (varying the random number generator seed and starting from 10 common different initial populations) and we collect the mean IGD value obtained. This metric computes the distance between a non-dominated front A and a reference set of points P which are assumed to belong to the Pareto set (we use only benchmark problems with a well known structure). The mathematical formulation of this metric is the following:

$$\mathrm{IGD}(A, P) = \frac{\sum_{p \in P} d(p, A)}{|P|},$$

where $d(p, A)$ is the minimum Euclidean distance between p and all the points in A. Low values of IGD are desirable, since this implies that the set A contains points near to **any** point in P. This means that the metric measures the accuracy, the uniformity and the extent of the front at the same time.

The comparison is made among the following algorithms:

- NSGA–II. This algorithm is widely accepted as state-of-the-art for general purpose GA. We set its parameters as described in the original paper [7].
- Only GA. This algorithm is the genetic part of metaHybrid. We obtain it setting as zero the probability for the SQP operator. Crossover and Mutation are tuned with the same parameters ad NSGA–II.
- Standard Hybrid. This version of metaHybrid does not use metamodels for computing derivatives, but it tries to approximate them by finite differences (forward differences). The parameters for the genetic part of this algorithm are the same as in "Only GA". The parameters of the SQP algorithm are taken from the paper presenting it [20], while its maximum number of iterations is dynamically modified as described in Section §2. The probability for the SQP operator is 0.015, we put $\epsilon = 0.001$.
- metaHybrid. We use the labels *SVD* and *RBF* to distinguish between the results obtained working with the two different metamodels. The GA and the SQP algorithms are set as in the previous instances.

5.1   ZDT4

This problem is well known to be a difficult test since it presents a large number (21^9) of local Pareto fronts. It involves two objective functions and ten input variables. We allowed 25000 evaluations for each run starting from an initial population of 75 randomly chosen individuals.

Although all the mathematical functions involved in the problem are infinitely many times differentiable, the objective space is very rough and therefore gradient information can be misleading. Indeed, as remarked in Fig. 1, the hybrid algorithm which uses finite differences obtained the poorest score. A second motivation for this behavior is the relative high number of input variables which forces that algorithm to waste a lot of evaluations for each gradient estimation.

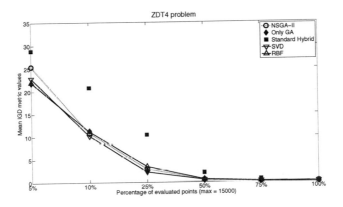

Fig. 1. Performance comparison on ZDT4 problem

On the contrary, the MetaHybrid algorithm shows performances very close to NSGA2, for instance. A non-hybrid genetic algorithm is surely more suited for this problem and NSGA2 is considered a state-of-the-art implementation. This suggests that the search direction offered by metamodels has a precision sufficiently high to capture the structure of the problem, but not so high to be confused by local roughnesses.

5.2   CEC '09 UP2

This problem is part of the CEC 09 Multi-Objective Evolutionary Algorithm competition [22]. We reduce the number of variables (10 instead of 30) and the total number of evaluations allowed (15000 instead of 300000). It is a continuous unconstrained problem involving two objective functions. The Pareto front is continuous.

The results are in line with the previous example. A bigger gap between the two pure genetic algorithms can be noticed. This failure of the new elitism approach is probably the cause also of the slight worsening in the performance of MetaHybrid. Notwithstanding this difficulty, both SVD and RBF schemes behave better than the "Standard Hybrid" and than the non-hybrid algorithm.

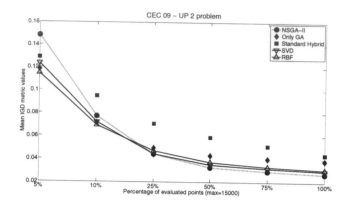

Fig. 2. Performance comparison on CEC '09 UP2 problem

5.3   Rotated OSY

The constrained optimization problem called OSY [15] involves six input variables, two objectives and six constraints. We decided to modify the original formulation in order to obtain a fairer test. The Pareto front for this problem is composed by five segments corresponding, in the variable space, to fixing at zero some of the input variables and to particular combination of the remaining ones in which only one at the time varies. Moreover, these combinations touch the variable bounds in many parts and it is possible to pass from a Pareto point to another one simply changing one single variable.

In order to avoid these unwanted features while maintaining the smooth mathematical structure, we work on a rotated and translated set of input variables $\mathbf{y}$ such that $A\mathbf{y} + (1, 1, 1, 1, 1, 1) = \mathbf{x}$. The matrix A applies a $\pi/6$ rotation to the first two variables, a $\pi/4$ rotation to the third and the fourth and a $\pi/3$ one to the last two. The old variable bounds are added to the problem as additional constraints, while we let $\mathbf{y}$ vary over all the range accessible through the rotation (i.e. in two dimensions, a rotated square would produce a diamond whose variables span a larger interval than the original ones). We allow 10000 evaluations for each run.

The IGD metric values reported in Fig. 3 show that pure genetic algorithms are still the best choice for this problem, but the applied rotation reduced the gap. We found promising that the "Only GA" implementation obtained better results than NSGA-II, since this confirms our studies on elitism.

5.4   CTP2

Deb, Pratap and Meyarivan in [9] propose a family of constrained problems which shares the same formulation. Modifying some parameters it is possible to obtain very different Pareto set shapes (and therefore problems with different kind of

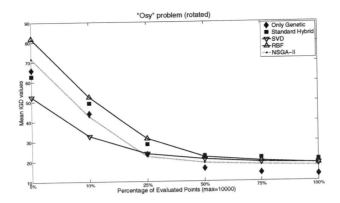

Fig. 3. Mean IGD metric values computed on the rotated version of OSY problem

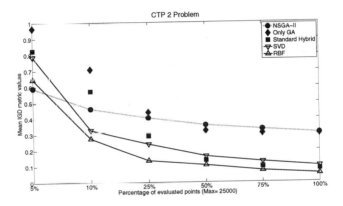

Fig. 4. Performance comparison on CTP2 problem

difficulties). The objective functions are always two and one single constraint draws the line of the Pareto set, which is fragmented in several pieces in the case of interest.

The hybrid implementations outperform the pure genetic ones on this example. The SQP algorithm confirm its ability in profitably handling constraints. In particular the RBF metamodels allowed a better comprehension of the problem than the SVD and even than the forward differences approximation.

5.5 Sym-Part

This problem has been introduced in [17] and it entered in the problems suite for the CEC '07 MOEA competition. It is a scalable unconstrained problem

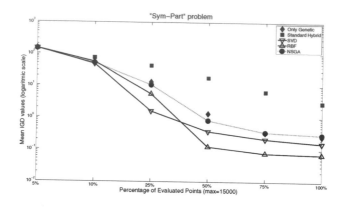

Fig. 5. Performance comparison on Sym-Part problem

involving two objective functions. We use 30 input variables for this test which are the main source of difficulties for this problem, since the Pareto front is continuous and convex.

The performance results for this benchmark are more similar to the CTP2 case than to the other unconstrained problems, with a small but significant exception. The "Standard Hybrid" exhibits low performances on this problem, while it was among the bests in the previous case. This is partially due to the high number of input variables. However, the Sym-Part problem highlights the benefits of the MetaHybrid approach, which outperforms both the classical GA and hybrid algorithms.

5.6   Lennard-Jones

Lennard-Jones atom clusters are an interesting real-world optimization problem. They involve only one objective function (the energy of the system), but the structure of the problem is rich, full of local minima, although the formulation is very smooth. The Lennard-Jones is a pair-wise potential which mimic the interatomic forces: it is strongly repulsive for too nearby atoms and it is weakly attractive for well separated ones. We study here the static problem, where we want to find the equilibrium solution, for a cluster of 10 atoms.

This problem is part of the CEC '11 competition on "Testing Evolutionary Algorythms on Real World Optimization Problems" [4] and we refer to the cited technical report for the formulation of the problem and the physical background. We do not need to modify any of the proposed algorithm, since they can work with single-objective problems as well. We modify the performance metric, since the IGD metric loses its meaning in this situation. We simply look at the best objective function value obtained at different steps during the optimization run. Since the global optimum is known (-28.422532), we plot in figure 6 the distance from that value. We allow 50000 evaluations for each run.

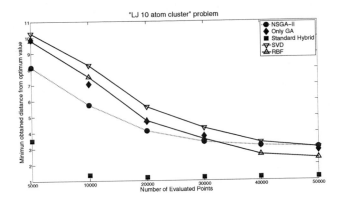

Fig. 6. Performance comparison on LJ problem

The performance of the "Standard Hybrid" implementation are the best ones for this benchmark. This reflects the smoothness of the problem. The RBF metamodels are the most precise approximation and the relative MetaHybrid algorithm outperforms all remaining implementation at the end of the runs.

6   Conclusions

We presented an hybrid genetic-SQP algorithm based on a steady-state evolution which uses metamodels in order to compute gradients. The algorithm is competitive as outlined in the performed tests, which cover a wide range of scenarios: single and multi-objective problems, constrained and unconstrained, smooth and rough objective space, continuous and disconnected Pareto fronts.

A validation on real-world problems is our first future task. However, the data collected are promising: the MetaHybrid algorithm can provide a sufficient precision (see the Lennard-Jones test). It can avoid local minima and it has good exploration properties (see ZDT4 and UP2 tests). The ratio between exploration and exploitation is balanced in such a way the algorithm is able to produce an accurate and extended approximated Pareto front as showed in the rotated OSY problem. Moreover, it is particularly efficient when working with constraints (like in the CTP2 problem) or when dealing with a considerable number of variables (as highlighted by the Sym-Part benchmark).

The proposed algorithm is hybrid under many points of view, since it is the sum of different components, but there is a common underlying idea: extracting as much information as possible from available data in order to request as less evaluations as possible. Indeed, we compute gradients through metamodels trained over the existing design database, we keep all non-dominated points in the parent population, we pass to elitism operator all iterates generated by the SQP algorithm, we judge the goodness of new SQP iterates with a filter made by old ones.

Not only data are intensively exploited: the steady-state evolution scheme and the multi-threading implementation of the SQP operator guarantee the full usage of the computational resources available. Further developments in this direction are planned: once the metamodels training is parallelized, we could use different methods for computing gradients at the same time using the most accurate approximation which can be identified through an iterative validation process, like in [16].

Another interesting research direction that will be explored in the next future is the use of gradient-based techniques for discrete (but not categorical) input variables. Categorical variables cannot support gradient information by definition, but a further hybridization could work effectively on mixed problem.

References

1. Aittokoski, T., Miettinen, K.: Efficient evolutionary method to approximate the Pareto optimal set in multiobjective optimization. In: EngOpt 2008 (2008)
2. Blum, C., Puchinger, J., Raidl, G.R., Roli, A.: A Brief Survey on Hybrid Meta-heuristics. In: Filipic, B., Silc, J. (eds.) Proceedings of BIOMA 2010 (2010) ISBN: 978-961-264-017-0
3. Buhmann, M.D.: Radial Basis Functions: Theory and Implementations. Cambridge University Press, Cambridge (2003)
4. Das, S., Suganthan, P.N.: Problem Definitions and Evaluation Criteria for CEC 2011 Competition on Testing Evolutionary Algorithms in Real World Optimization Problems. Technical Report (2010), http://www3.ntu.edu.sg/home/EPNSugan/
5. Deb, K.: Multi-objective optimization using evolutionary algorithms. Wiley, UK (2001)
6. Deb, K., Agraval, S.: Simulated binary crossover for continuous search space. Complex System 9, 115–148 (1995)
7. Deb, K., Agrawal, S., Pratap, A., Meyarivan, T.: A fast elitist non-dominated sort-ing genetic algorithm for multi-objective optimization: NSGA-II. KanGal Report, 200001 (2000)
8. Deb, K., Goel, T.: Controlled Elitist Non-dominated Sorting Genetic Algorithms for Better Convergence. KanGal Report, 200004 (2001)
9. Deb, K., Mathur, A.P., Meyarivan, T.: Constrained Test Problems for Multi-objective Evolutionary Optimization. In: Zitzler, E., Deb, K., Thiele, L., Coello Coello, C.A., Corne, D.W. (eds.) EMO 2001. LNCS, vol. 1993, pp. 284–298. Springer, Heidelberg (2001)
10. Fletcher, R., Leyffer, S.: Nonlinear programming without a penalty function. Math-ematical Programming 91, 239–269 (2002)
11. Gould, N.I.M., Toint, P.L.: SQP Methods for Large-Scale Nonlinear Programming. Invited Presentation at the 9th IFIP TC7 Conference on System Modelling and Optimization, Cambridge (1999)
12. Kumar, A., Sharma, D., Deb, K.: A hybrid multi-objective optimization proce-dure using PCX based NSGA-II and sequential quadratic programming. In: IEEE Congress on Evolutionary Computation, CEC 2007, pp. 3011–3018 (2008)
13. Miettinen, K.M.: Nonlinear Multiobjective Optimization. Kluvert Academic Pub-lisher, Boston (1999)
14. Nocedal, J., Wright, S.J.: Numerical Optimization. Springer, New York (1999)

15. Osyczka, A., Kundu, S.: A new method to solve generalized multicriteria oprimization problems using the simple genetic algorithm. Structural Optimization 10, 94–99 (1995)
16. Rigoni, E., Turco, A.: Metamodels for Fast Multi-objective Optimization: Trading Off Global Exploration and Local Exploitation. In: Deb, K., Bhattacharya, A., Chakraborti, N., Chakroborty, P., Das, S., Dutta, J., Gupta, S.K., Jain, A., Aggarwal, V., Branke, J., Louis, S.J., Tan, K.C. (eds.) SEAL 2010. LNCS, vol. 6457, pp. 523–532. Springer, Heidelberg (2010)
17. Rudolph, G., Naujoks, B., Preuß, M.: Capabilities of EMOA to Detect and Preserve Equivalent Pareto Subsets. In: Obayashi, S., Deb, K., Poloni, C., Hiroyasu, T., Murata, T. (eds.) EMO 2007. LNCS, vol. 4403, pp. 36–50. Springer, Heidelberg (2007)
18. Sharma, D., Kumar, A., Deb, K., Sindhya, K.: Hybridization of SBX based NSGA-II and sequential quadratic programming for solving multi-objective optimization problems. In: IEEE Congress on Evolutionary Computation, CEC 2007, pp. 3003–3010 (2008)
19. Strikwerda, J.: Finite Difference Schemes and Partial Differential Equations. SIAM, Philadelphia (2004)
20. Turco, A.: Adaptive Filter SQP. In: Blum, C., Battiti, R. (eds.) LION 4. LNCS, vol. 6073, pp. 68–81. Springer, Heidelberg (2010)
21. Turco, A., Kavka, C.: MFGA: A GA for Complex Real-World Optimization Problems. International Journal of Innovative Computing and Applications 3(1), 31–41 (2011)
22. Zhang, Q., Zhou, A., Zhao, S., Suganthan, P.N., Liu, W., Tiwari, S.: Multiobjective optimization Test Instances for the CEC 2009 Special Session and Competition. Techical Report CES–487 (2009)
23. Zitzler, E., Deb, K., Thieler, L.: Comparison of multiobjective evolutionary algorithms: Empirical results. IEEE Transactions on Ev. Comp. 8 (2000)
24. Zitzler, E., Thiele, L., Laumanns, M., Fonseca, C.M., da Fonseca, V.G.: Performance assessment of multiobjective optimizers: an analysis and review. IEEE Transactions on Evolutionary Computation 7(2), 117–132 (2003)

Designing Stream Cipher Systems Using Genetic Programming

Wasan Shaker Awad

Department of Information Systems
College of Information Technology
University of Bahrain
Sakheer, Bahrain
wasan_shaker@itc.uob.bh

Abstract. Genetic programming is a good technique for finding near-global optimal solutions for complex problems, by finding the program used to solve the problems. One of these complex problems is designing stream cipher systems automatically. Steam cipher is an important encryption technique used to protect private information from an unauthorized access, and it plays an important role in the communication and storage systems. In this work, we propose a new approach for designing stream cipher systems of good properties, such as high degree of security and efficiency. The proposed approach is based on the genetic programming. Three algorithms are presented here, which are simple genetic programming, simulated annealing programming, and adaptive genetic programming. Experiments were performed to study the effectiveness of these algorithms in solving the underlying problem.

1 Introduction

Encryption is an important mechanism for protecting sensitive information from an unauthorized access by transforming the information (plaintext) to another form which is unreadable (ciphertext). Nowadays, you can find many cipher systems of different types. However, cryptosystems (cipher systems) are commonly subdivided into block ciphers and stream ciphers. Stream ciphers are extremely fast and easy to implement. In addition, they usually have very minimal hardware resource requirements. Therefore stream ciphers are of great importance in applications where encryption speed is paramount and where area-constrained or memory constrained devices make it impractical to use block ciphers.

Designing good stream cipher automatically is a complex process. Therefore, this problem has been considered in this paper, and it can be formulated as follows:

- Given: Plaintext length in bits, which is the keystream length (size).
- Output: A keystream generator, which is the main component of stream cipher, that generates pseudorandom Binary sequence (keystream) of length size and fulfills the security and efficiency requirements.

C.A. Coello Coello (Ed.): LION 5, LNCS 6683, pp. 308–320, 2011.

Thus, the main purpose of this work is to present a new general automated approach for designing stream ciphers that satisfy the desired properties. The proposed approach is based on genetic programming (GP).

The problem considered here is the design automation of cipher systems. This problem has been considered by a number of researchers. For example, Genetic Algorithm (GA) has been used to find a set of rules of Cellular Automata (CA) suitable for cryptographic purposes [1]. Also, GA has been used for the construction of Boolean functions for cipher systems, such as block ciphers and stream ciphers [2]. The design of Boolean functions with properties of cryptographic significance is a hard task. Therefore, this problem has attracted a number of researchers [3]; they have proposed a GA-based method for finding Boolean functions which are mostly have high degree of nonlinearity. So far, a general automated method for designing stream ciphers is not known. However, this problem has been reviewed in more details by Awad [4].

Although GA (and GP) has gained many applications, it is reported that the simple GA suffers from many troubles such as getting stuck in a local minimum and parameters dependence [5]. There are many improvements have been proposed to enhance the performance of the GA, such as adaptive GA. Therefore, in this work, to avoid the problem of getting stuck in a local minimum and to preserve good individuals into the next generation, two algorithms are presented, in addition to simple GP (SGP), which are:

1. Simulated Annealing Programming (SAP)
2. Adaptive GP (AGP)

SAP is an integration of simulated annealing (SA) and GP. Many researchers explored the application of SA on many different types of problems, and it has been integrated with GA or GP in order to work on a population of individuals and to preserve good individuals into the next generation [6, 7, 8, 9].

AGP (or AGA) is a technique that dynamically adjusts selected control parameters, such as population size and genetic operation rates, during the course of evolving a problem solution [10]. That is because, one of the main problems related to GA is to find the optimal control parameter values that it uses, when a poor parameter setting is made for an evolutionary computation algorithm, the performance of the algorithm will be seriously degraded. Thus, different values may be necessary during the course of a run. A widely practiced approach to identify a good set of parameters for a problem is through experimentation. For these reasons, AGAs offering the most appropriate exploration and exploitation behavior. AGA has been studied by a number of researchers [11, 12].

2 Stream Cipher Systems

Every stream cryptosystem consists of two parts, which are [13]:

1. Keystream (random sequence bit) generator, and
2. Mixer (XOR for the binary sequences).

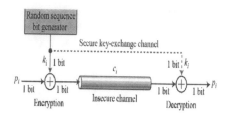

Fig. 1. The proposed enhance single point crossover in ATG

A keystream generator, which is the heart of stream ciphers, outputs a stream of bits (keystream) xored with a stream of plaintext bits to produce the stream of ciphertext, as shown in Fig. 1.

Currently, there are many stream cipher systems widely used in our day life that can be classified into:

1. Linear Feedback Shift Register (LFSR) based stream ciphers, in which a LFSR or nonlinear combination of LFSRS is used as keystream generator. Fig. 2 presents a LFSR of length five stages [13, 14, 15, 16, 17].
2. Nonlinear FSR (NLFSR), in which a nonlinear feedback function is used [13, 14, 15, 16, 17].
3. Feedback-with-Carry Shift Register (FCSR) [18].
4. (n,k)-NLFSR [19]
5. Cellular Automata (CA) [1].
6. Algebraic Shift Register [20].

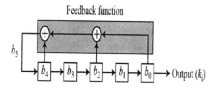

Fig. 2. The proposed enhance single point crossover in ATG

The stream cipher system's security depends entirely on the inside of keystream generator. The security of this generator can be analyzed in terms of randomness, linear complexity, and correlation immunity [21, 22, 23]. Thus, good keystream generators must have the following features:

1. They generate long period keystreams.
2. Their keystreams are random.
3. The generated keystreams are of large linear complexity.
4. They have high degrees of correlation immunity.

A binary sequence is said to be random if there is no obvious relationship between the individual bits of the sequence. Several research efforts exist in the literature for developing suites of tests for evaluating random number (Binary keystream) generators to be involved in stream ciphers [21, 22, 23]. In all these methodologies two criteria are used for the evaluation of the quality of random numbers obtained by using some generator in traditional applications such as simulation studies: uniform distribution and independence. The most important requirement imposed on random number generators is their capability to produce random numbers uniformly distributed in [0,1]; otherwise the application's results may be completely invalid. A number of statistical tests are applied to examine whether the pseudorandom number sequences are sufficiently random or not, which are frequency test, serial test, poker test, autocorrelation test and runs test.

1. Frequency Test: It calculates the number of ones and zeroes of the binary sequence and checks if there is no large difference.
2. Serial Test: The transition characteristics of a sequence such as the number 00, 01, 10 and 11 are evaluated. Ideally, it should be uniformly distributed within the sequence.
3. Poker Test: A N length sequence is segmented into blocks of M bits and the total number of segments is N/M. Within each segment, the integer value can vary from 0 to $m = 2M\text{-}1$. The objective of this test is to count the frequency of occurrence of each M length segment. Ideally, all the frequency of occurrences should be equal
4. Runs Test: A sequence is divided into contiguous stream of 1's that is referred as blocks and contiguous stream of 0's that is referred as gaps. If r_0^i is the number of gaps of length i, then half of the gaps will have length 1 bit, a quarter with length 2 bits, and an eighth with length 3 bits. If r_1^i is the number of blocks of length i, then the distribution of blocks is similar to the number of gaps.

Linear complexity is a well-known complexity measure in the theory of stream ciphers. Linear complexity of a keystream s is the length of the shortest LFSR which will produce the stream s, which is denoted by L(s). If the value of L(s) is L, then $2L$ consecutive bits can be used to reconstruct the whole sequence. Hence, to avoid the keystream reconstruction, the value of L should be large [24]. In order to obtain high linear complexity, several sequences can be combined in some nonlinear manner. The danger here is that one or more of the internal output sequences can be correlated with the combined keystream and attacked using linear algebra. A keysream generator has a higher degree of correlation immunity if there is no correlation between any internal output sequence and the combined keytream.

3   Genetic Programming and Simulated Annealing

One of the component methodologies of computational intelligence is evolutionary computation. There are number of evolutionary computation techniques, such as GA, GP, Cultured Algorithms, and Differential Evolution algorithms. Regardless of the technique used, evolutionary computation applications follow a similar procedure [25]:

1. Initialize the population.
2. Evaluate each individual in the population.
3. Select individuals.
4. Produce a new population by applying a number of operations on selected individuals.
5. loop to step 2 until some condition is met.

Automated design is an essential part of GP paradigm. GP receives a high level statement of a problem's requirements from the user and attempts to create a computer program that provides a solution for the problem. In this paper, the computer program to be created represents a keystream generator.

GP is the extension of the genetic model of learning the space of programs. These programs are expressed as trees. GP invented by John R. Koza in 1990s [26] which is regarded as an extension of GA [27, 28] attributed to John H. Holland [29]. Both techniques are identical in nature except for representation of individuals which in case of GP is parse trees based computer programs compared to fixed or variable length character strings in genetic algorithms. Representation is a major difference not only because it distinguishes the two techniques from each other but also because it greatly extends the problem handling capabilities of GP. It is one of the most promising domains independent and object oriented evolutionary computation techniques [30, 31]. GP is used mainly for design automation and automatic programming; such as the design of analog and digital circuits [32].

On the other hand, SA, which has been introduced by Kirkpatrik [33], is a general randomization technique for solving optimization problems; it is a recent technique for finding good solutions to a wide variety of combinatorial optimization problems. This technique can help to avoid the problem of getting stuck in a local minimum and to lead towards the globally optimum solution. It is inspired by the annealing process in metallurgy. At high temperatures, the molecules of liquid move freely with respect to one another. If the liquid is cooled slowly, thermal mobility is lost. In SA, the solution starts with a high temperature, and a sequence of trail vectors are generated until inner thermal equilibrium is reached. Once the thermal equilibrium is reached at a particular temperature, the temperature is reduced and a new sequence of moves will start. This process is continued until a sufficiently low temperature is reached, at which no further improvement in the objective function can be achieved. Thus, SA algorithm consists of: configurations, re-configuration technique, cost function, and cooling schedule [34, 35].

4   Simple Genetic Programming Method

This section is to describe the proposed SGP algorithm used for evolving keystream generators. The major steps for preparing GP for an application are [26]:

1. Determining the function library.
2. Determining the representation scheme.
3. Determining the fitness measure.

The description of these steps is given in the following sub sections along the proposed algorithm parameters.

4.1   Function Library

In GP, the structure under adaptation is a set of programs representing the candidate keystream generators. The keystream generators considered here are LFSR-based generators. Thus, the important basic function which is the shift register should be included. The function library used in this work is presented in table 1. The proposed function library is sufficient since:

1. It includes the LFSR function (SR), and there is no need to include other types of shift registers because for every shift register there is an equivalent LFSR.
2. Any combinational logical function can be expressed using (AND) and (XOR) only, that is because, any logical function can constructed from (AND), (OR), (NOT), and

$$\overline{x} = x \oplus 1 \tag{1}$$

$$x + y = \overline{\overline{x} \cdot \overline{y}} \tag{2}$$

4.2   Representation Scheme

The population chromosomes (programs), that represent candidate keystream generators, are strings of characters which are expressions represented using prefix polish notation. Fig. 3 shows the syntax of the population programs. These syntactic rules should be preserved during the generation of the initial population, and by the genetic operations. Therefore, strongly-typed GP [36] is adopted.

The initial states and feedback functions of the shift registers are represented as strings of the letters $'a'..'p'$. These letters represent the numbers 0..15. Thus, each letter is a sequence of four bits. The length of a LFSR is determined by the number of letters which are initially generated randomly. The number of these letters must be even, half of them for the initial state, and the second half for the feedback function. For example, if the number of these letters is eight letters, then four letters are used for the feedback function, thus, the length of LFSR is 16 bits (4×4). Furthermore, the first zeros of the feedback function are ignored. For example, consider the following LFSR: "SR abid", 'i' is the number $8 = (1000)_2$, then the first three zeros are ignored, and the length of this LFSR will be five bits $(1 + 4)$. Thus the feedback function will be (11100), or $g(x) = 1 + x + x^2 + x^5$.

Table 1. The function library

Symbols	Arty	Format	Description
SR	2	SRx	Shift register where x represents the feedback polynomial and initial state.
&	2	&xy	Bitwise AND operation between the two binary sequences x and y.
∧	2	∧xy	Bitwise XOR operation between the two binary sequences x and y.
X	0		Sequence of characters $'a'..'p'$, representing the numbers 0..15.
|	2	|xy	Bitwise OR operation between the two binary sequences x and y.

> S → SR X \| & S S \| ^ S S \| \| S S
>
> X → aX \| bX \| ... \| pX \| a \| b \| ... \| p

Fig. 3. The syntax rules of GP language

The following are examples of the chromosomes:

Chromosome: SRggbkbecdeh
Chromosome: $\wedge\wedge\&|S Rbpei S Rhoionm \wedge S Rlhhk\& S R fmcddiphhc S Rcgpjkg S Riech S Rkhji$
Chromosome: $\wedge S Rdcae S Ragojdfojfm$
Chromosome: $|\& S Rccga \wedge S Reehk\&|S Rpfdmingc \wedge S Rje S Rjmlidmbe S Rho S Rmhofoh$
Chromosome: $S Rlepjgc$

4.3   Fitness Function

The fitness value is a measurement of the goodness of the keystream generator, and it is used to control the application of the operations that modify a population. There are a number of metrics used to analyze keystream generators, which are keystream randomness, linear complexity and correlation immunity. Therefore, these metrics should be taken in our account in designing keystream generators, and they are in general hard to be achieved. The fitness value is calculated by generating the keystream after executing the program, and then the generated keystream is examined. The fitness function used to evaluate the chromosomes is to calculate at what percentage the chromosome satisfies the desired properties of the stream ciphers. Three factors are considered in the fitness evaluation of the chromosomes which are:

1. Randomness of the generated keystream.
2. Keystream period length.
3. Chromosome length.

Eq. (3) is used for the evaluation of keystream randomness using the frequency and serial tests, in which, nw is the frequency of w in the generated binary sequence. This function is derived from the fact that in the random sequence:

1. Probability (no) = Probability (n1), and
2. Probability (n01) = Probability (n11) = Probability (n10) = Probability (n00)

$$f_1 = |n_0 - n_1| + |n_{00} - \frac{size}{4}| + |n_{01} - \frac{size}{4}| + |n_{10} - \frac{size}{4}| + |n_{11} - \frac{size}{4}| \quad (3)$$

There is another randomness requirement which is: $1/2^i * n_r$ of the runs in the sequence are of length i, where n_r is the number of runs in the sequence. Thus, we have the following function:

$$f_2 = \sum_{i=1}^{M} |(\frac{1}{2^i} \times n_r) - n_i| \quad (4)$$

where M is maximum run length, and n_i is the desired number of runs of length i.

Another factor is considered in the evaluation of the fitness value which is the size of the candidate keystream generator (length of the chromosome). Thus, the fitness function used to evaluate the chromosome x will be as follows, where wt is a constant and $size$ is the keystream period length:

$$fit(x) = \frac{size}{1 + f_1 + f_2} + \frac{wt}{length(x)} \quad (5)$$

4.4   Algorithm Parameters

The parameters used in this work were set based on the experimental results, the parameter value that show the highest performance was chosen to be used in the implementation of the algorithm. Thus, the genetic operations used to update the population are 1-point crossover with probability pc=1.0 and mutation with probability pm=0.1. The selection strategy, used to select chromosomes for the genetic operations, is the 2- tournament selection. The old population is completely replaced by the new population which is generated from the old population by applying the genetic operations. Regarding the structure of each chromosome, the maximum chromosome length is 300 characters, and the maximum number of functions (except SR) is ten functions. The probability of the function SR is 0.5, and all other function are of probability 0.5. Finally, the maximum LFSR length is 20 bits. The run of GP is stopped after a fixed number of generations. The solution is the best chromosome of the last generation.

4.5   The Design Algorithm

The SGP algorithm for designing a keystream generator that meets the desired properties is illustrated in Algorithm 1.

Algorithm 1. SGP

1: Input : Keystream period length (size)
2: Output : LFSR-based keystream generator
3: Generate the initial population (pop) randomly
4: Evaluate pop
5: **while** not Max Number of generations **do**
6:     Generate a new population (pop1) by applying crossover and mutation
7:     Evaluate the fitness of the new generated chromosomes of pop_1
8:     Replace the old population by the new one, i.e.,$pop \leftarrow pop_1$
9: **end while**
10: Return the best chromosome of the last generation

---

5 Simulated Annealing Programming Method

The fitness function, chromosome representation, and the control parameters of SGP are also used in SAP. Algorithm 2 illustrates the process of SAP.

Algorithm 2. SAP

1: Input : Keystream period length (size)
2: Output : LFSR-based keystream generator
3: Generate the initial population (pop) randomly
4: Evaluate pop
5: $temp \leftarrow 250.$
6: **while** not Max Number of generations **do**
7:     Generate a new population (pop1) by applying crossover and mutation
8:     Evaluate the fitness of the new generated chromosomes of pop_1
9:     Calculate the averages of fitness values for pop and pop_1, av and av_1 respectively
10:    If $(av_1 > av)$ then replace the old population by the new one, i.e. $pop \leftarrow pop_1$
11:    Else
12:    Begin
13:    $e \leftarrow av - av_1$
14:    $Pr \leftarrow e/Temp$
15:    Generate a random number (rnd)
16:    If $(exp(-pr) > rnd)$ then $pop \leftarrow pop_1$
17:    EndElse
18:    EndIf
19:    $Temp \leftarrow Temp * 0.95$
20: **end while**
21: Return the best chromosome of the last generation

As shown in the algorithm, SA is the technique used for the construction of the keystream generators. The structure under adaption is the set of GP expressions, and the GA operations are used to update the population of expressions.

6   Adaptive Genetic Programming Method

The SGP algorithm has been modified to consider the dynamic setting of the algorithm parameters which are mutation and crossover rates. The concept of adapting crossover and mutation operators to improve the performance of GA has already been employed and studied by number of researchers. The goals with adaptive probabilities of crossover and mutation are to maintain the genetic diversity in the population and prevent the GAs to converge prematurely to local minima. Strinvivas [11] put forward the adaptive genetic algorithm, and its basic idea is to adjust pc and pm according to the individual fitness. In this paper, the mutation and crossover operation rates are adjusted adaptively based on the following formula [11]:

$$pc = \{ \begin{array}{ll} pc_1 - \frac{(pc_1 - pc_2)(f - f_{avg})}{f_{max} - f_{avg}} & f \geq f_{max} \\ pc_1 & f < f_{max} \end{array} \tag{6}$$

$$pm = \{ \begin{array}{ll} pm_1 - \frac{(pm_1 - pm_2)(f - f_{avg})}{f_{max} - f_{avg}} & f \geq f_{max} \\ pm_1 & f < f_{max} \end{array} \tag{7}$$

where f_{max} is the highest fitness value in the population; f_{avg} is the average fitness value in every population; f' is higher fitness value between two individuals; in addition, we set 1.0 for pc_1 and 0.7 for pc_2, and $pm_1 = 0.2$, and $pm_2 = 0.01$.

7   Results

This section presents the findings and results of the experiments carried out to demonstrate the effectiveness of the proposed methods for designing stream ciphers automatically. The experiments were carried out after implementing the proposed algorithms, that mentioned above, using C++ programming language. In all experiments, the keystream period length is 200 bits, and the population size is 100.

The researcher aimed at conducting the experiments, is to investigate the algorithm performances, and to make a comparison of the three algorithms: SGP, SAP, and AGP. Table 2 displays the obtained results. Results are obtained by running each algorithm 100 times for different values of maximum number of generations. The results shown in table 2 represent the average of the fitness values of the best chromosomes in 100 runs. According to the results, AGP and SAP are more effective than SGP in solving the underlying problem. They can evolve keystream generators that can generate keystreams of good statistical properties with large period lengths.

Table 3 presents the results of 20 runs of SAP and AGP. The values given in this table are the fitness values of best chromosomes, i.e. keystream generators evolved in each run. We can see that the highest fitness value in 20 runs is 50.66 which is the fitness value of the chromosome: SRphikje found by AGP.

Table 2. The comparison of the three algorithms for different values of maximum number of generations

Maximum Number of Generations	Average of fitness values		
	SGP	SAP	AGP
30	31.33	34.6199	34.9572
50	31.892	35.7214	35.9222
70	32.75	36.3741	36.42
90	32.865	35.3944	37.2937

Table 3. The best results of 20 runs of SAP and AGP

Run	Fitness Value of SAP	Fitness Value of AGP
1	34.4317	34.8774
2	33.5854	36.0652
3	40.3543	25.5107
4	29.9324	32.5907
5	49.3019	37.7025
6	35.7731	44.8094
7	35.7119	30.7139
8	43.9672	36.6298
9	28.4054	49.3019
10	42.6349	29.4603
11	25.0631	35.6179
12	36.0652	40.9058
13	34.02	35.9966
14	27.1358	44.8094
15	36.4571	41.0542
16	29.409	24.4168
17	35.5298	50.6605
18	33.5854	31.9766
19	40.9058	33.5854
20	35.6179	49.1889

The keystream generated (as dipcted bellow) by this generator is of period $length \geq 200$, and it passes the randomness tests considered in the fitness calculation. The keystrem is:

1111111000001101110100000000001010111000011010101010110 011110101110011001100010
00001100001000100010110101000101001011010011100110100110110001101111011100100011101
0001111110000100101111001011111110101101

Furthermore, according to the results of table 3 and by applying Wilcoxon signed-rank test, there is no significant difference in the performance of SAP and AGP

8 Conclusion

In this paper, a new approach for designing keystream generators automatically has been presented, which is a new promising direction for stream cipher design. It has been shown the capability of GP in designing the desired stream ciphers. Stream cipher design methods presented here can be used for evolving any generator that satisfies the given requirements, such as period length, and randomness. These requirements are expressed mathematically in the fitness function. Three algorithms have been designed and applied: SGP, SAP, and AGP. The numerical results have showed that the application of GP in stream cipher design is useful. Also, SAP and AGP methods are more effective than SGP, that is because, the performance of SGP algorithm has been improved by the dynamic setting of the algorithm parameters and by using SA with GP.

Based on the function library defined in this work, only LFSR-based keystream generators can be evolved. However, by changing the functions of the function library, the proposed approach can be used to evolve other types of stream ciphers.

The proposed automated approach in this study will save the time and effort of designing stream ciphers more than if using state-of-the-art techniques. It can be also regarded as a tool to serve the same purpose of designing good cipher systems. However, the results of the proposed algorithms can be improved by considering other factors, such as linear complexity, in the chromosome evaluation. In addition, it is useful to investigate the effectiveness of other evolutionary computation techniques.

References

1. Szaban, M., Seredynski, F., Bouvry, P.: Collective Behavior of Rules for Cellular Automata-Based Stream Ciphers. In: IEEE Congress on Evolutionary Computation, pp. 179–183 (2006)
2. Clark, A., Jacob, L.J.: Almost Boolean functions: the design of Boolean functions by spectral inversion. Computational Intelligence 20(3), 450–462 (2004)
3. Millan, W., Clark, A., Dawson, E.: An effective genetic algorithm for finding highly nonlinear Boolean functions. In: Proc. 1st Int. Conf. on Information and Communications Security, China, Beijing, pp. 149–158 (1997)
4. Awad, W.S.: The applications of GA in cryptology. Far East Journal of Experimental and Theoretical Artificial Intelligence 2(1), 59–76 (2008)
5. Eiben, A.E., Hinterding, R., Michalewic, Z.: Parameters control in evolutionary algorithms. IEEE Trans. Syst. Man Cybern. 16(1), 122–128 (1999)
6. Van Laarhoven, P.J.M., et al.: Simulated Annealing: Theory and applications. Reidel, Holland (1987)
7. Sadegheih: Sequence optimization and design of allocation using GA and SA. Applied Mathematics and Computation 186(2), 1723–1730 (2007)
8. Yuichiro, U., Mitsunori, M., Tomoyuki, H.: Simulated Annealing Programming Using Effective Subtrees. Doshisha Daigaku Rikogaku Kenkyu Hokoku 49(4), 205–209 (2009)
9. Miki, M., Hashimoto, M., Fujita, Y.: Program Search with Simulated Annealing. In: Proc. of the 9th Annual Conference on Genetic and Evolutionary Computation, London, England, pp. 1754–1754 (2007)
10. Sivanandam, S.N., Deepa, S.N.: Introduction to genetic algorithms. Springer, New York (2008)

11. Srinivas, M., Patnaik, L.M.: Adaptive Probabilities of Crossover and Mutation in Genetic Algorithms. IEEE Trans. Systems, Man and Cybernetics 24(4), 656–667 (1994)
12. Zhang, J., Hu, T.: Adaptive Genetic Algorithm Based on Population Diversity. Computer Engineering and Applications 9(1), 49–51 (2002)
13. Forouzan, B.A.: Cryptography and network security. McGraw-Hill, New York (2008)
14. Rueppel, R.A.: Analysis and Design of Stream Cipher. Springer, New York (1986)
15. Schneier, B.: Applied cryptography. John Wiley and Sons, New York (1996)
16. Golomb, S.W.: Shift Register Sequence. Holden-Day, San Francisco (1967)
17. Beker, P.F.: Cipher Systems. John Wiley, New York (1982)
18. Klapper, G.M.: Feedback shift registers, 2-adic span and combiners with memory. Journal of Cryptology 10(1), 111–147 (1997)
19. Dubrova, E., Teslenko, M., Tenhunen, H.: Analysis and Synthesis of (n,k)-Non-Linear Feedback Shift Registers. In: Proc. of the Conf. on Design, Automation and Test, Munich, Germany, pp. 1286–1290 (2008)
20. Goresky, M., Klapper, A.: Pseudonoise Sequence Based on Algebraic Feedback Shift Registers. IEEE Trans. Inf. Theory 52(4), 1649–1662 (2006)
21. Gustafson, H., et al.: A computer package for measuring the strength of encryption algorithm. Comp. and Sec. 14(1), 687–697 (1994)
22. Zeng, K., Yang, C., Rao, T.R.N.: Pseudorandom Bit Generator in Stream Cipher Cryptography. Comp. 2(24), 8–17 (1991)
23. L'ecuyer, P., Simard, R.: TestU01: A C library for empirical testing of random number generators. ACM Trans. Math. Softw. 33(4), 22–40 (2007)
24. Massey, J.L.: Shift register sequences and BCH decoding. IEEE Trans. on Inf. Theory IT 15(1), 122–127 (1976)
25. Eberhart, R., Shi, Y.: Computational Intelligence: concepts to implementation. Morgan Kaufmann, San Francisco (2008)
26. Koza, J.R.: Genetic programming. MIT Press, Cambridge (1992)
27. Goldberg, D.E.: Genetic algorithms in search, optimization, and machine learning. Addison-Wesley, New York (1989)
28. Mitchell, M.: An Introduction to Genetic Algorithm. MIT Press, Cambridge (1996)
29. Holland, J.H.: Adaptive in natural and artificial systems. University of Michigan, Ann Arbor (1975)
30. Hirsh, H., Banzhaf, W., Koza, J.R., Ryan, C., Spector, L., Jacob, C.: Genetic programming. IEEE Intelligent Systems 15(3), 74–84 (2000)
31. Koza, J.R., Keane, M.A., Streeter, M.: What's AI done for me lately? - genetic programming's human competitive results. IEEE Intelligent Systems 18(3), 25–31 (2003)
32. Koza, J.R.: Genetic Programming II: Automatic Discovery of Reusable Programs. MIT Press, Cambridge (1994)
33. Kirkpatrik, S., et al.: Optimization by simulated annealing. Science 220(4598), 671–680 (1983)
34. Yong, L., Lishan, K., Evans, D.J.: The annealing evolution algorithm as function optimizer. Parallel Computing 21(3), 389–400 (1995)
35. Cordon, O., et al.: An Inductive Query by Example Technique for Extended Boolean Queries Based on Simulated-Annealing Programming. In: The Proc. of 7th International ISKO Conference on Challenges in Knowledge Representation and Organization for the 21st Century, pp. 429–436. Integration of Knowledge Across Boundaries, Granada (2002)
36. Haynes, T., et al.: Strongly typed GP in evolving cooperation strategies. In: Proc. of the sixth Int. Conf. on GA, pp. 271–278. Morgan Kaufmann, San Francisco (1995)

GPU-Based Multi-start Local Search Algorithms

Thé Van Luong, Nouredine Melab, and El-Ghazali Talbi

INRIA Dolphin Project / Opac LIFL CNRS
40 avenue Halley, 59650 Villeneuve d'Ascq Cedex France
The-Van.Luong@inria.fr, {Nouredine.Melab,El-Ghazali.Talbi}@lifl.fr

Abstract. In practice, combinatorial optimization problems are complex and computationally time-intensive. Local search algorithms are powerful heuristics which allow to significantly reduce the computation time cost of the solution exploration space. In these algorithms, the multi-start model may improve the quality and the robustness of the obtained solutions. However, solving large size and time-intensive optimization problems with this model requires a large amount of computational resources. GPU computing is recently revealed as a powerful way to harness these resources. In this paper, the focus is on the multi-start model for local search algorithms on GPU. We address its re-design, implementation and associated issues related to the GPU execution context. The preliminary results demonstrate the effectiveness of the proposed approaches and their capabilities to exploit the GPU architecture.

Keywords: GPU-based metaheuristics, multi-start on GPU.

1 Introduction

Over the last years, interest in metaheuristics (generic heuristics) has risen considerably in the field of optimization. Indeed, plenty of hard problems in a wide range of areas including logistics, telecommunications, biology, etc., have been modeled and tackled successfully with metaheuristics. Local search (LS) algorithms are a class of metaheuristics which handle with a single solution iteratively improved by exploring its neighborhood in the solution space. Different parallel models have been proposed in the literature for the design and implementation of LSs [1]. The multi-start model consists in executing in parallel many LSs in an independent/cooperative manner. This mechanism may provide more effective, diversified and robust solutions.

Nevertheless, although LS methods have provided very powerful search algorithms, problems in practice are becoming more and more complex and CPU time-intensive and their resolution requires to harness more and more computational resources. In parallel, the recent advances in hardware architecture allow to provide such required tremendous computational power through GPU infrastructures. This new emerging technology is indeed believed to be extremely useful to speed up many complex algorithms. However, the exploitation of such computational infrastructures in metaheuristics is not straightforward.

C.A. Coello Coello (Ed.): LION 5, LNCS 6683, pp. 321–335, 2011.

Indeed, several scientific challenges mainly related to the hierarchical memory management or to the execution context have to be faced. The major issues are the efficient distribution of data processing between the CPU and the GPU, the thread synchronization, the optimization of data transfer between the different memories, the capacity constraints of these memories, etc. The main objective of our research work is to deal with such issues for the re-design of parallel metaheuristics models to allow solving of large scale optimization problems on GPU architectures. In [2,3], we have proposed to re-design the parallel evaluation of the neighborhood model for LSs on GPU. To go on this way, the main objective of this paper is to deal with the well-known multi-start model on GPU architectures where many LSs are executed in parallel.

We deal with the entire re-design of the multi-start model on GPU by taking into account the particular features related to both the LS process and the GPU computing. More exactly, we provide two different general schemes for building efficient multi-start LSs on GPU. The first scheme combines the multi-start model with the parallel evaluation of the neighborhood on GPU previously mentioned above. In the second scheme, the search process of each LS algorithm is fully distributed on GPU. The advantage of the full distribution of the search process on GPU is to reduce CPU/GPU memory copy latency. We will essentially focus on this approach throughout this paper.

Despite the fact that the second scheme for the multi-start model has already been applied in some previous works in the context of the tabu search on GPU [4,5], to the best of our knowledge, it has never been widely investigated in terms of 1) reproducibility for any other LS algorithm and 2) memory management. Indeed, the contribution of this paper is to provide a general methodology for the design of multi-start LSs on GPU applicable to any class of LS algorithms such as hill climbing, tabu search or simulated annealing. Furthermore, a particular focus is made on finding efficient associations between the different available memories and the data commonly used in the multi-start LS algorithms.

The remainder of the paper is organized as follows: on the hand, Section 2 highlights the principles of LS parallel models. On the other hand, a brief review of the GPU architecture is also depicted. Section 3 presents a methodology for the design and the implementation of parallel multi-start LS methods on GPU. The performance results obtained for the associated implementations are reported in Section 4. Finally, a discussion and some conclusions of this work are drawn in Section 5.

2   Parallel Local Search Algorithms and GPU Computing

2.1   Parallel Models of LS Algorithms

For non-trivial problems, executing the iterative process of a simple LS on large neighborhoods requires a large amount of computational resources. Consequently, a variety of algorithmic issues are being studied to design efficient LS heuristics. Parallelism arises naturally when dealing with a neighborhood, since each of the solutions belonging to it is an independent unit. Due to this, the performance of LS algorithms is particularly improved when running in parallel. Parallel design and implementation of metaheuristics have been studied as well on different architectures [6, 7, 8].

Basically, three major parallel models for LS heuristics can be distinguished: solution-level, iteration-level and algorithmic-level.

- *Solution-level Parallel Model.* A focus is made on the parallel evaluation of a single solution. Problem-dependent operations performed on solutions are parallelized. In that case, the function can be viewed as an aggregation of a given number of partial functions.
- *Iteration-level Parallel Model.* This model is a low-level Master-Worker model that does not alter the behavior of the heuristic. Exploration and evaluation of the neighborhood are made in parallel. At the beginning of each iteration, each parallel node manages some candidates and the results are returned back to the master. An efficient execution is often obtained particularly when the evaluation of each solution is costly.
- *Algorithmic-level Parallel Model.* Several LS algorithms are simultaneously launched for computing better and robust solutions. They may be heterogeneous or homogeneous, independent or cooperative, start from the same or different solution(s), configured with the same or different parameters.

The solution-level model is problem-dependent and does not present many generic concepts. In this paper, we will focus on the multi-start model which is an instantiation of the algorithmic-level model where LS algorithms are all homogeneous.

2.2   GPU Computing

GPUs have evolved into a highly parallel, multithreaded and many-core environment. Indeed, since more transistors are devoted to data processing rather than data caching and flow control, GPU is specialized for compute-intensive and highly parallel computation. A complete review of GPU architecture can be found in [9].

In general-purpose computing on graphics processing units, the CPU is considered as a host and the GPU is used as a device coprocessor. This way, each GPU has its own memory and processing elements that are separate from the host computer. Memory transfer from the CPU to the GPU device memory is a (a)synchronous operation which is time consuming. Bus bandwidth and latency between the CPU and the GPU can significantly decrease the performance of the search, so data transfers must be minimized.

Each processor device on GPU supports the single program multiple data (SPMD) model, i.e. multiple processors simultaneously execute the same program on different data. For achieving this, the concept of kernel is defined. The kernel is a function callable from the host and executed on the specified device by several processors in parallel.

This kernel handling is dependent of the general-purpose language. For instance, CUDA [10] or OpenCL [11] are parallel computing environments which provide an application programming interface. These toolkits introduce a model of threads which provides an easy abstraction for single-instruction and multiple-data (SIMD) architecture.

Regarding their spatial organization, threads are organized within so called thread blocks. A kernel is executed by multiple equally threaded blocks. Blocks can be organized into a one-dimensional or two-dimensional grid of thread blocks, and threads inside a block are grouped in a similar way. All the threads belonging to the same thread block will be assigned as a group to a single multiprocessor, while different thread blocks can be assigned to different multiprocessors.

From a hardware point of view, graphics cards consist of streaming multiprocessors, each with processing units, registers and on-chip memory. Since multiprocessors are used according to the SPMD model, threads share the same code and have access to different memory areas. Basically, the communication between the CPU host and its device is done through the global memory.

3   Design and Implementation of Multi-start Local Search Algorithms on GPU

With the recent advances in parallel computing particularly based on GPU computing, the multi-start model has to be re-visited from the design and implementation points of view. In this section, we propose multiple deployment schemes of the multi-start model for LS algorithms on GPU.

3.1   Multi-start Local Search Algorithms Based on the Iteration-Level

In [2, 3], we have proposed the design and the implementation of the parallel evaluation of the neighborhood (iteration-level) model for a single LS on GPU. That is the reason why, a natural way for designing multi-start LSs on GPU based on the iteration-level is to iterate the whole process (i.e. the execution of a single LS on GPU) to deal with as many LSs as needed (see Fig. 1). Indeed, in general, evaluating a fitness function for each neighbor is frequently the most costly operation of the LS. Therefore, in this scheme, task distribution is clearly defined: the CPU manages the whole sequential LS process for each LS algorithm and the GPU is dedicated only to the parallel evaluation of solutions.

Algorithm 2 gives the template of this model. The reader is referred to [2,3] for more details about the original algorithm. Basically, for each LS, the CPU first sends the number of expected neighbors to be evaluated to the GPU and then these solutions are processed on GPU. Regarding the kernel thread organization, as quoted above, a GPU is organized following the SPMD model, meaning that each GPU thread associated with one neighbor executes the same evaluation function kernel. Finally, results of the evaluation function are returned back to the host via the global memory.

This way, the GPU is used as a coprocessor in a synchronous manner. The time-consuming part i.e. the incremental evaluation kernel is calculated by the

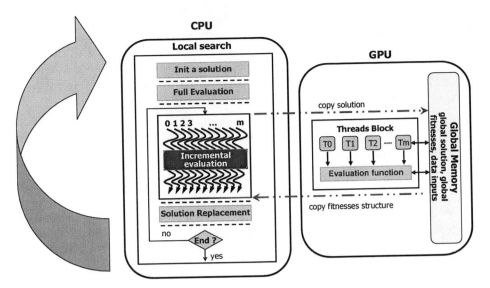

Fig. 1. Multi-start LS algorithms based on the parallel evaluation of the neighborhood on GPU (iteration-level). In this scheme, one thread is associated with one neighbor.

Algorithm 1. Multi-start local search algorithms template on GPU based on the iteration-level model

1: Allocate problem data inputs on GPU memory
2: Copy problem data inputs on GPU memory
3: Allocate a solution on GPU memory
4: Allocate a neighborhood fitnesses structure on GPU memory
5: Allocate additional solution structures on GPU memory
6: **for** $m = 1$ to $\#local\_searches$ **do**
7:     Choose an initial solution
8:     Evaluate the solution
9:     Specific LS initializations
10: **end for**
11: **repeat**
12:     **for** $m = 1$ to $\#local\_searches$ **do**
13:         Copy the solution on GPU memory
14:         Copy additional solution structures on GPU memory
15:         **for** each neighbor in parallel on GPU **do**
16:             Incremental evaluation of the candidate solution
17:             Insert the resulting fitness into the neighborhood fitnesses structure
18:         **end for**
19:         Copy back the neighborhood fitnesses structure on CPU memory
20:         Specific LS solution selection strategy on the neighborhood fitnesses structure

21:         Specific LS post-treatment
22:     **end for**
23:     Possible cooperation between the different solutions
24: **until** a stopping criterion satisfied

GPU and the rest is handled by the CPU. The advantage of this scheme resides in its highly parallel structure (i.e. an important number of generated neighbors to handle), leading to a significant multiprocessors occupancy of the GPU. However, depending on the number of LS algorithms, the main drawback of this scheme is that copying operations from the CPU to the GPU can become frequent and thus can lead to a significant performance decrease.

3.2 Design of Multi-start Local Search Algorithms Based on the Algorithmic-Level

A natural way for designing multi-start LSs on GPU is to parallelize the whole LS process on GPU by associating one GPU thread with one LS. This way, the main advantage of this approach is to minimize the data transfers between the host CPU memory and the GPU. Figure 2 illustrates this idea of this full distribution (algorithmic-level). In the rest of this paper, we will focus on this approach.

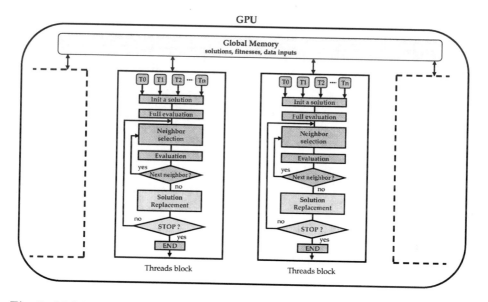

Fig. 2. Multi-start LS algorithms based on the full distribution of LSs on GPU (algorithmic-level). One thread is associated with one local search.

The details of the algorithm are given in Algorithm 1. First of all, at initialization stage, memory allocations on GPU are made: data inputs of the problem must be allocated and copied on GPU (lines 1 and 2). It is important to notice that problem data inputs (e.g. a matrix in the traveling salesman problem [12]) are a read-only structure and never change during all the execution of LS algorithms. Therefore, their associated memory is copied only once during all the execution. Second, a certain number of solutions corresponding to each LS must be allocated on GPU (line 3). Additional solution structures which are problem-dependent can also be allocated to facilitate the computation of incremental evaluation (line 4). Third, during the initialization of the different LS algorithms on GPU, each solution is generated and evaluated (from lines 5 to 9). Fourth, comes the algorithmic-level, in which the iteration process of each LS is performed in parallel on GPU (from lines 11 to 17). Since each neighbor is evaluated in a sequential manner on GPU, unlike the iteration-level scheme, there is no need to allocate and manipulate any neighborhood fitness structure. Fifth, an exchange of the best-so-far solutions could be made to accelerate the search process (line 18). In that case, operations on the global memory may be considered. Finally, the process is repeated until a stopping criterion is satisfied.

Algorithm 2. Multi-start local search algorithms template on GPU based on the algorithmic-level model

---

1: Allocate problem data inputs on GPU memory
2: Copy problem data inputs on GPU memory
3: Allocate #*local_searches* solutions on GPU memory
4: Allocate #*local_searches* additional solution structures on GPU memory
5: **for** each LS in parallel on GPU **do**
6:     Choose an initial solution
7:     Evaluate the solution
8:     Specific LS initializations
9: **end for**
10: **repeat**
11:     **for** each LS in parallel on GPU **do**
12:         **for** each neighbor **do**
13:             Incremental evaluation of the candidate solution
14:             Specific LS solution selection strategy
15:         **end for**
16:         Specific LS post-treatment
17:     **end for**
18:     Possible cooperation between the different solutions
19: **until** a stopping criterion satisfied

3.3   Memory Management of Multi-start Local Search Algorithms on the Algorithmic-Level

Memory Coalescing Issues. When an application is executed on GPU, each block of threads is split into SIMD groups of threads called *warps*. At any clock cycle, each processor of the multiprocessor selects a half-warp (16 threads) that is ready to execute the same instruction on different data. Global memory is conceptually organized into a sequence of 128-byte segments. The number of memory transactions performed for a half-warp will be the number of segments having the same addresses than those used by that half-warp. Fig. 3 illustrates an example of the memory management layer for a simple vector addition.

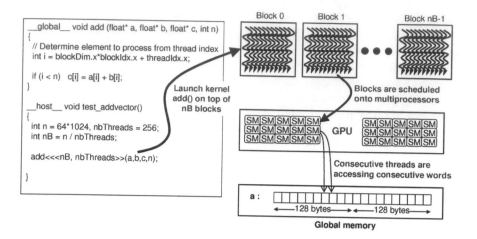

Fig. 3. An example of kernel execution for vector addition

For more efficiency, global memory accesses must be coalesced, which means that a memory request performed by consecutive threads in a half-warp is associated with precisely one segment. The requirement is that threads of the same warp must read global memory in an ordered pattern. If per-thread memory accesses for a single half-warp constitute a contiguous range of addresses, accesses will be coalesced into a single memory transaction. In the example of vector addition, memory accesses to the vectors a and b are fully coalesced, since threads with consecutive thread indices access contiguous words.

Otherwise, accessing scattered locations results in memory divergence and requires the processor to perform one memory transaction per thread. The performance penalty for non-coalesced memory accesses varies according to the size of the data structure. Regarding LS structures, coalescing is difficult when global memory accesses have a data-dependent unstructured pattern (especially for a permutation representation). As a result, non-coalesced memory accesses imply many memory transactions and it can lead to a significant performance decrease for LS methods.

Memory Organization. Optimizing the performance of GPU applications often involves optimizing data accesses which includes the appropriate use of the various GPU memory spaces. For instance, the use of texture memory is a solution for reducing memory transactions due to non-coalesced accesses. Texture memory provides a surprising aggregation of capabilities including the ability to cache global memory (separate from register, global, and shared memory). Regarding the data management on the different GPU memories, the following observations can be made whatever the used multi-start LS algorithm:

- **Global memory:** For each running LS on GPU (one thread), its associated solution is stored on the global memory. The same goes on for additional solution structures. This way, it ensures a global visibility among the different threads (LSs) during the entire search process for a possible cooperation. In a general way, all the data in combinatorial problems could be also associated with the global memory. However, as previously said, non-coalesced memory accesses may lead to a performance decrease. Therefore, the texture memory might be preferred since it can be seen as a relaxed mechanism for the threads to access the global memory. Indeed, the coalescing requirements do not apply to texture memory accesses.

- **Texture memory:** This read-only memory is adapted to LS algorithms since the problem inputs do not change during the execution of the algorithm. In most of optimization problems, problem inputs do not often require a large amount of allocated space memory. As a consequence, these structures can take advantage of the 8KB cache per multiprocessor of texture units. Indeed, minimizing the number of times that data goes through cache can increase significantly the efficiency of algorithms [13]. Moreover, cached texture data is laid out to give best performance for structures with 1D/2D access patterns such as matrices. The use of textures in place of global memory accesses is a completely mechanical transformation. Details of texture coordinate clamping and filtering is given in [14, 10].
- **Constant memory:** This memory is read only from kernels and is hardware optimized for the case where all threads read the same location. It might be used when the calculation of the evaluation function requires a common lookup table for all solutions (e.g. a decoder table for an indirect encoding on the job shop scheduling problem [15]).
- **Shared memory:** The shared memory is a fast memory located on the multiprocessors and shared by threads of each thread block. Since this memory area provides a way for threads to communicate within the same block, it might be used with the global memory in the context of a possible cooperation between different LS algorithms. In the case of the multi-start LS model, the type of shared information is the best-so-far solution found at each iteration of the search process.
- **Registers:** Among streaming processors, they are partitioned among the threads running on it and they constitute fast access memory. In the kernel code, each declared variable is automatically put into registers.
- **Local memory:** In a general way, additional structures such as declared array will reside in local memory. In fact, local memory resides in the global memory allocated by the compiler and its visibility is local to a thread (a LS).

Table 1 summarizes the kernel memory management in accordance with the different LS components. For the management of random numbers in SA, efficient techniques are provided in many books such as [16] to implement random generators on GPU. For deterministic multi-start LSs based on HC or TS, the random initialization of solutions might be done on CPU and then they can be copied on the GPU via the global memory to perform the LS process. This way, it ensures that the obtained results are the same as a multi-start LS performed on a traditional CPU. Regarding the management of the tabu list on GPU, since the list is particular to a TS execution, a natural mapping is to associate a tabu

Table 1. Kernel memory management. Summary of the different memories used in the multi-start LS algorithms on GPU.

Type of memory	LS structure
Texture memory	problem data inputs
Global memory	candidate solutions, additional candidate solution structures
Shared memory	possible solutions to exchange
Registers	additional LS variables
Local memory	additional LS structures
Constant memory	additional problem lookup tables

list to the local memory. However, since this memory has a limited size, large tabu lists should be associated with the global memory instead.

4   Experiments

To validate our approach, the multi-start model has been implemented on the quadratic assignment problem (QAP) on GPU using CUDA. The QAP arises in many applications such as facility location or data analysis. Let $A = (a_{ij})$ and $B = (b_{ij})$ be $n \times n$ matrices of positive integers. Finding a solution of the QAP is equivalent to finding a permutation $\pi = (1, 2, \ldots, n)$ that minimizes the objective function:

$$z(\pi) = \sum_{i=1}^{n} \sum_{j=1}^{n} a_{ij} b_{\pi(i)\pi(j)}$$

The problem has been implemented using a permutation representation. The chosen neighborhood for all the experiments is based on a 2-exchange operator ($\frac{n \times (n-1)}{2}$ neighbors). The incremental evaluation function has a time complexity of $O(n)$. The considered instances are the Taillard instances proposed in [17]. They are uniformly generated and are well-known for their difficulty.

Table 2. Used parameters for each particular LS

Tabu search	Simulated annealing
	geometric cooling schedule
tabu list size: $tl = \frac{n \times (n-1)}{16}$	initial temperature: $T_0 = 10000$
	threshold: $thr = 1$
iterations: $iters = 10000$	ratio: $r = 0.9$
	iterations: $iters = \frac{n \times (n-1)}{2}$
	equilibrium state: $T < thr$

The used configuration is an Intel Xeon 3GHz 2 cores with a GTX 280 (30 multiprocessors). From an implementation point of view, to build the CPU test code, the g++ compiler has been used with the -O2 optimization flag and SSE instructions. The specific parameters for each single LS algorithm are given in Table 2.

4.1   Measures of the Efficiency of Multi-start Algorithms Based on the Algorithmic-Level

In the next experiments, the effectiveness in terms of quality of solutions is not addressed here. Only execution times and acceleration factors are reported in comparison with a mono-core CPU. The objective is to evaluate the impact of a GPU implementation of multi-start algorithms based on the algorithmic-level (i.e. the full distribution of the search process on GPU) in terms of efficiency. For each multi-start algorithm, a standalone mono-core CPU implementation, a pure GPU one, and a GPU version using texture memory (GPU_{tex}) are considered. The number of LS algorithms of the multi-start model is set to 4096 which corresponds to a realistic scenario in accordance with the algorithm convergence. The average time has been measured in seconds for 30 runs. The standard deviation is not represented since its value is very low for each measured instance. The obtained results are reported in Table 3 for the different LS multi-start algorithms on GPU.

Table 3. Measures of the efficiency of the algorithmic-level on the QAP. The average time is reported in seconds for 30 executions, the number of LSs is fixed to 4096.

	tai30a	tai40a	tai50a	tai60a	tai80a	tai100a
HC CPU	5.48	17.18	44.56	88.32	302.43	810.39
HC GPU	$3.19_{\times 1.7}$	$7.44_{\times 2.3}$	$15.79_{\times 2.8}$	$30.06_{\times 2.9}$	$90.45_{\times 3.3}$	$224.51_{\times 3.6}$
HC GPUTex	$1.02_{\times 5.4}$	$2.96_{\times 5.8}$	$6.69_{\times 6.7}$	$12.52_{\times 7.1}$	$41.65_{\times 7.3}$	$103.59_{\times 7.8}$
TS CPU	335.57	725.39	1539.60	2439.86	6097.61	13004.76
TS GPU	$105.12_{\times 3.2}$	$207.12_{\times 3.5}$	$414.50_{\times 3.7}$	$655.32_{\times 3.7}$	$1544.32_{\times 3.9}$	$3222.01_{\times 4.0}$
TS GPUTex	$55.12_{\times 6.1}$	$105.65_{\times 6.9}$	$176.29_{\times 8.7}$	$262.31_{\times 9.3}$	$588.33_{\times 10.4}$	$1207.77_{\times 10.8}$
SA CPU	412.64	874.44	1672.63	2699.89	6807.88	13960.69
SA GPU	$115.32_{\times 3.6}$	$223.65_{\times 3.9}$	$422.32_{\times 4.0}$	$677.28_{\times 4.0}$	$1578.21_{\times 4.3}$	$3121.28_{\times 4.5}$
SA GPUTex	$72.25_{\times 5.7}$	$135.21_{\times 6.5}$	$205.74_{\times 8.1}$	$278.88_{\times 9.7}$	$609.78_{\times 11.2}$	$1161.52_{\times 12.0}$

Regarding the acceleration for a pure implementation on GPU based on HC (HC GPU), it varies between ×1.7 for the instance tai30a to ×3.6 for the last instance. In comparison with a pure CPU implementation, the obtained acceleration factors are positive but not impressive. Indeed, due to high misaligned accesses to global memories (flows and distances in QAP), non-coalescing memory reduces the performance of the implementation. Binding texture on global memory allows to overcome the problem (HC GPUTex). Indeed, from the instance tai30a, using texture memory starts providing significant acceleration factors (×5.4). GPU keeps accelerating the LS process as long as the size grows and the best results are obtained for the instance tai100a (×7.8).

Regarding the performance for the other multi-start algorithms (TS and SA based), similar observations can be made. Indeed, on the hand, the obtained speed-ups for the texture version of multi-start algorithms based on TS vary between ×6.1 to ×10.8. And on the other hand, they vary from ×5.7 to ×12.0

for the multi-start algorithms based on SA. In a general manner, the performance variation obtained with the different algorithms on GPU is in accordance with the algorithm complexity ($Complexity(SA) >= Complexity(TS) >> Complexity(HC)$).

The point to highlight in these experiments is that organizing data into cache such as texture memory clearly allows to improve the speed-ups in comparison with a standard GPU version where inputs are stored in the global memory.

4.2 Measures of the Efficiency of Large GPU-Based Implementations

Another experiment consists in measuring the impact in terms of efficiency by varying the number of LSs in the multi-start based on the algorithmic-level. In addition, we propose to compare this approach with the multi-start based on the iteration-level model (parallel evaluation of the neighborhood on GPU) presented in Section 3.1. For doing this, we propose to deal with the instance tai50a with the same parameters used before in the context of multi-start methods based on TS. The obtained results are depicted in Fig. 4 for the texture optimization.

For the algorithmic-level, one can notice that it starts providing a positive acceleration of ×1.7 from a number of 512 LSs (one thread per LS). From 1024 LSs, the acceleration factors are drastically improved until reaching ×8.7 for 4096 LSs. After that, the speed-up keeps improving slowly with the size increase. However, as one can see in Fig. 5, no significant difference can be made in terms of the quality of the solutions obtained for more than 168384 LSs. Therefore, since the execution is already time-consuming, it might not be relevant to perform more LSs.

Regarding a small number of running LSs, from 1 to 256 LSs, the multi-start for the algorithmic-level is clearly inefficient. This can be explained by the fact that since the number of threads is relatively small, the number of threads per block is not enough to fully cover the memory access latency.

Unlike the previous model, for the multi-start based on the iteration-level, the obtained speed-ups are quiet regular (from ×4.4 to ×5.1) whatever the number of running LSs. Indeed, since one thread is associated with one neighbor ($\frac{n \times (n-1)}{2}$ neighbors), during the kernel execution, there is enough threads to keep the GPU multiprocessors busy. However, as one can notice, the maximal performance of this scheme is quiet limited because of the multiple data copies between the CPU and the GPU (see [2] for an analysis of data transfers).

5   Discussion and Conclusion

Parallel metaheuristics such as the multi-start model allow to improve the effectiveness and robustness in optimization problems. Their exploitation for solving real-world problems is possible only by using a great computational power. High-performance computing based on GPU accelerators is recently revealed as an efficient way to use the huge amount of resources at disposal. However, the

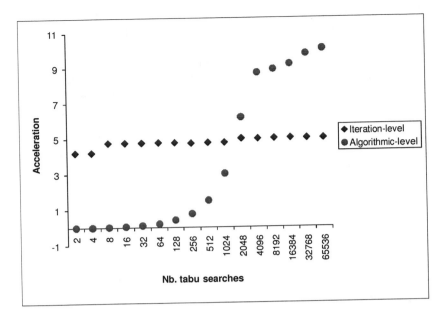

Fig. 4. Measures of the efficiency of the two multi-start approaches using the texture memory algorithmic-level approach in comparison with the iteration-level by varying the number of tabu searches (instance tai50a)

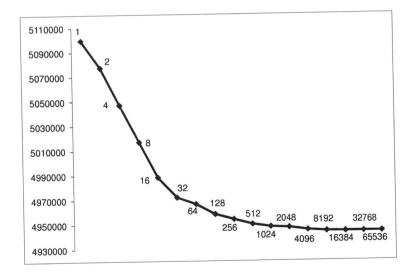

Fig. 5. Measures of the quality of the solutions for the multi-start model based on the algorithmic-level (tai50a). The average fitness is reported for 30 executions where each point represents a certain number of LSs.

exploitation of the multi-start model is not trivial and many issues related to the context execution and to the memory hierarchy of this architecture have to be considered.

In this paper, we have proposed a guideline to design and implement general GPU-based multi-start LS algorithms. The different concepts addressed throughout this paper takes into account popular LS algorithms such as HC, SA or TS. The designed and implemented approaches have been experimentally validated on a combinatorial optimization problem. To the best of our best of knowledge, multi-start parallel LS approaches have never been widely investigated so far.

The idea of our methodology is based on two natural schemes which exploit the GPU in a different manner. In the first scheme, the multi-start model is combined with the parallel evaluation of the neighborhood. The advantage of this scheme is to maximize the GPU in terms of multiprocessor occupancy. However, the performance of this scheme is limited due to the data transfers between the CPU and the GPU. To deal with this issue, we have particularly focused on the full distribution of the search process on GPU with the appropriate use of memory. Applying such mechanism with an efficient memory management allows to provide significant speed-ups (up to $\times 12$). However, this second scheme could also present some performance limitations when dealing with a small number of LS executions.

In a general manner, the two proposed schemes are complementary and their use strongly depends of the number of LSs to deal with. It would be interesting to test the performance of our approaches with some combinatorial optimization problems involving the use of different memories such as the constant and the shared memory.

Another perspective of this work is to combine the multi-start on GPU with a pure multi-core approach. Indeed, since this model has a high degree of parallelism, the CPU cores can also work in parallel in an independent manner. Moreover, since nowadays the actual configurations have 4 and 8 cores, instead of waiting the results back from the GPU, this computational power should be well-exploited in parallel to provide additional accelerations.

References

1. Talbi, E.G.: Metaheuristics: From design to implementation. Wiley, Chichester (2009)
2. Van Luong, T., Melab, N., Talbi, E.-G.: Local search algorithms on graphics processing units. A case study: The permutation perceptron problem. In: Cowling, P., Merz, P. (eds.) EvoCOP 2010. LNCS, vol. 6022, pp. 264–275. Springer, Heidelberg (2010)
3. Luong, T.V., Melab, N., Talbi, E.G.: Large neighborhood for local search algorithms. In: IPDPS. IEEE Computer Society, Los Alamitos (2010)
4. Zhu, W., Curry, J., Marquez, A.: Simd tabu search with graphics hardware acceleration on the quadratic assignment problem. International Journal of Production Research (2008)
5. Janiak, A., Janiak, W.A., Lichtenstein, M.: Tabu search on gpu. J. UCS 14(14), 2416–2426 (2008)

6. Alba, E., Talbi, E.G., Luque, G., Melab, N.: 4. Metaheuristics and Parallelism. In: Parallel Metaheuristics: A New Class of Algorithms, pp. 79–104. Wiley, Chichester (2005)
7. Zomaya, A.Y., Patterson, D., Olariu, S.: Sequential and parallel meta-heuristics for solving the single row routing problem. Cluster Computing 7(2), 123–139 (2004)
8. Melab, N., Cahon, S., Talbi, E.G.: Grid computing for parallel bioinspired algorithms. J. Parallel Distributed Computing 66(8), 1052–1061 (2006)
9. Ryoo, S., Rodrigues, C.I., Stone, S.S., Stratton, J.A., Ueng, S.Z., Baghsorkhi, S.S., Hwu, M.W.: Program optimization carving for gpu computing. J. Parallel Distributed Computing 68(10), 1389–1401 (2008)
10. NVIDIA: CUDA Programming Guide Version 3.0 (2010)
11. Group, K.: OpenCL 1.0 Quick Reference Card (2010)
12. Burkard, R.E., Deineko, V.G., Woeginger, G.J.: The travelling salesman problem on permuted monge matrices. J. Comb. Optim. 2(4), 333–350 (1998)
13. Bader, D.A., Sachdeva, V.: A cache-aware parallel implementation of the push-relabel network flow algorithm and experimental evaluation of the gap relabeling heuristic. In: Oudshoorn, M.J., Rajasekaran, S. (eds.) ISCA PDCS, ISCA, pp. 41–48 (2005)
14. Nickolls, J., Buck, I., Garland, M., Skadron, K.: Scalable parallel programming with cuda. ACM Queue 6(2), 40–53 (2008)
15. Dell'Amico, M., Trubian, M.: Applying tabu search to the job-shop scheduling problem. Ann. Oper. Res. 41(1-4), 231–252 (1993)
16. NVIDIA: GPU Gems 3. Chapter 37: Efficient Random Number Generation and Application Using CUDA (2010)
17. Taillard, É.D.: Robust taboo search for the quadratic assignment problem. Parallel Computing 17(4-5), 443–455 (1991)

Active Learning of Combinatorial Features for Interactive Optimization

Paolo Campigotto, Andrea Passerini, and Roberto Battiti

DISI - Dipartimento di Ingegneria e Scienza dell'Informazione
Università degli Studi di Trento
{campigotto,passerini,battiti}@disi.unitn.it
http://www.disi.unitn.it

Abstract. We address the problem of automated discovery of preferred solutions by an interactive optimization procedure. The algorithm iteratively learns a utility function modeling the quality of candidate solutions and uses it to generate novel candidates for the following refinement. We focus on combinatorial utility functions made of weighted conjunctions of Boolean variables. The learning stage exploits the sparsity-inducing property of 1-norm regularization to learn a combinatorial function from the power set of all possible conjunctions up to a certain degree. The optimization stage uses a stochastic local search method to solve a weighted MAX-SAT problem. We show how the proposed approach generalizes to a large class of optimization problems dealing with satisfiability modulo theories. Experimental results demonstrate the effectiveness of the approach in focusing towards the optimal solution and its ability to recover from suboptimal initial choices.

1 Introduction

The field of combinatorial optimization focussed in the past mostly on solving well defined problems, where the function $f(x)$ to optimize is given, either in a closed form, or as a simulator which can be interrogated to deliver f values corresponding to inputs, possibly with some noise leading to stochastic optimization. One therefore distinguishes two separated phases, a first one related to defining the problem through appropriate consulting, knowledge elicitation, modeling steps, and a second one dedicated to solving the problem either optimally, in the few cases when this is possible, or approximately, in most real-world cases leading to NP-hard problems.

Unfortunately the above picture is not realistic in many application scenarios, where *learning* about the problem definition goes hand in hand with delivering a set of solutions of improving quality, as judged by a decision maker (DM) responsible for selecting the final solution. In particular, this holds in the context of multi-objective optimization, where one aims at maximizing at the same time a set of functions f_1, ..., f_n. Multi-objective optimization, when cast in the language of machine learning, is a paradigmatic case of lack of information, where only some relevant building blocks (*features*) are initially given as the individual function f_i's, but their combination into a *utility function* modeling the

C.A. Coello Coello (Ed.): LION 5, LNCS 6683, pp. 336–350, 2011.

preferences of the DM is not given and has to be learnt by interacting with the DM [1]. Dealing with human DM, characterized by limited patience and bounded rationality, demands for some form of strategic production of candidates to be evaluated (query learning), and requires to account for the possible mistakes and dynamical evolution of her preferences (learning about concrete possibilities may lead somebody to change his/her initial objectives and evaluations). A further complication is related to the difficulty of delivering quantitative judgments by the DM, who is often better off in ranking possibilities more than in delivering utility values. The interplay of optimization and machine learning has been advocated in the past for example in the Reactive Search Optimization (RSO) context, see [2,3] also for an updated bibliography and [4] for an application of RSO in the context of multi-objective optimization.

In this work, we focus on a setting in which the optimal utility function is both *unknown* and *complex* enough to prevent exhaustive enumeration of possible solutions. We start by considering combinatorial utility functions expressed as weighted combinations of terms, each term being a conjunction of Boolean features. A typical scenario would be a house sale system suggesting candidate houses according to their characteristics, such as "the kitchen is roomy", "the house has a garden", "the neighbourhood is quiet". The task can be formalized as a weighted MAX-SAT problem, a well-known formalization which allows to model a large number of real-world optimization problems. However, in the setting we consider here the underlying utility function is unknown and has to be jointly and interactively learned during the optimization process.

Our method consists of an iterative procedure alternating a search phase and a model refinement phase. At each step, the current approximation of the utility function is used to guide the search for optimal configurations; preference information is required for a subset of the recovered candidates, and the utility model is refined according to the feedback received. A set of randomly generated examples is employed to initialize the utility model at the first iteration.

We show how to generalize the proposed method to more complex utility functions which are combinations of *predicates* in a certain theory of interest. A standard setting is that of scheduling, where solutions could be starting times for each job, predicates define time constraints for related jobs, and weights specify costs paid for not satisfying a certain set of constraints. The generalization basically consists of replacing satisfiability with satisfiability modulo theory [5] (SMT). SMT is a powerful formalism combining first-order logic formulas and theories providing interpretations for the symbols involved, like the theory of arithmetic for dealing with integer or real numbers. It has received consistently increasing attention in recent years, thanks to a number of successful applications in areas like verification systems, planning and model checking.

Experimental results on both weighted MAX-SAT and MAX-SMT problems demonstrate the effectiveness of our approach in focusing towards the optimal solutions, its robustness as well as its ability to recover from suboptimal initial choices.

This manuscript is organized as follows: Section 2 introduces the algorithm for the SAT case. Section 3 introduces SMT and its weighted generalization and shows how to adapt our algorithm to this setting. Related works are discussed in Section 4. Section 5 reports the experimental evaluation for both SAT and SMT problems. A discussion including potential research directions concludes the paper.

2   Overview of Our Approach

Candidate configurations are n dimensional Boolean vectors $\mathbf{x}$ consisting of *catalog* features. The only assumption we make on the utility function is its sparsity, both in the number of features (from the whole set of catalog ones) and in the number of terms constructed from them. We rely on this assumption in designing our optimization algorithm.

The candidate solutions are obtained by applying a stochastic local search (SLS) algorithm that searches the Boolean vectors maximizing the weighted sum of the terms of the learnt utility model. At each iteration, the algorithm chooses between a random and a greedy move with probability wp and $1 - wp$, respectively. A greedy move consists of flipping one of the variables leading to the maximum increase in the sum of the weights of the satisfied terms (if improving moves are not available, the least worsening move is accepted). The main difference w.r.t the "standard" weighted SLS algorithms consists of the DNF rather than CNF representation, which we believe to be a more natural choice when modeling combined effects of multiple non-linearly related features. Since switching from disjunctive to conjunctive normal form representations may involve an exponential increase in the size of the Boolean formula, we implemented a method that operates on formulae represented as a weighted linear sum of terms.

The candidate solutions generated by the optimizer during the search phase are first sorted by their predicted score values and then shuffled uniformly at random. The first $s/2$ configurations are selected, where s is the number of the random training examples generated at the initialization phase. The evaluation of the selected configurations completes the generation of the new training examples.

The refinement of the utility model consists of learning the weights of the terms, discarding the terms with zero weight. In the following, we assume that the available feedback consists of a quantitative score. We thus learn the utility function by performing regression over the set of the Boolean vectors. Adapting the method to other forms of feedback, such as ranking of sets of solutions, is straightforward as will be discussed in Section 6. We address the regression task by the Lasso [6]. The Lasso is an appropriate choice on problem domains with many irrelevant features, as its 1-norm regularization can automatically select input features by assigning zero weights to the irrelevant ones. Feature selection is crucial for achieving accurate prediction if the underlying model is sparse [7].

Let $D = (\mathbf{x}_i, y_i)_{i=1...m}$ the set of m training examples, where $\mathbf{x}_i$ is the Boolean vector and y_i its preference score. The learning task is accomplished by solving the following lasso problem:

```
1.   procedure interactive_optimization
2.   ┌   input: set of the catalog variables
3.   │   output: configuration optimizing the learnt utility function
4.   │   /* Initialization phase */
5.   │   initialize training set D by selecting s configurations uniformly at random;
6.   │   get the evaluation of the configurations in D;
7.   │   while (termination_criterion)
8.   │   ┌   /* Learning phase */
9.   │   │   Based on D, select terms and relative weights for current
10.  │   │   weighted MAX-SAT formulation (Eq. 1);
11.  │   │   /* Optimization phase */
12.  │   │   Get new configurations by optimizing current weighted MAX-SAT
13.  │   │   formulation;
14.  │   │   /* Training examples selection phase */
15.  │   └   Select s/2 configurations, get their evaluation and add them to D;
16.  └   return configuration optimizing the learnt weighted MAX-SAT formulation
```

Fig. 1. Pseudocode for the interactive optimization algorithm

$$\min_{\mathbf{w}} \sum_{i=1}^{m} (y_i - \mathbf{w}^T \cdot \Phi(\mathbf{x}_i))^2 + \lambda \|\mathbf{w}\|_1 \tag{1}$$

where the mapping function Φ projects sample vectors to the space of all possible conjunctions of up to d Boolean variables. The learnt function $f(\mathbf{x}) = \mathbf{w}^T \cdot \Phi(\mathbf{x})$ will be used as the novel approximation of the utility function. A new iteration of our algorithm can now take place. The pseudocode of our algorithm is in Fig. 1.

Note that dealing with the explicit projection Φ in Eq. 1 is tractable only for a rather limited number of catalog features and size of conjunctions d. This will typically be the case when interacting with a human DM. A possible alternative consists of directly learning a non-linear function of the features, without explicitly projecting them to the resulting higher dimensional space. We do this by kernel ridge regression [8] (Krr), where 2-norm regularization is used in place of 1-norm. The resulting dual formulation can be kernelized into:

$$\boldsymbol{\alpha} = (K + \lambda I)^{-1} \mathbf{y}$$

where K and I are the kernel and identity matrices respectively and λ is again the regularization parameter. The learnt function is a linear combination of kernel values between the example and each of the training instances: $f(\mathbf{x}) = \sum_{i=1}^{m} \alpha_i K(\mathbf{x}, \mathbf{x}_i)$. We employ a Boolean kernel [9] which implicitly considers all conjunctions of up to d features:

$$K_B(\mathbf{x}, \mathbf{x}') = \sum_{l=1}^{d} \binom{\mathbf{x}^T \cdot \mathbf{x}'}{l}$$

With the lasso, the function $\Phi(\cdot)$ maps the Boolean variables to all possible terms of size up to d. This allows for an explicit representation of the learnt utility function f as a weighted combination of the selected Boolean terms. On the other hand, in the kernel ridge regression case terms are only implicitly represented via the Boolean kernel K_B. In both cases, the value of the learnt function f is used to guide the search of the SLS algorithm. In the following, the two proposed approaches are referred as the *Lasso* and the *Krr* algorithms. As will be shown in the experimental section, the sparsity-inducing property of the *Lasso* allows it to consistently outperform *Krr*. The problem of addressing more complex scenarios, possibly involving non-human DM, where we can not afford an explicit projection, will be discussed in Section 6.

3   Satisfiability Modulo Theory

In the previous section, we assumed our optimization task could be cast into a *propositional* satisfiability problem. However, many applications of interest require or are more naturally described in more expressive logics as first-order logic (FOL), involving quantifiers, functions and predicates. In these cases, one is usually interested in validity of a FOL formula with respect to a certain *background theory* T fixing the interpretation of (some of the) predicate and function symbols. A general purpose FOL reasoning system such as Prolog, based on the resolution calculus, needs to add to the formula a conjunction of all the axioms in T. This is, for instance, the standard setting we consider in inductive logic programming when verifying whether a certain hypothesis covers an example given the available background knowledge. Whenever the cost of including such additional background theory is affordable, our algorithm can be applied rather straightforwardly.

Unfortunately, adding all axioms of T is not viable for many theories of interest: consider for instance the theory of *arithmetic*, which restricts the interpretation of symbols such as $+, \geq, 0, 5$. A more efficient alternative consists of using *specialized* reasoning methods for the background theory of interest. The resulting problem is known as *satisfiability modulo theory* (SMT)[5] and has drawn a lot of attention in recent years, guided by its applicability to a wide range of real-world problems. Among them, consider, for example, problems arising in formal hardware/software verification or in real-time embedded systems design. Popular examples of useful theories include various theories of arithmetic over reals or integers such as linear or difference ones. Linear arithmetic considers $+$ and $-$ functions alone, applied to either numerical constants or variables, plus multiplication by a numerical constant. Difference arithmetic is a fragment of linear arithmetic limiting legal predicates to the form $x - y \leq c$, where x, y are variables and c is a numerical constant. Very efficient procedures exists for checking satisfiability of difference logic formulas [10]. A number of theories have been studied apart from standard arithmetic ones (e.g., the theory of bit-vector arithmetic to model machine arithmetic).

```
1.  procedure SMT-solver(φ)
2.     φ' = α(φ)
3.     while (true)
4.        (r,M) ← SAT(φ')
5.        if r = unsat then return unsat
6.        (r,J) ← T-Solver(β(M))
7.        if r = sat then return sat
8.        C ← ⋁_{l∈J} ¬α(l)
9.        φ' ← φ' ∧ C
```

Fig. 2. Pseudocode for a basic lazy SMT-solver

3.1   Satisfiability Modulo Theory Solvers

The most successful SMT solvers can be grouped into the two main approaches named *eager* and *lazy*. The eager approach consists of developing theory-specific and efficient translators which translate a query formula into an equisatisfiable propositional one, much like compilers do when optimizing the code generated from a high-level program. Lazy approaches, on the other hand, work by building efficient *theory solvers*, inference systems specialized on a theory of interest. These solvers are integrated as submodules into a generic SAT solver. In the rest of the paper we will focus on this latter class of SMT solvers, which we integrated in our optimization algorithm. The simplest approach for building a lazy SMT-solver consists of alternating calls to the satisfiability and the theory solver respectively, until a solution satisfying both solvers is retrieved or the problem is found to be unsatisfiable. Let φ be a formula in a certain theory T, made of a set of n predicates $A = \{a_1, \ldots, a_n\}$. A mapping α maps φ into a propositional formula $\alpha(\varphi)$ by replacing its predicates with propositional variables $p_i = \alpha(a_i)$. The inverse mapping β replaces propositional variables with their corresponding predicates, i.e., $\beta(p_i) = a_i$. For example, consider the following formula in a non-linear theory T:

$$(\cos(x) = 3 + \sin(y)) \wedge (z \leq 8) \tag{2}$$

Then, $p_1 = \alpha(\cos(x) = 3 + \sin(y))$ and $p_2 = \alpha(a_i \leq 8)$. Note that the truth assignment $p_1 = true, p_2 = false$ is equivalent to the statement $(\cos(x) = 3 + \sin(y)) \wedge (z > 8)$ in the theory T.

Figure 2 reports the basic form [11] of an SMT algorithm. $\text{SAT}(\varphi)$ calls the SAT solver on the φ instance, returning a pair (r, M), where r is **sat** if the instance is satisfiable, **unsat** otherwise. In the former case, M is a truth assignment satisfying φ. $\text{T-Solver}(S)$ calls the theory solver on the formula S and returns a pair (r, J), where r indicates if the formula is satisfiable. If $r =$ **unsat**, J is a *justification* for S, i.e any unsatisfiable subset $J \subset S$. The next iteration calls the SAT solver on an extended instance accounting for this justification.

State-of-the-art solvers introduce a number of refinements to this basic strategy, by pursuing a tighter integration between the two solvers. A common underlying idea is to prune the search space for the SAT solver by calling the theory solver on partial assignments and propagating its results. Finally, combination methods exist to jointly employ different theories, see [12] for a basic procedure.

3.2 Weighted MAX-SMT

Weighted MAX-SMT generalizes SMT problems much like weighted MAX-SAT does with SAT ones. While a body of works exist addressing weighted MAX-SAT problems, the former generalization has been tackled only recently and very few solvers have been developed [13,14,15]. The simplest formulation consists of adding a cost to each or part of the formulas to be jointly satisfied, and returning the assignment of variables minimizing the sum of the costs of the unsatisfied clauses, or a satisfying assignment if it exists. The following is a "weighted version" of Eq. 2:

$$5 \cdot (\cos(x) = 3 + \sin(y)) + 12 \cdot (z \leq 8) \tag{3}$$

where 5 and 12 are the cost of the violation of the first and the second predicate, respectively.

Generalizing, consider a true utility function f expressed as a weighted sum of terms, where a term is the conjunction of up to d predicates defined over the variables in the theory T. The set of all n possible predicates represents the search space S of the MAX-SAT solver integrated in the MAX-SMT solver. Our approach learns an approximation $\hat{f}$ of f and gets one of its optimizers $\mathbf{v}$ from the MAX-SMT solver. The optimizer (and in general each candidate solution in the theory T) identifies an assignment $\mathbf{p}^* = (p_1^*, \ldots, p_n^*)$ of Boolean values ($p_i^* = \{\mathbf{true}, \mathbf{false}\}$) to the predicates in S. The DM is asked for a feedback on the candidate solution $\mathbf{v}$ and returns a possibly noisy quantitative score $s \approx f(\mathbf{v})$. The pair $(\mathbf{p}^*, s)$ represents a new training example for our approach. In order to obtain multiple training examples, we optimize again $\hat{f}$ with the additional *hard*[1] constraint generated by the disjunction of all the terms of $\hat{f}$ unsatisfied by $\mathbf{p}^*$. For example, let t_1 and t_5 be the terms of $\hat{f}$ unsatisfied by $\mathbf{p}^*$, then the hard constraint becomes:

$$(t_1 \vee t_5)$$

If $\mathbf{p}^*$ satisfies all the terms of $\hat{f}$, i.e., $\hat{f}(\mathbf{p}^*) = 0$, the additional *hard* constraint generated is

$$(\neg p_1^* \vee \neg p_2^* \ldots \vee \neg p_n^*)$$

which excludes $\mathbf{p}^*$ from the feasible solutions set of $\hat{f}$. The generation of the training examples is iterated till the desired number of examples have been created or the hard constraints generated made the MAX-SMT problem unsatisfiable.

The learning component of our algorithm is then re-trained, including in the training set the new collected examples and the approximation of the true utility function is refined. A new optimization phase can now take place (see Fig. 1).

---

[1] *Hard* constraints do not have a cost, and they have to be satisfied. On the contrary, the terms with a cost, which may or may not be satisfied, are called *soft* constraints.

The mechanism creating the training examples is motivated by the tradeoff between the selection of good solutions (w.r.t. the current approximation of the true utility function) and the diversification of the search process.

4   Related Works

Active learning is a hot research area and a broad range of different approaches has been proposed (see [16] for a review). The simplest and most common framework is that of *uncertainty sampling*: the learner queries the instances on which it is least certain. However, the ultimate goal of a recommendation or optimization system is selecting the best instance(s) rather than correctly modeling the underlying utility function. The query strategy should thus tend to suggest good candidate solutions and still learn as much as possible from the feedback received. Typical areas where research on this issue is quite popular are single- and multi-objective interactive optimization [1] and information retrieval [17]. The need to trade off multiple requirements in this active learning setting is addressed in [18] where the authors consider relevance, diversity and density in selecting candidates. Note that our approach relies on query *synthesis* rather than selection, as *de-novo* candidate solutions are generated by the SLS algorithm. Nonetheless, our diversification strategies are very simple and could be significantly improved by taking advantage of the aforementioned literature.

Choosing relevant features according to their weight within the learnt model is a common selection strategy (see e.g. [19]). When dealing with implicit feature spaces as in kernel machines, the problem can be addressed by introducing a hyper-parameter for each input feature, like a feature-dependent variance for Gaussian kernels [20]. Parameters and hyper-parameters (or their relaxed real-valued version) are jointly optimized trying to identify a small number of relevant features. One-norm regularization [6] has the advantage of naturally inducing sparsity in the set of selected features. Approaches also exist [21] which directly address the combinatorial problem of zero-norm optimization.

A large body of recent work exists for developing interactive approaches [1] to multiobjective optimization. A common approach consists of modeling the utility function as a linear combination of objectives, and iteratively updating its weights trying to match the DM requirements. Our algorithm allows to deal with complex non-linear interactions between (Boolean) objectives and, thanks to the SMT extension, can be applied to a wide range of optimization problems.

Very recent works in the field of constraint programming [22] define the user preferences in terms of *soft* constraints and introduce constraint optimization problems where the data are not completely known before the solving process starts. In particular, the work in [22] introduces an elicitation strategy for soft constraint problems with missing preferences, with the purpose of finding the solution preferred by the DM asking to reveal as few preferences as possible. Despite the common purpose, this approach is different from ours. A major difference regards the preference elicitation problem considered. In [22] decision variables and soft constraints are assumed to be known in advance and the information uncertainty consists only of missing preference values. On the other hand, our settings assume sparsity of the utility function, both in the number of features (from the whole set of catalog features) and in the selection of the terms constructed from them. Furthermore, our technique is robust to imprecise information from the DM, modeled in terms of inaccurate preference scores for the candidate solutions. Even if interval-valued constraints [23] have been introduced to handle uncertainty in the evaluations of the DM, the experiments in [22] do not consider the case of inconsistent preference information. Finally, while the technique in [22] combines branch and bound search with preference elicitation and the adoption of local search algorithms is matter of research, our approach works straightforwardly with both incomplete and complete search techniques.

5   Experimental Results

The following empirical evaluation demonstrates the versatility and the efficiency of our approach for the weighted MAX-SAT and the weighted MAX-SMT problems. The MAX-SMT tool used for the experiments is the "Yices" solver [13].

5.1   Weighted MAX-SAT

The *Lasso* and the *Krr* algorithms were tested over a benchmark of randomly generated utility functions according to the triplet (*number of features, number of terms, max term size*), where *max term size* is the maximum allowed number of Boolean variables per term. We generate functions for: $\{(5,3,3), (6,4,3), (7,6,3), (8,7,3), (9,8,3), (10,9,3)\}$. Each utility function has two terms with maximum size. Terms weights are integers selected uniformly at random in the interval $[-100,0) \cup (0,100]$. We consider as *gold* standard solution the configuration obtained by optimizing the true utility function.

The number of catalog features is 40. The maximum size of terms is assumed to be known. The walk probability parameter of the SLS algorithm wp is set to 0.2. Furthermore, the score values of the training examples are affected by Gaussian noise, with mean 0 and standard deviation 10.

We run a set of experiments for $10, 20, \ldots 100$ initial training examples, for the *Lasso* and the *Krr* versions of the algorithm. Results are expressed in terms of the quality of the learnt utility function (Fig. 3) and of the approximation of the *gold* solution (Fig. 4). Each point of the curves in the Fig. 3 and 4 is the mean and the median values, respectively, over 400 runs with different random seeds.

Fig. 3 shows the quality of the learnt utility function, in terms of the root mean squared error (rmse) between the true and the predicted values for a benchmark of 1000 test examples. A better approximation is generated by the *Lasso* algorithm for all the considered true utility functions. Furthermore, while increasing the number of training examples, a faster improvement is observed for the *Lasso* w.r.t. the *Krr* algorithm. Consider, for example, the case of nine terms. With 40 training examples, the performance of *Krr* is within 10 units from the value observed for the *Lasso* method. When 100 examples are employed, the

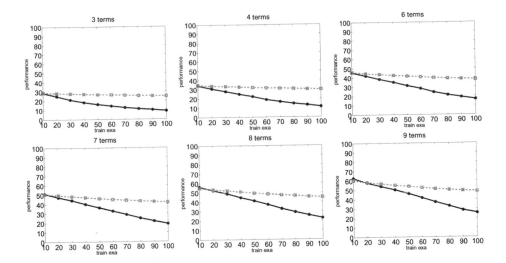

Fig. 3. Quality of the learnt utility function for an increasing number of training examples observed for the algorithms at the first iteration. The y-axis reports the root mean squared error between the true and the predicted values for a benchmark of 1000 test examples. The x-axis contains the number of training examples. The solid blue and the dashed green lines show the performance of the *Lasso* and the *Krr* algorithms, respectively. See text for details.

mean rmse of the *Lasso* algorithm is less than value 30, while the performance of the *Krr* method does not increase beyond value 50.

The superior performance of the *Lasso* algorithm is confirmed by the experiments in Fig. 4, reporting the quality of the *best* configuration at the different iterations for an increasing number of initial training examples. The *best* configuration is the configuration optimizing the current approximation of the true utility function. Its quality is measured in terms of the approximation error w.r.t. the gold solution.

Considering the simplest problems with three and four terms, the performance of *Krr* is comparable with the results obtained by *Lasso*, except at the first iteration of *Krr* in the case of four terms true utility functions, where the gold solution is not identified even with 100 initial training examples.

However, the *Lasso* approach outperforms the *Krr* results when the true utility function includes at least six terms. First, note that the *Lasso* algorithm succeeds in exploiting its active learning strategy, and converges rather quickly to the optimal solution when enough iterations are provided. At the first iteration its approximation error is above 40 even when 30 training examples are used. At the third iteration, the *Lasso* algorithm identifies the gold standard solution, when at least 60 training examples are available. On the other hand, for true utility functions with more than seven terms *Krr* fails to improve over its suboptimal solution when increasing the number of examples and iterations. As a consequence, the *Krr* algorithm does not identify the gold solution, even

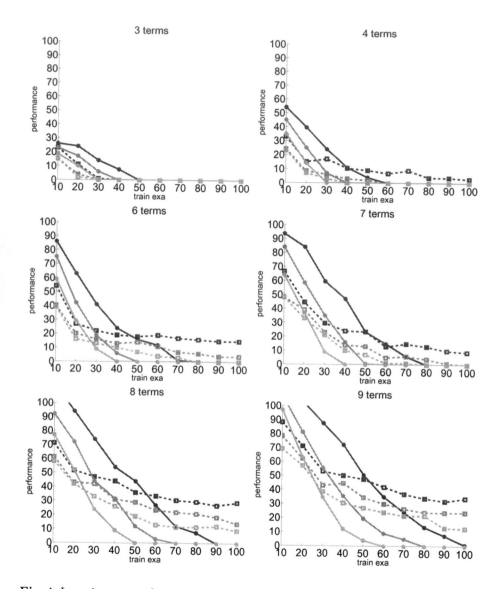

Fig. 4. Learning curves for an increasing number of training examples observed for the two algorithms at different iterations. The y-axis reports the solution quality, while the x-axis contains the number of training examples. The dashed lines refer to the *Krr* algorithm, while the solid lines are for the *Lasso* algorithm. Furthermore, red, green and cyan colors show the performance of the algorithms at the first, the second and the third iteration, respectively. See text for details.

in the case of 100 training examples. However, when very few training examples are available, the *Krr* algorithm reaches a better approximation than *Lasso*.

5.2   Weighted MAX-SMT

SMT is a hot research area [11]. However, MAX-SMT techniques are very recent and there are no well established publicly available benchmarks for weighted MAX-SMT problems. Existing results [14] indicate that MAX-SMT solvers can efficiently address real-world problems.

In this work, we modeled a scheduling problem as a MAX-SMT problem. In detail, a set of five jobs must be scheduled over a given period of time. Each job has a fixed known duration, the constraints define the overlap of two jobs or their non-concurrent execution. The true utility function is generated by selecting uniformly at random weighed terms over the constraints. The solution of the problem is a schedule assigning a starting date to each job and minimizing the cost, where the cost of the schedule is the sum of the weights of the violated terms of the true utility function. The temporal constraints are expressed by using the difference arithmetic theory. In detail, let s_i and d_i, with $i = 1 \ldots 5$, be the starting date and the duration of the i-*th* job, respectively. If s_i is scheduled before s_j, the constraint expressing the overlap of the two jobs is $s_j - s_i < d_i$, while their non-concurrent execution is encoded by $s_j - s_i \geq d_i$ Note that there are 40 possible constraints for a set of 5 jobs. The maximum size of the terms of the true utility function is three and it is assumed to be known. Their weights are distributed uniformly at random in the range $[1, 100]$. Similarly to the MAX-SAT case, the experimental setting includes Gaussian noise (with mean 0 and standard deviation 10) affecting the cost values of the training examples.

Fig. 5 depicts the performance of the Lasso algorithm for the cases of 3, 4, 6, 7, 8, 9 terms in the true utility function. The y-axis reports the solution quality measured in terms of deviation from the gold solution, while the x-axis contains the number n of training examples at the first iteration. At the following iterations, $n/2$ examples are added to the training set (see Sec. 2). Each point of the curves is the median value over 400 runs with different random seeds.

As expected, the learning problem becomes more challenging while increasing the number of terms. However, the results for the scheduling problem are promising: our approach identifies the gold standard solution in all the cases. In detail, less than 40 examples are required to identify the gold solution at the second iteration. At the third iteration our algorithm needs only 20 training examples for convergence to the gold solution.

Finally, note that the approach based on Krr does not maintain an explicit representation of the learnt utility function, and therefore a direct extension to SMT problems is not possible for the current MAX-SMT solvers which tightly integrate SAT and theory solvers as discussed in Section 3.

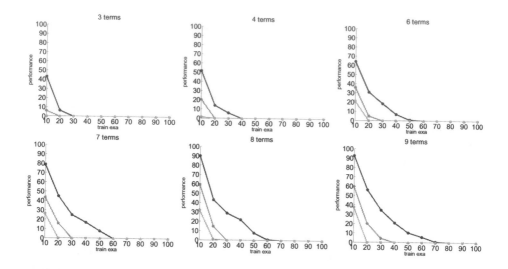

Fig. 5. Learning curves observed at different iterations of the Lasso algorithm while solving the scheduling problem. The y-axis reports the solution quality, while the x-axis contains the number of training examples. Red, green and cyan colors show the performance of the algorithm at the first, the second and the third iteration, respectively. See text for details.

6   Discussion

We presented an interactive optimization strategy for combinatorial problems over an unknown utility function. The algorithm alternates a search phase using the current approximation of the utility function to generate candidate solutions, and a refinement phase exploiting feedback received to improve the approximation. One-norm regularization is employed to enforce sparsity of the learned function. An SLS algorithm addresses the weighted MAX-SAT problem resulting from the search phase. We show how to adapt the approach to a large class of relevant optimization problems dealing with satisfiability modulo theories. Experimental results on both weighted MAX-SAT and MAX-SMT problems demonstrate the effectiveness of our approach in focusing towards the optimal solutions, its robustness as well as its ability to recover from suboptimal initial choices.

The algorithm can be generalized in a number of directions. The availability of a quantitative feedback is not necessarily straightforward, especially when a human DM is involved in the loop. A more affordable request is often that of ranking sets of candidates according to preference. Our setting can be easily adapted to this setting by replacing the squared error loss in the learning stage with appropriate ranking losses. The simplest solution consists of formulating it as correctly ordering each pair of instances as done in support vector ranking, and applying 1-norm SVM [24]. More complex ranking losses have been proposed in the literature (see for instance [25]), especially to increase the importance of correctly ranking the best solutions, and could be combined with 1-norm regularization.

Our experimental evaluation is focused on small-scale problems, typical of an interaction with a human DM. In principle, when combined with appropriate SMT solvers, our approach could be applied to larger real-world optimization problems, whose formulation is only partially available. In this case, a local search algorithm rather than a complete solver will be used during the optimization stage, as showed in the experiments on the weighted MAX-SAT instances. However, the cost of requiring an explicit representation of all possible conjunction of predicates (even if limited to the unknown part) would rapidly produce an explosion of computational and memory requirements. One option is that of resorting to an implicit representation of the function to be optimized, like the one we used in the *Krr* algorithm. Kernelized versions of zero-norm regularization [26] could be tried in order to enforce sparsity in the projected space. However, the lack of an explicit formula would prevent the use of all the efficient refinements of SMT solvers, based on a tight integration between SAT and theory solvers. A possible alternative is that of pursuing an incremental feature selection strategy and iteratively solving increasingly complex approximations of the underlying problem. We are currently investigating both research directions.

Finally, we are also considering larger preference elicitation problems, with both known hard constraints limiting the set of feasible solutions and unknown user preferences. This setting allows us to address many real-world scenarios. In the house sale system, for instance, the hard constraints could define the available house types or locations, and the preferences of the DM would drive the search within the set of feasible solutions.

References

1. Branke, J., Deb, K., Miettinen, K., Słowiński, R. (eds.): Multiobjective Optimization: Interactive and Evolutionary Approaches. Springer, Heidelberg (2008)
2. Battiti, R., Brunato, M., Mascia, F.: Reactive search and intelligent optimization. Springer, Heidelberg (2008)
3. Battiti, R., Brunato, M.: Reactive search optimization: Learning while optimizing. In: Gendreau, M., Potvin, J.Y. (eds.) Handbook of Metaheuristics, 2nd edn. Int. Series in Op. Res. & Man. Sci., vol. 146, pp. 543–571. Springer Science, Heidelberg (2010)
4. Battiti, R., Campigotto, P.: Reactive Search Optimization: Learning While Optimizing. An Experiment in Interactive Multi-Objective Optimization. In: VIII Metaheur. Int. Conf. (MIC 2009), Germany. LNCS, Springer, Heidelberg (2009)
5. Barrett, C., Sebastiani, R., Seshia, S.A., Tinelli, C.: Satisfiability modulo theories. In: Handbook of Satisfiability, pp. 825–885. IOS Press, Amsterdam (2009)
6. Tibshirani, R.: Regression shrinkage and selection via the lasso. Journal of the Royal Statistical Society, Series B 58, 267–288 (1996)
7. Friedman, J., Hastie, T., Rosset, S., Tibshirani, R.: Discussion of boosting papers. Annals of Statistics 32, 102–107 (2004)

8. Suanders, C., Gammerman, A., Vovk, V.: Ridge regression learning algorithm in dual variables. In: ICML 1998 (1998)
9. Khardon, R., Roth, D., Servedio, R.: Efficiency versus convergence of boolean kernels for on-line learning algorithms. Journal of Artif. Int. Res. 24(1), 341–356 (2005)
10. Nieuwenhuis, R., Oliveras, A.: DPLL(T) with exhaustive theory propagation and its application to difference logic. In: Etessami, K., Rajamani, S.K. (eds.) CAV 2005. LNCS, vol. 3576, pp. 321–334. Springer, Heidelberg (2005)
11. de Moura, L., Bjorner, N.: Satisfiability modulo theories: An appetizer. In: Oliveira, M.V.M., Woodcock, J. (eds.) SBMF 2009. LNCS, vol. 5902, pp. 23–36. Springer, Heidelberg (2009)
12. Nelson, G., Oppen, D.C.: Simplification by cooperating decision procedures. ACM Trans. Program. Lang. Syst. 1(2), 245–257 (1979)
13. Dutertre, B., de Moura, L.: A Fast Linear-Arithmetic Solver for DPLL(T). In: Ball, T., Jones, R.B. (eds.) CAV 2006. LNCS, vol. 4144, pp. 81–94. Springer, Heidelberg (2006)
14. Nieuwenhuis, R., Oliveras, A.: On sat modulo theories and optimization problems. In: In Theory and App. of Sat. Testing. LNCS, pp. 156–169. Springer, Heidelberg (2006)
15. Cimatti, A., Franzén, A., Griggio, A., Sebastiani, R., Stenico, C.: Satisfiability modulo the theory of costs: Foundations and applications. In: Esparza, J., Majumdar, R. (eds.) TACAS 2010. LNCS, vol. 6015, pp. 99–113. Springer, Heidelberg (2010)
16. Settles, B.: Active learning literature survey. Technical Report Computer Sciences Technical Report 1648, University of Wisconsin-Madison (2009)
17. Radlinski, F., Joachims, T.: Active exploration for learning rankings from clickthrough data. In: 13th ACM SIGKDD International Conference on Knowledge Discovery and Data Mining (KDD 2007), pp. 570–579. ACM Press, New York (2007)
18. Xu, Z., Akella, R., Zhang, Y.: Incorporating diversity and density in active learning for relevance feedback. In: Amati, G., Carpineto, C., Romano, G. (eds.) ECIR 2007. LNCS, vol. 4425, pp. 246–257. Springer, Heidelberg (2007)
19. Guyon, I., Weston, J., Barnhill, S., Vapnik, V.: Gene selection for cancer classification using support vector machines. Machine Learning 46(1-3), 389–422 (2002)
20. Chapelle, O., Vapnik, V., Bousquet, O., Mukherjee, S.: Choosing multiple parameters for support vector machines. Machine Learning 46(1-3), 131–159 (2002)
21. Kaizhu, H., Irwin, K., Michael, R.: Direct Zero-Norm Optimization for Feature Selection. In: IEEE International Conference on Data Mining, pp. 845–850 (2008)
22. Gelain, M., Pini, M.S., Rossi, F., Venable, K.B., Walsh, T.: Elicitation strategies for soft constraint problems with missing preferences: Properties, algorithms and experimental studies. Artif. Intell. 174(3-4), 270–294 (2010)
23. Gelain, M., Pini, M.S., Rossi, F., Venable, K.B., Wilson, N.: Interval-valued soft constraint problems. Annals of Mat. and Art. Int. 58, 261–298 (2010)
24. Zhu, J., Rosset, S., Hastie, T., Tibshirani, R.: 1-norm Support Vector Machines. In: Neural Information Processing Systems. MIT Press, Cambridge (2003)
25. Chakrabarti, S., Khanna, R., Sawant, U., Bhattacharyya, C.: Structured learning for non-smooth ranking losses. In: 14th ACM SIGKDD International Conference on Knowledge Discovery and Data Mining, KDD 2008, pp. 88–96. ACM, New York (2008)
26. Weston, J., Elisseeff, A., Schölkopf, B., Tipping, M.: Use of the zero norm with linear models and kernel methods. Journal of Mach. Learn. Res. 3, 1439–1461 (2003)

A Genetic Algorithm Hybridized with the Discrete Lagrangian Method for Trap Escaping

Madalina Raschip and Cornelius Croitoru

"Al.I.Cuza" University of Iasi, Romania
{mionita,croitoru}@info.uaic.ro

Abstract. This paper introduces a genetic algorithm enhanced with a trap escaping strategy derived from the dual information presented as discrete Lagrange multipliers. When the genetic algorithm is trapped into a local optima, the Discrete Lagrange Multiplier method is called for the best individual found. The information provided by the Lagrangian method is unified, in the form of recombination, with the one from the last population of the genetic algorithm. Then the genetic algorithm is restarted with this new improved configuration. The proposed algorithm is tested on the winner determination problem. Experiments are conducted using instances generated with the combinatorial auction test suite system. The results show that the method is viable.

1 Introduction

Genetic algorithms (GAs) are powerful optimization techniques working with populations of individuals which are improved each iteration using specific operators. When dealing with difficult real-world problems, they may be trapped into local optima. Different methods for solving this problem were developed in literature. Maintaining the population diversity is a preemptive way. A straightforward procedure is to increase the mutation rate after a change has been detected. The loss of diversity is dependent on the selection intensity. Scaling techniques address the problem for fitness-proportionate selection schemes. Niching methods assist the selection procedure in order to reduce the effect of the genetic drift caused by this [1]. The island model [2], the random immigrants [3], restarting [4] are other examples of techniques used for preserving diversity. The restarting techniques are used inside genetic algorithms when some threshold is reached (local convergence is detected typically when no progress has been made for a long time). The current run is terminated and the algorithm is restarted with a new seed. Recently, restarting techniques have been applied to complete algorithms based on backtracking for constraint satisfaction problems, including the satisfiability problem [5]. They yield good performance improvements.

Recent publications describe hybrid approaches which often lead to faster and more robust algorithms for hard optimization problems [6]. The traditional methods come in two distinct flavors: heuristic search algorithms which find a satisfactory even if not necessarily optimal solution and exact algorithms which

C.A. Coello Coello (Ed.): LION 5, LNCS 6683, pp. 351–363, 2011.

guaranty for finding a provably optimal solution. Hybrid methods were developed in order to borrow ideas from both sources.

In particular, the hybridization of metaheuristics with (integer) linear programming (LP) techniques have proven to be feasible and useful in practice [7]. The two complementary techniques benefit from the synergy. The information provided by the LP-relaxed solutions could be exploited inside metaheuristics for creating promising initial solutions, inside repairing procedures, or to guide local improvement [7]. Approaches which use dual variables and the relations between primal and dual variables are also present in literature. For example, in [8] the shadow prices of the relaxed Multi-constrained Knapsack problem are used by a genetic algorithm inside a repairing procedure. Ratios based on the shadow prices give the likeliness of the items to be included in a solution. In [9] a primal-dual variable neighborhood search for the simple plant location problem is presented. After a primal feasible solution is obtained using a variable neighborhood decomposition search, a dual solution which exploits the complementary slackness conditions is created. The dual solution is transformed into an exact solution and used to derive a good lower bound and to strengthen next a Branch and Bound algorithm. The Lagrange multipliers could also be used inside metaheuristics. For example, in [11] an approach that combines Lagrangian decomposition with local search based metaheuristics, like variable neighborhood descent method, was proposed for the design of the last mile in fiber optic networks.

In [10] a hybrid technique based on duality information is proposed in order to escape from local optima. When the evolutionary algorithm reaches a local trap, the method leads the search out of a local optima. It constructs the appropriate dual relaxed space and improves it. The evolutionary algorithm is then restarted with a new population of primal individuals generated using the information from the dual solutions. The method was applied for determining the winner in combinatorial auctions.

The new approach presented here is based on the ideas from [10], but it uses a Lagrange Multiplier method inside the genetic algorithm. The Discrete Lagrange Multiplier (DLM) method is started with the initial solution equal to the best individual from the genetic algorithm when this is stuck into a local optima. The DLM method is a general search method based on the Lagrange multipliers (the dual solutions). The Lagrange method will give a new solution to be used by the GA in order to escape from the local optima. In contrast to local search methods that restart from a new starting point when are trapped into a local optima, the new method moves the search out of a local optima in a direction provided by the recombination with the DLM solution. In the traditional methods based on restarts, breaks in the trajectory are made. The new method escapes from a local optima in a continuous trajectory. The advantage is given by the fact that the optima may be in the vicinity of the already found local optima.

The method was tested for the winner determination problem (WDP) from the combinatorial auction field. In combinatorial auctions, multiple distinct items are sold simultaneously and the bidders may bid on combination of items [12].

The valuation for a combination of items is not necessarily equal to the sum of the individual items. This expressiveness can lead to more efficient allocations, as the applications in many real-world problems has demonstrated [13]. The problem of determining the winners is computational complex (NP-complete and inapproximable) [14].

The paper is organized as follows. Section 2 presents the new approach. In the following section the winner determination problem together with the application of the general scheme for the WDP is described. Next, the experimental results on the generated instances are shown. Finally, conclusions are drawn.

2   The Hybrid Method

The method follows the ideas from [10]. When the genetic search reaches a local trap, the approach runs the Discrete Lagrange Multiplier method, having as starting point the best individual from the (primal) genetic algorithm. Because of time constraints, the DLM method runs only for the best individual.

The Discrete Lagrange Multiplier method is the discrete version of the continuous Lagrange Multipliers method, which uses difference equations instead of using differential calculus. The method searches for saddle points in discrete neighborhoods. It performs ascents in the original-variable subspace and descents in the Lagrange-multiplier subspace. When the search reaches a local optima, the DLM method uses the Lagrange multipliers to lead the search out of the local optima.

The DLM algorithm helps the genetic algorithm to escape from the local optima. The approach restarts the genetic algorithm with the initial configuration modified: the last population of the genetic algorithm is recombined with the solution provided by the DLM method. By using the past experience, in the form of the last population of individuals, and the new information, which continues the previous direction of search and which is resulted from a DLM run, the algorithm is able to improve its future performance. These steps are iterated several times, or until an optimum solution is found. The scheme of the algorithm is presented in Figure 1.

In contrast to [10], the new scheme does not need to transform a primal solution into a dual one. The dual solutions (the Lagrangian multipliers) are initialized greedy and their values are changed 'online' in accordance with the modified primal solutions. Another benefit of the new scheme is that the DLM algorithm provides a feasible solution, constructed from the Lagrangian saddle point found, to be used in restarting the evolutionary algorithm. In the previous approach, the new primal solutions after restarting are constructed in a greedy manner from the dual ones.

In [15] a framework based on genetic algorithms and a constrained simulated annealing method was proposed for solving discrete constrained optimization problems. The simulated annealing technique provides initial solutions for the genetic algorithm and could be replaced by the DLM method. The purpose of using the DLM method in our approach is to escape from local optima.

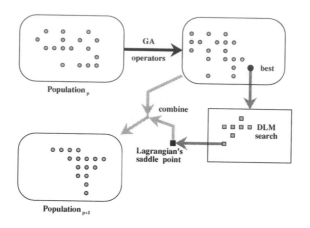

Fig. 1. The GA hybridized with the DLM method scheme

3 The Hybrid Method Applied to WDP

3.1 Winner Determination

An auctioneer has a set of goods, $M = \{1, 2, ..., m\}$ to sell. The buyers (bidders) submit a set of bids, $\mathcal{B} = \{B_1, ..., B_n\}$. A bid is a tuple $B_j = (S_j, p_j)$ where $S_j \subseteq M$ is a set of goods and p_j is a price. The winner determination problem is to label the bids as winning or losing so as to maximize the auctioneer's revenue (the sum of the accepted bid prices) under the constraint that each good is allocated to at most one bid. The problem can be formulated as an Integer Linear Programming problem as follows:

$$max \sum_{j=1}^{n} p_j x_j$$

$$s.t. \sum_{j | i \in S_j} x_j \leq 1, \forall i = 1, 2, ..., m \qquad \text{(WDP)}$$

$$x_j \in \{0, 1\}, \forall j = 1, 2, ..., n$$

$x_j = 1$ if bid j with price p_j is selected in the solution and $x_j = 0$ otherwise. The definition assumes the free disposal case, i.e. not all items need to be covered. If there is no free disposal, an equality is used in the constraint formulation.

Different methods for solving the problem were developed. Complete methods based on the Branch and Bound procedure [16] or linear programming [17] were designed. Stochastic methods like stochastic local search [18], simulated annealing [19] and genetic algorithms [20] have also been applied for solving the problem. A heuristic method based on the Lagrangian relaxation with subgradient optimization is proposed in [21]. The heuristic methods compare well with CPLEX or with other exact algorithms.

3.2   The Discrete Lagrangian Method for WDP

The Lagrange Multiplier methods have been developed for continuous constrained optimization problems. For a minimization problem, they do descents in the original space and ascents in the Lagrange-multiplier space. Equilibrium is reached when an optimal solution is found.

The Discrete Lagrangian method (DLM) is a global search method which works on discrete values. It was initially proposed for solving satisfiability problems [22]. A Lagrangian function determines the search direction. The method escapes from a local optima by using the information provided by the Lagrange multipliers.

Define the problem $WDP(\lambda)$:

$$max \sum_{j=1}^{n} p_j x_j + \sum_{i=1}^{m} [\lambda_i (1 - \sum_{j|i \in S_j} x_j)] \qquad (WDP(\lambda))$$

$$s.t. x_j \in \{0, 1\}, \forall j = 1, 2, ..., n$$

for any Lagrangian multiplier vector $\lambda = (\lambda_1, ..., \lambda_m)$ such that $\lambda_i \geq 0$, for all $i = 1, 2, ..., m$, as the discrete Lagrangian formulation of WDP. The formulation where $x_j \in [0, 1]$ is the classical (continuous) Lagrangian formulation. The Discrete Lagrangian function is defined as:

$$L(x, \lambda) = \sum_{j=1}^{n} p_j x_j + \sum_{i=1}^{m} [\lambda_i (1 - \sum_{j|i \in S_j} x_j)]$$

DLM searches for a saddle-point for the problem $WDP(\lambda)$. A saddle-point (x^*, λ^*) of $L(x, \lambda)$ satisfies the following condition:

$$L(x^*, \lambda) \leq L(x^*, \lambda^*) \leq L(x, \lambda^*)$$

for all λ sufficiently close to λ^* and for all x whose Hamming distance between x^* and x is 1.

The pseudo-code of the algorithm is given next (Algorithm 1). Note that the solution x could be unfeasible because is a solution for the problem $WDP(\lambda)$.

The step of updating the Lagrange multipliers is detailed next. Denote by $s(\lambda) = (s_i(\lambda)), \forall i = 1, 2, ..., m$ the subgradient vector.

$$s_i(\lambda) = 1 - \sum_{j|i \in S_j} x_j(\lambda)$$

The Lagrange multiplier λ^k (at iteration k) can be computed from λ^{k-1} using the following formula:

$$\lambda_i^k = \lambda_i^{k-1} - step\_size \frac{LB - L(x, \lambda)}{||s(\lambda^{k-1})||^2} s_i(\lambda^{k-1}) \qquad (1)$$

Algorithm 1. DLM_WDP(x_init)

---

initialize the solution x (if exists x_init then $x = x\_init$ else set x random, in the Lagrange space)

initialize the Lagrange multipliers λ (greedy)

$step\_size = 1$

while x is not a solution **do**

    find the first (or best) neighbor, x' of x (at distance 1)

    if exists x' **then**

      replace x with x'

    else

      update Lagrange multipliers($step\_size$)

    end if

    if after no consecutive iterations the best solution doesn't change **then**

      $step\_size/ = 2$

    end if

end while

return the best feasible solution

---

where LB is the best lower bound found so far, and $||s(\lambda)||$ is the norm of the subgradient vector.

$$||s(\lambda)|| = \sqrt{\sum_{i=1}^{m} s_i(\lambda)^2}$$

The Lagrange multiplier λ can be interpreted as the price for the items. The subgradient $s_i(\lambda)$ denotes the stock of item i. When an item is out of stock, i.e. $s_i(\lambda) < 0$ (more bids request item i), the price for the item i, λ_i is increased. Otherwise, if the item is not allocated $s_i(\lambda) = 1$, the price for the item is lowered down. When $s_i(\lambda) = 0$, we have balanced the supply and the demand, so the price of the item is not changed.

3.3 The Scheme of the Hybrid Algorithm for WDP

The genetic algorithm starts with a population of individuals, possible solutions to the WDP problem. The individuals are evolved to better solutions by using a selection scheme and specific operators like mutation and crossover.

An individual is encoded by a permutation of bids. A solution is constructed according to the permutation. A *first-fit* algorithm is used to decode such a permutation into a feasible solution. It starts with an empty allocation and it considers each bid in the order determined by the permutation. A bid is included in the solution if it satisfies the restrictions together with the previous selected bids. This representation ensures feasibility of the children. A disadvantage is that the search space becomes larger because the same solution can be encoded by multiple permutations.

The scheme of the hybrid algorithm is presented bellow (Algorithm 2). The initial population is generated randomly. The fitness function is equal to the

Algorithm 2. PDLMGA()

init population
while stopping condition not met **do**
    while not trapped into a local optima **do**
        selection
        apply operators
        local optimization (use best dual relaxed solution)
        keep best in population
    end while
    get best from population
    bestDLM ← DLM(best)
    recombine population with bestDLM
end while

objective function of the WDP problem, that is the auctioneer's outcome. A fitness proportional selection scheme is used, as well as the standard permutation operators, namely uniform order based crossover and swap mutation.

After the application of the operators, each solution is improved using a local optimization step. The same optimization method as in [10] is used. An unsatisfied bid is selected greedily to be added to the solution. The bid i with the largest shadow surplus value $p_i / \sum_{j \in S_i} y_j$ is considered. The dual prices y_j of the best dual relaxed solution, found at a previous step in DLM is used. The positions of the new bid and the first bid from permutation in conflict with are swapped. The assignment of bids is renewed. If the value of the new chromosome is better, the algorithm continues for a number of iterations; otherwise the optimization method stops.

The genetic algorithm uses the elitism mechanism; at each iteration the best solution is kept in population. The algorithm iterates for a number of steps, or until a local optima is reached.

When the genetic algorithm is stuck into a local optima, the DLM algorithm is called. The primal starting point for DLM is the best individual from the GA. All individuals from the last iteration of the GA are recombined using the crossover operator with the solution found by DLM, transformed into a feasible one. For constructing a feasible solution from a saddle point, the selected bids are sorted in decreasing order of the reduced profit, $(p_i - \sum_{j \in S_i} y_j)/|S_i|$. The genetic algorithm is restarted for a number of steps.

4   Experiments

4.1   Experimental Settings

The method was tested on instances from the CATS test suite [23]. Each distribution models a realistic scenario. For example, the arbitrary distribution simulates the auction of various electronic components; the regions distribution simulates the auction of radio spectrum rights; etc. Problems from each of the main distributions were generated: arbitrary, matching, paths, regions and scheduling.

Instances with a variable number of bids and items were generated. The number of items ranges from 40 to 400 and the number of bids ranges from 50 to 2000. Ten problem instances were drawn from each distribution.

The optimal solutions were determined using a mixed integer linear programming solver [24]. If the solver could not give a solution in a reasonable amount of time, the approximation algorithm ALPH [25] was considered. The ALPH algorithm first runs an approximation algorithm on the linear programming relaxation of the problem. Then a hill-climbing algorithm improves the order of the bids determined early. The ALPH heuristic was run with a small value for the approximation error parameter of the linear programming phase $\epsilon = 0.01$. For eight instances (out of ten generated) from the arbitrary distribution the LP solver was unable to find an exact solution. Four instances from the regions distribution were not solved by the LP solver.

The new method, denoted by PDLMGA, uses a population size of 500 individuals, a crossover probability of 0.6 and a mutation probability of 0.02. The maximum number of iterations is set to 500 and the number of consecutive iterations without no change of the best was equal to 75. The number of restarts is set to five. To avoid increasing the execution time, the DLM algorithm used the step of finding the *first* best neighbor for the current solution. The algorithm was stopped after 1500 maximum iterations and the number of consecutive iterations without no change of the primal best was set to 30.

The PDLMGA method was compared against the stochastic algorithm ALPH, the stochastic local search approach, Casanova [18] and the previous approach, the PDGA algorithm [10]. In [25] it was shown that ALPH runs faster than CPLEX on large problem instances. The ALPH algorithm was run with the parameter ϵ equal with 0.2 (the same value as in the experiments from [25]). The value of the approximation error is greater than the value used in the process of finding 'the optimum'. The Casanova algorithm adds at each step unsatisfied bids in a greedy or a random way depending on a specific probability, the 'walk' probability. Within the profit, the age of a bid is also considered in the greedy selection. The algorithm was tested with the walk probability of 0.2 and the novelty probability of 0.02. The θ_r parameter was 0 (no soft restarting strategy). The maximum number of steps from Casanova was equal to the product of the number of individuals and the number of iterations. The number of independent searches from Casanova was equal to the number of restarts from the genetic algorithm. The PDGA algorithm has the same settings as the PDLMGA.

4.2   Results

Table 1 displays the results obtained for CATS instances with 'varsize' bids. The results are averaged over 20 independent runs for each problem instance, except for the ALPH algorithm. As measure of comparison we used the gap from optimum which is equal to the difference between the optimum value and the value of the objective function for the solution found, divided by the optimum value. The Wilcoxon Signed-Rank non-parametric test is conducted. The test is done for two approaches: the PDLMGA and one algorithm from ALPH/Casanova/PDGA.

Table 1. The average gap (in percents) for CATS instances

Distribution	ALPH	Casanova	PDGA	PDLMGA
arbitrary	2.3	10.8	8	**3**
matching	0.3	5	**4.8**	6.3
paths	0.3	12.1	6.9	**2.2**
regions	-1	11.4	4.6	**1.3**
scheduling	0.2	2.2	0.7	**0.4**

Table 2. The mean and the standard deviation for ten instances of the paths data set

Instance (goods,bids)	optimal	Casanova mean (stdev)	PDGA mean (stdev)	PDLMGA mean (stdev)
(219,1132)	60.71	51.04 (1.11)	51.54 (0.89)	57.77 (0.89)
(61,1198)	26.17	23.31 (0.37)	25.3 (0.17)	25.87 (0.11)
(51,279)	23.73	21.07 (0.58)	23.4 (0.15)	23.66 (0.06)
(302,185)	27.94	25.47 (0.49)	27.83 (0.11)	27.88 (0.09)
(159,1028)	45.7	39.39 (0.78)	40.73 (0.97)	44.33 (0.36)
(129,1913)	41.54	35.24 (0.57)	37.87 (0.8)	40.43 (0.24)
(44,1208)	17.39	15.58 (0.19)	16.9 (0.17)	17.24 (0.08)
(117,970)	43.31	38.53 (0.64)	39.7 (0.61)	42.03 (0.46)
(189,1332)	55.18	47.83 (0.80)	47.41 (1.06)	52.85 (0.42)
(90,789)	38.42	34.37 (0.41)	36.77 (0.27)	37.81 (0.21)

The Null hypothesis was that there is no difference in the performances of the two algorithms. p-values below 0.05 were considered to be statistically significant. In the cases where the differences are significant, the winner is marked in bold.

The best solutions are provided by the ALPH algorithm (note that ALPH is a specially constructed algorithm for WDP). The new approach gives statistically better results than the Casanova algorithm and the PDGA algorithm for almost all distribution, except the matching data set. For five problems (out of eight) from the arbitrary distributions and two problems (out of four) from the regions distribution which are not solved by the LP-solver, the PDLMGA found better 'optima' than ALPH.

The mean and the standard deviation of the best fitness value found by the genetic algorithms and Casanova for the instances of the paths distribution are shown in Table 2. The best means are provided by the PDLMGA approach. The small values of the standard deviation of the algorithms show that the algorithms are robust and find good solutions consistently.

Because the new approach is compared against the stochastic local search algorithm, the time costs need to be considered. The local search approaches are usually often orders of magnitudes faster. Figure 2 presents the time values (in seconds) for the two algorithms[1].

---

[1] Computer settings: 2GHz Pentium single core processor, 1 GB RAM.

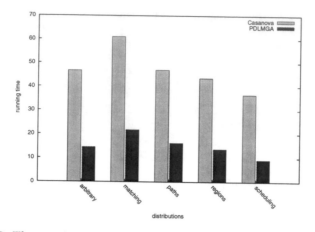

Fig. 2. The running time of the Casanova and PDLMGA approaches

In our case, the genetic algorithm runs faster than the Casanova approach. Note that the GA uses a mechanism for early stopping in case of premature convergence, while Casanova has not included such a mechanism. The DLM method is simple and runs faster, as you can see from Table 3. Table 3 shows how much time the new approach it spends for running the DLM method.

Table 3. The percent of time, from the total time of PDLMGA, spent for running DLM

Distribution	arbitrary	matching	paths	regions	scheduling
time%	0.21	0.27	0.31	0.19	0.29

How much does improve the new method? Next we analyze the amount of the improvement achieved by the new method. We compared the new approach with:

- a random restart algorithm, without the local optimization step (rrGA), and
- the same algorithm, only that the DLM method is started with a random initial solution (rPDLMGA).

Table 4 shows the differences of the gaps (in percents) between the PDLMGA and the two previous considered algorithms. The matching data set uses extensively the dual information, when compared with rrGA. Starting the DLM with a random solution has a weak influence for this data set. For the arbitrary, paths and regions distributions the information gain is also evident in both cases. The scheduling distribution is easy when solving with a GA.

The number of restarts. In Table 5 the gap versus the number of restarts used in the PDLMGA for the paths distribution is represented. For each case

Table 4. The differences of the gaps (average and standard deviation) in percents between the considered algorithms and the PDLMGA

Distribution	rrGA	rPDLMGA
arbitrary	4.3 (0.09)	3.1 (0.04)
matching	13.5 (0.11)	1 (0.05)
paths	6.2 (0.06)	3.2 (0.02)
regions	3.2 (0.05)	1.7 (0.02)
scheduling	0.7 (0.04)	0 (0)

Table 5. The number of restarts vs. the gap for the paths distribution. For each configuration the running time is shown (in seconds).

restarts	1	3	5	7
gap	5.5	3	2.2	1.9
time	5.6	11.8	16.5	21.2

the time requirements are also presented. As expected, the accuracy of the algorithm improves when using a larger number of restarts. A trade-off between performance and computation costs must be kept.

Evaluation on more difficult problems. Further experiments on instances with a larger number of bids were made. Problems from the matching distribution with 1000 bids and from paths distribution with 10000 and 20000 bids (and 256 items) were considered. The same parameters were kept for the algorithms, as in the experiments with smaller instances. The results are presented in Table 6. The Wilcoxon Signed-Rank non-parametric test is conducted on pairs of algorithms, except for ALPH. In cases where the differences are significant (at the level 0.05) the winner is marked in bold.

Table 6. The average gap (in percents) for larger CATS instances

Distribution	instances	ALPH	Casanova	PDGA	PDLMGA
matching (10000,256)	5	0.4	16.4	**8.2**	18.4
paths (10000,256)	10	0.3	20.1	10.6	**4.7**
paths (20000,256)	5	0.2	21.2	8.8	**4.8**

The generated instances appear not to be so difficult for the approximative methods. Casanova is most influenced by the increasing size of the problems. The solution quality of PDLMGA is superior to the one returned by Casanova and PDGA for the large paths distributions. For the matching data set, the PDGA seems to be the best alternative, when comparing the algorithms. ALPH again finds better solutions.

For difficult problem instances approximative methods are preferred to the classic ones. In experiments, Casanova outperformed the CASS algorithm, a deterministic approach, on large problem instances [18]. The results found by the new approach compared favorably to simpler genetic algorithms and some other local search techniques, like Casanova.

5   Conclusion

The paper investigates the development of a novel hybrid algorithm by combining techniques from Evolutionary Computing and Integer Programming areas. The new hybrid evolutionary algorithm uses the dual information in the form of Lagrange multipliers to escape from a local optima. The method was applied for an important problem from the combinatorial auction realm. It was tested on different types of problem instances and the obtained allocations are very close to the optimal solutions. Although at a first sight, the algorithm seems to be complex and time consuming, it is fast enough to run in less than a minute problems with tens of thousands of bids.

Comparisons to other trap escaping strategies are necessary. Applications on other optimization problems is mandatory for the future work.

References

1. Mahfoud, S.: Niching methods for genetic algorithms. University of Illinois at Urbana-Champaign (1996)
2. Starkweather, T., Whitley, D., Mathias, K.: Optimization using distributed genetic algorithms. In: Schwefel, H.P., Männer, R. (eds.) PPSN 1990. LNCS, vol. 496, pp. 176–185. Springer, Heidelberg (1991)
3. Cobb, H.G., Grefenstette, J.F.: Genetic algorithms for tracking changing environments. In: Proceedings of the 5th International Conference on Genetic Algorithms, pp. 523–530 (1993)
4. Fukunaga, A.S.: Restart scheduling for genetic algorithms. In: Eiben, A.E., Bäck, T., Schoenauer, M., Schwefel, H.-P. (eds.) PPSN 1998. LNCS, vol. 1498, pp. 357–366. Springer, Heidelberg (1998)
5. Gomes, C., Selman, B., Crato, N., Kautz, H.: Heavy-tailed phenomena in satisfiability and constraint satisfaction problems. Journal of Automated Reasoning 24(1/2), 67–100 (2000)
6. Raidl, G.: A Unified View on Hybrid Metaheuristics. In: Almeida, F., Blesa Aguilera, M.J., Blum, C., Moreno Vega, J.M., Pérez Pérez, M., Roli, A., Sampels, M. (eds.) HM 2006. LNCS, vol. 4030, pp. 1–12. Springer, Heidelberg (2006)
7. Raidl, G., Puchinger, J.: Combining (Integer) Linear Programming Techniques and Metaheuristics for Combinatorial Optimization. In: Hybrid Metaheuristics, An Emerging Approach to Optimization. SCI, vol. 114, pp. 31–62 (2008)
8. Pfeiffer, J., Rothlauf, F.: Analysis of Greedy Heuristics and Weight-Coded EAs for Multidimensional Knapsack Problems and Multi-Unit Combinatorial Auctions. In: Proceedings of the 9th Conference on Genetic and Evolutionary Computation, p. 1529 (2007)
9. Hansen, P., Brimberg, J., Mladenović, N., Urosević, D.: Primal-dual variable neighbourhood search for the simple plant location problem. INFORMS Journal on Computing 19(4), 552–564 (2007)

10. Raschip, M., Croitoru, C.: A New Primal-Dual Genetic Algorithm: Case Study for the Winner Determination Problem. In: Cowling, P., Merz, P. (eds.) EvoCOP 2010. LNCS, vol. 6022, pp. 252–263. Springer, Heidelberg (2010)
11. Leitner, M., Raidl, G.: Lagrangian Decomposition, Metaheuristics, and Hybrid Approaches for the Design of the Last Mile in Fiber Optic Networks. In: Blesa, M.J., Blum, C., Cotta, C., Fernández, A.J., Gallardo, J.E., Roli, A., Sampels, M. (eds.) HM 2008. LNCS, vol. 5296, pp. 158–174. Springer, Heidelberg (2008)
12. de Vries, S., Vohra, R.: Combinatorial auctions: A survey. INFORMS Journal on Computing 15(3), 284–309 (2000)
13. Rassenti, S.J., Smith, V.L., Bulfin, R.L.: A combinatorial auction mechanism for airport time slot allocation. Bell J. of Economics 13, 402–417 (1982)
14. Rothkopf, M., Pekec, A., Harstad, R.: Computationally manageable combinatorial auctions. Management Science 44(8), 1131–1147 (1998)
15. Wah, B.W., Chen, Y.X.: Constrained genetic algorithms and their applications in nonlinear constrained optimization. In: Evolutionary Optimization. International Series in Operations Research and Management Science, vol. 48(IV), pp. 253–275 (2003)
16. Sandholm, T., Suri, S., Gilpin, A., Levine, D.: CABoB: a fast optimal algorithm for combinatorial auctions. In: Proceedings of the International Joint Conferences on Artificial Intelligence, pp. 1102–1108 (2001)
17. Nisan, N.: Bidding and Allocation in Combinatorial Auctions. In: Proceedings of the ACM Conference on Electronic Commerce, pp. 1–12 (2000)
18. Hoos, H.H., Boutilier, C.: Solving combinatorial auctions using stochastic local search. In: Proceedings of the 17th National Conference on Artificial Intelligence, pp. 22–29 (2000)
19. Guo, Y., Lim, A., Rodrigues, B., Zhu, Y.: Heuristics for a bidding problem. Computers and Operations Research 33(8), 2179–2188 (2006)
20. Boughaci, D., Benhamou, B., Drias, H.: A memetic algorithm for the optimal winner determination problem. Soft Computing 13(8-9), 905–917 (2009)
21. Guo, Y., Lim, A., Rodrigues, B., Tang, J.: Using a Lagrangian Heuristic for a Combinatorial Auction Problem. In: Proceedings of the 17th IEEE International Conference on Tools with Artificial Intelligence, pp. 99–103 (2005)
22. Shang, Y., Wah, B.: A Discrete Lagrangian-Based Global-Search Method for Solving Satisfiability Problems. Journal of Global Optimization 12, 61–99 (1998)
23. Leyton-Brown, K., Pearson, M., Shoham, Y.: Towards a Universal Test Suite for Combinatorial Auction Algorithms. In: Proceedings of the ACM Conference on Electronic Commerce, pp. 66–76 (2000)
24. Berkelaar, M.: "lp_solve - version 5.5", Eindhoven University of Technology, http://sourceforge.net/projects/lpsolve/
25. Zurel, E., Nisan, N.: An Efficient Approximate Allocation Algorithm for Combinatorial Auctions. In: Proceedings of the ACM Conference on Electronic Commerce, pp. 125–136 (2001)

Greedy Local Improvement of SPEA2 Algorithm to Solve the Multiobjective Capacitated Transshipment Problem

Nabil Belgasmi[1,2], Lamjed Ben Said[1,3], and Khaled Ghedira[1,3]

University of Tunis
[1] Research Unit Strategies for Optimizing Information and knowledge (SOIE)
[2] Higher School of Computer Sciences (ENSI), Campus Universitaire de La Manouba, 2010
[3] Higher Institute of Management (ISG Tunis) Bouchoucha, Le Bardo, 2000
Belgasmi.nabil@gmail.com,
{lamjed.bensaid,khaled.ghedira}@isg.rnu.tn

Abstract. We consider a multi-location inventory system where inventory choices at each location are centrally coordinated through the use of lateral Transshipments. This cooperation between different locations of the same echelon level often leads to cost reduction and service level improvement. However, when some locations face embarrassing storage capacity limits, inventory sharing through transshipment may cause undesirable lead time. In this paper, we propose a more realistic multiobjective transshipment model which optimizes three conflicting objectives: (1) minimizing the aggregate cost, (2) maximizing the fill rate and (3) minimizing the transshipment lead time, in the presence of different storage capacity constraints. We improve the performance of the well-known evolutionary multiobjective algorithm SPEA2 by adequately applying a multiobjective quasi-gradient local search to some candidate solutions that have lower density estimation. The resulting hybrid evolutionary algorithm outperforms NSGA-II and the original SPEA2 in both spread and convergence. It is also shown that lateral transshipments constitute an efficient inventory repairing mechanism in a wide range of system configurations.

Keywords: Evolutionary multiobjective optimization, local search, simulation, inventory management.

1 Introduction

In the past, research in operations management focused on single-firm analysis. Its goal was to provide managers in practice with suitable tools to improve the performance of their firm by calculating optimal inventory quantities, among others. Nowadays, business decisions are dominated by the globalization of markets and increased competition among firms. Further, more and more products reach the customer through supply chains that are composed of independent firms. Following these trends, research in operations management has shifted its focus from single-firm analysis to multi-firm analysis, in particular to improving the efficiency and performance of supply chains under centralized control. The proactive use of transshipments is an example of such coordination.

C.A. Coello Coello (Ed.): LION 5, LNCS 6683, pp. 364–378, 2011.

Referred to as physical pooling of inventories, the transshipment has been widely used in practice to reduce cost and improve customer service [8]. It is usually recognized as the monitored movement of material among locations at the same echelon. It affords a valuable mechanism for correcting the discrepancies between the locations' observed demand and their on-hand inventory. Subsequently, transshipments may reduce costs and improve service without increasing the system-wide inventories [7].

The study of multi-location models with transshipments is an important contribution for mathematical inventory theory as well as for inventory practice. The idea of lateral transshipments is not new. The first study dates back to the sixties. The two-location-one-period case with linear cost functions was considered by [1]. The N-location-one-period model was studied by Krishnan [15] where the cost parameters are the same for all locations. Non-negligible replenishment lead times and transshipment lead times were incorporated among stocking locations to the multi-location model in [11]. The effect of lateral transshipment on the service levels in a two-location-one-period model was studied in [21]. There is a considerable amount of Supply Chain Management studies in the last past decades. Some papers provided interesting surveys. Pokharel [17] indicates that various objectives could be considered for strategic decision making on Supply Chain Network: (1) increasing service level, (2) decreasing warehouse costs, (3) decreasing total fixed and variable costs, (4) decreasing lead time (order processing and supply lead times), (5) consolidating supplier base, (6) increasing supplier reliability, (7) increasing capacity utilization and (8) increasing total quality of supply. In the same work, it was developed a two-objective decision-making model for the choice of suppliers and warehouses for a supply chain network design.

In most of the mentioned researches, transshipment lead times were assumed to be negligible despite its direct impact on service levels. Moreover, storage capacity at all system location was assumed to be unlimited. These are two noticeable limitations of the existent works.

In this study, we, first, incorporate storage capacity constraints into the traditional transshipment model which leads to a better modeling of real-world situations. Secondly, we propose a multiobjective transshipment model which minimizes the aggregate cost and transshipment lead times while maximizing the global fill rate subject to several predefined storage capacity constraints. We incorporate a greedy local search into the evolutionary algorithm SPEA2 and compare its performance to NSGA-II.

The remainder of this paper is organized as follows. In section 2, we formulate the multiobjective transshipment model. In section 3, we give a brief description of the multiobjective evolutionary optimization, and we present the hybrid SPEA2 algorithm (H-SPEA2). In section 4, we show our experimental results. In section 5, we state our concluding remarks.

2   Model

2.1   Problem Description

We consider the following real life problem where we have n stores selling a single product. The stores may differ in their cost and demand parameters. The system inventory is reviewed periodically. At the beginning of the period, and long before the demands realization, replenishments take place in store i to increase the stock level up to S_i. The storage capacity of each location is limited to $S_{max,i}$. In other way, the replenishment quantities should not exceed $S_{max,i}$ inventory units. This may be due to expensive fixed holding costs, or to the limited physical space of the stores. Thus, the inventory level of store i will be always less or equal to $min(S_i, S_{max,i})$. After the replenishment, the observed demands D_i which represents the only uncertain event in the period are totally or partially satisfied depending on the on-hand inventory of local stores. However, some stores may be run out of stock while others still have unsold goods. In such situation, it will be possible to move these goods from stores with surplus inventory to stores with still unmet demands. This is called lateral transshipment within the same echelon level. It means that stores in some sense share the stocks. The set of stores holding inventory I^+ can be considered as temporary suppliers since they may provide other stores at the same echelon level with stock units. Let τ_{ij} be the transshipment cost of each unit sent by store i to satisfy a one-unit unmet demand at store j. The transshipment lead time of 1 unit transferred from i to j is equal to L_{ij}. After the end of the transshipment process, if store i still has a surplus inventory, it will be penalized by a per-unit holding cost of h_i. If store j still has unmet demands, it will be penalized by a per-unit shortage cost of p_j. Fixed cost transshipment costs are assumed to be negligible in our model. It was proved in [8] that, in the absence of fixed costs, if transshipments are made to compensate for an actual shortage and not to build up inventory at another store, there exists an optimal base stock policy S^* for all possible stationary policies. To see the effect of the fixed costs on a two-location model formulation, see [21].

The following notation is used in our model formulation:

n	Number of stores
S_i	Order quantities for store i
S	Vector of order quantities, $S = (S_1, S_2, ..., S_n)$ (Decision variable)
$S_{max,i}$	Maximum storage capacity of store i
S_{max}	Vector of storage capacities, $S_{max} = (S_{max,1}, S_{max,2}, ..., S_{max, n})$
D_i	Demand realized at i
D	Vector of demands, $D = (D_1, D_2, ..., D_n)$
h_i	Unit inventory holding cost at i
p_j	Unit penalty cost for shortage at j
τ_{ij}	Unit cost of transshipment from i to j
T_{ij}	Amount transshipped from i to j
L_{ij}	Unit transshipment lead time from i to j
I^+	Set of stores with surplus inventory (before transshipment)
I^-	Set of stores with unmet demands (before transshipment)

2.2 Modeling Assumptions

Mainly three assumptions are made in this study to simplify the model. Some assumptions can be relaxed in further researches.

- **Assumption 1 (Lead time):** All transshipment lead times are both positive and deterministic. The case of stochastic transshipment lead times is under investigation.
- **Assumption 2 (Demand):** Customers' demands at each store could be fulfilled partially either by the local available inventory or by the shipped quantities that may come from other stores. For example, if a customer orders 100 units, and finally gets only 30 units, its demand could not be cancelled. The customer should accept the partial fulfillment of his demand.
- **Assumption 3 (Replenishment policy):** At the beginning of every period, replenishments take place to increase inventory position of store i up to S_i.

2.3 Model Formulation

2.3.1 Cost Function

Since inventory choices in each store are centrally coordinated, it would be a common interest among the stores to minimize aggregate cost. At the end of the period, the system cost is given by (1):

$$C(S,D) = \sum_{i \in I^+} h_i(S_i - D_i) + \sum_{j \in I^-} p_j(D_j - S_j) - K(S,D) \tag{1}$$

The first and the second term on the right hand side of (1) can be respectively recognized as the total holding cost and shortage cost before the transshipment. However, the third term is recognized as the aggregate transshipment profit since every unit shipped from i to j decreases the holding cost at i by h_i and the shortage cost at j by p_j. However, the total cost is increased by τ_{ij} because of the transshipment cost. Due to the complete pooling policy, the optimal transshipment quantities T_{ij} can be determined by solving the following linear programming problem (2):

$$K(S,D) = \max_{T_{ij}} \sum_{i \in I^+} \sum_{j \in I^-} (h_i + p_j - \tau_{ij}) T_{ij} \tag{2}$$

$$\sum_{j \in I^-} T_{ij} \leq S_i - D_i \quad , \forall i \in I^+ \tag{3}$$

$$\sum_{i \in I^+} T_{ij} \leq D_j - S_j \quad , \forall j \in I^- \tag{4}$$

$$T_{ij} \geq 0 \tag{5}$$

In (2), problem K can be recognized as the maximum aggregate income due to the transshipment. T_{ij} denotes the optimal quantity that should be shipped from i to fill unmet demands at j. Constraints (3) and (4) say that the shipped quantities cannot exceed the available quantities at store i and the unmet demand at store j. Since demand is stochastic, the aggregate cost function is built as a stochastic programming model which is formulated in (6). The objective is to minimize the expected aggregate cost with respect to storage capacity constraints that may exist in some locations.

$$\min_{S} E(C(S,D)) = \min_{S} (C^{BT}(S) - K^{TR}(S)) \tag{6}$$

Where C^{BT} denotes the expected cost before the transshipment, called Newsvendor cost, and K^{TR} denotes the expected aggregate income due to the transshipment. This decomposition shows the important relationship between both the Newsvendor and the transshipment problem. By setting very high transshipment costs, i.e. $\tau_{ij} > h_i + p_j$, no transshipments will occur. Problem K^{TR} will then return zero.

2.3.2   Fill Rate Function

One of the most important performance measures of inventory distribution systems is the fill rate at the lowest echelon stocking locations. The fill rate is equivalent to the proportion of the satisfied demand. We extend the fill rate formulation given in [21] to n locations model. Let F be the aggregate fill rate measure after the transshipment realization:

$$F(S,D) = \frac{\sum_j \min(D_j, S_j + \sum_i T_{ij})}{\sum D_j} \tag{7}$$

Notice that the whole system fill rate would be maximized if we order very large quantities at the beginning of every period (without exceeding the local storage capacities). However, this may results in global holding cost increase. If we order very little quantities S_i, we certainly avoid holding costs, but the different locations will often be unable to satisfy customers' demands. This badly affects the system fill rate. Thus, we need to find good solutions taking into account the balance among costs and service level.

2.3.3   Lead Time Function

The fill rate measure is widely used service criteria to evaluate the performance of inventory distribution systems. However, it does not take into account the lead times caused by the transshipment process. In other words, we can have a perfect fill rate value while making customers waiting for long time. In our attempt to integrate the lead time in our Transshipment model, and following [16], we suggest this aggregate performance measure:

$$LT(S,D) = \sum_{i,j} L_{ij} T_{ij} \quad , \forall i \in I^+, j \in I^- \tag{8}$$

2.4   Objective Functions Estimation

The considered objective functions are stochastic because of the demand randomness modeled by the continuous random variables D_i with known joint distributions. The stochastic nature of the problem leads us to compute the expected values of each objective function. In addition, an analytical tractable expression for problem K given in (2) exists only in the case of a generalized two-location problem or N-location with identical cost structures [13]. In both cases, the open linear programming problem K has an analytical solution. But in the general case (many locations with different cost structures), we can use any linear programming method to solve problem K. In this study, we used the Simplex Method. The most common method to deal with noise or randomness is re-sampling or re-evaluation of objective values [3]. With the re-sampling method, if we evaluate a solution S for N times, the estimated objective value is obtained as in equation below and the noise is reduced by a factor of $N^{1/2}$. For this purpose, draw N random scenarios $D^1,...,D^N$ independently from each other (in our problem, a scenario D^k is equivalent to a vector demand $D^k=(D^1_1,...,D^N_N)$. A sample estimate of $f(S)$, noted $E(f(S,D))$, is given by:

$$E[f(S,D)] \approx \overline{f}(S) = \frac{1}{N}\sum_{k=1}^{N} f(S,D^k) \Rightarrow \overline{\sigma} = \sqrt{Var[\overline{f}(S)]} \approx \frac{\sigma}{\sqrt{N}}$$

3 Evolutionary Multiobjective Optimization

Most real world problems have several (usually conflicting) objectives to be satisfied. A general multiobjective optimization problem has the following form:

$$\min \left[f_1(S), f_2(S),..., f_k(S) \right]$$

Subject to the m inequality constraints and the p equality constraints:

$$g_i(S) \geq 0, \quad i = 1,2,...,m$$

$$h_i(S) = 0, \quad i = 1,2,...,p$$

The most popular approach to handle multiobjective problems is to find a set of the best alternatives that represent the optimal tradeoffs of the problem. After a set of such trade-off solutions are found, a decision maker can then make appropriate choices. In a simple optimization problem, the notion of optimality is simple. The best element is the one that realizes the minimum (or the maximum) of the objective function. In a multiobjective optimization problem, the notion of optimality is not so obvious. In other words, there is no solution that is the best for all criteria, but there exists a set of solutions that are better than other solutions in all the search space, when considering all the objectives. This set of solutions is known as the optimal solutions of the Pareto set or nondominated solutions. This is the most commonly adopted notion of optimality. We say a vector of decision variable S^* is Pareto optimal if there does not exist another S such that:

$$\begin{cases} f_i(S) \leq f_i(S^*), & \forall i = 1,2,...,k \\ and \\ f_j(S) < f_j(S^*), & for\ at\ least\ one\ j \end{cases}$$

In other words, this definition says that S^* is Pareto optimal if there exists no feasible vector of decision variable S that would decrease some criterion without causing a simultaneous increase in at least one criterion. This concept almost always gives not a single solution, but rather a set of solutions called the Pareto optimal set. The plot of the objective functions whose nondominated vectors are in the Pareto optimal set is called the Pareto front.

Many performance measures were designed either to evaluate the quality of a given Pareto front (unary metric) or to compare two nondominated sets (binary metric). In this study, we focus on two well-known unary indicators: Hypervolume and Spread. The Hypervolume quality indicator computes the volume covered by a nondominated set of solutions (the region of objective space dominated by the obtained Pareto front). Higher values of Hypervolume are preferred. The Spread metric is a diversity indicator that measures the extent of spread achieved among the obtained solutions. This metric takes a zero value for an ideal distribution. Before applying it, the objective function values must be normalized.

Evolutionary algorithms are population based metaheuristics that operate on a set of individuals in order to find trade-off solutions as most as possible. This noticeable characteristic make them the most adapted to solve multi-objective optimization problems. Several works have been done in this field. In most cases, as set out in [9], genetic algorithms are defined to be not enough effective because the crossover and mutation operators do not allow to intensify the search sufficiently. The mutation operator is typically expected to make a slight modification to an individual. Its role is to promote the diversification of individuals while the selection role is to conserve the best of them. Evolutionary algorithms researchers have suggested several approaches to overcome the weakness of these search methods and improve their performance by increasing their convergence rate and solutions diversity. One promising approach is *Hybridization*. The most common and effective technique is to incorporate local search (LS) into evolutionary algorithms. The local search operator replaces or follows the mutation operator, and then helps to intensify the research in various areas pointed by the genetic mechanisms: selection and crossover, we call this type of hybrid algorithms Memetic Algorithms (MAs) [15]. When designing multi-objective MAs, we face several design issues. Most of them are evoked in single objective case [6]. These issues can be summarized as follows:

- How to incorporate LS method into MOEAs?

- How to generate neighborhood?

- How long does the LS take?

- How often LS should be performed?

- How to select solutions for LS?

- What is the replacement strategy?

- How to maintain population diversity?

We can classify MAs on the basis of the used LS method type:

- *Gradient based schemes*: This type of MAs incorporates LS methods that exploit gradient information. For example, two versions of the NSGA-II [5] algorithm were hybridized with the sequential quadratic programming (SQP): SBX-NSGA-II [14and PCX-NSGA-II [19]. All these works have shown good results in terms of convergence and CPU time.

- *Neighborhood based schemes*: Here, MAs integrate LS methods that explore solutions neighborhoods without using gradient information such that MOGLS [10], PHC-NSGA-II [2] and M-PAES [12].

In this study, we used a neighborhood based scheme which integrates a greedy local search in the main loop of the SPEA2 algorithm that will be presented in the next section.

3.1 SPEA2: Brief Description

Many multiobjective evolutionary algorithms have been proposed in the last few years. Comparative studies have shown for large number of test cases that, among all major multiobjective EAs, Strength Pareto Evolutionary Algorithm (SPEA2) is clearly superior. The key results of the comparison [22] were: (1) SPEA2 performs better than SPEA on all test problems and (2) SPEA2 and NSGA-II show the best overall performance. But in higher dimensional spaces, SPEA2 seems to have advantages over PESA [4] and NSGA-II. In addition, it was proven that SPEA2 is less sensitive to noisy function evaluations since it saves the non-dominated solutions in an archive. At the beginning of the SPEA2 optimization process, an initial population is generated randomly respecting the different local storage constraints (S_i is less than $S_{max,i}$). In our multi-location problem, an individual is a base stock decision $S = (S_1, S_2,...,S_n)$ consisting of n genes S_i. At each generation, all the individuals are evaluated. A finc-grained fitness assignment strategy is used to perform individuals' evaluation. It incorporates Pareto dominance and density information (respectively $R(i)$ and $D(i)$ according to [22]). The density function $D(i)$ can be recognized as a crowding measure computed at a solution i. In other words, good individuals are the less dominated and the well spaced ones. Good individuals are conserved in an external set (archive). This is called the environmental selection. If the archive is full, a truncation operator is used to determine which individuals should be removed from the archive. The truncation operator is based on the distance of the k-th nearest neighbor computation method [20]. In other words, an individual is removed if it has the minimum distance to the other individuals. This mechanism preserves the diversity of the optimal Pareto front. The archived individuals participate in the creation of new individuals for the coming generations. These steps are repeated for a fixed number of generations. The resulting optimal Pareto front is located in the archive.

3.2 SPEA2 with a Greedy Local Search

Here is the description of SPEA2 hybridized with a greedy local search in order to improve the spread and the convergence of the resulting Pareto front. The local search phase introduces two additional parameters: *NLS* and *SNS* corresponding respectively to the Number of individuals on which the Local Search will be applied, and, the Size of the Neighborhood Set to be generated for each selected individual. SPEA2 fitness function $F(i)$ of a new generated solution i is based on two important components: the raw fitness $R(i)$ and the density $D(i)$. The raw fitness $R(i)$ provides a sort of niching mechanism based on the concept of Pareto dominance whereas the density function $D(i)$ indicates whether a solution is located in a crowded area or no. After evaluating each new individual, SPEA2 constructs the elitist archive of the next generation $t+1$ which consists of the best individuals obtained during the optimization process. At this step, our greedy local search selects the least crowded individuals according to the already computed density and samples its neighborhood using a polynomial quasi-gradient mutation. The idea behind the proposed multiobjective quasi-gradient mutation is quite simple. If a search direction "d" applied to an individual X leads to a degradation in all the objective functions, then the opposite direction "-d" may be more interesting than "d" in finding new good individuals. That is, let Y be the image of X by the translation of vector "d". Three cases are possible: (a) Y dominates X, then Y is returned; (b) Y is equivalent to X, then Y is kept; (c) Y is dominated by X, that is $F_i(Y=X+d) > F_i(X)$ for all objective functions (minimization problem). Thus, Y becomes equal to X - d. The obtained set of locally generated individuals is then evaluated and tested against each selected parents which may replaced by its offspring if this latter dominates it. Thus, the local search tries to guide the search toward less explored regions of the objective space. As noticed, there are no additional computation efforts except those related to the sorting procedure (2.d.a) and to the evaluations of the sampled neighborhood set. In this study, we used both SBX crossover operator [19] and polynomial mutation [18] since the considered problem is continuous, with real-coded decision variables.

The main loop of the hybrid algorithm H-SPEA2 is described as follow:

Input: N_p (population size), N_A (archive size), T (number of generations), NLS and SNS.

Output: A (nondominated set)

1) Initialize Population

  1.a) Create an initial population P_0

  1.b) Create empty external set A_0 ("archive")

2) For t = 0 to T

  2.a) Evaluate fitness of each individual in P_t and A_t

  2.b) Copy all nondominated individuals in P_t and A_t to A_{t+1}. If the A_{t+1} size exceeds archive size N_A reduce A_{t+1} using truncation operator. If the A_{t+1} size is less than archive size then use dominated individuals in P_t and A_t to fill A_{t+1}.

  2.c) Local search phase

- Sort the nondominated individuals in A_{t+1} with respect to their density estimation values already computed by SPEA2 in step 2.a) in order to select NLS worst individuals which have the lowest density values.

- **For each selected individual X:**

  - Generate SNS new individuals by applying successive calls to the proposed mutation **MO_QuasiGradient_Mutation(X)**

  - Add the new individuals to the mating pool.

> **2.d)** Perform Binary Tournament Selection with replacement on A_{t+1} to fill the mating pool.
>
> **2.e)** Apply crossover and mutation to the mating pool and update A_{t+1}
>
> **End FOR**

Pseudo-code of the proposed multiobjective quasi-gradient mutation:

> **Function** *MO_QuasiGradient_Mutation(X):Y*
>
> ---
>
> **Input:** An evaluated individual X
>
> **Output:** A new individual Y
>
> ---
>
> **1.** Apply polynomial mutation on individual X to get a new individual Y (use a small distribution index value)
>
> **2.** Let "d" be the direction from X to Y in the decision space. That is, d := Y - X.
>
> **3.** Evaluate the individual Y.
>
> **4.** If Y is dominated by X, then let Y be the image of X by the translation of vector "-d"; (Y:=X-d); Then evaluate the new Y.
>
> **5.** Return Y.

4 Optimization Results

In this section, we report on our numerical study. We consider a four-location system. For all system settings (table 1), we maintain the same costs and demand structure: Shortage cost = \$4, holding cost = \$2, transshipment costs = \$0.5, demands are random variables uniformly distributed over the interval [0, 200] and Lead time = 5. We only vary the storage capacity of some stores in order to analyze the resulting system response.

Table 1. Four systems with different storage capacities. (Infinity) means that storage capacity is unlimited. (0) means that it is not possible to hold inventory.

	C-0	C-1	C-2	C-3
S(max,1)	infinity	100	100	100
S(max,2)	infinity	infinity	0	0
S(max,3)	infinity	infinity	infinity	0
S(max,4)	infinity	infinity	infinity	infinity

Table 2. SPEA2, NSGAII and H-SPEA2 settings

Parameters	SPEA2	NSGAII	H-SPEA2
Population size	50	200	50
Archive size	200	-	200
Max evaluations	50000	50000	50000
Crossover probability	0,90	0,90	0,90
Mutation probability	0,25	0,25	0,25
(NLS, SNS)	-	-	(5, 20)
Crossover distribution index	10	10	10
Mutation distribution index	20	20	20

4.1 Cost vs. Fill Rate Problem

Figure 1 illustrates the Pareto fronts of Cost/Fill Rate problem (C/F), when C-3 system and H-SPEA2, SPEA2 and NSGA-II are considered. Non-dominated solutions are well spread over the entire Pareto front obtained by H-SPEA2 optimizer.

Table 3 proves also that our hybrid algorithm outperforms SPEA2 and NSGA-II in term of spread. The system can achieve high fill rate level (95%) while ensuring a low cost value ($350). However, increasing the fill rate up to (100%) affects considerably the cost ($700).This is due to increasing cost of transshipment resulting from the frequent inventory transfer from stores 1 and 4 to stores 2 and 3 to repair their embarrassing storage limits ($S_{max.2}=0$ and $S_{max.3}=0$).

The resulting Pareto fronts are also well spread since we obtained good Spread metric values. We conclude that our hybrid SPEA2 outperforms NSGA-II and SPEA2 in both convergence and spread. It intensifies the search in the regions where nondominated individuals suffer from lower density levels, while both NSGA-II and SPEA2 include density information or crowding distance implicitly into the fitness assignment strategy.

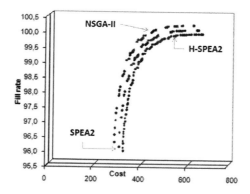

Fig. 1. Cost vs. Fill rate non-dominated sets of system C-3 instance using H-SPEA2, SPEA2 and NSGA-II

Table 3. Comparison of all systems with respect to Hypervolume and Spread metrics for the two-objective problem *(Cost/Fill Rate)*. Best results are in bold.

Cost vs. Fill Rate		Hypervolume		Spread	
		Mean	Variance (10^{-8})	Mean	Variance (10^{-2})
C-0	NSGA-II	0,9951	15,560	0,7502	0,1863
	SPEA2	0,9952	8,050	0,7893	0,2093
	H-SPEA2	**0,9972**	6,235	**0,5326**	0,1064
C-1	NSGA-II	0,9953	11,556	0,7729	0,1880
	SPEA2	0,9952	11,787	0,7918	0,1881
	H-SPEA2	**0,9954**	11,775	**0,5144**	0,1641
C-2	NSGA-II	0,9919	181,076	0,6288	1,1755
	SPEA2	0,9926	69,145	0,6536	1,7226
	H-SPEA2	**0,9941**	9,962	**0,4122**	1,0036
C-3	NSGA-II	0,9946	4,355	0,8064	0,0142
	SPEA2	**0,9952**	3,104	0,7951	0,0204
	H-SPEA2	0,9951	9,941	**0,5066**	0,0120

4.2 Cost vs. Lead Time

This section deals with the bi-objective problem that minimizes both aggregate cost and transshipment lead time. Figure 2 illustrates the Pareto front of the problem when system C-3 and H-SPEA2 are considered. In figure 2, we notice that the Pareto front is very dense, and nondominated solutions are well spread and diversified. The cost values vary from \$350 to \$1600, while lead time varies from 0 to 900 time unit. To achieve the lowest (highest) cost value, lead time would be considerable (null). This proves that costs and lead time are very conflicting. The decision maker may think about providing the system with a sufficient number of transporting vehicles so that inventory transfers occur simultaneously. When lead times are negligible or null, the cost reaches its highest value. In fact, "no lead time" means that there are not unsold units neither unmet demands. This happens only in the case of (a) large ordered replenishment quantities (all demands are satisfied by on-hand stock) or (b) very high transshipment costs. Since in our numerical examples all unit transshipment profits are positive ($\tau_{ij} < h_i + p_j$, see formulae (2)), "no lead time" is explained by ordering large replenishment quantities.

For system C-3, only store 4 will be able to hold such quantities. It will be considered as an emergency inventory provider, while stores 2 and 3 are its important "virtual" customers. According to table 4, the resulting Hypervolume values are less than those of table 3 where very high values (>94%) prove that fill rate and cost are not very conflicting. H-SPEA2 outperforms SPEA2 and NSGA-II in both Hypervolume and spread.

Table 4. Comparison of the 4 systems with respect to Hypervolume and Spread metrics for *"Cost/Lead time"* problem. Best results are in bold.

Cost vs. Lead time		Hypervolume		Spread	
		Mean	Variance (10^{-3})	Mean	Variance (10^{-2})
C-0	NSGA-II	0,6289	0, 250711	0,5358	0, 29433
	SPEA2	**0,6308**	0, 519017	0,5713	0, 43205
	H-SPEA2	0,6207	0, 213418	**0,4110**	0, 12304
C-1	NSGA-II	0,5813	0, 145691	0,5952	0, 19771
	SPEA2	0,6013	0, 091115	0,6452	0, 09991
	H-SPEA2	**0,7001**	0, 012015	**0,4410**	0, 01302
C-2	NSGA-II	**0,6392**	0, 111331	0,4105	0, 42436
	SPEA2	0,6371	0, 265278	0,4767	0, 19322
	H-SPEA2	0,6344	0, 101132	**0,3395**	0, 24770
C-3	NSGA-II	0,6946	0, 148475	0,3812	0, 15594
	SPEA2	0,7241	0, 091144	0,4108	0, 17441
	H-SPEA2	**0,8410**	0, 160311	**0,2175**	0, 09861

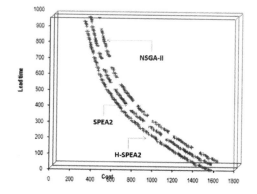

Fig. 2. Cost vs. Lead time non-dominated sets of C-3 instance using H-SPEA2, SPEA2 and NSGA-II

5   Conclusions

This research proposes a multiobjective model for the multi-location transshipment problem with local storage capacity constraints. The model incorporates optimization of the aggregate cost; fill rate and transshipment lead time. The SPEA2 algorithm was improved by a greedy local search that uniformly explores the neighborhood of candidate solutions having poor density in objective space. There are no extra density calculations. We only reuse the density values provided by SPEA2 fitness assignment strategy. Many instances of the problem were solved. Different Pareto fronts were successfully generated in relatively short computation time. Based on Hypervolume and Spread metrics, H-SPEA2 is shown to be better than SPEA2 and NSGA-II. In short, the main contribution of this work is an effective use of a simple local search to improve the exploration and exploitation capabilities of SPEA2. In addition, experiments indicated that the transshipment is very interesting: it guaranties high service level even when holding extra inventory is not allowed for some stores.

References

1. Aggarwal, S.P.: Inventory control aspect in warehouses. In: Symposium on Operations Research. Indian National Science Academy, New Delhi (1967)
2. Bechikh, S., Belgasmi, N., Said, L.B., Ghédira, K.: PHC-NSGA-II: A Novel Multi-objective Memetic Algorithm for Continuous Optimization. In: Proceedings of the 2008 20th IEEE International Conference on Tools with Artificial Intelligence, ICTAI, November 03-05, vol. 01, pp. 180–189. IEEE Computer Society, Washington, DC (2008)
3. Beyer, H.-G.: Evolutionary algorithms in noisy environments: Theoretical issues and guidelines for practice. Computer Methods in Applied Mechanics and Engineering 186(2-4), 239267 (2000)
4. Corne, D., Knowles, J.D., Oates, M.J.: The Pareto Envelope-Based Selection Algorithm for Multi-objective Optimisation. In: Deb, K., Rudolph, G., Lutton, E., Merelo, J.J., Schoenauer, M., Schwefel, H.-P., Yao, X. (eds.) PPSN 2000. LNCS, vol. 1917, pp. 839–848. Springer, Heidelberg (2000)
5. Deb, K., Argawal, S., Pratap, A., Meyarivan, T.: A fast and elitist multi-objective genetic algorithm: NSGA-II. IEEE Transaction on Evolutionary Computation 6(2), 182–197 (2002)
6. El-Mihoub, T.A., Hopgood, A.A., Nolle, L., Battersby, A.: Hybrid genetic algorithms: A review. Engineering Letters 3(2), 124–137 (2006)
7. Herer, Y., Rashit, A.: Policies in a general two-location infinite horizon inventory system with lateral stock transshipments. Department of Industrial Engineering, Tel Aviv University (1999b)
8. Herer, Y.T., Tzur, M., Yücesan, E.: The multi-location transshipment problem (Forthcoming in IIE Transactions) (2005)
9. Hoos, H.H., Stützle, T.: Stochastic local search: Foundations and Applications. Morgan Kaufmann Publishers, San Francisco (2005)
10. Jaszkiewicz, A.: Genetic local search for multiple objective combinatorial optimization, Technical Report RA-014/98, Institute of Computing Science, Poznan University of Technology (1998)
11. Jonsson, H., Silver, E.A.: Analysis of a Two-Echelon Inventory Control System with Complete Redistribution. Management Science 33, 215–227 (1987)
12. Knowles, J., Corne, D.: M-PAES: A memetic algorithm for multiobjective optimization. In: Congress on Evolutionary Computation, Piscataway, New Jersey, vol. 1, pp. 325–332 (2000)
13. Krishnan, K.S., Rao, V.R.K.: Inventory control in N warehouses. J. Industrial Engineering 16(3), 212–215 (1965)
14. Kumar, A., Sharma, D., Deb, K.: A hybrid multi-Objective optimization procedure using PCX based NSGA-II and sequential quadratic programming. In: Special Session & Competition on Performance Assessment of Multi-Objective Optimization Algorithms, CEC 2007, Singapore, pp. 25–28 (2007)

15. Moscato, P.: On evolution, search, optimization, genetic algorithms and martial arts: Towards memetic algorithms. Caltech Concurrent Computation Program, C3P Report 826 (1989)
16. Pan, A.: Allocation of order quantity among suppliers. Journal of Purchasing and Materials Management 25(3), 36–39 (1989)
17. Pokharel, S.: A two objective model for decision making in a supply chain. International Journal of Production Economics 111(2), 378–388 (2008)
18. Raghuwanshi, M.M., Kakde, O.G.: Survey on multiobjective evolutionary and real coded genetic algorithms. In: Proceeding of the 8th Asia Pacific Symposium on Intelligent and Evolutionary Systems, vol. 11, pp. 150–161 (2004)
19. Sharma, D., Kumar, A., De, K., Sindhya, K.: Hybridization of SBX based NSGA-II and sequential quadratic programming for solving multiobjective optimization problems. In: Special Session & Competition on Performance Assessment of Multi-Objective Optimization Algorithms, CEC 2007, Singapore, pp. 25–28 (2007)
20. Silverman, B.W.: Density estimation for statistics and data analysis. Chapman and Hall, London (1986)
21. Tagaras, G.: Effects of pooling on the optimization and service levels of two-location inventory systems. IIE Trans. 21(3), 250–257 (1989)
22. Zitzler, E., Laumanns, M., Thiele, L.: SPEA2: Improving the strength Pareto Evolutionary Algorithm for Multiobjective Optimization. In: Evolutionary Methods for Design, Optimisation, and Control, Barcelona, Spain, pp. 19–26 (2002)

Hybrid Population-Based Incremental Learning Using Real Codes

Sujin Bureerat

Department of Mechanical Engineering, Faculty of Engineering, Khon Kaen University,
40002, Thailand
sujbur@kku.ac.th

Abstract. This paper proposes a hybrid evolutionary algorithm (EA) dealing with population-based incremental learning (PBIL) and some efficient local search strategies. A simple PBIL using real codes is developed. The evolutionary direction and approximate gradient operators are integrated to the main procedure of PBIL. The method is proposed for single objective global optimization. The search performance of the developed hybrid algorithm for box-constrained optimization is compared with a number of well-established and newly developed evolutionary algorithms and meta-heuristics. It is found that, with the given optimization settings, the proposed hybrid optimizer outperforms the other EAs. The new derivative-free algorithm can maintain outstanding abilities of EAs.

Keywords: Population-Based Incremental Learning, Approximate Gradient, Evolutionary Direction, Meta-Heuristics, Evolutionary Algorithms.

1 Introduction

Evolutionary algorithms (EAs) or meta-heuristic search algorithms are commonly known as alternative optimizers to classical mathematical programming (MP) or gradient-based optimizers. Using EAs is advantageous over MP since they are simple to implement, more robust, capable of tackling global optimization, and derivative-free. Nevertheless, the methods have some unavoidable disadvantages since they have a low convergence rate and require a large number of function evaluations to achieve optimum results. With no guarantee of convergence, the optimum results obtained from using EAs are usually classified as near optima. EAs also have a complete lack of search consistency since, with multiple simulation runs, they are unlikely to find the same optimum point. As a result of the attractiveness of their ability to tackle almost all kinds of optimization problems and the aforementioned advantages, many researchers and engineers have invested considerable effort to improve and develop evolutionary optimizers. The target is to retain their outstanding abilities and alleviate their drawbacks. From genetic algorithms (GA) to the countless number of EAs presently being used in a wide variety of real world applications, only those EAs with a high searching performance are receiving considerable attention.

Several methods have been used to enhance the search performance of EAs. One of the most popular and efficient strategies is the use of EAs in combination with

C.A. Coello Coello (Ed.): LION 5, LNCS 6683, pp. 379–391, 2011.
© Springer-Verlag Berlin Heidelberg 2011

surrogate models [1, 2]. This hybridization approach is said to be well-established and successfully implemented on a variety of real world applications. Another approach is the integration between EAs and their variants. Since the weak and strong points of EA operators have been thoroughly investigated, their proper combinations can also improve EA performance [3].

This paper proposes a hybrid evolutionary algorithm dealing with population-based incremental learning and some efficient local search strategies. The method is developed to deal with single objective global optimization. A simple PBIL using real codes is detailed. An evolutionary direction operator and an approximate gradient are integrated with the main procedure of PBIL. The search performance of the developed hybrid algorithm while solving 35 box-constrained optimization problems is compared with a number of well-established and newly developed EAs and meta-heuristics. It is found that the proposed optimizer can be regarded as one of the best evolutionary optimizers.

The paper is organized as follows. The following section 2 gives the details of population-based incremental learning using real codes, evolutionary direction and approximate gradient operators, and the hybrid algorithm. Section 3 provides the testing functions for performance comparison. Section 4 shows the comparative results and assessments of the EAs performance. The paper is concluded in section 5.

2 Hybrid Algorithm

2.1 Population-Based Incremental Learning

The population-based incremental learning was first developed by Baluja as an alternative search algorithm to genetic algorithm [4]. Unlike most traditional EAs, PBIL uses the so-called probability vector to estimate a binary population. The method accomplishes optimization search by improving the probability vector iteratively. The real-code variants of PBIL have been developed [5-6] but they seem to be less popular than the original binary-code PBIL. In this paper, we propose PBIL using real code, which exploits the probability matrix similar to the histogram PBIL (PBIL$_H$) in [6]. Given that the box-constrained optimization problem is of the form:

$$\text{Min } f(\mathbf{x}) \tag{1}$$

$$L_i \le x_i \le U_i; \ i = 1, \ldots, n$$
$$\mathbf{x} \in R^n$$

where $\mathbf{x}$ is the vector of design variables size $n \times 1$, f is an objective function, L_i are the lower bounds of $\mathbf{x}$, and U_i are the upper bounds of $\mathbf{x}$.

The probability matrix ($\mathbf{P}$ or P_{ij} size $n \times m$) is proposed to deal with real design parameters in such a way that the feasible range $[L_i, U_i]$ of a design variable is divided into m sections. The element P_{ij} determines the probability that the i-th element of $\mathbf{x}$ will be placed in the range $[L_i + (j-1)\delta_i, L_i + j\delta_i]$ where $\delta_i = (U_i - L_i)/m$. Generation of a real-code population can be carried out in a similar manner as with binary PBIL. A real-code PBIL search starts with a probability matrix $\mathbf{P}$ where all elements values are assigned as $1/m$. An initial population according to $\mathbf{P}$ is then created with their corresponding objectives being evaluated while the best individual $\mathbf{x}^{best}$ is detected. The probability matrix is then updated based upon $\mathbf{x}^{best}$ as

$$P'_{ij} = (1 - L_{R,j})P^{old}_{ij} + L_{R,j} \tag{2}$$

where

$$L_{R,j} = 0.5\exp(-(j-r)^2). \tag{3}$$

The element r is determined in such a way that x_i^{best} is placed in the range $[L_i + (r-1)\delta_i, L_i + r\delta_i]$. The learning rate L_R in Eq. 3 is set to prevent premature convergence. In order to preserve the condition $\sum_{j=1}^{T} P_{ij} = 1$, the i-th row of $\mathbf{P'}$ is normalized as:

$$P''_{ij} = (1/\sum_{j=1}^{m} P'_{ij})P'_{ij} \tag{4}$$

As a result, the finally updated P_{ij} is in Eq. 4. The probability matrix and $\mathbf{x}^{best}$ are iteratively improved until the termination condition is met. The pseudo-code of real-code PBIL is given in Fig. 1 where t is a generation number, N_G is the total number of generations, and N_P is the population size.

Input: N_G, N_P, n, m, L_R
Output. $\mathbf{x}^{best}$, f^{best}
Initialization: $P_{ij} = 1/m$, $\delta_i = (U_i - L_i)/m$, $\mathbf{x}^{best}(0) = \{\}$.
1: For $t = 1$ to N_G
2:      Generate a real code population $\mathbf{X}(t)$ from P_{ij}
      2.1: For $i = 1$ to n
      2.2:   For $j = 1$ to m
      2.3:   Randomly generate $N_P.P_{ij}$ elements of x_i in the interval
                $[L_i + (j-1)\delta_i, L_i + j\delta_i]$.
      2.4:   End
      2.5:   Randomly permute the positions of N_P elements of x_i.
      2.6:   Put the N_P values of x_i in the i-th row of the population matrix $\mathbf{X}(t)$.
      2.7: End
3:      Evaluate $\mathbf{f}(t) = \text{fun}(\mathbf{X}(t))$.
3:      Find new $\mathbf{x}^{best}(t)$ from $\mathbf{X} \cup \mathbf{x}^{best}(t-1)$.
4:      Update P_{ij} based on the current $\mathbf{x}^{best}$.
      4.1: For $i = 1$ to n
      4.2:   Find r such that $x_i^{best} \in [L_i + (r-1)\delta_i, L_i + r\delta_i]$.
      4.3:   Update P_{ij} using Eq. 2 and 3.
      4.4: End
5: End

Fig. 1. Algorithm for real code population-based incremental learning

2.2 Evolutionary Direction Recombination

The evolutionary direction operator was proposed in [7], which is the modification of the work in [8]. It can be thought of as a special kind of evolutionary recombination. One operation requires three randomly selected individuals from the current population to produce a pair of children. Let the three individuals be $\mathbf{x}_1$, $\mathbf{x}_2$, and $\mathbf{x}_3$ where $\mathbf{x}_1$ has the best (minimum) objective among them. An evolutionary direction is computed as:

$$s = (\mathbf{x}_1 - \mathbf{x}_2) + (\mathbf{x}_1 - \mathbf{x}_3) + \mathbf{c} \tag{5}$$

where $c_i = \varepsilon.randn$, $randn$ is a normally distributed random number with mean zero and standard deviation one, and ε is a small number to be specified (default value is 0.05). The random vector $\mathbf{c}$ is used to prevent a premature convergence. The new individuals as the product of this evolutionary operator can be obtained as

$$\mathbf{y}_1 = \mathbf{x}_1 + rand.\lambda(t)\mathbf{s} \tag{6}$$

$$\mathbf{y}_2 = \mathbf{x}_1 - rand.\lambda(t)\mathbf{s}$$

where $rand \in [0,1]$ is a uniform random number, and $\lambda(t)$ is the maximum step length at the t-th generation. The value of λ is set to have greater value earlier. As the optimization progresses, it becomes smaller. In this work, the maximum step length is set to be

$$\lambda(t) = \exp\left(\frac{6.2146}{N_G - 1} - 0.6931\right)\exp\left(-\frac{6.2146}{N_G - 1}t\right), \tag{7}$$

which means $\lambda(1) = 0.5$ and $\lambda(N_G) = 0.001$.

This strategy is set for local search. Note that the solutions will be treated to satisfy the bound constraints before performing function evaluation. Fig. 2 illustrates the evolutionary operation given that $\mathbf{c}$ in Eq. 5 is set to be a zero vector for simplicity. In Fig. 2 a) and b), the solid line arrow is the search direction for $\mathbf{y}_1$ whereas the dashed arrow is the search direction for $\mathbf{y}_2$. In Fig.2 a), it is shown that the offspring $\mathbf{y}_1$ is better than $\mathbf{y}_2$. However, in Fig. 2 b), $\mathbf{y}_2$ has the possibility to be better than $\mathbf{y}_1$ and for this reason the + and − signs are used in Eq. 6.

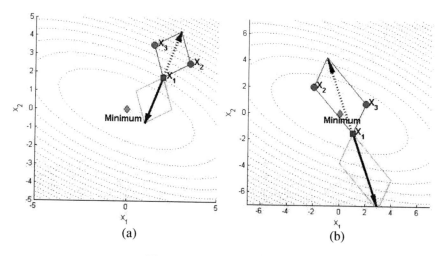

Fig. 2. Evolutionary directions

2.3 Approximate Gradient

The approximate gradient is estimated from the members of a population and their corresponding objectives [7]. The approximate gradient is calculated exploiting the relation of the directional derivative. The directional derivative of a function f in the direction of s at a point x in R^n space, denoted by df/ds can be expressed as:

$$\mathbf{u}_s^T \nabla f = \frac{df}{ds} \approx \frac{f(\mathbf{x}) - f(\mathbf{x} + \Delta \mathbf{x})}{\Delta s} \tag{8}$$

where $\mathbf{u}_s$ is a unit vector of s. With the current population size $n \times N_P$ $\{\mathbf{x}_1, ..., \mathbf{x}_{NG}\}$, their objectives $\{f_1, ..., f_{NG}\}$, the current best solution $\mathbf{x}^{best}$, and its objective f^{best}, the approximate gradient of f at the point $\mathbf{x}^{best}$ can be computed as:

$$\begin{bmatrix} (\mathbf{x}^{best} - \mathbf{x}_1)^T \\ \vdots \\ (\mathbf{x}^{best} - \mathbf{x}_R)^T \end{bmatrix} \nabla f = \mathbf{A} \nabla f = \begin{Bmatrix} f^{best} - f_1 \\ \vdots \\ f^{best} - f_R \end{Bmatrix} = \mathbf{b} . \tag{9}$$

The first R individuals that are closest to $\mathbf{x}^{best}$ are chosen to approximate the gradient. The value R is set to be greater than n in order to prevent matrix singularity. Since the matrix $\mathbf{A}$ is not a square matrix, Eq. 9 can be solved using the pseudo-inverse operation; thus, the approximate gradient is termed pseudo-gradient in [7]. Note that pseudo-inverse algorithms are available in both free and commercial software such as SCILAB and MATLAB. Since ∇f is used for local search rather than global search, it is also useful to apply a quadratic interpolation to enhance the search efficiency. Based on the steepest descent method, the search direction is set to be $s = -\nabla f$. Two solutions extended from $\mathbf{x}^{best}$ along the search direction are found as follows:

$$\mathbf{z}_1 = \mathbf{x}^{best} + \beta_1.\mathbf{s} \tag{10}$$

$$\mathbf{z}_2 = \mathbf{x}^{best} + \beta_2.\mathbf{s}$$

where $\beta_1 \neq \beta_2$ are two randomly generated numbers in the range of $(0,1]$. The third solution for this process can be determined by applying a quadratic interpolation technique. We can assume that the objective function along the search direction is a quadratic function of the variable β as

$$f(\mathbf{x}^{best} + \beta.\mathbf{s}) = C_1\beta^2 + C_2\beta + C_3. \tag{11}$$

The quadratic function coefficients can be found by solving

$$\begin{bmatrix} 0 & 0 & 1 \\ \beta_1^2 & \beta_1 & 1 \\ \beta_2^2 & \beta_2 & 1 \end{bmatrix} \begin{Bmatrix} C_1 \\ C_2 \\ C_3 \end{Bmatrix} = \begin{Bmatrix} f^{best} \\ f(\mathbf{z}_1) \\ f(\mathbf{z}_2) \end{Bmatrix} \tag{12}$$

where $f(\mathbf{z}_1)$ and $f(\mathbf{z}_2)$ are the objective function values at the points $\mathbf{z}_1$ and $\mathbf{z}_2$ respectively. Then, the third individual can be found

$$\mathbf{z}_3 = \mathbf{x}^{best} + \beta_3.\mathbf{s} \qquad (13)$$

where $\beta_3 = -C_2/2/C_1$.

This is equivalent to performing the Powell line search method for one step. In cases that $C_1 = 0$, which implies that the objective could be a linear function, t_3 is generated at random. Fig. 3 shows the process of evaluating $\mathbf{z}_1$, $\mathbf{z}_2$, and $\mathbf{z}_3$ of the approximate gradient operator. The solution $\mathbf{z}_3$ is treated to be inside the bounds before performing function evaluation. For $\mathbf{z}_1$ and $\mathbf{z}_2$, if they are located outside the feasible region, they will be discarded from the optimization process. Nevertheless, their function evaluations are counted to the total number of function evaluations.

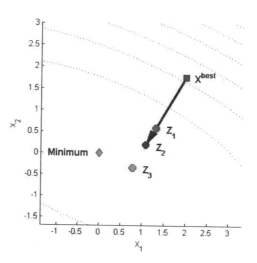

Fig. 3. Step length determination of an approximate gradient operator

2.4 Hybrid Algorithm

The algorithm of the hybrid PBIL is given in Fig. 4. Initially, the hybrid algorithm starts with an initial probability matrix of PBIL, an initial real-code population, and an initial best design solution $\mathbf{x}^{best}$. $N_P/2$ individuals as the first sub-population are then created according to the probability matrix. An approximate gradient is estimated while three new solutions are created from this operation. The rest of the solutions (approximately $N_P/2 - 3$ solutions) to fill in the current population are created by using the evolutionary direction operator. After combining the three sub-populations, the function evaluation is performed. Afterwards, the best individual is detected and used to update the probability matrix and compute an approximate gradient direction. The process is repeated until the maximum number of iterations is reached. In this hybrid algorithm, three sub-populations are created in parallel but the operators share information during the search since updating a probability matrix and computing an approximate gradient require $\mathbf{x}^{best}$ for the operation.

Input: $N_G, N_P, n, m, L_R, \varepsilon$
Output: $\mathbf{x}^{best}, f^{best}$
Initialization: $P_{ij} = 1/m, \delta_i = (U_i - L_i)/m, \mathbf{X}(0), \mathbf{f}(0), \mathbf{x}^{best}(0)$
1: For $t = 1$ to N_G
2:      Generate a sub-population $\mathbf{X}_1$ ($N_P/2$ solutions) from P_{ij} following the computational steps 2.1-2.7 in Fig. 1.
   4: Compute an approximate gradient using Eq. 9.
3:      Generate 3 individuals $\mathbf{X}_2$ from an approximate gradient operator.
4:      Generate $N_P/2 - 3$ individuals $\mathbf{X}_3$ from an evolutionary direction operator.
3:      Combine $\mathbf{X}(t) = \mathbf{X}_1 \cup \mathbf{X}_2 \cup \mathbf{X}_3$, and evaluate $\mathbf{f}(t) = \text{fun}(\mathbf{X}(t))$.
3:      Find new $\mathbf{x}^{best}(t)$ from $\mathbf{X} \cup \mathbf{x}^{best}(t-1)$.
4:      Update P_{ij} based on the current $\mathbf{x}^{best}$ using the computational steps 4.1 - 4.4 in Fig. 1.
5: End

Fig. 4. Algorithm of the Hybrid PBIL

3 Testing Functions

In order to examine the searching performance of the proposed hybrid algorithm, 35 testing functions of box-constrained optimization are posed as detailed in Table 1. With the exception of F_0 [9], all of the functions are taken from [10] and [11] where their expressions and more details can be tracked back from their given names. Evolutionary optimizers used to compare with the hybrid algorithm are as follows:

Real-code ant colony optimization (ACO) [10]: The parameters used for computing the weighting factor and the standard deviation in the algorithm are set to be ξ 1.0 and $q = 0.2$ respectively.

Charged system search (CSS) [12]: The number of solutions in the charge memory is $0.2N_P$. The charged moving considering rate and the parameter PAR are set to be 0.75 and 0.5 respectively.

Differential evolution (DE) [13]: DE step size, crossover probability, and refresh iterations are set as 0.8, 0.5, and 10 respectively. The DE/rand/1/bin strategy is used.

Continuous tabu search (TS) [14]: The sizes of tabu list and promising list are set to be $2N_P$ and $0.5N_P$ respectively.

Fireworks algorithm (FA) [15]: The number of fireworks for each generation is $0.25N_P$, the limit of sparks created with algorithm 1 is N_P, the amount of sparks created with algorithm 2 is $0.25 N_P$, the floor parameter for rounding the amount of sparks created with algorithm 1 is 0.004, the ceiling parameter for rounding the amount of sparks created with algorithm 1 is 0.8, and the maximum explosion amplitude is $0.5(U_i - L_i)$.

Binary-code genetic algorithm (GA) [9]: The crossover and mutation probabilities are 1.0 and 0.1 respectively.

Particle swarm optimization [16]: The starting inertia weight, ending inertia weight, cognitive learning factor, and social learning factor are assigned as 0.5, 0.01, 0.5 and 0.5 respectively.

Simulated annealing (SA) [17]: During an optimization run, an annealing temperature is reduced exponentially 10 times from the value of 10 to 0.001. On each loop, $2n$ children are created by means of mutation to be compared with their parent.

Continuous scatter search (SS) [11]: The BLX-α recombination method is used. The number of high-quality solutions in the reference set is $0.25N_P$, and the number of diverse solutions in the reference set is $0.25N_P$.

Binary PBIL (PB) [4]: The learning rate, mutation shift, and mutation probability are set as 0.5, 0.2, and 0.05 respectively.

The two algorithms from this paper are real-code PBIL (PR) as detailed in subsection 2.1, and real code PBIL in combination with the evolutionary direction and approximate gradient operators (HPR). The number of columns of the probability matrix is set to be $10n$. Each method is used to solve each optimization problem 30 runs starting with the same initial population. The best results the methods can search for are taken as near optimum solutions. The number of iterations is set to be $10n$ whereas the population size is $7n$. In cases of EAs and meta-heuristics that use different search strategies such as fireworks algorithm, simulated annealing, continuous scatter search, and charged system search, the number of iterations and population size may not be the same values as previously mentioned but they will use the same total number of function evaluations i.e. $10n \times 7n$ evaluations.

Table 1. Testing functions

Function no., details	$[L,U]^n$	Function no., details	$[L,U]^n$
1, B_2 [10]	$[-50,100]^2$	18, Griewangk [10]	$[-5.12,5.12]^{10}$
2, Beale [11]	$[-4.5,4.5]^2$	19, Perm [11]	$[-15,15]^{15}$
3, Booth [11]	$[-10,10]^2$	20, Perm0 [11]	$[-15,15]^{15}$
4, Easom [10]	$[-100,100]^2$	21, Cigar [10]	$[-3,7]^{20}$
5, Goldstein & Price [10]	$[-2,2]^2$	22, Diagonal plane [10]	$[0.5,1.5]^{20}$
6, Martin & Gaddy [10]	$[-20,20]^2$	23, Dixon & Price [11]	$[-10,10]^{20}$
7, Matyas [11]	$[-5,10]^2$	24, Levy(n) [11]	$[-10,10]^{20}$
8, Penny & Linfield* [9]	$[-5,5]^2$	25, Powell(n) [11]	$[-4,5]^{20}$
9, Powersum [11]	$[0,2]^2$	26, Rastrigin(n) [10]	$[-2.56,5.12]^{20}$
10, Branin [10]	$[-5,15]^2$	27, Rosenbrock [10]	$[-5,5]^{20}$
11, Shubert [11]	$[-10,10]^2$	28, Sum Squares [11]	$[-5,10]^{20}$
12, Six Hump Camel Back [11]	$[-5,10]^2$	29, Schwefel [11]	$[-500,500]^{20}$
		30, Trid(n) [11]	$[-400,400]^{20}$
13, Colville [11]	$[-10,10]^4$	31, Ackley(n) [11]	$[-15,30]^{30}$
14, Hartmann 3,4 [10]	$[0,4]^4$	32, Ellipsoid [10]	$[-3,7]^{30}$
15, Shekel [10]	$[0,10]^4$	33, Plane [10]	$[-0.5,1.5]^{30}$
16, Zakharov [10]	$[-5,10]^5$	34, Sphere [10]	$[-3,7]^{30}$
17, Hartmann 6,4 [10]	$[0,6]^6$	35, Tablet [10]	$[-3,7]^{30}$

* $f_8(\mathbf{x}) = 0.5(x_1^4 - 16x_1^2 + 5x_1) + 0.5(x_2^4 - 16x_2^2 + 5x_2)$

4   Comparison Results

For each testing function, each method will produce 30 near optimum values. The average value of 30 near optimum values obtained from using 12 EAs is found and normalized using the relation

Table 2. Comparative results by normalized function values

Fn No.	ACO	CSS	DE	TS	FA	GA	PSO	SA	SS	PB	PR	HPR
1	0.000	0.035	0.040	0.016	0.050	0.869	0.018	0.032	0.105	1.000	0.019	0.005
2	0.020	0.520	1.000	0.130	0.220	0.265	0.000	0.148	0.542	0.302	0.431	0.058
3	0.014	0.056	0.166	0.005	0.056	1.000	0.015	0.353	0.413	0.778	0.014	0.000
4	0.829	1.000	0.830	0.553	0.997	0.550	0.283	0.988	1.000	0.912	0.529	0.000
5	0.015	1.000	0.439	0.000	0.090	0.836	0.520	0.718	0.687	0.841	0.024	0.144
6	0.001	0.051	0.068	0.003	0.037	0.400	0.001	0.083	0.089	1.000	0.005	0.000
7	0.036	0.109	0.293	0.006	0.099	0.500	0.006	0.501	0.578	1.000	0.016	0.001
8	0.085	0.074	0.051	0.516	0.284	0.159	1.000	0.000	0.193	0.185	0.190	0.188
9	0.009	0.030	0.132	0.002	0.043	0.440	0.119	1.000	0.153	0.658	0.002	0.002
10	1.000	0.076	0.516	0.016	0.149	0.421	0.163	0.069	0.290	0.755	0.041	0.000
11	0.934	0.520	0.862	0.165	0.315	0.346	0.539	0.116	1.000	0.570	0.340	0.000
12	0.038	0.801	0.476	0.273	0.363	0.536	0.998	0.337	0.749	1.000	0.125	0.000
13	0.239	0.186	1.000	0.031	0.319	0.329	0.141	0.154	0.337	0.344	0.165	0.000
14	0.000	1.000	0.002	0.005	0.006	0.036	0.301	0.439	0.023	0.203	0.084	0.004
15	0.052	0.000	0.821	0.623	0.885	0.634	0.454	0.920	0.377	1.000	0.772	0.573
16	0.033	0.017	0.412	0.001	0.020	0.632	0.339	0.180	0.205	1.000	0.006	0.000
17	0.000	1.000	0.051	0.237	0.129	0.191	0.908	0.001	0.531	0.405	0.251	0.247
18	0.704	0.000	1.000	0.163	0.175	0.076	0.150	0.161	0.015	0.040	0.173	0.097
19	0.000	0.462	0.000	0.546	0.619	0.027	0.890	0.000	0.267	0.081	1.000	0.417
20	0.000	0.000	0.002	0.000	0.000	0.000	0.000	1.000	0.000	0.001	0.000	0.000
21	0.000	0.000	0.036	0.002	0.026	0.338	0.179	0.001	0.199	1.000	0.043	0.000
22	0.000	0.014	0.000	0.006	0.000	0.001	1.000	0.000	0.057	0.022	0.022	0.000
23	0.009	0.000	0.114	0.000	0.013	0.166	0.308	0.003	0.022	1.000	0.042	0.007
24	0.030	0.023	0.088	1.000	0.886	0.059	0.320	0.000	0.042	0.212	0.403	0.408
25	0.047	0.000	0.147	0.006	0.011	0.420	0.712	0.005	0.049	1.000	0.061	0.001
26	1.000	0.386	0.588	0.601	0.485	0.176	0.477	0.000	0.470	0.293	0.722	0.379
27	0.076	0.011	0.759	0.000	0.081	0.481	1.000	0.004	0.117	0.860	0.153	0.087
28	0.001	0.000	0.042	0.002	0.024	0.149	0.810	0.001	0.157	1.000	0.036	0.000
29	1.000	0.266	0.468	0.449	0.485	0.130	0.827	0.000	0.795	0.223	0.679	0.507
30	0.110	0.333	0.080	0.429	0.000	0.410	1.000	0.454	0.698	0.515	0.332	0.203
31	0.279	0.074	0.519	0.323	0.264	0.537	1.000	0.000	0.642	0.889	0.331	0.079
32	0.004	0.000	0.041	0.001	0.036	0.119	1.000	0.000	0.179	0.452	0.031	0.004
33	0.000	0.000	0.000	0.000	0.000	0.000	0.000	0.000	0.000	0.000	0.000	0.000
34	0.007	0.000	0.065	0.000	0.009	0.106	1.000	0.000	0.187	0.319	0.016	0.000
35	0.000	0.000	0.004	0.008	0.001	0.030	1.000	0.000	0.025	0.054	0.232	0.002
total	6.573	8.043	11.113	6.118	7.177	11.370	17.477	7.669	11.192	19.915	7.291	3.414

* PB = binary PBIL, PR = real code PBIL, HPR =Hybrid real-code PBIL

$$\bar{f}_i = \left| \frac{f_i - f_{min}}{f_{max} - f_{min}} \right| \tag{14}$$

where f_{min} is the average near optimum value of the best method, and f_{max} is the average near optimum value of the worst method. By using Eq. 12, the best method will have $\bar{f} = 0$ whereas the worst method will have $\bar{f} = 1$. This relative comparison is given in Table 2 where each value in the table stands for a $\bar{f}$ value. From the results, the overall top five best performers are the proposed hybrid PBIL algorithm, continuous tabu search, real-code ant colony optimization, fireworks algorithm, real-code PBIL, and simulated annealing. Clearly, there is no absolute best method. The proposed hybrid approach is slightly ahead of the second best TS and the third best ACO. Charged system search and simulated annealing are the two best methods for large scale objective functions having one optimum. The continuous tabu search uses the longest computational time for each optimization run. The hybrid PBIL takes a slightly longer time than the real-code PBIL. Among the PBIL variants, real-code PBIL outperforms its binary-code counterpart. The searching performance of PBIL is improved when integrated with the evolutionary direction and approximate gradient operators. This ranking is made to show the convergence rate of EAs, which means the hybrid PBIL has a high convergence rate when compared to the other EAs with the given optimization parameters and conditions.

An alternative EA performance assessment is given in Table 3-4. Table 3 shows the ranking of the 12 implemented evolutionary algorithms for the first test function. Firstly the performance matrix size 12×12, whose elements are full of zeros, is generated. Then, the results obtained from method I and method J are compared using the statistical t-test at 95% confidence level. In cases that the mean objective function value from method I is significantly different from that obtained from method J, an element of the performance matrix is modified. The element at row I and column J of the matrix is changed to be one if the mean value obtained from method I is higher; otherwise, element at row J and column I is changed to be one. Having a complete performance matrix, the values on each column are summed up. The algorithm having that highest score (ACO in Table 3) is considered the best method while the method having the lowest total value is the worst for solving this test function.

Table 3. Performance matrix and ranking score using t-test: Function number 1

EAs	ACO	CSS	DE	TS	FA	GA	PSO	SA	SS	PB	PR	HPR
ACO	0	0	0	0	0	0	0	0	0	0	0	0
CSS	1	0	0	0	0	0	0	0	0	0	0	1
DE	1	0	0	1	0	0	1	0	0	0	1	1
TS	1	0	0	0	0	0	0	0	0	0	0	1
FA	1	0	0	1	0	0	1	0	0	0	1	1
GA	1	1	1	1	1	0	1	1	1	0	1	1
PSO	1	0	0	0	0	0	0	0	0	0	0	0
SA	1	0	0	1	0	0	0	0	0	0	0	1
SS	1	0	0	1	0	0	1	0	0	0	1	1
PB	1	1	1	1	1	0	1	1	1	0	1	1
PR	1	0	0	0	0	0	0	0	0	0	0	1
HPR	1	0	0	0	0	0	0	0	0	0	0	0
Total	11	2	2	6	2	0	5	2	2	0	5	9
Ranking	1	6	6	3	6	11	4	6	6	11	4	2

Having determined the performance matrices of all the test functions, the best method for each design problem will have a score as 1 whereas the worst has 12 as given in Table 4. After summing up the scores of all the testing functions, the top five EAs and meta-heuristics are: the hybrid PBIL, continuous tabu search, real-code ant colony optimization, simulated annealing, and charged system search. The order of the top five methods is slightly different from that in the first comparison. Among the

PBIL versions, real-code PBIL outperforms binary PBIL, the worst method in this study. The hybrid PBIL is superior to the real-code PBIL, which means the inclusion of an evolutionary direction and an approximate gradient helps enhance PBIL search performance. Charged system search is the best method for larger scale testing functions having one optimum solution while the proposed hybrid approach is the best for multi-modal small scale functions.

Table 4. Comparative results by ranking

Function No.	ACO	CSS	DE	TS	FA	GA	PSO	SA	SS	PB	PR	HPR
1	1	6	6	3	6	11	4	6	6	11	4	2
2	2	9	9	4	6	6	1	5	9	6	9	3
3	3	6	8	2	6	11	3	8	10	11	3	1
4	6	6	6	4	6	4	1	6	6	6	3	1
5	1	8	6	1	4	8	7	8	8	8	1	4
6	2	6	6	3	6	11	3	6	6	12	5	1
7	3	6	8	2	6	9	3	9	9	9	3	1
8	3	3	2	11	5	5	12	1	5	5	5	5
9	4	4	7	2	4	10	7	11	7	11	2	1
10	11	3	9	2	6	9	7	3	8	11	3	1
11	10	5	10	3	4	5	5	2	10	5	5	1
12	2	7	7	5	5	7	7	4	7	7	3	1
13	7	4	12	1	7	7	3	4	7	7	4	1
14	1	12	2	3	3	6	10	10	6	9	8	3
15	1	1	8	5	10	5	4	10	3	12	8	5
16	6	4	9	2	4	11	9	7	7	12	3	1
17	1	12	3	5	4	5	11	1	10	9	5	5
18	11	1	12	7	7	4	7	6	2	3	7	4
19	3	7	1	9	10	4	11	1	6	5	12	7
20	4	1	4	2	4	4	4	12	4	4	4	2
21	3	1	7	5	6	11	9	4	9	12	8	2
22	4	8	1	7	4	6	12	1	11	9	9	1
23	4	1	9	1	5	10	11	3	7	11	8	5
24	3	2	6	11	11	4	8	1	4	7	8	8
25	6	1	9	3	5	10	11	3	6	11	6	2
26	12	5	9	9	6	2	6	1	6	3	11	4
27	4	2	10	1	5	9	10	2	5	10	7	7
28	3	1	8	5	6	9	11	3	9	11	7	2
29	12	3	5	5	5	2	10	1	10	3	9	5
30	2	5	2	7	1	8	12	8	11	8	5	4
31	4	2	8	6	4	8	12	1	10	11	6	2
32	4	1	7	3	7	9	12	2	10	11	6	4
33	1	11	1	1	1	1	1	1	1	1	12	1
34	5	1	8	4	5	9	12	3	10	11	7	2
35	3	1	6	6	4	8	12	2	8	10	11	5
total	152	156	231	150	188	248	268	156	253	292	217	104

5 Conclusions and Discussion

The real-code PBIL is developed to deal with box-constrained optimization. The hybridization of the real-code PBIL with the evolutionary directions and approximate gradient is proposed. The new method is derivative-free and capable of maintaining the outstanding advantages of traditional EAs e.g. global optimization. From the comparative results, it is shown that the proposed hybrid approach is one of the best EAs. In fact, it is the overall best method based on the assessment in this work. The main real-code PBIL is used for global search while the evolutionary direction and approximate gradient are efficient for local search. Nevertheless, it should be noted that the comparative results rely on specific optimization settings such as crossover and mutation probabilities of GA, and number of fireworks in FA. The future work will be the implementation of an approximate gradient, an evolutionary direction, and some other efficient evolutionary operators for constrained optimization problems, and multiobjective optimization.

Acknowledgments.   The author is grateful of the support from the Thailand Research Fund (TRF). Many thanks are also directed to my colleague, Peter Warr, for his careful proofreading.

References

1. Farina, M., Amato, P.: Linked Interpolation-Optimization Strategies for Multicriteria Optimization Problems. Soft Computing 9, 54–65 (2005)
2. Srisoporn, S., Bureerat, S.: Geometrical Design of Plate-Fin Heat Sinks Using Hybridization of MOEA and RSM. IEEE Transactions on Components and Packaging Technologies 31, 351–360 (2008)
3. Kaveh, A., Talatahari, S.: Particle Swarm Optimizer, Ant Colony Strategy and Harmonic Search Scheme Hybridized for Optimization of Truss Structures. Computer and Structures 87, 1245–1287 (2009)
4. Baluja, S.: Population-Based Incremental Learning: a Method for Integrating Genetic Search Based Function Optimization and Competitive Learning. Technical Report CMU_CS_95_163, Carnegie Mellon University (1994)
5. Sebag, M., Ducoulombier, A.: Extending Population-Based Incremental Learning to Continuous Search Spaces. In: Eiben, A.E., Bäck, T., Schoenauer, M., Schwefel, H.-P. (eds.) PPSN 1998. LNCS, vol. 1498, pp. 418–427. Springer, Heidelberg (1998)
6. Yuan, B., Gallagher, M.: Playing in Continuous Spaces: Some Analysis and Extension of Population-Based Incremental Learning. In: CEC 2003, CA, USA, pp. 443–450 (2003)
7. Bureerat, S., Cooper, J.E.: Evolutionary Optimisation Using Evolutionary Direction and Pseudo-Gradient. In: 1st ASMO UK/ISSMO, Ilkley, UK, pp. 81–87 (1999)
8. Yamamoto, K., Inoue, O.: New Evolutionary Direction Operator for Genetic Algorithms. AIAA 33, 1990–1993 (1995)
9. Lindfield, G., Penny, J.: Numerical Methods Using MATLAB. Ellis Horwood, England (1995)
10. Socha, K., Dorigo, M.: Ant Colony Optimization for Continuous Domains. European Journal of Operational Research 185, 1155–1173 (2008)
11. Herrera, F., Lozano, M., Molona, D.: Continuous Scatter Search: An Analysis of the Integration of Some Combination Methods and Improvement Strategies. European Journal of Operational Research 169, 450–476 (2006)
12. Kaveh, A., Talatahari, S.: A Novel Heuristic Optimization Method: Charged System Search. Acta Mechanica 213, 267–289 (2010)
13. Storn, R., Price, K.: Differential Evolution - A Simple and Efficient Adaptive Scheme for Global Optimization over Continuous Spaces. Technical Report TR-95-012. International Computer Science Institute, Berkeley, CA (1995)

14. Teh, Y.S., Rangaiah, G.P.: Tabu Search for Global Optimization of Continuous Functions with Application to Phase Equilibrium Calculations. Computers and Chemical Engineering 27, 1665–1679 (2003)

15. Tan, Y., Zhu, Y.: Fireworks Algorithm for Optimization. In: Tan, Y., Shi, Y., Tan, K.C. (eds.) ICSI 2010. LNCS, vol. 6145, pp. 355–364. Springer, Heidelberg (2010)

16. Reyes-Sierra, M., Coello Coello, C.A.: Multi-objective Particle Swarm Optimizers: a Survey of the State-of-the-Art. Int. J. of Computational Intelligence Research 2, 287–308 (2006)

17. Bureerat, S., Limtragool, J.: Structural Topology Optimisation Using Simulated Annealing with Multiresolution Design Variables. Finite Element in Analysis and Design 44, 738–747 (2008)

Pareto Autonomous Local Search

Nadarajen Veerapen and Frédéric Saubion

LERIA, Université d'Angers, 49045 Angers, France
{nadarajen.veerapen,frederic.saubion}@univ-angers.fr

Abstract. This paper presents a study for the dynamic selection of operators in a local search process. The main purpose is to propose a generic autonomous local search method which manages operator selection from a set of available operators, built on neighborhood relations and neighbor selection functions, using the concept of Pareto dominance with respect to quality and diversity. The latter is measured using two different metrics. This control method is implemented using the COMET language in order to be easily introduced in various constraint local search algorithms. Focusing on permutation-based problems, experimental results are provided for the QAP and ATSP to assess the method's effectiveness.

1 Introduction

Metaheuristics are now widely adopted as efficient solving methods for combinatorial optimization and constraint satisfaction problems. Nevertheless, these approaches often require a fair amount of knowledge of the problem as well as of the solving method. A recent development has been to consider building generic high level control strategies in an effort to make optimization techniques easier to use [4].

Focusing on local search (LS) techniques, a good LS algorithm [9] should explore the search space effectively in the quest for the optimum solution. This involves balancing two generally diverging objectives: intensification (converging towards a local optimum) with diversification (suitably sampling different areas of the search space). The effectiveness of those two strategies is largely dependent on the chosen neighborhood structure(s). This balance can be controlled by means of basic operations (i.e., moves) that are applied along the search process. Therefore, an increasing number of works now attempt at building more autonomous algorithms [8]. Of course this trend has been explored for LS algorithms in the context of *Reactive Search* [1], based on seminal works such as reactive tabu [2] or adaptive simulated annealing [12]. For instance in [10], an adaptive LS uses several neighborhood relations. Nevertheless, as recently mentioned in [14], most LS algorithms handle diversity and quality as two opposite objectives and thus use alternate stages of diversification and intensification, sometimes in a supervised way and focus most of the time on the quality of the current incumbent solution, but may introduce prohibition mechanisms to avoid local optima trapping. Agreeing with the remarks of [14], we believe that more coordination can be achieved between these two objectives, which can be

C.A. Coello Coello (Ed.): LION 5, LNCS 6683, pp. 392–406, 2011.
© Springer-Verlag Berlin Heidelberg 2011

assessed by the target quality/diversity balance fluctuating in response to the state of the search process.

Recent works in evolutionary algorithms provide new techniques for adaptive operator selection. *Compass* [16] evaluates the performance of an operator as a scalarization of fitness improvement from parent to offspring, variation in quality and execution time. In [6], a *Dynamic Multi-Armed Bandit* is used to select the operator that maximizes a sum of two quantities, the first one representing the performance of the operator and the second ensuring that an operator is selected an infinite number of times. Using the *Compass* principles, an adaptive local search algorithm has been presented in [20]. In these works the performance of an operator is defined w.r.t. a static target balance between quality versus diversity.

In this paper, we first consider a generic algorithmic model for local search as a selection process of move operators from a set of available ones, which combine a neighborhood relation and a neighbor selection within this neighborhood. Then the purpose of the algorithm is to choose and apply a operator on the current incumbent solution to progressively build a search path. Therefore, our attempt is twofold : 1) to introduce a new compromise between quality and diversity in the search and 2) to provide a control framework that is able to use general purpose operators for a wide range of problems in order to provide optimization facilities to non expert users by relieving them from algorithm design and tuning.

We present local search control features for solving permutation problems, i.e. those whose configuration can be modeled as permutations. This general framework allows us to define various operators by combining basic permutation neighborhoods and selectors.

At each step of the search, the operators are selected according to the Pareto dominance principle, computed w.r.t. the recorded performance in intensification and diversification of each operator. Moreover, since our purpose is to provide a generic development framework for local search users, our control features are inserted in COMET [24], which is a language dedicated to the design of local search algorithms with constraint handling facilities. In order to outline the generality of our controller, we then test our implementation on two well known permutation problems: the Quadratic Assignment Problem and the Asymmetric Traveling Salesman Problem.

The rest of this paper is organized in 4 sections. Section 2 establishes the definitions to deal with neighborhoods, neighborhood selectors and operators for permutation problems. In Section 3 we present the control framework, two distance metrics and the Pareto selection method. This is followed by test protocol and results in Section 4. Finally, Section 5 ends with concluding remarks and some possibilities for further investigation.

2   Neighborhood, Selectors and Operators

The purpose of this section is to provide a formal description of the permutation based problems and their associated operators. In [21], the authors propose such a formal review of different neighborhoods and they define distances associated to these neighborhoods. As mentioned above, since our goal is to dynamically manage operators according to their behavior and properties we are thus particularly interested in such metrics. Nevertheless, in [21], the authors deal with single operator methods and the metrics that could be used to assess the diversity of a local search path are indeed fully dependent of the operator.

Here, we aim at providing a generic and simple description of the neighborhood and the operators that can be useful to define new operators and to manage their application according to their impact on the search process. Our purpose is also to provide a framework to compare neighborhoods and selectors in a multi operator local search procedure.

2.1 General Definitions

In this section, our purpose is to clearly define the neighborhood and the selection of the neighbor and thus the operators, together with the different notions associated to the search process.

Neighborhood. Let S be the search space of candidate solutions. A neighbor relation is an irreflexive binary relation $\mathcal{N} \subseteq S^2$ over the search space. In most cases, the relation is also symmetric.

Search Paths. Given a neighbor relation $\mathcal{N}$ we define the set of search paths as $\mathcal{P}_{\mathcal{N}} = \{s_1 \cdots s_n \in S^* | \forall i > 1, (s_{i-1}, s_i) \in \mathcal{N}\}$, where S^* classically denotes the set of words constructed over S . Therefore, any pair (s, s') of elements of S, such that $(s, s') \in \mathcal{N}^+$,[1] defines an equivalence class over the set $\mathcal{P}_{\mathcal{N}}$, which corresponds to all the paths that link s to s'. We may denote this subset by $\mathcal{P}_{\mathcal{N}}/(s, s')$. In most of the cases, the neighborhood should be complete, i.e. $\forall s, s' \in S, \mathcal{P}_{\mathcal{N}}/(s, s') \neq \emptyset$.

Distances. The neighbor relation actually defines the declarative structure of the search space. We may thus define the distance between s and s' as $d_{\mathcal{N}}(s, s') = min_{p \in \mathcal{P}_{\mathcal{N}}/(s,s')}|p|$, where $|p|$ is the classic word length. By definition, we impose $d_{\mathcal{N}}(s, s) = 0$. Note that we may require $\mathcal{N}$ to be symmetric if we want d to be a distance.

Combining Neighborhoods. In order to express more complex neighborhood structures, we denote $\mathcal{N} \circ \mathcal{N}'$ the composition and $\mathcal{N} \cup \mathcal{N}'$ the union, which are the most commonly used neighborhood constructors. A neighborhood composed with itself is denoted by $\mathcal{N}^2$ and $\mathcal{N}^{n+1} = \mathcal{N} \circ \mathcal{N}^n$.

Search Landscape. Turning now to the search landscape, we first introduce an ordering relation $<$ over S that corresponds to the order induced by the fitness function of the problem. Note that we consider here only minimization problems, which is general enough.

---

[1] $\mathcal{N}^+$ is the transitive closure of $\mathcal{N}$.

Operational Landscape. We now have to introduce the operational structure of local search in order to move through the neighborhood relation.

In this context, a selector is a function that performs a selection over a neighborhood, eventually guided by the ordering $<$ and is defined as $\sigma : \mathcal{S} \times 2^{\mathcal{S}^2} \mapsto \mathcal{S}$ (here the selection returns only one neighbor), such that $(s, \sigma(s, \mathcal{N})) \in \mathcal{N}^=$ (the reflexive closure of $\mathcal{S}$ to include identity). An operator is then defined by a pair $(\mathcal{N}, \sigma)$.

Again, we consider the paths induced by an operator

$$\mathcal{P}_o = \bigcup_{n>1} \{s_1 \cdots s_n \in \mathcal{S}^* | o = (\mathcal{N}, \sigma), \forall i > 1, s_i = \sigma(s_{i-1}, \mathcal{N})\}$$

In order to simplify the notation, we use $o(s) = \sigma(s, \mathcal{N})$ when $o = (\mathcal{N}, \sigma)$ since o can be viewed as a function on $\mathcal{S}$. We denote $o \circ o'$ the composition between operators, o^2 the composition of o with itself and $o^{n+1} = o \circ o^n$.

Here, we should note that we only have the inclusion $\mathcal{P}_o \subseteq \mathcal{P}_\mathcal{N}$, since some neighborhood paths cannot be necessarily constructed by the operators as soon as it includes a selection process among the neighbors. Moreover, if there exists a path in $\mathcal{P}_o$ from s to s', there does not necessarily exist a path from s' to s. Therefore, due to this non symmetric aspect of operators it is not obvious to use a simple distance over the paths created by the operators. Now we may handle multiple neighborhoods local search by composing or joining neighborhood relations.

2.2   Permutations

We now focus on permutations which correspond to the encoding that we will use in our problems. Our purpose is to propose a comprehensive view of the possible operators that could be used in this context.

Let $\Pi(n)$ be the search space, i.e. set of all permutations of the set $\{0, 1, \ldots, n-1\}$. If $\pi \in \Pi(n)$ and $0 \le i \le n-1$, then π_i denotes element i in π.

As described in [21] we may use a set of basic neighborhood relations induced by the basic possible permutations.

Swap $\mathcal{N}_S$ $(s, s') \in \mathcal{N}_S$ iff $s = (\pi_0, \ldots, \pi_i, \pi_{i+1}, \ldots, \pi_{n-1})$
and $s' = (\pi_0, \ldots, \pi_{i+1}, \pi_i, \ldots, \pi_{n-1})$ for some i.

Exchange $\mathcal{N}_E$ $(s, s') \in \mathcal{N}_E$ iff $s = (\pi_0, \ldots, \pi_{i-1}, \pi_i, \pi_{i+1}, \ldots, \pi_{j-1}, \pi_j, \pi_{j+1}, \ldots, \pi_{n-1})$
and $s' = (\pi_0, \ldots, \pi_{i-1}, \pi_j, \pi_{i+1}, \ldots, \pi_{j-1}, \pi_i, \pi_{j+1}, \ldots, \pi_{n-1})$ for some i and some j.

Insertion $\mathcal{N}_I$ $(s, s') \in \mathcal{N}_I$ iff $s = (\pi_0, \ldots, \pi_{i-1}, \pi_i, \pi_{i+1}, \ldots, \pi_{j-1}, \pi_j, \pi_{j+1}, \ldots, \pi_{n-1})$
and $s' = (\pi_0, \ldots, \pi_{i-1}, \pi_{i+1}, \ldots, \pi_{j-1}, \pi_i, \pi_j, \pi_{j+1}, \ldots, \pi_{n-1})$ for some i and some j.

Edge Exch. $\mathcal{N}_{EE}$ $(s, s') \in \mathcal{N}_{EE}$ iff $s = (\pi_0, \ldots, \pi_{i-1}, \pi_i, \pi_{i+1}, \ldots, \pi_{j-1}, \pi_j, \pi_{j+1}, \ldots, \pi_{n-1})$
and $s' = (\pi_0, \ldots, \pi_i, \pi_j, \pi_{j-1}, \ldots, \pi_{i+1}, \pi_{j+1}, \ldots, \pi_{n-1})$ for $i+1 < j$.

It is easy to see that the neighborhood constructed by $\mathcal{N}_S$ can also be constructed by $\mathcal{N}_E$ and $\mathcal{N}_I$. Therefore, ordering relations can be defined to classify the neighborhood in order to highlight the relationships between the distances they induce (see [21] for more details).

We may now propose several classic selection functions in order to build operators.

Random σ_R such that $\sigma(s, \mathcal{N})$ is any randomly chosen s' such that $(s, s') \in \mathcal{N}$

Best Improve $\sigma_B I$ such that $\sigma(s, \mathcal{N})$ is a minimal element s' according to the order $<$, such that $(s, s') \in \mathcal{N}$

Best Improve k $\sigma_B I k$ such that $\sigma(s, \mathcal{N})$ is an uniformly selected element $s' \in K$, K being the set of k-best elements according to the order $<$, such that $(s, s') \in \mathcal{N}$

Improve σ_I such that $\sigma(s, \mathcal{N})$ is any element s' such that $(s, s') \in \mathcal{N}$ and $s' < s$.

Tournament k $\sigma_T k$ such that $\sigma(s, \mathcal{N})$ is an element s' such that K is a subset of k elements that are in relation with s in $\mathcal{N}$ and s' is the best of these k elements.

3   Operator Control for Local Search

The aim of our method is to select from a given set of operators the appropriate one to apply at each iteration (Fig. 1). This requires evaluating the efficiency of the operators based on their previous behavior and selecting one which is capable of advancing the search process, either for intensification or diversification purposes.

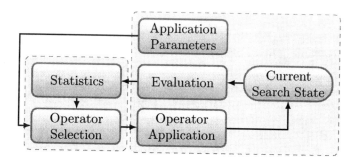

Fig. 1. Overview

Our objective is to have an approach as generic as possible for solving permutation problems. As such, our solving method involves four distinct modules.

- Permutation problem definition and search process
- Path (sliding window) Manager.
- Operator Manager.
- Operator Store (An operator is a neighborhood to which is associated a selector).

For the user, solving a new permutation problem only involves defining a procedure to read the instance data and specifying the objective function and constraints. Optionally, the user can add new operators to the Operator Store and provide a method to generate initial solutions other than random ones.

Our implementation is written in COMET [24]. We believe that this constraint-based local search language provides interesting avenues for our work because of its focus on making local search easier and it already provides simple mechanisms for manipulating neighborhoods and selectors. Our program builds on those intuitive features and could be considered a plug-in for COMET. Indeed, the one line instructions in the main loop of the general algorithmic outline (Algorithm 1) each correspond to one line calls to our plug-in. We think that this genericity and simplicity (modulo a minimum amount of knowledge needed to define the problem) allied to the inherent ease-of-use of COMET is a good step in empowering end-users of optimization software.

define problem as instance of Permutation Problem
add operators to Operator Store
initialize Path Manager
initialize Operator Manager
$s \leftarrow$ initial solution
$s^* \leftarrow s$
repeat
    $op \leftarrow$ select Operator
    $s \leftarrow op(s)$
    Update Path Manager with s
    Update Operator Manager with measures of s
    if s *is better than* s^* **then** $s^* \leftarrow s$
until *end condition reached*
return s^*

Algorithm 1. General algorithmic outline

Our approach is meant to be as generic as possible but has some shortcomings. In practice (for operators more complex than simple exchanges), the user is required to provide the function to compute the delta in evaluating a candidate solution since it is dependent on the objective function.

3.1 Metrics

As mentioned in the introduction, an important issue in the control is to assess the balance diversification/intensification by means of metrics that can evaluate the efficiency of an operator w.r.t. the visited search path in order to choose the next move. We propose here to handle simultaneously two criteria, quality and diversity, to manage this balance as a compromise.

Quality. Quality is measured directly using the objective function. The relative change in quality when applying an operator op to a solution s is given by

$$\Delta Q = \frac{eval(op(s)) - eval(s)}{eval(s) + 1}$$

Distance. Diversity is a natural concept when considering populations of solutions in evolutionary algorithms. This is less intuitive in local search which produces one solution at each iteration. This notion has been investigated for instance in [22] and in [13]. We could consider the diversity of the path of the search (the sequence of solutions already found) or a sliding window of this path. Instead we choose to try to measure the difference between the path and the current candidate solution $c = op(s)$. We propose two different perspectives: first, how different the path is compared to c at the variable level; second, how far c is from the path in terms of the numbers of operations between them.

The L_1 (Manhattan) distance between two vectors p and q is defined as $d_1(p, q) = \sum_{i=1}^{n} |p_i - q_i|$. We use a simple metric measuring the L_1 distance between representations of the candidate solution and the centroid of the path. The dimensions of the points representing the solutions are the binary variables, $x_{a,b}$, the value of which is 1 in the candidate solution, where $x_{a,b} = 1$ implies that variable a is assigned value b for assignment problems or that a is followed by b for ordering problems. The centroid therefore corresponds to the frequencies of $x_{a,b} = 1$ in the path.

More formally, let $X = \{1, \ldots, n\}$ such that there exists a bijection g from X to the set of variables representing the solutions where Y is the domain of these variables. Let $f_s : X \to Y$ be the function representing the values assigned to the variables of solution s. Let $P_{i,j}$ be the path from iteration i through j, $i \leq j$. Then

$$d_1^P(c, P_{i,j}) = \frac{1}{n} \times \sum_{k=1}^{n} \left(1 - \frac{occ(P_{i,j}, x_{k, f_c(k)})}{|P_{i,j}|} \right)$$

where $occ(P_{i,j}, x_{a,b})$ returns the number times $x_{a,b} = 1$ in $P_{i,j}$.

Next, we use the basic neighborhood distance presented in Section 2 and the idea is to compare the effective search path with the optimal path that may have been built with this reference neighborhood relation.

Thus we define distance

$$d_{\mathcal{N}}^P(p_k, P_{i,j}) = \sum_{l=i}^{j} \frac{|P_{l,k}|}{d_{\mathcal{N}}(p_l, p_k)} \quad , i \leq j \leq k$$

Using a sufficiently simple operator, $d_{\mathcal{N}}^P$ can be used to evaluate the exploratory characteristics of more complex operators. We suggest the use of the simple exchange operator, $\mathcal{N}_E$, for this purpose. An algorithm to compute $d_{\mathcal{N}_E}$ is presented in [21].

3.2   Operator Selection

We now provide some insights on the selection process that is used to choose the move operator at each search step. Given two vectors u and v of equal cardinality p and considering a maximization problem, u *dominates* v if

$u_k \geq v_k, \forall k \in \{1, \ldots, p\}$ with at least one strict inequality. This is often referred to as Pareto dominance.

We consider the population of two-dimensional vectors representing the performance of each operator. In this paper the performance corresponds to the average ΔQ and d^P over an independent sliding window of length m for each operator. The initial performance of each operator is calculated by applying each of them once to the initial solution. If an operator has not been used in the last m iterations, the sliding window for this operator will not be empty: it will contain at least one element (and at most m elements) computed before those m last iterations.

The operator to use at each iteration of the algorithm is selected by fair random choice, that is with a probability proportional to its utility value [18]. We define the utility value of an operator as the number of operators which it dominates to which we add an ϵ to ensure a non-zero utility value.

4   Experiments

We test our method on the QAP and the ATSP. These problems were chosen because their solutions are easy to model as a single array of variables. The *Quadratic Assignment Problem* (QAP) models the problem of finding a minimum cost allocation of facilities into locations, taking the costs as the sum of all possible distance-flow products [15]. The *Asymmetric Traveling Salesman Problem* (ATSP) involves finding a minimum weight Hamiltonian tour in a directed graph [7]. Initial solutions for the QAP are randomly generated and the nearest neighbor construction heuristic is used for the ATSP. In these experiments, we use 10 operators:

O1 $(\sigma_I, \mathcal{N}_E)$, the first-improving exchange between two variables.

O2 $(\sigma_B I, \mathcal{N}_E)$, the best exchange between two variables.

O3 $(\sigma_B I5, \mathcal{N}_E)$, random choice among the 5-best exchanges between two variables.

O4 $(\sigma_B I, \mathcal{N}_E)^2$, two consecutive best exchanges between two variables. The variables exchanged in the first step are forbidden in the second.

O5 $(\sigma_B I, \mathcal{N}_E)^3$, three consecutive best exchanges between two variables. The variables exchanged in previous steps are forbidden in the following steps.

O6 $(\sigma_T 3!, \mathcal{N}_E^2)$, best exchange between 3 randomly chosen variables.

O7 $(\sigma_T 4!, \mathcal{N}_E^3)$, best exchange between 4 randomly chosen variables.

O8 $(\sigma_T 5!, \mathcal{N}_E^4)$, best exchange between 5 randomly chosen variables.

O9 $(\sigma_T 6!, \mathcal{N}_E^5)$, best exchange between 6 randomly chosen variables.

O10 $(\sigma_R, \mathcal{N}_E^3)$, three consecutive random exchanges between two variables.

As described below, these 10 operators provide very poor results for the ATSP. We therefore add operator O11. The 3-opt [7] move $(\sigma_B I, \mathcal{N}_{EE}^2)$ involves selecting the best solution obtained by breaking 3 edges and rebuilding new edges in such a way that no sub-path is reversed.

In these experiments, we focus on different neighborhoods built on $\mathcal{N}_E$, which seems to be a good intermediate level of neighborhood. As in above, other neighborhoods could also be used (indeed they can also be expressed in terms of $\mathcal{N}_E$). Further works could investigate larger sets of combinations as it has been done for evolutionary algorithms in [17].

4.1   Experimental Protocol

The test instances used are from QAPLIB [3] for the QAP and from TSPLIB [19] for the ATSP. Each (algorithm, instance) pair is replicated 30 times. The sliding window length is arbitrarily set to 100 and $\epsilon = 1$ for the selection process. All runs were allowed a maximum of 40 000 iterations. We use the non-parametric paired Wilcoxon signed-rank test [5,23]. Given two algorithms A and B, the null hypothesis is: the medians of the distribution of solutions generated by A and B are equal. It is rejected with a confidence level of 95%.

4.2   Results and Discussion

Table 1 shows the results for the QAP and Table 2 those of the ATSP. The average percentage difference between the best known value (BKV) and the fair random choice for the following utility values: Uniform distribution, Quality, number of Pareto dominated solutions using distance d_1^P (ParDom d_1^P) and distance $d_{\mathcal{N}_E}^P$ (ParDom $d_{\mathcal{N}_E}^P$). The results for Robust Tabu Search for the QAP (RoTS) are also provided for comparison (of course more recent works on LS obtain better results than RoTS, e.g. [11]). Our purpose here is just to provide a simple baseline, reimplemented in COMET, and to show that our method, using a non optimized set of operators may achieve interesting results. The best results for each instance are indicated in bold font (RoTS results are not considered because they are better or equal in all but two instances). For Table 2, column ParDom10 contains the results when using only the same 10 operators used for the QAP with distance d_1^P.

For the QAP ParDom d_1^P and ParDom $d_{\mathcal{N}_E}^P$ manage to share most of the bold font results between the two of them. However, based on the Wilcoxon test, ParDom d_1^P seems to be the best algorithm when compared to uniform selection (p-value 0.02, when comparing the means) and quality-proportional selection (p-value 0.08). In contrast, results for ParDom $d_{\mathcal{N}_E}^P$ are not statistically significant. The results thus appear to show that ParDom d_1^P is better than ParDom $d_{\mathcal{N}_E}^P$ (although the null hypothesis cannot be rejected when they are compared to one another).

For the ATSP, using only the 10 operators that were used for the QAP is not effective. This is easy to explain because none of them take the cyclic nature of ATSP solutions into consideration. Adding the 3-opt operator produces a marked improvement. With only one ATSP-specific operator, the population of operators remains very biased against the ATSP. This results in a stronger improvement than with the QAP when comparing the 3 non-trivial selection methods with uniform selection. Here, ParDom d_1^P and ParDom $d_{\mathcal{N}_E}^P$ share the best results

Table 1. Experimental Results for the QAP

Instance	BKV	Uniform	Quality	ParDom d_1^P	ParDom $d_{\mathcal{N}_E}^P$	RoTS
bur26a	5426670	0.000244	0.001629	**0.000000**	0.004015	0.000000
bur26c	5426795	0.000061	**0.000000**	0.000059	0.000002	0.000000
bur26f	3782044	**0.000000**	0.000000	**0.000000**	0.000000	0.000000
chr25a	3796	11.790306	10.353003	10.189673	**9.381454**	7.093783
els19	17212548	**0.000000**	0.000000	**0.000000**	0.000000	0.000000
kra30a	88900	**0.470416**	0.488939	0.499888	0.730034	0.067267
kra30b	91420	0.110698	0.124335	**0.063881**	0.098666	0.023408
nug20	2570	**0.000000**	0.000000	0.000000	0.000000	0.000000
nug30	6124	0.12279556	0.091444	0.057478	**0.050947**	0.014370
sko42	15812	0.163167	0.148832	**0.090817**	0.115608	0.029598
sko49	23386	0.266655	0.194703	**0.186265**	0.193962	0.125203
sko56	34458	0.212781	**0.196955**	0.229497	0.292762	0.118753
tai30a	1818146	1.131385	1.178607	0.794332	**0.633736**	0.512898
tai35a	2422002	1.538266	1.391353	0.943254	**0.745479**	0.762013
tai50a	4941410	1.847374	1.815764	1.377229	**1.363935**	1.391181
tai30b	637117113	0.150888	0.107800	**0.103892**	0.129518	0.026246
tai50b	458821517	**0.173836**	0.186702	0.269760	0.537427	0.150598
wil50	48816	0.076696	**0.074429**	0.079400	0.090216	0.053425

between them but with a significant proportion for ParDom $d_{\mathcal{N}_E}^P$. In terms of statistical significance both Pareto dominance selections over 11 operators are indeed better than uniform and quality selection. When comparing ParDom d_1^P to ParDom $d_{\mathcal{N}_E}^P$, the null hypothesis can be rejected. This strongly shows that ParDom $d_{\mathcal{N}_E}^P$ is the better one for the ATSP.

Fig. 2(a) shows the cumulative frequency of applications of operators for an arbitrary run of the QAP and Fig. 3 shows the same for the ATSP. One can observe that the operators are clearly separated into two groups: one whose frequency is higher than average (0.1) and one lower. Closer examination reveals that the former is the group which improves quality the most while the latter is the one which perturbs the solutions the most without improving quality.

The operator which is selected the most is the one which has managed to consistantly improve the solution while modifying a number of variables of the solution at the same time over its last 100 applications. As can be seen in Fig. 2(a), operator O5 performs very well at the start of the search. Its performance then drops off but gradually increases back to the level of the other best performing operators, with the search stagnating at the end. This illustrates the fact that one operator is not always the best during the whole duration of the search and that it is important for the selection mechanism to be influenced by the stage of the search. In contrast, in Fig. 3, operator O11 always remains the most selected operator. This is to be expected since it is the only one which is specific to the ATSP and therefore consistantly outperforms the other operators. The operators which are good for quality in the QAP remain of interest for the ATSP.

Table 2. Experimental Results for the ATSP

Instances	BKV	ParDom10	Uniform	Quality	ParDom d_1^P	ParDom $d_{\mathcal{N}_E}^P$
br17	39	**0.000000**	**0.000000**	**0.000000**	0.000000	0.000000
p43	5620	0.202847	0.009490	0.002372	0.001779	**0.000593**
ry48p	14422	4.309620	0.661721	0.347155	0.204086	**0.168955**
ft53	6905	12.608255	1.108858	0.517982	0.186338	**0.172822**
ft70	38673	5.891276	0.749791	0.455787	0.080849	**0.048268**
ftv33	1286	7.550544	**0.000000**	**0.000000**	0.000000	0.000000
ftv35	1473	5.489930	0.495587	0.072415	0.067889	**0.031681**
ftv38	1530	6.141612	0.718954	0.429194	0.305011	**0.259259**
ftv44	1613	9.070056	0.725356	0.378177	**0.237652**	0.252118
ftv47	1776	11.006006	0.478604	0.191441	0.138889	**0.114489**
ftv55	1608	14.195688	0.972222	0.213516	0.136816	**0.093284**
ftv64	1839	17.130687	1.386623	0.781222	**0.554649**	0.580025
ftv70	1950	17.042735	1.540171	0.919658	**0.635897**	0.637607
ftv90	1579	25.429597	2.180705	1.253958	0.975301	**0.821195**
ftv100	1788	24.571216	2.839299	1.498881	1.168904	**1.047726**
ftv110	1958	31.089547	4.375213	2.667688	2.378277	**2.311883**
ftv120	2166	25.386273	3.464143	2.368421	2.136042	**2.132964**
ftv130	2307	22.831961	4.838896	2.781390	2.417281	**2.265569**
ftv140	2420	32.836088	4.720386	3.286501	3.004132	**2.965565**
ftv150	2611	32.370739	5.581514	3.993361	**3.150772**	3.292481
ftv160	2683	35.242887	5.998261	3.467511	3.473723	**3.334576**
ftv170	2755	33.393829	5.929825	3.680581	**3.097816**	3.553539
kro124p	36230	18.522587	2.327813	1.358451	1.150520	**1.055479**
rbg323	1326	7.986425	0.072901	0.012569	**0.000000**	**0.000000**
rbg358	1163	9.203210	0.005732	**0.000000**	**0.000000**	**0.000000**
rbg403	2465	1.150778	**0.000000**	**0.000000**	**0.000000**	**0.000000**
rbg443	2720	1.455882	**0.000000**	**0.000000**	**0.000000**	**0.000000**

To explain why quality-improving operators are selected more often, we can notice that modifying a solution to make it better also requires modifying its variables and thus the distance from the last solution. Quality-improving operators are therefore more likely to dominate operators whose sole action is to cause perturbations in the solution.

Fig. 2(b) displays the number of operators dominated by each operator over a subset of the search in Fig. 2(a) (shaded region). It can be observed that there are times in the search where one operator dominates almost all others, thus having the highest chance of being selected. This increased probability is reflected in the cumulative frequency graph. We can observe that in the few hundred iterations prior to iteration 7 000, no operator is considered to be much better than the others. This, in effect, reduces the selection process to a simple uniform selection. A more discriminating selection then emerges as different operators seem more suited to the following portion of the search.

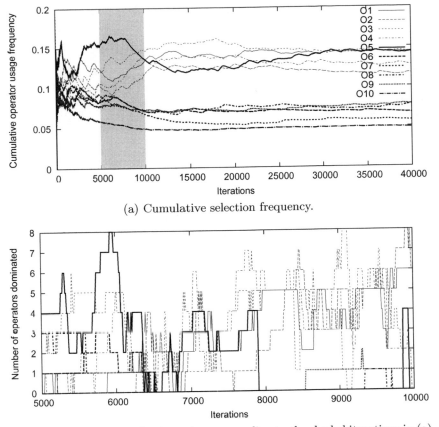

(a) Cumulative selection frequency.

(b) Number of operators dominated corresponding to the shaded iterations in (a).

Fig. 2. QAP tai50a sample run

Another interesting observation, especially in Fig. 3, is that a marked increase in the usage frequency of an operator often implies the opposite for another operator.

Fig. 4 is a snapshot of part the search on a QAP instance. It features the best value as well as the current value of the objective function plotted alongside the distance of the new solution from the path. It highlights that the control, managing the compromise between quality and diversity, is able to escape from local optima but also to reach good solutions. The correlation between our distance measure and the quality also appears clearly.

Although the detailed results are not reported in this paper, we can note that if we add 5 clones of an operator that does nothing, the performance gap widens between the proposed selection method and uniform selection. If we only use the best operator for intensification and the best operator for diversification and set ϵ to 0.1 (1 being too similar to uniform selection), the results are worse than for 10 operators.

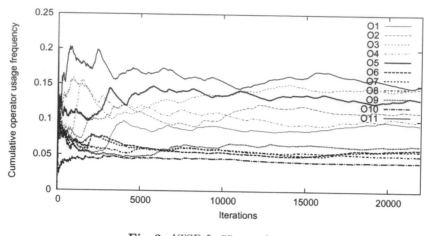

Fig. 3. ATSP ftv55 sample run

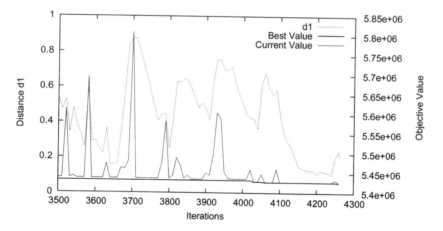

Fig. 4. Final part of the search for one run of QAP bur26a

5   Conclusion

In this paper we have presented a generic method, which manages local search operators for permutation problems, as well as two metrics to measure the distance of a solution from a subsection of the search path. This method was implemented in COMET with the objective of making it generic and easy to use from an end-user's perspective. The numerical results of the tests run on the QAP and ATSP have shown that the algorithm was effective but very much open to many improvements.

At present, the selection process favors the operators which maintain a cooperative balance between intensification and diversification, i.e., the operators in the middle of the Pareto curve. The next milestone in our work will be to introduce a reactive element to the process. Another important question to investigate is the handling of restarts (and not only small perturbation moves as used in this paper), either as an "external" restart upon reaching some condition or as an "internal" restart, that is, as an operator in itself. Obviously, testing the genericity of the approach on additional problems is also required, in particular problems with constraints and ones which require the exploration of non-feasible solutions.

Acknowledgements. This work was supported by Microsoft Research through its PhD Scholarship Programme.

References

1. Battiti, R., Brunato, M., Mascia, F.: Reactive Search and Intelligent Optimization. Springer, Heidelberg (2008) (incorporated)
2. Battiti, R., Tecchiolli, G.: The reactive tabu search. Informs Journal On Computing 6(2), 126–140 (1994)
3. Burkard, R.E., Karisch, S.E., Rendl, F.: QAPLIB – a quadratic assignment problem library. Journal of Global Optimization 10(4), 391–403 (1997)
4. Burke, E., Kendall, G., Newall, J., Hart, E., Ross, P., Schulenburg, S.: Hyper-Heuristics: an emerging direction in modern search technology. In: Handbook of Metaheuristics, pp. 457–474 (2003)
5. Chiarandini, M., Paquete, L., Preuss, M., Ridge, E.: Experiments on metaheuristics: Methodological overview and open issues. Technical Report DMF-2007-03-003, The Danish Mathematical Society (2007)
6. DaCosta, L., Fialho, Á., Schoenauer, M., Sebag, M.: Adaptive operator selection with dynamic multi-armed bandits. In: Proceedings of the 10th Annual Conference on Genetic and Evolutionary Computation, pp. 913–920. ACM, Atlanta (2008)
7. Gutin, G., Punnen, A.P.: The traveling salesman problem and its variations. Springer, Heidelberg (2002)
8. Hamadi, Y., Monfroy, E., Saubion, F.: What Is Autonomous Search? In: Hybrid Optimization: The Ten Years of CPAIOR. Springer, Heidelberg (2010)
9. Hoos, H., Stützle, T.: Stochastic Local Search: Foundations & Applications. Morgan Kaufmann Publishers Inc., San Francisco (2004)
10. Hu, B., Raidl, G.R.: Variable neighborhood descent with self-adaptive neighborhood-ordering. In: Proc. of the 7th EU Meeting on Adaptive, Self-Adaptive and Multilevel Metaheuristics (2006)
11. Hussin, M.S., Stützle, T.: Hierarchical iterated local search for the quadratic assignment problem. In: Blesa, M.J., Blum, C., Di Gaspero, L., Roli, A., Sampels, M., Schaerf, A. (eds.) HM 2009. LNCS, vol. 5818, pp. 115–129. Springer, Heidelberg (2009)
12. Ingber, L.: Adaptive simulated annealing (ASA): lessons learned. Control and Cybernetics 25, 33–54 (1996)
13. Linhares, A.: The structure of local search diversity. In: Math 2004: Proceedings of the 5th WSEAS International Conference on Applied Mathematics, pp. 1–5. World Scientific and Engineering Academy and Society (WSEAS), Stevens Point (2004)
14. Linhares, A., Yanasse, H.H.: Search intensity versus search diversity: a false trade off? Applied Intelligence 32(3), 279–291 (2010)

15. Loiola, E.M., de Abreu, N.M.M., Boaventura-Netto, P.O., Hahn, P., Querido, T.:
    A survey for the quadratic assignment problem. European Journal of Operational
    Research 176(2), 657–690 (2007)
16. Maturana, J., Saubion, F.: A compass to guide genetic algorithms. In: Rudolph, G.,
    Jansen, T., Lucas, S., Poloni, C., Beume, N. (eds.) PPSN 2008. LNCS, vol. 5199,
    pp. 256–265. Springer, Heidelberg (2008)
17. Maturana, J., Lardeux, F., Saubion, F.: Autonomous operator management for
    evolutionary algorithms. Journal of Heuristics (2010)
18. Nareyek, A.: Choosing search heuristics by non-stationary reinforcement learning.
    In: Metaheuristics: Computer Decision-Making, pp. 523–544. Kluwer Academic
    Publishers, Dordrecht (2004)
19. Reinelt, G.: TSPLIB - a traveling salesman problem library. Informs Journal On
    Computing 3(4), 376–384 (1991)
20. Robet, J., Lardeux, F., Saubion, F.: Autonomous control approach for local search.
    In: Stützle, T., Birattari, M., Hoos, H.H. (eds.) SLS 2009. LNCS, vol. 5752, pp.
    130–134. Springer, Heidelberg (2009)
21. Schiavinotto, T., Stützle, T.: A review of metrics on permutations for search land-
    scape analysis. Comput. Oper. Res. 34(10), 3143–3153 (2007)
22. Sidaner, A., Bailleux, O., Chabrier, J.J.: Measuring the spatial dispersion of evolu-
    tionary search processes: Application to walksat. In: Collet, P., Fonlupt, C., Hao,
    J.-K., Lutton, E., Schoenauer, M. (eds.) EA 2001. LNCS, vol. 2310, pp. 77–90.
    Springer, Heidelberg (2002)
23. Sprent, P.: Applied Nonparametric Statistical Methods. Chapman & Hall, London
    (1989)
24. Van Hentenryck, P., Michel, L.: Constraint-Based Local Search. The MIT Press,
    Cambridge (2005)

Transforming Mathematical Models Using Declarative Reformulation Rules

Antonio Frangioni[1] and Luis Perez Sanchez[2]

[1] Dipartimento di Informatica, Università di Pisa, Polo Universitario della Spezia,
Via dei Colli 90, 19121 La Spezia, Italy
`frangio@di.unipi.it`
[2] Dipartimento di Informatica, Università di Pisa, Largo B. Pontecorvo 3,
56127 Pisa, Italy
`perez@di.unipi.it`

Abstract. Reformulation is one of the most useful and widespread activities in mathematical modeling, in that finding a "good" formulation is a fundamental step in being able so solve a given problem. Currently, this is almost exclusively a human activity, with next to no support from modeling and solution tools. In this paper we show how the reformulation system defined in [13] allows to automatize the task of exploring the formulation space of a problem, using a specific example (the Hyperplane Clustering Problem). This nonlinear problem admits a large number of both linear and nonlinear formulations, which can all be generated by defining a relatively small set of general Atomic Reformulation Rules (ARR). These rules are not problem-specific, and could be used to reformulate many other problems, thus showing that a general-purpose reformulation system based on the ideas developed in [13] could be feasible.

1 Introduction

It is a striking discovery that while the term *reformulation* is ubiquitous in mathematics (e.g. [4, 9, 14, 16]), there are few formal definitions and theoretical characterizations of the concept. Some are limited to *syntactic* reformulations, i.e., those that can be obtained by application of algebraic rewriting rules to the elements of a given model [11]. These reformulations are capable of exploiting *syntactical structure* of the model, such as presence of particular algebraic terms in parts of its algebraic description [7]. While being very relevant, these do not include all transformations that have shown to be of practical use.

Indeed, oftentimes reformulations are based on nontrivial theorems which link the properties of two seemingly very different structures. Some notable examples are the equivalent representations of a polyhedron in terms of extreme points and faces (which underpins a number of important approaches such as decomposition methods, and has many relevant special cases such as the path formulation and the arc formulation of flows [1]) and the equivalence between the optimal solution value of a convex problem and that of its dual. These reformulations require a higher view of the concept of *structure* of a model, i.e., a *semantic structure*

C.A. Coello Coello (Ed.): LION 5, LNCS 6683, pp. 407–422, 2011.

which considers the mathematical properties of the entire represented mathematical objects as opposed to these of small parts of their algebraic description; we therefore refer to them as *semantic* reformulations. Proper definitions of reformulation capable of capturing this concept are thin on the ground.

For instance, an attempt was made in [15] by demanding that a bijection exists between the feasible regions of the two models and that one objective function is obtained by applying a monotonic univariate function to the other, which are extremely strict conditions. A view based on complexity theory was proposed in [2], but since it requires a polynomial time mapping between the problems it already cuts off a number of well-known reformulation techniques where the mapping is pseudo-polynomial [6] or even exponential in theory [3, 5, 8], but quite effective in practice. Only recently a wider attempt at formalizing the definition of formulation has been done which covers several techniques such as reformulation based on the preservation of the optimality information, changes of variables, narrowing, approximation and relaxation [11, 12].

However, a general formal definition of reformulation is not enough; the aim is to identify *classes of reformulation rules* for which automatic search in the formulation space is possible. In this sense, syntactic reformulations, being somewhat more limited in scope and akin to *rewriting systems*, may prove to have stronger properties that allow more efficient specialized search strategies. Yet, defining appropriate more general classes of semantic reformulations is also necessary in order for the system to be able to cover a large enough set of possible reformulations.

In this paper we showcase the modeling capabilities of the I-DARE (Intelligence-Driven Automatic Reformulation Engine) system developed in [13] by using a specific example (the Hyperplane Clustering Problem). This nonlinear problem admits a large number of both linear and nonlinear formulations, which can all be generated by defining a relatively small set of general *Atomic Reformulation Rules* (ARR) on a set of properly defined *structures* described in §2.

The ARRs are a key component of the I-DARE reformulation system (I-DARE(t)) [13]; it informally defines a reformulation rule based on the fact that we can transform structure A into B if and only if A's input is transformable into B's input, and B's output is transformable into A's output. ARRs are defined between two structures; in the I-DARE system, structures are classes that are derived from the hierarchy in Figure 1 where d_LeafProblem_C represents the atomic structures, and d_Block_C represents the structures that are composed of other structures. The composition of structures is controlled by the arrangement of the sub-structure's shared variables. I-DARE exploits the power of a declarative language (in particular $\mathcal{F}$LORA-2 [17]) for the definition of the structures and of the ARRs.

ARRs are divided in two classes, Algebraic ARRs (ARR$^\Sigma$) and Algorithmic ARRs (ARR$^\mathcal{A}$). The ARR$^\Sigma$s defines the transformation of the input and output using solely algebraic operation, whereas the ARR$^\mathcal{A}$s need the intervention of an algorithmic approach for reformulation the input and/or output. In this paper, for space reasons, we only concentrate on the former. Further, we will not define formally the concept of ARR, which is described in details in [13]. The aim

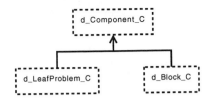

Fig. 1. I-DARE(lib) hierarchy

here is to show that a relatively small set of general (algebraic) ARRs suffice for producing a large number of both linear and nonlinear formulations for the problem. These ARRs are not problem-specific, and could be used to reformulate many other problems, thus showing that a general-purpose reformulation system based on the ideas developed in [13] could be feasible.

2 Structures

One of the main I-DARE potentialities is the capacity of declaring and relating structures that contain a specific semantic value. In this section we will focus on creating a set of global structures that will allow us to build models by combining them.

For instance we may declare some simple structures just to define a binary variable (BV), continuous variable (CV), relation and a constant.

```
d_SingleBV_C :: d_LeafProblem_C   d_SingleCV_C :: d_LeafProblem_C   d_Relation_C :: d_LeafProblem_C   d_Constant_C :: d_LeafProblem_C
[                                 [                                 [                                [
   args -> [ v = d_var ]            args -> [ v = d_var ]            args -> [ rel = d_rel]          args -> [c = d_constant]
].                                ].                                ].                               ].
```

We may also define, for example a vector of continuous variables,

```
d_VectorCV_C :: d_LeafProblem_C
[
   dim_var -> [D] ,
   args    -> [ v = d_vector(d_var , [D]) ]
].
```

Considering more complex structures, we can create for instance a product between a CV and a BV,

```
d_ProdBC_C :: d_Block_C
[
   ids   -> [bin        , cont       ] ,
   subsC -> [d_SingleBV_C , d_SingleCV_C],
   link  -> [([X],d_all) , ([Y], d_all)]
   rplR  -> [bin = 1, cont = 1]
].
```

Moreover we can declare a structure to represent a semi-continuous expression, like $f * x$, where f is a continuous structure (i.e. using only CVs) and x is a BV.

```
d_SemiContinuous_C :: d_Block_C
[
   ids   -> [ct        , bv         ] ,
   subsC -> [d_Component_C , d_SingleBV_C] ,
   link  -> [([X], d_all) , ([Y], d_all)] ,
   rplR  -> [ct = 1, bv = 1]
].
```

Considering operators like $|\cdot|$ (absolute value), we can create further structures. For instance the following leftmost structure represents $|\sum_i v_i c_i|$, where v_i is a CV and c_i is a constant, and the rightmost represents its non-vectorial version.

```
d_VAbs_C :: d_LeafProblem_C
[
  dim_var -> [D],
  args -> [
              v = d_vector(d_var, [D]),
              c = d_vector(d_constant, [D])
          ]
].
```

```
d_SAbs_C :: d_LeafProblem_C
[
  args -> [
              v = d_var,
              c = d_constant
          ]
].
```

Structures representing specific collections of constraints and/or optimization problems can also be defined, like Linear Programs (d_LP_C); Mixed-Integer Linear Programs (d_MILP_C); Semi-Assignment Constraints (d_SemiAssign_C), and Complementary Constraints (d_ProdCC_C) defined by $xy = 0$ where $x, y \geq 0$ are CVs.

```
d_LP_C :: d_LeafProblem_C
[
  dim_var -> [cols, cons],
  args -> [
    x    = d_vector(d_var, [cols]),
    c    = d_vector(d_constant, [cols]),
    A    = d_vector(d_constant, [cons, cols]),
    b    = d_vector(d_constant, [cons]),
    rels = d_vector(d_rel, [cons]),
    dir  = d_direction
  ]
].
```

```
d_MILP_C :: d_LeafProblem_C
[
  dim_var -> [cons, colsR, colsI],
  args -> [
    xr   = d_vector(d_var, [colsR]),
    xi   = d_vector(d_var, [colsI]),
    cr   = d_vector(d_constant, [colsR]),
    ci   = d_vector(d_constant, [colsI]),
    Ar   = d_vector(d_constant, [cons, colsR]),
    Ai   = d_vector(d_constant, [cons, colsI]),
    b    = d_vector(d_constant, [cons]),
    rels = d_vector(d_rel, [cons]),
    dir  = d_direction
  ]
].
```

```
d_SemiAssign_C :: d_LeafProblem_C
[
  dim_var -> [D],
  args -> [
              v = d_vector(d_var, [D])
          ]
].
```

```
d_ProdCC_C :: d_LeafProblem_C
[
  args -> [
              x = d_var,
              y = d_var
          ]
].
```

Beside those specific structures we can define a structure to represent a general constraint f =</=/>= c, where c is a constant, and f can be any component. Likewise we could define a minimization objective function,

```
d_Constraint_C :: d_Block_C
[
  ids   -> [expr        , rel         , c],
  subsC -> [d_Component_C, d_Relation_C, d_Constant_C],
  link  -> [([X], d_all) , ([], d_all) , ([], d_all)],
  rplR  -> [expr=1, rel = 1, c = 1]
].
```

```
d_OFMin_C :: d_Block_C
[
  ids   -> [expr         ],
  subsC -> [d_Component_C],
  link  -> [([X], d_all) ],
  rplR  -> [expr = 1     ]
].
```

Note that in d_Constraint_C, d_Relation_C and d_Constant_C are helper structures to put a single relation and/or a constant inside a block. Also, observe that if expr (as well as rel and c) has free indices, they must be equal to the free indices in the constraint. Therefore no internal replication is allowed (also in the case of d_OFMin_C).

2.1 Compositions

Once we have the single structures we may want to compose them to obtain more complex structures. The following structure combines two structures that share a set of variables,

```
d_Composition_C  ::  d_Block_C
[
  ids    -> [p1              , p2             ],
  subsC  -> [d_Component_C ,  d_Component_C  ],
  link   -> [([X,Y], d_all), ([X,Z], d_all)],
  rplR   -> [p1 = 1, p2 = 1]
].
```

Observe that both substructures share a set of variables (x) and have independent sub-sets of variables (Y and Z).

Another composition case can be based on the internal replication of a substructure.

```
d_IndComposition_C  ::  d_Block_C
[
  ids    -> [s             ],
  subsC  -> [d_Component_C],
  link   -> [([X], d_all) ]
].
```

Notice that the internal structure s can be replicated inside of d_IndComposition_C, implying that each replication will have an independent set of variables. Therefore, the substructures are completely separable. This fact will prove useful during reformulations, while integrating narrowings of d_IndComposition_C. We can specify a general behavior by saying that d_IndComposition_C will sum all isolated terms and concatenate all constraints.

3 Creating a Model

In this section we propose the representation of a Hyperplane Clustering Problem (HCP) using an I-DARE model. In a HCP we have a set of points $p = \{p_i \mid i \in M\} \in \mathbb{R}^D$ and we want to find the set of N hyperplanes $w = \{w_{j1}x_1 + \ldots + w_{jd}x_d = w_j^0 \mid j \in N\} \in \mathbb{R}^D$ and an assignment of points to hyperplanes such that the distances from the hyperplanes to their assigned points are minimized. HCP can be algebraically defined by the following MINLP,

$$\min \sum_{i \in M} \sum_{j \in N} |w_j p_i - w_j^0| x_{ij} \tag{3.1}$$

$$s.t. \sum_{j \in N} x_{ij} = 1 \quad \forall i \in M \tag{3.2}$$

$$\sum_{k \in D} |w_{jk}| = 1 \quad \forall j \in N \tag{3.3}$$

$$w \in \mathbb{R}^{N \times D}, \quad w^0 \in \mathbb{R}^N, \quad x \in \{0,1\}^{M \times N}$$

Note HCP has a parameter $p \in \mathbb{R}^{M \times D}$, and dimensions $N, M, D \subset \mathbb{N}$.

To model HCP we will use a combination of the previously specified structures. Note that (3.1) is an objective function containing products between absolute values and BVs; (3.2) is a semi-assignment; and (3.3) is a constraint containing absolute value operations. Hence, we can build the following model.

Dimensions, indices and Properties

```
d_dimension(D).  d_dimension(N).  d_dimension(M).
d_index(i, M).   d_index(j, N).   d_index(k, D).

p  :  d_constant.
p  :  d_property
[
    dims -> [M, D]
].

w  :  d_var.
w  :  d_property
[
    dims -> [N, D]
].
```

```
w0  :  d_var.
w0  :  d_property
[
    dims -> [D]
].

x  :  d_var.
x  :  d_property
[
    dims  -> [M, N].
    lower -> 0.
    upper -> 1
].
```

Structures

```
vabsof : d_VAbs_C
[
    args -> [
        v = $([$(w(j,k),[k]), w0(j)]),
        c = $([$(p(i,k),[k]), 1])
    ] //freeinds i,j
].

bvof : d_SingleBV_C
[
    args -> [
        v = x(i,j)
    ] //freeinds i,j
].

semiof : d_SemiContinuous_C
[
    subs  -> [absof  , bvof],
    subVP -> [[(w,w0)], [x]],
    freel -> [i,j]
].

indof : d_IndComposition_C
[
    subs  -> [semiof  ],
    subVP -> [[(w,w0,x)]]
].
```

```
of : d_OFMin_C
[
    subs  -> [indof],
    subVP -> [[(w,w0,x)]]
].

semiac : d_SemiAssign_C
[
    args -> [
        v = $(x(i,j),[j])
    ] //freeinds i
].

sabsc : d_SAbs_C
[
    args -> [
        v = w(j,k),
        c = 1,
    ] //freeinds j,k
].

rel : d_Relation_C
[
    args -> [rel = '=']
].
```

```
c : d_Constant_C
[
    args -> [c = 1]
].

indc1 : d_IndComposition_C
[
    subs  -> [sabsc],
    subVP -> [[(w)]],
    freel -> [j]
].

constraint : d_Constraint_C
[
    subs  -> [indc1, rel , c],
    subVP -> [[w], [[]], [[]]],
    freel -> [j]
].

indc2 : d_IndComposition_C
[
    subs  -> [semiac],
    subVP -> [[(x)]]
].
```

```
indc3 : d_IndComposition_C
[
    subs  -> [constraint],
    subVP -> [[(w)]]
].

cmpdc : d_Composition_C
[
    subs  -> [indc2, indc3],
    subVP -> [[[], x], [[], w]]
].

fcmp : d_Composition_C
[
    subs  -> [of, cmpdc],
    subVP -> [[(x,w), w0],
              [(x,w), []]]
].

HCP : d_Formulation
[
    root       -> fcmp,
    dimensions -> [D,M,N],
    indices    -> [i,j,k],
    properties -> [w,w0,p,x]
].
```

The diagram in Figure 2 shows the HCP formulation by representing only the name and class of the structures used, plus the relations between them:

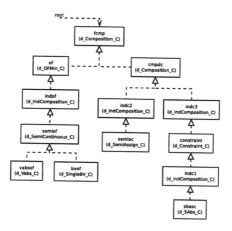

Fig. 2. HCP Formulation

4   Reformulations

In this section we will introduce some of the reformulations that can be created based on the previously defined structures. One usual goal when reformulating nonlinear problems is to remove the nonlinear elements, e.g. by adding the proper additional variables and constraints. We will use the classes d_MILP_C and d_LP_C as the main goals in the ARRs to be presented herein. Most of the reformulation rules exposed in this section were extracted from [10].

In some the cases, the generated MILP and LP will have no objective function (i.e. the cost is constant), so we will not specify the direction parameter, because it is irrelevant. In other cases, when integrating two MILPs, for instance, we will use the fact that the d_direction type is evaluated as 1 if equal to min and -1 if equal to max. So depending on the unified direction we want to produce, we will transform the cost constants of the objective function.

4.1   ProdBC to MILP

A product between a BV b and a CV $x \in [0..U]$, can be substituted by a continuous variable $w \in [0..U]$ and the constraints: $w - Ub \leq 0$, $w - x \leq 0$ and $x + Ub - w \leq U$. So we can build the following ARR$^\Sigma$.

```
d_ProdBC_to_MILP_ARR : d_ARR_Algebraic
[
    A -> d_ProdBC_C(?_, ?_),
    B -> d_MILP_C,
    indexA -> [1=(i,i1), 2=j],
    indexD -> [],
    dimRel  -> [colsI=1, colsR=2, cons=3],
    arg_map -> [
                  B..ci = 0,
                  B..cR = $([0, 1]),
                  B..Ai = $([
                              [ $(-1*_up(A..cont..v), [i1,i]) ],
                              [ $( 0, [i1,i]) ],
                              [ $( _up(A..cont..v) , [i1,i]) ]
                            ]),
                  B..Ar = $([
                              [ $(_cs([1->j=1, 0]), [i1,j]) ],
                              [ $(_cs([1->j=1, -1]), [i1,j]) ],
                              [ $( cs([1->j=0, -1]), [i1,j]) ]
                            ]),
                  B..rels = '=<',
                  B..b   = $([0, 0, _up(A..cont..v)]),
                  B..xi = lower(0),
                  B..xi = upper(1),
                  B..xr = lower(0),
                  B..xr = upper(_up(A..cont..v)),
                  B..xr = [v=1, aux=1]
                ],
    ans_map -> [
                  A..bin..v  = B..xi,
                  A..cont..v = B..xr(v)
                ]
]
```

Note that the objective function of the generated MILP has a non-zero constant for the variable that must substitute bx, and the rest of the constants are 0. The utility of this objective function constants will be seen when reformulating d_OFMin_C and d_Constraint_C. Most of the ARRs proposed in the rest of the section are written in a similar way to the one previously exposed. Therefore, we will avoid the specification of the ARR subclass (in I-DARE(t) language) due to space limitations (for the complete set of ARR consult [13]).

4.2 SAbs to Composition

If we consider a structure involving a term $|pv|$ (d_SAbs_C, p is a constant and v is a CV), this term can be reformulated so that it is differentiable, by adding two CVs $t^+, t^- \in [0.. + \infty]$; replacing $|pv|$ by $t^+ + t^-$; and adding the constraints $pv - t^+ - t^- = 0$ and $t^+t^- = 0$. This reformulation involves a linear substructure, plus a complementary constraint ($xy = 0$). So we can define an ARR^Σ that transforms d_SAbs into a composition between a d_LP_C and a d_ProdCC_C. Notice that the substitution of $|pv|$ may be expressed by defining the c constants in d_LP_C with 0 for v and 1 for t^+ and t^-.

4.3   VAbs to LP

Considering now a term $|\sum_i p_i v_i|$ we can apply a similar reformulation to the one defined in the previous section. However in this case we will consider that the term is inside a minimization function (the same way can be done for d_SAbs_C). In this case, the complementary constraint can be eliminated because we are minimizing $t^+ + t^-$, so due to the function's direction, at a global optimum, one of t^+ or t^- will have value zero. Therefore implying the complementary constraint. In this case we used a condition inside the ARR^Σ indicating that A must have a parent d_OFMin_C inside the block's tree, thus it must be inside a minimization function.

4.4   SemiContinuous to MILP

If we manage to narrow a d_SemiContinuous_C until the point of knowing that it has an d_LP_C inside, then we can easily transform d_SemiContinuous_C into a d_MILP_C. Assume the LP has the form (leftmost equation)

$$\min c^T x$$
$$\text{s.t. } Ax = b$$
$$x_i \in [0..B_i]$$

$$\min \sum_i c_i x_i y$$
$$\text{s.t. } Ax = b$$
$$x_i \in [0..B_i], \quad y \in \{0,1\}$$

$$\min \sum_i w_i$$
$$\text{s.t. } Ax = b$$
$$w_i - B_i y \le 0 \quad \forall(i)$$
$$w_i - x_i \le 0 \quad \forall(i)$$
$$x_i + B_i y - w_i \le B_i \quad \forall(i)$$
$$w_i, x_i \in [0..B_i], \quad y \in \{0,1\}$$

then the fact of multiplying this LP by a BV y (only in the objective function) creates the following MINLP (previous center equation). This MINLP can be reformulated into a MILP by applying the same mechanism used for d_ProdBC_to_MILP_ARR (cf. §4.1). We may add a CV $w_i \in [0..B_i]$ to substitute each product $x_i y$, and then add the constraints $w_i - B_i y \le 0$, $w_i - x_i \le 0$ and $x_i + B_i y - w_i \le B_i$. Resulting in the MILP present in the rightmost part of the previous equations.

4.5   ProdCC to MILP

When in presence of a complementary constraint $xy = 0$, we can substitute it by the following MILP constraints, $x - Mz \le 0$ and $y + Mz \le M$, where $z \in \{0,1\}$

and M is a sufficiently large number. Since d_ProdCC_C represents a constraint, the generated MILP will have no objective function.

4.6   SemiAssign to MILP

The semi-assignment constraint $\sum_i y_i = 1$, has trivial transformation into a MILP with no CVs.

4.7   Constraint to MILP

Having a d_Constraint_C with its substructure narrowed to a MILP, allows us to transform the whole constraint structure into a MILP. We will assume that the objective function $(\sum_i c_i x_i)$ of the inner MILP will represent, regardless of its direction, a last row of the LHS matrix of the new generated MILP. This last row is obtained by combining $\sum_i c_i x_i$ with the d_Relation_C and d_Constant_C substructures of d_Constraint_C. Therefore the resulting MILP will include all constraints of the inner MILP plus $\sum_i c_i x_i$ d_rel d_constant. Notice that an ARR^{Σ} to reformulate d_Constraint_C(d_LP_C, ?_, ?_) into d_LP_C can be created in an analogous way.

4.8   OFMin to MILP

The reformulation of a d_OFMin_C with the inner structure narrowed to a MILP is even more direct that the d_Constraint_C case, because the objective function is left as it is, except for the sign transformation depending on the inner MILP direction. Again in this case the reformulation from d_OFMin(d_LP_C) to d_LP_C can be done in an analogous way.

4.9   IndComposition to MILP

The d_IndComposition_C structure with the inner structure narrowed to MILP, can be reformulated into a single MILP, by mixing the inner replicated structures. For instance if the inner MILP has a free index j then each $MILP^j$ has an independent set of variables with respect to the other $MILP^{j'}$, with $j \neq j'$. Therefore the resulting MILP can be composed as shown in Figure 3.

The c^j constants will be multiplied by the direction of $MILP^j$ in order to unify the objective function to a minimization. Applying this composition we

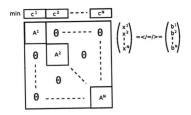

Fig. 3. Independent Composition of N MILP subproblems

can define the following ARR^Σ to reformulate a d_IndComposition_C(d_MILP_C) into a single d_MILP_C. We could define a similar reformulation to integrate several d_LP_C into a single d_LP_C.

4.10   Composition to MILP

When the composition of two structures, with shared variables (d_Composition_C), has both substructures narrowed to MILP, it can be reformulated into a single MILP. The main difficulty in this case are the common variables, for instance assume we have an inner $MILP^1$ with variables x, y and another inner $MILP^2$ with variables x, z (note that x are the shared variables), then to integrate both of them into a single MILP we need to,

- create the objective function $\min(d^1 c_x^1 + d^2 c_x^2)x + c_y^1 y + c_z^2 z$, and
- create the constraints $A_x^1 x + A_y^1 y \leq / = / \geq b^1$ and $A_x^2 x + A_z^2 z \leq / = / \geq b^2$.

where

- d^1 and d^2 are the directions of $MILP^1$ and $MILP^2$, respectively;
- c_x^1 and $_c x^2$ are the costs related with the shared variables of $MILP^1$ and $MILP^2$, respectively;
- c_y^1 and $_c z^2$ are the costs related with the independent variables of $MILP^1$ and $MILP^2$, respectively;
- A_x^1 and A_x^2 are the LHS matrices related with the shared variables of $MILP^1$ and $MILP^2$, respectively;
- A_y^1 and A_z^2 are the LHS matrices related with the independent variables of $MILP^1$ and $MILP^2$, respectively; and
- b^1 and b^2 are the RHS vectors of $MILP^1$ and $MILP^2$, respectively.

The diagram in Figure 4 is a representation of this composition.

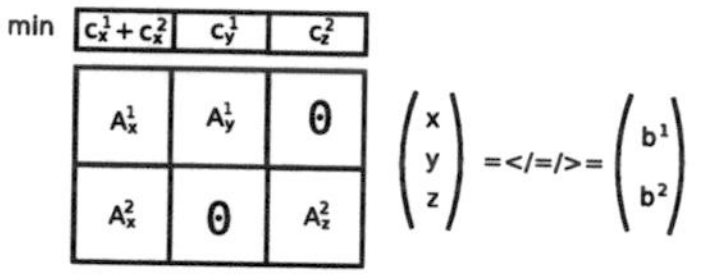

$$\min \begin{array}{|c|c|c|} \hline c_x^1 + c_x^2 & c_y^1 & c_z^2 \\ \hline \end{array}$$

$$\begin{array}{|c|c|c|} \hline A_x^1 & A_y^1 & \theta \\ \hline A_x^2 & \theta & A_z^2 \\ \hline \end{array} \begin{pmatrix} x \\ y \\ z \end{pmatrix} = </=/>= \begin{pmatrix} b^1 \\ b^2 \end{pmatrix}$$

Fig. 4. Composition of two MILP subproblems with shared variables

By using this integration mechanism we can define the ARR^Σ to reformulate d_Composition_C(d_MILP_C, d_MILP_C) into d_MILP_C. Other combinations of d_MILP_C and d_LP_C as substructures of d_Composition_C can conduct to similar ARR^Σ to treat those cases. We only have to be careful with the resulting structure, that it is always d_MILP_C except for the case when both substructures are d_LP_C (in that case the generated structure must be d_LP_C).

5   Applying the ARR^Σs to HCP

Taking the HCP formulation we defined in §3, we could apply a combination of the previously defined ARR^Σs until finally obtain a MILP formulation. To show

how the HCP formulation is modified by the application of the ARR^{Σ} we will use the HCP algebraic formulation combined with the graphical representation, pointing out the latest reformulation applied. To do so, we will dim all the model except for the structure being transformed, and the new structure obtained will have a gray background color (instead of white). We will start from the original HCP formulation (see Figure 5).

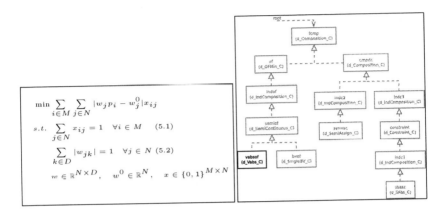

$$\min \sum_{i \in M} \sum_{j \in N} |w_j p_i - w_j^0| x_{ij}$$
$$s.t. \quad \sum_{j \in N} x_{ij} = 1 \quad \forall i \in M \quad (5.1)$$
$$\sum_{k \in D} |w_{jk}| = 1 \quad \forall j \in N \quad (5.2)$$
$$w \in \mathbb{R}^{N \times D}, \quad w^0 \in \mathbb{R}^N, \quad x \in \{0,1\}^{M \times N}$$

Fig. 5. HCP non-linear formulation

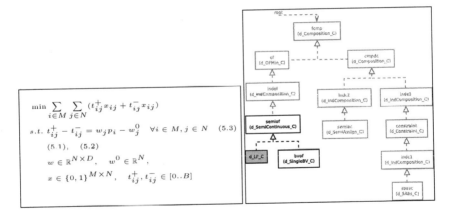

$$\min \sum_{i \in M} \sum_{j \in N} (t_{ij}^+ x_{ij} + t_{ij}^- x_{ij})$$
$$s.t. \quad t_{ij}^+ - t_{ij}^- = w_j p_i - w_j^0 \quad \forall i \in M, j \in N \quad (5.3)$$
$$(5.1), \quad (5.2)$$
$$w \in \mathbb{R}^{N \times D}, \quad w^0 \in \mathbb{R}^N,$$
$$x \in \{0,1\}^{M \times N}, \quad t_{ij}^+, t_{ij}^- \in [0..B]$$

Fig. 6. Transforming VAbs to LP

First we apply the ARR^{Σ} d_VAbs_to_LP_oncond_OFMin_ARR ($\S 4.3$) to the structure vabsof in the formulation (see Figure 6). A new structure of class d_LP_C substitutes the structure vabsof, even if in the actual reformulated model vabsof is exchanged with the track structure _tr(vabsof, d_LP_C). To keep the example simple we will only show the tail of the track structures.

We can now reformulate semicof by applying the ARR^{Σ} d_SemiContinuous_LP_SingleBV_to_MILP_ARR (cf. $\S 4.4$), see Figure 7. Observe that semicof meets the criteria for this reformulation, since it has a substructure of class d_LP_C and another of class d_SingleBV_C.

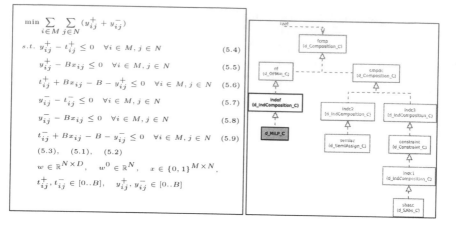

$$\min \sum_{i \in M} \sum_{j \in N} (y_{ij}^+ + y_{ij}^-)$$

$$s.t. \quad y_{ij}^+ - t_{ij}^+ \leq 0 \quad \forall i \in M, j \in N \qquad (5.4)$$

$$y_{ij}^+ - Bx_{ij} \leq 0 \quad \forall i \in M, j \in N \qquad (5.5)$$

$$t_{ij}^+ + Bx_{ij} - B - y_{ij}^+ \leq 0 \quad \forall i \in M, j \in N \qquad (5.6)$$

$$y_{ij}^- - t_{ij}^- \leq 0 \quad \forall i \in M, j \in N \qquad (5.7)$$

$$y_{ij}^- - Bx_{ij} \leq 0 \quad \forall i \in M, j \in N \qquad (5.8)$$

$$t_{ij}^- + Bx_{ij} - B - y_{ij}^- \leq 0 \quad \forall i \in M, j \in N \qquad (5.9)$$

$$(5.3), \quad (5.1), \quad (5.2)$$

$$w \in \mathbb{R}^{N \times D}, \quad w^0 \in \mathbb{R}^N, \quad x \in \{0,1\}^{M \times N},$$

$$t_{ij}^+, t_{ij}^- \in [0..B], \quad y_{ij}^+, y_{ij}^- \in [0..B]$$

Fig. 7. Transforming semi-continuous LP to MILP

Since d_IndComposition_C has a substructure of type d_MILP_C, then we can apply the ARR$^\Sigma$ d_IndComposition_MILP_to_MILP_ARR (cf. §4.9). Notice that the MILP has the same free indices semicof had in the original model ($i \in M, j \in N$), so this reformulation will integrate the $\|M\| * \|N\|$ replications of the inner MILP. Moreover, after doing this we can apply the ARR$^\Sigma$ d_OFMin_MILP_to_MILP_ARR (cf. §4.8), since of has d_MILP_C has its inner structure, see Figure 8.

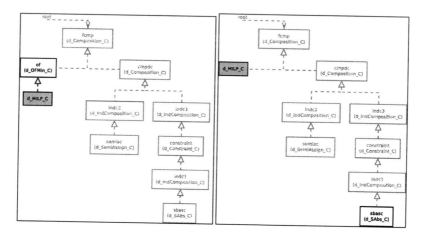

Fig. 8. IndComposition to MILP and OFMin to MILP

Figure 9 moves to the constraints part, staring by reformulating sabsc using ARR$^\Sigma$ d_SAbs_to_Composition_LP_ProdCC_ARR (cf. §4.2).

Although in this case the complexity of the model augmented a little bit, this will allow us to simplify it further by applying ARR$^\Sigma$ d_ProdCC_to_MILP_ARR (cf. §4.5) to the d_ProdCC_C structure class, that can be seen in Figure 10.

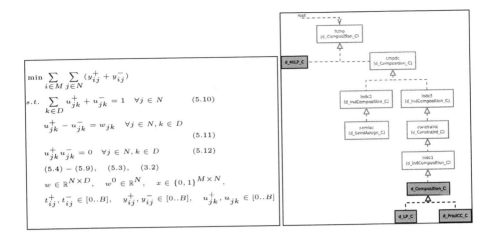

Fig. 9. Transforming SAbs to Composition

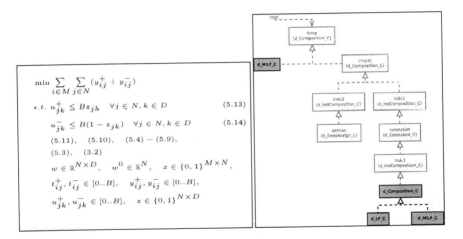

Fig. 10. Transforming ProdCC to MILP

Observe that at this point the algebraic representation is in MILP form. However, the formulation still have to undergo some other reformulations to be completely transformed into a d_MILP_C, see Figure 11.

Note how in this example the reformulations are applied only when the narrowing requisites are met. Only at that point the corresponding ARR$^\Sigma$ can be applied to transform the structure. Thanks the the deductive power of $\mathcal{F}$LORA-2, the system easily detects which ARRs it can apply to a certain (maybe intermediate) formulation, allowing the creation of all possible reformulations.

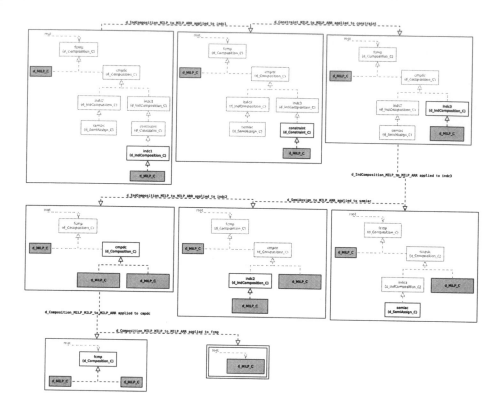

Fig. 11. Rest of the reformulations

6   Discussion

This paper show, with a relatively simple example, how the I-DARE system allows to automatically produce a large set of reformulations of a given mathematical model based on a small set of general structures and Automatic Reformulation Rules. This system matches the capabilities of the framework envisioned in [11, 12], which covers a large number of real-life problems and reformulation techniques. However, our system also allows to deal with algorithmic reformulation rules that are out of reach for frameworks based exclusively on algebraic techniques, and it makes explicit use of the concept of *structure* to allow exploiting reformulation rules based on the semantic (as opposed to purely syntactic) meaning of each block. Our system also provide explicit algorithmic notions for its definition of reformulation, exploiting the power of declarative languages, unlike e.g. that of [15]. On the other hand, [2] manages the idea of mapping functions; while in theory it has the same power that our reformulation system has, we propose a reformulation system defined over a precise modeling language, that allows us to algorithmically and algebraically deduce reformulations. I-DARE(t) offers a way of determining which structures can be reformulated and how they will be reformulated, obtaining at the end of the process valid formulations and data ready to be given to the solvers.

As the example shows, a small set of structures and reformulation rules produces a large set of possible formulations. One of the main design goals of I-DARE(t) is extensibility, i.e., the fact that one can easily define new structures and reformulation rules to cover all kinds of algebraic reformulations [11]. By doing so in a general way, i.e., defining reformulation rules for general models rather than for specific applications, the system can then exploit reformulations developed for a specific model for entirely different classes of problems. This means that a system like I-DARE could act as a central repository for reformulation techniques, allowing more effective sharing of these ideas between researchers and practitioners and fostering a positive feedback loop whereby researchers in reformulation techniques find a much wider audience for their work, while practitioners have access to sophisticated reformulation techniques that they would be unlikely to develop (or even use) themselves. We believe that such a system could have a substantial positive impact both on the research in reformulation techniques and, possibly more importantly, on the practice of the solution of mathematical models.

References

[1] Ahuja, R.K., Magnanti, T.L., Orlin, J.B.: Network Flows: Theory, Algorithms and Applications. Prentice Hall, Englewood Cliffs (1993)

[2] Audet, C., Hansen, P., Jaumard, B., Savard, G.: Links between linear bilevel and mixed 0-1 programming problems. Journal of Optimization Theory and Applications 93(2), 273–300 (1997)

[3] Ben Amor, H., Desrosiers, J., Frangioni, A.: On the Choice of Explicit Stabilizing Terms in Column Generation. Discrete Applied Mathematics 157(6), 1167–1184 (2009)

[4] Bjorkqvist, J., Westerlund, T.: Automated reformulation of disjunctive constraints in minlp optimization. Computers and Chemical Engineering 23, S11–S14 (1999)

[5] Desaulniers, G., Desrosiers, J., Solomon, M.M. (eds.): Column generation. Springer, Heidelberg (2005)

[6] Frangioni, A., Gendron, B.: 0-1 Reformulations of the Multicommodity Capacitated Network Design Problem. Discrete Applied Mathematics 157(6), 1229–1241 (2009)

[7] Frangioni, A., Gentile, C.: SDP Diagonalizations and Perspective Cuts for a Class of Nonseparable MIQP. Operations Research Letters 35(2), 181–185 (2007)

[8] Frangioni, A., Scutellà, M.G., Necciari, E.: A Multi-exchange Neighborhood for Minimum Makespan Machine Scheduling Problems. Journal of Combinatorial Optimization 8, 195–220 (2004)

[9] Judice, J., Mitra, G.: Reformulation of mathematical programming problems as linear complementarity problems and investigation of their solution methods. Journal of Optimization Theory and Applications 57(1), 123–149 (1988)

[10] Liberti, L.: Reformulation techniques in mathematical programming, in preparation. Thèse d'Habilitation à Diriger des Recherches, Université Paris IX

[11] Liberti, L.: Reformulations in mathematical programming: Definitions and systematics. RAIRO-RO 43(1), 55–86 (2009)

[12] Liberti, L., Cafieri, S., Tarissan, F.: Reformulations in mathematical programming: a computational approach. In: Abraham, A., Hassanien, A.-E., Siarry, P., Engelbrecht, A. (eds.) Foundations of Computational Intelligence. SCI, vol. 3, pp. 153–234. Springer, Berlin (2009)

[13] Sanchez, L.P.: Artificial Intelligence Techniques for Automatic Reformulation and Solution of Structured Mathematical Models. PhD thesis, University of Pisa (2010)

[14] Sherali, D., Adams, W.P.: A Reformulation-Linearization Technique for Solving Discrete and Continuous Nonconvex Problems. Kluwer Academic Publishers, Dodrecht (1999)

[15] Sherali, H.: Personal communication (2007)

[16] van Roy, T.J., Wolsey, L.A.: Solving mixed integer programming problems using automatic reformulation. Operations Research 35(1), 45–57 (1987)

[17] Yang, G., Kifer, M., Wan, H., Zhao, C.: Flora-2: User's Manual

Learning Heuristic Policies –
A Reinforcement Learning Problem

Thomas Philip Runarsson

School of Engineering and Natural Sciences
University of Iceland
tpr@hi.is

Abstract. How learning heuristic policies may be formulated as a reinforcement learning problem is discussed. Reinforcement learning algorithms are commonly centred around estimating value functions. Here a value function represents the average performance of the learned heuristic algorithm over a problem domain. Heuristics correspond to actions and states to solution instances. The problem of bin packing is used to illustrate the key concepts. Experimental studies show that the reinforcement learning approach is compatible with the current techniques used for learning heuristics. The framework opens up further possibilities for learning heuristics by exploring the numerous techniques available in the reinforcement learning literature.

1 Introduction

The current state of the art in search techniques concentrate on problem specific systems. There are many examples of effective and innovative search methodologies which have been adapted for specific applications. Over the last few decades, there has been considerable scientific progress in building search methodologies and customizing these methodologies. This has usually been achieved through hybridization with problem specific techniques for a broad scope of applications. This approach has resulted in effective methods for intricate real world problem solving environments and is commonly referred to as heuristic search. At the other extreme an exhaustive search could be applied without a great deal of proficiency. However, the search space for many real world problems is too large for an exhaustive search, making it too costly. Even when an effective search method exists, for example mixed integer programming, real world problems frequently do not scale well, see eg. [6] for a compendium of so-called NP optimization problems. In such cases heuristics offer an alternative approach to complete search.

In optimization the goal is to search for instances x, from a set of instances $\mathcal{X}$, which maximize a payoff function $f(x)$ while satisfying a number of constraints. A typical search method starts from an initial set of instances. Then, iteratively, search operators are applied locating new instances until instances with the highest payoff are reached. The key ingredient to any search methodology is thus the structure or representation of the instances x and the search operators that

C.A. Coello Coello (Ed.): LION 5, LNCS 6683, pp. 423–432, 2011.

manipulate them. The aim of developing automated systems for designing and selecting search operators or heuristics is a challenging research objective. Even when a number of search heuristics have been designed, for a particular problem domain, the task still remains of selecting those heuristics which are most likely to succeed in generating instances with higher payoff. Furthermore, the success of a heuristic will depend on a particular case in point and the current instance when local search heuristics are applied. For this reason additional heuristics may be needed to guide and modify the search heuristics in order to produce instances that might otherwise not be created. These additional are so-called meta-heuristics. Hyper-heuristics are an even more general approach where the space of the heuristics themselves is searched [4].

A recent overview on methods of automating the heuristic design process is given in [2,5]. In general we can split the heuristic design process into two parts; the first being the actual heuristic h or operator used to modify or create instance[1] $x \in \mathcal{X}$, the second part being the heuristic policy $\pi(\phi(x), h)$, the probability of selecting h, where $\phi(x)$ are features of instance x, in the simplest form $\phi(x) = x$. Learning a heuristic h can be quite tricky for many applications. For example, for a designed heuristic space $h \in \mathcal{H}$ there may exist heuristics that create instances $x \notin \mathcal{X}$ or where the constraints are not satisfied. For this reason most of the literature in automating the heuristic design process is focused on learning heuristic policies [15,12,3], although sometimes not explicitly stated.

The main contribution of this paper is on how learning heuristics can be put in a reinforcement learning framework. The approach is illustrated for the bin packing problem. The focus is on learning a heuristic policy and the actual heuristics will be kept as simple and intuitive as possible. In reinforcement learning policies are found directly or indirectly, via a value functions, using a scheme of reward and punishment. To date only a handful of examples [15,11,1,10] exist on applying reinforcement learning for learning heuristics. However, ant system have also many similarities to reinforcement learning and can be thought of as learning a heuristic policy, see [7,8]. Commonly researchers apply reinforcement learning only to a particular problem instance, not to the entire problem domain as will be attempted here.

The literature of reinforcement learning is rich in applications which can be posed as Markov decision processes, even partially observable ones. Reinforcement learning methods are also commonly referred to as approximate dynamic programming [13], since commonly approximation techniques are used to model policies. Posing the task of learning heuristic within this framework opens up a wealth of techniques for this research domain. It also may help formalize better open research questions, such as how much *human expertise* is required for the design of a satisfactory heuristic search method, for a given problem domain $f \in \mathcal{F}$?

The following section illustrates how learning heuristics may be formulated as a reinforcement learning problem. This is followed by a description of the bin-packing problem and a discussion of commonly used heuristic for this task.

---

[1] So called construction heuristics versus local search heuristics.

Section 4 illustrated how temporal difference learning can be applied to learning heuristic policies for bin packing and the results compared with classical heuristics as well as those learned using genetic programming in [12]. Both off-line and on-line bin packing are considered. The paper concludes with a summary of main results.

2   Learning Heuristics – A Reinforcement Learning Problem

In *heuristic search* the goal is to search for instances x, which maximize some payoff or objective $f(x)$ while satisfying a number of constraints set by the problem. A typical search method starts from an initial set of instances. Then, iteratively, heuristic operators h are applied locating new instances until instances with the highest payoff are reached. The key ingredients to any *heuristic search methodology* is thus; the structure or representation of the instances x, the heuristic $h \in \mathcal{H}$, the heuristic policy, π, and payoff $f(x)$. Analogously, it is possible to conceptualise heuristic search in the reinforcement learning framework [14] as pictured below. Here the characteristic features of our instance $\phi(x)$ is synonymous to a *state* in the reinforcement learning literature and likewise the heuristic h to an *action*. Each iteration of the search heuristic is denoted by t. The reward must be written as follows·

$$f(x) = \sum_{t=0}^{T} c(x_t) \qquad (1)$$

where T denotes the final iteration, found by some termination criteria for the heuristic search. For many problems one would set $c(x_t) = 0$ for all $t < T$ and then $c(x_T) = f(x_T)$. For construction heuristics T would denote the iteration for when the instance has been constructed completely. For some problems, the objective $f(x)$ can be broken down into a sum as shown in (1). One such example is the bin packing problem. Each time a bin new bin needs to be opened a reward of $c(x) = -1$ is given else 0.

It is the search agent's responsibility to update its heuristic policy based on the feedback from the particular problem instance $f \in \mathcal{F}$ being searched. Once the search has been terminated the environment is updated with a new problem instance sampled from $\mathcal{F}$. This way a new learning episode is initiated. This makes the heuristic learning problem noisy. The resulting policy learned, however, is one that maximizes the average performance over the problem domain, that is

$$\max_{\pi} \frac{1}{|\mathcal{F}|} \sum_{f \in \mathcal{F}} f(x_T^{(f)}) \qquad (2)$$

where $x_T^{(f)}$ is the solution found by the learned heuristic policy for problem f. The average performance over the problem domain corresponds to the so called value function in reinforcement learning. Reinforcement learning algorithms are commonly centred around estimating value functions.

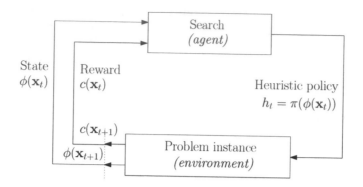

Fig. 1. Learning heuristic search as a reinforcement learning problem

3   Bin Packing

Given a bin of size $\bar{W}$ and a list of items of sizes $w_1, w_2, \ldots, w_n$, each item must be in exactly one bin,

$$\sum_{i=1}^{n} z_{i,j} = 1, \quad j = 1, \ldots, m \tag{3}$$

where $z_{i,j}$ is 1 if item i is in bin j. The bins should not overflow, that is

$$\sum_{j=1}^{m} z_{i,j} <= \bar{W} x_j, \quad i = 1, \ldots, n \tag{4}$$

where x_j is 1 when bin j is used else 0. The objective is to minimize the number of bins used,

$$\min_{z,x} \sum_{j=1}^{m} x_j \tag{5}$$

The number of decision variables are therefore $(n+1)m$ binary variables, where m is an upper estimate on the number of bins needed. The bin packing problem is a combinatorial NP-hard problem. Problem instance can be generated quite easily from this problem domain by randomly sampling the weights w from some known distribution. Previous studies, see for example [12], have used a discrete uniform distribution $U(w_{\min}, w_{\max})$ and kept the number of items n fixed. Clearly one would expect that different distributions of w would result in different heuristic policies. However, hand crafted heuristics found in the literature often do not take into account the weight distribution.

There are two heuristic approaches to solving the bin packing problem, one is on-line in nature and the other off-line. In the on-line case on must pack each weight in the order in which they arrive, that is w_1 first, then w_2 and so on.

In the off-line case the order does not matter, in essence you have been given all the weights at the same time. The common on-line heuristics include first-fit (FF) and best-fit. First-fit places the next item to be packed into the first bin j with with sufficient residual capacity or gap. The best fit searches for the bin with the smallest but sufficient capacity. Both methods have a worst case number of bins needed of $17/10\mathrm{OPT}(\boldsymbol{w}) + 2$, where OPT is the optimal number of bins. An off-line version of FF is first fit decreasing (FFD), where the weights have been placed in a non-increasing order. Using this new order the largest unpacked item is always packed into the first possible bin. A new bin is opened when needed and all bins stay open. The number of bins used by FFD is at most $11/9\mathrm{OPT}(\boldsymbol{w}) + 6/9$. A modification of FFD [9] also exists and numerous other variations may be found in the literature.

4   Illustrative Example Using Bin-Packing

Now the techniques described above are illustrated for the bin-packing problem. We consider the problem domain $\mathcal{F}$ where $n = 100$ and items $w \sim U(w_{\min}, w_{\max}) = U(20, 80)$ are to be packed into bins of size $\bar{W} = 150$. Both on-line and off-line approaches to bin packing will be studied. As with most reinforcement learning methods a value function will be approximated. The value function approximates the expected value of the solutions found $f(\boldsymbol{x}_T)$ over the domain $\mathcal{F}$. For example, in bin packing the aim is to minimize the number of bins used, and so the value function approximates the mean number of bins used by the heuristic search algorithm for the entire problem domain. The optimal policy π is the one that is greedy with respect to this value function. There are in principle two types of value functions, so-called value function V^π and heuristic-state value function Q^π. A policy greedy with respect to the heuristic-state value function is the optimal policy, defined as follows,

$$h_t^* = \operatorname*{argmax}_{h \in \mathcal{H}} Q^\pi(\phi(\boldsymbol{x}_t), h), \tag{6}$$

however, for a state value function a one step lookahead must be performed

$$h_t^* = \operatorname*{argmax}_{h \in \mathcal{H}} V^\pi(\phi(\boldsymbol{x}_{t+1}^{(h)})) \tag{7}$$

where $\phi(\boldsymbol{x}_{t+1}^{(h)})$ is the resulting (post-heuristic) state when heuristic h is applied to solution instance $\boldsymbol{x}_t$. The reinforcement learning algorithm applied here is known as temporal difference learning. The learned policy is one that minimizes the mean number of bins used for the problem domain. The temporal difference learning formula is simply

$$V(\phi(\boldsymbol{x}_t)) = V(\phi(\boldsymbol{x}_t)) + \alpha\big(V(\phi(\boldsymbol{x}_{t+1})) + c(\boldsymbol{x}_{t+1}) - V(\phi(\boldsymbol{x}_t))\big) \tag{8}$$

and

$$Q(\phi(\boldsymbol{x}_t), h_t) = Q(\phi(\boldsymbol{x}_t), h_t) + \alpha\big(Q(\phi(\boldsymbol{x}_{t+1}), h_{t+1}) + c(\boldsymbol{x}_{t+1}) - Q(\phi(\boldsymbol{x}_t), h_t)\big) \tag{9}$$

where $0 < \alpha < 1$ is a step size parameter which needs to be tuned.

The heuristic h for bin packing will usually be simply the assignment of a weight w_i to a particular bin j given that there is sufficient capacity. Two heuristics are illustrated below. In one approach the heuristic chooses in what order the weights should be assigned to a bin, but the learned heuristic policy decides in which bin the item should be placed. Another approach uses the learned heuristic policy to select the weight to be assigned but the heuristic selects the bin to place the weight in. Both of these are so-called construction heuristics. Each iteration step t corresponds to a weight being assigned and so $T = n$. The cost occurred at each iteration t can be -1 when a new bin is opened else 0. Alternatively the cost can be at all times zero, but at the terminal iteration the negative number of bins opened or even the negative mean number of gaps created.

4.1   On-Line Bin Packing

In on-line bin packing the items have independent and identically distributed (IID) weights sampled from the distribution $U(20, 80)$. The heuristic simply selects one item after the other. If the item will not fit in any open bin the heuristic will open a new bin and place the item there. If the item fits in more than one open bin then the learned heuristic policy is used to select the most appropriate bin. Having decided on the heuristic one must select an appropriate state description for the solution instance x. In this case the current total weight W_j of a bin j under consideration seems appropriate. The post-heuristic state would then simply be $W_j + w_i$, where we are considering placing the next weight w_i in bin j. This state description, which is simply the content of a single bin, is clearly not rich enough to predict how many bins will be opened in the future. Nevertheless, one may be able to predict the final gap, i.e. $(\bar{W} - W_j)$, for the bins. The cost function in this case would be the mean gap created. The temporal difference (TD) learning scheme would then be as follows:

$$V(W_j) = V(W_j) + \alpha\big(V(W_j + w_i) - V(W_j)\big) \tag{10}$$

and once all bins have been packed a final update is performed for all bins opened as

$$V(W_j) = V(W_j) + \alpha\big((\bar{W} - W_j) - V(W_j)\big) \tag{11}$$

The heuristic policy is one which is greedy with respect to V, i.e. the bin chosen for item w_i is

$$j = \operatorname*{argmin}_{j, W_j + w_i \leq \bar{W}} V(W_j + w_i) \tag{12}$$

The noise needed to drive the learning process is created by generating new problem instances or items to be packed within each new episode of the temporal difference learning algorithm[2]. Figure 2 shows the value function learned using the TD algorithm, the expected gap versus bin weight for on-line packing.

---

[2] Learning parameter $\alpha = 0.001$.

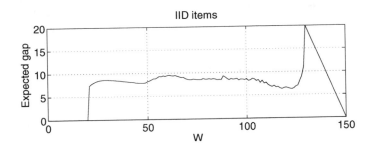

Fig. 2. Expected gap as a function of bin weight for on-line packing

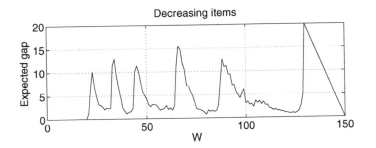

Fig. 3. Expected gap as a function of bin weight for off-line packing

There we can see that one should avoid leaving gaps of around size 10 to 20, this seems reasonable since the smallest item weight is 20 and so it becomes impossible to fill this gap. The policy learned is, therefore, very specific to the distribution of weights being packed, as one may expect. The learned heuristic policy is specialized for the problem domain in question.

4.2 Off-Line Bin Packing

The simplest off-line heuristic approach to bin-packing is FFD. We can repeat the exercise in the last section by ordering the items to be packed in a decreasing order. Now the distribution is no longer IID and the predicted gap, shown in figure 3, is completely different. The regions of smaller expected gaps seen are now smaller than those in figure 2 and as a result the performance of the learned heuristic is better. The mean number of bins packed now is 35.01, or a savings of one bin on average.

We now illustrate how the reinforcement learning framework is able to implement the histogram-matching approach in [12] for off-line packing. In [12] a part of the heuristic value function is found using genetic programming (GP). The problem instance domain is extended in their study to include $\bar{W} = 150, 75, 300$

and weight distributions $U(20, 80), U(1, 150), U(30, 70), U(1, 80)$ when $\bar{W} = 150$, for example. In this case the heuristic selects the bin with the smallest gap and the learned heuristic policy is defined as follows:

$$w^* = \operatorname*{argmin}_{w}\{GP(w)\}\big(g_t(w) - o_t(w)\big) \tag{13}$$

where $g_t(w)$ is the number of gaps of size w and $o_t(w)$ the number of unpacked items of size w. Initially there are no gaps ($g_0 = 0$) and o_0 is simply a histogram of the items weights to be packed. Integer weights are assumed such that the maximum number of bins needed for the gap histogram (g) is $\bar{W}$ and w_{max} for the items (o). New gaps are created once bins are filled and when new ones are opened. The $GP(w)$ function in the above formulation is part of the heuristic value function discovered by a genetic program and

$$GP(w) = \left\{ \frac{w_{\max} + w_{\min} + w}{\bar{W}} + 10^{-4} \right\} \tag{14}$$

was found to be very robust [12]. The same state description will now be used to learn the decision value function using temporal difference learning. Here Q values are used, where the decisions made are the weights assigned to a bin. In [12] the bins are selected in such a manner that the bin with the smallest gap is selected, i.e. best fit. However, this gap is only available as long as $g_t(w_i) > o_t(w_i)$. The same heuristic strategy for selecting a bin is used here. The temporal difference (TD) formulation is as follows:

$$Q(s_t, w_t) = Q(s_t, w_t) + \alpha\big(Q(s_{t+1}, w_{t+1}) + c_{t+1} - Q(s_t, w_t)\big) \tag{15}$$

were the state $s_t = g_t(w_t) - o_t(w_t)$ and c_{t+1} is 1 if a new bin was opened else 0. The value of a terminal state is zero as usual, i.e. $Q(s_{n+1}, \cdot) = 0$. The weight selected follows then the policy

$$w_t = \operatorname*{argmin}_{w, o_t(w) > 0} Q(s_t, w) \tag{16}$$

The value function now tells us the expected number of bins that will be opened given the current state s_t and taking decision w_t, at iteration t while following the heuristic policy π. The number of bins used is, therefore, $\sum_{i=1}^{n} c_i$. However, for more general problems the cost of a solution is not known until the complete solution has been built. So an alternative formulation is to have no cost during search ($c_t = 0, t = 1, \ldots, n$) and only at the final iteration give a terminal cost which is the number of bins used, i.e. $c_{n+1} = m$.

Figure 4 below shows the moving average number of bins used as a function of learning episodes[3]. The noise is the result from generating new problem instance at each episode. When the performance of this value function is compared with

---

[3] $\alpha = 0.01$, slightly larger than before.

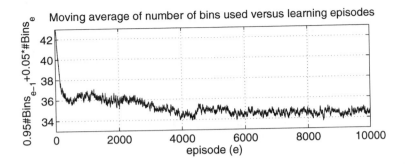

Fig. 4. Moving average number of bins used as a function of learning episodes. Each new episode generates a new problem instance, hence the noise.

the one in [12] on 100 test problems, no statistical difference in the mean number of bins used is observed, $\mu_{GP} = 34.27$ and $\mu_{TD} = 34.53$. These results improve on the off-line approach of first-fit above.

5   Summary and Conclusions

The challenging task of heuristic learning was put forth as a reinforcement learning problem. The heuristics are assumed to be designed by the human designer and correspond to the actions in the usual reinforcement learning setting. The states represent the solution instances and the heuristic policies learned decide when these heuristics should be used. The simpler the heuristic the more challenging it becomes to learn a policy. At the other extreme a single powerful heuristic may be used and so no heuristic policy need be learned (only one action possible).

The heuristic policy is found indirectly by approximating a value function, whose value is the expected performance of the algorithm over the problem domain as a function of specific features of a solution instance and the applied heuristics. It is clear that problem domain knowledge will be reflected in the careful design of features, such as we have seen in the histogram-matching approach to bin packing, and in the design of heuristics. The machine learning side is devoted to learning heuristic policies.

The exploratory noise needed to drive the reinforcement learning is introduced indirectly by generating completely new problem instance at each learning episode. This very different from the reinforcement learning approaches commonly seen in the literature for learning heuristic search, where usually only a single problem (benchmark) instance is considered. One immediate concern, which needs to be addressed, is the level of noise encountered during learning for when a new problem instance is generated at each new episode. Although the bin packing problems tackled in this paper could be solved, figure 4 shows that convergence may also be an issue. One possible solution to this may be to correlate the instances generated.

References

1. Bai, R., Burke, E.K., Gendreau, M., Kendall, G., McCollum, B.: Memory length in hyper-heuristics: An empirical study. In: IEEE Symposium on Computational Intelligence in Scheduling, SCIS 2007, pp. 173–178. IEEE, Los Alamitos (2007)
2. Burke, E.K., Hyde, M.R., Kendall, G., Ochoa, G., Ozcan, E., Woodward, J.R.: Exploring hyper-heuristic methodologies with genetic programming. Computational Intelligence, 177–201 (2009)
3. Burke, E.K., Hyde, M.R., Kendall, G., Woodward, J.: A genetic programming hyperheuristic approach for evolving two dimensional strip packing heuristics. IEEE Transactions on Evolutionary Computation (2010) (to appear)
4. Burke, E.K., Kendall, G.: Search methodologies: introductory tutorials in optimization and decision support techniques. Springer, Heidelberg (2005)
5. Burker, E.K., Hyde, M., Kendall, G., Ochoa, G., Özcan, E., Woodward, J.R.: A classification of hyper-heuristic approaches. In: Handbook of Metaheuristics, pp. 449–468 (2010)
6. Crescenzi, P., Kann, V.: A compendium of NP optimization problems. Technical report, http://www.nada.kth.se/~viggo/problemlist/compendium.html (accessed September 2010)
7. Dorigo, M., Gambardella, L.: A study of some properties of Ant-Q. In: Ebeling, W., Rechenberg, I., Voigt, H.-M., Schwefel, H.-P. (eds.) PPSN 1996. LNCS, vol. 1141, pp. 656–665. Springer, Heidelberg (1996)
8. Dorigo, M., Gambardella, L.M.: Ant colony system: A cooperative learning approach to the traveling salesman problem. IEEE Transactions on Evolutionary Computation 1(1), 53–66 (2002)
9. Floyd, S., Karp, R.M.: FFD bin packing for item sizes with uniform distributions on $[0, 1/2]$. Algorithmica 6(1), 222–240 (1991)
10. Meignan, D., Koukam, A., Créput, J.C.: Coalition-based metaheuristic: a self-adaptive metaheuristic using reinforcement learning and mimetism. Journal of Heuristics, 1–21
11. Nareyek, A.: Choosing search heuristics by non-stationary reinforcement learning. Applied Optimization 86, 523–544 (2003)
12. Poli, R., Woodward, J., Burke, E.K.: A histogram-matching approach to the evolution of bin-packing strategies. In: IEEE Congress on Evolutionary Computation, CEC 2007, pp. 3500–3507 (September 2007)
13. Powell, W.B.: Approximate Dynamic Programming: Solving the curses of dimensionality. Wiley-Interscience, Hoboken (2007)
14. Sutton, R.S., Barto, A.G.: Reinforcement learning: An introduction. The MIT Press, Cambridge (1998)
15. Zhang, W., Dietterich, T.G.: A Reinforcement Learning Approach to Job-shop Scheduling. In: Proceedings of the Fourteenth International Joint Conference on Artificial Intelligence, pp. 1114–1120. Morgan Kaufmann, San Francisco (1995)

Continuous Upper Confidence Trees

Adrien Couëtoux[1,2], Jean-Baptiste Hoock[1], Nataliya Sokolovska[1],
Olivier Teytaud[1], and Nicolas Bonnard[2]

[1] TAO-INRIA, LRI, CNRS UMR 8623,
Université Paris-Sud, Orsay, France
[2] Artelys, 12 rue du Quatre Septembre Paris, France

Abstract. Upper Confidence Trees are a very efficient tool for solving Markov Decision Processes; originating in difficult games like the game of Go, it is in particular surprisingly efficient in high dimensional problems. It is known that it can be adapted to continuous domains in some cases (in particular continuous action spaces). We here present an extension of Upper Confidence Trees to continuous stochastic problems. We (i) show a deceptive problem on which the classical Upper Confidence Tree approach does not work, even with arbitrarily large computational power and with progressive widening (ii) propose an improvement, termed double-progressive widening, which takes care of the compromise between variance (we want infinitely many simulations for each action/state) and bias (we want sufficiently many nodes to avoid a bias by the first nodes) and which extends the classical progressive widening (iii) discuss its consistency and show experimentally that it performs well on the deceptive problem and on experimental benchmarks. We guess that the double-progressive widening trick can be used for other algorithms as well, as a general tool for ensuring a good bias/variance compromise in search algorithms.

1 Introduction

Monte-Carlo Tree Search [3] is now widely accepted as a great tool for high-dimensional games [9] and high-dimensional planning [10]; its most well known variant is Upper Confidence Trees [7]. It is already adapted to continuous domains [12,11], but not for arbitrary stochastic transitions; this paper is devoted to this extension.

In section 2, we will present Progressive Widening (PW), a classical improvement of UCT in continuous or large domains. We will see that PW is not sufficient for ensuring a good behavior in the most general setting; a simple but not trivial modification, termed double-PW, is proposed and validated. Experiments (section 3) will show that this modification makes UCT for Markov Decision Processes compliant with high-dimensional continuous domains with arbitrary stochastic transition.

In all the paper, $\#E$ denotes the cardinal of a set E.

C.A. Coello Coello (Ed.): LION 5, LNCS 6683, pp. 433–445, 2011.

2   Progressive Widening for Upper Confidence Trees

Progressive strategies have been proposed in [4,2] for tackling problems with
big action spaces; they have been theoretically analyzed in [13], and used for
continuous spaces in [11,12]. We will here (i) define a variant of progressive
widening (section 2.1), (ii) show why it can't be directly applied in some cases
(section 2.2), (iii) define our version (section 2.3).

2.1   Progressive Widening

Consider an algorithm, choosing between options $O = \{o_1, o_2, \ldots, o_n, \ldots\}$ at
several time steps. More formally, this is as follows:

$R_0 = 0$
for $t = 1, t = 2, t = 3, \ldots$ **do**
    Choose an option $o_{(t)} \in O$.
    Test it: get a reward r_t.
    Cumulate the reward: $R_t = R_{t-1} + r_t$.
end for

The goal is to design the "Choose" method so that the cumulated reward
increases as fast as possible. An option (terminology of bandits) is equivalent
to an action (terminology of reinforcement learning) or a move (terminology of
games).

Many papers have been published on such problems, in particular around
upper confidence bounds [8,1]. Upper Confidence Bounds, in its simplest version,
proceeds as follows:

Upper confidence bound algorithm with parameter k.
$R_0 = 0$
for $t = 1, t = 2, t = 3, \ldots$ **do**
    Choose an option $o_{(t)} \in O$ maximizing $score_t(o)$ defined as follows:
    $totalReward_t(o) = \sum_{1 \leq l \leq t-1, o_l = o} r_l$
    $nb_t(o) = \sum_{1 \leq l \leq t-1, o_l = o} 1$
    $score_t(o) = \frac{totalReward_t(o)}{nb_t(o)+1} + k_{ucb}\sqrt{\log(t)/(nb_t(o)+1)}$ $\quad$ ($+\infty$ if $nb_t(o) = 0$)
    Test it: get a reward r_t.
    Cumulate the reward: $R_t = R_{t-1} + r_t$.
end for

Variants of the score function are termed "bandit algorithms"; there are plenty of variants of the score formula; this is essentially independent of the aspects investigated in this paper.

A trouble in many mathematical works around such problems is that the set O is usually assumed small in front of the number of iterations. More precisely, the behavior of the algorithm above is trivial for $t \leq \#O$. [14] proposed the use of a constant s such that $nb_t(o) = 0 \Rightarrow score_t(o) = s$; this is the so-called First Play Urgency algorithm. There are other specialized efficient tools for bandits used in "trees" such as rapid action value estimates [6,5]; however these tools assume some sort of homogenity between the actions at various time steps. [3,13,2] proposed progressive strategies for big/infinite sets of arms. The principle is as follows for some constants $C > 0$ and $\alpha \in]0, 1[$ (as it is independent of the algorithm used for choosing an option, within a given pool of possible options, we do not explicitly write a score function as above):

Progressive widening with constants $C > 0$ and $\alpha \in]0, 1[$.
$R_0 = 0$
for $t = 1, t = 2, t = 3, \ldots$ **do**
    Let $k = \lceil Ct^\alpha \rceil$.
    Choose an option $o_{(t)} \in \{o_1, \ldots, o_k\}$.
    Test it: get a reward r_t.
    Cumulate the reward: $R_t = R_{t-1} + r_t$.
end for

The key point is that the chosen option is restricted to have index $\leq k$; the complete set $O = \{o_1, o_2, \ldots\}$ is not allowed. This algorithm has the advantage that it is anytime: we do not have to know in advance at which value of t the algorithm will be stopped. [3] applied it successfully in the very efficient CrazyStone implementation of Monte-Carlo Tree Search [4]. Upper Confidence Tree (or Monte-Carlo Tree Search) is not a simple setting as above: when applying an option, we reach a new state; one can think of Monte-Carlo Tree Search (or UCT) as having one bandit in each possible state s of the reinforcement learning problem, for choosing between (infinitely many) options $o_1(s), o_2(s), \ldots, o_n(s), \ldots$. The algorithm is as follows, for a task in which all the reward is obtained in the final state[1]. The last line of the algorithm (returning the most simulated action from S) is often surprising for people who are not used to MCTS; it is known as much better than choosing the action with best expected reward.

---

[1] This assumption (that the reward is null except in the final state) simplifies the writing, but is not necessary for the work presented here.

Progressive Widening (PW) applied in state s with constants $C > 0$ and $\alpha \in]0, 1[$.

Input: a state s.

Output: an action.

Let $nbVisits(s) \leftarrow nbVisits(s) + 1$

and let $t = nbVisits(s)$

Let $k = \lceil Ct^\alpha \rceil$.

Choose an option $o_{(t)}(s) \in \{o_1(s), \ldots, o_k(s)\}$ maximizing $score_t(s, o)$ defined as follows:

$totalReward_t(s, o) = \sum_{1 \le l \le t-1, o_l(s)=o} r_l(s)$

$nb_t(s, o) = \sum_{1 \le l \le t-1, o_l=o} 1$

$score_t(s, o) = \frac{totalReward_t(s,o)}{nb_t(s,o)+1} + k_{ucb}\sqrt{\log(t)/(nb_t(s,o) + 1)}$

$\qquad (+\infty \text{ if } nb_t(o) = 0)$

Test it: get a state s'.

UCT algorithm with progressive widening

Input: a state S, a time budget.

Output: an action a.

Initialize: $\forall s, nbSims(s) = 0$

while Time not elapsed **do**

    // starting a simulation.

    $s = S$.

    while s is not a terminal state **do**

        Apply progressive widening in state s for choosing an option o.

        Let s' be the state reached from s when choosing action o.

        $s = s'$

    end while

    // the simulation is over; it started at S and reached a final state.

    Get a reward $r = Reward(s)$ // s is a final state, it has a reward.

    For all states s in the simulation above, let $r_{nbVisits(s)}(s) = r$.

end while

Return the action which was simulated most often from S.

It is important to keep in mind that the progressive widening algorithm is applied in each visited state; some states might be visited only once, or never, and some other states are visited very often. MCTS with progressive widening or progressive strategies is the only version of MCTS which works in continuous action spaces [12,11]; however, it was applied only with the property that applying a given action a in a given state s can lead to finitely many states only. We will see in section 2.2 that this methodology (the algorithm above) does not work as is in the case in which there is a null probability of reaching twice the same state when applying the same action in the same state (i.e. typically it does not work for stochastic transitions with continuous support).

2.2 Why It Does Not Work as Is for Randomized Transitions in Continuous Domains

We have presented UCT with progressive widening. In this section we will show why it is not sufficient for a consistent behavior (i.e. for a convergence toward maximum expected reward) in some cases, in particular when the transitions are stochastic and never lead twice to the state - one can think of the case of a Gaussian additive noise, or any other noise such that states can be reached only once.

Let us assume now that we have an infinite (discrete or continuous) domain of options. This is not too much a trouble for progressive widening: if $\alpha \in]0, 1[$, and if the o_t are a good approximation of the set of possible actions (typically, in continuous domains, we assume that the set $\{o_i; i \geq 1\}$ is dense in the set of actions and the reward has some smoothness properties), then asymptotically, good actions are explored, and all these explored actions are sampled infinitely often [13].

Let us now consider what happens if we have randomized transitions. Assume that for a state s and an action a, we can reach infinitely many transitions. Consider such a state s, and assume that we visit it infinitely often; we would like the algorithms to have two characteristics:

1. infinitely many actions $(a_1, a_2, a_3, \ldots)$ will be explored (for reducing the bias due to the choice of actions);
2. all states that can be reached from s are themselves explored infinitely often (for reducing the variance due to random exploration).

We do not have a mathematical proof that these two requirements are enough, but they look quite reasonable: in order to approximate a continuous set of actions, and unless we have an efficient pruning to a finite set of actions, we will have to explore infinitely many actions; and if we consider only finitely many possible consequences of an action whereas the real support is infinite we will miss important facts and it is hard to believe that the algorithm can be consistent.

For classical score functions, progressive widening will ensure the first property. But the second property will not be ensured, as in continuous domains with stochastic transitions, nothing ensures that we will reach twice the same state, whenever we play infinitely often a given action a in a given state s. This will be illustrated on the Trap problem later.

The following section is devoted to proposing a solution to this problem.

2.3 Proposed Solution: Double Progressive Widening

Section 2.1 has presented the known form of progressive widening, and section 2.2 has shown that in some cases it does not work (namely, when there are pairs (s, a) such that, with probability one, applying a in s infinitely often does not lead to visiting the following states infinitely often). In this section, we propose the use of a second form of progressive widening in MCTS, as follows:

Double Progressive Widening (DPW) applied in state s with constants $C > 0$ and $\alpha \in]0,1[$.
Input: a state s.
Output: a state s'.
Let $nbVisits(s) \leftarrow nbVisits(s) + 1$
and let $t = nbVisits(s)$
Let $k = \lceil Ct^\alpha \rceil$.
Choose an option $o_{(t)}(s) \in \{o_1(s), \ldots, o_k(s)\}$ maximizing $score_t(s,o)$ defined as follows:

$$totalReward_t(s,o) = \sum_{1 \leq l \leq t-1, o_l(s)=o} r_l(s)$$
$$nb_t(s,o) = \sum_{1 \leq l \leq t-1, o_{(l)}(s)=o} 1$$
$$score_t(s,o) = \frac{totalReward_t(s,o)}{nb_t(s,o)+1} + k_{ucb}\sqrt{\log(t)/(nb_t(s,o)+1)} \quad (+\infty \text{ if } nb_t(o)=0)$$

Let $k' = \lceil Cnb_t(s, o_{(t)}(s))^\alpha \rceil$
if $k' > \#Children_t(s, o_{(t)}(s))$ // progressive widening on the random part **then**
　Test option $o_{(t)}(s)$; get a new state s'
　if $s' \notin Children_t(s, o_{(t)})$ **then**
　　$Children_{t+1}(s, o_{(t)}) = Children_t(s, o_{(t)}) \cup \{s'\}$
　else
　　$Children_{t+1}(s, o_{(t)}) = Children_t(s, o_{(t)})$
　end if
else
　$Children_{t+1}(s, o_{(t)}) = Children_t(s, o_{(t)})$
　Choose s' in $Children_t(s, o_{(t)})$ // s' is chosen with probability $nb_t(s, o, s')/nb_t(s, o)$
end if

UCT algorithm with DPW
Input: a state S.
Output: an action a.
Initialize: $\forall s, nbSims(s) = 0$
while Time not elapsed **do**
　// starting a simulation.
　$s = S$.
　while s is not a terminal state **do**
　　Apply DPW in state s for choosing an option o.
　　Let s' be the state given by DPW.
　　$s = s'$
　end while
　// the simulation is over; it started at S and reached a final state.
　Get a reward $r = Reward(s)$ // s is a final state, it has a reward.
　For all states s in the simulation above, let $r_{nbVisits(s)}(s) = r$.
end while
Return the action which was simulated most often from S.

This algorithm is not so intuitive, for the second progressive widening part. The idea is as follows:

– If k' is large enough, we consider adding one more child to the pool of visited children: we simulate a transition and get a state s'. If we get an already visited child, then we go to this child; otherwise, we create a new child.

– If k' is not large enough, then we sample one of the previously seen children. As they are not necessarily equally likely, we select a child proportionally to the number of times it has been generated.

The algorithm has been designed with a "consistency" objective in mind, which is twofolds:

– Infinite visiting: we want that if a node is visited infinitely often, then we generate infinitely many children, and each of these children is itself visited infinitely often. By induction, this property ensures that all created nodes are visited infinitely often. Progressive widening and the UCB formula (or many other formulas in fact) ensure this property.

– Propagation: the average reward of any node visited infinitely often converges to a limit and this limit (for a non-terminal node) is the average reward corresponding to its children which have best asymptotic average reward. This property is ensured by the careful sampling in the progressive widening.

3   Experiments

In section 3.1 we present a deceptive problem designed specifically for pointing out the inconsistency of the classical PW approach. In section 3.2 we treat a more real problem.

3.1   Trap Problem

In this section we present the toy problem, aimed at being (i) deceptive for the simple progressive widening (ii) as simple as possible. We provide our experimental results as well.

Problem Description. This problem has been designed to clearly illustrate the weakness of the simple progressive widening. In this problem, one has to make two successive decisions, in order to maximize the reward. As we will see, the optimal policy is to make a risky move at the first step, in order to be able to obtain the maximum reward on the second (and last) step. The state will be denoted x, and is initialized at $x_0 = 0$. At each time step t the decision is denoted $d_t \in [0, 1]$. Let $R > 0$ be the noise amplitude at each time step. At a time step t, given the current state x_t and a decision d_t, we have:

$$x_{t+1} = x_t + d_t + R \times Y,$$

Y being a random variable following a uniform distribution on $[0, 1]$.

The trap problem relies on five positive real numbers: the high reward h, the average reward a, the initial ramp length l, and the trap width w. The high reward will be given if and only if we cross the trap, otherwise we obtain 0. If we stay on the initial ramp, we get the average reward. We thus define the reward function $r(\cdot)$ as follows:

$$r(x) = \begin{cases} a \text{ if } & x < l \\ 0 \text{ if } l < x < l + w \\ h \text{ if } & x > l + w \end{cases}$$

The objective is to maximize $r(x_0) + r(x_1)$, the cumulated reward. The shape of the reward function is shown in Fig.1.

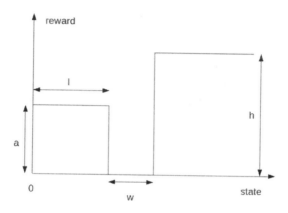

Fig. 1. Shape of the reward function: Trap problem

Experimental Results. We compare simple progressive widening and double progressive widening on the trap problem. In our experiments, we used the following settings: $a = 70$, $h = 100$, $l = 1$, $w = 0.7$, $R = 0.01$. With these parameters, the optimal behavior is to have the first decision $d_0 \in [0.7, 1]$ and $d_1 \geq 1.7 - d_0$. If one makes optimal decisions, one has an expected reward of $r^* = 170$. That is the reward toward which the Double progressive widening version of Monte Carlo Tree Search converges. However, the Simple progressive widening version does not reach this optimal reward. Worse, as we increase the computation time, it becomes less efficient, converging toward a local optimum, 140.

The mean values of the rewards are shown in Fig. 2 and the medians of the rewards are shown in Fig. 3. Each point is computed according to 100 simulations.

3.2   The Power Management Problem

In this section we present a real world problem. We show our experimental results for various settings of the power management problem.

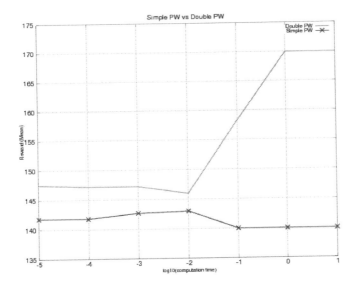

Fig. 2. Mean of the reward, for the trap problem with $a = 70$, $h = 100$, $l = 1$, $w = 0.7$, $R = 0.01$. The estimated standard deviations of the rewards are $STD_{DPW} = [13.06, 12.88, 12.88, 12.06, 14.70, 0, 0]$ for Double PW and $STD_{SPW} = [7.16, 7.16, 8.63, 9.05, 0, 0, 0]$ for Simple PW - the differences are clearly significant, where STD means standard deviation.

Problem Description. The experimental setup is an energy stock management problem. We have finitely many energy stocks (nuclear stocks, water stocks), each of them can be used to produce electricity; we can also produce electricity with classical thermal plants, that are more costly. The problem is to find the right tradeoff between

– using stocks now (in order to save up money), with the risk that later we might have peaks of demands, leading to very high costs if we do not have enough stocks.
– keeping stocks for later (in order to avoid the trouble above), with the risk that we might have too much in a stock if there is no big peak of demand.

Also, even for a fixed amount of water used from the stocks, we have to decide which stock we want to use. In particular, stocks above a given level are lost (because we have to get rid of water when the level is too high). All stocks are not equivalent: some of them have stronger inflows than others, and the used part of some stocks is transfered to other stocks whereas others are not or not to the same. One can think of a graph of reservoirs, water used in a given stock being forwarded to another stock given by the graph. In our implementation, the demand is a function of the time, determined in advance. The inflows, however, follow a lognormal distribution. Hence, they take different values from one simulation to the other.

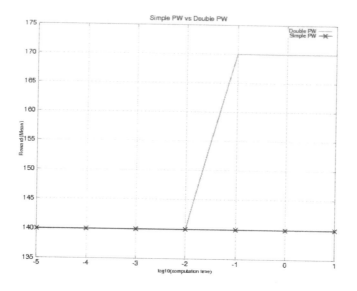

Fig. 3. Median of the reward, for the trap problem with $a = 70$, $h = 100$, $l = 1$, $w = 0.7$, $R = 0.01$

The code of the problem can be found in `http://www.lri.fr/~couetoux/stock.cpp` or requested by email.

Small Size Experiments. We consider here 2 stocks only and 5 time steps. We compare the Q-learning algorithm from the Mash project `http://mash-project.eu/`, our progressive widening Monte-Carlo Tree Search approach, a greedy algorithm only maximizing the short term, and a blind planner optimizing a sequence of decisions regardless of stock levels. Results are presented in Fig. 4. We plot the median values of cumulated reward as a function of time. It is easy to see that the Simple and Double PW MCTS achieve the best performance, compared to Blind, Greedy, and Q-learning approaches. In this particular (power management) problem, decisions are strongly associated with stock levels (states). The performance of the Blind is poor, since it makes illegal decisions rather often. Our implementation of the Q-learning suffers from the same phenomenon.

Bigger Size Experiments. We here switch to 6 stocks and 21 time steps, corresponding to 3 time steps per day, one week, with an expected increase of demand at some point during the week. Results are presented in Fig. 5. Note, that the conclusion is the same as for the small scale problem, described above: the proposed Simple and Double PW MCTS are very competitive compared to other tested methods. We did not include the results of the Q-learning. In this

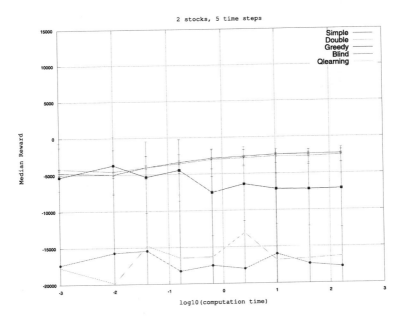

Fig. 4. The power management problem. Median values of cumulated reward with 2 stocks and 5 time steps.

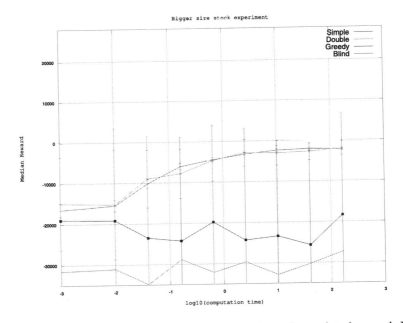

Fig. 5. The power management problem. Median values of cumulated reward. Experiments with 6 stocks and 21 time steps.

setting, the Q-learning obtained rather poor rewards. For the sake of clarity of results of other approaches, we do not show the performance of the Q-learning on Fig 5.

4   Conclusion

We have modified progressive widening in order to make it compliant with continuous domains with general noise. Experimentally, the "double-PW" modification was very efficient on deceptive problems aimed at pointing out the weaknesses of simple PW; we conjecture that for some problems, both versions are roughly equivalent, and for some problems the double PW is much better. On the other hand, on a realistic problem, the modification had disappointingly little effect. The formal proof of the consistency of the double PW (i.e. the convergence to the optimal reward for wide classes of Markov Decision Processes) has not been given and is the main further work.

References

1. Auer, P.: Using confidence bounds for exploitation-exploration trade-offs. The Journal of Machine Learning Research 3, 397–422 (2003)
2. Chaslot, G., Winands, M., Uiterwijk, J., van den Herik, H., Bouzy, B.: Progressive Strategies for Monte-Carlo Tree Search. In: Wang, P., et al. (eds.) Proceedings of the 10th Joint Conference on Information Sciences (JCIS 2007), pp. 655–661. World Scientific Publishing Co. Pte. Ltd., Singapore (2007)
3. Coulom, R.: Efficient Selectivity and Backup Operators in Monte-Carlo Tree Search. In: Ciancarini, P., van den Herik, H.J. (eds.) Proceedings of the 5th International Conference on Computers and Games, Turin, Italy (2006)
4. Coulom, R.: Computing elo ratings of move patterns in the game of go. In: Computer Games Workshop, Amsterdam, The Netherlands (2007)
5. Finnsson, H., Björnsson, Y.: Simulation-based approach to general game playing. In: AAAI 2008: Proceedings of the 23rd National Conference on Artificial Intelligence, pp. 259–264. AAAI Press, Menlo Park (2008)
6. Gelly, S., Silver, D.: Combining online and offline knowledge in UCT. In: ICML 2007: Proceedings of the 24th International Conference on Machine Learning, pp. 273–280. ACM Press, New York (2007)
7. Kocsis, L., Szepesvári, C.: Bandit based monte-carlo planning. In: Fürnkranz, J., Scheffer, T., Spiliopoulou, M. (eds.) ECML 2006. LNCS (LNAI), vol. 4212, pp. 282–293. Springer, Heidelberg (2006)
8. Lai, T., Robbins, H.: Asymptotically efficient adaptive allocation rules. Advances in Applied Mathematics 6, 4–22 (1985)
9. Lee, C.-S., Wang, M.-H., Chaslot, G., Hoock, J.-B., Rimmel, A., Teytaud, O., Tsai, S.-R., Hsu, S.-C., Hong, T.-P.: The Computational Intelligence of MoGo Revealed in Taiwan's Computer Go Tournaments. IEEE Transactions on Computational Intelligence and AI in Games (2009)
10. Nakhost, H., Müller, M.: Monte-carlo exploration for deterministic planning. In: Boutilier, C. (ed.) IJCAI, pp. 1766–1771 (2009)

11. Rolet, P., Sebag, M., Teytaud, O.: Optimal active learning through billiards and upper confidence trees in continous domains. In: Proceedings of the ECML Conference (2009)
12. Rolet, P., Sebag, M., Teytaud, O.: Optimal robust expensive optimization is tractable. In: Gecco 2009, p. 8. ACM, Montréal (2009)
13. Wang, Y., Audibert, J.-Y., Munos, R.: Algorithms for infinitely many-armed bandits. In: Advances in Neural Information Processing Systems, vol. 21 (2008)
14. Wang, Y., Gelly, S.: Modifications of UCT and sequence-like simulations for Monte-Carlo Go. In: IEEE Symposium on Computational Intelligence and Games, Honolulu, Hawaii, pp. 175–182 (2007)

Towards an Intelligent Non-stationary Performance Prediction of Engineering Systems

David J.J. Toal and Andy J. Keane

Computational Engineering and Design Group, School of Engineering Sciences,
University of Southampton, Southampton, U.K., SO17 1BJ
djjt@soton.ac.uk

Abstract. The analysis of complex engineering systems can often be expensive thereby necessitating the use of surrogate models within any design optimization. However, the time variant response of quantities of interest can be non-stationary in nature and therefore difficult to represent effectively with traditional surrogate modelling techniques. The following paper presents the application of partial non-stationary kriging to the prediction of time variant responses where the definition of the non-linear mapping scheme is based upon prior knowledge of either the inputs to, or the nature of, the engineering system considered.

Keywords: Performance Prediction, Non-Stationary Kriging.

1   Introduction

Surrogate modelling strategies are often employed in the design optimization of engineering systems as the cost of the computational simulations involved prohibit direct optimization[1]. Similarly, determining the performance of a system may require a transient analysis of that system over a period of time. A predictor of the time variant response throughout a design space would therefore reduce the overall cost of such analyzes within any design optimization.

While the variation of the quantities of interest throughout the design space may be relatively stationary the variation of these quantities within the time domain may be non-stationary as the inputs to the system are changed throughout the cycle. A transient thermo-mechanical analysis of an engine, for example, may involve the variation of prescribed temperatures and pressures over time.

The following paper demonstrates the application of partial non-stationary kriging to the prediction of the time variant response of an engineering system throughout a design space. This strategy assumes that whilst the time variant response is non-stationary the response with respect to variations in the system's design is stationary. A non-linear mapping scheme[2] is then employed within the time domain to map the non-stationary response to one which can be approximated by a stationary correlation function. The mapping scheme is represented by a piecewise linear variation in a density function which consists of several controlling knots. The placement of these knots within the time domain can reflect the inputs, and changes, within a simulation.

C.A. Coello Coello (Ed.): LION 5, LNCS 6683, pp. 446–449, 2011.

The approach of Romero et al.[3], which considers observed responses as time-correlated spatial processes could be considered as an alternative to the method proposed here. However, Romero et al. assume that data is available at discrete time steps whereas the current method is more general in nature.

2   Partial Non-stationary Kriging

Partial non-stationary kriging is a combination of stationary and non-stationary kriging. A number of variables within the model are assumed to be stationary and the remaining are assumed to be non-stationary. Given a black box function this distinction can be difficult to determine but for the purposes of time variant response prediction we assume that only time is non-stationary.

The assumption that the objective functions of two designs, x_i and x_j are similar when close together, can be modelled statistically by assuming that the correlation between two random variables, $Y(x_i)$ and $Y(x_j)$ is given by,

$$\text{Corr}[Y(x_i), Y(x_j)] = \exp\left(-\sum_{l=1}^{d} 10^{\theta^{(l)}} |x_i^{(l)} - x_j^{(l)}|^{p^{(l)}}\right.$$
$$\left. - \sum_{m=d|1}^{d+e} 10^{\theta^{(l+1)}} |f(x_i^{(m)}) - f(x_j^{(m)})|^{p^{(m)}}\right), \quad (1)$$

where d defines the number of variables assumed stationary and e defines the number assumed non-stationary. The hyperparameters θ and p determine the rate of correlation decrease and the degree of smoothness of the response respectively. The non-linear mapping is defined by,

$$f(x^{(l)}) = \int_{0}^{x^{(l)}} g(x')dx' , \quad (2)$$

where the univariate density function, $g(x)$, is represented by a piecewise linear function defined by $K + 1$ knots of density function value 10^{η_k} and position ζ_k.

The kriging hyperparameters θ and p as well as the magnitudes of the density function are unknowns and can be determined via a maximization of the concentrated log-likelihood function[4],

$$\phi = -\frac{n}{2}\ln(\hat{\sigma}^2) - \frac{1}{2}\ln(|R|) , \quad (3)$$

with the mean, $\hat{\mu}$, and variance, $\hat{\sigma}^2$, given by,

$$\hat{\mu} = \frac{1^T R^{-1} y}{1^T R^{-1} 1} \quad \text{and} \quad \hat{\sigma}^2 = \frac{1}{n}(y - 1\hat{\mu})^T R^{-1}(y - 1\hat{\mu}) , \quad (4)$$

where R denotes the correlation matrix defined by Equation 1. Although this optimization can prove costly due to the $O(n^3)$ factorization of R, the cost can be reduced and the efficiency of the optimization improved through the consideration of an adjoint of the concentrated likelihood function[5]. Predictions are made using the standard kriging predictor[4] and Equation 1.

3   Engine Casing Temperature Response Prediction

Consider the intercasing section from an aero engine shown in Figure 1(a). The parameterization of this geometry permits modifications to the casing thickness, flange thickness and height and the thrust linkage setting angle. A thermal analysis of 20 different designs, defined by a random latin hypercube, is carried out over an arbitrary operating cycle. The prescribed temperature on the inner casing surface varies linearly from T_1 to T_2, see Figure 1(b), over a cycle.

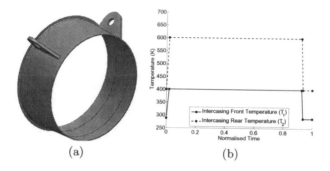

(a)   (b)

Fig. 1. Compressor intercasing geometry (a) and cycle definition (b)

The data from each of the 20 simulations is used to construct a partial non-stationary kriging model of the variation of mean temperature with changes to the design of the intercasing. As previously noted, the time domain is modelled as non-stationary with the design variables modelled as stationary. The non-linear mapping of the time domain, however, requires an appropriate parameterization of the density function. In this instance, the knots are placed in accordance with the changes to the simulation inputs i.e. the ramp points of Figure 1(b) determine where the knots are placed in the time domain. In this case knots are placed at every ramp point with an additional two knots in between.

During each thermal analysis the mean casing temperature is recorded at every time step. As there are approximately 32 time steps per analysis the surrogate model is constructed from 640 data points. Naturally the additional data in the time domain increases the cost of the correlation matrix factorization. This is countered by retaining only those temperatures corresponding to the ramp points of Figure 1(b) and a random subset in between. This produces something akin to a random sampling plan through the time domain. Using this strategy the effort required during the SQP optimization of the hyperparameters can be reduced. By considering a subset of only 15 time steps the cost of the hyperparameter optimization for this example can be reduced by 80% from 34 minutes to approximately 6.5 minutes. Given an optimized set of hyperparameters the complete dataset can then be used in the predictor.

Figure 2 demonstrates the accuracy of a partial non-stationary prediction of the time variant response of an unsampled design. The partial non-stationary

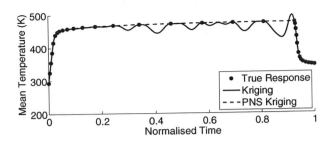

Fig. 2. Predictions of the time variant response of an unsampled intercase design

kriging model predicts the true response extremely well giving a r^2 correlation of 0.9998, and a root mean squared error of 0.48. Conversely a traditional stationary kriging prediction results in a r^2 of 0.9016 and RMSE of 12.43.

4 Conclusions

A partial non-stationary kriging strategy incorporating prior knowledge of simulation inputs to help define the non-linear mapping scheme has been presented and applied to the prediction of the time variant response of a compressor intercasing throughout a design space. The presented technique has been demonstrated to be more accurate than stationary kriging and can be modified further to predict the time variant response of other engineering systems.

Acknowledgments. The research leading to these results has received funding from the European Community's Seventh Framework Programme (FP7/2007-2013) under grant agreement no. 234344 (www.crescendo-fp7.eu).

References

1. Lim, D., Jin, Y., Ong, Y.S., Sendhoff, B.: Generalizing Surrogate-Assisted Evolutionary Computation. IEEE Transactions on Evolutionary Computation 14(3), 329–355 (2010)
2. Xiong, Y., Chen, W., Apley, D., Ding, X.: A Non-Stationary Covariance-Based Kriging Method for Metamodelling in Engineering Design. International Journal for Numerical Methods in Engineering 71(6), 733–756 (2007)
3. Romero, D.A., Amon, C., Finger, S., Verdinelli, I.: Multi-Stage Bayesian Surrogates for the Design of Time-Dependent Systems. In: Proceedings of DETC 2004, Salt Lake City, Utah, USA, September 28-October 2 (2004)
4. Jones, D.: A Taxonomy of Global Optimization Methods Based on Response Surfaces. Journal of Global Optimization 21(4), 345–383 (2001)
5. Toal, D.J.J., Bressloff, N.W., Keane, A.J., Holden, C.M.E.: The Development of a Hybridized Particle Swarm for Kriging Hyperparameter Tuning. Engineering Optimization 43(6), 675–699 (2011)

Local Search for Constrained Financial Portfolio Selection Problems with Short Sellings

Luca Di Gaspero[1], Giacomo di Tollo[2], Andrea Roli[3], and Andrea Schaerf[1]

[1] DIEGM, Università degli Studi di Udine, via delle Scienze 208,
I-33100, Udine, Italy
{l.digaspero,schaerf}@uniud.it

[2] LERIA, Université d'Angers en Pays-de-Loire, 2, Boulevard Lavoisier,
F-49045 Angers Cedex 01, France
giacomodt@gmail.com

[3] DEIS, *Alma Mater Studiorum* Università di Bologna, via Venezia 52,
I-47023 Cesena, Italy
andrea.roli@unibo.it

1 Introduction

The Portfolio Selection Problem [7] is amongst the most studied issues in finance. In this problem, given a universe of assets (shares, options, bonds, ...), we are concerned in finding out a portfolio (i.e., which asset to invest in and by how much) which minimizes the risk while ensuring a given minimum return. In the most common formulation it is required that all the asset shares have to be non-negative. Even though this requirement is a common assumption behind theoretical approaches, it is not enforced in real-markets, where the presence of short positions (i.e., assets with negative shares corresponding to speculations on falling prices) is intertwined to long positions (i.e., assets with positive shares).

Realistic portfolio selection under short selling is recently receiving more attention amongst scholars (see, e.g., [4]), but often the computational phase is not discussed properly in terms of strategies used and models at hand; furthermore most approaches are aimed in determining single risk/return points [5].

The aim of this paper is instead to draw out the whole efficient frontier in presence of realistic short selling constraints. We extend our previous metaheuristic approach for Portfolio Selection [3] to the case of short sellings and we propose a new set of benchmark instances, constructed from real-world market data. Our solver favorably compares with a Mixed Quadratic Programming formulation of the problem solved with IBM ILOG CPLEX 12.2.

2 The Portfolio Selection Problem with Short Sellings

A common hypothesis in financial theory is that information about future asset prices is contained in their current and historical prices, so that returns can be treated as stochastic variables. In the most common approach, all information about return realization and deviation risk are described by the return expected value and its variance.

C.A. Coello Coello (Ed.): LION 5, LNCS 6683, pp. 450–453, 2011.

Given a target return R and a set of n assets $\mathcal{A} = \{1, \ldots, n\}$, each of them is characterized by an expected return r_i and any pair of assets (i, j) has associated the covariance of expected returns denoted by σ_{ij}. In this setting, the formal statement of the Portfolio Selection Problem is the following:

$$\min \sum_{i=1}^{n} \sum_{j=1}^{n} \sigma_{ij} x_i x_j \tag{1}$$

subject to

$$\sum_{i=1}^{n} r_i x_i \geq R \tag{2}$$

$$\sum_{i=1}^{n} x_i = 1 \tag{3}$$

In the equations above, $x_i \in \mathbb{R}$ is the proportion of money invested in asset i. Positive values of x_i are classical investments, while negative values represent short sellings.

By solving the problem for a set of values of R it is possible to estimate the *efficient frontier* (called EF). In this work, we construct the EF by solving the problem for 100 equally distributed values of R. The investor can then choose the portfolio depending on specific risk/return requirements.

Since we are allowing short sellings, additional constraints on the portfolio in case of short sellings are imposed by law. For instance, US *regulation T* imposes to warrant the investor position with a *collateral*, to provide against the case in which the price of the sold asset rises instead of falling. In our model this is represented by introducing a risk-free asset $n + 1$ (i.e., whose variance and covariances are zero), whose return is r_{n+1}. The investment in the risk-free asset must be no less than a proportion γ of the overall sum of the short positions.

$$x_{n+1} \geq -\gamma \cdot \sum_{i=1}^{n} \min\{0, x_i\} \tag{4}$$

Furthermore, it is also imposed a limit in the total (short and long) exposure:

$$\sum_{i=1}^{n+1} |x_i| \leq 2 \tag{5}$$

Additional constraints are added to this basic formulation to encompass practical behaviors. For example, for facilitating the portfolio management and reducing its management costs, the number of assets in the portfolio should be limited. To this aim, for each asset we introduce an integer variable z_i, which is equal to 1 if the asset is in the long part of the portfolio, -1 if it is in the short part, and 0 otherwise. The constraint can be expressed as follows:

$$\sum_{i=1}^{n} |z_i| \leq k \tag{6}$$

Another constraint can be introduced to limit the share invested in single assets, e.g., to avoid excessive exposure to a specific asset or to avoid the cost of administrating very small portions of assets. Therefore, on each asset we are imposing a minimum and maximum proportion: (ε_i and δ_i respectively) allowed to be held/sold for each asset in a portfolio, so that either $x_i = 0$ or $\varepsilon_i \leq |x_i| \leq \delta_i$ ($i = 1 \ldots n$); in other words, the portion of the portfolio for a specific asset must obey the following inequality:

$$\varepsilon_i z_i \leq x_i \leq \delta_i z_i \tag{7}$$

The introduction of these constraints makes the problem a Mixed Integer Quadratic Program, which is NP-complete [1]. Moreover, it is worth to observe that in our model the collateral asset $n+1$ is not counted in the number of assets that compose the portfolio (6) nor it is bounded by constraint (7).

3 Local Search

The solution algorithm we propose is an extension of a solution method proposed for the long-only Portfolio Selection Problem [3]. In this solver local search works on the search space composed by assignments to the integer variables z_i, thus selecting the assets to be included in the portfolio and in which of its parts (long or short). Instead, the asset proportions x_i are determined by solving a Quadratic Programming (QP) subproblem that models the optimal assignment of proportions to the selected assets.

The neighborhood relation is the set union of the basic moves that manage the insertion, deletion, or replacement of an asset either in the long or in the short part. That is, a move consists in changing the value of one z_i variable, or swapping the value of two of them.

We implemented a steepest descent (SD) strategy. At each step, the algorithm searches for the best solution in the whole neighborhood and it stops when no further improvement is possible. Other local search strategies are currently under investigation but preliminary results show that SD outperforms them.

In the current implementation, the Local Search part is developed in C++ using EASYLOCAL++ [2], while QP is solved by means of the Fortran QPB routine available from the Galahad library [6].

4 Experiments

We experimented our techniques on several instances obtained from real stock markets. The instances are available and fully described at http://satt.diegm.uniud.it/portfolio/. To the purpose of comparing the efficacy of our technique, we have also developed an exact solver for the full MIQP problem formulation using the IBM ILOG CPLEX C++ library (rel. 12.2).

For space limits we report only the results on one instances, namely the biggest one in the dataset. The exact solver was stopped after 24 hours of computation

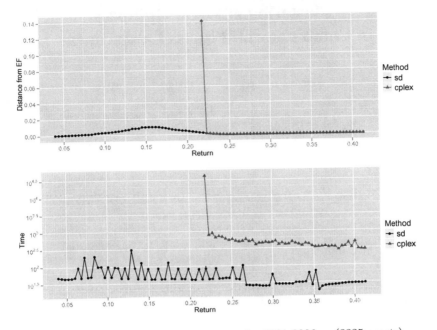

Fig. 1. Results on us_nasdaq_composite-2001-2006-m (2235 assets)

and it was able to compute only 50 frontier points. Results are shown in Figure 1. Our method clearly outperforms the exact implementation in terms of running times (1-2 orders of magnitude less) and it is consistently able to obtain the optimal results. This behavior is confirmed by the experiments on all the remaining instances.

References

[1] Bienstock, D.: Computational study of a family of mixed-integer quadratic programming problems. Mathematical Programming 74, 121–140 (1996)

[2] Di Gaspero, L., Schaerf, A.: EasyLocal++: An object-oriented framework for flexible design of local search algorithms. Software — Practice & Experience 33(8), 733–765 (2003)

[3] Di Gaspero, L., Di Tollo, G., Roli, A., Schaerf, A.: Hybrid metaheuristics for constrained portfolio selection problem. Quantitative Finance (2010) ISSN 1469-7688 (published online), doi:10.1080/14697680903460168

[4] Shannon, E., Johnson, G., Vikram, S.: An empirical analysis of 130/30 strategies: Domestic and international 130/30 strategies add value over long-only strategies. The Journal of Alternative Investments (2007)

[5] Gilli, M., Schumann, E., di Tollo, G., Cabej, G.: Constructing long-short portfolios with the omega ratio. Technical Report 08-34, Swiss Finance Institute (2008)

[6] Gould, N.I.M., Orban, D., Toint, P.L.: GALAHAD, a library of thread-safe fortram 90 packages for large-scale nonlinear optimization. ACM Transactions on Mathematical Software 29(4), 353–372 (2003)

[7] Markowitz, H.: Portfolio selection. Journal of Finance 7(1), 77–91 (1952)

Clustering of Local Optima
in Combinatorial Fitness Landscapes

Gabriela Ochoa[1], Sébastien Verel[2], Fabio Daolio[3], and Marco Tomassini[3]

[1] School of Computer Science, University of Nottingham, Nottingham, UK
[2] INRIA Lille - Nord Europe and University of Nice Sophia-Antipolis, France
[3] Information Systems Department, University of Lausanne, Lausanne, Switzerland

Abstract. Using the recently proposed model of combinatorial landscapes: *local optima networks*, we study the distribution of local optima in two classes of instances of the *quadratic assignment problem*. Our results indicate that the two problem instance classes give rise to very different configuration spaces. For the so-called real-like class, the optima networks possess a clear modular structure, while the networks belonging to the class of random uniform instances are less well partitionable into clusters. We briefly discuss the consequences of the findings for heuristically searching the corresponding problem spaces.

1 Introduction

We have recently introduced a model of combinatorial landscapes: *Local Optima Networks* (LON) [1,2], which allows the use of complex network analysis techniques [3] for studying fitness landscapes and problem difficulty in combinatorial optimization. The model, inspired by work in the physical sciences on energy surfaces[4], is based on the idea of compressing the information given by the whole problem configuration space into a smaller mathematical object which is the graph having as vertices the local optima and as edges the possible transitions between them. This characterization of landscapes as networks has brought new insights into the global structure of the landscapes studied. Moreover, some network features have been found to correlate and suggest explanations for search difficulty on the studied domains. Our initial work considered binary search spaces and the NK family of abstract landscapes [1,2]. Recently, we have turned our attention to more realistic combinatorial spaces (permutation spaces), specifically, the *Quadratic Assignment Problem* (QAP) [5]. In this article, we focus on a particular characteristic of the optima networks using the QAP, namely, the manner in which local optima are distributed in the configuration space. Several questions can be raised. Are they uniformly distributed, or do they cluster in some nonhomogeneous way? If the latter, what is the relation between objective function values within and among different clusters and how easy is it to go from one cluster to another? Knowing even approximate answers to some of these questions would be very useful to further characterize the difficulty of a class of problems and also, potentially, to devise new search heuristics or variation to known heuristics that take advantage of this information. This short paper starts to address some of these questions. The sections below summarize our methodology and preliminary results.

C.A. Coello Coello (Ed.): LION 5, LNCS 6683, pp. 454–457, 2011.

2   Methodology

The Quadratic Assignment Problem: The QAP is a combinatorial problem in which a set of facilities with given flows has to be assigned to a set of locations with given distances in such a way that the sum of the product of flows and distances is minimized. A solution to the QAP is generally written as a permutation π of the set $\{1, 2, ..., n\}$. The cost associated with a permutation π is: $C(\pi) = \sum_{i=1}^{n} \sum_{j=1}^{n} a_{ij} b_{\pi_i \pi_j}$, where n denotes the number of facilities/locations and $A = \{a_{ij}\}$ and $B = \{b_{ij}\}$ are referred to as the distance and flow matrices, respectively. The structure of these two matrices characterizes the class of instances of the QAP problem. For the statistical analysis conducted here, the two instance generators proposed in [6] for the multi-objective QAP were adapted for the single-objective QAP. The first generator produces uniformly random instances where all flows and distances are integers sampled from uniform distributions. The second generator produces flow entries that are non-uniform random values. The instances produced have the so called "real-like" structure since they resemble the structure of QAP problems found in practical applications. For the purpose of community detection, 200 instances were produced and analyzed with size 9 for the random uniform class, and 200 of size 11 for the real-like instances class. Problem size 11 is the largest one for which an exhaustive sample of the configuration space was computationally feasible in our implementation.

Local Optima Networks: In order to define the local optima network of the QAP instances, we need to provide the definitions for the nodes and edges of the network. The vertexes of the graph can be straightforwardly defined as the local minima of the landscape. In this work, we select small QAP instances such that it is feasible to obtain the nodes exhaustively by running a best-improvement local search algorithm from every configuration (permutation) of the search space. The neighborhood of a configuration is defined by the pairwise exchange operation, which is the most basic operation used by many meta-heuristics for QAP. This operator simply exchanges any two positions in a permutation, thus transforming it into another permutation. The neighborhood size is thus $|V(s)| = n(n - 1)/2$. The edges account for the transition probability between basins of attraction of the local optima. More formally, the edges reflect the total probability of going from basin b_i to basin b_j, which is the average over all $s \in b_i$ of the transition probabilities to solutions $s' \in b_j$. The reader is referred to [5] for a more detailed exposition.

We define a *Local Optima Network* (LON) as being the graph $G = (S^*, E)$ where the set of vertices S^* contains all the local optima, and there is an edge $e_{ij} \in E$ with weight $w_{ij} = p(b_i \rightarrow b_j)$ between two nodes i and j iff $p(b_i \rightarrow b_j) > 0$. Notice that since each maximum has its associated basin, G also describes the interconnection of basins.

The study of LONs for the QAP instances [5], showed that the networks are dense. Indeed, they are complete or almost complete graphs, which is inconvenient for cluster detection algorithms. Therefore, we opted for filtering out the networks edges keeping the more likely transitions (which are the most relevant for heuristic search). In filtering, we first replace the directed graph by an undirected one ($w_{ij} = \frac{w_{ij} + w_{ji}}{2}$), and then suppress all edges that have w_{ij} smaller than the value making the α-quantile ($\alpha = 0.05$

in experiments) in the weights distribution. Such a less dense network provides a coarser but clearer view of the fitness landscape backbone, and can be used for minima cluster analysis.

3   Results and Discussion

Clusters or communities in networks can be loosely defined as being groups of nodes that are strongly connected between them and poorly connected with the rest of the graph. Community detection is a difficult task, but today several good approximate algorithms are available [7]. Here we use two of them: (i) a method based on greedy modularity optimization, and (ii) a spin glass ground state-based algorithm, in order to double check the community partition results. Figure 1 shows the modularity score (Q) distribution calculated for each algorithm/instance-class. In general, the higher the value of Q of a partition, the crisper the community structure [7]. The plot indicates that the two instance classes are well separated in terms of Q, and that the community detection algorithm does not seem to have any influence on such a result.

The modularity measurements (Fig. 1) indicate that real-like instances have significantly more minima cluster structure than the class of random uniform instances of the QAP problem. This can be appreciated visually by looking at Fig. 2 where the community structures of the LON of two particular instances are depicted. Although these are the two particular cases with the highest Q values of their respective classes, the trends observed are general. For the real-like instance (Fig. 2, left) one can see that groups of minima are rather recognizable and form well separated clusters (encircled with dotted lines), which is also reflected in the high corresponding modularity value $Q = 0.79$. Contrastingly, the right plot represents a case drawn from the class of random uniform instances. The network has communities, with a $Q = 0.53$, although they are hard to represent graphically, and thus are not shown in the picture.

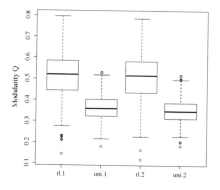

Fig. 1. Boxplots of the modularity score Q on the y-axis with respect to class problem (rl stands for real-like and uni stands for random uniform) and community detection algorithm (1 stands for fast greedy modularity optimization and 2 stands for spin glass search algorithm).

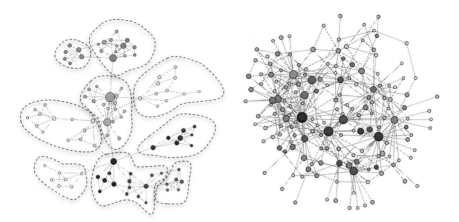

Fig. 2. Community structure of the filtered LONs for two selected instances: real-like (Left); uniform (Right). Node sizes are proportional to the corresponding basin size. Darker colors mean better fitness. The layout has been produced with the R interface to the *igraph* library.

Our analysis so far, considers only small instances, and even in this case, the local optima networks show an interesting modular structure. We argue that for larger instances, the modular structure will also be present or even increased. In order to study larger instances, we are currently exploring adequate sampling algorithms. Our results may have consequences in the design of effective heuristic search algorithms. For example, on the random uniform instances a simple local heuristic search, such as hill-climbing, should be sufficient to quickly find satisfactory solutions since they are homogeneously distributed. In contrast, in the real-like case they are much more clustered in regions of the search space. This leads to more modular optima networks and using multiple parallel searches, or large neighborhood moves would probably be good strategies. These ideas clearly deserve further investigation.

References

1. Tomassini, M., Vérel, S., Ochoa, G.: Complex-network analysis of combinatorial spaces: The NK landscape case. Phys. Rev. E 78(6), 066114 (2008)
2. Vérel, S., Ochoa, G., Tomassini, M.: Local optima networks of NK landscapes with neutrality. IEEE Trans. on Evol. Comp. (2010) (to appear)
3. Doye, J.P.K.: The network topology of a potential energy landscape: a static scale-free network. Phys. Rev. Lett. 88, 238701 (2002)
4. Newman, M.E.J.: The structure and function of complex networks. SIAM Review 45, 167–256 (2003)
5. Daolio, F., Vérel, S., Ochoa, G., Tomassini, M.: Local optima networks of the quadratic assignment problem. In: IEEE Congress on Evolutionary Computation, CEC 2010, pp. 3145–3152. IEEE Press, Los Alamitos (2010)
6. Knowles, J., Corne, D.: Instance generators and test suites for the multiobjective quadratic assignment problem. In: Fonseca, C.M., Fleming, P.J., Zitzler, E., Deb, K., Thiele, L. (eds.) EMO 2003. LNCS, vol. 2632, pp. 295–310. Springer, Heidelberg (2003)
7. Fortunato, S.: Community detection in graphs. Physics Reports 486, 75–174 (2010)

Multi-Objective Optimization with an Adaptive Resonance Theory-Based Estimation of Distribution Algorithm: A Comparative Study

Luis Martí, Jesús García, Antonio Berlanga, and José M. Molina

Universidad Carlos III de Madrid, Group of Applied Artificial Intelligence
Av. de la Universidad Carlos III, 22. Colmenarejo, Madrid 28270, Spain
{lmarti,jgherrer}@inf.uc3m.es, {aberlan,molina}@ia.uc3m.es
http://www.giaa.inf.uc3m.es/

Abstract. The introduction of learning to the search mechanisms of optimization algorithms has been nominated as one of the viable approaches when dealing with complex optimization problems, in particular with multi-objective ones. One of the forms of carrying out this hybridization process is by using multi-objective optimization estimation of distribution algorithms (MOEDAs). However, it has been pointed out that current MOEDAs have a intrinsic shortcoming in their model-building algorithms that hamper their performance.

In this work we argue that error-based learning, the class of learning most commonly used in MOEDAs is responsible for current MOEDA underachievement. We present adaptive resonance theory (ART) as a suitable learning paradigm alternative and present a novel algorithm called multi-objective ART-based EDA (MARTEDA) that uses a Gaussian ART neural network for model-building and an hypervolume-based selector as described for the HypE algorithm. In order to assert the improvement obtained by combining two cutting-edge approaches to optimization an extensive set of experiments are carried out. These experiments also test the scalability of MARTEDA as the number of objective functions increases.

1 Introduction

Multi-objective optimization has received a lot of attention by the evolutionary computation community leading to *multi-objective evolutionary algorithms* (MOEAs) (cf. [1]). A *multi-objective optimization problem* (MOP) can be expressed as the problem in which a set of M *objective functions* $f_1(\boldsymbol{x}), \ldots, f_M(\boldsymbol{x})$ with should be jointly optimized;

$$\min \ \boldsymbol{F}(\boldsymbol{x}) = \langle f_1(\boldsymbol{x}), \ldots, f_M(\boldsymbol{x}) \rangle \, ; \ \boldsymbol{x} \in \mathcal{D} \, ; \tag{1}$$

where $\mathcal{D} \subseteq \mathbb{R}^n$ is known as the *feasible set* and could be expressed as a set of restrictions over the decision set, $\mathbb{R}^n$. The image set of $\mathcal{D}$ produced by function vector $\boldsymbol{F}(\cdot)$, $\mathcal{O} \subseteq \mathbb{R}^M$, is called *feasible objective set* or criterion set (see [2] for details on notation).

C.A. Coello Coello (Ed.): LION 5, LNCS 6683, pp. 458–472, 2011.

The solution to this problem is a set of trade-off points. The adequacy of a solution can be expressed in terms of the *Pareto dominance relation* [3]. The solution of (1) is the Pareto-optimal set, $\mathcal{D}^*$; which is the subset of $\mathcal{D}$ that contains all elements of $\mathcal{D}$ that are not dominated by other elements of $\mathcal{D}$. Its image in objective space is called Pareto-optimal front, $\mathcal{O}^*$.

There is a class of MOPs that are particularly appealing because of their inherent complexity: the so-called *many-objective problems* [4]. These are problems with a relatively large number of objectives. It has been shown that "established" approaches fail to yield adequate solutions because of the exponential relation between the dimension of the objective space and the amount of resources, in particular population size, required to solve the problem correctly. Although somewhat counterintuitive and hard to visualize for a human decision maker, these problems are not uncommon in real-life engineering practice. For example, [5] details some relevant real problems of this type.

Many-objective problems have been addressed from three main fronts:

1. the design of better fitness assignment (selection) functions;
2. the use of objective reduction strategies, and;
3. application of better search (variation) methods

There has been has been a relatively large body of work on the first two issues. For example, it has been shown that the use performance indicators and some forms of relaxed Pareto dominance for the fitness assignment task allows the resulting algorithm to cope with higher dimension problems (cf. [6,7,8]). Similarly, some works have focused on the reduction of the amount of objectives to a minimum by eliminating redundant or irrelevant objectives (cf. [9,10,11]).

The third direction remains to be properly explored. Here, a viable approach is to employ cutting-edge evolutionary algorithms that could effectively deal with high-dimensional problems more efficiently.

The incorporation of learning as part of the search processes has been nominated as a viable way of dealing with that third issue [12]. There are some approaches that perform this task by providing hybrid evolutionary/machine learning method, like, for example, the *learnable evolution model* (LEM) [13]. However, these efforts seem to have been concentrated on single-objective optimization (c. f. [14,15]).

Another form of carrying out this task is to resort to *estimation of distribution algorithms* (EDAs) [16]. This is because of EDAs capacity of learning the problem structure. EDAs replace the application of evolutionary operators with the creation of a statistical model of the fittest elements of the population in a process known as model-building. This model is then sampled to produce new elements. Nevertheless, the so-called *multi-objective EDAs* (MOEDAs) [17] have not live up to their a priori expectations. This is can be attributed to the fact that most MOEDAs have limited themselves to transforming single-objective EDAs into a multi-objective formulation by including an existing multi-objective fitness assignment function.

This straightforward extrapolation has prompted the existence of a number of shortcomings en current MOEDAs. We have recognized three of them, in particular, those derived from the incorrect treatment of outstanding but isolated elements of the population (outliers); the loss of population diversity, and that too much computational effort is being spent on finding an optimal population model.

The performance issue of current MOEDAs has been traced back to the their underlying learning paradigm: the dataset-wise error minimization learning, or error-based learning, for short [18]. This class of learning, in different forms, is shared by most machine learning algorithms. It implies that model is tuned in order to minimize a global error measured across the dataset. In this type of learning isolated data is not taken into account because of their little contribution to the overall error and therefore they do not take an active part of learning process. This assertion is in part supported by the fact that most the approaches that had a better performance in comparative experiments like [18] do not exactly conform to the error-based scheme. That is why, other learning paradigms should be assessed.

Adaptive resonance theory (ART) [19] is a theory of human cognition that has seen a realization as a family of neural networks. It relies on a learning scheme denominated match-based learning and on intrinsic topology self-organization. These features make it interesting as a case study as model-building approach. Match-based learning equally weights isolated and clustered data [20], and, therefore, the algorithm does not disregard outliers. Similarly, self-organization makes possible the on-the-fly determination the model complexity required to correctly represent the data set, thus eliminating the need of an external algorithm for that task.

In this work we argue that error-based learning, the class of learning most commonly used in MOEDAs is responsible for current MOEDA underachievement. We discuss in detail ART-based learning as a viable alternative and present a novel algorithm called *multi-objective ART-based EDA* (MARTEDA) that uses a Gaussian ART neural network [21] for model-building and an hypervolume-based selection as described for the hypervolume estimation algorithm for multiobjective optimization (HypE) [8]. We experimentally show that thanks to MARTEDA's novel model-building approach and an indicator-based population ranking the algorithm it is able to outperform similar MOEDAs and MOEAs. Elements of MARTEDA have been discussed in some preliminary works [22], but it has not yet been presented in detail.

The remaining part of the work proceeds as we discuss the model-building issue. Following that we describe the Gaussian ART network that is used as start point for our model-building algorithm. Subsequently, MARTEDA is introduced, describing how the HypE selection and Gaussian ART are blended together in a MOEDA framework. Section 5 presents and discusses the results of the comparative experiments involving MARTEDA and a selection of other current state-of-the-art algorithms dealing with a set of community accepted problems. These problems are configured with an progressive number of objectives (3, 6, 9 and 12) in order to assess the performance of our proposal in the context of many-objective optimization. Finally, some conclusive remarks and future lines of research are outlined.

2   The Model-Building Issue

Notwithstanding the diverse efforts dedicated to providing usable model-building methods for EDAs the nature of the problem itself has received relatively low attention. An analysis of the results yielded by current multi-objective EDAs and their scalability with regard to the number of objective leads the identification of certain issues that might be hampering the obtention of substantially better results with regard to other evolutionary approaches.

Data outliers issue is a good example of insufficient comprehension of the nature of the model-building problem. In machine-learning practice, outliers are handled as noisy, inconsistent or irrelevant data. Therefore, outlying data is expected to have little influence on the model or just to be disregarded.

However, that behavior is not adequate for model-building. In this case, is known beforehand that all elements in the data set should be take into account as they represent newly discovered or candidate regions of the search space and therefore must be explored. Therefore, these instances should be at least equally represented by the model and perhaps even reinforced.

Another weakness of most MOEDAs (and most EDAs, for that matter) is the loss of population diversity. This is a point that has already been made, and some proposals for addressing the issue have been laid out [23,24,25,26,27]. This loss of diversity can be traced back to the above outliers issue of model-building algorithms.

The incorrect treatment of outliers and the loss of population diversity can be attributed the error-based learning approaches that take place in the most MOEDAs. Error-based learning is rather common in machine learning algorithms. It implies that model topology and parameters are tuned in order to minimize a global error measured across the learning data set. This type of learning isolated data is not taken into account because of their little contribution to the overall error and therefore they do not take an active part of learning process. In the context of many problems this behavior makes sense, as isolated data can be interpreted as spurious, noisy or invalid data.

That is not the case of model-building, as we have already argued. In model-building all data is equally important and, furthermore, isolated data might have a bigger significance as they represent unexplored zones of the current optimal search space. This assessment is supported by the fact that most the approaches that had a better performance do not follow the error-based scheme, like the k-means algorithm, randomized leader algorithm and the growing neural gas network [18]. That is why, perhaps another classes of learning, like instance-based learning or match-based learning would yield a sizable advantage.

3   Model Building with Adaptive Resonance Theory

Adaptive Resonance Theory (ART) neural networks are capable of fast, stable, on-line, unsupervised or supervised, incremental learning, classification, and prediction following a match-based learning scheme [19]. Match-based learning is complementary to error-based learning. During training, ART networks adjust previously-

learned categories in response to familiar inputs, and create new categories dynamically in response to inputs different enough from those previously seen. A
vigilance test allows to regulate the maximum tolerable difference between any
two input patterns in a same category. It has been pointed out that ART networks
are not suitable for some classes of classical machine-learning applications [20],
however, what is an inconvenience in that area is a feature in our case.

3.1   Gaussian ART for Model-Building

There are many variations of ART networks. Among them, the Gaussian ART
[21] is most suitable for model-building since it capable of handling continuous
data. The result of applying Gaussian ART is a set of nodes each representing
a local Gaussian density. These nodes can be combined as a Gaussian mixture
that can be used to synthesize new individuals.

Gaussian ART creates classes of similar inputs. A *match tracking mechanism*
induces the creation of more specific classes when the prediction of the network
differs from the expected output at some degree.

Gaussian ART has a layer of afferent or input nodes, F1, and a classification
layer, F2. The F2 layer stores classes of inputs. Its activation is a combined
measure of the similarity of the input and the prototype of each class, and the
size of the given class.

For the model-building task we have modified the original formulation of the
network to make it more suited for the task. When an input $x \in \mathbb{R}^n$ is presented
to the input layer it is propagated to the F2 layer. F2 has N^* units, with N of
them committed. Each committed unit models a local density of the input space
using Gaussian receptive fields with mean μ_j and standard deviation σ_j. A unit
is activated if it satisfies the match criterion. That is, the match function,

$$G_j = \exp\left(-\frac{1}{2}\sum_{i=1}^{n}\left(\frac{x_i - \mu_{ji}}{\sigma_{ji}}\right)^2\right), \, j = 1, \ldots, N, \tag{2}$$

must be greater than the F2 vigilance parameter, ρ; according to this, the input
strength of a unit is computed as

$$g_j = \begin{cases} \frac{\eta_j}{\prod_{i=1}^{n}\sigma_{ji}}G_j, & \text{if } G_j > \rho \\ 0 & \text{otherwise} \end{cases}, \rho > 0, \tag{3}$$

where η_j is a measure of the unit a priori activation probability. This is different
from the original Gaussian ART network where only one unit was allowed be
active after an input presentation.

After the presentation of an input, if no F2 unit is active, then an uncommitted
unit must be committed. The task of detecting when an input is not sufficiently
coded in F2 is accomplished by the F2 gain control, G, that fires if no committed
units are active. The signal

$$\Gamma = \begin{cases} 1 \text{ if } \max_{j=1,\ldots,N} g_j = 0 \\ 0 \text{ otherwise} \end{cases} \tag{4}$$

is used to commit an uncommitted unit.

The activation of each unit is then calculated normalizing the unit's input strength,

$$v_j = \frac{g_j}{\sum_{l=1}^{N} g_l}.$$ (5)

As other ART networks, this model is an on-line learning neural network. Therefore, all adaptation processes have local rules. In F2, μ_j and σ_j are updated using a learning rule based on the gated steepest descent learning rule. η_j is updated to represent the cumulative category activation,

$$\eta_j(t+1) = \eta_j(t) + v_j,$$ (6)

and, therefore, the amount of training that has taken place in the jth unit. The use of η_j equally weights inputs over time with the intention to measure their sample statistics.

Learning the first and second moments of the input is

$$\mu_{ji}(t+1) = \left(1 - \eta_j^{-1} v_j\right) \mu_{ji}(t) + \eta_j^{-1} v_j x_i,$$ (7)

$$\lambda_{ji}(t+1) = \left(1 - \eta_j^{-1} v_j\right) \lambda_{ji}(t) + \eta_j^{-1} v_j x_i^2.$$ (8)

The standard deviation,

$$\sigma_{ji}(t+1) = \sqrt{\lambda_{ji}(t+1) - \mu_{ji}(t+1)^2},$$ (9)

is calculated using (7) and (8).

Gaussian ART is initialized with all units uncommitted ($N = 0$). Learning takes place in active ($v_j > 0$) F2 units following (7)–(9). However if no F2 units becomes active an uncommitted unit is committed and therefore N is incremented. The new unit is indexed by N and initialized with $v_N = 1$, $\eta_N = 0$. Learning will proceed as usual but a constant γ_i^2 will be added to each λ_{Ni} to set $\sigma_{Ni} = \gamma_i$. The value of γ_i has a direct impact on the quality of learning. A larger γ_i slows down learning in its corresponding input feature but warranties a more robust convergence.

The local Gaussian densities resulting from the described algorithm can be combined to synthesize a Gaussian mixture. This Gaussian mixture is then used can be used by the EDA to generate new individuals.

4   Multi-Objective ART-Based EDA

The multi-objective ART-based EDA (MARTEDA) is a MOEDA that uses the previously described Gaussian ART network as its model-building algorithm. Although it intends to deal with the issues raised by the previous discussion, it was also designed with scalability in mind, since it is expected to cope with many-objective problems. It also exhibits an elitist behavior, as it has proved itself a very advantageous property. Finally, thanks to the combination of fitness assignment and model-building it promotes diversity preservation.

MARTEDA maintains a population, $\mathcal{P}_t$, of n_{pop} individuals; where t is a given iteration. The algorithm's workflow is similar to other EDAs. It starts with a random initial population $\mathcal{P}_0$ of individuals.

At a given iteration t the algorithm determines the set $\hat{\mathcal{P}}_t$ containing the best $\lfloor \alpha |\mathcal{P}_t| \rfloor$ elements.

$$\left|\hat{\mathcal{P}}_t\right| = \lfloor \alpha |\mathcal{P}_t| \rfloor = \lfloor \alpha n_{\text{pop}} \rfloor . \tag{10}$$

Different selection strategies can be applied. However, indicator-based selection seems to have a superior performance in complex and many-objective problems. The hypervolume-based selection have many theoretical features, like being the only indicator that have the properties of a metric and the only to be strictly Pareto monotonic [28] but has the drawback of being computationally intensive to compute.

A lot of research has focused on improving the computational complexity of this indicator [29, 30, 31, 32]. The exact computation of the algorithm has been shown to be #**P**-hard [33] in the number of objectives. #**P** problems are the analogous of **NP** for counting problems [34]. Therefore, all algorithms calculating a hypervolume must have an exponential runtime with regard to the number of objectives if **P**≠**NP**, something that seems to be true [35].

The HypE algorithms attempt to circumvent this problem by estimating the value of the hypervolume by means of a Monte Carlo simulation. The detailed description of this procedure is out of the scope of this paper, and, therefore we invite the interested reader to consult the corresponding paper.

A Gaussian ART network is then trained using $\hat{\mathcal{P}}_t$ as its training data set. In order to have a controlled relation between size of $\hat{\mathcal{P}}_t$ and the maximum size of the network, N_{max}, these two sizes are bound by the rate $\gamma \in (0, 1]$,

$$N_{\text{max}} = \left\lceil \gamma \left| \hat{\mathcal{P}}_t \right| \right\rceil = \lceil \gamma \lfloor \alpha n_{\text{pop}} \rfloor \rceil . \tag{11}$$

The trained GNG network is a model of $\hat{\mathcal{P}}_t$. The network can be interpreted as a Gaussian mixture, as explained in the previous section. Therefore it can be used to sample new individuals. In particular, $\lfloor \omega |\mathcal{P}_t| \rfloor$ new individuals are synthesized.

The local Gaussian densities resulting from the described algorithm can be combined to synthesize the Gaussian mixture with parameters $\boldsymbol{\Theta}$,

$$P(\boldsymbol{x}|\boldsymbol{\Theta}) = \frac{1}{N} \sum_{i=1}^{N} P(\boldsymbol{x}|\boldsymbol{\mu}_i, \boldsymbol{\sigma}_i) . \tag{12}$$

Each Gaussian density is formulated as

$$P(\boldsymbol{x}|\boldsymbol{\mu}_i, \boldsymbol{\sigma}_i) = \frac{1}{(2\pi)^{n/2}|\boldsymbol{\Sigma}_i|^{1/2}} \exp\left(-\frac{1}{2}(\boldsymbol{x} - \boldsymbol{\mu}_i)^\top \boldsymbol{\Sigma}_i^{-1}(\boldsymbol{x} - \boldsymbol{\mu}_i)\right), \tag{13}$$

with the covariance matrices Σ_i defined as a diagonal matrix with its non-zero elements set to the values of the deviations σ_i. The Gaussian mixture can be used by the EDA to generate new individuals. These new individuals are created by sampling the $P(x|\Theta)$. The generation of randomly distributed numbers that follow a given distribution has been dealt in depth by many authors. In our case, we applied the Box-Muller transformation [36].

Each one of these individuals substitute a randomly selected ones from the section of the population not used for model-building $\mathcal{P}_t \setminus \hat{\mathcal{P}}_t$. The set obtained is then united with the best elements, $\hat{\mathcal{P}}_t$, to form the population of the next iteration $\mathcal{P}_{t+1}$. Some other substitution strategies could be used in this step. For example, the new individuals could substitute the worst individuals of $\mathcal{P}_t \setminus \hat{\mathcal{P}}_t$. We have chosen the previously described approach because it promotes diversity and avoids stagnation.

Iterations are repeated until a given stopping criterion is met. The output of the algorithm is a subset of $\mathcal{P}_t$ that contains the non-dominated solutions, $\mathcal{P}_t^*$.

5   Experimental Study

The results of the experiments involving MARTEDA, some current state-of-the-art MOEDAs and MOEAs in a selection of current community-accepted problems are reported in this section.

The Walking Fish Group (WFG) problem toolkit [37] is a toolkit for creating complex synthetic multi-objective test problems. The WFG test suite exceeds the functionality of previous existing test suites. These include: non-separable problems, deceptive problems, a truly degenerate problem, a mixed shape Pareto front problem, problems scalable in the number of position related parameters, and problems with dependencies between position- and distance-related parameters. The WFG test suite provides a better form of assessing the performance of optimization algorithms on a wide range of different problems.

From the set of nine problems WFG4 to WFG9 were selected because of the simplicity of their Pareto-optimal front that lies on the first orthant of a unit hypersphere. This decision was also caused by the high computational cost of the experiments being carried out and by the length restriction imposed upon this contribution. Each problem was configured with 3, 6, 9 and 12 objective functions. For all cases the decision space dimension was fixed to 30.

Besides applying MARTEDA to the aforementioned problems some other MOEDAs and MOEAs are also assessed in order to provide a comparative ground for the tests. One algorithm is of particular interest, the MONEDA [38] algorithm. This approach was previously proposed by the authors to deal with the model-building issue of MOEDAs and MARTEDA is supposed to be an improvement over it. However, as MONEDA used the less-performing NSGA-II selection, we have also tested MONEDA with the HypE selection, in order to have some basis for comparison.

Besides MONEDA, we also tested the naïve MIDEA [39], and MrBOA [40] MOEDAs and the SMS-EMOA [41], HypE [8] and NSGA-II [1] MOEAs. One

Parameters:
▷ γ, initial deviations.
▷ n_{pop}, population size.
▷ $\alpha \in (0, 1]$, selection percentile.
▷ $\omega \in (0, 1]$, substitution percentile.

Algorithm:
$t \leftarrow 0$.
Randomly generate the initial population $\mathcal{P}_0$ with n_{pop} individuals.

repeat
  Sort population $\mathcal{P}_t$ using the HypE+ ranking algorithm.
  Extract first $\alpha|\mathcal{P}_t|$ elements the sorted $\mathcal{P}_t$ to $\hat{\mathcal{P}}_t$.
  A Gaussian ART with $\hat{\mathcal{P}}_t$ as training data set.
  Sample $\lfloor \omega|\mathcal{P}_t| \rfloor$ from the network.
  Substitute randomly selected individuals of $\mathcal{P}_t \setminus \hat{\mathcal{P}}_t$ with the new individuals to produce $\mathcal{P}'_t$.
  $\mathcal{P}_{t+1} = \hat{\mathcal{P}}_t \cup \mathcal{P}'_t$.
  $t = t + 1$.
until end condition = true
Determine the set of non-dominated individuals of $\mathcal{P}_t$, $\mathcal{P}^*_t$.
return $\mathcal{P}^*_t$ as the algorithm's solution.

Fig. 1. Algorithmic representation of MARTEDA

of the purposes of this study is to assess the parameter robustness of the algorithms. That is why the same parameter values have been for all problems, only increasing the population size as the number of objectives grows. For each problem/dimension pair each algorithm was executed 30 times.

The quality of the solutions is determined by the use of the hypervolume indicator [42].

The stochastic nature of evolutionary algorithms prompts the use of statistical tools in order to reach a valid judgement of the quality of the solutions and how different algorithms compare with each other. Box plots [43] are one of such representations and have been repeatedly applied in our context. Although box plots allows a visual comparison of the results and, in principle, some conclusions could be deduced out of them. Nevertheless, in order to reach a substantiated judgement it is necessary go beyond reporting the descriptive statistics of the performance indicators. For this task is required to carry out a set of statistical inferences that would support any judgements made from the data.

The statistical validity of the judgment of the results calls for the application of statistical hypothesis tests. It has been previously remarked by different authors that the Mann-Whitney-Wilcoxon U test [44] is particularly suited for experiments in the context of multi-objective evolutionary optimization [42]. This test is commonly used as a non-parametric method for testing equality of population medians. In our case we performed pair-wise tests on the significance of the difference of the indicator values yielded by the executions of the algorithms. A significance level, α, of 0.05 was used for all tests.

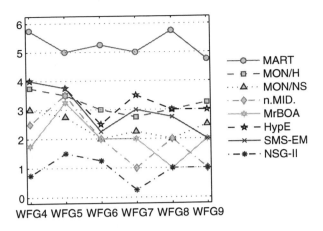

Fig. 2. Mean values of the performance index of MARTEDA (MART), MONEDA with HypE (MON/H) or NSGA-II selection (MON/NS), naïve MIDEA (n.MID), MrBOA, HypE, SMS-EMOA (SMS-EM) and NSGA-II (NSG-II) across the different problems, P_p ().

The visual analysis of the results is rather difficult as it implies cross-examining and comparing the results presented separately. That is why we decided to adopt a more integrative representation such as the one proposed in [45]. That is, for a given set of algorithms $A_1, \ldots, A_K$, a set of P test problem instances $\Phi_{1,m}, \ldots, \Phi_{P,m}$, configured with m objectives, the function $\delta(\cdot)$ is defined as

$$\delta(A_i, A_j, \Phi_{p,m}) = \begin{cases} 1 \text{ if } A_i \gg A_j \text{ solving } \Phi_{p,m} \\ 0 \text{ in other case} \end{cases}, \tag{14}$$

where the relation $A_i \gg A_j$ defines if A_i is significantly better than A_j when solving the problem instance $\Phi_{p,m}$, as computed by the statistical tests previously described.

Relying on $\delta(\cdot)$, the performance index $P_{p,m}(A_i)$ of a given algorithm A_i when solving $\Phi_{p,m}$ is then computed as

$$P_{p,m}(A_i) = \sum_{j=1; j \neq i}^{K} \delta(A_i, A_j, \Phi_{p,m}). \tag{15}$$

This index intends to summarize the performance of each algorithm with regard to its peers.

Figs. 2 and 3 exhibit the results computing the performance indexes grouped by problems and dimensions.

Fig. 2 represents the mean performance indexes yielded by each algorithm when solving each problem in all of its configured objective dimensions,

$$\bar{P}_p(A_i) = \frac{1}{|\mathcal{M}|} \sum_{m \in \mathcal{M}} P_{p,m}(A_i). \tag{16}$$

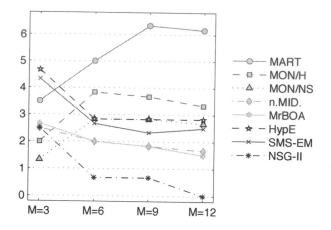

Fig. 3. Mean values of the performance index across the different space dimensions, $\bar{P}_m$. See Fig. 2 for a description of the acronyms.

It is worth noticing that MARTEDA has better overall results with respect to the other algorithms in all problems. As it could be expected, the use of indicator-based selection in MONEDA has yielded better results than the original MONEDA. Indicator-based MONEDA and the indicator-based MOEAs have a similar performance. It can be hypothesized that these results can be biased by the three objective problems, having dramatic differences in their results with respect to the rest of the dimensions considered.

This situation is clarified in Fig. 3, which presents the mean values of the index computed for each dimension

$$\bar{P}_m\left(A_i\right) = \frac{1}{P}\sum_{p=1}^{P} P_{p,m}\left(A_i\right) . \tag{17}$$

In this case MARTEDA is shown to clearly outperform the rest in more than three dimensions. Still, another important conclusion can be extracted. For more than three objectives, the MOEDAs that attempt to tackle the model-building issue (MONEDA and MARTEDA) and that also exploit indicator-based selection have outperformed the rest of the methods. This is very important, as it transcends the particular results of a given algorithm but instead casts some light on what should be the proper trend of development in this field.

Finally, the above experiments lead us to hypothesize that thanks to the treatment of the outliers in the model-building data-set, the MARTEDA approach manages to overcome the difficulties that hampers the rest of the methods. Another important result is that MARTEDA was able to yield good results across a varied set of problems without tuning its parameters in every case. This implies that MARTEDA has a certain degree of robustness regarding its parameters.

6   Final Remarks

In this paper we have explored the model-building issue of MOEDAs and the requirements it imposes on its supporting learning paradigm. We put forward adaptive resonance theory as a alternative learning paradigm. Based on it, we introduced a novel algorithm called multi-objective ART-based EDA (MARTEDA) that uses a Gaussian ART neural network for model-building and the hyper volume-based selection described for the HypE algorithm. We showed that by using this novel model-building approach and an indicator-based population ranking the algorithm is able to outperform similar MOEDAs and MOEAs.

Still, the main conclusion of this work is that we provide strong evidences that further research must be dedicated to the model-building issue in order to make current MOEDAs capable of dealing with complex multi-objective problems with many objectives. In spite of the fact that obviously further studies are necessary, these extensive experiments have provided solid ground for the use of MARTEDA in a real-world application context.

Acknowledgements

This work was supported by projects CICYT TIN2008-06742-C02-02/TSI, CICYT TEC2008-06732-C02-02/TEC, CAM CONTEXTS (S2009/TIC-1485) and DPS2008-07029-C02-02.

References

1. Coello Coello, C.A., Lamont, G.B., Van Veldhuizen, D.A.: Evolutionary Algorithms for Solving Multi-Objective Problems. In: Genetic and Evolutionary Computation, 2nd edn. Springer, New York (2007)
2. Miettinen, K.: Nonlinear Multiobjective Optimization. International Series in Operations Research & Management Science, vol. 12. Kluwer, Norwell (1999)
3. Pareto, V.: Cours D'Économie Politique. F. Rouge, Lausanne (1896)
4. Purshouse, R.C., Fleming, P.J.: On the evolutionary optimization of many conflicting objectives. IEEE Transactions on Evolutionary Computation 11(6), 770–784 (2007)
5. Stewart, T., Bandte, O., Braun, H., Chakraborti, N., Ehrgott, M., Göbelt, M., Jin, Y., Nakayama, H., Poles, S., Di Stefano, D.: Real-world applications of multiobjective optimization. In: Branke, J., Deb, K., Miettinen, K., Słowiński, R. (eds.) Multiobjective Optimization. LNCS, vol. 5252, pp. 285–327. Springer, Heidelberg (2008)
6. Wagner, T., Beume, N., Naujoks, B.: Pareto-, aggregation-, and indicator-based methods in many-objective optimization. In: Obayashi, S., Deb, K., Poloni, C., Hiroyasu, T., Murata, T. (eds.) EMO 2007. LNCS, vol. 4403, pp. 742–756. Springer, Heidelberg (2007)
7. Bader, J., Deb, K., Zitzler, E.: Faster hypervolume-based search using Monte Carlo sampling. In: Beckmann, M., Künzi, H.P., Fandel, G., Trockel, W., Basile, A., Drexl, A., Dawid, H., Inderfurth, K., Kürsten, W., Schittko, U., Ehrgott, M., Naujoks, B., Stewart, T.J., Wallenius, J. (eds.) Multiple Criteria Decision Making for Sustainable Energy and Transportation Systems. LNEMS, vol. 634, pp. 313–326. Springer, Berlin (2010)

8. Bader, J., Zitzler, E.: HypE: An Algorithm for Fast Hypervolume-Based Many-Objective Optimization. TIK Report 286, Computer Engineering and Networks Laboratory (TIK), ETH Zurich (2008)
9. Deb, K., Saxena, D.K.: Searching for Pareto–optimal solutions through dimensionality reduction for certain large–dimensional multi–objective optimization problems. In: 2006 IEEE Conference on Evolutionary Computation (CEC 2006), pp. 3352–3360. IEEE Press, Piscataway (2006)
10. Brockhoff, D., Zitzler, E.: Dimensionality reduction in multiobjective optimization: The minimum objective subset problem. In: Waldmann, K.H., Stocker, U.M. (eds.) Operations Research Proceedings 2006, pp. 423–429. Springer, Heidelberg (2007)
11. Brockhoff, D., Saxena, D.K., Deb, K., Zitzler, E.: On handling a large number of objectives a posteriori and during optimization. In: Knowles, J., Corne, D., Deb, K. (eds.) Multi–Objective Problem Solving from Nature: From Concepts to Applications. Natural Computing Series, pp. 377–403. Springer, Heidelberg (2008)
12. Corne, D.W.: Single objective = past, multiobjective = present,??? = future. In: Michalewicz, Z. (ed.) 2008 IEEE Conference on Evolutionary Computation (CEC), Part of 2008 IEEE World Congress on Computational Intelligence (WCCI 2008). IEEE Press, Piscataway (2008)
13. Michalski, R.S.: Learnable evolution model: Evolutionary processes guided by machine learning. Machine Learning 38, 9–40 (2000)
14. Sheri, G., Corne, D.W.: The simplest evolution/learning hybrid: LEM with KNN. In: IEEE World Congress on Computational Intelligence, pp. 3244–3251. IEEE Press, Hong Kong (2008)
15. Sheri, G., Corne, D.W.: Learning-assisted evolutionary search for scalable function optimization: LEM(ID3). In: IEEE World Congress on Computational Intelligence. IEEE Press, Barcelona (2010)
16. Lozano, J.A., Larrañaga, P., Inza, I., Bengoetxea, E. (eds.): Towards a New Evolutionary Computation: Advances on Estimation of Distribution Algorithms. Springer, Heidelberg (2006)
17. Pelikan, M., Sastry, K., Goldberg, D.E.: Multiobjective estimation of distribution algorithms. In: Pelikan, M., Sastry, K., Cantú-Paz, E. (eds.) Scalable Optimization via Probabilistic Modeling: From Algorithms to Applications. SCI, pp. 223–248. Springer, Heidelberg (2006)
18. Martí, L., García, J., Berlanga, A., Coello Coello, C.A., Molina, J.M.: On current model-building methods for multi-objective estimation of distribution algorithms: Shortcommings and directions for improvement. Technical Report GIAA2010E001, Grupo de Inteligencia Artificial Aplicada, Universidad Carlos III de Madrid, Colmenarejo, Spain (2010)
19. Grossberg, S.: Studies of Mind and Brain: Neural Principles of Learning, Perception, Development, Cognition, and Motor Control. Reidel, Boston (1982)
20. Sarle, W.S.: Why statisticians should not FART. Technical report, SAS Institute, Cary, NC (1995)
21. Williamson, J.R.: Gaussian ARTMAP: A neural network for fast incremental learning of noisy multidimensional maps. Neural Networks 9, 881–897 (1996)
22. Martí, L., García, J., Berlanga, A., Molina, J.M.: Moving away from error-based learning in multi-objective estimation of distribution algorithms. In: Branke, J., Alba, E., Arnold, D., Bongard, J., Brabazon, A., Butz, M.V., Clune, J., Cohen, M., Deb, K., Engelbrecht, A., Krasnogor, N., Miller, J., O'Neill, M., Sastry, K., Thierens, D., Vanneschi, L., van Hemert, J., Witt, C. (eds.) GECCO 2010: Proceedings of the 12th Annual Conference on Genetic and Evolutionary Computation, pp. 545–546. ACM Press, New York (2010)

23. Ahn, C.W., Ramakrishna, R.S.: Multiobjective real-coded Bayesian optimization algorithm revisited: Diversity preservation. In: GECCO 2007: Proceedings of the 9th Annual Conference on Genetic and Evolutionary Computation, pp. 593–600. ACM Press, New York (2007)
24. Shapiro, J.: Diversity loss in general estimation of distribution algorithms. In: Runarsson, T.P., Beyer, H.-G., Burke, E.K., Merelo-Guervós, J.J., Whitley, L.D., Yao, X. (eds.) PPSN 2006. LNCS, vol. 4193, pp. 92–101. Springer, Heidelberg (2006)
25. Yuan, B., Gallagher, M.: On the importance of diversity maintenance in estimation of distribution algorithms. In: GECCO 2005: Proceedings of the 2005 Conference on Genetic and Evolutionary Computation, pp. 719–726. ACM Press, New York (2005)
26. Peña, J.M., Robles, V., Larrañaga, P., Herves, V., Rosales, F., Pérez, M.S.: GA-EDA: Hybrid evolutionary algorithm using genetic and estimation of distribution algorithms. In: Orchard, B., Yang, C., Ali, M. (eds.) IEA/AIE 2004. LNCS (LNAI), vol. 3029, pp. 361–371. Springer, Heidelberg (2004)
27. Zhang, Q., Sun, J., Tsang, E.: An evolutionary algorithm with guided mutation for the maximum clique problem. IEEE Transactions on Evolutionary Computation 9(2), 192–200 (2005)
28. Zitzler, E., Thiele, L., Laumanns, M., Fonseca, C.M., Grunert da Fonseca, V.: Performance assessment of multiobjective optimizers: An analysis and review. IEEE Transactions on Evolutionary Computation 7(2), 117–132 (2003)
29. While, L., Hingston, P., Barone, L., Huband, S.: A faster algorithm for calculating hypervolume. IEEE Transactions on Evolutionary Computation 10(1), 29–38 (2006)
30. Fonseca, C.M., Paquete, L., López-Ibánez, M.: An improved dimension–sweep algorithm for the hypervolume indicator. In: 2006 IEEE Congress on Evolutionary Computation (CEC 2006), pp. 1157–1163 (2006)
31. Beume, N., Rudolph, G.: Faster S–metric calculation by considering dominated hypervolume as Klee's measure problem. In: Kovalerchuk, B. (ed.) Proceedings of the Second IASTED International Conference on Computational Intelligence, pp. 233–238. IASTED/ACTA Press (2006)
32. Beume, N.: S–metric calculation by considering dominated hypervolume as Klee's measure problem. Evolutionary Computation 17(4), 477–492 (2009); PMID: 19916778
33. Bringmann, K., Friedrich, T.: Approximating the volume of unions and intersections of high–dimensional geometric objects. Computational Geometry 43(6-7), 601–610 (2010)
34. Papadimitriou, C.M.: Computational Complexity. Addison-Wesley, Reading (1994)
35. Deolalikar, V.: P≠NP. Technical report, Hewlett Packard Research Labs, Palo Alto, CA, USA (2010)
36. Box, G.E.P., Muller, M.E.: A note on the generation of random normal deviates. Annals of Mathematical Statistics 29, 610–611 (1958)
37. Huband, S., Hingston, P., Barone, L., While, L.: A review of multiobjective test problems and a scalable test problem toolkit. IEEE Transactions on Evolutionary Computation 10(5), 477–506 (2006)

38. Martí, L., García, J., Berlanga, A., Molina, J.M.: Introducing MONEDA: Scalable multiobjective optimization with a neural estimation of distribution algorithm. In: Keizer, M., Antoniol, G., Congdon, C., Deb, K., Doerr, B., Hansen, N., Holmes, J., Hornby, G., Howard, D., Kennedy, J., Kumar, S., Lobo, F., Miller, J., Moore, J., Neumann, F., Pelikan, M., Pollack, J., Sastry, K., Stanley, K., Stoica, A., Talbi, E.G., Wegener, I. (eds.) GECCO 2008: 10th Annual Conference on Genetic and Evolutionary Computation, pp. 689–696. ACM Press, New York (2008); EMO Track "Best Paper" Nominee
39. Bosman, P.A.N., Thierens, D.: The naive MIDEA: A baseline multi–objective EA. In: Coello Coello, C.A., Hernández Aguirre, A., Zitzler, E. (eds.) EMO 2005. LNCS, vol. 3410, pp. 428–442. Springer, Heidelberg (2005)
40. Ahn, C.W.: Advances in Evolutionary Algorithms. In: Theory, Design and Practice. Springer, Heidelberg (2006) ISBN 3-540-31758-9
41. Beume, N., Naujoks, B., Emmerich, M.: SMS–EMOA: Multiobjective selection based on dominated hypervolume. European Journal of Operational Research 181(3), 1653–1669 (2007)
42. Knowles, J., Thiele, L., Zitzler, E.: A tutorial on the performance assessment of stochastic multiobjective optimizers. TIK Report 214, Computer Engineering and Networks Laboratory (TIK), ETH Zurich (2006)
43. Chambers, J., Cleveland, W., Kleiner, B., Tukey, P.: Graphical Methods for Data Analysis. Wadsworth, Belmont (1983)
44. Mann, H.B., Whitney, D.R.: On a test of whether one of two random variables is stochastically larger than the other. Annals of Mathematical Statistics 18, 50–60 (1947)
45. Bader, J.: Hypervolume-Based Search for Multiobjective Optimization: Theory and Methods. PhD thesis, ETH Zurich, Switzerland (2010)

Multi-Objective Differential Evolution with Adaptive Control of Parameters and Operators

Ke Li[1], Álvaro Fialho[2], and Sam Kwong[1]

[1] Department of Computer Science, City University of Hong Kong, Hong Kong
`jerryli3@student.cityu.edu.hk`, `cssamk@cityu.edu.hk`
[2] LIX, École Polytechnique, Palaiseau, France
`fialho@lix.polytechnique.fr`

Abstract. Differential Evolution (DE) is a simple yet powerful evolutionary algorithm, whose performance highly depends on the setting of some parameters. In this paper, we propose an adaptive DE algorithm for multi-objective optimization problems. Firstly, a novel tree neighborhood density estimator is proposed to enforce a higher spread between the non-dominated solutions, while the Pareto dominance strength is used to promote a higher convergence to the Pareto front. These two metrics are then used by an original replacement mechanism based on a three step comparison procedure; and also to port two existing adaptive mechanisms to the multi-objective domain, one being used for the autonomous selection of the operators, and the other for the adaptive control of DE parameters CR and F. Experimental results confirm the superior performance of the proposed algorithm, referred to as Adap-MODE, when compared to two state-of-the-art baseline approaches, and to its static and partially-adaptive variants.

Keywords: Multi-Objective Optimization, Differential Evolution, Tree Neighborhood Density, Parameter Control, Adaptive Operator Selection.

1 Introduction

Differential Evolution, proposed by Storn and Price [15], is a popular and efficient population-based, direct heuristic for solving global optimization problems in continuous search spaces. The main benefits brought by DE are its simple structure, ease of use, fast convergence speed and robustness, which enables it to be widely applied to many real-world applications. For the generation of new solutions (trial vectors), each individual (target vector) is combined with others by means of different forms of weighted sums (mutation strategies). Originally, in case the newly generated solution has a better fitness value than its corresponding parent, it replaces its parent in the population for the next generation. The aim of these iterations is basically to find a proper direction for the search process towards the optimum, by following the quality distribution of the solutions in the current population.

C.A. Coello Coello (Ed.): LION 5, LNCS 6683, pp. 473–487, 2011.

One of the possible application domains of DE are the Multi-objective Optimization Problems (MOPs), which exist everywhere in real-world applications, such as engineering, financial, and scientific computing. The main difficulty in these cases lies in providing a way to compare the different solutions, as the involved multiple criteria might compete with one another, besides possibly not being directly comparable. Multi-Objective Evolutionary Algorithms (MOEAs) tackle this issue by searching for the set of optimal trade-off solutions, the so-called Pareto optimal set: the aim is not only to approach the Pareto optimal front as closely as possible, but also to find solutions that are distributed over the Pareto optimal front as uniformly as possible, in order to better satisfy all the different objectives considered. Needless to say, to be applied to MOPs, the DE original scheme needs to be adapted according to the mentioned aims.

Many different types of DE variants proposed to tackle MOPs can be found in the literature, such as GDE3 [12], and DEMO [17]. We refer the reader to [2] for a recent comprehensive survey of DE, including its application to MOPs. But the performance of DE largely depends on the definition of some parameters. Besides the crossover rate CR, and the mutation scaling factor F, there is the need of choosing which mutation strategies, from the many available ones, should be used for the generation of new solutions, and at which rate each of the chosen strategies should be applied. The setting of these parameters is usually a crucial and very time-consuming task: the optimal values for them do not only depend on the problem at hand, but also on the region of the search space that is being explored by the current population, while solving the problem. Following the intuition of the Exploration versus Exploitation (EvE) balance, exploration tends to be more beneficial in the early stages of the search (consequently more exploratory mutation strategies, high values for F and CR), while more exploitation should be promoted when getting closer to the optimum (respectively, more fine-tuning operators, and a smaller value for F).

A prominent paradigm to automate the setting of these parameters on-line, i.e., while solving the problem, is the so-called Adaptive parameter control. It constantly adapts the values of the parameters based on feedbacks received from the search process. Some algorithms have been recently proposed for the on-line adaptation of CR and F, and for the autonomous control of which of the strategies should be applied at each instant of the search, the latter being commonly referred to as Adaptive Operator Selection (AOS). Some DE algorithms using adaptive methods can be found in the literature, such as SaDE [16], JADE [21], jDE [1] and ISADE [11]. Regarding DE for MOPs, there also exists some pioneering works, such as JADE2 [20] and OW-MOSaDE [10]. However, to the best of our knowledge, the employment of both adaptive parameter control of CR and F, and adaptive operator (mutation strategy) selection, is still relatively scarce in the domain of MOPs.

In this work, we employ an adaptive parameter control of CR and F slightly different from the one employed by the JADE method [21], which adapts their values based on the recent success rate of the search process; and an AOS mechanism inspired from the PM-AdapSS-DE method [9], which uses the Probability

Matching mechanism to select between the available mutation strategies, based on the normalized relative fitness improvements brought by their recent applications. The main contribution of this work lies in the porting of these adaptive methods to the multi-objective domain. More specifically, a novel method is proposed to partially evaluate the fitness of the solutions, referred to as Tree Neighborhood Density (TND) estimator. The aggregation of the TND with the Pareto Dominance Strength (brought from the SPEA2 [23] method) is the information used by the AOS mechanism to keep its operator preferences up-to-date, and by a novel replacement mechanism based on a three-step comparison scheme. Lastly, the output of this replacement mechanism defines the success rates used for the adaptive parameter control of CR and F. The resulting algorithm, referred to as Adaptive Multi-Objective DE (Adap-MODE), is assessed in the light of a set of multi-objective benchmark functions, and shows to achieve significantly better results than other state-of-the-art approaches (NSGA-II [4] and GDE3 [12]) and than its static and partially-adaptive variants in most of the cases.

The remainder of this paper is organized as follows. Firstly, the background and some related work are briefly reviewed in Section 2. Then, our proposed algorithm is described in detail in Section 3. After that, some experimental results are analyzed in Section 4. Finally, Section 5 concludes this paper and gives possible directions for further work.

2 Related Work

The performance of an Evolutionary Algorithm (EA) strongly depends on the setting of some of its parameters. Section 2.1 will briefly overview the different ways of doing parameter setting in EAs, focusing on the kind of approach used in this work, referred to as Adaptive Parameter Control. Then, Section 2.2 will survey more specifically the Adaptive Operator Selection (AOS) paradigm.

2.1 Parameter Setting in Evolutionary Algorithms

There are different ways of doing parameter setting in EAs, as acknowledged by the well-known taxonomy proposed by Eiben et al. in [6]. In the higher level, there is the separation between Parameter Tuning and Parameter Control methods. Parameter Tuning methods set the parameters off-line, based on statistics over several runs; besides being computationally expensive, it provides a single parameter setting, that remains static during all the run. Parameter Control methods continuously adapt the parameters on-line, i.e., while solving the problem; these methods are further sub-divided into three branches, as follows.

The Deterministic methods adapt the parameter values according to predefined (deterministic) rules; but the definition of these rules already defines a complex optimization problem *per se*, besides hardly adapting to different problems. The Self-Adaptive methods adapt the parameter values *for free*, by encoding them within the candidate solution and letting the evolution take care of their control; in this case, however, the search space of the parameter values is aggregated to that of the problem, what might significantly increase the overall complexity of the search process. Lastly, the Adaptive methods control the parameter values based on feedback received from the previous search steps of the current optimization process.

In this work, we use an adaptive method very similar to the one proposed in the JADE algorithm [21], which controls the values of DE crossover rate CR and mutation scaling factor F based on the recent success rate (more details in Section 3.4). Another example of adaptive method proposed for the same aim is the SaDE [16] algorithm. Furthermore, another kind of adaptive method is also used in our algorithm, the AOS, surveyed in the following.

2.2   Adaptive Operator Selection

A recent paradigm, referred to as Adaptive Operator Selection (AOS), proposes the autonomous control of which operator (or mutation strategy in the case of DE) should be applied at each instant of the search, while solving the problem, based on their recent performance. A general AOS method usually consists of two components: the Credit Assignment scheme defines how each operator should be rewarded based on the impacts of its recent applications on the search progress; and the Operator Selection mechanism decides which of the available operators should be applied next, according to their respective empirical quality estimates, which are built and constantly updated by the rewards received. Each of these components will now be briefly overviewed in turn.

Credit Assignment

The most common way of assessing the impact of an operator application is the fitness improvement achieved by the offspring generated by its application, with respect to a baseline individual. In [9], the fitness improvement with respect to its parent is considered, while [3] use as baseline individual the best individual of the current population.

Based on this impact assessment, different ways of assigning credit to the operators can be found, in addition to the common average of the recent fitness improvements. In [19], a statistical technique rewards the operators based on their capability of generating outlier solutions, arguing that rare but highly beneficial improvements might be more important than frequent small improvements. Along the same line, in [8] each operator is rewarded based on the extreme (or maximal) fitness improvement recently achieved by it. In the quest for a more robust rewarding, in [7] a rank-based scheme is proposed. In multi-modal problems, however, the diversity is also important; in [14], both diversity variation and fitness improvement are combined to evaluate the operator application.

Operator Selection

The Operator Selection mechanism usually keeps an empirical quality estimate for each operator, built by the received rewards, which is used to guide its selection. The most popular method for Operator Selection is referred to as Probability Matching (PM) [18]: basically, the probability of selecting each operator is proportional to its empirical quality estimate with respect to the others; this is the method used in this work, more details in Section 3.3.

Other more complex Operator Selection methods worth to be mentioned are: the Adaptive Pursuit (AP) [18], originally proposed for learning automata, employs a winner-takes-all strategy to enforce a higher exploitation of the best operator; and the Dynamic Multi-Armed Bandit (DMAB) [8], which tackles the Operator Selection problem as yet another level of the Exploration vs. Exploitation dilemma, efficiently exploiting the current best operator, while minimally exploring the others, inspired from the multi-armed bandit paradigm.

3   Adaptive Multi-Objective DE

The general framework of the proposed adaptive Differential Evolution (DE) algorithm for multi-objective problems is illustrated in Fig. 1. As can be seen, it is divided into three modules. In the middle, there is the main cycle of the DE algorithm, represented here by only three steps for the sake of brevity: once after every generation, the fitness (see Section 3.1) of each offspring is evaluated by the sum of its Pareto Dominance (PD) strength and its Tree Neighborhood Density (TND). While the PD enforces convergence towards the Pareto front, the TND promotes diversification between the non-dominated solutions. These two measures are separately used by the Replacement mechanism, that decides which of the individuals should be maintained for the next generation by means of an original three-step comparison procedure (Section 3.2).

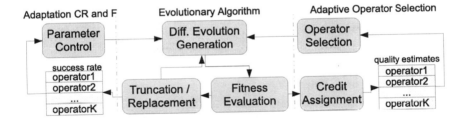

Fig. 1. The framework of the proposed adaptive Differential Evolution algorithm

Two adaptive mechanisms are employed in parallel. On the right side, there is the AOS module, inspired from the PM-AdapSS-DE algorithm [9]. And on the left side, there is the Adaptive Parameter Control module slightly modified from the JADE algorithm [21]. Both adaptive mechanisms are described, respectively, in Sections 3.3 and 3.4. Although these are adaptive mechanisms brought from the literature, it is worth noting that in this work they are originally ported to the multi-objective domain, by receiving inputs based on the special aggregation between the PD and the novel TND measures.

3.1   Fitness Evaluation

In multi-objective optimization, the aims of the search can be said to be two-fold. On the one hand, the solutions found should approach as much as possible to the Pareto front. On the other hand, the non-dominated solutions should be distributed over the Pareto front as uniformly as possible, in order to have satisfiable solutions for all the different objectives. In this work, we use the Pareto Dominance (PD) strength proposed in the SPEA2 algorithm [23] to enforce the first issue (convergence). For the second issue, we propose a novel measure to promote spread between the non-dominated solutions, referred to as the Tree Neighborhood Density (TND). The fitness of each individual is assessed by an aggregation of these two criteria, as described in the following.

Pareto Dominance Strength
In order to calculate the Pareto Dominance (PD) strength, we use the mechanism proposed in the SPEA2 algorithm [23]. The only difference is that the external archive to store elite individuals is not implemented here. Briefly, a strength value $S(i)$ is assigned to each individual i in the population P, representing the number of solutions it dominates. If solely based on this criterion, the fitness of each individual i, referred to as $PD(i)$ here, would be calculated as:

$$PD(i) = \sum_{j \in P, j \succ i} S(j) \tag{1}$$

i.e., the sum of the strengths of all the individuals that dominate individual i. Intuitively, the smaller the better, with $PD(i) = 0$ corresponding to a non-dominated solution; whereas a large $PD(i)$ means that the individual i is dominated by many others.

Tree Neighborhood Density
As previously mentioned, the Tree Neighborhood Density (TND) is a novel estimation proposed to enforce a higher level of spread between the non-dominated solutions. For the sake of a clearer discussion, some definitions and terminologies are firstly given as follows.

Definition 1 (Tree crowding density). *Let T be a minimum spanning tree connecting all the individuals of population P. For any individual i in P, let d_i be the degree of i in T, i.e., the number of edges of T connected to i; and let these edges be $\{l_{i,1}, l_{i,2}, \ldots, l_{i,d_i}\}$. The tree crowding density of i is estimated as:*

$$T_{crowd}(i) = \sum_{j=1}^{d_i} l_{i,j}/d_i \tag{2}$$

Definition 2 (Tree neighborhood). *Let $r_i = \max\{l_{i,1}, l_{i,2}, \ldots, l_{i,d_i}\}$. A circle centered in individual i, with radius r_i, is defined as the tree neighborhood of i.*

Definition 3 (Membership of individual on the tree neighborhood).
Let the Euclidean distance between individuals i and j be denoted as $dist_{i,j}$. The individual j is considered as a member of the tree neighborhood of i if and only if $dist_{i,j} \leq r_i$ (denoted as $i \triangleright_T j$).

Based on these definitions, the calculation procedure for the Tree Neighborhood Density (TND) is implemented as follows:

1. The Euclidean distance between each individual of the population P with the other $NP - 1$ individuals is calculated;
2. A minimal spanning tree T connecting all individuals is generated;
3. The tree crowding density for each individual i in T is assessed, and the corresponding tree neighborhood is generated;
4. For each individual i, the degrees of the individuals pertaining to its tree neighborhood are summed:

$$sumdegrees(i) = \sum_{j \in U} d_j, \text{ where } U = \{j | j \in P, i \triangleright_T j\} \tag{3}$$

5. Then, the Tree Neighborhood Density of individual i is calculated as:

$$TND(i) = \frac{\sum_{j \in U}(1/Tcrowd_j)}{sumdegrees(i)} \tag{4}$$

6. Finally, the TND values of all individuals are normalized:

$$nTND(i) = \frac{TND(i) - TND_{min}}{TND_{max} - TND_{min}}. \tag{5}$$

where $nTND(i)$ is the normalized TND of individual i, and TND_{max} and TND_{min} indicate, respectively, the maximum and minimum TND in the current population.

In the same way as for the PD measure, the smaller TND the better. The underlying motivation for its proposal can be explained as follows. The whole set of solutions in the population can be regarded as a connected graph, with the Euclidean minimum spanning tree of this graph being an optimized structure that reflects the distribution of the solutions of the current population in the search space. Then, for a given individual, the corresponding neighborhood can be defined by the other individuals connected to it, and finally, the crowdedness of this neighborhood can be said to represent its density.

Aggregated Fitness Evaluation
Based on the aforementioned discussion, the fitness value (to be minimized) of each individual i is calculated as the sum of both criteria:

$$f(i) = PD(i) + nTND(i) \tag{6}$$

It is worth noting that only the TND measure is normalized between 0 and 1. Hence, evolution proceeds by firstly minimizing PD, i.e., approaching the Pareto front; and then, as soon as some non-dominated solutions (i.e., with $PD = 0$) are found, $nTND$ becomes significant in the fitness evaluation, and a higher spread between the non-dominated solutions is promoted.

3.2   Replacement Mechanism

At each generation, each of the NP parental solutions is used to generate other NP offspring solutions. In the original DE algorithm, the offspring replaces its parent in the next generation if it has a better fitness value. In the case of multi-objective optimization, a different replacement mechanism is needed in order to incorporate the already mentioned properties of this kind of problem. To this aim, a three-step comparison method is proposed in this work, as follows.

Starting from the mixed population of size $2 \times NP$, containing the NP parental and the NP offspring individuals, firstly, the Pareto dominance relationship is considered: each pair (parent, offspring) is compared at a time, and the dominated one is immediately rejected.

In case the mixed population is still bigger than NP, the replacement mechanism proceeds to the second step, which uses the non-dominated sorting method proposed in the NSGA-II algorithm [4]. Briefly, at each round, the non-dominated individuals of the mixed population are chosen to survive to the next generation, and are removed from the mixed population. This is done iteratively up to the completion of the population for the next generation (i.e., NP chosen individuals after the first and second steps), or until there are no less than NP individuals with assigned rank values in the population.

If there are still individuals to be filtered for the next generation, the third step finally considers the TND values. At each iteration, the individual that has the lowest TND (i.e., the most crowded individual) is maintained, until the exact number of individuals for the completion of the new population is achieved.

3.3   Adaptive Operator Selection

As surveyed in Section 2.2, to implement the AOS paradigm, there is the need of defining two elements, the Credit Assignment and the Operator Selection mechanisms. The approaches used in this work will be now detailed in turn.

Credit Assignment: Normalized Relative Fitness Improvement
The Credit Assignment scheme is inspired from the one used in the PM-AdapSS-DE algorithm [9]; the differences are the use of a different and normalized calculation of the relative fitness improvements (which showed to perform better after some preliminary experiments) and in the already described fitness evaluation, specially designed for multi-objective optimization.

The impact of each operator application i is evaluated as the normalized relative fitness improvement η_i achieved by it, measured as:

$$\eta_i = \frac{|pf_i - cf_i|}{|f_{best} - f_{worst}|} \tag{7}$$

where f_{best} (respectively f_{worst}) is the fitness value of the best (respectively the worst) solution in the current population; pf_i and cf_i are the fitness values of the (parent) target vector and its offspring, respectively. As in [9], in case no improvement is achieved i.e., $pf_i - cf_i \geq 0$, η_i is set to zero.

All the normalized relative fitness improvements achieved by the application of operator (mutation strategy in this case) $a \in \{1, \ldots, K\}$ during each generation g are stored in a specific set R_a. Following [9], at the end of each generation g, a unique credit (or reward) is assigned to each operator, calculated as the average of all the normalized relative fitness improvements achieved by it:

$$r_a(g) = \sum_{i=1}^{|R_a|} \frac{R_a(i)}{|R_a|} . \tag{8}$$

Operator Selection: Probability Matching
The Operator Selection mechanism used is the Probability Matching (PM) [18]. Formally, let the strategy pool be denoted by $S = \{s_1, \ldots, s_K\}$ where $K > 1$. The probability vector $P(g) = \{p_1(g), \ldots, p_K(g)\}(\forall t : p_{min} \leq p_i(g) \leq 1; \sum_{i=1}^{K} p_i(g) = 1)$ represents the selection probability of each operator at generation g. At the end of every generation, the PM technique updates the probability $p_a(g)$ of each operator a based on the received reward $r_a(g)$, as follows. Firstly, the empirical quality estimate $q_a(g)$ of operator a at generation g is updated as [18]:

$$q_a(g+1) = q_a(g) + \alpha \left[r_a(g) - q_a(g) \right] \tag{9}$$

where $\alpha \in (0, 1]$ is the adaptation rate; the selection probability is updated as:

$$p_a(t+1) = p_{min} + (1 - K \cdot p_{min}) \frac{q_a(g+1)}{\sum_{i=1}^{K} q_i(g+1)}. \tag{10}$$

where $p_{min} \in (0, 1)$ is the minimal selection probability value of each operator, used to ensure that all the operators have a minimal chance of being selected. The rationale for this minimal exploration is that the operators that are currently performing badly might become useful at a further moment of the search [18].

3.4   Adaptive Parameter Control of CR and F

The parameter adaptation method used here is similar to that used in the JADE algorithm [21]. Let CR_i^a denote the crossover rate for the individual i using operator $a \in \{1, \ldots, K\}$. At each generation, CR_i^a is independently generated according to a normal distribution with mean μ_{CR}^a and standard deviation 0.1:

$$CR_i^a = norm(\mu_{CR}^a, 0.1) \tag{11}$$

being regenerated whenever it exceeds 1. All successful crossover rates at generation g for operator a are stored in a specific set denoted as S_{CR}^a. The mean μ_{CR}^a is initialized to a user defined value and updated after each generation as:

$$\mu_{CR}^a = (1 - c) \cdot \mu_{CR}^a + c \cdot mean(S_{CR}^a) \tag{12}$$

where c is a constant and $mean(S_{CR}^a)$ is the arithmetic mean of values in S_{CR}^a. An analogous adaptation mechanism is used for the scaling factor F_i^a. After some preliminary experiments, a difference with respect to the JADE algorithm [21] at this point is that the mean value μ_F^a is calculated by the root-mean-square of the values in S_F^a, instead of Lehmer mean.

4   Performance Comparison

In this section, three different empirical comparisons are presented. Firstly, the proposed Adap-MODE is compared with two state-of-the-art MOEAs, namely, NSGA-II [4] and GDE3 [12]. Then, in order to assess the benefits brought by the combined use of both adaptive parameter control modules, Adap-MODE is compared with four static variants, each using one of the four mutation strategies and a fixed values for control parameters ($CR = 0.5, F = 1.0$). Lastly, we compare Adap-MODE with its "partially-adaptive" variants, namely, the same MODE but using only AOS (and $CR = 0.5, F = 1.0$), and the same MODE but using only the adaptive parameter control of CR and F (and the mutation strategies being uniformly selected). This latter is done in order to evaluate the gain achieved by the combination of both modules, compared with each of the modules being independently applied.

4.1   Experimental Settings

For the sake of a fair empirical comparison, the parameters of the two state-of-the-art MOEAs are set as in the respective original papers. For the NSGA-II [4], $\eta_c = \eta_m = 20, p_c = 0.9, p_m = 1/D$, with D representing the dimension of the problem; and for GDE3 [12], $CR = 0.5, F = 1.0$. For the parameters of the proposed Adap-MODE method, the PM adaptation rate is set to $\alpha = 0.3$ and minimal probability $p_{min} = 0.05$, as in [9]; and the parameter c for the adaptive parameter control of CR and F is set to 0.1, as in [21], with CR and F being both initialized to 0.2. Lastly, the DE population size is set to 100.

In this work, the AOS mechanism implemented in Adap-MODE is used to select between the following four DE mutation strategies: (1) DE/rand/1/bin, (2) DE/current-to-rand/1/bin, (3) DE/rand/2/bin, and (4) DE/rand-to-best/2/bin. These are the same strategies used in some previous works [16,9]; no theoretical or empirical analysis was preliminary performed for their choice. It is worth highlighting that the AOS scheme is generic: any other set of mutation strategies could be considered here.

In order to compare the performance of the proposed and baseline approaches, ZDT [22] and DTLZ [5] test suites are considered as benchmark functions. The maximum number of generations is set to 300 for ZDT, and to 500 for DTLZ.

Two assessment metrics are used to quantitatively evaluate the performance of each algorithm at the end of each run, averaged over 50 runs. The Uniform Assessment (UA) metric [13] is used to evaluate the spread of the solutions, while the Hyper-Volume (HV) [24] is a comprehensive performance indicator. Generally, for the values of both UA and HV, the larger the better.

4.2   Experimental Results

The comparative results, for each of the are presented in Tables 1 to 3. Following the central limit theorem, we assume that the sample means are normally distributed; therefore, the paired t-test statistical test at 95% confidence level

Table 1. Comparative results of NSGA-II, GDE3 and Adap-MODE

		NSGA-II	GDE3	Adap-MODE	S
ZDT1	UA	4.433e-1/3.56e-2	2.359e-1/4.42e-2	**8.080e-1/1.62e-2**	†
	HV	3.65960/3.00e-4	3.65990/3.55e-4	**3.66193/3.15e-5**	†
ZDT2	UA	4.391e-1/4.68e-2	2.551e-1/4.98e-2	**8.069e-1/1.89e-2**	†
	HV	3.32618/3.21e-4	3.32673/3.06e-4	**3.32853/4.19e-5**	†
ZDT3	UA	4.252e-1/4.49e-2	2.069e-1/4.17e-2	**7.660e-1/1.98e-2**	†
	HV	4.80650/5.13e-2	4.81433/1.95e-4	**4.81463/4.81e-4**	†
ZDT4	UA	4.173e-1/4.69e-2	2.403e-1/4.64e-2	**8.055e-1/1.85e-2**	†
	HV	3.65413/4.04e-3	3.63033/1.84e-1	**3.66201/5.33e-4**	†
ZDT6	UA	4.529e-1/4.86e-2	2.226e-1/4.67e-2	**7.896e-1/2.27e-2**	†
	HV	3.03090/1.51e-3	3.04029/2.67e-4	**3.04183/1.62e-5**	†
DTLZ1	UA	3.742e-1/4.44e-2	5.256e-1/3.58e-2	**8.246e-1/1.48e-2**	†
	HV	0.967445/1.95e-3	0.965469/9.45e-4	**0.973582/2.75e-4**	†
DTLZ2	UA	3.688e-1/3.78e-2	4.868e-1/3.30e-2	**8.236e-1/1.84e-2**	†
	HV	7.33017/2.70e-2	7.31392/9.05e-3	**7.40523/1.14e-2**	†
DTLZ3	UA	3.353e-1/7.92e-2	4.857e-1/4.09e-2	**8.304e-1/1.72e-2**	†
	HV	6.41853/1.80e+0	5.85267/2.37e+0	**7.32465/5.76e-1**	†
DTLZ4	UA	**4.404e-1/9.19e-2**	2.532e-1/3.91e-2	2.654e-1/2.99e-2	‡
	HV	6.90792/7.55e-1	5.46000/1.10e+0	**7.02943/5.46e-1**	†
DTLZ5	UA	3.930e-1/4.63e-2	4.379e-1/3.85e-2	**7.866e-1/1.82e-2**	†
	HV	6.10048/1.42e-3	6.08543/1.83e-3	**6.10548/4.40e-3**	
DTLZ6	UA	2.939e-1/5.19e-2	2.652e-1/4.33e-2	**7.759e-1/2.18e-2**	†
	HV	5.86932/7.09e-2	6.10187/2.12e-3	**6.10732/4.88e-3**	†
DTLZ7	UA	4.102e-1/3.96e-2	4.491e-1/3.70e-2	**7.723e-1/1.86e-2**	†
	HV	13.15151/8.55e-2	13.19772/9.29e-2	**13.46486/7.43e-2**	†

is adopted to compare the significance between two competing algorithms, with the † indicating that Adap-MODE is significantly better than all its competitors in the corresponding Table, and ‡ representing that the best competitor significantly outperforms Adap-MODE. Moreover, the best results for each metric on each problem function are highlighted in **boldface**.

Starting with the comparison between Adap-MODE and the two state-of-the-art MOEAs, namely NSGA-II and GDE3, the results are presented in Table 1. These results clearly show that Adap-MODE is the best choice when compared to its competitors: it achieves the best results in 23 out of the 24 performance metrics, performing significantly better in 22 of them. The only exception is for the UA metric in the DTLZ4 problem, in which NSGA-II wins. It is worth noting that Adap-MODE performs around two times better than its competitors w.r.t. the uniformity metric UA in most of the functions, what might be largely attributed to the use of the proposed tree neighborhood density estimator by the fitness assignment.

Table 2 compares the performance of Adap-MODE with four static variants of it, each using one of the four available mutation strategies, without any adaptive parameter control. From these results, it becomes clear that there is no single

Table 2. Comparative results of Adap-MODE and its pure versions, following the same order of the problems as in Table 1

	Str.1	Str.2	Str.3	Str.4	Adap-MODE	S
UA	7.9e-1/1.9e-2	7.9e-1/1.9e-2	7.4e-1/2.6e-2	4.1e-1/5.2e-2	**8.1e-1/1.6e-2**	†
HV	3.662/3.4e-5	3.662/3.4e-5	3.656/2.2e-3	1.902/3.9e-1	**3.662/3.1e-5**	
UA	7.9e-1/1.5e-2	8.0e-1/1.9e-2	7.2e-1/2.7e-2	3.6e-1/6.3e-2	**8.1e-1/1.9e-2**	
HV	3.328/3.4e-5	**3.328/3.9e-5**	3.319/4.6e-3	1.905/2.4e-1	3.328/4.2e-5	‡
UA	**7.7e-1/2.1e-2**	7.5e-1/2.7e-2	5.1e-1/8.0e-2	3.4e-1/2.7e-2	7.6e-1/1.9e-2	
HV	**4.815/6.4e-5**	4.814/1.6e-3	4.775/1.3e-2	1.781/3.4e-1	4.814/4.8e-4	‡
UA	**8.1e-1/1.7e-2**	8.0e-1/1.6e-2	8.0e-1/1.9e-2	3.3e-1/3.9e-2	8.0e-1/1.8e-2	
HV	3.636/1.0e-1	3.662/3.8e-5	3.649/8.6e-2	0.0/0.0	**3.662/5.3e-4**	
UA	7.9e-1/1.9e-2	8.1e-1/2.0e-2	**8.2e-1/1.9e-2**	7.6e-1/4.6e-2	7.9e-1/2.2e-2	‡
HV	3.042/1.7e-5	3.042/2.4e-5	**3.042/1.5e-5**	3.041/3.1e-3	3.042/1.6e-5	
UA	**8.3e-1/2.1e-2**	8.2e-1/1.7e-2	8.2e-1/1.5e-2	4.7e-1/4.3e-2	8.2e-1/1.5e-2	
HV	0.97/1.0e-3	0.97/5.6e-4	0.969/7.1e-4	0.0/0.0	**0.973/2.7e-4**	†
UA	8.1e-1/1.8e-2	8.0e-1/2.0e-2	8.0e-1/1.9e-2	7.9e-1/2.5e-2	**8.2e-1/1.8e-2**	
HV	7.348/1.4e-2	7.337/1.4e-2	7.335/7.8e-3	7.303/8.8e-3	**7.405/1.1e-2**	†
UA	8.0e-1/1.8e-2	3.5e-1/3.9e-2	3.4e-1/3.2e-2	4.0e-1/3.9e-2	**8.3e-1/1.7e-2**	†
HV	6.538/2.0	0.0/0.0	0.0/0.0	0.0/0.0	**7.324/5.7e-1**	†
UA	2.5e-1/3.8e-2	2.4e-1/3.3e-2	2.5e-1/3.0e-2	2.3e-1/3.0e-2	**2.6e-1/2.9e-2**	†
HV	5.58/1.1	6.639/4.8e-1	6.359/7.7e-1	5.971/1.1	**7.029/5.4e-1**	†
UA	7.4e-1/2.2e-2	7.2e-1/2.3e-2	7.2e-1/2.1e-2	7.3e-1/2.8e-2	**7.8e-1/1.8e-2**	†
HV	6.073/3.4e-3	6.067/3.4e-3	6.065/3.9e-3	6.052/5.3e-3	**6.105/4.4e-3**	†
UA	7.9e-1/2.0e-2	7.9e-1/2.1e-2	**7.9e-1/1.7e-2**	7.6e-1/2.5e-2	7.7e-1/2.2e-2	‡
HV	6.107/4.4e-3	6.106/4.2e-3	**6.108/5.4e-3**	5.764/1.0	6.107/4.9e-3	
UA	7.6e-1/1.8e-2	7.7e-1/1.6e-2	7.4e-1/2.2e-2	5.1e-1/1.3e-1	**7.7e-1/1.8e-2**	
HV	13.412/5.6e-2	13.427/4.8e-2	13.346/7.3e-2	7.735/3.7	**13.46/7.4e-2**	†

mutation strategy that is the best over all the functions. For example, for the ZDT2 function, strategy 2 is the best in terms of HV, while strategy 1 is the winner for ZDT3. It is also worth noting that strategy 4 performs worst, while strategies 1 and 3 are the most competitive ones. This kind of situation motivates the use of the AOS paradigm. And indeed, Adap-MODE remains the best option in most of the functions, while achieving very similar performance in others.

The last comparative results, shown in Table 3, presents the performance of Adap-MODE compared with its "partially"-adaptive variants, one using only the AOS, and the other using only the adaptive parameter control of CR and F. From these results, it is not clear which of the adaptive modules is the most beneficial for the performance of Adap-MODE: at some functions, the "AOS only" method is better than the "parameter control only" one, while in others the opposite occurs. But these results clearly demonstrate that the combined use of both adaptive modules is better than their sole use, what is shown by the fact that Adap-MODE significantly outperforms them in most of functions, in terms of both UA and HV.

Table 3. Comparative results of Adap-MODE, Adap-MODE with AOS only and Adap-MODE with parameter control only

		CR/F(fixed)+AOS	CR/F(adapt.)+Unif.OS	Adap-MODE	S
ZDT1	UA	7.860e-1/2.08e-2	7.851e-1/2.42e-2	**8.080e-1/1.62e-2**	†
	HV	3.66162/2.97e-4	3.66066/2.69e-4	**3.66193/3.15e-5**	†
ZDT2	UA	7.809e-1/2.02e-2	7.793e-1/1.71e-2	**8.069e-1/1.89e-2**	†
	HV	3.32840/3.13e-4	3.32612/5.27e-4	**3.32853/4.19e-5**	†
ZDT3	UA	7.538e-1/2.83e-2	7.487e-1/1.52e-2	**7.660e-1/1.98e-2**	†
	HV	4.81448/1.18e-3	4.81228/1.18e-3	**4.81463/4.81e-4**	
ZDT4	UA	**8.127e-1/2.30e-2**	7.486e-1/6.12e-2	8.055e-1/1.85e-2	
	HV	3.64150/1.43e-1	3.65409/4.26e-2	**3.66201/5.33e-4**	†
ZDT6	UA	7.626e-1/2.34e-2	**8.078e-1/2.34e-2**	7.896e-1/2.27e-2	‡
	HV	3.04179/3.22e-5	3.04183/4.93e-5	**3.04183/1.62e-5**	
DTLZ1	UA	**8.247e-1/1.80e-2**	8.200e-1/1.73e-2	8.246e-1/1.48e-2	
	HV	0.969925/5.41e-4	0.917842/1.25e-1	**0.973582/2.75e-4**	†
DTLZ2	UA	8.096e-1/2.01e-2	8.224e-1/1.56e-2	**8.236e-1/1.84e-2**	
	HV	7.33762/1.10e-2	7.40368/9.20e-3	**7.40523/1.14e-2**	
DTLZ3	UA	6.365e-1/1.44e-1	8.289e-1/1.42e-2	**8.304e-1/1.72e-2**	
	HV	7.13704/3.70e-1	4.59535/2.92e+0	**7.32465/5.76e-1**	†
DTLZ4	UA	2.092e-1/3.32e-2	9.814e-2/4.33e-3	**2.654e-1/2.99e-2**	†
	HV	6.78321/6.03e-1	4.66216/1.09e+0	**7.02943/5.46e-1**	†
DTLZ5	UA	7.334e-1/2.26e-2	7.792e-1/1.95e-2	**7.866e-1/1.82e-2**	†
	HV	6.07005/3.69e-3	**6.10649/3.78e-3**	6.10548/4.40e-3	
DTLZ6	UA	7.739e-1/2.32e-2	**7.876e-1/2.00e-2**	7.759e-1/2.18e-2	‡
	HV	**6.10841/5.67e-3**	6.10640/4.14e-3	6.10732/4.88e-3	
DTLZ7	UA	7.621e-1/1.83e-2	7.634e-1/1.70e-2	**7.723e-1/1.86e-2**	†
	HV	13.42436/6.19e-2	13.43145/7.25e-2	**13.46486/7.43e-2**	†

5 Conclusion

In this paper, we propose a new DE algorithm for multi-objective optimization that uses two adaptive mechanisms in parallel: the Adaptive Operator Selection mechanism, to control which operator should be applied at each instant of the search; and the Adaptive Parameter Control, that adapts the values of the DE parameters CR and F while solving the problem. A tree neighborhood density estimator is proposed and, combined with the Pareto dominance strength measure, is used in order to evaluate the fitness of each individual. Additionally, a novel replacement mechanism is proposed, based on a three-step comparison procedure. As a consequence, the adaptive methods employed by the proposed algorithm, inspired from recent literature, are originally ported to the multi-objective domain.

Numerical experiments demonstrate that the proposed Adap-MODE is capable of efficiently adapting to the characteristics of the region that is currently being explored by the algorithm, by efficiently selecting appropriate operators and their corresponding parameters. Adap-MODE is shown to outperform two state-of-the-art MOEAs, namely NSGA-II [4] and GDE3 [12], in most of the functions. It also performs significantly better, in most of the functions, than the same MODE with static parameters, and than the partially-adaptive variants using each of the two adaptive modules.

But there is still a lot of space for improvements. Firstly, for the fitness evaluation, more sophisticated schemes to control the balance between both convergence and spread could be analyzed. Regarding the AOS implementation, other schemes have already shown to perform better than PM in the literature and should also be analyzed in the near future, such as the Adaptive Pursuit [18] and the Dynamic Multi-Armed Bandit [8]; a more recent work, that also use bandits, reward the operators based on ranks [7], thus achieving a much higher robustness w.r.t. different benchmarking situations. In the same way, there are different alternatives for the adaptive parameter control of CR and F that could be further explored.

Another issue that deserves further exploration is related to the (hyper) parameters of the adaptive modules. In the case of Adap-MODE, the AOS requires the definition of the adaptation rate α and the minimum probability p_{min}, while the adaptive parameter control requires the setting of c. In this work, these parameters were set as in the original references, but further analysis of their sensitivity should be done. Ideally, Adap-MODE and the other methods used as baseline should also be all compared again, after a proper off-line tuning phase. Another important baseline would be the same MODE with off-line tuned CR, F, and mutation application rates.

Lastly, the extra computational time resulting from the use of these adaptive schemes should be further analyzed; although it is true to say that, in real-world problems, the fitness evaluation is usually the most computationally expensive step, all the rest becoming negligible.

Acknowledgement

This work is supported by Hong Kong RGC GRF Grant 9041353(CityU 115408).

References

1. Brest, J., Greiner, S., Boskovic, B., Mernik, M., Zumer, V.: Self-adapting control parameters in differential evolution: A comparative study on numerical benchmark problems. IEEE Trans. Evol. Comput. 10, 646–657 (2006)
2. Das, S., Suganthan, P.N.: Differential evolution – a survey of the state-of-the-art. IEEE Trans. Evol. Comput. (in press)
3. Davis, L.: Adapting operator probabilities in genetic algorithms. In: Proc. ICGA, pp. 61–69 (1989)
4. Deb, K., Pratap, A., Agarwal, S., Meyarivan, T.: A fast and elitist multiobjective genetic algorithm: NSGA-II. IEEE Trans. Evol. Comput. 6, 182–197 (2002)
5. Deb, K., Thiele, L., Laummans, M., Zitzler, E.: Scalable test problems for evolutionary multiobjective optimization. In: Abraham, A., et al. (eds.) Evolutionary Multiobjective Optimization, pp. 105–145. Springer, Heidelberg (2005)

6. Eiben, A.E., Hinterding, R., Michalewicz, Z.: Parameter control in evolutionary algorithms. IEEE Trans. Evol. Comput. 3, 124–141 (1999)
7. Fialho, Á., Ros, R., Schoenauer, M., Sebag, M.: Comparison-based adaptive strategy selection with bandits in differential evolution. In: Schaefer, R., Cotta, C., Kołodziej, J., Rudolph, G., et al. (eds.) PPSN XI. LNCS, vol. 6238, pp. 194–203. Springer, Heidelberg (2010)
8. Fialho, Á., Da Costa, L., Schoenauer, M., Sebag, M.: Extreme value based adaptive operator selection. In: Rudolph, G., Jansen, T., Lucas, S., Poloni, C., Beume, N. (eds.) PPSN 2008. LNCS, vol. 5199, pp. 175–184. Springer, Heidelberg (2008)
9. Gong, W., Fialho, A., Cai, Z.: Adaptive strategy selection in differential evolution. In: Branke, J., et al. (eds.) Proc. GECCO. ACM, New York (2010)
10. Huang, V.L., Zhao, S.Z., Mallipeddi, R., Suganthan, P.N.: Multi-objective optimization using self-adaptive differential evolution algorithm. In: Proc. CEC, pp. 190–194. IEEE, Los Alamitos (2009)
11. Jia, L., Gong, W., Wu, H.: An improved self-adaptive control parameter of differential evolution for global optimization. In: Cai, Z., Li, Z., Kang, Z., Liu, Y. (eds.) Computational Intelligence and Intelligent Systems. CCIS, vol. 51, pp. 215–224. Springer, Heidelberg (2009)
12. Kukkonen, S., Lampinen, J.: GDE3: The third evolution step of generalized differential evolution. In: Proc. CEC, pp. 443–450. IEEE, Los Alamitos (2005)
13. Li, M., Zheng, J., Xiao, G.: Uniformity assessment for evolutionary multi-objective optimization. In: Proc. CEC, pp. 625–632. IEEE, Los Alamitos (2008)
14. Maturana, J., Lardeux, F., Saubion, F.: Autonomous operator management for evolutionary algorithms. J. Heuristics (2010)
15. Price, K.V.: An introduction to differential evolution. In: Corne, D., et al. (eds.) New Ideas in Optimization, pp. 79–108. McGraw-Hill, New York (1999)
16. Qin, A.K., Huang, V.L., Suganthan, P.N.: Differential evolution algorithm with strategy adaptation for global numerical optimization. IEEE Trans. Evol. Comput. 13, 398–417 (2009)
17. Robič, T., Filipič, B.: DEMO: Differential evolution for multiobjective optimization. In: Coello Coello, C.A., Hernández Aguirre, A., Zitzler, E. (eds.) EMO 2005. LNCS, vol. 3410, pp. 520–533. Springer, Heidelberg (2005)
18. Thierens, D.: An adaptive pursuit strategy for allocating operator probabilities. In: Beyer, H.-G., et al. (eds.) Proc. GECCO, pp. 1539–1546. ACM, New York (2005)
19. Whitacre, J., Pham, T., Sarker, R.: Use of statistical outlier detection method in adaptive evolutionary algorithms. In: Proc. GECCO, pp. 1345–1352. ACM, New York (2006)
20. Zhang, J., Sanderson, A.C.: Self-adaptive multi-objective differential evolution with direction information provided by archived inferior solutions. In: Proc. CEC, pp. 2806–2815. IEEE, Los Alamitos (2008)
21. Zhang, J., Sanderson, A.C.: JADE: Adaptive differential evolution with optional external archive. IEEE Trans. Evol. Comput. 13, 945–958 (2009)
22. Zitzler, E., Deb, K., Thiele, L.: Comparison of multiobjective evolutionary algorithms: Empirical results. Evol. Comput. 8, 173–195 (2000)
23. Zitzler, E., Laumanns, M., Thiele, L.: SPEA2: Improving the strength pareto evolutionary algorithm for multiobjective optimization. In: Giannakoglou, K.C., et al. (eds.) Evolutionary Methods for Design, Optimisation and Control with Application to Industrial Problems, pp. 95–100. CIMNE (2002)
24. Zitzler, E., Thiele, L.: Multiobjective evolutionary algorithms: a comparative case study and the strength pareto approach. IEEE Trans. Evol. Comput. 3, 257–271 (1999)

Distribution of Computational Effort in Parallel MOEA/D

Juan J. Durillo[1], Qingfu Zhang[2], Antonio J. Nebro[1], and Enrique Alba[1]

[1] Department Lenguajes y Ciencias de la Computación, University of Málaga, Spain
{durillo,antonio,eat}@lcc.uma.es
[2] The School of Computer Science and Electronic Engineering, University of Essex, Wivenhoe Park, Colchester, CO4 3SQ, U.K.
qzhang@essex.ac.uk

Abstract. MOEA/D is a multi-objective optimization algorithm based on decomposition, which consists in dividing a multi-objective problem into a number of single-objective sub-problems. This work presents two variants, called pMOEA/Dv1 and pMOEA/Dv2, of a new parallel model of MOEA/D that have been developed under the observation that different sub-problems may require different computational effort, and thus, demand different number of evaluations. Our interest in this paper is to analyze whether the proposed models are able of outperforming the MOEA/D in terms of the quality of the computed fronts. To cope with this issue, our proposals have been evaluated using a benchmark composed of eight problems and the obtained results have been compared against MOEA/D-DE, an extension of the original MOEA/D where new individuals are generated by an operator taken from differential evolution. Our experiments show that some configurations of pMOEA/Dv1 and pMOEA/Dv2 have been able to compute fronts of higher quality than MOEA/D-DE in many of the evaluated problems, giving room for further research in this line.

1 Introduction

A multi-objective optimization problem (MOP) requires to reconcile several conflicting objectives. A solution is Pareto optimal if any improvement in one objective leads to deterioration in at least one other objective. The Pareto front of a MOP is the set of all the Pareto optimal solutions in the objective space. In many applications, a decision marker would like to have a good approximation of the Pareto front for selecting their preferred tradeoff solutions.

Multi-objective Evolutionary Algorithms (MOEAs) aim at finding a number of Pareto optimal solutions to approximate the Pareto front in a single run. Many MOEAs have been developed during the last twenty years. Most of current popular MOEAs are Pareto dominance based, in which the fitness of an individual solution is mainly determined by dominance relationships with other individuals. Very recently, some effort has been made to develop other MOEA paradigms. MOEA/D [8] is such an example. It decomposes a MOP into a number of single objective optimization sub-problems. The objective of each sub-problem is

C.A. Coello Coello (Ed.): LION 5, LNCS 6683, pp. 488–502, 2011.
© Springer-Verlag Berlin Heidelberg 2011

a (linear or nonlinear) weighted aggregation of all the individual objectives in the MOP. Neighborhood relations among these sub-problems are defined based on the distances among their aggregation weight vectors. Each sub-problem is optimized in MOEA/D by using information mainly from its neighboring sub-problems. The MOEA/D framework has been studied and used with success for dealing with a number of multi-objective problems [8][5][9].

For continuous multi-objective problems, three different versions of MOEA/D have been designed, namely, MOEA/D with SBX operator (MOEA/D-SBX), MOEA/D with DE operators (MOEA/D-DE), and MOEA/D with Dynamic Resource Allocation (MOEA/D-DRA). In MOEA/D-SBX [8], SBX crossover and polynomial mutation operators are used for generating new solutions. A solution is allowed to mate only with its neighbors and a new solution could replace any old solutions in its neighborhood if it is better than them. In MOEA/D-DE [5], differential evolution operators are used for generating new solutions. To encourage diversity, two extra measures are used in MOEA/D-DE. One is that a new solution is allowed to replace only a small number of old solutions, the other is that a solution, with a very low probability, can mate with any other solution in the population. Aiming at further improvement, MOEA/D-DRA [9] assigns different amounts of computational effort to different sub-problems according to their utilities, which are estimated during the search.

A parallel version of MOEA/D-DE, called pMOEA/D, has been suggested in [7] and linear speedup has been observed when using up to 8 cores. In this work, based on the ideas of dynamic resource allocation and pMOEA/D, we propose two new parallel models of MOEA/D. The major purpose of our work is to study whether or not those models outperform MOEA/D-DE on a set of benchmark problems.

To cope with this issue, we first present a new parallel model consisting on the use of several disjoint partitions of the population. This model is similar to the one we previously proposed in [7] or the one proposed by Branke et al. in [1]. However, while in this last work the divisions are made in the search space, in our proposal, as well as in [7], the search space is the same as for the original MOP in each partition, being only modified the objectives to compute.

Although in this paper we have considered MOEA/D-DE, our approach is easily applicable to any version of MOEA/D. Then, taken this model as starting point we propose two different versions, called pMOEA/Dv1 and pMOEA/Dv2, that are based on different mechanisms for distributing the computational effort among the different partitions employed. The experiments carried out in this work will show that it is possible to improve the results obtained by MOEA/D, giving room to open new research lines in that direction.

The contributions of this paper are the following:

- We propose a new parallel model for MOEA/D.
- Two versions of that model, named pMOEA/Dv1 and pMOEA/Dv2, have been developed by considering different ways of balancing the computational effort.
- We have compared the behavior of those versions over a benchmark of bi-objective problems belonging to the LZ09 family [5].

The rest of the paper is organized as follows. First, we describe MOEA/D in the next section. The parallel model for MOEA/D is presented in Section 3. In Section 4, we extend this model by incorporating a mechanism for balancing the computational effort among different sub-problems. Sections 5 and 6 are devoted to describing the experiments we have performed and to analyzing the obtained results. Finally, the conclusions and lines of future work are presented in 7.

2   Sequential MOEA/D

Consider the following multi-objective optimization problem:

$$\text{minimize } F(x) = (f_1(x), \ldots, f_m(x)) \tag{1}$$
$$\text{subject to} \qquad x \in \Omega$$

MOEA/D employs an aggregation approach to decompose (1). In principle, any aggregation approach works. In this paper, we make use of the Tchebycheff aggregation approach [6], where the scalar optimization problems (i.e., sub-problems) are in the form

$$\text{minimize } g^{te}(x|\lambda, z^*) = \max_{1 \le i \le m}\{\lambda_i|f_i(x) - z_i^*|\} \tag{2}$$
$$\text{subject to} \qquad x \in \Omega$$

where $z^* = (z_1^*, \ldots, z_m^*)^T$ is the reference point, i. e., $z_i^* = \max\{f_i(x)|x \in \Omega\}$. $\lambda = (\lambda_1, \ldots, \lambda_m)$ is a weight vector, i.e. $\lambda_i \ge 0$ for all $i = 1, \ldots, m$ and $\sum_{i=1}^m \lambda_i = 1$. For each Pareto optimal point x^* there exists a weight vector λ such that x^* is the optimal solution of (2) and each optimal solution of (2) is a Pareto optimal solution of (1). Therefore, one is able to obtain different Pareto optimal solutions by solving a set of single objective optimization problem defined by the Tchebycheff approach with different weight vectors.

MOEA/D chooses a set of uniformly distributed weight vectors $\lambda^1, \ldots, \lambda^N$, and decompose (1) into N single objective optimization subproblems. The objective of the i-th subproblem is $g^{te}(x|\lambda^i, z^*)$.

During the search, MOEA/D maintains:

- a population of N points $x^1, \ldots, x^N \in \Omega$, where x^i is the current solution to the i-th subproblem;
- $FV^1, \ldots, FV^N$, where FV^i is the F-value of x^i, i.e., $FV^i = F(x^i)$ for each $i = 1, \ldots, N$;
- $z = (z_1, \ldots, z_m)^T$, where z_i is the best value found so far for objective f_i, z is used to substitute z_i^* in computing g^{te} during the search.

For each $i = 1, \ldots, N$, set $B(i) = \{i_1, \ldots, i_T\}$ where $\lambda^{i_1}, \ldots, \lambda^{i_T}$ are the T closest weight vectors to λ^i in terms of Euclidean distance. The neighborhood of x_i is $\{x_k|k \in B(i)\}$.

For each $i = 1, \ldots, N$ at each generation, MOEA/D does the following:

Step 1. *Selection of Mating/Update Range*: Uniformly randomly generate a number $rand$ from $(0, 1)$. Then set

$$P = \begin{cases} B(i) & \text{if } rand < \delta, \\ \{1, \ldots, N\} & \text{otherwise.} \end{cases}$$

where $0 < \delta < 1$ is a prefixed control parameter.

Step 2. *Reproduction*: Randomly select three current solutions from $\{x^k | k \in P\}$. Apply genetic operators on them to generate a new solution y, and reply y if y is not feasible. Compute $F(y)$.

Step 3. *Update of z*: For each $j = 1, \ldots, m$, if $z_j > f_j(y)$, then set $z_j = f_j(y)$.

Step 4. *Update of Solutions*: Set $c = 0$ and then do

   While $\{c = n_r$ or P is empty$\}$

     1. Randomly pick an index j from P.

     2. If $g(y|\lambda^j, z) \leq g(x^j|\lambda^j, z)$, then set $x^j = y$, $FV^j = F(y)$.

     3. Remove j from P and set $c = c + 1$.

   End of While Loop

   where n_r is a predetermined control parameter.

More details about this MOEA/D variant can be found in [5], a DE operator and a polynomial mutation operator are used in Step 2 as genetic operator in [5].

3   A Parallel Model of MOEA/D

The parallel model we propose here for MOEA/D consists in defining a number of partitions of the whole population, as in [7]. Each partition is composed by a number of different sub-problems and is to be evolved in parallel. As we indicated in previous section, each sub-problem is defined by a given weight vector, λ, which has as many components as objectives in the target MOP. The closer the vectors, the closer the solutions they define into the Pareto front. In order to define the partitions, we sort these vectors by mean of one of their components, and we assign the first closest (N/Number of sub-problems) weights to the first partition, the next closest (N/Number of sub-problems) to the second partition, and so on. The idea is that each partition computes a small region of the Pareto front. Fig. 1 depicts an example of how this method works for three partitions.

The main difference between this approach and [7] lies in the way in which neighborhood are defined. In the latter, they are defined taking into account the whole population, as in the original MOEA/D algorithm. As a consequence, neighborhoods could be defined across different partitions. In our approach, the neighborhood of each solution is defined by considering only sub-problems that belong to the same partition. Our motivation for doing so is that we are interested in isolating each partition in order to better analyze its behavior.

Additionally, this model reduces the number of possible concurrent access to the same solution in comparison with pMOEA/D. While in that algorithm a sub-problem can be accessed by different processors (considering that every partition is assigned by a different processor), in the model proposed here each

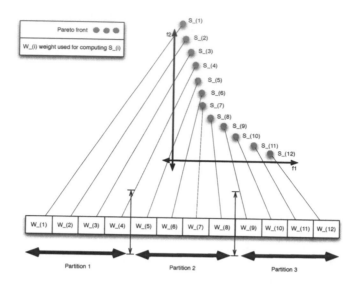

Fig. 1. Distribution of the Pareto front into different partitions. Notation: S_(i) and W_(j) refer to sub-problem$_i$ and weight$_j$, respectively.

sub-problem is accessed only by one. Thus, under the assumption of considering an enough number of sub-problems, we can see our model as a set of independent MOEA/D instances executing in each partition, and, hence, working each of them in finding a part of the Pareto front. It is worth to notice that this model can easily be implemented in a parallel system by just defining as many partitions as available cores/processors and then assigning each of those partitions to a different core/processor. Finally, as done in MOEA/D-DE, in a small number of cases (10% of the total number of performed evaluations) a partition could select a sub-problem from the whole population (i.e., from other partitions). This mechanism allows partitions to share information, and it can be seen as a migration of solutions.

Fig. 2 includes an example of Pareto front computed by this model when solving the LZ09_F2 problem. In this figure, we can see different sub-fronts, represented by a different symbols. Each of these sub-fronts has been generated by a different partition (in this example we have made use of only four partitions). As we see, all the partitions have succeeded in converging towards the optimal Pareto front (authors unfamiliar with this problem, please refer to [5]).

4   Extending the Proposed Parallel Model

In the previous section, we have proposed a parallel model of MOEA/D and we have shown that it was possible to compute an accurate Pareto front by using it. As in the original MOEA/D algorithm, in this parallel model all the sub-problems are considered as having the same degree of difficulty. Hence, the same

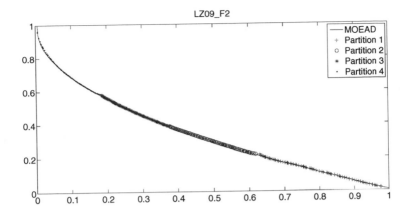

Fig. 2. Computed front for the LZ09_F2 problem with a partitioned MOEA/D

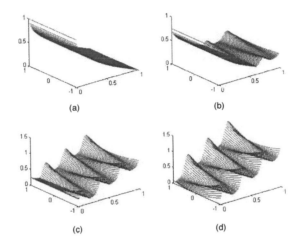

Fig. 3. Landscapes of different sub-problems defined for the LZ09_F2 problem

computational effort is allocated to each of them. However, this is not a realistic approximation. An example supporting this claim is depicted in Fig.3. This figure shows the landscape of four different sub-problems using four different weights in the LZ09_F2 problem. More specifically, these landscapes show the value that the objective function associated to each sub-problem takes for every combination of the two first decision variables defining the LZ09_F2. For simplicity, the rest of decision variables has been fixed to the optimal value for each sub-problem.

Focusing on the analysis of the figure, we can observe that not all the landscapes have the same degree of difficulty. For example, the landscape represented in (a) could be the simplest one and it could be solved in a fast manner by a gradient-based method. As long as we move from left to right and from top to bottom, it is possible to see how these landscapes presents a higher number of peaks and local optima. This way, one may think that sub-problem (d), for example, would require a higher computational effort than problem (a) for being solved. As a consequence, it is reasonable to think that by balancing the computational effort among the different sub-problems it would be possible to reduce the number of required evaluations for solving the problem. The idea is to avoid performing more function evaluations in those sub-problems which have already converged towards their optimal solution, and to perform a higher number of evaluations in the rest of sub-problems, as done in MOEA/D-DRA [9]. In this last algorithm the number of performed evaluations in each sub-problem depended on how fast that sub-problem converged towards their optimal solution. Our aim is to extend the parallel model proposed before for taking advantage of this fact, in such a way that different computational efforts can be allocated to different partitions depending on their behavior.

Thus, the success of our proposal lies in an accurate balance of the computational effort; however, this information is not known beforehand. Our approach has consisted in determining in running time which partition is performing better or worse than others, and to adapt dynamically the evaluations performed in each partition. In order to accomplish this idea, we need:

- A way of measuring the advance of each partition with respect to the others, and
- a way of dynamically assigning different computational efforts to different partitions.

Regarding to the first point, we have considered in this work an adaptation of the method used in MOEA/D-DRA, where the authors define a way of quantifying the utility of each point based on the differences in the fitness in different iterations (Equation 3). In this work, we have adapted it for measuring the utility of one partition instead of a single point. Specifically, we have considered that the utility of a partition in a given iteration, t, is given by the mean of the utility of the points included into that partition.

$$utility(t) = \begin{cases} 1 & \text{if} \Delta(t-1) \leq 0.001; \\ (0.95 + 0.05 * \frac{\Delta(t-1)}{0.001} * utility(t-1)) & \text{otherwise} \end{cases}$$

$$\Delta(t) = \frac{g(x^{(t-1)}|\lambda, z^*) - g(x^{(t)}|\lambda, z^*)}{g(x^{(t-1)}|\lambda, z^*)} \tag{3}$$

where $x^{(t)}$ refers to decision variables at iteration t.

For dealing with the second issue we have considered two different alternatives:

1. to dynamically change the number of evaluations performed in each partition, and
2. to dynamically change the size of some partitions.

These alternatives give rise to two different versions of our parallel model, namely pMOEA/Dv1 and pMOEA/Dv2 respectively.

4.1 pMOEA/Dv1

Let us start describing the first alternative. The idea beyond this version is very simple: based on their utility, some partitions are executed for a higher number of evaluations than others. The procedure for doing so is defined as follows:

- Step 1. Evolve each partition by using a given number of evaluations.
- Step 2. Compute the utility of each partition.
- Step 3. Distribute the evaluations between partitions based on the utility of each particle (the higher the utility of a given partition, the higher the number of evaluations assigned to that partition).
- Step 4. Perform the number of evaluations assigned to each partition.
- Step 5. If the number of maximum evaluations has not been performed, go to Step 2.

The underlaying idea of this approach consists in giving more chances to those partitions that show a good utility. Once a partition has converged, its utility should be close to zero (no improvements of the fitness are possible for that partition), and, as a consequence the rest of evaluations are mainly distributed among the rest of partitions. A drawback of this approach is that a partition could also get stuck in a local minimum, reporting no utility. In this case, the communication mechanism among the partitions described before is the only way of escaping from that situation.

For implementing this approach there are two key factors which should be taken into account:

- How many evaluations should be performed among all the partitions before recomputing their utility, and
- how to distribute these evaluations among the partitions.

In this paper, the number of evaluations has been determined empirically after a set of preliminary experiments by using pMOEA/Dv1 with four partitions. Considering a total number of 150,000 function evaluations, these experiments showed that the overall best results were obtained when 30,000 function evaluations were performed (steps 1 and 4) before recomputing their utility. In order to determine how to distribute those evaluations, we have proceed as follows:

- Step 1. The utility of each partition i, u_i, has been determined.
- Step 2. The total utility, u_{total}, has been defined as $\sum_i^N u_i$, being N the number of partitions.
- Step 3. A probability of being selected, p_i, is assigned to each partition, with p_i proportional to the contribution of u_i to u_{total}.
- Step 4. Based on its probability a partition is selected.
- Step 5. A number of 1,000 function evaluations is allocated to the selected partition.
- Step 6. If the total number of function of evaluations has not reached the maximum (30,000 in our case), goto Step 4.

4.2   pMOEA/Dv2

The second alternative consists in dynamically changing the size of the partitions. In particular, whenever the algorithm detects that a partition has no utility for several consecutive iterations, the size of that partition is augmented by taking individuals from other partitions. The way in which a partition size is increased is depicted in Fig. 4. The procedure consists in taking those sub-problems whose weight vectors are the closest to the weight vectors defining the sub-problems already included in the partition.

It is worth noting that in this alternative there are two cases in which a partition is augmented:

- When partition is stuck in a local minimum. In this case, the effect of this mechanism is twofold:
  - We increase the diversity of the stuck partition by adding new sub-problems, thus increasing the chances of escaping from that local minimum.
  - We decrease the number of sub-problems of other partitions, thus letting them (as with the first alternative) to converge faster.
- When all the solutions in this partition have converged. This way, new added sub-problems could benefit from the already known solutions.

In this approach, there are also two key issues which should be considered in order to implement it:

- When to consider that a partition should be augmented, and
- how many new individuals are added to that partition.

As with pMOEA/Dv1, in this paper those parameters have been fixed after carrying out a preliminary set of experiments. In particular, these experiments showed that the best results were obtained when a partition is increased in 20 individuals if it had no utility for five consecutive iterations.

5   Experimentation

In this section, we evaluate the two parallel versions of MOEA/D, comparing them against the original algorithm.

For assessing the performance of algorithms we have used the following quality indicators: hypervolume (I_{HV}) [10], and additive epsilon indicator ($I_{\epsilon+}^1$) [4]. The former indicator measures the convergence and diversity of an approximation to the Pareto front of a problem while the latter indicator only measures convergence. In the case of the $I_{\epsilon+}^1$ indicator, the lower the value, the better the quality of the results. Conversely, for the I_{HV} indicator, higher values of the indicator mean approximations to the Pareto front of better quality.

All the algorithms evaluated in this work have been implemented using jMetal [3], a framework aimed multi-objective optimization with metaheuristics. As our intention in this work is to evaluate the effectiveness of our models, we have considered here a sequential implementation of pMOEA/Dv1 and pMOEA/Dv2, simulating a concurrent behavior.

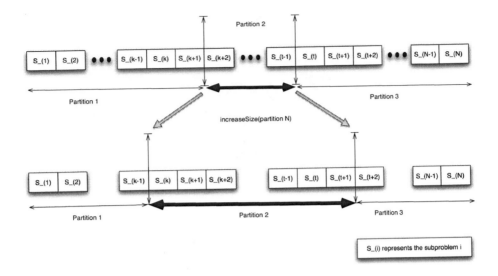

Fig. 4. Partitions with variable size

5.1 Configuration

In this work, we have used the same parameter settings proposed in [8] for all the evaluated algorithms:

- Population size: 300
- Stopping condition: 150,000 function evaluations
- Weight vectors: we have used the values provided in [9]
- Control parameters in DE: $CR = 1.0$, $F = 0.5$
- Polynomial mutation operators: $p_m = 1/n$ (n is the number of decision variables), distribution index $= 20$
- Rest of parameters: $T = 20$, $\delta = 0.9$, $n_r = 2$

For the parallel models proposed here, we have considered a number of partitions ranging between 2 and 8.

5.2 Benchmark

For evaluating our proposals, we have considered as a benchmark the bi-objective problems belonging to the LZ09 benchmark [5]. These problems are labelled as LZ09_F1, LZ09_F2, ..., LZ09_F9 but LZ09_F6, which has three objectives.

6 Analysis of the Results

This section is devoted to analyzing the obtained results. For each combination of algorithm and problem we have made 100 independent runs, and we report

Table 1. Median and IQR of the $I_{\epsilon+}^1$ indicator for pMOEA/Dv1 in the considered benchmark

Problem	MOEA/D $\tilde{x}_{IQR}$	pMOEA/Dv1 (2) $\tilde{x}_{IQR}$	pMOEA/Dv1 (4) $\tilde{x}_{IQR}$	pMOEA/Dv1 (8) $\tilde{x}_{IQR}$
LZ09_F1	$4.06e-03_{7.0e-04}$	$4.27e-03_{6.6e-04}$	$3.98e-03_{6.7e-04}$	$5.66e-03_{1.7e-03}$
LZ09_F2	$1.03e-02_{4.0e-03}$	$9.49e-03_{3.7e-03}$	$9.48e-03_{5.2e-03}$	$1.22e-02_{6.9e-03}$
LZ09_F3	$1.64e-01_{2.1e-01}$	$5.81e-02_{2.1e-01}$	$4.83e-02_{2.1e-01}$	$4.57e-02_{2.4e-01}$
LZ09_F4	$5.69e-02_{2.3e-02}$	$3.10e-02_{1.8e-02}$	$1.97e-02_{1.2e-02}$	$2.66e-02_{2.2e-02}$
LZ09_F5	$7.89e-02_{2.4e-02}$	$7.28e-02_{2.4e-02}$	$7.37e-02_{3.2e-02}$	$8.77e-02_{2.1e-01}$
LZ09_F7	$3.33e-02_{3.4e-02}$	$2.56e-02_{1.7e-02}$	$2.86e-02_{9.3e-03}$	$9.85e-02_{7.3e-02}$
LZ09_F8	$1.85e-01_{1.0e-01}$	$2.71e-01_{2.0e-01}$	$3.54e-01_{1.1e-01}$	$2.84e-01_{1.5e-01}$
LZ09_F9	$1.61e-02_{1.6e-02}$	$1.09e-02_{9.3e-03}$	$1.30e-02_{5.1e-03}$	$1.80e-02_{8.3e-03}$

Table 2. Statistical comparison between each configuration of pMOEA/Dv1 for the $I_{\epsilon+}^1$ indicator

	pMOEA/Dv1 (2)	pMOEA/Dv1 (4)	pMOEA/Dv1 (8)
MOEA/D	$--\nabla\nabla-\nabla\blacktriangle\nabla$	$--\nabla\nabla-\nabla\blacktriangle-$	$\blacktriangle\blacktriangle-\nabla-\blacktriangle\blacktriangle-$
pMOEA/Dv1 (2)		$\nabla--\nabla--\blacktriangle-$	$\blacktriangle\blacktriangle--\blacktriangle\blacktriangle-\blacktriangle$
pMOEA/Dv1 (4)			$\blacktriangle\blacktriangle-\blacktriangle-\blacktriangle\nabla\blacktriangle$

the median, $\tilde{x}$, and the interquartile range, IQR, as measures of location (or central tendency) and statistical dispersion, respectively. The best result for each problem has a gray colored background. For the sake of a better understanding, we have also used a clear grey background to indicate the second best result; this way, we can visualize at a glance the most salient techniques.

Let us start by analyzing the results obtained by the first approach. Tables 1 and 3 summarize the values obtained in the $I_{\epsilon+}^1$ and I_{HV} indicators, respectively. In these tables, MOEA/D refers to MOEA/D-DE, and pMOEA/Dv1 (X) refers to our approach, being X the number of partitions employed (as we mentioned before, ranging from 2 to 8).

Focusing on the $I_{\epsilon+}^1$ indicator (Table 1), we observe that pMOEA/Dv1 has obtained the best results in this indicator when using two and four partitions. Actually, it has yielded the best values in three problems, and the second best in other two when two partitions are used, and three best values and four second best value when the number of partition is four. This way, these results indicate that it is possible to improve the convergence towards the Pareto optimal front when using up to four partitions.

For each pair of variants we have also analyzed if the differences between them are statistically confident. To cope with this issue, we have applied the Wilcoxon rank sum test, a non-parametric statistical hypothesis test, which allows us to make pairwise comparisons between algorithms to know about the significance of the obtained data [2]. A confidence level of 95% (i.e., significance level of 5% or p-value under 0.05) has been used in all the cases, which means that the differences are unlikely to have occurred by chance with a probability of 95%. In each cell, the eight considered MOPs are represented with a symbol. Three different symbols are used:"–" indicates that there hast not been statistical significance between the algorithms, "▲" means that the algorithm in the row has yielded better results than the algorithm in the column with statistical confidence, and

Table 3. Median and IQR of the I_{HV} indicator for pMOEA/Dv1 in the considered benchmark

Problem	MOEA/D $\tilde{x}_{IQR}$	pMOEA/Dv1 (2) $\tilde{x}_{IQR}$	pMOEA/Dv1 (4) $\tilde{x}_{IQR}$	pMOEA/Dv1 (8) $\tilde{x}_{IQR}$
LZ09_F1	$6.64e-01_{9.0e-05}$	$6.64e-01_{9.9e-05}$	$6.64e-01_{2.4e-04}$	$6.64e-01_{3.1e-04}$
LZ09_F2	$6.61e-01_{6.4e-04}$	$6.61e-01_{6.6e-04}$	$6.60e-01_{6.4e-04}$	$6.59e-01_{2.3e-03}$
LZ09_F3	$6.22e-01_{5.6e-02}$	$6.54e-01_{5.0e-02}$	$6.56e-01_{4.2e-02}$	$6.54e-01_{6.4e-02}$
LZ09_F4	$6.56e-01_{2.6e-03}$	$6.60e-01_{1.7e-03}$	$6.61e-01_{8.0e-04}$	$6.59e-01_{2.5e-03}$
LZ09_F5	$6.48e-01_{4.4e-03}$	$6.49e-01_{5.9e-03}$	$6.48e-01_{1.1e-02}$	$6.44e-01_{5.1e-02}$
LZ09_F7	$6.60e-01_{7.9e-03}$	$6.61e-01_{3.4e-03}$	$6.60e-01_{1.5e-03}$	$6.43e-01_{2.4e-02}$
LZ09_F8	$5.58e-01_{6.0e-02}$	$4.81e-01_{1.6e-01}$	$4.05e-01_{1.2e-01}$	$4.45e-01_{1.5e-01}$
LZ09_F9	$3.27e-01_{2.1e-03}$	$3.27e-01_{1.2e-03}$	$3.26e-01_{1.3e-03}$	$3.24e-01_{1.3e-03}$

Table 4. Statistical comparison between each configuration of pMOEA/Dv1 for the I_{HV} indicator

	pMOEA/Dv1 (2)	pMOEA/Dv1 (4)	pMOEA/Dv1 (8)
MOEA/D	▲ ▲ ▽ ▽ − ▽ ▲ −	▲ ▲ ▽ ▽ − − ▲ −	▲ ▲ − ▽ ▲ ▲ ▲ ▲
pMOEA/Dv1 (2)		− ▲ − ▽ − − ▲ ▲	▲ ▲ − ▲ ▲ ▲ ▲ ▲
pMOEA/Dv1 (4)			▲ ▲ − ▲ ▲ ▲ − ▲

Table 5. Median and IQR of the I_c^1 Indicator for pMOEA/Dv2 in the considered benchmark

Problem	MOEA/D $\tilde{x}_{IQR}$	pMOEA/Dv2 (2) $\tilde{x}_{IQR}$	pMOEA/Dv2 (4) $\tilde{x}_{IQR}$	pMOEA/Dv2 (8) $\tilde{x}_{IQR}$
LZ09_F1	$4.00e-03_{7.0e-04}$	$4.15e-03_{3.0e-04}$	$4.35e-03_{4.8e-04}$	$4.28e-03_{5.0e-01}$
LZ09_F2	$1.03e-02_{4.0e-03}$	$6.90e-03_{1.7e-03}$	$7.98e-03_{4.6e-03}$	$8.48e-03_{1.9e-03}$
LZ09_F3	$1.64e-01_{2.1e-01}$	$1.10e-02_{3.2e-03}$	$4.86e-02_{6.5e-02}$	$4.18e-02_{4.3e-02}$
LZ09_F4	$5.69e-02_{2.3e-02}$	$4.69e-02_{1.9e-02}$	$4.34e-02_{1.7e-02}$	$4.49e-02_{2.1e-02}$
LZ09_F5	$7.89e-02_{2.4e-02}$	$6.53e-02_{2.6e-02}$	$7.57e-02_{1.5e-02}$	$7.27e-02_{2.2e-02}$
LZ09_F7	$3.33e-02_{3.4e-02}$	$6.87e-02_{7.4e-02}$	$8.43e-02_{1.2e-01}$	$6.19e-02_{6.5e-02}$
LZ09_F8	$1.85e-01_{1.0e-01}$	$2.49e-01_{7.9e-02}$	$2.75e-01_{1.6e-01}$	$2.92e-01_{2.2e-01}$
LZ09_F9	$1.61e-02_{1.6e-02}$	$9.37e-03_{5.6e-03}$	$1.70e-02_{1.1e-02}$	$1.80e-02_{9.0e-02}$

"▽" is used when the algorithm in the column has been statistically better than the algorithm in the row.

Table 2 summarizes the results of that statistical analysis for the epsilon indicator. This table confirms the results described above. In fact, we can observe that by using up to four partitions, pMOEA/Dv1 has equaled or improved the results of the original algorithm in all the problems but one (LZ09_F8). We can also observe that the configuration with eight partitions has been worse than the others in most of the cases.

Regarding to the I_{HV} indicator, we observe in Table 3 that pMOEA/Dv1 configured with two partitions has been the most salient algorithm in the comparison, obtaining either the best or second best value in practically all the problems. In this case, with more than four partitions the approach has not been able to outperform the original MOEA/D.

Focusing on the results of the statistical tests (Table 4), pMOEAD/v1 (2) has been statistically better than MOEA/D in two out of the eight problems, and statistically worse in other two (in the rest of the cases there are not statistical differences between them). As long as the number of partitions increase, the results have been of lower quality than those obtained by MOEA/D.

Table 6. Statistical comparison between each configuration of pMOEA/Dv2 for the $I^1_{\epsilon+}$ indicator

	pMOEA/Dv2 (2)	pMOEA/Dv2 (4)	pMOEA/Dv2 (8)
MOEA/D	– ▽ ▽ ▽ ▲ ▲ ▽	▲ ▽ ▽ ▽ – ▲ ▲ –	▲ ▽ ▽ ▽ – – ▲ –
pMOEA/Dv2 (2)		– ▲ ▲ – ▲ – – ▲	– ▲ ▲ – ▲ – ▲ ▲
pMOEA/Dv2 (4)			– – – – – – – –

Table 7. Median and IQR of the I_{HV} indicator for pMOEA/Dv2 in the considered benchmark

Problem	MOEA/D $\tilde{x}_{IQR}$	pMOEA/Dv2 (2) $\tilde{x}_{IQR}$	pMOEA/Dv2 (4) $\tilde{x}_{IQR}$	pMOEA/Dv2 (8) $\tilde{x}_{IQR}$
LZ09_F1	$6.64e-01_{9.0e-05}$	$6.64e-01_{9.6e-05}$	$6.64e-01_{9.6e-05}$	$6.64e-01_{7.4e-05}$
LZ09_F2	$6.61e-01_{6.4e-04}$	$6.62e-01_{4.0e-04}$	$6.62e-01_{1.2e-03}$	$6.61e-01_{8.6e-04}$
LZ09_F3	$6.22e-01_{5.6e-02}$	$6.61e-01_{1.3e-02}$	$6.57e-01_{1.5e-02}$	$6.58e-01_{9.8e-03}$
LZ09_F4	$6.56e-01_{2.6e-03}$	$6.58e-01_{3.1e-03}$	$6.59e-01_{3.3e-03}$	$6.57e-01_{3.1e-03}$
LZ09_F5	$6.48e-01_{4.4e-03}$	$6.51e-01_{5.5e-03}$	$6.48e-01_{4.9e-03}$	$6.47e-01_{7.7e-03}$
LZ09_F7	$6.60e-01_{7.9e-03}$	$6.51e-01_{2.0e-02}$	$6.47e-01_{3.4e-02}$	$6.52e-01_{1.8e-02}$
LZ09_F8	$5.58e-01_{6.0e-03}$	$4.99e-01_{7.8e-02}$	$5.02e-01_{7.7e-02}$	$4.75e-01_{9.8e-02}$
LZ09_F9	$3.27e-01_{2.1e-03}$	$3.27e-01_{1.4e-03}$	$3.27e-01_{1.1e-03}$	$3.24e-01_{2.3e-02}$

Table 8. Statistical comparison between each configuration of pMOEA/Dv2 for the I_{HV} indicator

	pMOEA/Dv2 (2)	pMOEA/Dv2 (4)	pMOEA/Dv2 (8)
MOEA/D	▲ ▽ ▽ ▽ ▽ ▲ ▲ ▽	– ▽ ▽ – ▲ ▲ –	▲ – ▽ – – ▲ ▲ ▲
pMOEA/Dv2 (2)		– – ▲ – ▲ – – –	▲ ▲ ▲ – ▲ – – ▲
pMOEA/Dv2 (4)			▲ ▲ – ▲ – – – ▲

Let us analyze pMOEA/Dv2 now. In this case, the data summarizing the experiments are included in tables 5, 7, 6, and 8. We will refer to this version as pMOEA/Dv2 (X), where X indicates the number of partitions.

Proceeding as before, we start by analyzing the results of the $I^1_{\epsilon+}$ indicator, summarized in Table 5. In this case, the configuration with two partitions has been the most salient of the comparison (best value of the indicator in four out of the eight analyzed problems, and second best value in other two cases).

If we pay attention to the statistical analysis (Table 6), we observe that the alternative with two partition has been better than MOEA/D in five out of the eight evaluated problems and it has obtained worse values in only two out of these eight problems. When four partitions are used, our approach has improved the results of the MOEA/D in three problems and worsened the results in other three ones. It is worth mentioning that increasing the number of partitions in more than four has not provided any advantage.

The results of the I_{HV} indicator are summarized Table 7. In this case the best and second best value are mainly distributed between the configurations making use of two and four partitions. Attending to the comparison between pairs (included in Table 8), the conclusions are similar to the previous analyzed cases: pMOEA/Dv2 has been able to outperform original MOEA/D in various problems when is configured up to four partitions.

Summarizing this section, we have shown that by distributing the computational effort it has been possible to improve the results obtained by MOEA/D in some problems, in particular the convergence of the obtained Pareto front approximations. Regarding to the number of partitions, we have also observed that using more than four partitions has not been convenient. An explanation for this behavior could be that each partition is focused on computing a very small part of the Pareto front and, as a consequence, the diversity on these partitions is very small.

7   Conclusions and Future Work

In this work, we have proposed a new parallel model of MOEA/D, consisting in dividing the population of MOEA/D into a number of disjoint partitions. Under the evidence that some points of the Pareto front require a higher computational effort than others, we have designed two extensions of this model by including a mechanism for balancing the computational effort. This mechanism consists in performing a higher number of evaluations in some partitions in the parallel search of parts of the Pareto front.

Those two new versions (using a number of partitions between 2 and 8) have been evaluated using the LZ09 benchmark. For assessing the quality of the results we have made use of the $I_{\epsilon+}^1$ and the I_{HV} indicators.

Our experiments have shown that our proposals have been able to improve the results of MOEA/D in many cases, particularly in terms of convergence, when a number of partitions up to four has been used. On the other hand, there are a few problems where the proposed schemes for balancing the computational effort have resulted in worse Pareto approximations.

This work is a first approximation to the complex issue of dynamically adjusting the search effort, in the context of the modern multi-objective metaheuristic MOEA/D, with the idea of giving more computing power to those sub-problems which have to find more complex solutions in the Pareto front. Consequently, there is big room for carrying out research on it. Some ideas are hybridizing the two proposed versions and analyzing other schemes of balancing the computational effort among the different partitions.

Acknowledgments. This work has been partially funded by the "Consejería de Innovación, Ciencia y Empresa", Junta de Andalucía under contract P07-TIC-03044 DIRICOM project, http://diricom.lcc.uma.es and the Spanish Ministry of Science and Innovation and FEDER under contract TIN2008-06491-C04-01 (the M* project). Juan J. Durillo is supported by grant AP-2006-03349 from the Spanish Ministry of Education and Science.

References

1. Branke, J., Schmeck, H., Deb, K., Reddy, M.,, S.: Parallelizing multi-objective evolutionary algorithms: cone separation. In: Congress on Evolutionary Computation, CEC 2004, vol. 2, pp. 1952–1957 (2004)
2. Demšar, J.: Statistical Comparisons of Classifiers over Multiple Data Sets. J. Mach. Learn. Res. 7, 1–30 (2006)

3. Durillo, J.J., Nebro, A.J., Alba, E.: The jmetal framework for multi-objective optimization: Design and architecture. In: Proceedings of the IEEE 2010 Congress on Evolutionary Computation, pp. 4138–4325 (2010)
4. Knowles, J., Thiele, L., Zitzler, E.: A Tutorial on the Performance Assessment of Stochastic Multiobjective Optimizers. Technical Report 214, Computer Engineering and Networks Laboratory (TIK), ETH Zurich (2006)
5. Li, H., Zhang, Q.: Multiobjective optimization problems with complicated pareto sets, MOEA/D and NSGA-II. IEEE Transactions on Evolutionary Computation 2(12), 284–302 (2009)
6. Miettinen, K.: Nonlinear Multiobjective Optimization. Kluwer, Norwell (1999)
7. Nebro, A.J., Durillo, J.J.: A study of the parallelization of the multi-objective metaheuristic MOEA/D. In: Blum, C., Battiti, R. (eds.) LION 4. LNCS, vol. 6073, pp. 303–317. Springer, Heidelberg (2010)
8. Zhang, Q., Li, H.: MOEA/D: A multi-objective evolutionary algorithm based on decomposition. IEEE Transactions on Evolutionary Computation 1(6), 712–731 (2007)
9. Zhang, Q., Zhou, A., Li, H.: The performance of a new version of MOEA/D on cec09 unconstrained mop test instances. Technical Report CES-491, School of CS & EE, University of Essex (2009)
10. Zitzler, E., Thiele, L.: Multiobjective Evolutionary Algorithms: A Comparative Case Study and the Strength Pareto Approach. IEEE Transactions on Evolutionary Computation 3(4), 257–271 (1999)

Multi Objective Genetic Programming for Feature Construction in Classification Problems

Mauro Castelli, Luca Manzoni, and Leonardo Vanneschi

Università degli Studi di Milano-Bicocca,
Dipartimento di Informatica, Sistemistica e Comunicazione (DISCo),
20126 Milan, Italy
{mauro.castelli,luca.manzoni,vanneschi}@disco.unimib.it

Abstract. This work introduces a new technique for features construction in classification problems by means of multi objective genetic programming (MOGP). The final goal is to improve the generalization ability of the final classifier. MOGP can help in finding solutions with a better generalization ability with respect to standard genetic programming as stated in [1]. The main issue is the choice of the criteria that must be optimized by MOGP. In this work the construction of new features is guided by two criteria: the first one is the entropy of the target classes as in [7] while the second is inspired by the concept of margin used in support vector machines.

1 Introduction

Genetic programming (GP) has been successfully applied in problems of different domain. In particular genetic programming has been widely used in those problems characterized by a high dimensionality of the space of the features. In this kind of problems common machine learning techniques are not able to find a good approximation of the global optimum due to the complexity of the space of the features. To overcome this problem, features space reduction can be used; typically the idea is to consider only the most relevant features in the original set of features and to use these features to guide the search [6,3]. This technique is quite simple and there are several straightforward methods to choose the most relevant features [5]. The most important problem is that these techniques do not consider the interaction between features and, in many cases, they can only represent linear relations between variables (features). The use of MO optimization to derive near optimal feature extraction was proposed in [12].

It is important to underline that the problem of features selection is that of finding a subset of the original features of a dataset while in the features construction problem the focus is the definition of new features starting from the original ones.

Existing features extraction and construction methods can be divided in two main classes: the wrapper approach in which the final learner is used as an indicator for the appropriateness of the constructed features and the non-wrapper

C.A. Coello Coello (Ed.): LION 5, LNCS 6683, pp. 503–506, 2011.

approach in which the process of feature construction is performed as a preprocessing phase. This work falls in the second category and the main aim is to
build a set of new features that can provide a better classification performance
with respect to the original set of features.

Genetic Programming has been used as a features construction method [11,4]
especially in problems characterized by a high dimensionality of the space of the
features. In [7] the authors proposed the use of GP for features construction in
classification problems and the experimental results clearly show that this approach is effective in improving the classification accuracy. The idea is to extend
and improve the work proposed in [7] by using a multi objective optimization
approach.

2   Methods

In the proposed work a well known multi objective optimization algorithm called
NSGA-II [2] is used. The main issue in using multi-optimization techniques is
to find a way to combine the fitness values given by all the chosen criteria. The
majority of the multi-optimization algorithms uses in some way the concept of
Pareto set. The NSGA and NSGA-II [2] algorithms also share the same basic idea. For details about the accurate definition of the NSGA and NSGA-II
algorithm the reader is referred to [2].

Regarding the use of $MOGP$ for features construction the idea is to use GP
to build new variables starting from the original set of features. Every individual in the population represents a candidate feature built up by combining the
original features. In particular the terminal set is composed of the original set of
features while the function set contained the four binary operators $+$, $-$, $*$, and
$/$ (protected as in [8]). Each GP run produces an individual that represents the
best new feature to classify instances of a certain target class c. If the problem
consists of n target classes, n $MOGP$ run are performed and at each run the
best tree represents a new feature.

In $MOGP$ the search process is guided by two different criteria. The first
one is the entropy of a class while the second criteria is the average difference
between the distribution of a class and the distributions of the other classes.
To evaluate the defined criteria it is necessary to build the class distribution.
To build the distribution of a particular class we consider the mean and the
standard deviation of a class with respect to a candidate new feature (that is a
tree). Assuming that the class data follow a normal distribution, it is possible
to determine the limit of the distribution by the following formula:

$$\mu - 3\sigma \leq x_c \leq \mu + 3\sigma$$

where x_c is the value of the candidate new feature for an instance of a class c
and μ and σ are mean and standard deviation of the class with respect to the
candidate new feature. Having the class distribution for a particular class c, it
is possible to measure the level of uncertainty of the class interval. The level of
uncertainty can be measured by means of the entropy. The concept of entropy is

widely used in information theory where entropy is a measure of the uncertainty associated with a random variable. The term by itself in this context usually refers to the Shannon entropy [10], which quantifies, in the sense of an expected value, the information contained in a message, usually in bits. Equivalently, the Shannon entropy is a measure of the average information content one is missing when one does not know the value of the random variable. In our application entropy for the interval I can be calculated as follows:

$$\text{Entropy}(I) = \sum_{c \in C} -p_I(c) \log_2 (p_I(c))$$

where $p_I(c)$ denotes the probability for an instance of the class c to belong to the interval I.

The use of entropy in features construction problem is also proposed in [7]. In this work a second criterion that must guide genetic programming through the search process is proposed. In particular, entropy could be a useful criteria to construct new features that are able to produce better generalization performances. The problem is that entropy does not consider the distribution of other classes. So the algorithm builds features with the lowest entropy value but does not consider the distance from the other class distributions. This is an important point, especially if we consider that test data does not follow exactly the same distribution represented by training data. Hence it may happen that features extracted considering only entropy values are not able to classify test instances with a good accuracy. Considering the distance between the distribution of the current considered class c and the previously defined distributions can help in building features with a high discriminative power.

The idea is that given two individual t_1 and t_2 and built the distributions for a certain class c_1 if the entropy values derived from distributions built up using t_1 and t_2 are closer, we must prefer the distribution with the highest distance from all the other class distributions. This can intuitively improve classification performances and can also help in reducing the number of misclassified instances. The second criterion is in some way equivalent to the maximization of the margin when support vector machines are used. While the usage of entropy is useful in dividing the search space, the usage of the second criterion maximizes the margin between different area of the search space.

To calculate the degree of difference between two distributions d_1 and d_2 the cumulative distribution function (cdf) is considered. In probability theory and statistics, the cumulative distribution function describes the probability that a real-valued random variable X with a given probability distribution will be found at a value less than x. Having the *cdf* of d_1 and d_2 it is possible to compare the two distributions and the result obtained by the comparison of the *cdf* values can be confirmed by a Kolmogorov-Smirnov test (K-S test) [9]. This test is a form of minimum distance estimation used as a nonparametric test of equality of probability distributions. The Kolmogorov-Smirnov statistic quantifies the distance between the empirical distribution function of the sample and the cumulative distribution function of the reference distribution, or between the empirical distribution functions of two samples. For further details the reader is referred to [9].

So the *GP* search is guided by the minimization of the entropy and by the maximization of the margin (that is the difference between the distribution of the considered class and other classes distributions).

3   Conclusions

The use of multi-objective optimization can aid *GP* in finding solutions with a better generalization ability as reported in [1]. In features construction problems the use of MO optimization can also help in finding new features that can be useful in creating a robust model for the considered classification problem.

References

1. Castelli, M., Manzoni, L., Silva, S., Vanneschi, L.: A comparison of the generalization ability of different genetic programming frameworks. In: WCCI 2010: Proceedings of IEEE World Congress on Computational Intelligence. Springer, Heidelberg (2010)
2. Deb, K., Pratap, A., Agarwal, S., Meyarivan, T.: A fast elitist multi-objective genetic algorithm: Nsga-ii. IEEE Transactions on Evolutionary Computation 6, 182–197 (2000)
3. Kohavi, R., John, G.H.: Wrappers for feature subset selection. Artif. Intell. 97(1-2), 273–324 (1997)
4. Krawiec, K.: Genetic programming-based construction of features for machine learning and knowledge discovery tasks. Genetic Programming and Evolvable Machines 3(4), 329–343 (2002)
5. Lee, C., Lee, G.G.: Information gain and divergence-based feature selection for machine learning-based text categorization. Inf. Process. Manage. 42(1), 155–165 (2006)
6. Neshatian, K., Zhang, M.: Genetic programming and class-wise orthogonal transformation for dimension reduction in classification problems. In: O'Neill, M., Vanneschi, L., Gustafson, S., Esparcia Alcázar, A.I., De Falco, I., Della Cioppa, A., Tarantino, E. (eds.) EuroGP 2008. LNCS, vol. 4971, pp. 242–253. Springer, Heidelberg (2008)
7. Neshatian, K., Zhang, M., Johnston, M.: Feature construction and dimension reduction using genetic programming. In: Orgun, M.A., Thornton, J. (eds.) AI 2007. LNCS (LNAI), vol. 4830, pp. 160–170. Springer, Heidelberg (2007)
8. Poli, R., Langdon, W.B., McPhee, N.F.: A field guide to genetic programming (2008), http://lulu.com, http://www.gp-field-guide.org.uk
9. Pollard, J.H.: A handbook of numerical and statistical techniques. Cambridge University Press, Cambridge (1977)
10. Shannon, C.E.: A mathematical theory of communication. SIGMOBILE Mob. Comput. Commun. Rev. 5(1), 3–55 (2001)
11. Smith, M.G., Bull, L.: Genetic programming with a genetic algorithm for feature construction and selection. Genetic Programming and Evolvable Machines 6(3), 265–281 (2005)
12. Zhang, Y., Rockett, P.: Domain-independent feature extraction for multi-classification using multi-objective genetic programming. Pattern Analysis & Applications 13, 273–288 (2010), 10.1007/s10044-009-0154-1

Sequential Model-Based Optimization for General Algorithm Configuration

Frank Hutter, Holger H. Hoos, and Kevin Leyton-Brown

University of British Columbia, 2366 Main Mall, Vancouver BC, V6T 1Z4, Canada
{hutter,hoos,kevinlb}@cs.ubc.ca

Abstract. State-of-the-art algorithms for hard computational problems often expose many parameters that can be modified to improve empirical performance. However, manually exploring the resulting combinatorial space of parameter settings is tedious and tends to lead to unsatisfactory outcomes. Recently, automated approaches for solving this *algorithm configuration* problem have led to substantial improvements in the state of the art for solving various problems. One promising approach constructs explicit regression models to describe the dependence of target algorithm performance on parameter settings; however, this approach has so far been limited to the optimization of few numerical algorithm parameters on single instances. In this paper, we extend this paradigm for the first time to general algorithm configuration problems, allowing many categorical parameters and optimization for sets of instances. We experimentally validate our new algorithm configuration procedure by optimizing a local search and a tree search solver for the propositional satisfiability problem (SAT), as well as the commercial mixed integer programming (MIP) solver CPLEX. In these experiments, our procedure yielded state-of-the-art performance, and in many cases outperformed the previous best configuration approach.

1 Introduction

Algorithms for hard computational problems—whether based on local search or tree search—are often highly parameterized. Typical parameters in local search include neighbourhoods, tabu tenure, percentage of random walk steps, and perturbation and acceptance criteria in iterated local search. Typical parameters in tree search include decisions about preprocessing, branching rules, how much work to perform at each search node (*e.g.*, to compute cuts or lower bounds), which type of learning to perform, and when to perform restarts. As one prominent example, the commercial mixed integer programming solver IBM ILOG CPLEX has 76 parameters pertaining to its search strategy [1]. Optimizing the settings of such parameters can greatly improve performance, but doing so manually is tedious and often impractical.

Automated procedures for solving this *algorithm configuration* problem are useful in a variety of contexts. Their most prominent use case is to optimize parameters on a training set of instances from some application ("offline", as part of algorithm development) in order to improve performance when using the algorithm in practice ("online"). Algorithm configuration thus trades human time for machine time and automates a task that would otherwise be performed manually. End users of an algorithm can also apply

C.A. Coello Coello (Ed.): LION 5, LNCS 6683, pp. 507–523, 2011.
© Springer-Verlag Berlin Heidelberg 2011

algorithm configuration procedures (*e.g.*, the automated tuning tool built into CPLEX versions 11 and above) to configure an existing algorithm for high performance on problem instances of interest.

The algorithm configuration problem can be formally stated as follows: given a parameterized algorithm A (the *target algorithm*), a set (or distribution) of problem instances I and a cost metric c, find parameter settings of A that minimize c on I. The cost metric c is often based on the runtime required to solve a problem instance, or, in the case of optimization problems, on the solution quality achieved within a given time budget. Various automated procedures have been proposed for solving this algorithm configuration problem. Existing approaches differ in whether or not explicit models are used to describe the dependence of target algorithm performance on parameter settings.

Model-free algorithm configuration methods are relatively simple, can be applied out-of-the-box, and have recently led to substantial performance improvements across a variety of constraint programming domains. This research goes back to the early 1990s [2, 3] and has lately been gaining momentum. Some methods focus on optimizing numerical (*i.e.*, either integer- or real-valued) parameters (see, *e.g.*, [4, 5]), while others also target categorical (*i.e.*, discrete-valued and unordered) domains [6, 7, 8, 9]. The most prominent configuration methods are the racing algorithm F-RACE [5] and our own iterated local search algorithm PARAMILS [7, 8]. A recent competitor is the genetic algorithm GGA [9]. F-RACE and its extensions have been used to optimize various high-performance algorithms, including iterated local search and ant colony optimization procedures for timetabling tasks and the travelling salesperson problem [6, 5]. Our own group has used PARAMILS to configure highly parameterized tree search [10] and local search solvers [11] for the propositional satisfiability problem (SAT), as well as several solvers for mixed integer programming (MIP), substantially advancing the state of the art for various types of instances. Notably, by optimizing the 76 parameters of CPLEX—the most prominent MIP solver—we achieved up to 50-fold speedups over the defaults and over the configuration returned by the CPLEX tuning tool [1].

While the progress in practical applications described above has been based on model-free optimization methods, recent progress in model-based approaches promises to lead to the next generation of algorithm configuration procedures. *Sequential model-based optimization (SMBO)* iterates between fitting models and using them to make choices about which configurations to investigate. It offers the appealing prospects of interpolating performance between observed parameter settings and of extrapolating to previously unseen regions of parameter space. It can also be used to quantify importance of each parameter and parameter interactions. However, being grounded in the "black-box function optimization" literature from statistics (see, e.g., [12]), SMBO has inherited a range of limitations inappropriate to the automated algorithm configuration setting. These limitations include a focus on deterministic target algorithms; use of costly initial experimental designs; reliance on computationally expensive models; and the assumption that all target algorithm runs have the same execution costs. Despite considerable recent advances [13, 14, 15], all published work on SMBO still has three key limitations that prevent its use for general algorithm configuration tasks: (1) it only supports numerical parameters; (2) it only optimizes target algorithm performance for single instances; and (3) it lacks a mechanism for terminating poorly performing target algorithm runs early.

The main contribution of this paper is to remove the first two of these SMBO limitations, and thus to make SMBO applicable to general algorithm configuration problems with many categorical parameters and sets of benchmark instances. Specifically, we generalize four components of the SMBO framework and—based on them—define two novel SMBO instantiations capable of general algorithm configuration: the simple model-free Random Online Adaptive Racing (ROAR) procedure and the more sophisticated Sequential Model-based Algorithm Configuration (SMAC) method. These methods do not yet implement an early termination criterion for poorly performing target algorithm runs (such as, e.g., PARAMILS's adaptive capping mechanism [8]); thus, so far we expect them to perform poorly on some configuration scenarios with large captimes. In a thorough experimental analysis for a wide range of 17 scenarios with small captimes (involving the optimization of local search and tree search SAT solvers, as well as the commercial MIP solver CPLEX), SMAC indeed compared favourably to the two most prominent approaches for general algorithm configuration: PARAMILS [7, 8] and GGA [9].

The remainder of this paper is structured as follows. Section 2 describes the SMBO framework and previous work on SMBO. Sections 3 and 4 generalize SMBO's components to tackle general algorithm configuration scenarios, defining ROAR and SMAC, respectively. Section 5 experimentally compares ROAR and SMAC to the existing state of the art in algorithm configuration. Section 6 concludes the paper.

2   Existing Work on Sequential Model-Based Optimization (SMBO)

Model-based optimization methods construct a regression model (often called a *response surface model*) that predicts performance and then use this model for optimization. *Sequential* model-based optimization (SMBO) iterates between fitting a model and gathering additional data based on this model. In the context of parameter optimization, the model is fitted to a training set $\{(\boldsymbol{\theta}_1, o_1), \ldots, (\boldsymbol{\theta}_n, o_n)\}$ where parameter configuration $\boldsymbol{\theta}_i = (\theta_{i,1}, \ldots, \theta_{i,d})$ is a complete instantiation of the target algorithm's d parameters and o_i is the target algorithm's observed performance when run with configuration $\boldsymbol{\theta}_i$. Given a new configuration $\boldsymbol{\theta}_{n+1}$, the model aims to predict its performance o_{n+1}.

SMBO has its roots in the statistics literature on experimental design for global continuous ("black-box") function optimization. Most notable is the efficient global optimization (EGO) algorithm by Jones et al. [12], which is, however, limited to optimizing continuous parameters for noise-free functions (*i.e.*, the performance of deterministic algorithms). Bartz-Beielstein et al. [13] were the first to use the EGO approach to optimize algorithm performance. Their sequential parameter optimization (SPO) toolbox–which has received considerable attention in the evolutionary algorithms community–provides many features that facilitate the manual analysis and optimization of algorithm parameters; it also includes an automated SMBO procedure for optimizing numerical parameters on single instances. We studied the components of this automated procedure, demonstrated that its intensification mechanism mattered most, and improved it in our SPO$^+$ algorithm [14]. In [15], we showed how to reduce the

---

Algorithm Framework 1: Sequential Model-Based Optimization (SMBO)

$\mathbf{R}$ keeps track of all target algorithm runs performed so far and their performances (*i.e.*, SMBO's training data $\{([\boldsymbol{\theta}_1, \boldsymbol{x}_1], o_1), \ldots, ([\boldsymbol{\theta}_n, \boldsymbol{x}_n], o_n)\}$), $\mathcal{M}$ is SMBO's model, $\vec{\Theta}_{new}$ is a list of promising configurations, and t_{fit} and t_{select} are the runtimes required to fit the model and select configurations, respectively.

Input : Target algorithm A with parameter configuration space Θ; instance set Π; cost metric $\hat{c}$

Output : Optimized (incumbent) parameter configuration, $\boldsymbol{\theta}_{inc}$

1 $[\mathbf{R}, \boldsymbol{\theta}_{inc}] \leftarrow Initialize(\Theta, \Pi)$

2 **repeat**

3     $[\mathcal{M}, t_{fit}] \leftarrow FitModel(\mathbf{R})$

4     $[\vec{\Theta}_{new}, t_{select}] \leftarrow SelectConfigurations(\mathcal{M}, \boldsymbol{\theta}_{inc}, \Theta)$

5     $[\mathbf{R}, \boldsymbol{\theta}_{inc}] \leftarrow Intensify(\vec{\Theta}_{new}, \boldsymbol{\theta}_{inc}, \mathcal{M}, \mathbf{R}, t_{fit} + t_{select}, \Pi, \hat{c})$

6 **until** *total time budget for configuration exhausted*

7 **return** $\boldsymbol{\theta}_{inc}$

---

overhead incurred by construction and use of response surface models via approximate GP models. We also eliminated the need for a costly initial design by interleaving randomly selected parameters throughout the optimization process, and by exploiting the fact that different algorithm runs take different amounts of time. The resulting time-bounded SPO variant, TB-SPO, is the first practical SMBO method for parameter optimization given a user-specified time budget. Although it was shown to significantly outperform PARAMILS in certain cases, it is still limited to the optimization of numerical algorithm parameters on single problem instances.

In Algorithm Framework 1, we give pseudocode for the time-bounded SMBO framework of which TB-SPO is an instantiation: in each iteration, it fits a model, selects a list of promising parameter configurations and performs target algorithm runs on (a subset of) these, until a given time bound is reached. This time bound is related to the combined overhead due to fitting the model and selecting promising configurations. In the following, we generalize the components of this algorithm framework, extending its scope to tackle general algorithm configuration problems with many categorical parameters and sets of benchmark instances.

3 Random Online Aggressive Racing (ROAR)

In this section, we first generalize SMBO's *Intensify* procedure to handle multiple instances, and then introduce ROAR, a very simple model-free algorithm configuration procedure based on this new intensification mechanism.

3.1 Generalization I: An Intensification Mechanism for Multiple Instances

A crucial component of any algorithm configuration procedure is the so-called *intensification* mechanism, which governs how many evaluations to perform with each configuration, and when to trust a configuration enough to make it the new current best known configuration (the *incumbent*). When configuring algorithms for sets of

---

Procedure 2: Intensify($\vec{\Theta}_{new}$, θ_{inc}, $\mathcal{M}$, $\mathbf{R}$, $t_{intensify}$, Π, $\hat{c}$)

$\hat{c}(\theta, \Pi')$ denotes the empirical cost of θ on the subset of instances $\Pi' \subseteq \Pi$, based on the runs in $\mathbf{R}$; *maxR* is a parameter, set to $2\,000$ in all our experiments

---

Input : Sequence of parameter settings to evaluate, $\vec{\Theta}_{new}$; incumbent parameter setting, θ_{inc}; model, $\mathcal{M}$; sequence of target algorithm runs, $\mathbf{R}$; time bound, $t_{intensify}$; instance set, Π; cost metric, $\hat{c}$

Output: Updated sequence of target algorithm runs, $\mathbf{R}$; incumbent parameter setting, θ_{inc}

1 **for** $i := 1, \dots, length(\vec{\Theta}_{new})$ **do**
2 $\quad$ $\theta_{new} \leftarrow \vec{\Theta}_{new}[i]$
3 $\quad$ **if** $\mathbf{R}$ *contains less than* $maxR$ *runs with configuration* θ_{inc} **then**
4 $\quad\quad$ $\Pi' \leftarrow \{\pi' \in \Pi \mid \mathbf{R}$ contains less than or equal number of runs using θ_{inc} and π' than using θ_{inc} and any other $\pi'' \in \Pi\}$
5 $\quad\quad$ $\pi \leftarrow$ instance sampled uniformly at random from Π'
6 $\quad\quad$ $s \leftarrow$ seed, drawn uniformly at random
7 $\quad\quad$ $\mathbf{R} \leftarrow$ ExecuteRun($\mathbf{R}$, θ_{inc}, π, s)
8 $\quad$ $N \leftarrow 1$
9 $\quad$ **while** *true* **do**
10 $\quad\quad$ $S_{missing} \leftarrow \langle$instance, seed$\rangle$ pairs for which θ_{inc} was run before, but not θ_{new}
11 $\quad\quad$ $S_{torun} \leftarrow$ random subset of $S_{missing}$ of size $\min(N, |S_{missing}|)$
12 $\quad\quad$ **foreach** $(\pi, s) \in S_{torun}$ **do** $\mathbf{R} \leftarrow$ ExecuteRun($\mathbf{R}$, θ_{new}, π, s)
13 $\quad\quad$ $S_{missing} \leftarrow S_{missing} \setminus S_{torun}$
14 $\quad\quad$ $\Pi_{common} \leftarrow$ instances for which we previously ran both θ_{inc} and θ_{new}
15 $\quad\quad$ **if** $\hat{c}(\theta_{new}, \Pi_{common}) > \hat{c}(\theta_{inc}, \Pi_{common})$ **then break**
16 $\quad\quad$ **else if** $S_{missing} = \emptyset$ **then** $\theta_{inc} \leftarrow \theta_{new}$; **break**
17 $\quad\quad$ **else** $N \leftarrow 2 \cdot N$
18 $\quad$ **if** *time spent in this call to this procedure exceeds* $t_{intensify}$ *and* $i \geq 2$ **then break**
19 **return** $[\mathbf{R}, \theta_{inc}]$

---

instances, we also need to decide which instance to use in each run. To address this problem, we generalize TB-SPO's intensification mechanism. Our new procedure implements a variance reduction mechanism, reflecting the insight that when we compare the empirical cost statistics of two parameter configurations across multiple instances, the variance in this comparison is lower if we use the same N instances to compute both estimates.

Procedure 2 defines this new intensification mechanism more precisely. It takes as input a list of promising configurations, $\vec{\Theta}_{new}$, and compares them in turn to the current incumbent configuration until a time budget for this comparison stage is reached.[1] In each comparison of a new configuration, θ_{new}, to the incumbent, θ_{inc}, we first perform an additional run for the incumbent, using a randomly selected $\langle$instance, seed$\rangle$ combination. Then, we iteratively perform runs with θ_{new} (using a doubling scheme) until either θ_{new}'s empirical performance is worse than that of θ_{inc} (in which case we reject θ_{new}) or we performed as many runs for θ_{new} as for θ_{inc} and it is still at least as good as θ_{inc} (in which case we change the incumbent to θ_{new}). The $\langle$instance, seed$\rangle$ combi-

---

[1] If that budget is already reached after the first configuration in $\vec{\Theta}_{new}$, one more configuration is used; see the last paragraph of Section 4.3 for an explanation why.

nations for θ_{new} are sampled uniformly at random from those on which the incumbent has already run. Similar to the FOCUSEDILS algorithm [7, 8], θ_{inc} and θ_{new} are always compared using only instances on which they have both been run. However, every comparison in Procedure 2 is based on a *different* randomly selected subset of instances and seeds, while FOCUSEDILS's Procedure "better" uses a fixed ordering to which it can be very sensitive.

3.2   Defining ROAR

We now define Random Online Aggressive Racing (ROAR), a simple model-free instantiation of the general SMBO framework (see Algorithm Framework 1).[2] This surprisingly effective method selects parameter configurations uniformly at random and iteratively compares them against the current incumbent using our new intensification mechanism. We consider ROAR to be a racing algorithm, because it runs each candidate configuration only as long as necessary to establish whether it is competitive. It gets its name because the set of candidates is selected at *random*, each candidate is accepted or rejected *online*, and we make this online decision *aggressively*, before enough data has been gathered to support a statistically significant conclusion. More formally, as an instantiation of the SMBO framework, ROAR is completely specified by the four components *Initialize*, *FitModel*, *SelectConfigurations*, and *Intensify*. *Initialize* performs a single run with the target algorithm's default parameter configuration (or a random configuration if no default is available) on an instance selected uniformly at random. Since ROAR is model-free, its *FitModel* procedure simply returns a constant model which is never used. *SelectConfigurations* returns a single configuration sampled uniformly at random from the parameter space, and *Intensify* is as described in Procedure 2.

4   Sequential Model-Based Algorithm Configuration (SMAC)

In this section, we introduce our second, more sophisticated instantiation of the general SMBO framework: Sequential Model-based Algorithm Configuration (SMAC). SMAC can be understood as an extension of ROAR that selects configurations based on a model rather than uniformly at random. It instantiates *Initialize* and *Intensify* in the same way as ROAR. Here, we discuss the new model class we use in SMAC to support categorical parameters and multiple instances (Sections 4.1 and 4.2, respectively); then, we describe how SMAC uses its models to select promising parameter configurations (Section 4.3).

4.1   Generalization II: Models for Categorical Parameters

The models in all existing SMBO methods of which we are aware are limited to numerical parameters. In this section, we discuss the new model class SMAC uses to also handle *categorical* parameters.

---

[2] We previously considered random sampling approaches based on less powerful intensification mechanisms; see, *e.g.*, RANDOM* defined in [15].

SMAC's models are based on random forests [16], a standard machine learning tool for regression and classification.[3] Random forests are collections of regression trees, which are similar to decision trees but have real values (here: target algorithm performance values) rather than class labels at their leaves. Regression trees are known to perform well for categorical input data; indeed, they have already been used for modeling the performance of heuristic algorithms (e.g., [18, 19]). Random forests share this benefit and typically yield more accurate predictions [16]; they also allow us to quantify our uncertainty in a given prediction. We construct a random forest as a set of B regression trees, each of which is built on n data points randomly sampled with repetitions from the entire training data set $\{(\boldsymbol{\theta}_1, o_1), \ldots, (\boldsymbol{\theta}_n, o_n)\}$. At each split point of each tree, a random subset of $\lceil d \cdot p \rceil$ of the d algorithm parameters is considered eligible to be split upon; the split ratio p is a parameter, which we left at its default of $p = 5/6$. A further parameter is n_{min}, the minimal number of data points required to be in a node if it is to be split further; we use the standard value $n_{min} = 10$. Finally, we set the number of trees to $B = 10$ to keep the computational overhead small.[4] We compute the random forest's predictive mean $\mu_{\boldsymbol{\theta}}$ and variance $\sigma_{\boldsymbol{\theta}}^2$ for a new configuration $\boldsymbol{\theta}$ as the empirical mean and variance of its individual trees' predictions for $\boldsymbol{\theta}$.

Model fit can often be improved by transforming the cost metric. In this paper, we focus on minimizing algorithm runtime. Previous work on predicting algorithm runtime has found that logarithmic transformations substantially improve model quality [20] and we thus use log-transformed runtime data throughout this paper; that is, for runtime r_i, we use $o_i = \ln(r_i)$. (SMAC can also be applied to optimize other cost metrics, such as the solution quality an algorithm obtains in a fixed runtime; other transformations may prove more efficient for other metrics.) However, we note that in some models such transformations implicitly change the cost metric users aim to optimize [17]. We avoid this problem in our random forests by computing the prediction in the leaf of a tree by "untransforming" the data, computing the user-defined cost metric, and then transforming the result again.

4.2   Generalization III: Models for Sets of Problem Instances

There are several possible ways to extend SMBO's models to handle multiple instances. Most simply, one could use a fixed set of N instances for every evaluation of the target algorithm run, reporting aggregate performance. However, there is no good fixed choice for N: small N leads to poor generalization to test data, while large N leads to a

---

[3] In principle, other model families can also be used. Notably, one might consider Gaussian processes (GPs), or the projected process (PP) approximation to GPs we used in TB-SPO [15]. Although GPs are canonically defined only for numerical parameters, they can be extended to categorical parameters by changing the kernel function. We defined such a kernel function, based on the weighted Hamming distance between two parameter configurations. However, there is a more significant obstacle to using GPs to support general algorithm configuration: response variable transformations distort the GP cost metric, which is particularly problematic for multi-instance models. Further information, including the definition of the weighted Hamming distance kernel function, can be found in the extended version of this paper [17].

[4] An optimization of these three parameters might improve performance further. We plan on studying this in the context of an application of SMAC to optimizing its own parameters.

prohibitive N-fold slowdown in the cost of each evaluation. (This is the same problem faced by the PARAMILS instantiation BASICILS(N) [7].) Instead, we explicitly integrate information about the instances into our response surface models. Given a vector of *features* x_i describing each training problem instance $\pi_i \in \Pi$, we learn a joint model that predicts algorithm runtime for combinations of parameter configurations and instance features. We then aggregate these predictions across instances.

Instance Features. Existing work on empirical hardness models [21] has demonstrated that it is possible to predict algorithm runtime based on features of a given problem instance. Most notably, such predictions have been exploited to construct portfolio-based algorithm selection mechanisms, such as SATzilla [20]. For SAT instances in the form of CNF formulae, we used 126 features including features based on graph representations of the instance, an LP relaxation, DPLL probing, local search probing, clause learning, and survey propagation. For MIP instances we computed 39 features, including features based on graph representations, an LP relaxation, the objective function, and the linear constraint matrix. Both sets of features are detailed in the extended version of this paper [17]. To reduce the computational complexity of learning, we applied *principal component analysis* (see, *e.g.*, [22]), to project the feature matrix into a lower-dimensional subspace spanned by the seven orthogonal vectors along which it has maximal variance. For new domains, for which no features have yet been defined, SMAC can still be applied with an empty feature set or simple domain-independent features, such as instance size or the performance of the algorithm's default setting (which, based on preliminary experiments, seems to be a surprisingly effective feature). Note that in contrast to per-instance approaches, instance features are only needed for the *training* instances: the end result of algorithm configuration is a single parameter configuration that is used without a need to compute features for test instances.

Predicting Performance Across Instances. So far, we have discussed models trained on pairs (θ_i, o_i) of parameter configurations θ_i and their observed performance o_i. Now, we extend this data to include instance features. Let x_i denote the vector of features for the instance used in the ith target algorithm run. Concatenating parameter values, θ_i, and instance features, x_i, into one input vector yields the training data $\{([\theta_1, x_1], o_1), \dots, ([\theta_n, x_n], o_n)\}$. From this data, we learn a model that takes as input a parameter configuration θ and predicts performance across all training instances. To achieve this, we do not need to change random forest construction: all input dimensions are handled equally, regardless of whether they refer to parameter values or instance features. The prediction procedure changes as follows: within each tree, we first predict performance for the combinations of the given configuration and each instance; next, we combine these predictions with the user-defined cost metric (*e.g.*, arithmetic mean runtime across instances); finally, we compute means and variances across trees.

4.3 Generalization IV: Using the Model to Select Promising Configurations in Large Mixed Numerical/Categorical Configuration Spaces

The *SelectConfiguration* component in SMAC uses the model to select a list of promising parameter configurations. To quantify how promising a configuration θ is, it uses the model's predictive distribution for θ to compute its *expected* positive *improvement*

$(EI(\theta))$ [12] over the best configuration seen so far (the *incumbent*). $EI(\theta)$ is large for configurations θ with low predicted cost and for those with high predicted uncertainty; thereby, it offers an automatic tradeoff between exploitation (focusing on known good parts of the space) and exploration (gathering more information in unknown parts of the space). Specifically, we use the $E[I_{exp}]$ criterion introduced in [14] for log-transformed costs; given the predictive mean μ_θ and variance σ_θ^2 of the log-transformed cost of a configuration θ, this is defined as

$$EI(\theta) := E[I_{exp}(\theta)] = f_{min}\Phi(v) - e^{\frac{1}{2}\sigma_\theta^2 + \mu_\theta} \cdot \Phi(v - \sigma_\theta), \qquad (1)$$

where $v := \frac{\ln(f_{min}) - \mu_\theta}{\sigma_\theta}$, Φ denotes the cumulative distribution function of a standard normal distribution, and f_{min} denotes the empirical mean performance of θ_{inc}.[5]

Having defined $EI(\theta)$, we must still decide how to identify configurations θ with large $EI(\theta)$. This amounts to a maximization problem across parameter configuration space. Previous SMBO methods [13, 14, 15] simply applied random sampling for this task (in particular, they evaluated EI for 10 000 random samples), which is unlikely to be sufficient in high-dimensional configuration spaces, especially if promising configurations are sparse. To gather a set of promising configurations with low computational overhead, we perform a simple multi-start local search and consider all resulting configurations with locally maximal EI.[6] This search is similar in spirit to PARAMILS [7, 8], but instead of algorithm performance it optimizes $EI(\theta)$ (see Equation 1), which can be evaluated based on the model predictions μ_θ and σ_θ^2 without running the target algorithm. More concretely, the details of our local search are as follows. We compute EI for all configurations used in previous target algorithm runs, pick the ten configurations with maximal EI, and initialize a local search at each of them. To seamlessly handle mixed categorical/numerical parameter spaces, we use a randomized one-exchange neighbourhood, including the set of all configurations that differ in the value of exactly one discrete parameter, as well as four random neighbours for each numerical parameter. In particular, we normalize the range of each numerical parameter to [0,1] and then sample four "neighbouring" values for numerical parameters with current value v from a univariate Gaussian distribution with mean v and standard deviation 0.2, rejecting new values outside the interval [0,1]. Since batch model predictions (and thus batch EI computations) for a set of N configurations are much cheaper than separate predictions for N configurations, we use a best improvement search, evaluating EI for all neighbours at once; we stop each local search once none of the neighbours has larger EI. Since SMBO sometimes evaluates many configurations per iteration and because batch EI computations are cheap, we simply compute EI for an additional 10 000 randomly-sampled configurations; we then sort all 10 010 configurations in descending order of EI. (The ten results of local search typically had larger EI than all randomly sampled configurations.)

---

[5] In TB-SPO [15], we used $f_{min} = \mu(\theta_{inc}) + \sigma(\theta_{inc})$. However, we now believe that setting f_{min} to the empirical mean performance of θ_{inc} yields better performance overall.

[6] We plan to investigate better mechanisms in the future. However, we note that the best problem formulation is not obvious, since we desire a *diverse* set of configurations with high EI.

Having selected this list of 10 010 configurations based on the model, we interleave randomly-sampled configurations in order to provide unbiased training data for future models. More precisely, we alternate between configurations from the list and additional configurations sampled uniformly at random. Since *Intensify* always compares at least two configurations against the current incumbent, at least one randomly sampled configuration is evaluated in every iteration of SMBO. In finite configuration spaces, thus, each configuration has a positive probability of being selected in each iteration. In combination with the fact that *Intensify* increases the number of runs used to evaluate each configuration unboundedly, this allows us to prove that SMAC (and ROAR) eventually converge to the optimal configuration when using consistent estimators of the user-defined cost metric.[7] The proof is very simple and uses the same arguments as a previous proof about FocusedILS (see [8]); we omit it here and refer the reader to the extended version of this paper [17].

5   Experimental Evaluation

We now compare the performance of SMAC, ROAR, TB-SPO [15], GGA [9], and PARAMILS (in particular, FOCUSEDILS 2.3) [8] for a range of configuration scenarios that involve minimizing the runtime of SAT and MIP solvers. In principle, our ROAR and SMAC methods also apply to optimizing other cost metrics, such as the solution quality an algorithm can achieve in a fixed time budget; we plan on studying their empirical performance for this case in the near future.

5.1   Experimental Setup

Configuration scenarios. We considered a diverse set of 17 algorithm configuration problem instances (so-called *configuration scenarios*) that had been used previously to analyze PARAMILS [8, 1] and TB-SPO [15].[8] These scenarios involve the configuration of the local search SAT solver SAPS (4 parameters), the tree search solver SPEAR (26 parameters), and the most widely used commercial mixed integer programming (MIP) solver, IBM ILOG CPLEX (76 parameters); references for these algorithms, as well as details on their parameter spaces, are given in the extended version of this paper [17]. In all 17 configuration scenarios, we terminated target algorithm runs at $\kappa_{max} = 5$ seconds, the same per-run captime used in previous work for these scenarios. In previous work, we have also applied PARAMILS to optimize MIP solvers with very large per-run captimes (up to $\kappa_{max} = 10\,000s$), and obtained better results than the CPLEX tuning tool [1]. We believe that for such large captimes, an adaptive capping mechanism, such as the one implemented in ParamILS [8], is essential; we are currently working on

---

[7] This proof does not cover continuous parameters, since they lead to infinite configuration spaces; in that case, we would require additional smoothness assumptions to prove convergence.

[8] All instances we used are available at http://www.cs.ubc.ca/labs/beta/Projects/AAC

integrating such a mechanism into SMAC.[9] In this paper, to study the remaining components of SMAC, we only use scenarios with small captimes of 5s. In order to enable a fair comparison with GGA, we changed the optimization objective of all 17 scenarios from the original PAR-10 (*penalized average runtime*, counting timeouts at κ_{max} as $10 \cdot \kappa_{max}$, which is not supported by GGA) to simple average runtime (PAR-1, counting timeouts at κ_{max} as κ_{max}). [10] However, one difference remains: we minimize the runtime reported by the target algorithm, but GGA can only minimize its own measurement of target algorithm runtime, including (sometimes large) overheads for reading in the instance.

Parameter transformations. Some numerical parameters naturally vary on a non-uniform scale (*e.g.*, a parameter θ with an interval $[100, 1600]$ that we discretized to the values $\{100, 200, 400, 800, 1600\}$ for use in PARAMILS). We transformed such parameters to a domain in which they vary more uniformly (*e.g.*, $\log(\theta) \in [\log(100), \log(1600)]$), un-transforming the parameter values for each call to the target algorithm.

Comparing configuration procedures. We performed 25 runs of each configuration procedure on each configuration scenario. For each such run r_i, we computed *test performance t_i* as follows. First, we extracted the incumbent configuration θ_{inc} at the point the configuration procedure exhausted its time budget; SMAC's overhead due to the construction and use of models were counted as part of this budget. Next, in an offline evaluation step using the same per-run cutoff time as during training, we measured the mean runtime t_i across 1 000 independent test runs of the target algorithm parameterized by θ_{inc}. In the case of multiple-instance scenarios, we used a test set of previously unseen instances. For a given scenario, this resulted in test performances $t_1, \ldots, t_{25}$ for each configuration procedure. We report medians across these 25 values, visualize their variance in boxplots, and perform a Mann-Whitney U test to check for significant differences between configuration procedures. We ran GGA through HAL [23], using parameter settings recommended by GGA's author, Kevin Tierney, in e-mail communication: we set the population size to 70, the number of generations to 100, the number of runs to perform in the first generation to 5, and the number of runs to perform in the last generation to 70. We used default settings for FOCUSEDILS 2.3, including aggressive capping. We note that in a previous comparison [9] of GGA and FOCUSEDILS, capping was disabled in FOCUSEDILS; this explains its poor performance there and its better performance here.

---

[9] In fact, preliminary experiments for configuration scenario CORLAT (from [1], with $\kappa_{max} = 10\,000$s) highlight the importance of developing an adaptive capping mechanism for SMAC: *e.g.*, in one of SMAC's run, it only performed 49 target algorithm runs, with 15 of them timing out after $\kappa_{max} = 10\,000$s, and another 3 taking over 5\,000 seconds each. Together, these runs exceeded the time budget of 2 CPU days (172\,800 seconds), despite the fact that all of them could have safely been cut off after less than 100 seconds. As a result, for scenario CORLAT, SMAC performed a factor of 3 worse than PARAMILS with $\kappa_{max} = 10\,000$s. On the other hand, SMAC can sometimes achieve strong performance even with relatively high captimes; *e.g.*, on CORLAT with $\kappa_{max} = 300$s, SMAC outperformed PARAMILS by a factor of 1.28.

[10] Using PAR-10 to compare the remaining configurators, our qualitative results did not change.

With the exception of FOCUSEDILS, all of the configuration procedures we study here support numerical parameters without a need for discretization. We present results both for the mixed numerical/categorical parameter space these methods search, and— to enable a direct comparison to FOCUSEDILS—for a fully discretized configuration space.

Computational environment. We conducted all experiments on a cluster of 55 dual 3.2GHz Intel Xeon PCs with 2MB cache and 2GB RAM, running OpenSuSE Linux 11.1. We measured runtimes as CPU time on these reference machines.

5.2   Experimental Results for Single Instance Scenarios

In order to evaluate our new general algorithm configuration procedures ROAR and SMAC one component at a time, we first evaluated their performance for optimizing the continuous parameters of SAPS and the mixed numerical/categorical parameters of SPEAR on single SAT instances; multi-instance scenarios are studied in the next section. To enable a comparison with our previous SMBO instantiation TB-SPO, we used the 6 configuration scenarios introduced in [15], which aim to minimize SAPS's runtime on 6 single SAT-encoded instances, 3 each from quasigroup completion (QCP) and small world graph colouring (SWGCP). We also used 5 similar new configuration scenarios, which aim to minimize SPEAR's runtime for 5 single SAT-encoded instances, 2 from a hard distribution of IBM bounded model checking (IBM) and 3 from software verification (SWV). For more information and references for these instances, please see [17]. The time budget for each algorithm configuration run was 30 CPU minutes, exactly following [15].

The model-based approaches SMAC and TB-SPO performed best in this comparison, followed by ROAR, FOCUSEDILS, and GGA. Table 1 shows the results achieved by each of the configuration procedures, for both the full parameter configuration space (which includes numerical parameters) and the discretized version we made for use with FOCUSEDILS. For the special case of single instances and a small number of all-numerical parameters, SMAC and TB-SPO are very similar, and both performed best.[11] While TB-SPO does not apply in the remaining configuration scenarios, our more general SMAC method achieved the best performance in all of them. ROAR performed well for small but not for large configuration spaces: it was among the best (*i.e.*, best or not significantly different from the best) in most of the SAPS scenarios (4 parameters) but only for one of the SPEAR scenarios (26 parameters). Both GGA and FOCUSEDILS performed slightly worse than ROAR for the SAPS scenarios, and slightly (but statistically significantly) worse than SMAC for most SPEAR configuration scenarios. Figure 1 visualizes each configurator's 25 test performances for all scenarios. We note that SMAC

---

[11] In fact, in 1 of the 6 scenarios for which TB-SPO is applicable, it performed better than SMAC. This is because for all-numerical parameters, projected process (PP) models performed better than random forest (RF) models. In further experiments (not reported here, see [17]), we evaluated a version of SMAC based on PP instead of RF models; its median performance was slightly better than TB-SPO's, but the two were statistically indistinguishable in all 6 scenarios. With categorical parameters, SMAC performed better with the RF models we use here.

Table 1. Comparison of algorithm configuration procedures for optimizing parameters on single problem instances. We performed 25 independent runs of each configuration procedure and report the median of the 25 test performances (mean runtimes across 1 000 target algorithm runs with the found configurations). We bold-faced entries for configurators that are not significantly worse than the best configurator for the respective configuration space, based on a Mann-Whitney U test. The symbol "—" denotes that the configurator does not apply for this configuration space.

Scenario	Unit	Full configuration space					Discretized configuration space				
		SMAC	TB-SPO	ROAR	F-ILS	GGA	SMAC	TB-SPO	ROAR	F-ILS	GGA
SAPS-QCP-MED	$[\cdot 10^{-2}s]$	4.70	**4.58**	4.72	—	6.28	**5.27**	—	**5.25**	5.50	6.24
SAPS-QCP-Q075	$[\cdot 10^{-1}s]$	2.29	**2.22**	2.34	—	2.74	**2.87**	—	2.92	2.91	2.98
SAPS-QCP-Q095	$[\cdot 10^{-1}s]$	1.37	**1.35**	1.55	—	1.75	**1.51**	—	1.57	1.57	1.95
SAPS-SWGCP-MED	$[\cdot 10^{-1}s]$	**1.61**	**1.63**	1.70	—	2.48	2.54	—	2.58	2.57	2.71
SAPS-SWGCP-Q075	$[\cdot 10^{-1}s]$	**2.11**	2.48	2.32	—	3.19	**3.26**	—	3.38	3.55	3.55
SAPS-SWGCP-Q095	$[\cdot 10^{-1}s]$	**2.36**	2.69	2.49	—	3.13	**3.65**	—	3.79	3.75	3.77
SPEAR-IBM-Q025	$[\cdot 10^{-1}s]$	**6.24**	—	6.31	—	6.33	**6.21**	—	6.30	6.31	6.30
SPEAR-IBM-MED	$[\cdot 10^{0}s]$	**3.28**	—	3.36	—	3.35	**3.16**	—	3.38	3.47	3.84
SPEAR-SWV-MED	$[\cdot 10^{-1}s]$	**6.04**	—	6.11	—	6.14	**6.05**	—	6.14	6.11	6.15
SPEAR-SWV-Q075	$[\cdot 10^{-1}s]$	**5.76**	—	5.88	—	5.83	**5.76**	—	5.89	5.88	5.84
SPEAR-SWV-Q095	$[\cdot 10^{-1}s]$	**8.38**	—	8.55	—	8.47	**8.42**	—	8.53	8.58	8.49

and ROAR often yielded more robust results than FOCUSEDILS and GGA: for many scenarios some of the 25 FOCUSEDILS and GGA runs did very poorly.

Our new SMAC and ROAR methods were able to explore the full configuration space, which sometimes led to substantially improved performance compared to the discretized configuration space PARAMILS is limited to. Comparing the left *vs* the right side of Table 1, we note that the SAPS discretization (the same we used to optimize SAPS with PARAMILS in previous work [7, 8]) left substantial room for improvement when exploring the full space: roughly 1.15-fold and 1.55-fold speedups on the QCP and SWGCP instances, respectively. GGA did not benefit as much from being allowed to explore the full configuration space for the SAPS scenarios; however, in one of the SPEAR scenarios (SPEAR-IBM-MED), it did perform 1.15 times better for the full space (albeit still worse than SMAC).

5.3   Experimental Results for General Multi-instance Configuration Scenarios

We now compare the performance of SMAC, ROAR, GGA, and FOCUSEDILS on six general algorithm configuration tasks that aim to minimize the mean runtime of SAPS, SPEAR, and CPLEX for various sets of instances. These are the 5 BROAD configuration scenarios used in [8] to evaluate PARAMILS's performance, plus one further CPLEX scenario, and we used the same time budget of 5 hours per configuration run.

Overall, SMAC performed best in this comparison: as shown in Table 2 its performance was among the best (*i.e.*, statistically indistinguishable from the best) in all 6 configuration scenarios, for both the discretized and the full configuration spaces. Our simple ROAR method performed surprisingly well, indicating the importance of the intensification mechanism: it was among the best in 2 of the 6 configuration scenarios for either version of the configuration space. However, it performed substantially worse than the best approaches for configuring CPLEX—the algorithm with the largest configuration space; we note that ROAR's random sampling approach lacks the guidance offered by either FOCUSEDILS's local search or SMAC's response surface model.

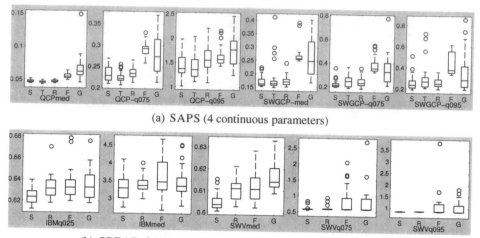

(a) SAPS (4 continuous parameters)

(b) SPEAR (26 parameters; 12 of them continuous and 4 integral)

Fig. 1. Visual comparison of configuration procedures' performance for setting SAPS and SPEAR's parameters for single instances. For each configurator and scenario, we show boxplots for the 25 test performances underlying Table 1, for the full configuration space (discretized for FOCUSEDILS). 'S' stands for SMAC, 'T' for TB-SPO, 'R' for ROAR, 'F' for FOCUSEDILS, and 'G' for GGA.

GGA performed slightly better for optimizing CPLEX than ROAR, but also significantly worse than either FOCUSEDILS or SMAC. Figure 2 visualizes the performance each configurator achieved for all 6 scenarios. We note that—similarly to the single instance cases—the results of SMAC were often more robust than those of FOCUSEDILS and GGA.

Although the performance improvements achieved by our new methods might not appear large in absolute terms, it is important to remember that algorithm configuration is an optimization problem, and that the ability to tease out the last few percent of improvement often distinguishes good algorithms. We expect the difference between configuration procedures to be clearer in scenarios with larger per-instance runtimes. In order to handle such scenarios effectively, we believe that SMAC will require an adaptive capping mechanism similar to the one we introduced for PARAMILS [8]; we are actively working on integrating such a mechanism with SMAC's models.

As in the single-instance case, for some configuration scenarios, SMAC and ROAR achieved much better results when allowed to explore the full space rather than FOCUSED-ILS's discretized search space. Speedups for SAPS were similar to those observed in the single-instance case (about 1.15-fold for SAPS-QCP and 1.65-fold for SAPS-SWGCP), but now we also observed a 1.17-fold improvement for SPEAR-QCP. In contrast, GGA actually performed worse for 4 of the 6 scenarios when allowed to explore the full space.

Table 2. Comparison of algorithm configuration procedures for benchmarks with multiple instances. We performed 25 independent runs of each configuration procedure and report the median of the 25 test performances (mean runtimes across 1 000 target algorithm runs with the found configurations on a test set disjoint from the training set). We bold-face entries for configurators that are not significantly worse than the best configurator for the respective configuration space. We also list performance of the default configuration, and of the configuration found by the CPLEX tuning tool (see [1]); note that on the test set this can be worse than the default.

Scenario	Unit	Default	CPLEX Tuning Tool	Full configuration space				Discretized configuration space			
				SMAC	ROAR	F-ILS	GGA	SMAC	ROAR	F-ILS	GGA
Saps-QCP	$[\cdot 10^{-1} s]$	11.8	—	**7.05**	7.52	—	7.84	**7.65**	**7.65**	**7.62**	**7.59**
Saps-SWGCP	$[\cdot 10^{-1} s]$	25.0	—	**1.77**	1.8	—	2.82	**2.94**	3.01	**2.91**	3.04
Spear-QCP	$[\cdot 10^{-1} s]$	3.27	—	**1.65**	1.84	—	2.21	**1.93**	2.01	2.08	2.01
Spear-SWGCP	$[\cdot 10^{0} s]$	1.62	—	**1.16**	**1.16**	—	1.17	**1.16**	**1.16**	1.18	1.18
CPLEX-Regions100	$[\cdot 10^{-1} s]$	7.40	8.60	**3.45**	6.67	—	4.37	**3.50**	7.23	**3.23**	3.98
CPLEX-MIK	$[\cdot 10^{0} s]$	4.87	3.56	**1.20**	2.81	—	3.42	**1.24**	3.11	2.71	3.32

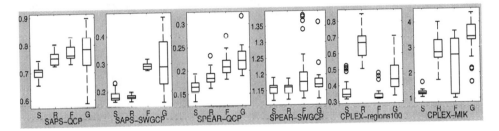

Fig. 2. Visual comparison of configuration procedures for general algorithm configuration scenarios. For each configurator and scenario, we show boxplots for the runtime data underlying Table 2, for the full configuration space (discretized for FOCUSEDILS). 'S' stands for SMAC, 'R' for ROAR, 'F' for FOCUSEDILS, and 'G' for GGA.

6   Conclusion

In this paper, we extended a previous line of work on sequential model-based optimization (SMBO) to tackle general algorithm configuration problems. SMBO had previously been applied only to the optimization of algorithms with numerical parameters on single problem instances. Our work overcomes both of these limitations, allowing categorical parameters and configuration for sets of problem instances. The four technical advances that made this possible are (1) a new intensification mechanism that employs blocked comparisons between configurations; an alternative class of response surface models, random forests, to handle (2) categorical parameters and (3) multiple instances; and (4) a new optimization procedure to select the most promising parameter configuration in a large mixed categorical/numerical space.

We presented empirical results for the configuration of two SAT algorithms (one local search, one tree search) and the commercial MIP solver CPLEX on a total of 17 configuration scenarios with small per-run captimes for each target algorithm run. Overall, our new SMBO procedure SMAC yielded statistically significant—albeit sometimes small—improvements over all of the other approaches on several configuration scenarios, and never performed worse. In contrast to FOCUSEDILS, our new methods are also able to search the full (non-discretized) configuration space, which led to further substantial improvements for several configuration scenarios. We note that our new intensification mechanism enabled even ROAR, a simple model-free approach, to perform better than previous general-purpose configuration procedures in many cases; ROAR only performed poorly for optimizing CPLEX, where good configurations are sparse. SMAC yielded further improvements over ROAR and—most importantly—also state-of-the-art performance for the configuration of CPLEX.

In future work, we plan to improve SMAC to better handle configuration scenarios with large per-run captimes for each target algorithm run; specifically, we plan to integrate PARAMILS's adaptive capping mechanism into SMAC, which will require an extension of SMACs models to handle the resulting partly *censored* data. While in this paper we aimed to find a single configuration with overall good performance, we also plan to use SMAC's models to determine good configurations on a per-instance basis. Finally, we plan to use these models to characterize the importance of individual parameters and their interactions, and to study interactions between parameters and instance features.

Acknowledgements

We thank Kevin Murphy for many useful discussions on the modelling aspect of this work. Thanks also to Chris Fawcett and Chris Nell for help with running GGA through HAL, to Kevin Tierney for help with GGA's parameters, and to James Styles and Mauro Vallati for comments on an earlier draft of this paper. We gratefully acknowledge support from a postdoctoral research fellowship by the Canadian Bureau for International Education (FH), support from NSERC through HH's and KLB's respective discovery grants, and from the MITACS NCE through a seed project grant.

References

[1] Hutter, F., Hoos, H.H., Leyton-Brown, K.: Automated configuration of mixed integer programming solvers. In: Lodi, A., Milano, M., Toth, P. (eds.) CPAIOR 2010. LNCS, vol. 6140, pp. 186–202. Springer, Heidelberg (2010)

[2] Minton, S., Johnston, M.D., Philips, A.B., Laird, P.: Minimizing conflicts: A heuristic repair method for constraint-satisfaction and scheduling problems. AIJ 58(1), 161–205 (1992)

[3] Gratch, J., Dejong, G.: Composer: A probabilistic solution to the utility problem in speed-up learning. In: Proc. of AAAI 1992, pp. 235–240 (1992)

[4] Adenso-Diaz, B., Laguna, M.: Fine-tuning of algorithms using fractional experimental design and local search. Operations Research 54(1), 99–114 (2006)

[5] Birattari, M., Yuan, Z., Balaprakash, P., Stützle, T.: F-race and iterated F-race: an overview. In: Empirical Methods for the Analysis of Optimization Algorithms. Springer, Berlin (2010)

[6] Birattari, M., Stützle, T., Paquete, L., Varrentrapp, K.: A racing algorithm for configuring metaheuristics. In: Proc. of GECCO 2002, pp. 11–18 (2002)

[7] Hutter, F., Hoos, H.H., Stützle, T.: Automatic algorithm configuration based on local search. In: Proc. of AAAI 2007, pp. 1152–1157 (2007)

[8] Hutter, F., Hoos, H.H., Leyton-Brown, K., Stützle, T.: ParamILS: an automatic algorithm configuration framework. JAIR 36, 267–306 (2009)

[9] Ansótegui, C., Sellmann, M., Tierney, K.: A gender-based genetic algorithm for the automatic configuration of algorithms. In: Gent, I.P. (ed.) CP 2009. LNCS, vol. 5732, pp. 142–157. Springer, Heidelberg (2009)

[10] Hutter, F., Babić, D., Hoos, H.H., Hu, A.J.: Boosting Verification by Automatic Tuning of Decision Procedures. In: Proc. of FMCAD 2007, pp. 27–34 (2007)

[11] KhudaBukhsh, A., Xu, L., Hoos, H.H., Leyton-Brown, K.: SATenstein: Automatically building local search SAT solvers from components. In: Proc. of IJCAI 2009 (2009)

[12] Jones, D.R., Schonlau, M., Welch, W.J.: Efficient global optimization of expensive black box functions. Journal of Global Optimization 13, 455–492 (1998)

[13] Bartz-Beielstein, T., Lasarczyk, C., Preuss, M.: Sequential parameter optimization. In: Proc. of CEC 2005, pp. 773–780. IEEE Press, Los Alamitos (2005)

[14] Hutter, F., Hoos, H.H., Leyton-Brown, K., Murphy, K.P.: An experimental investigation of model-based parameter optimisation: SPO and beyond. In: Proc. of GECCO 2009 (2009)

[15] Hutter, F., Hoos, H.H., Leyton-Brown, K., Murphy, K.P.: Time-bounded sequential parameter optimization. In: Blum, C., Battiti, R. (eds.) LION 4. LNCS, vol. 6073, pp. 281–298. Springer, Heidelberg (2010)

[16] Breiman, L.: Random forests. Machine Learning 45(1), 5–32 (2001)

[17] Hutter, F., Hoos, H.H., Leyton-Brown, K.: Sequential model-based optimization for general algorithm configuration (extended version). Technical Report TR-2010-10, UBC Computer Science (2010), http://www.cs.ubc.ca/~hutter/papers/10-TR-SMAC.pdf

[18] Bartz-Beielstein, T., Markon, S.: Tuning search algorithms for real-world applications: A regression tree based approach. In: Proc. of CEC 2004, pp. 1111–1118 (2004)

[19] Baz, M., Hunsaker, B., Brooks, P., Gosavi, A.: Automated tuning of optimization software parameters. Technical Report TR2007-7, Univ. of Pittsburgh, Industrial Engineering (2007)

[20] Xu, L., Hutter, F., Hoos, H.H., Leyton-Brown, K.: SATzilla: portfolio-based algorithm selection for SAT. JAIR 32, 565–606 (2008)

[21] Leyton-Brown, K., Nudelman, E., Shoham, Y.: Empirical hardness models: Methodology and a case study on combinatorial auctions. Journal of the ACM 56(4), 1–52 (2009)

[22] Hastie, T., Tibshirani, R., Friedman, J.H.: The Elements of Statistical Learning, 2nd edn. Springer Series in Statistics. Springer, Heidelberg (2009)

[23] Nell, C., Fawcett, C., Hoos, H.H., Leyton-Brown, K.: HAL: A framework for the automated analysis and design of high-performance algorithms. In: LION-5 (to appear, 2011)

Generalising Algorithm Performance in Instance Space: A Timetabling Case Study

Kate Smith-Miles and Leo Lopes

School of Mathematical Sciences, Monash University, Victoria 3800, Australia
{kate.smith-miles,leo.lopes}@sci.monash.edu.au

Abstract. The ability to visualise how algorithm performance varies across the feature space of possible instance, both real and synthetic, is critical to algorithm selection. Generalising algorithm performance, based on learning from a subset of instances, creates a "footprint" in instance space. This paper shows how self-organising maps can be used to visualise the footprint of algorithm performance, and illustrates the approach using a case study from university course timetabling. The properties of the timetabling instances, viewed from this instance space, are revealing of the differences between the instance generation methods, and the suitability of different algorithms.

Keywords: Algorithm Selection, Timetabling, Hardness Prediction, Phase Transition, Combinatorial optimisation, Instance Difficulty.

1 Introduction

Understanding the performance of optimisation algorithms for a class of problems involves studying the behaviour of the algorithms across an instance space defined by some measurable features or characteristics of the instances. The properties of the instances that we study to test the power of an algorithm are often defined by the source of the instances. Typically, instances used for testing optimisation algorithms either come from real world optimisation problems or they are synthetically generated via some instance generation procedure. However, it is often challenging to synthetically generate instances that are real-world-like [1]. If we intend to develop algorithms for the purpose of applying them to tackle real-world problems though, we need to understand the performance of the algorithms in the part of the feature space corresponding to the real-world instances. How do we visualise if a set of synthetic instances are similar to a small set of real-world instances? In addition, we may be interested to discover the types of instances that result in an algorithm excelling or failing, regardless of whether those instances are real-world-like.

All of these concerns are addressed by developing an understanding of the generalisation performance of algorithms across instance space. Corne and Reynolds [2] have noted that when claiming a certain algorithm performance it is important to make clear the boundaries of that performance in instance space. The performance of the algorithm on studied instances can be generalised to

C.A. Coello Coello (Ed.): LION 5, LNCS 6683, pp. 524–538, 2011.
© Springer-Verlag Berlin Heidelberg 2011

unseen instances in the same region of instance space. They introduced the idea of a "footprint" in instance space as a means to visualise the generalisation region, and noted that "understanding these footprints, how they vary between algorithms and across instance space dimensions, may lead to a future platform for wiser algorithm-choice decisions" [2]. Using two features at a time, the footprints of algorithm performance were shown in an instance space defined by two features of the problem under study (a scheduling problem and a vehicle routing problem). For problems with more than two significant features that define classes of instances though, a different visualisation approach will be needed.

In this paper we explore further these ideas of footprints and generalisation of algorithm performance in high dimensional instance spaces. We propose the use of self-organising feature maps to visualise a high-dimensional feature space as a two-dimensional map, where the generalisation footprint of algorithm performance can be clearly seen. We have previously shown how these maps can be used to visualise the relationship between features of instances and algorithm performance [3]. Using a case study of course timetabling, we demonstrate how this view of the instance space is also revealing of the relationship between the instance generation method and the resulting features of the instances. Consequently, the limitations of synthetic instance generation methods when trying to create real-world-like instances can be explained. We consider the performance of two highly competitive algorithms across three classes of instances: real-world Udine timetabling instances [4], synthetically generated instances [5], and instances that have been iteratively refined via a learning process to resemble a seed set of real-world instances [6]. We seek to understand the regions of instance space where each algorithm is superior to the other, and whether the footprint of the algorithm's strong performance includes the region of real-world instances. By providing this kind of analysis of algorithm performance we hope to enable context-specific advice to be given on algorithm selection, particularly in practical real-world settings.

The remainder of this paper is as follows: In Section 2 we describe the timetabling meta-data for our case study. In particular, we describe the three classes of instances of course timetabling and how they were generated, we provide a comprehensive list of features of the timetabling problem that we will use to define the instance space (both from the timetabling characteristics and features of the underlying graph colouring problem), we discuss the two algorithms in our algorithm portfolio, and how we measure the performance of the algorithms on the instances. Once our meta-data has been defined in this way, the analysis of the relationship between features of the instances and algorithm performance can begin. In Section 3 we present some data mining approaches to understanding the relationships in the meta-data. We begin with a self-organising feature map to visualise the high-dimensional feature space as a two-dimensional map of the instance space. The footprints of each algorithm and the regions corresponding to the types of instances seen in practise are shown. We also partition the instance space using a decision tree to provide rules describing the differences between the classes of instances, and the differences between the performance of the two algorithms in terms of the features of the instances. In Section 4 we discuss these findings and draw conclusions.

2   Course Timetabling

Timetabling is better described as a class of problems, rather than a single problem type. This research focuses on the Udine Timetabling problem, also known as Curriculum-based Course Timetabling (heretofore CTT) problem. CTT was used for track 3 of the 2007 International Timetabling Competition (ITC2007). Our choice of CTT as a case study was motivated by several factors: the existence of instance generators as well as real-world instances; and access to two of the top five search procedures from ITC2007.

In the interest of space we describe the problem only briefly here. A detailed description of the problem can be found in [7].

CTT arises because students follow specific tracks along the coursework that leads to a degree. For example: a typical first semester engineering curriculum could be Calculus, Linear Algebra, Physics, Introduction to Computing, and Introduction to a subject of study, such as Mechanical Engineering (ME) or Chemical Engineering (ChE). Lectures for Introduction to ME can be scheduled at the same time as those for Introduction to ChE, but not at the same time as those for the first four courses. The instructor assigned to each course must also be available at the times for each lecture.

The set of conflicts within a particular instance of the CTT can be described as a conflict graph $G(V, E)$, where V is a set of vertices corresponding to "events" that need to be timetabled, and E is a set of edges connecting any two vertices when the events cannot occur at the same time. We can describe conflicts in this manner for both teachers and the curriculum.

There are also soft constraints, which incur penalties when they are violated, but do not invalidate the solution. If ME has 40 first-year students and ChE has 30, then the first four courses should be taught in a room with capacity for at least 70 students, while Introduction to ME and Introduction to ChE can be taught in rooms with capacity for 40 and 30 students respectively. All lectures for the same course should ideally be in the same room, and should be distributed throughout the week. Finally, the lectures from each curriculum should run consecutively.

This description is sufficient to describe the meta-data for our experiment using the framework in [8].

- The *problem space* $\mathcal{P}$ is the union of three sets of instances: the original 21 from the competition [4]; a set of 4500 obtained using the generator in [5]; and another set of 4500 from [6]. The latter set was generated specifically to be differentiating of performance and similar to the original 21 instances, after our previous work showed that the 21 instances were not particularly discriminating of two highly competitive solvers, and that the random generator created instances that were dissimilar to real instances [9]. After excluding from $\mathcal{P}$ instances whose optimal solution violated a hard constraint (this was proved using an integer programming model), 8199 instances remained.

- The *performance space* $\mathcal{Y}$ is the sum of violations of soft constraints after 600s of computational work.
- The *algorithm space* $\mathcal{A}$ comprises two solvers: Algorithm A (TSCS[1]) is a Tabu Search over a weighed constraint satisfaction problem written in C++. Algorithm B (SACP [10]) is a constraint propagation code combined with Simulated Annealing written in Java. Of the 8199 instances in $\mathcal{P}$, 3694 resulted in draws; on 2409 instances TSCS won; and on 2096 instances SACP won.
- The *feature space* $\mathcal{F}$ is summarised in Table 1. In addition to straightforward features (like number of events), we use features related to landmarking [13] (obtained by running the DSATUR algorithm [14], which is optimal for bipartite graphs); features related to the conflicts, thus related to the underlying Graph Colouring problem and features that come from the application (Timetabling).

3   Visualising Instance Space

Now that we have assembled the meta-data for course timetabling based on two competitive algorithms, we are in a position to analyse the meta-data with a view to understanding the instance space and its properties. We seek to identify the various types of instances within the instance space and to understand the effect of instance generation method on the properties of the instances. We also seek to visualise the generalisation footprint of each algorithm's performance behaviours, and to determine the parts of instance space where one algorithm dominates the other. In order to visualise the high dimensional instance space (defined by the set of 32 candidate features in Table 1) we will be employing self-organising maps that produce a topologically-preserved mapping to a two dimensional space.

3.1   Self-Organising Feature Maps

Self-Organising Feature Maps (SOFMs) are the most well known unsupervised neural network approach to clustering. Their advantage over traditional clustering techniques such as the k-means algorithm lies in the improved visualisation capabilities resulting from the two-dimensional map of the clusters. Often patterns in a high dimensional input space have a very complicated structure, but this structure is made more transparent and simple when they are clustered in a lower dimensional feature space. Kohonen [15] developed SOFMs as a way of automatically detecting strong features in large data sets. SOFMs find a mapping from the high dimensional input space to low dimensional feature space, so the clusters that form become visible in this reduced dimensionality. They can be viewed as an approximation to a nonlinear generalisation of principle component analysis.

---

[1] The 3rd-placed solver in ITC 2007, by Astuta, Nonobe, and Irabaki. The authors have not published a paper on this solver.

Table 1. All features used in the meta-data. For features that are computed for every node, both the mean and standard deviation of the resulting distribution are used.

Feature name	Description
Size related features: those that define the dimension of the problem (3 features).	
Number of Courses	Number of courses, independently of how many lectures are in each course.
Number of Events	Sum of lectures across all courses.
Number of Rooms	Total number of rooms available.
Landmarking features: obtained from landmarking the instance by running the DSATUR algorithm [11] (2 features).	
DSATUR Solution	Upper bound on the number of colours
DSATUR Colour Sum	Sum of colour values over all nodes (DSATUR tries to minimise this quantity)
Graph Colouring features: from each of the conflict graphs $G(V, E)$, where V is a course, and E is a conflict between two courses, generated by: the curricula; the teacher availability; and the combination of both constraints (21 features):	
Edge Density	$\frac{\lvert E \rvert}{(\lvert V \rvert - 1)^2}$
Node Clustering Index[12] mean and standard deviation	For each node $v \in V$, the edge density of the graph induced by v and its immediate neighbours.
Unweighted Event Degree mean and standard deviation	The degree of each node v.
Weighted Event Degree mean and standard deviation	The sum of the enrolments of all neighbours of v.
Timetabling features: features that come from the constraints unique to timetabling, as opposed to conflicts, which are more closely related to Graph Colouring (6 features) .	
Slack	Total seats in all the rooms - Total seats required by all the courses.
One Room events	Number of events that will only fit in one room
Event Size mean and standard deviation	Number of students in each course
Room Options mean and standard deviation	The number of rooms into which each course can fit without penalty.

The architecture of the SOFM is a feed-forward neural network with a single layer of neurons arranged into a rectangular array. Figure 1 depicts the architecture with n inputs connected via weights to a 3×3 array of 9 neurons. The number of neurons used in the output layer is determined by the user.

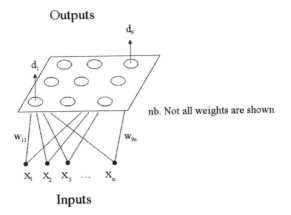

Fig. 1. Architecture of Self-Organising Feature Map

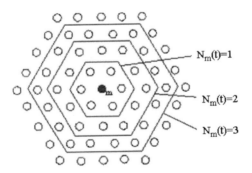

Fig. 2. Varying neighbourhood sizes around winning neuron m

When an input pattern is presented to the SOFM, each neuron calculates how similar the input is to its weights. The neuron whose weights are most similar (minimal distance d in input space) is declared the winner of the competition for the input pattern, the weights of the winning neuron are strengthened to reflect the outcome, and the learning is shared with neurons in the neighbourhood of the winning neuron. This creates a process of global competition, followed by local cooperation. Figure 2 provides an example of how a neighbourhood N_m can be defined around a winning neuron m. Initially the neighbourhood size around a winning neuron is allowed to be quite large to encourage the regional response to inputs. As the learning proceeds however, the neighbourhood size is slowly decreased so that the response of the network becomes more localised.

The localised response, which is needed to help clearly differentiate distinct input patterns, is also encouraged by varying the amount of learning received by each neuron within the winning neighbourhood. The winning neuron receives the most learning at any stage, with neighbours receiving less the further away they are from the winning neuron. If we denote the size of the neighbourhood around winning neuron m at time t by $N_{m(t)}$, then the amount of learning that every neuron i within the neighbourhood of m receives is determined by:

$$c = \alpha(t)e^{-\frac{\|r_i - r_m\|}{\sigma^2(t)}} \tag{1}$$

where $r_i - r_m$ is the physical distance (number of neurons) between neuron i and the winning neuron m. The two functions $\alpha(t)$ and $\sigma^2(t)$ are used to control the amount of learning each neuron receives in relation to the winning neuron. These functions are usually slowly decreased over time. The amount of learning is greatest at the winning neuron (where $i = m$ and $r_i = r_m$) and decreases the further away a neuron is from the winning neuron, as a result of the exponential function. Neurons outside the neighbourhood of the winning neuron receive no learning.

Like all neural network models, the learning algorithm for the SOFM follows the basic steps of presenting input patterns, calculating neuron outputs, and updating weights. The weight update rule, for all neurons within the neighbourhood of the winning neuron m for a given input pattern x_i is:

$$w_{ji}(t + 1) = w_{ji}(t) + c[x_i - w_{ji}(t)]$$

with c as defined by equation (1). For neurons outside the neighbourhood of the winning neuron, $c = 0$. The initialisation stage involves setting the weights to small random values, setting the initial neighbourhood size $N_m(0)$ to be large (but less than the number of neurons in the smallest dimension of the array), and setting the values of the parameter functions to be between 0 and 1. The algorithm iterates through all of the input patterns repeatedly, with diminishing neighbourhood size and decaying functions $\alpha(t)$ and $\sigma^2(t)$ each time, until eventual convergence of the weights.

3.2   Visualising the Instance Space

In order to determine the features most likely to be predictive of algorithm performance, a correlation analysis was employed. Wherever a feature had a correlation greater than 0.7 with the performance metric of either algorithm, it was selected for inclusion as a feature for the SOFM. The selected features were:

- From the graph built from both curriculum and teacher conflicts:
  - the minimum colours and the colour sum (from the DSATUR algorithm); the clustering index; the edge density; the mean and standard deviation of the unweighted event degree;
- From the graph built only from curriculum conflicts; and from the graph built only from teacher conflicts:

- the edge density; the mean and standard deviation of the unweighted event degree;
- Timetabling features:
  - The mean and standard deviation of Event Size and Room Options; slack; the number of one room events, courses, events, and rooms.

The instance space is therefore characterised by a set of 8199 course timetabling instances, each defined by a set of 21 features related to both the properties of the timetabling environment and the underlying graph colouring problem.

All features were normalised to the range [0,1] using variance. The software package Viscovery SOMine [16] was used to generate the SOFM, using a rectangular map of approximate ratio 100:52 based on the dimensions of the plane spanned by the two largest eigenvectors of the correlation matrix of the features (i.e. the first two principal components of the correlation matrix). The final map contains 2030 neurons arranged in 58 rows and 35 columns. 48 complete presentations of all 8199 instances were required to achieve convergence, with a decay factor of 0.5 applied to the functions $\alpha(t)$ and $\sigma^2(t)$. The initial neighbourhood size was 7. While these values were chosen arbitrarily based on past experience, experimentation with different values showed that the resulting maps were quite robust.

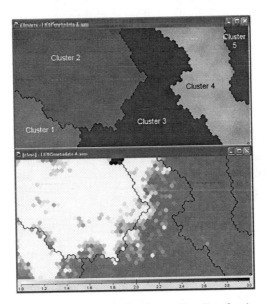

Fig. 3. Five clusters in instance space (top) and the distribution of the three classes of instances across instance space (bottom). The real-world Udine instances are shown as black, the synthetic instances as grey, and the refined synthetic instances as white

Figure 3 shows that there are five natural clusters of instances in the 21-dimensional feature space when projected onto a two-dimensional map of instance space. Instances that belong to the same cluster are similar (according to

Euclidean distance in 21-dimensional feature space) to each other, and significantly different from other instances in other clusters. The lower map in Figure 3 shows the location of the three classes of instances across the instance space. We find the small set of real-world Udine instances (shown as black regions) all located in the top-centre of the map (top right corner of cluster 2). Clusters 3, 4 and 5 contain predominantly the synthetic instances (shown as grey regions of the lower map) generated from the synthetic generator [5], and are not in the same region as the real-world instances. The instances that we have modified to be more "real-world-like" [6] (shown as white region on the map) surround the Udine instances and are therefore quite similar based on their features, but more diverse.

In order to determine which features make an instance more real-world like, we can inspect the distribution of features across the map. A subset of the features relating to the timetabling environment are shown in Figure 4, and some of the features relating to the underlying graph colouring problem are shown in Figure 5. Here we see that one of the main differences between the Udine instances and our real-world-like instances is the mean and standard deviation of the degree of the teacher conflict graph (significantly smaller in the Udine instances). In addition, the mean and standard deviation of the event size is significantly smaller for the Udine instances. Thus we have obtained some immediate feedback on how to make our real-world-like instances more similar to the Udine instances.

The synthetic instances [5] are clearly quite different in distribution from the Udine instances. The main observations about these differences are revealed in the map by considering the boundary separating clusters 2 and 3, which correlates quite closely with the distribution of colorsum, the number of courses and mean room options.

It should be noted that no information about the class of instance was used to generate the clusters, only features of the timetabling problem and the underlying graph colouring problem. Yet the three classes of instances are clearly seen as quite distinguishable in this instance space. We now examine the performance of the two algorithms across the instance space with a view to visualising the footprint of their generalisation.

3.3   Visualising the Footprints of Algorithm Performance

Once the clusters have formed based on similarity of features of the instances, we can now superimpose additional information such as the performance of algorithms on those instances. The penalty of each algorithm for instances across the map is shown in the top row of Figure 6, with Algorithm A (TSCS) shown on the left and Algorithm B (SACP) on the right. Visually, these two algorithms appear to be very competitive with each other, producing low penalty solutions to instances in cluster 1, high penalty solutions to the difficult instances at the bottom of cluster 3, and similar penalties to each other across the map. The difference in penalty (Algorithm A minus Algorithm B) is shown in the lower left of Figure 6, and reveals that there is little difference in the performance of the algorithms on the synthetic instances in clusters 3, 4 and 5. Only instances in

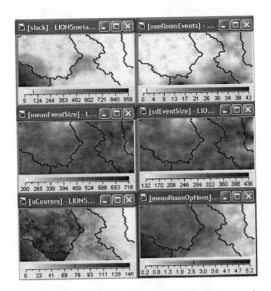

Fig. 4. The distribution of timetabling based features across instance space (white represents a minimal value of the feature and black represents a maximal value of the feature)

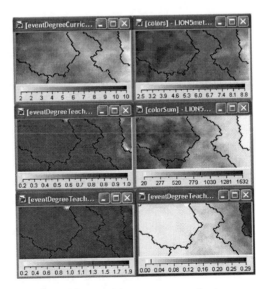

Fig. 5. The distribution of graph based features across instance space (white represents a minimal value of the feature and black represents a maximal value of the feature)

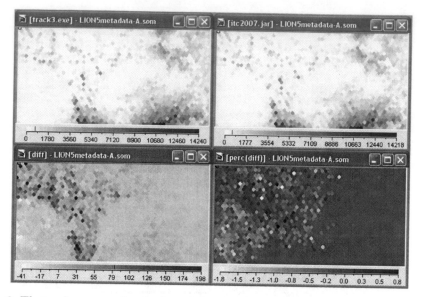

Fig. 6. The performance of each algorithm across instance space. The penalty obtained by Algorithm A (TSCS) is shown top left, and Algorithm B (SACP) is shown top right. The difference in penalty between Algorithm A and B is shown bottom left (penalty of Algorithm A minus penalty of Algorithm B), and the relative difference is shown bottom right. White represents a minimal value and black represents a maximal value.

clusters 1 and 2 provide the opportunity for each algorithm to show its relative power. Since a difference of 1 penalty point is less meaningful for a high penalty solution than a low penalty solution, we also show the relative difference in algorithm performance on the bottom right of Figure 6. The relative difference is calculated as the ratio of the difference in penalty to the mean penalty for each instance, which reduces the impact of small differences in high penalty solutions.

Clearly, the boundary between cluster 2 and 3 provides some kind of partition across instance space to separate those instances that elicit identical performance from both algorithms from those instances that present unique challenges to each algorithm. It is also clear that there are some instances in clusters 1 and 2 where Algorithm A outperforms Algorithm B, and others where the reverse is true. Unfortunately, the regions where one algorithm clearly outperforms the other are not so well defined. There is a region spreading from the top left corner of the map diagonally down through cluster 2 where the relative difference is large and positive (i.e. Algorithm A has a much higher penalty than Algorithm B, and therefore Algorithm B is the superior algorithm for such instances). Returning to the features however, it is difficult to see a single feature that explains this diagonal pattern (possibly slack, colorsum and number of courses, but the superior performance of Algorithm B does not continue into cluster 1 where colorsum and number of courses are also high). Likewise, the superior performance of Algorithm A in cluster 1 (and a small region just over the boundary into cluster 2) is not well explained by any of the features of the instances.

So while we have features that are clearly capable of distinguishing between the classes of instances (real world versus synthetic), and whether the instances will be discriminating of algorithm performance (or just elicit a tied outcome), it would appear that the selected features are not ideal for completely explaining the conditions under which Algorithm A outperforms Algorithm B and vice versa. It is possible that this is a problem with the chosen method of analysis (self organising maps), and so we now employ other machine learning methods to see if we can automate the discovery of relationships.

3.4   Partitioning the Instance Space via Decision Trees

Decision trees can be very powerful tools for elucidating rules that can help explain performance differences between algorithms. However, extraordinary care must be taken in designing the experiment, especially to ensure class balance and avoid bias.

In these experiments, we perform training on a random subset of each class of cardinality equal to the cardinality of the class with the fewest elements to help control bias.

Node	competition	real-world-like	synthetic
1. Root	21	21	21
2. Teacher Clustering Index Mean< 0.4	21 (100%)	0	0
3. Teacher Clustering Index Mean≥ 0.4	0	21 (50%)	21 (50%)
4. One Room Events < 2.5	0	20 (87%)	3 (13%)
5. One Room Events ≥ 2.5	0	1 (5%)	18 (95%)

Fig. 7. A decision tree that describes features that can be used to determine the origin of the instance. Unfortunately, these features do not help predict algorithm performance.

Figure 7 summarises a decision tree experiment run on a subset of $\mathcal{P}$ obtained by randomly sampling 21 instances (the number of real instances from the ITC2007 competition) from each of the *synthetic* and *real-world-like* instances, then trying to separate all three types of instances. While the clustering index of the teacher conflict graph can be easily used to separate the real instances from the synthetic ones, such separation is inconsequential, at least if the goal is to compare these two solvers. The tree in Figure 8 , obtained by randomly sampling an equal number of instances on which each solver won or there was a draw, illustrates that while it is possible to learn some aspects of the relationships between the features and which solver will win, the most important features are distinct from those in Figure 7.

Another word of caution illustrated by Figure 8 is that simply counting wins can be a dangerous way to decide which solver is superior to the other. These data show that in problems in which the DSATUR colour sum is low, there is not a significant difference between the solvers, and that in the remaining instances,

Node	SACP wins	Tie	TSCS wins
1. Root	2096	2096	2096
2. DSATUR colour sum $\leq$ 393	339 (17%)	1486 (76%)	126 (7%)
3. DSATUR colour sum $>$ 393	1757 (41%)	610 (14%)	1970 (45%)
4. Slack$<$ 112	1337 (63%)	216 (10%)	567 (27%)
5. Slack $\geq$ 112	420 (19%)	394 (18%)	1403 (63%)

Fig. 8. A decision tree that describes features that can be used to determine which algorithm will win. The results are insightful but inconclusive.

SACP wins on those instances in which there is little slack. This tree, therefore, supports the argument that SACP is the better solver on hard instances, since it provides evidence that SACP wins where there is little slack and heuristics have difficulty reducing conflicts. In contrast, if only raw performance data excluding feature information is used, a statistically strong conclusion would be reached that TSCS wins more often overall in $\mathcal{P}$, and that the difference in mean performance is greater than 0. This conclusion would hold even in the subset of $\mathcal{P}$ composed exclusively of real-world-like instances.

It is up to the reader of a specific research paper to decide whether or not it is important to them that a solver wins on a subset of instances that is harder. However, having that information – as opposed to relying on statistics over an entire instance set – is valuable independent of any subjective consideration. Furthermore, the information is supported by a repeatable (and challengeable), concrete experiment.

4   Conclusions

In this paper we have shown, through a case study of university course timetabling, that data mining techniques like self-organising feature maps and decision trees can be used to explore the high-dimensional feature space that defines an instance space. Specifically, the instance space can be visualised with a view to understanding the applicability of synthetic instance generators to real-world instances, and examining the generalisation footprint of algorithm performance. For our case study, we have utilised a comprehensive set of features based on both timetabling and graph colouring properties. We have demonstrated that these features create an instance space where the differences between real-world and synthetically generated instances are readily visualised. The effectiveness of different algorithms can then be superimposed across the instance space and the footprint can be visualised. For our chosen algorithms we have been able to partition the instance space to separate instances that elicit tied performance from these two highly competitive algorithms, from those instances where one algorithm outperforms the other. The chosen features have proven to be insufficient, however, for discovering the properties of instances that make SACP outperform TSCS, or vice versa. The footprint of both algorithms, where they performs well, includes the Udine real-world instances (which is why both algorithms performed well in the competition), but as we move away from the Udine instances in the instance space, we find some regions where one algorithm dominates. The boundaries of these regions are less well defined based on the current features. It remains for future research to consider additional features of instances that could distinguish between these two competitive algorithms, and the relationship between landscape metrics [17,18,19] and algorithm performance should also be considered.

The ability to generate instances that are both discriminating of algorithm performance and real-world-like is critical for progress in understanding the strengths and weaknesses of various algorithms, and ensuring that the right algorithm is being selected to avoid deployment failures for practical applications [9]. Analysis of the kind presented in this paper provides a starting point to examine the characteristics of a set of instances, and enables feedback into the instance generation process [6] to develop a meaningful set of instances to drive future research developments.

References

1. Hill, R., Reilly, C.: The effects of coefficient correlation structure in two-dimensional knapsack problems on solution procedure performance. Management Science, 302–317 (2000)
2. Corne, D., Reynolds, A.: Optimisation and Generalisation: Footprints in Instance Space. In: Schaefer, R., Cotta, C., Kołodziej, J., Rudolph, G. (eds.) PPSN XI. LNCS, vol. 6238, pp. 22–31. Springer, Heidelberg (2010)
3. Smith-Miles, K., van Hemert, J., Lim, X.: Understanding TSP Difficulty by Learning from Evolved Instances. In: Blum, C., Battiti, R. (eds.) LION 4. LNCS, vol. 6073, pp. 266–280. Springer, Heidelberg (2010)
4. McCollum, B., Schaerf, A., Paechter, B., McMullan, P., Lewis, R., Parkes, A.J., Gaspero, L.D., Qu, R., Burke, E.K.: Setting the research agenda in automated timetabling: The second international timetabling competition. INFORMS Journal on Computing 22(1), 120–130 (2010)
5. Burke, E.K., Mareček, J., Parkes, A.J., Rudová, H.: Decomposition, reformulation, and diving in university course timetabling. Computers & Operations Research 37(3), 582–597 (2010)
6. Lopes, L., Smith-Miles, K.: Generating applicable synthetic instances for branch problems, under review (2011)
7. Gaspero, L.D., McCollum, B., Schaerf, A.: The second international timetabling competition (itc-2007): Curriculum-based course timetabling (track 3). Technical report, DIEGM, University of Udine (2007)
8. Rice, J.: The Algorithm Selection Problem. Advances in Computers 15, 65–117 (1976)
9. Lopes, L., Smith-Miles, K.: Pitfalls in Instance Generation for Udine Timetabling. In: Blum, C., Battiti, R. (eds.) LION 4. LNCS, vol. 6073, pp. 299–302. Springer, Heidelberg (2010)
10. Müller, T.: Itc2007 solver description: A hybrid approach. In: Proceedings of the Seventh PATAT Conference (2008)
11. Culberson, J., Luo, F.: Exploring the k-colorable landscape with iterated greedy. In: Cliques, Coloring, and Satisfiability: Second DIMACS Implementation Challenge, pp. 245–284 (1996)

12. Beyrouthy, C., Burke, E., Landa-Silva, D., McCollum, B., McMullan, P., Parkes, A.: Threshold effects in the teaching space allocation problem with splitting. European Journal of Operational Research (EJOR) (2008) (under review)
13. Pfahringer, B., Bensusan, H., Giraud-Carrier, C.: Meta-learning by landmarking various learning algorithms. In: Proceedings of the Seventeenth International Conference on Machine Learning Table of Contents, pp. 743–750. Morgan Kaufmann Publishers Inc., San Francisco (2000)
14. Wood, D.: An algorithm for finding a maximum clique in a graph. Operations Research Letters 21(5), 211–217 (1997)
15. Kohonen, T.: Self-organized formation of topologically correct feature maps. Biological Cybernetics 43(1), 59–69 (1982)
16. SOMine, V.: Eudaptics software Gmbh
17. Knowles, J., Corne, D.: Towards landscape analyses to inform the design of a hybrid local search for the multiobjective quadratic assignment problem. In: Soft Computing Systems: Design, Management and Applications, pp. 271–279 (2002)
18. Bierwirth, C., Mattfeld, D., Watson, J.: Landscape regularity and random walks for the job-shop scheduling problem. In: Gottlieb, J., Raidl, G.R. (eds.) EvoCOP 2004. LNCS, vol. 3004, pp. 21–30. Springer, Heidelberg (2004)
19. Schiavinotto, T., Stützle, T.: A review of metrics on permutations for search landscape analysis. Comput. Oper. Res. 34(10), 3143–3153 (2007)

A Hybrid Fish Swarm Optimisation Algorithm for Solving Examination Timetabling Problems

Hamza Turabieh[1] and Salwani Abdullah[2]

[1] Computer Science Department, Faculty of Science and Information Technology
Zarka University, Jordan
turabieh@zp.edu.jo
[2] Data Mining and Optimisation Research Group (DMO)
Center for Artificial Intelligence Technology,
Universiti Kebangsaan Malaysia, 43600 Bangi, Selangor, Malaysia
salwani@ftsm.ukm.my

Abstract. A hybrid fish swarm algorithm has been proposed to solve exam timetabling problems where the movement of the fish is simulated when searching for food inside water (refer as a search space). The search space is categorised into three categories which are crowded, not crowded and empty areas. The movement of fish (where the fish represents the solution) is determined based on a Nelder-Mead simplex search algorithm. The quality of the solution is enhanced using a great deluge algorithm or a steepest descent algorithm. The proposed hybrid approach is tested on a set of benchmark examination timetabling problems in comparison with a set of state-of-the-art methods from the literature. The experimental results show that the proposed hybrid approach is able to produce promising results for the test problem.

Keywords: Exam Timetabling, Fish Swarm Algorithm.

1 Introduction

Timetabling problems present a challenging problem area for researchers across both operational research and artificial intelligence. This kind of problems can be classified as scheduling problems. This concept can be defined based on Fox and Sadeh-Koniecpol [15] as follows:

"Scheduling selects among alternative plans, and assigns resources and times to each activity so that they obey the temporal restrictions of activities and the capacity limitations of a set of shared resources."

The examination timetabling problem is a popular problem in the academic world for schools or higher educational institutes which are concerned with allocating exams into a limited number of time slots (periods) subject to a set of constraints so that no students should sit for two or more exams at the same time and the scheduled exams must not exceed the room capacity.

C.A. Coello Coello (Ed.): LION 5, LNCS 6683, pp. 539–551, 2011.
© Springer-Verlag Berlin Heidelberg 2011

Up to date, many approaches have been introduced to solve examination timetabling problems. Carter [9] categorised these approaches into four types: sequential methods, cluster methods, constraint-based methods and generalised search. Petrovic and Burke [26] added the following categories: hybrid evolutionary algorithms, meta-heuristics, multi-criteria approaches, case based reasoning techniques, hyper-heuristics and adaptive approaches.

Graph colouring heuristics methods seem to be the earliest algorithms applied in this problem, followed by stochastic search methods such as simulated annealing, tabu search, genetic algorithm etc., that are considered as meta-heuristic approaches. In general, these approaches can be classified into two categories i.e. single-based approach and population-based approach. Single-based approach such as simulated annealing [2], large neighbourhood search [1] and tabu search [17] in general works on a single solution and try to find a better solution in a solution space. On the other hand, population-based approaches start with many solutions and then try to obtain optimal solution(s) in the whole search space. The most common population-based approaches are evolutionary algorithms [11], ant colony algorithms [12], artificial immune systems [22] etc. In this work, we deal with a hybrid population-based approach that combine the good properties of a single-based approach (a great deluge algorithm in this case) and a population-based approach (i.e. fish swarm algorithm) to solve uncapacitated examination timetabling problem. We try to create a balance between the exploration (generated by the fish swarm algorithm) and exploitation (generated by the great deluge algorithm). These algorithms are discussed in details in Section 3 and Section 4.2, respectively. Interested readers can refer to [20], [23], [29] [30], and [31] for a comprehensive survey on timetabling.

The rest of the paper is organised as follows. The next section describes the examination timetabling problem in details. Section 3 presents a fish swarm algorithm followed by the hybrid approach in Section 4. Experimental results of comparing the proposed hybrid approach with other algorithms from the literature are reported and discussed in Section 5. Finally, Section 6 concludes this paper with brief concluding comments and discussion on the future work.

2 Uncapacitated Examination Timetabling Problem

In this work, the formulations of examination timetabling problem have been adapted from the description presented in Burke et al. [6]. The problem consists of:

- E_i is a collection of N examinations ($i=1,...,N$).
- T is the number of timeslots.
- $C=(c_{ij})_{NxN}$ is the conflict matrix where each record, denoted by c_{ij} ($i,j \in \{1,...,N\}$), represents the number of students taking exams i and j.
- M is the number of students.
- t_k ($1 \leq t_k \leq T$) specifies the assigned timeslots for exam k ($k \in \{1,...,N\}$.

The specifications of the uncapacitated examination timetabling problems are shown in Table 1 as taken from [29].

Table 1. Specifications of the uncapacitated examination timetabling problem

Datasets	Number of timeslots	Number of examinations	Number of students	Conflict matrix density
car91	32	543	18419	0.14
car92	35	682	16925	0.13
ear83I	24	190	1125	0.29
hec92I	18	81	2823	0.42
kfu93	20	461	5349	0.06
lse91	18	381	2726	0.06
sta83I	13	139	611	0.14
tre92	23	261	4360	0.18
uta92I	35	622	21267	0.13
ute92	10	184	2750	0.08
yor83I	21	181	941	0.27

Only one hard constraint is considered for uncapacitated exam timetabling which is no students should be required to sit two examinations simultaneously.

In this problem, we formulate an objective function which tries to spread out students' exams throughout the exam period (Expression (1)) that is treated as a soft constraint.

$$\min \frac{\sum_{i=1}^{N-1} F(i)}{M} \tag{1}$$

Where:

$$F(i) = \sum_{j=i+1}^{N} c_{ij} \cdot proximity \ (t_i, t_j) \tag{2}$$

$$proximity \ (t_i, t_j) = \begin{cases} 2^5 / 2^{|t_i - t_j|} & \text{if } 1 \le |t_i - t_j| \le 5 \\ 0 & \text{otherwise} \end{cases} \tag{3}$$

Subject to:

$$\sum_{i=1}^{N-1} \sum_{j=i+1}^{N} c_{ij} \cdot \lambda(t_i, t_j) = 0 \quad \text{where} \quad \lambda(t_i, t_j) = \begin{cases} 1 & \text{if } t_i = t_j \\ 0 & \text{otherwise} \end{cases} \tag{4}$$

Equation (2) presents the cost for an exam i which is given by the proximity value multiplied by the number of students in conflict. Equation (3) represents a proximity value between two exams [8]. Equation (4) represents a clash-free requirement so that no student is asked to sit two exams at the same time. The clash-free requirement is considered to be a hard constraint.

3 The Fish Swarm Optimisation Algorithm

Studying the behavior of fish through searching for food attracts a lot of researchers due to its ability to find foods in a large searching area. In this section, we present the idea of fish swarming algorithm, that simulates the behavior of a fish inside water (search space) while they are searching for food. Fish swarm algorithm has been successfully applied on many optimisation problem such as in [14], [16], [18], [19], [21], [32], and [33]. In this work, the fish swarm algorithm works on selected solutions rather than all solutions, with an aim to reduce the computational time. We will use the words "fish" and "solution" interchangeably throughout the paper. The basic point of a fish swarm algorithm is the visual scope (area). We categorised three possible cases for the visual scope for Sol_i (where Sol represents a solution) as in Figure 1:

- Empty visual: no solution is closed to Sol_i.
- Crowded visual: many solutions are closed to Sol_i.
- Not crowded visual: a few solutions are closed to Sol_i.

A visual scope for a selected solution Sol_i can be represented as the scope of closed solutions. The determination of the number of solutions that are closed to Sol_i is based on the number of solutions inside the visual scope. The visual is used to determine the closeness of two solutions which is based on the distance (solution quality) between two solutions (i.e. $f(x') - f(x)$). Note that in this work, if the distance is less or equal to 10, then the solution x' is closed to x. The category of the visual scope is determined based on the following mechanism:

$$\frac{Number\ of\ closed\ solutions}{size\ of\ population} \leq \theta$$

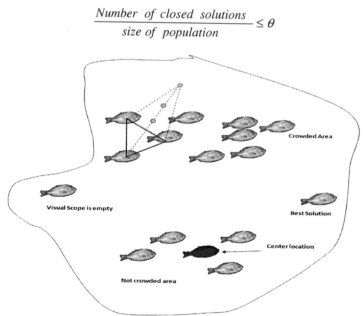

Fig. 1. A representation of solutions in the search space

where θ is set to 0.5. Note that all the constant values used in this algorithm are obtained through some preliminary experiments. If the number of the closed solution is more than the size of the visual scope, then the visual scope is considered as a crowded area. The size of visual scope is calculated as:

$$\frac{Size\ of\ Population}{2} + 1$$

i.e. in other words, the visual scope is considered as a crowded area if the number of the closed solution inside this visual scope is at least greater than half of the population size.

The simulation behavior of the fish swarm algorithm can be described in three steps i.e. swarming, chasing and searching behaviors which are based on the category of the visual scope as shown in Figure 2. If the visual scope is a crowded area then a searching behavior is employed; if the visual scope is a not crowded area then a swarming or a chasing behavior is applied; if the visual scope is empty, then a chasing behavior is applied. The details of the simulation behavior of the fish swarm algorithm that is employed in this work are discussed as below:

```
Input
m: size of population
Set Iteration =1
Initialization Phase
    (x₁, x₂, …, xₘ) ⟵ Construction();
Improvement Phase
While termination condition is not satisfied do
    Select a solution using Roulette Wheel Selection (RWS) Solᵢ
    Set best solution as Sol_best;
    Compute the visual for Solᵢ
    if  visual scope is an empty area then
        Solᵢ' = Steepest descent(Solᵢ)
    else
        if visual scope is a crowded area then
            Apply a multi-decay rate great deluge to obtain Solᵢ'
            as in Fig 5.
        else
            Central Location = compute the central point of visual
            if  Central Location is better than Solᵢ  then
                Estimated quality = Central Location;
                Apply a standard great deluge (swarming
                behavior) as in [13] to obtain Sol_iA;
            else
                Estimated quality  =  Best Solution, Sol_best;
                Apply a standard great deluge (chasing behavior)
                as in [13] to obtain Sol_iA;
            end if
            if  Sol_best < Solᵢ  then
                Estimated quality  =  Best Solution, Sol_best;
                Apply a standard great deluge (chasing behavior)
                as in [13] to obtain Sol_iB;
            else
                Sol_iB = Steepest descent(Solᵢ)
            end if
            Solᵢ' = min{Sol_iA , Sol_iB}
        end if
    end if
    Solᵢ'' = min{Solᵢ, Solᵢ'} // Update solution
    Iteration++
end while
```

Fig. 2. The pseudo code for the fish swarm algorithm

A.   Swarming or Chasing behavior

A fish (solution) swarms towards a central point (which is an average of the *visual scope*) of the search area. This behavior is applied if and only if the central point (in terms of the quality of the solution) is better than the selected solution Sol_i. Otherwise, a fish will chase a best solution so far. These behaviors are represented by a great deluge algorithm (see Section 4.2) where a central point or a best solution is treated as an estimated quality, respectively. However, if the *visual scope* is empty, a chasing behavior that employed a steepest descent algorithm is employed.

B.   Searching behaviour

In this behavior, a Nelder-Mead simplex algorithm [21] (as in Section 4.1) is used to determine the movement directions of the fish. There are three directions called a Contraction-External (CE), a Reflection (R) and an Expansion (E). These movement directions are later to be used to intelligently control the multi decay rate in the great deluge algorithm. Note that the details on the multi decay rate great deluge algorithm are discussed in Section 4.2.

4   The Hybrid Approach

The main issue of hybridising a fish swarm algorithm and the great deluge (or steepest descent) algorithm is to combine the advantages these algorithms. Before the great deluge is employed, a Nelder-Mead simplex method is used to calculate the expected location (estimated quality value) of the selected solution in the population. Nelder-Mead simplex method is a very efficient local search method procedure but its convergence is extremely sensitive to the selected starting point, thus a great deluge algorithm is hybridised in order to control the convergence by using a multi decay rate that control the level of accepting a worse solution during the search process. The following two subsections will describe Nelder-Mead simplex algorithm and a multi decay rate great deluge algorithm, respectively.

4.1   Nelder-Mead Simplex Algorithm

Nelder-Mead simplex algorithm has been proposed by Nelder and Mead [25] which is a local search method designed for unconstrained optimisation without using gradient information. This algorithm tries to find an approximation of a local optimum of a problem with N variables. The method is based on the theory of simplex, which is a special polytope of $N + 1$ vertices in N dimensions. The operations of this method rescale the simplex based on the local behavior of the function by using four basic procedures: reflection, expansion, contraction external and shrinkage. In this work only three values have been estimated based on Nelder-Mead simplex algorithm which they are as follows:

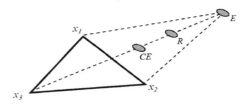

Fig. 3. Nelder-Mead simplex search algorithm

- Contraction_External(CE) = $\lfloor \frac{x_1 + x_2}{2} \rfloor - [\frac{[\frac{x_1 + x_2}{2}] + x_3}{2}]$

- Reflection(R) = Contraction_External $- [\frac{[\frac{x_1 + x_2}{2}] + x_3}{2}]$

- Expansion(E) = Reflection $- [\frac{[\frac{x_1 + x_2}{2}] + x_3}{2}]$

The three expected locations (estimated values) will be used within the multi decay rate great deluge algorithm in order to control the great deluge decay rate in accepting a worse solution. Figure 3 represents the three possible new locations for the x_3 in the search space (i.e. at CE, R or E).

4.2 Multi Decay Rate Great Deluge Algorithm

In this work, we applied a great deluge algorithm [13] that can intelligently control the decay rate based on the estimated value calculated using Nelder-Mead simplex algorithm. The great deluge algorithm always accepts a better solution, and a worse solution is accepted in order to escape from local minimum. The acceptance of the worse solution is controlled by the decay rate that later will affect the performance of searching behavior. Figure 4 illustrates the changing of the decay rate at different estimated values.

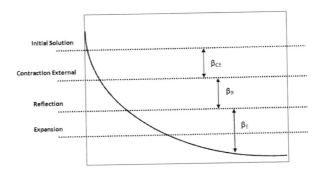

Fig. 4. Changing decay rate at different estimated values (Contraction-External, Reflection and Expansion)

Figure 5 represents the pseudo code of the proposed multi decay rate great deluge algorithm.

```
Calculate estimated qualities based on Nelder-Mead Simplex Algorithm
Contraction-External (CE), Reflection (R), and Expansion (E)) (see
Section 4.1);
Calculate the force decay rate, βCE = Contraction_External/NumOfIte;
Calculate the force decay rate, βR = Reflection/NumOfIte;
Calculate the force decay rate, βE = Expansion/NumOfIte;
Set initial decay rate as Contraction_External (CE) i.e. β = βCE;
Set level = βCE;
Set maximum number of iteration, NumOfIte;
Iteration ← 1;
do while (iteration < NumOfIte)
     Define neighbourhood (N1 and N2) of  Soli by randomly assigning
     exam to a valid timeslot to generate a new solution called Soli*;
     Calculate f(Soli*);
     if (f(Soli*) < f(Solbest)) where Solbest represents the best solution
     found so far
          Soli ← Soli*;
          Solbest ← Soli*;
     else
          if (f(Soli*)≤ level)
          Soli← Soli*;
     end if
     if   Solbest< Reflection (R)
          β = βR;
     else
          if Solbest< Expansion (E)
          β = βE;
     end if
     level = level - β;
     Iteration++
end while
```

Fig. 5. The pseudo code for the multi decay rate great deluge algorithm

Two different neighbourhood structures and their explanation are outlined as follows:
N1: Select two exams at random and swap timeslots.
N2: Choose a single exam at random and move to a new random feasible timeslots.

5   Simulation Results

In order to test our proposed algorithm, we employed it over uncapacitated examination timetabling datasets that was introduced by Carter et al. [8]. The details of datasets can also be found in [29]. The proposed algorithm was programmed using Matlab and simulations were performed on the Intel Pentium 4 2.33 GHz computer. The parameter settings used in this work are shown in Table 2.

Table 2. Parameters setting

Parameter	Iteration	Population size	GD-iteration	Steepest descent iteration	Visual
Value	1000000	50	2000	2000	10

Table 3. Results

Instance	Best	Avg.	Median	Q1	Q3
car91	4.81	4.93	4.90	4.84	4.92
car92	4.11	4.33	4.19	4.17	4.53
ear83I	36.10	36.54	36.47	36.33	35.58
hec92I	10.95	11.34	11.01	11.00	11.36
kfu93	13.21	13.43	13.60	13.33	13.75
lse91	10.20	10.54	10.65	10.51	10.93
sta83I	159.74	159.93	159.82	159.81	159.96
tre92	8.00	8.45	8.36	8.35	8.51
uta92I	3.32	3.62	3.53	3.33	3.70
ute92	26.17	26.77	26.65	26.55	26.81
yor83I	36.23	36.55	36.42	36.33	36.71

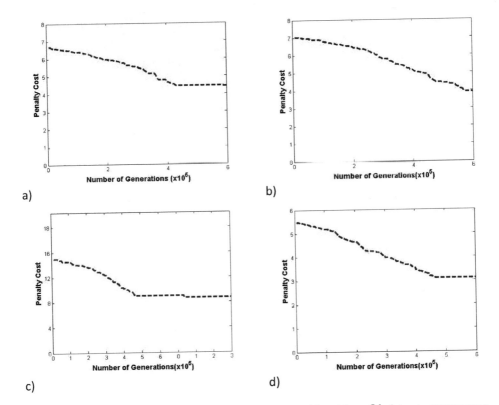

Fig. 6. Convergence using fish swarm optimization algorithm (a) car91 dataset convergence; (b) car92 dataset convergence; (c) tre92 dataset convergence; (d) ute92 dataset convergence.

The experiment carried out in this section attempts to space out students' examination throughout the examination period. Termination condition is the number of generations (i.e. 1,000,000 generations). The best, average, median together with the upper and lower quartiles results are reported in Table 3 which is out of 11 runs.

From Table 3, we can see that our approach is able to obtain good enough results since the difference between best and average for all datasets is in between (0.08 to 0.35) except for ear83I. We believe the algorithm is robust since the difference between best and average is small and also the difference between the upper and lower quartiles is in between (0.03 to 0.44), again except for ear83I. This shows that the solutions obtained are close to each other at the solution search space.

Figures 6 (a), (b), (c) and (d) show the behavior of our algorithm when exploring the search space on car91, car92, tre92 and ute92 datasets, respectively. The x-axis represents the number of generations, while the y-axis represents the overall penalty cost. These graphs demonstrate how the hybrid algorithm explores the search space. Note that the timetable quality is measured by taking the average penalty per student. The curves show that the algorithm begins with initial solutions and rapidly improves the quality of the timetable. For most of the cases (except for car92), the graphs show no improvement on the quality of solutions at about 200000 iterations (for $tre92$) and 400000 iterations (for car91 and ute92). For the car92 dataset, the algorithm is still able to reduce the penalty cost even after 1000000 iterations. This shows that given extra processing time, the quality of solution on certain datasets might be able to be enhanced further.

Table 4. Comparison of our results with other approaches in the literature

Instance	Our approach	Merlot et al. [24]	Casey and Thompson [10]	Côté et al. [11]	Yang and Petrovic [34]	Turabieh and Abdullah [3]
car91	4.81	5.1	5.4	5.2	4.50	4.80
car92	4.11	4.3	4.2	4.2	3.93	4.10
ear83I	36.10	35.1	34.2	34.2	33.7	34.92
hec92I	10.95	10.6	10.2	10.2	10.83	10.73
kfu93	13.21	13.5	14.2	14.2	13.82	**13.00**
lse91	10.20	10.5	14.2	11.2	10.35	10.01
sta83I	159.74	157.3	134.9	157.2	158.35	158.26
tre92	8.00	8.4	8.2	8.2	7.92	**7.88**
uta92I	3.32	3.5	-	3.2	3.14	3.20
ute92	26.17	25.1	25.2	25.2	25.39	26.11
yor83I	36.23	37.4	37.2	36.2	36.35	36.22

Instance	Caramia et al. [7]	Abdullah and Burke [1]	Qu and Burke [28]	Qu et al. [27]	Burke et al. [5]	Burke and Bykov [4]
car91	6.6	4.1	5.16	5.11	4.6	**4.42**
car92	6.0	4.8	4.16	4.32	3.9	**3.74**
ear83I	**29.3**	36.0	35.86	35.56	32.8	32.76
hec92I	**9.2**	10.8	11.94	11.62	10.0	10.15
kfu93	13.8	15.2	14.79	15.18	13.0	12.96
lse91	**9.6**	11.9	11.15	11.32	10.0	9.83
sta83I	158.2	159.0	159	158.88	**156.84**	157.03
tre92	9.4	8.5	8.6	8.52	7.9	**7.75**
uta92I	3.5	3.6	3.59	3.21	3.2	**3.06**
ute92	24.4	26.0	28.3	28	24.8	24.82
yor83I	36.2	36.2	41.81	40.71	**34.9**	34.84

Table 4 shows the comparison of our final results in terms of penalty cost compared to other recent published results in the literature. The best results are presented in bold. Our algorithm is capable to find a good enough feasible timetables for all eleven cases. Even though, our algorithm is not able to obtain new best results, but from Table 4 we can see that the algorithm works reasonably well across all datasets and we did not perform the worse in any of the datasets, even the complexity of the datasets are different (based on the conflict density value). We believe this is because the capability of the proposed hybrid approach that can make a balance between exploration and exploitation during the search process that help to minimise the objective function values and give competitive results for the uncapacitated examination timetabling problem compared to other algorithm in the literature.

6 Conclusion and Future Work

In this paper, a simulation of fish swarm algorithm has been applied on uncapacitated exam timetabling problems. The search space has been categorised into three categories, which are crowded, not crowded and empty areas. The movement of each fish (solution) is based on its location in the search space. Intelligent decay rate have been proposed for great deluge algorithm based on the hybridisation with the Nelder–Mead simplex algorithm. The obtained results are good enough and we strongly believe that categorising the search space enhances the exploration process while intelligent controlling the decay rate enhances the exploitation process. Our future work will aim to test this algorithm on International Timetabling Competition dataset (ITC2007) for both exam and course timetabling problems.

References

[1] Abdullah, S., Burke, E.K.: A Multi-start large neighbourhood search approach with local search methods for examination timetabling. In: International Conference on Automated Planning and Scheduling (ICAPS 2006), Cumbria, UK, pp. 334–337 (2006)

[2] Abdullah, S., Shaker, K., McCollum, B., McMullan, P.: Dual sequence simulated annealing with round-robin approach for university course timetabling. In: Cowling, P., Merz, P. (eds.) EvoCOP 2010. LNCS, vol. 6022, pp. 1–10. Springer, Heidelberg (2010)

[3] Turabieh, H., Abdullah, S.: An integrated hybrid approach to the examination timetabling problem. OMEGA (2011), doi:10.1016/j.omega.2010.12.005

[4] Burke, E.K., Bykov, Y.: Solving exam timetabling problems with the flex-deluge algorithm. In: Burke, E.K., Rudová, H. (eds.) PATAT 2007. LNCS, vol. 3867, pp. 370–372. Springer, Heidelberg (2007) ISBN: 80-210-3726-1

[5] Burke, E.K., Eckersley, A.J., McCollum, B., Petrovic, S., Qu, R.: Hybrid variable neighbourhood approaches to university exam timetabling. European Journal of Operational Research 206, 46–53 (2010)

[6] Burke, E.K., Kingston, J., de Werra, D.: Applications to timetabling. In: Gross, J., Yellen, J. (eds.) Handbook of Graph Theory, pp. 445–474. Chapman Hall/CRC Press (2004)

[7] Caramia, M., Dell'Olmo, P., Italiano, G.F.: New algorithms for examination timetabling. In: Näher, S., Wagner, D. (eds.) WAE 2000. LNCS, vol. 1982, pp. 230–241. Springer, Heidelberg (2001)

[8] Carter, M.W., Laporte, G., Lee, S.: Examination timetabling: Algorithmic strategies and applications. Journal of the Operational Research Society 47(3), 373–383 (1996)

[9] Carter, M.W.: A survey of practical applications of examination timetabling algorithms. Operations Research 34(2), 193–202 (1986)

[10] Casey, S., Thompson, J.: GRASPing the examination scheduling problem. In: Burke, E.K., De Causmaecker, P. (eds.) PATAT 2002. LNCS, vol. 2740, pp. 232–244. Springer, Heidelberg (2003)

[11] Côté, P., Wong, T., Sabourin, R.: A hybrid multi-objective evolutionary algorithm for the uncapacitated exam proximity problem. In: Burke, E.K., Trick, M.A. (eds.) PATAT 2004. LNCS, vol. 3616, pp. 294–312. Springer, Heidelberg (2005)

[12] Dowsland, K., Thompson, J.: Ant colony optimization for the examination scheduling problem. Journal of the Operational Research Society 56(4), 426–438 (2005)

[13] Dueck, G.: New Optimization Heuristics. The great deluge algorithm and the record-to-record travel. Journal of Computational Physics 104, 86–92 (1993)

[14] Fernandes, E.M.G.P., Martins, T.F.M.C., Rocha, A.M.A.C.: Fish Swarm Intelligent Algorithm for Bound Constrained Global Optimization. In: Proceedings of the International Conference on Computational and Mathematical Methods in Science and Engineering, CMMSE 2009, June 30 , July 1-3 (2009)

[15] Fox, M.S., Sadeh-Koniecpol, N.: Why is scheduling so difficult? A csp perspective. In: Proceedings of the European Conference on Artificial Intelligence, pp. 754–767 (1990)

[16] Gao, S., Yang, J.Y.: Swarm intelligence algorithms and applications. China Waterpower Press, Beijing (2006)

[17] Gaspero, L.D., Schaerf, A.: Tabu search techniques for examination timetabling. In: Burke, E., Erben, W. (eds.) PATAT 2000. LNCS, vol. 2079, pp. 104–117. Springer, Heidelberg (2001)

[18] Jiang, M., Mastorakis, N., Yuan, D., Lagunas, M.A.: Image segmentation with improved artificial fish swarm algorithm. In: Mastorakis, N., Mladenov, V., Kontargyri, V.T. (eds.) Proceedings of the European Computing Conference. Lecture Notes in Electrical Engineering, vol. 28, pp. 133–138. Springer, Heidelberg (2009) ISBN: 978-0-387-84818-1

[19] Jiang, M., Wang, Y., Pfletschinger, S., Lagunas, M.A., Yuan, D.: Optimal multiuser detection with artificial fish swarm algorithm. In: Huang, D.-S., Heutte, L., Loog, M. (eds.) ICIC 2007. CCIS, vol. 2, pp. 1084–1093. Springer, Heidelberg (2007)

[20] Lewis, R.: A survey of metaheuristic-based techniques for university timetabling problems. OR Spectrum 30(1), 167–190 (2008)

[21] Li, X.L., Shao, Z.J., Qian, J.X.: An optimizing method based on autonomous animate: fish swarm algorithm. System Engineering Theory and Practice 11, 32–38 (2002)

[22] Malim, M.R., Khader, A.T., Mustafa, A.: Artificial immune algorithms for university. In: Burke, E.K., Rudová, H. (eds.) PATAT 2007. LNCS, vol. 3867, pp. 234–245. Springer, Heidelberg (2007)

[23] McCollum, B.: A perspective on bridging the gap between theory and practice in university timetabling. In: Burke, E.K., Rudová, H. (eds.) PATAT 2007. LNCS, vol. 3867, pp. 3–23. Springer, Heidelberg (2007)

[24] Merlot, L.T.G., Boland, N., Hughes, B.D., Stuckey, P.J.: A Hybrid Algorithm for the Examination Timetabling Problem. In: Burke, E.K., De Causmaecker, P. (eds.) PATAT 2002. LNCS, vol. 2740, pp. 207–231. Springer, Heidelberg (2003)

[25] Nelder, J.A., Mead, R.: A simplex method for function minimization. Computer Journal 7, 308–313 (1965)

[26] Petrovic, S., Burke, E.K.: Chapter 45: University timetabling. In: Leung, J. (ed.) Handbook of Scheduling: Algorithms Models and Performance Analysis. CRC Press, Boca Raton (2004)

[27] Qu, R., Burke, E.K., McCollum, B.: Adaptive automated construction of hybrid heuristics for exam timetabling and graph colouring problems. European Journal of Operational Research (EJOR) 198(2), 392–404 (2009)

[28] Qu, R., Burke, E.K.: Hybridisations within a graph based hyper-heuristic framework for university timetabling problems. Journal of Operational Research Society (JORS) 60, 1273–1285 (2009)

[29] Qu, R., Burke, E.K., McCollum, B., Merlot, L.T.G., Lee, S.Y.: A survey of search methodologies and automated system development for examination timetabling. Journal of scheduling, 55–89 (2009)

[30] Sadeh, N., Kaujnunn, M.: Micro-opportunistic scheduling: The micro-boss factory scheduler. In: Intelligent Scheduling, pp. 99–135. Morgan Kaufmann, San Francisco (1994)

[31] Schaerf, A.: A survey of automated timetabling. Artificial Intelligence Review 13(2), 87–127 (1999)

[32] Wang, C.-R., Zhou, C.-L., Ma, J.-W.: An improved artificial fish swarm algorithm and its application in feed-forward neural networks. In: Proceedings of the Fourth International Conference on Machine Learning and Cybernetics, pp. 2890–2894 (2005)

[33] Wang, X., Gao, N., Cai, S., Huang, M.: An Artificial Fish Swarm Algorithm Based and ABC Supported QoS Unicast Routing Scheme in NGI. In: Min, G., Di Martino, B., Yang, L.T., Guo, M., Rünger, G. (eds.) ISPA Workshops 2006. LNCS, vol. 4331, pp. 205–214. Springer, Heidelberg (2006)

[34] Yang, Y., Petrovic, S.: A Novel Similarity Measure for Heuristic Selection in Examination Timetabling. In: Burke, E.K., Trick, M.A. (eds.) PATAT 2004. LNCS, vol. 3616, pp. 247–269. Springer, Heidelberg (2005)

The Sandpile Mutation Operator for Genetic Algorithms

C.M. Fernandes[1,2], J.L.J. Laredo[1], A.M. Mora[1], A.C. Rosa[2], and J.J. Merelo[1]

[1] Department of Architecture and Computer Technology, University of Granada, Spain
[2] LaSEEB-ISR-IST, Technical Univ. of Lisbon (IST)
{cfernandes,acrosa}@laseeb.org
{jjmerelo,juanlu.jimenez,amorag77}@gmail.com

Abstract. This paper describes an alternative mutation control scheme for Genetic Algorithms (GAs) inspired by the Self-Organized Criticality (SOC) theory. The strategy, which mimics a SOC system known as *sandpile*, is able to generate mutation rates that, unlike those given by other methods of adaptive parameter control, oscillate between very low values and cataclysmic mutations. In order to attain the desired behaviour, the sandpile is not just attached to a GA; it is also modified in an attempt to link its rates to the stage of the search, i.e., the fitness distribution of the population. Due to its characteristics, the sandpile mutation arises as a promising candidate for efficient and yet simple and context-independent approach to dynamic optimization. An experimental study confirms this assumption: a GA with sandpile mutation outperforms a recently proposed SOC-based GA for dynamic optimization. Furthermore, the proposed method does not increase traditional GAs' parameter set.

1 Introduction

Many industrial applications have dynamic components that lead to variations of the fitness function — i.e., the problem is defined by a time-varying fitness function — and Genetic Algorithms (GAs) [1] characteristics make them candidate tools to solve this class of problems. However, issues like genetic diversity, premature convergence, exploration-exploitation balance and re-adaptation, may require, in dynamic optimization, rather different approaches. Self-Organized Criticality (SOC) [3] can provide solutions to the difficulties intrinsic to non-stationary problems.

SOC describes a property of complex systems that consists of a critical state formed by self-organization at the border of order and chaos. While *order* in this context means that the system is working in a predictable regime where small disturbances have only local impact, *chaos* is an unpredictable state sensitive to initial conditions or small disturbances. One of the characteristics of SOC is that small disturbances can lead to the so-called *avalanches*, i.e., events that are spread spatially or temporally through the system. Such events occur independently of the initial state and the same perturbation may lead to small or large avalanches, which show a power-law proportion between their size and quantity. This means that large (catastrophic) events may hit the system from time to time and reconfigure it.

When combined with a GA, SOC systems can introduce large amounts of genetic novelty into the population, periodically, in an unsupervised and non-deterministic manner, as shown in [14]. In fact, SOC has already been applied to Evolutionary

C.A. Coello Coello (Ed.): LION 5, LNCS 6683, pp. 552–566, 2011.
© Springer-Verlag Berlin Heidelberg 2011

Computation in the past [6, 12, 13, 14]. The present work, which follows a different approach, describes a mutation operator for binary GAs based on a SOC model called *sandpile* [3] and investigates its performance on dynamic optimization problems. The *sandpile mutation* is able to evolve periods of low mutation rate values punctuated by macro mutation peaks and is a promising candidate to deal with dynamic problems. Previous results [9] confirm these assumptions. In this paper, we present an enhanced version of the method and follow a different experimental methodology.

The paper is structured as follows. A state-of-the-art review is provided in Section 2 and the sandpile model and the sandpile mutation are introduced in Section 3. Section 4 describes the experiments and discusses the results. Finally, Section 5 concludes the paper and discusses future lines of research.

2 SOC in Evolutionary Computation

Strategies for controlling GAs' parameters are usually divided into three categories [8]: deterministic, adaptive and self-adaptive. *Deterministic methods*, which change the values according to deterministic rules, may be useful when developing GAs for problems of which the characteristics are known. However, they are not robust and usually do not maintain the performance when switching to different problems or even different instances. In these situations, *adaptive methods* are more suitable since the variation depends indirectly on the problem and the search stage. However, adaptive control requires strategies that may depend, for instance, on population genotypes, phenotypes or fitness. *Self-adaptive methods* follow the same intuition that led to GAs, by allowing the values to evolve together with the solutions to the problem. Nevertheless, and according to Bäck *et al.* [2], this method, when applied to binary GAs, may slow down the convergence of the standard GA or deteriorate the best fitness values. Besides, self-adaptive GAs enlarge the search space by codifying the parameters in the chromosome.

Another approach is possible. SOC may be used in GAs for controlling the parameter values, diversity or population size and possibly overcoming the difficulties inherent to other control methods. Previous works suggest that the task is feasible and, in some situations, may improve the algorithms' performance. *Extremal Optimization* [4], for instance, is an optimization algorithm based on SOC that evolves a single solution to the problem by means of local search and modification. By plotting the fitness of the solution, it is possible to observe distinct stages of evolution, where improvement is disturbed by brief periods of dramatic decrease in the quality of the solution.

In the realm of Evolutionary Computation, Krink *et al.* [10] proposed two control schemes — later extended to cellular GAs [11] — based on the sandpile. The model's equations are computed offline in order to obtain the "power-law values", which are then used during the run to control the number of individuals that will be replaced by randomly generated solutions (SOC mass extinction model) or the mutation probability of the algorithm (SOC mutation model).

Tinós and Yang [12] were also inspired by SOC to create a sophisticated *Random Immigrants GA* (RIGA) [11], called *Self-Organized Random Immigrants GA* (SORIGA). The rules of SORIGA's dynamics are the following. In each generation, the algorithm replaces the worst individual of the population and its neighbors (determined by the individuals' indexes in the population) by r_r random solutions.

Because this strategy by itself does not guarantee that the system exhibits SOC behavior — the new chromosomes are quickly replaced by the fittest chromosomes in the population —, the random solutions are stored in a subpopulation and the chromosomes from the main population are not allowed to replace the new individuals. By plotting the extent of extinction events, which shows a power-law proportion between the size and their frequency, the authors argue that the model exhibits SOC [14].

Our proposal differs from previous approaches. Power-law values are not previously computed, like in [12] and [13]. This feature may be very important when tackling dynamic problems, because large avalanches can be linked (online) to changes in the environment. In addition, SORIGA, the closest method to the one presented in this paper, gives new genetic material to the population by inserting r_r new chromosomes in each generation, while the sand pile mutation may completely reconfigure the population's alleles in only one generation. The tests in Section 4 demonstrate that the GA with sandpile mutation is able to outperform SORIGA.

3   The Sandpile Model and the Sandpile Mutation

In 1987, Bak *et al.* [3] identified the SOC phenomenon in a model called the *sandpile*, a cellular automaton where each cell of the lattice keeps a value that corresponds to the slope of the pile. In its simplest form, the sand pile is a linear lattice of L sites $(x_1, x_2, \ldots x_L)$ where "sand" is randomly dropped, one grain at a time. The number of grains deposited on site x_j is represented by the function $h(x_j)$, which may be referred as the height of the pile. The grains accumulate in the lattice as long as the height difference between adjacent sites does not exceed a threshold value. If that happens, the grain topples from site j to the adjacent sites, and if the height difference between the following adjacent also exceeds the threshold, then the grain topples again. The toppling only stops when the grain reaches a site were the slope does not exceed the value defined as threshold. Considering the whole system, it may be stated that the toppling stops when the pile reaches the equilibrium state.

The process can be generalized to two dimensions (the case that matters for this paper). Grains of sand are randomly dropped on the lattice where they pile up and increment the values of the slopes $z(x, y)$. Then, if the slope at site (x, y) is bigger than critical z, the grains are distributed by its neighboring sites (a *von Neumann* neighborhood is considered here. If one of those sites also exceeds the threshold value, the avalanche continues.

If the lattice is previously driven (initialized) to a critical state, we then see avalanches of all sizes, from a single tumble to events that reconfigure almost the entire pile. The likelihood of an avalanche is in power-law proportion to the size of the event, and avalanches are known to occur at all size scales. Large avalanches are very rare while small ones appear very often. Without any fine-tuning of parameters, the system evolves to a non-equilibrium critical state: SOC.

A two-dimensional sandpile model can be constructed with simple rules. In this straightforward design, the number of grains of sand in a cell (x, y) characterizes its state. The update rule states that if a cell has fours grains of sand in it, it loses four, and from each of its four immediate neighbor cell (*von Neumman* neighborhood) with four or more grains in it, it gains one. The sandpile mutation uses this description of the two-dimensional model in order to evolve self-regulated mutation rates.

Sand pile Mutation

for g grains **do**
    drop grain at random within the bounds of the lattice $h(x, y) \rightarrow h(x, y) + 1$
    if $h(x, y) \geq h_c$ and $random(0, 1.0) > f_n$
        mutate (flips the bit with probability 0.5)
        avalanche $h(x, y) \rightarrow h(x, y) - 4$
                     $h(x \pm 1, y) \rightarrow h(x \pm 1, y) + 1$
                     $h(x, y \pm 1) \rightarrow h(x, y \pm 1) + 1$
    and **update** lattice h recursively

f_n: normalized fitness associated with solution over which the grain has been dropped

Fig. 1. Pseudo-code of the sand pile mutation

First, the GA's population is linked to a $N \times L$ lattice with $N = 1, ... n$ and $L = 1, ... l$, where n is the population size and l is the chromosome length. For instance, the first gene of the first chromosome is linked to the $(1,1)$, the second gene of the first chromosome is linked to cell $(1,2)$ and so on. Then, the sandpile is initialized so that it is near critical state when the GA starts. This is done by running the algorithm of the model, without linking it to the GA, until the rate of dropping sand is approximately equal to the rate at which the sand is falling of the sides of the table. Then, in each generation, the individuals are selected, recombined and evaluated (no mutation at this stage). After that, the solutions are ranked according to their fitness, each individual is mapped into the lattice and g grains are randomly dropped on the lattice thus incrementing the cells' values $h(x, y)$. When a cell reaches the critical value $h_c = 4$, an avalanche occurs if a value randomly generated from a uniform distribution between 0 and 1.0 is higher than the normalized fitness of the individual associated with the cell. This way, fitter individuals have less chances of being mutated. After a first avalanche, the neighboring cells are recursively updated and the avalanche may proceed through the lattice. (See the pseudo-code in Fig. 1.)

The sandpile mutation has one restriction that is not present in the sandpile model: if a cell is already involved in an avalanche, and the recursive nature of the process has not allowed it to complete its sequence, then the cell is ignored. This restriction eliminates hypothetical avalanche cycles and several mutations of the same gene. One more details must be referred: if a cell reaches h_c but there is no mutation (due to the fitness test), then the grain is discarded.

The critical issue here is the f_n value. Please note that the chromosomes are evaluated before being mutated by the sand pile scheme. This is the only way to assure that fitness values influence the mutation, but then the next selection stage is working with values that do not correspond to the current genotype. A possible solution could be to re-evaluate the entire population after the mutation stage, but this would double the computational effort. The first version of the sandpile mutation [15] assumed the apparent drawback of selecting individuals with outdated fitness values and defined f_n as:

$$f_n(j) = \frac{fitness(j) - worstFitness}{bestFitness - worstFitness}. \tag{1}$$

where j is the index of the chromosome associated with the cell (x, y), $worstFitness$ is the lowest fitness in the population and $bestFitness$ is the highest. However, and although the results published in [9] were quite promising, further tests [10] showed that the algorithm's efficiency when compared to other algorithms was not as noticeable as expected after the preliminary tests. A modified version was then tested, improving consistently the first sandpile mutation:

$$f_n(j) = \frac{parentsFitness(j) - worstFitness}{bestFitness - worstFitness}. \tag{2}$$

where $parentsFitness(j)$ is the average fitness of chromosome j parents. This way, the sandpile mutation acts right after the new population is created by selection and crossover, and before evaluation. Since fitter parents have more chance to generate fitter offspring, this approach may be a good approximation to the original idea.

Another modification was made when handling the exception $bestFitness = worstFitness$. When this equality occurs the GA may have fully converged, and therefore it may be more suited to set $f_n(j) = 0$ for all j, meaning that the mutation surely occurs after an avalanche. This way, the sand pile is "open" for massive mutations. Finally, to avoid extra difficulties in testing the algorithm, the mutation type is set to *flip the bit with probability 0.5* (i.e., in case of mutation, the allele flips with a probability of 0.5). A bit-flip mutation (i.e., the alleles flip with a probability of 1.0) could bias the results towards the algorithm in some test functions, like the trap functions, for instance. The global and local optima of these problems are, respectively, strings of 1s and strings of 0s. If a population has converged to the local optimum, a massive bit-flip mutation can lead the population to the global optimum, thus taking advantage of the problem's structure.

In the taxonomy of evolutionary solutions to dynamic optimization, a *Generational GA with the sand pile mutation* (GGA$_{SM}$) may be classified as a diversity maintenance strategy [6], along with the already referred RIGA and SORIGA. The *Elitism-based Immigrants GA* (EIGA) is another diversity maintenance strategy, recently proposed in [16]. EIGA is a very simple strategy that in every generation replaces a fraction r_i of the population by mutated copies of the best solution of the previous generation (with mutation probability p_m^i). The results reported in [16] situate it as a state-of-the-art GA for dynamic optimization. For that reason, EIGA was introduced in the test set, together with SORIGA and a standard Generational GA (GGA).

There are other types of evolutionary approaches to dynamic optimization, such as *memory* schemes [5], *reaction to changes* [7] and *multi-population approaches* [6], but for now these are left out of the experimental setup since they usually rely on a different premises (*reactive* algorithms, for instance, require that changes are easy to detect) or are more suited for a specific type of dynamics (memory, for instance, loses efficiency when the changes are not cyclic).

4 Test Set and Results

The experiments were conducted on dynamic versions of trap functions, royal road R1 problem and $0 - 1$ knapsack problem. Therefore, we have quasi-deceptive functions (order-3 traps), deceptive functions (order-4 traps), non-deceptive functions (Royal Road) and constrained combinatorial problems (knapsack) as base-functions for constructing the non-stationary problems.

4.1 Functions

The knapsack version used in these experiments is described in [15]. The Royal Road R1 function is defined by:

$$f(\vec{x}) = \sum_{i=1}^{q} c_s \delta_s(\vec{x}) \tag{3}$$

where q is the number of schemata $S = \{s_1, ..., s_q\}$, $\delta s(\vec{x})$ is set as 1 if $\vec{x}$ is an instance of S and 0 otherwise, and $c_s = 8$ for all s; a 64-bit string was used and each schema is composed of 8 contiguous bits.

A trap function is a piecewise-linear function defined on *unitation* (the number of ones in a binary string) that has two distinct regions in the search space, one leading to a global optimum and other leading to the local optimum. Depending on its parameters, traps may be deceptive or not. The traps in these experiments are defined by:

$$F(u(\vec{x}) = \begin{cases} k, & if \ u(\vec{x}) = k \\ k - 1 & u(\vec{x}), & otherwise \end{cases} \tag{4}$$

where $u(\vec{x})$ is the unitation function and k is the problem size (and also the fitness of the global optimum). With these definitions, order-3 traps are in the region between deceptive and non-deceptive, while order-4 traps are deceptive. For this study, 30-bit and 40-bit problems were designed by juxtaposing ten order-3 and order-4 traps.

4.2 Methodology

The test environment proposed in [15] was used to create a dynamic experimental setup based on the functions described above. With this problem generator, it is possible to construct dynamic problems with different degrees of severity and speed — i.e., the extent of the changes and the frequency of the changes, respectively — using stationary base-functions with binary variables. This is accomplished by applying a binary mask to the solutions, thus shifting the fitness landscape. The generator has two parameters that control the severity of the changes and their frequency: ρ is a value between 0 and 1.0 that controls the severity of changes and τ defines the number of generations between changes. In this paper, we use the number of evaluations $\varepsilon = \tau \times n$ between changes, where n is the population size. Since, without the population size value, τ does not give enough information on the computational cost required between each change, we think that using ε is more suited for discussing the performance of GAs on dynamic optimization problems.

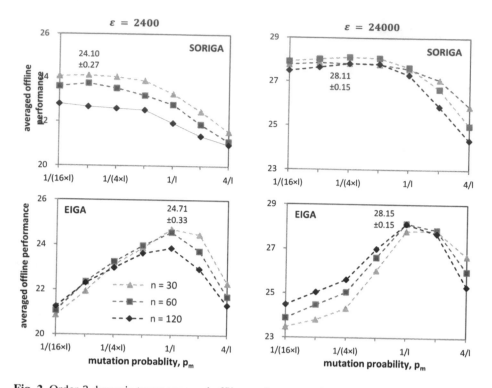

Fig. 2. Order-3 dynamic traps: averaged offline performance with different n (population size). Best value (and standard deviation) in each graph is shown. SORIGA's r_r (number of immigrants introduced in the population) is set to **3**. EIGA's r_i (ratio of mutated copies of the best solution that are introduced in the population) is set to **0.2**, and $p_m^i = p_m$.

The experiments were designed by setting, for each of the stationary base-functions, $\varepsilon = (1200, 2400, 24000, 48000)$ and $\rho = (0.05, 0.3, 0.6, 0.95)$, thus making 16 different dynamic scenarios of each type of problem. Every run covered 10 periods of changes. For each experiment, 30 independent runs were executed with the same 30 random seeds. This methodology is similar to the one in [14], in which SORIGA is tested with 10 periods of changes and the ε of the fastest scenario is set to 1200.

We are particularly interested in the GAs' performance when varying the mutation probability, because diversity maintenance strategies may shift the optimal mutation probability values (i.e., the p_m values that maximize the performance). Therefore, it is of extreme importance to test the GAs under a reasonable range of p_m values, so that the results don't become biased towards some of the approaches. A similar proceeding was conducted when tuning the parameter the g of GGA$_{SM}$, which was set to several values in the range $(n \times l)/32$ to $n \times l$, where n is the population size and l is the chromosome dimension. The population size also affects the performance of the GAs, not only in stationary problems, but also in dynamic environments. In this study, the algorithms were tested with $n = 30, 60$ and 120.

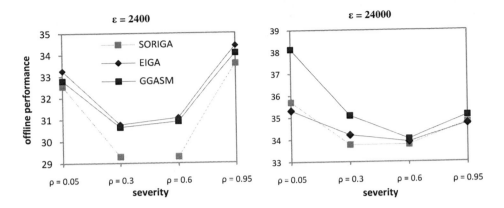

Fig. 3. Order-4 dynamic traps: offline performance. $n = 60$. GGA$_{SM}$: $g = (n \times l)/8$ if $\varepsilon = 2400$; $g = (n \times l)/32$ if $\varepsilon = 24000$. EIGA: $p_m = 1/l$ if $\varepsilon = 2400$; $p_m = 2/l$ if $\varepsilon = 24000$. SORIGA: $p_m = 1/(16 \times l)$ if $\varepsilon = 2400$; $p_m = 1/(4 \times l)$ if $\varepsilon = 24000$.

Table 1. Kolmogorov-Smirnov tests with 0.05 level of significance. + signs when GGA$_{SM}$ is significantly better than the specified GA, signs when GGA$_{SM}$ is significantly worst, and ≈ signs when the differences are not statistically significant (i.e., the null hypothesis is not rejected). Order 3 and Knapsack: $n = 30$; order-4 and Royal Road: $n = 60$; see [10] for g and $p_m$values.

	ρ →	$\varepsilon = 1200$				$\varepsilon = 2400$				$\varepsilon = 24000$				$\varepsilon = 48000$			
		.05	.3	.6	.95	.05	.3	.6	.95	.05	.3	.6	.95	.05	.3	.6	.95
order-3	GGA	≈	−	−	≈	≈	−	−	≈	≈	≈	+	+	≈	≈	+	+
	SORIGA	+	≈	≈	+	+	≈	≈	+	+	+	+	+	+	+	+	+
	EIGA	≈	−	−	≈	≈	≈	≈	≈	+	+	+	+	+	+	+	+
order-4	GGA	−	≈	≈	−	≈	≈	≈	≈	+	+	+	+	+	+	+	+
	SORIGA	≈	+	+	+	≈	+	+	+	+	+	+	+	+	+	+	+
	EIGA	−	≈	≈	−	≈	≈	≈	−	+	+	+	≈	+	+	+	+
R. Road	GGA	+	≈	≈	≈	≈	+	≈	≈	≈	+	+	+	+	+	+	+
	SORIGA	+	≈	−	−	+	≈	−	−	+	≈	−	−	+	+	≈	−
	EIGA	≈	≈	+	≈	+	+	+	≈	+	+	+	+	+	+	+	+
Knapsack	GGA	−	−	+	+	−	≈	+	+	+	+	≈	−	+	+	+	≈
	SORIGA	≈	+	+	+	≈	+	+	+	+	+	+	+	+	+	+	+
	EIGA	−	−	+	+	−	−	+	+	+	+	−	−	+	+	+	−

Uniform crossover was chosen in order to avoid taking advantage of the trap function building blocks tight linkage. Every algorithm in the test set uses binary tournament. Preliminary tests demonstrated that a high crossover probability together with an elitist strategy maximize the performance of the GA, therefore, we use GAs with

2-elitism and $p_c = 1.0$. The *offline performance*, as defined in [15] (best-of-generation fitness values averaged over the total number of runs and over the data gathering period), is used to evaluate the GAs:

$$\overline{F_{BG}} = \frac{1}{G} \times \sum_{i=1}^{G} \left(\frac{1}{R} \times \sum_{j=1}^{R} F_{BG_{ij}} \right) \tag{5}$$

where G is the number of generations, R is the number of runs and $F_{BG_{ij}}$ is the best-of-generation fitness of generation i of run j of a GA on a specific problem. The output of the experiments generated a large amount of data that cannot be entirely analyzed and described in this paper. The analysis is thus limited to some fundamental issues.

4.3 Results

As expected, different algorithms require different p_m values in order to maximize their performance. Fig. 2 shows EIGA and SORIGA's offline performance in order-3 traps, averaged over the results in scenarios with different severity. First, it is clear that SORIGA's optimal performance values are worst than EIGA, although the difference is not statistically significant when $\varepsilon = 24000$. Those optimal values are attained with different mutation probability values: while SORIGA attains the best performance with p_m values in the range $1/(16 \times l)$ to $1/(4 \times l)$], EIGA's best performance is attained with $p_m = 1/l$. In addition, the graphics show that the performance also depends on the population size: small populations (30 and 60 individuals) are more efficient. Similar behavioral patterns have been observed in the other problems. These results demonstrate that experimental studies involving diversity maintenance GAs for dynamic optimization problems must include preliminary tests covering a wide range of mutation probability values and different population size values.

Another relevant outcome of the experiments conducted for this investigation is illustrated by **Fig. 3**, which shows the results attained by the best configurations in order-4 traps. In general, GGA$_{SM}$ outperforms the other GAs when frequency of

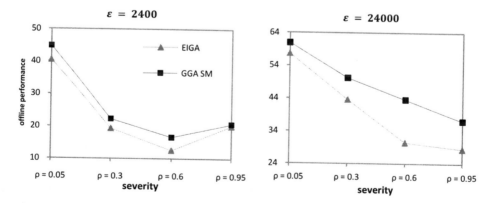

Fig. 4. Dynamic royal road R1: offline performance. $n = 60$. GGA$_{SM}$: $g_r = (n \times l)/32$; EIGA: $p_m = 1/l$.

changes is lower. In faster environments, the algorithm is not so efficient, although, in this case (and also in the knapsack problem), it still outperforms SORIGA. As for EIGA, it is in general better than GGA$_{SM}$ in the fastest scenarios ($\varepsilon = 1200$), except in the Royal Road function, in which the GA with the sandpile mutation is always better or at least equivalent to EIGA. Please note that the optimal p_m may also depend on the frequency of change, as seen in **Fig. 3**.

Table 1 summarizes the experiments and illustrates the previous comments by displaying the non-parametric statistical tests performed on the offline performance values attained by the best configurations of each algorithm in each type of problem. For each scenario, GGA$_{SM}$'s result is compared with the other GA's offline performance using a paired Kolmogorov-Smirnov test with 0.05 level of significance. The null hypothesis states that the datasets from which the offline performance and the standard deviation are calculated are drawn from the same distribution. In general, GGA$_{SM}$ is better than SORIGA and it is at least competitive with EIGA (in slower scenarios it clearly outperforms EIGA).

The exception is the behavior of SORIGA and GGA$_{SM}$ in Royal Road dynamic problems. In this case, SORIGA is the best GA when severity is high, while the sandpile mutation outperforms the other algorithms when severity is low. On the other hand, GGA$_{SM}$ attains better or equivalent results as EIGA in every Royal Road dynamic problem. As shown in Fig. 4, the GA with the sandpile mutation outperforms EIGA in most of the Royal Road scenarios, even when $\varepsilon = 2400$. These results demonstrate that the sandpile mutation may improve a generational GA's performance on dynamic optimization problems with different characteristics, as well as two state-of-the-art GAs for dynamic optimization, especially when the period between changes is not very small. The next section analyses the mutation rates' distribution of GGA$_{SM}$ and tries to shed some light on the working mechanisms of the sandpile mutation.

Table 2. Order-4 traps. Mutation rate median values. $g = (n \times l)/32$.

	$\varepsilon = 2400$	$\varepsilon = 24000$
$\rho = 0.05$	0.0011	0.0007
$\rho = 0.3$	0.0021	0.0011
$\rho = 0.6$	0.0023	0.0014
$\rho = 0.95$	0.0010	0.0009

4.4 Mutation Rate Analysis

As demonstrated in the previous section, the optimal p_m may vary according to the dynamics of the problem. The same happens with the parameter g, which defines the number of grains that are dropped over the sandpile in each generation. Being g a parameter of the new algorithm, it is important to, at least, give some hints on how it must be set. In our experiments, optimal g lies between $(n \times l)/32$ and $(n \times l)/2$, depending on the problem, and for now we haven't devised any rule to avoid testing and hand-tuning the parameter. This complicates the GA's parameter tuning, but the same happens with p_m in a standard GA. A positive trait, though, is that previous results [10] suggest that GAs are less sensitive to g than to p_m.

This section investigates how the mutation rates vary during the run, and if their distribution somehow reflects the type of dynamics. For that purpose, in each generation, the population before and after the sandpile mutation operator (i.e., before and after all the g grains are dropped) is compared. The *mutation rate* in generation t is then defined as the ratio between the number of alleles that flipped and the size of the sandpile, as defined by Equation 6, where $m(i,j) = 1$ if the gene i of the chromosome j has mutated, and 0 otherwise (n is the population size and l is the chromosome length). Percentage is given by $m_r(t) \times 100$.

$$m_r(t) = \frac{\sum_{i=1}^{n}\sum_{j=1}^{l} m(i,j)}{(n \times l)} \tag{6}$$

If we compute the median values of the mutation rates over 30 independent runs and over the data gathering period, an interesting pattern shows up — see **Error! Not a valid bookmark self-reference.**. The median varies with the severity of the changes. The scenarios with $\rho = 0.05$ and $\rho = 0.95$ have similar values, but if one takes a closer look into the mutation rate values during the run, it is clear that low and high severity give rise to a rather different behavior. **Fig. 5** shows the mutation rate values during 400 generations of a GGA_{SM} run on order-4 scenarios with $\varepsilon = 24000$. When comparing the curves of , it is clear that the dynamics of the mutation is different for each case. With $\rho = 0.95$, there is more activity in the macro-mutation region.

Fig. 6 addresses the same issue with a different perspective, by plotting the log-log of the mutation rates and their abundance. We see that the shapes of the log-log are different: with $\rho = 0.05$ there is more activity in the medium range (1~5%), while

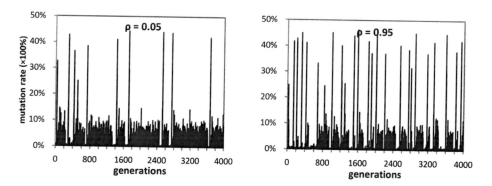

Fig. 5. Order-4 dynamic trap problems. GGA_{SM} online mutation rate. Population size: $n = 60$. Grain rate: $g_r = (n \times l)/32$. $\varepsilon = 24000$.

$\rho = 0.95$ shows more activity at higher rates ($> 30\%$). In general, and as expected, the graphics show that low rates arise more frequently than high rates. Such a distribution was the main objective of this work.

The previous experiment suggests that the sandpile may be adapting the mutation rates' distribution to the severity of the changes. Another experiment was designed in order to investigate if there are differences in the distribution when varying the period between changes while maintaining the other parameters. For that purpose, the maximum number of evaluations was fixed at 600000, and three problems were constructed, by setting ρ to *random* (that is, ρ is randomly generated from a uniform distribution between 0 and 1.0 at the arrival of a new change) and ε to 6000, 24000 and 120000. The resulting distributions, in Fig. 7, show that one of the effects of increasing ε is an attenuation of the activity in the high range ($\sim$30%), and a decrease of smaller rates ($\sim$3%). That is, like severity, ε seems to affect the distribution of GGA_{SM} mutation rates.

These experiments give some hints on the reasons why GGA_{SM} is able to improve other GAs performance on some dynamic problems. As intended, the sandpile mutation is able to evolve slow and medium mutation rates punctuated by mutation

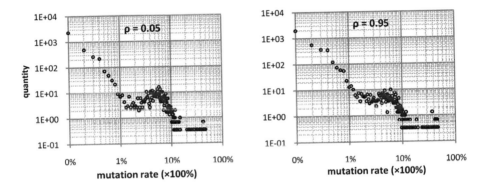

Fig. 6. Logarithm of the mutation rates abundance plotted against their values

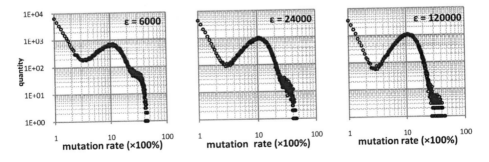

Fig. 7. Mutation rates distribution. Order-3 traps; $\rho = random$; $g = (n \times l)/16$

bursts that reconfigure almost 50% of the alleles in the population. These macro-mutations can help the population to escape full convergence or local optima, thus giving the GA a chance to track the moving optimum. However, when the problems are fast (i.e.,there are only a few generations between changes), the sandpile may lack the necessary time between each change for that state in which large avalanches are likely to occur, and therefore it is not able to improve standard GAs' performance. As a matter of fact, in the experiments conducted for these investigations, neither SORIGA nor EIGA were able to clearly outperform GGA in the low and medium severity scenarios of the fastest problems ($\varepsilon = 1200$), a result that suggests that a standard evolutionary search may be the best choice to tackle such kind of problems.

These last experiments, although they shed some light on the working mechanisms of the sandpile mutation, need to be extended in order fully understand the self-regulated behaviour of the sandpile mutation.

5   Conclusions and Future Work

This paper describes an online mutation rate adjustment strategy for GAs based on the SOC theory. Since dynamic optimization requires a particular balance between exploration and exploitation to provide the algorithm with means to track the optimum when the function changes, new strategies need to be devised in order to deal with this issue. Due to its characteristics, SOC is a promising candidate for improving traditional GAs abilities to deal with dynamic problems. Our approach uses SOC at the mutation level, providing the GA with a self-regulated mutation rate that reflects the search stage and the dynamics of the problem.

The algorithm was compared with two state-of-the art approaches to dynamic optimization — one of them, SORIGA, is also based on SOC. The sandpile GA demonstrated to outperform SORIGA consistently over the entire test set, except in the Royal Road low severity problems. When compared to EIGA, an efficient algorithm for dynamic optimization recently proposed, the sandpile mutation showed to improve its performance on lower frequency scenarios, while being competitive in faster problems. The paper also provides an analysis of the distribution of the mutation rates throughout the run. The algorithm seems to be able to adapt the distribution to the type of problem. Finally, parameter g replaces p_m; therefore, the size of the GAs' parameter space does not increase.

Future research will focus on the mutation rates' distribution; metrics are needed in order to understand how the rates vary with the type of problem and the genetic diversity during the run should be investigated so that the working mechanisms of the operator can be understood. In addition, an in-depth comparison with EIGA is in course, so that the potential of the sandpile can be fully acknowledged. In particular, we intend to study it in even faster scenarios and increase the periods of changes to 50, as in [16].

It is also of great importance to study the behavior of the sandpile mutation with problems (such as the knapsack) in which the variables forming the building-blocks are not near-encoded, or, alternatively, to devise a sandpile-like model in which the avalanches do not spread trough adjacent cells. Finally, the structure of the sandpile itself may be affecting the performance of the operator, and other topologies (such as a torus lattice) should be tested.

Acknowledgements. This work has been partially funded by FCT, Ministério da Ciência e Tecnologia, his Research Fellowship SFRH/BPD/66876/2009, also supported by FCT (ISR/IST plurianual funding) through the POS_Conhecimento Program. In addition, this paper has also been funded in part by the Spanish MICYT projects NoHNES (TIN2007-68083) and TIN2008-06491-C04-01, and the Junta de Andalucía P06-TIC-02025 and P07-TIC-03044.

References

1. Bäck, T.: Evolutionary Algorithms in Theory and Practice. Oxford University Press, Oxford (1996)
2. Bäck, T., Eiben, A.E., van der Vart, N.A.L.: An Empirical Study on GAs "Without Parameters". In: Schoenauer, M., et al. (eds.) Proceedings of the 6th International Conference on Parallel Problem Solving from Nature (PPSN VI), pp. 315–324. Springer, London (2000)
3. Bak, P., Tang, C., Wiesenfeld, K.: Self-organized criticality: an explanation of 1/f noise. Physical Review of Letters 59, 381–384 (1987)
4. Boettcher, S., Percus, A.G.: Optimization with extremal dynamics. Complexity 8(2), 57–62 (2003)
5. Branke, J.: Memory enhanced evolutionary algorithms for changing optimization problems. In: Proc. of the 1999 Congress on Evolutionary Computation, pp. 1875–1882. IEEE Press, Los Alamitos (1999)
6. Branke, J.: Evolutionary optimization in dynamic environments. Kluwer Academic Publishers, Norwell (2002)
7. Cobb, H.G.: An investigation into the use of hypermutation as an adaptive operator in GAs having continuous, time-dependent nonstationary environments. Tech. Report AIC-90-001, Naval Research Laboratory, Washington, USA (1990)
8. Eiben, A.E., Hinterding, R., Michalewicz, Z.: Parameter Control in Evolutionary Algorithms. IEEE Trans. on Evolutionary Computation 3(2), 124–141 (1999)
9. Fernandes, C.M., Merelo, J.J., Ramos, V., Rosa, A.C.: A Self-Organized Criticality Mutation Operator for Dynamic Optimization Problems. In: Proc. of the 2008 Genetic and Evolutionary Computation Conference, pp. 937–944. ACM, New York (2008)
10. Fernandes, C.M.: Diversity-enhanced GAs for dynamic optimization. Ph.D Thesis, Tech. U. Lisbon (2009),
   http://geneura.ugr.es/pub/tesis/PhD-CFernandes.pdf
11. Grefenstette, J.J.: Genetic algorithms for changing environments. In: Parallel Problem Solving from Nature II, pp. 137–144. North-Holland, Amsterdam (1992)
12. Krink, T., Rickers, P., René, T.: Applying self-organized criticality to Evolutionary Algorithms. In: Deb, K., Rudolph, G., Lutton, E., Merelo, J.J., Schoenauer, M., Schwefel, H.-P., Yao, X. (eds.) PPSN 2000. LNCS, vol. 1917, pp. 375–384. Springer, Heidelberg (2000)
13. Krink, T., Thomsen, R.: Self-Organized Criticality and mass extinction in Evolutionary Algorithms. In: Proceedings of the 2001 IEEE Congress on Evolutionary Computation (CEC 2001), vol. 2, pp. 1155–1161. IEEE Press, Los Alamitos (2001)
14. Tinós, R., Yang, S.: A self-organizing RIGA for dynamic optimization problems. Genetic Programming and Evolvable Machines 8, 255–286 (2007)

15. Yang, S., Yao, X.: Experimental study on PBIL algorithms for dynamic optimization problems. Soft Computing 9(11), 815–834 (2005)
16. Yang, S.: Genetic Algorithms with Memory- and Elitism-Based Immigrants in Dynamic Environments. Evolutionary Computation 16(3), 385–416 (2008)

Self-adaptation Techniques Applied to Multi-Objective Evolutionary Algorithms

Saúl Zapotecas Martínez*, Edgar G. Yáñez Oropeza,
and Carlos A. Coello Coello**

CINVESTAV-IPN (Evolutionary Computation Group)
Departamento de Computación
México D.F. 07300, MÉXICO
saul.zapotecas@gmail.com,
eyanez@computacion.cs.cinvestav.mx,
ccoello@cs.cinvestav.mx

Abstract. In spite of the success of evolutionary algorithms for dealing with multi-objective optimization problems (the so-called multi-objective evolutionary algorithms (MOEAs)), their main drawback is the fine-tuning of their parameters, which is normally done in an empirical way (using a trial-and-error process for each problem at hand), and usually has a significant impact on their performance. In this paper, we present a self-adaptation methodology that can be incorporated into any MOEA, in order to allow an automatic fine-tuning of parameters, without any human intervention. In order to validate the proposed mechanism, we incorporate it into the NSGA-II, which is a well-known elitist MOEA and we analyze the performance of the resulting approach. The results reported here indicate that the proposed approach is a viable alternative to self-adapt the parameters of a MOEA.

1 Introduction

The design of mechanisms that allow to automate the fine-tuning of the parameters of an evolutionary algorithm (EA) has been subject of a considerable amount of research throughout the years [1,2]. When dealing with optimization problems having several (often conflicting) objectives (the so-called multi-objective optimization problems), the fine-tuning of parameters gets even more complicated, since we aim to converge to a set of solutions (the so-called Pareto optimal set). Because of such complexity, the design of online and self-adaptation mechanisms have been scarce within the multi-objective evolutionary algorithms (MOEAs) literature (see for example [3,4,5]).

The main goal of this work is to define a multi-objective evolutionary algorithm that does not require any user-defined parameters. In order to achieve such a goal,

* The first author acknowledges support from CINVESTAV-IPN and CONACyT to pursue graduate studies at CINVESTAV-IPN.
** The third author acknowledges support from CONACyT project number 103570.

C.A. Coello Coello (Ed.): LION 5, LNCS 6683, pp. 567–581, 2011.
© Springer-Verlag Berlin Heidelberg 2011

we define different techniques to self-adapt the main parameters of a well-known MOEA (the NSGA-II [6]). The resulting approach is then validated using 12 test problems taken from the specialized literature. Results are compared with respect to those obtained with the original NSGA-II. As will be seen, the obtained results are very competitive and indicate that the proposed approach can be a viable alternative to automate the fine-tuning of parameters of a MOEA.

The remainder of this paper is organized as follows. In Section 2, we present the previous related work reported in the specialized literature. In Section 3, we describe in detail our proposed self-adaptation approach. In Section 4, we validate our proposed approach using standard test problems and performance measures reported in the specialized literature. Finally, in Section 5 we present our conclusions and provide some possible paths for future research.

2   Previous Related Work

The interest in reducing the number of parameters of a MOEA, has been studied by relatively few researchers. Apparently, the first attempt to self-adapt the parameters of a MOEA was the one reported by Kursawe [7]. His proposal was to provide individuals in the population of a MOEA with a set of step lengths for each objective function. The aim of Kursawe's work, however, was to be able to deal with dynamic environments rather than automating the fine-tuning of parameters of a MOEA.

Other authors have only focused on the self-adaptation of a single operator. For example, Büche et al. [8] proposed to use Kohonen's self-organizing maps to adapt the step length of a MOEA's mutation operator.

Tan et al. [9] proposed the incrementing multi-objective evolutionary algorithm (IMOEA) with adopts an adaptive population size whose value is computed based on the online discovered trade-off surface and the desired population distribution density. IMOEA relies on a convergence metric that is based on Pareto dominance and a performance measure called "progress ratio", which was proposed by Van Veldhuizen [10]. Additionally, IMOEA also incorporates dynamic niching (i.e., the user does not need to define a niche radius for performing fitness sharing).

Kumar and Rockett [11] proposed the Pareto converging genetic algorithm (PCGA). This MOEA uses a systematic approach based on Pareto rank histograms for assessing convergence towards the Pareto front.

Abbass [3] proposed the self-adaptive Pareto differential evolution (SPDE) which extends a MOEA called Pareto differential evolution (PDE) [12] with self-adaptive crossover and mutation operators. In SPDE, both the crossover and the mutation rates are treated as additional decision variables which are added to the chromosomic string and are affected by the evolutionary process.

Zhu and Leung [13] proposed a parallel multi-objective genetic algorithm which is implemented in an island model and has an asynchronous self-adjustable mechanism. This mechanism adopts certain information about the current status of each island and uses it to focus the search effort towards non-overlapping regions of the search space.

Toscano and Coello [4] proposed the micro genetic algorithm 2 (μGA^2) which is a parameterless version of the micro genetic algorithm (μGA) for multi-objective optimization previously introduced by the same authors [14]. The new approach adopts several self-adaptation mechanisms to select the type of encoding (binary or real-numbers), and the type of crossover operator (from several available). For this sake, it executes several μGAs in parallel and performs a comparison of their results. The μGA^2 also incorporates a mechanism based on a performance measure in order to decide when to stop iterating.

Martí et al. [15] proposed a mechanism that gathers information about the solutions obtained so far. This information is accumulated and updated using a discrete Kalman filter and is used to decide when to stop a MOEA.

Zielinski and Laur [16] proposed a mechanism for self-adapting three important parameters in a multi-objective particle swarm optimization: inertia, the cognitive component and the social component. The proposed mechanism is based on a design of experiments technique called evolutionary operation (EVOP). The authors adopt analysis of variance in a two-level factorial design [17] (i.e., two values are considered for each parameter being self-adapted) to determine the effect of each combination of parameters. The information obtained from the analysis of variance allows to determine how should the parameters be modified. The approach defines a measure of "success" based on Pareto dominance, which is used to guide the search.

Trautmann et al. [5] proposed a new convergence criteria for MOEAs. This mechanism consists of analyzing the performance of a MOEA through its iterative process with respect to three well-known performance measures: generational distance [10], hypervolume [18] and spread [6]. In this way, if there is not a significant variance of these performance measures, it is possible to conclude that the MOEA has converged to the real Pareto front and the evolutionary process is consequently stopped.

None of these previous approaches, however, constitutes a full proposal of a self-adaptation framework for MOEAs, which is precisely what we introduce here, with certain specific mechanisms specifically tailored for the NSGA-II.

3   Our Proposed Approach

Our approach consists of two phases. In the first of them, an analysis of variance (ANOVA) [19] of a MOEA, using a certain set of test problems and performance measures is undertaken. This analysis is meant to provide us with the set of parameters to which the MOEA under study is most sensitive. In our study, we adopted the NSGA-II as our baseline MOEA, but any other state-of-the-art MOEA could be used as well (e.g., SPEA2 [18]).

In the second phase of our proposed approach (called here NSGA-II$_{self\_adap}$), we introduce some specific self-adaptation techniques that are used to automatically tune the values of the most sensitive parameters identified in the first phase.

Next, we will provide a summary of the results obtained from our ANOVA and will also describe the self-adaptation mechanisms that we propose to use.

3.1 Phase 1: Sensitivity Analysis

As indicated before, in order to define the parameters to which the NSGA-II is most sensitive, we performed an analysis of variance. For this analysis, we adopted five problems taken from the Zitzler-Deb-Thiele (ZDT) [20] and from the Deb-Thiele-Laumanns-Zitzler (DTLZ) test suites [21]. The problems were selected in such a way that different features were covered (e.g., non-convexity, disconnected Pareto fronts, etc.) using two and three objectives. The problems chosen for the study are presented next.

- **ZDT3:** The true Pareto front of this problem is disconnected (in two dimensions), consisting of several noncontiguous convex parts.
- **ZDT4:** This problem contains 21^9 false Pareto fronts and, therefore, tests the ability of a MOEA to deal with multifrontality.
- **DTLZ5:** The true Pareto front of this problem is a curve formed by a set of well-distributed solutions.
- **DTLZ6:** The true Pareto front of this problem is unimodal, biased, with a many-to-one mapping and is hard to converge to it.
- **DTLZ7:** The true Pareto front of this problem is disconnected (in three dimensions).

In Table 1, we show the parameters and the values that we used for the ANOVA. For each test problem, we performed 20 independent runs using each of the possible combinations of parameters from those indicated in Table 1.

For evaluating the performance of each set of parameters, we used two performance measures: inverted generational distance ($\mathcal{IGD}$) [10] and the multiplicative unary ϵ-indicator ($\mathcal{I}_\epsilon$) [22] (using the true Pareto front of each problem).

The analysis of results led us to conclude that both the crossover rate and the crossover type (for binary encoding) could take a fixed value, since no variation of these parameters had significant effect on the performance of the NSGA-II. Thus, we decided to adopt a crossover rate $P_c = 0.7$ and two-points crossover for binary encoding. Our study indicated that these values produced the best overall performance for the NSGA-II.

Table 1. Analyzed parameters

Parameter	Values
Population size	100, 200 and 500
Number of generations	100, 200 and 500
Crossover rate	0.5, 0.7 and 1.0
Mutation rate	0.001, 0.1 and 0.3
Encoding	Real and binary
Crossover type	Two-point and uniform crossover for binary encoding
	SBX and uniform crossover for real numbers encoding
Mutation type	Uniform mutation for binary encoding
	Parameter-based and boundary mutation for real numbers encoding

3.2   Self-adaptation of Parameters

The analysis indicated that the variation of the other parameters of the NSGA-II had a greater impact on performance and, therefore, we incorporated them into our proposed self-adaptation scheme. In Fig. 1, we show the general scheme of our self-adaptive MOEA and the corresponding details are presented next.

Initially, a population of 100 individual is randomly generated. For each individual of the population, the type of encoding to be adopted (real or binary) is randomly assigned. The mutation rate and the individual's chromosome are also randomly initialized using the corresponding encoding. For the individuals with real numbers encoding, it is necessary to define, in a random way, the type of mutation and crossover to be used. This is unnecessary when using binary encoding, as was indicated before (see Section 3.1), since fixed values and operators are adopted in that case.

In this work, we assume that all the test problems use real numbers for their decision variables. When using binary encoding, a decoding is evidently needed to transform the binary numbers of each chromosome into real numbers (an accuracy of eight decimal places is adopted in that case). After doing this, the ranking mechanism of the original NSGA-II is applied.

The tournament selection adopted in our case is different from that of the original NSGA-II, because parents are only selected from individuals that have the same encoding. This way, appropriate crossover and mutation operators are applied to individuals having the same encoding. The specific type of crossover and mutation to be applied are chosen from those available (see Sections 3.4 and 3.5) for each type of encoding. The details about the use of these genetic operators are provided in Sections 3.4 and 3.5, respectively.

Once the offspring population is obtained, both the parents and the offspring populations are merged with the purpose of selecting from them to the best individuals for the next generation. For this task, the crowding comparison

```
 1.  t = 0
 2.  Initialize the population;
 3.  Encode the individuals;
 4.  Evaluate the population;
 5.  Rank the Population;
 6.  while (there are no improvements according to the hypervolume) do
 7.      Select the parents // using the same encoding between the individuals;
 8.      Perform crossover;
 9.      Encode the offspring population;
10.      Evaluate the offspring population;
11.      Join the parent and offspring population;
12.      Perform the elitism procedure;
13.      Performed the Inheritance-Fertilization procedure;
14.      if (t ≥ 100) then
15.          Perform a hypervolume analysis
16.          Add/remove individuals
17.      end if
18.      t = t + 1;
19. end while
```

Fig. 1. Our proposed self-adaptation techniques coupled to the NSGA-II

operator of the original NSGA-II is adopted to generate a total ordering of the individuals, so that the best half is selected [6].

Our approach introduces an additional step called the inheritance-fertilization procedure. This is a mechanism that we propose for diversifying the population. This procedure is applied at each generation and its details are discussed in Section 3.6.

Finally, the stopping criterion is defined using the hypervolume performance measure [23]. Specifically, what we do is to check if there is a change in the hypervolume value of the individuals in the population. If no significant change is detected after several iterations, then the MOEA is stopped. The details of this mechanism are discussed in Section 3.7.

3.3   The Individual

In evolutionary algorithms, the individual is commonly represented by a single chromosomic string (i.e., haploids are normally adopted). However, in our proposed approach, we adopt diploids, since we simultaneously encode the individual in binary and real-numbers representation. This is a pragmatic solution to deal with the encoding of each individual, since in our approach, the type of representation could change during the self-adaptation process.

In our case, an individual includes the following elements: the type of encoding (real numbers or binary), the decision variables of the problem, the mutation rate, the type of crossover, the type of mutation, the parents and the fertility. Additionally, each individual also has the parameters from the original NSGA-II (the rank, which relates to Pareto dominance and the crowding distance value, which relates to diversity). Fig. 2 shows the parameters contained in each individual. Since each individual has two possible representations (real numbers or binary), the decision variables and the rates of the operators will be encoded and initialized using the corresponding representation. The type of crossover indicates the crossover operator that was used to generate that individual. This operator will also be used to decide which type of crossover will be used in case the individual is selected for breeding. Similarly, the type of mutation refers to the specific mutation operator that will be applied on the individual that contains it. Since the type of crossover and the type of mutation are already fixed for binary encoding, these parameters are not included in an individual. The

| Type of encoding |
| Decision variables |
| Mutation rate |
| Type of crossover |
| Type of mutation |
| Parents |
| Fertility |
| Crowding distance |
| Rank |

Fig. 2. Definition of each individual in our proposed approach

parameters called *parents* and *fertility* are used in the inheritance-fertilization procedure which will be explained below.

3.4   Crossover Operator

Since the tournament takes place only among individuals with the same type of encoding, the parents selected for breeding will also have the same encoding among themselves.

As indicated before, when using binary encoding, two-points crossover is always adopted in the traditional way [24]. When using real numbers encoding, we have five types of crossover operators available: (1) Simulated Binary Crossover (SBX) [25], (2) simple crossover [26], (3) uniform crossover [27], (4) intermediate crossover [28] and (5) two-points crossover [24].

In order to choose the type of operator to be applied to each pair of individuals, we employ a probabilistic event using a probability $p = 0.9$. If this event returns true, we use the crossover type of the best parent (in terms of its rank). If this event returns false, then we employ another probabilistic event, but using a probability $p = 0.5$. If this second event returns true again, we choose the type of crossover that was adopted to generate the best parent (in terms of its rank). Otherwise, we choose the type of crossover of the other parent. If both parents have the same rank, we choose the type of crossover in a random manner between them.

The mutation rate is encoded (in binary or as a real number) in the chromosomic string. Thus, the mutation rate can be affected by the crossover operator. When using real numbers encoding, the crossover operator is applied using a probability $p = 0.5$ for two-points, simple and uniform crossover. For intermediate recombination, we adopt $k = 0$. SBX is applied as suggested in [6].

Finally, each child generated by the crossover operator inherits the type of encoding and the type of crossover from its best parent (in terms of rank).

3.5   Mutation Operator

For binary encoding, the mutation rate is defined within the interval $(0.001, 0.3)$ and is also encoded in the chromosome. Mutation is applied to the decision variables first, and then to the mutation rate as well. Then, the type of encoding is mutated (or not) using a probability $p = 0.5$. If the type of encoding changes (binary $\mapsto$ real) then the decision variables and the mutation rate are represented using real numbers.

Since there are different types of crossover and mutation operators available (for real-numbers representation), if an individual changes its encoding from binary to real numbers, then we need to define new values for the type of operators to be adopted. In our case, we define such values in a random way.

Since the range of the mutation rate is different for each encoding (in real numbers encoding, the mutation rate is in the range $(1/L, 0.5)$, where L is the number of decision variables), we use a linear mapping to transform the mutation rate from one encoding to the other (i.e., $(0.001, 0.3) \mapsto (1/L, 0.5)$). The mutation rate defined for real numbers encoding ensures that at least one decision variable will get mutated. It also guarantees that more than 50% of the decision variables will be mutated. For perturbing the type of crossover and mutation to be adopted, we perform a similar mapping. Here, we use a mapping defined by $(0.001, 0.3) \mapsto (1/8, 0.8)$.

If the type of crossover or mutation has to be changed then the new types are defined in a random way. Finally, as in the binary case, an individual can change its type of encoding (real↦binary). In this case, the decision variables and mutation rate would be transformed to their equivalent binary representation. The type of crossover and mutation are removed because they are both fixed for binary encoding.

3.6   Inheritance-Fertilization Operator

When the NSGA-II selects the population for the next generation, the parent and offspring populations are merged. The inheritance-fertilization operator identifies the parents and offspring that have been selected to constitute the following generation. Thus, each child has information about who were his parents and viceversa. Once the parents and children have been identified, the mechanism detects parents which have not produced children that had been selected during a certain number of generations (in this work we used a gap of five generations). If this is the case, the parameters of this individual are perturbed.

The mutation of the parameters of each parent is performed according to its encoding, as was indicated before. The same applies to the perturbation of the type of encoding, crossover and mutation (see Section 3.5).

When using the inheritance-fertilization operator, the decision variables are not perturbed. However, the type of encoding can be modified. The aim of this operator is to maintain diversity in the population. The underlying assumption of this operator is that if the children generated by the parents selected in previous generations are not good (in terms of their ranking), is because the genetic operators are not working properly. Thus, they must be modified so that better results can be achieved and that is precisely what the operator does.

3.7   Stopping Criterion and a Varying Population Size

We adopted the hypervolume performance measure [29] to detect when the algorithm has converged (i.e., when no further improvement is found) and we use that as the stopping criterion of our approach. The hypervolume (also known as the S metric or the Lebesgue Measure) of a set of solutions measures the size of the portion of objective space that is dominated by those solutions collectively.

The number of generations and the size of the population play an important role in MOEAs. However, it is well-known that it is unnecessary to have an extremely large population to perform a better search. It is possible to use a modest population size, as long as we have a good mechanism to maintain diversity and we run the MOEA during a sufficiently large number of generations [3]. These aspects were taken into account for the design of the strategy that is explained next.

Initially, the population size is set to 100 individuals. After 100 generations, we start applying hypervolume at each generation. If after 30 generations, there is no improvement in the hypervolume, then 100 new individuals (randomly generated) are added to the population. If some improvement is detected, then the counter is reset so that we start counting again 30 more generations, and repeat this process until no improvement is detected.

Once the 100 new individuals have been added, we continue with the second phase at which we run our MOEA during 20 more generations and check again for improvements in the hypervolume. If no improvement is detected, then we generate 300 new (random) individuals. On the contrary, if some improvement is detected, then, we reduce the population size from 200 to 100. We keep the best half, using Pareto ranking and the crowding comparison operator of the original NSGA-II. Then, the counter is reset again and the search continues, aiming to find 30 consecutive generations without any improvement, before adding 100 new individuals.

Once the population reaches 500 individuals, we enter the third stage. At that point, we run the MOEA for 40 generations. If no improvement in the hypervolume is detected, we consider that the algorithm has converged and we stop the execution of our MOEA. However, if there is an improvement in the hypervolume during these 40 generations, we remove 300 individuals from the population using the same procedure indicated before. In this case, the counter is reset to 30 generations, so that we try to obtain 60 consecutive generations without any improvements in the hypervolume before stopping the execution of the MOEA. Since the hypervolume requires a reference vector, we use the same in all cases, to avoid any errors in its calculation. The complete process is graphically depicted in Fig. 3.

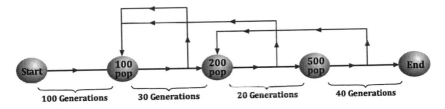

Fig. 3. Graphical illustration of the stopping criterion and the adaptive population size mechanisms

4   Experimental Results

In order to validate the performance of our proposed approach, we compared its results with respect to those obtained by the original NSGA-II using twelve problems taken form ZDT [20] (ZDT1, ZDT2, ZDT3, ZDT4 and ZDT6) and the DTLZ [21] (DTLZ1, DTLZ2, DTLZ3,DTLZ4, DTLZ5, DTLZ6 and DTLZ7) test suites. We adopted three performance measures to assess our results: Inverted Generational Distance ($\mathcal{IGD}$) [10], the multiplicative unary ϵ-indicator ($\mathcal{I}_\epsilon$) [22] and Spread ($\mathcal{S}$) [6].

4.1   Experimental Setup

We performed 20 independent runs per problem per approach. Since our approach does not require any extra parameters, we define only the parameters for the NSGA-II: crossover probability $P_c = 0.7$, mutation probability $P_{mr} = 0.1$. For the genetic operators (SBX and PBM) we used a crossover index $\eta_c = 1$ and a mutation index $\eta_m = 50$. The parameters presented above, were used because the ANOVA of the NSGA-II showed a better performance when adopting them. Additionally, we established a populations size $N = 500$ which is precisely the number of solutions that our proposed approach reports at the end of each run.

Since the stopping criteria used for our approach does not have a fixed number of generations, in order to define the number of generations to be performed by the original NSGA-II we experimented with two different criteria:

1. **Average number of evaluations:** In this case, we used the average (over all runs) number of objective function evaluations performed by our self-adaptive approach to set the number of generations[1] of the NSGA-II.
2. **Average number of generations:** In this case, we used instead the average number of generations (over all runs) performed.

In Tables 2 and 3, we show the average of number of objective function evaluations and the average of the number of generations in which our proposed approach was stopped.

Table 2. Number of generations for the ZDT test suite

Problem	ZDT1	ZDT2	ZDT3	ZDT4	ZDT6
Evaluations Average	238	241	353	611	858
Generations Average	630	643	960	1676	2016

Table 3. Number of generations for the DTLZ test suite

Problem	DTLZ1	DTLZ2	DTLZ3	DTLZ4	DTLZ5	DTLZ6	DTLZ7
Evaluations Average	138	115	170	106	90	99	117
Generations Average	379	324	497	289	248	250	306

Thus, in order to obtain the number of generations during which the original NSGA-II would run, we used either the average of the number of fitness function evaluations (this variant was called NSGA-II$_{eval}$) or the average of the number of generations (this variant was called NSGA-II$_{gen}$).

---

[1] Knowing the total number of objective function of evaluations and the population size, it is straightforward to obtain the total number of generations.

4.2 Discussion of Results

The results obtained by NSGA-II$_{eval}$, NSGA-II$_{gen}$ and NSGA-II$_{self\_adap}$ (our self-adaptive approach) are summarized in Tables 4 to 9. Each table displays both, the mean and the standard deviation (σ) of each performance measure, for each test problem. The best results are shown in **boldface**.

Table 4. Results of $\mathcal{IGD}$ for the ZDT test suite

	$NSGA-II_{eval}$		$NSGA-II_{gen}$		$NSGA-II_{self\_adap}$	
	average	(σ)	average	(σ)	average	(σ)
$ZDT1$	0.000058	(0.000002)	0.000058	(0.000002)	**0.000056**	**(0.000001)**
$ZDT2$	0.000059	(0.000002)	0.000059	(0.000002)	**0.000058**	**(0.000003)**
$ZDT3$	0.000132	(0.000007)	0.000132	(0.000006)	**0.000127**	**(0.000007)**
$ZDT4$	**0.000083**	**(0.000004)**	0.000084	(0.000004)	0.000093	(0.000004)
$ZDT6$	0.002230	(0.000191)	0.002230	(0.000191)	**0.000018**	**(0.000002)**

Table 5. Results of $\mathcal{I}_\epsilon$ for the ZDT test suite

	$NSGA-II_{eval}$		$NSGA-II_{gen}$		$NSGA-II_{self\_adp}$	
	average	(σ)	average	(σ)	average	(σ)
$ZDT1$	1.002235	(0.000322)	**1.002152**	**(0.000225)**	1.002214	(0.000373)
$ZDT2$	1.001938	(0.000440)	1.001967	(0.000317)	**1.001792**	**(0.000346)**
$ZDT3$	1.001863	(0.000393)	**1.001830**	**(0.000226)**	1.002024	(0.000609)
$ZDT4$	1.001709	(0.000316)	**1.001074**	**(0.000324)**	1.001016	(0.000420)
$ZDT6$	1.140147	(0.011069)	1.140147	(0.011069)	**1.001409**	**(0.000171)**

Table 6. Results of $\mathcal{S}$ for the ZDT test suite

	$NSGA-II_{eval}$		$NSGA-II_{gen}$		$NSGA-II_{self\_adp}$	
	average	(σ)	average	(σ)	average	(σ)
$ZDT1$	0.65008	(0.040356)	0.663761	(0.034663)	**0.542132**	**(0.029852)**
$ZDT2$	0.666593	(0.037943)	0.662773	(0.035102)	**0.539112**	**(0.025966)**
$ZDT3$	0.746819	(0.036011)	0.732614	(0.030007)	**0.612249**	**(0.021919)**
$ZDT4$	**0.556536**	**(0.019120)**	0.578010	(0.041313)	0.801464	(0.044060)
$ZDT6$	**0.788426**	**(0.024690)**	0.796272	(0.024848)	0.856744	(0.046337)

Table 7. Results of $\mathcal{IGD}$ for the DTLZ test suite

	$NSGA-II_{eval}$		$NSGA-II_{gen}$		$NSGA-II_{self\_adp}$	
	average	(σ)	average	(σ)	average	(σ)
$DTLZ1$	0.005666	(0.006462)	**0.001938**	**(0.003133)**	0.002161	(0.003941)
$DTLZ2$	0.000172	(0.000004)	**0.000168**	**(0.000004)**	0.000174	(0.000005)
$DTLZ3$	0.250904	(0.098975)	**0.055570**	**(0.032845)**	0.012097	(0.013000)
$DTLZ4$	0.000547	(0.000009)	**0.000542**	**(0.000008)**	0.000558	(0.000030)
$DTLZ5$	**0.000015**	**(0.000002)**	0.000023	(0.000002)	0.000037	(0.000013)
$DTLZ6$	0.015200	(0.002084)	0.002186	(0.000930)	**0.000094**	**(0.000034)**
$DTLZ7$	0.000817	(0.000125)	0.000579	(0.000062)	**0.000389**	**(0.000016)**

$\mathcal{IGD}$ **performance measure.** In Tables 4 and 7, we can see that our proposed approach outperforms the NSGA-II in most of the test problems adopted with respect to $\mathcal{IGD}$. However, in five of the twelve adopted problems (ZDT4, DTLZ1, DTLZ2, DTLZ4 and DTLZ5), our algorithm was outperformed by the original

Table 8. Results of $\mathcal{I}_\epsilon$ for the DTLZ test suite

	$NSGA-II_{eval}$		$NSGA-II_{gen}$		$NSGA-II_{self\_adp}$	
	average	(σ)	average	(σ)	average	(σ)
$DTLZ1$	1.1904444	(0.181980)	**1.066070**	**(0.082606)**	1.079437	(0.123285)
$DTLZ2$	1.049461	(0.006330)	1.047081	(0.006184)	**1.046300**	**(0.005485)**
$DTLZ3$	10.176407	(3.861924)	2.667639	(0.900274)	**1.396945**	**(0.380866)**
$DTLZ4$	1.042668	(0.004336)	1.040609	(0.004801)	**1.039888**	**(0.004084)**
$DTLZ5$	**1.001418**	**(0.000338)**	1.001947	(0.000312)	1.002976	(0.001318)
$DTLZ6$	1.732627	(0.082842)	1.103957	(0.040262)	**1.008537**	**(0.004132)**
$DTLZ7$	1.081822	(0.021139)	1.057326	(0.011657)	**1.026261**	**(0.003132)**

Table 9. Results of S for the DTLZ test suite

	$NSGA-II_{eval}$		$NSGA-II_{gen}$		$NSGA-II_{self\_adp}$	
	average	(σ)	average	(σ)	average	(σ)
$DTLZ1$	0.974938	(0.190992)	0.576810	(0.087172)	**0.546923**	**(0.160001)**
$DTLZ2$	0.429743	(0.022495)	**0.429741**	**(0.017543)**	0.490002	(0.034583)
$DTLZ3$	1.188861	(0.085993)	0.955311	(0.117838)	**0.931206**	**(0.336853)**
$DTLZ4$	**0.412915**	**(0.019106)**	0.424239	(0.018287)	0.453508	(0.036211)
$DTLZ5$	**0.523047**	**(0.092002)**	0.717033	(0.018537)	0.746702	(0.018931)
$DTLZ6$	0.972776	(0.067786)	**0.782302**	**(0.025047)**	0.804278	(0.013795)
$DTLZ7$	0.478671	(0.038158)	**0.440922**	**(0.030344)**	0.531645	(0.021466)

NSGA-II. Evidently, the original NSGA-II is not significantly better than our proposed approach (the NSGA-II$_{self\_adap}$), for these specific test problems.

$\mathcal{I}_\epsilon$ **performance measure.** In Tables 5 and 8, we can see that our proposed approach outperforms the NSGA-II in seven of the twelve test problems adopted with respect to $\mathcal{I}_\epsilon$. Although for ZDT1, ZDT3, ZDT4, DTLZ1 and DTLZ5 the original NSGA-II is better, our algorithm is not significantly worse.

S **performance measure.** Tables 6 and 9 show that our proposed approach was outperformed by the NSGA-II in most of the test problems adopted (seven out of twelve) with respect to spread. However, we do not consider this to be a major drawback, since our self-adaptation mechanisms were focused on convergence rather than on spreading solutions along the Pareto front and in terms of convergence, we found better results in most cases. We believe that the use of additional individuals in the population, in order to maintain diversity, is the main reason why our proposed approach does not reach the same quality of results with respect to spread as the original NSGA-II.

5   Conclusions and Future Research

In this paper, we have presented self-adaptation mechanisms for a MOEA, aiming to have an approach that does not require any manual fine-tuning of its parameters. It is important to emphasize, however, that the parameters are not removed. Instead, we use mechanisms that automatically define them using information gathered during the search, so that no user intervention is required.

Our results indicate that our proposed approach is able to outperform the original NSGA-II in several test problems, with respect to convergence, with the advantage of not requiring any empirical fine-tuning of parameters. Thus, we believe that our proposal can be a viable alternative for end-users who want to apply an out-of-the-box NSGA-II in a certain application, without having much knowledge about evolutionary computation techniques.

Since self-adaptation mechanisms are, in general, hard to define (particularly in the context of MOEAs), in order to simplify things, we tailored most of the mechanisms described here to the specific selection scheme and density estimator adopted by the NSGA-II. However, as part of our future work, we are interested in defining more general versions of some of the self-adaptation mechanisms introduced here, so that they are applicable to more than one MOEA. We are interested in strengthening our algorithm through a finer tuning of its parameters. We aim to achieve this by using more complicated functions such as the Walking-Fish-Group (WFG) [30] and CEC'2009 [31] test problems. Additionally, we are also interested in adding self-adaptation mechanisms that improve the spread of solutions produced by our approach. In that regard, the use of archiving techniques may be useful.

References

1. Eiben, Á.E., Hinterding, R., Michalewicz, Z.: Parameter Control in Evolutionary Algorithms. IEEE Transactions on Evolutionary Computation 3(2), 124–141 (1999)
2. Lobo, F.G., Lima, C.F., Michalewicz, Z. (eds.): Parameter Setting in Evolutionary Algorithms. Springer, Berlin (2007) ISBN 978-3-540-69431-1
3. Abbass, H.A.: The Self-Adaptive Pareto Differential Evolution Algorithm. In: Congress on Evolutionary Computation (CEC 2002), vol. 1, pp. 831–836. IEEE Service Center, Piscataway (2002)
4. Toscano Pulido, G., Coello Coello, C.A.: The Micro Genetic Algorithm 2: Towards Online Adaptation in Evolutionary Multiobjective Optimization. In: Fonseca, C.M., Fleming, P.J., Zitzler, E., Deb, K., Thiele, L. (eds.) EMO 2003. LNCS, vol. 2632, pp. 252–266. Springer, Heidelberg (2003)
5. Trautmann, H., Ligges, U., Mehnen, J., Preuß, M.: A Convergence Criterion for Multiobjective Evolutionary Algorithms Based on Systematic Statistical Testing. In: Rudolph, G., Jansen, T., Lucas, S., Poloni, C., Beume, N. (eds.) PPSN 2008. LNCS, vol. 5199, pp. 825–836. Springer, Heidelberg (2008)
6. Deb, K., Pratap, A., Agarwal, S., Meyarivan, T.: A Fast and Elitist Multiobjective Genetic Algorithm: NSGA–II. IEEE Transactions on Evolutionary Computation 6(2), 182–197 (2002)
7. Kursawe, F.: A Variant of Evolution Strategies for Vector Optimization. In: Schwefel, H.-P., Männer, R. (eds.) PPSN 1990. LNCS, vol. 496, pp. 193–197. Springer, Heidelberg (1991)
8. Büche, D., Guidati, G., Stoll, P., Koumoutsakos, P.: Self-Organizing Maps for Pareto Optimization of Airfoils. In: Guervós, J.J.M., Adamidis, P.A., Beyer, H.-G., Fernández-Villacañas, J.-L., Schwefel, H.-P. (eds.) PPSN 2002. LNCS, vol. 2439, pp. 122–131. Springer, Heidelberg (2002)
9. Tan, K., Lee, T., Khor, E.: Evolutionary Algorithms with Dynamic Population Size and Local Exploration for Multiobjective Optimization. IEEE Transactions on Evolutionary Computation 5(6), 565–588 (2001)

10. Veldhuizen, D.A.V.: Multiobjective Evolutionary Algorithms: Classifications, Analyses, and New Innovations. PhD thesis, Department of Electrical and Computer Engineering. Graduate School of Engineering. Air Force Institute of Technology, Wright-Patterson AFB, Ohio (1999)
11. Kumar, R., Rockett, P.: Improved Sampling of the Pareto-Front in Multiobjective Genetic Optimizations by Steady-State Evolution: A Pareto Converging Genetic Algorithm. Evolutionary Computation 10(3), 283–314 (2002)
12. Abbass, H.A., Sarker, R., Newton, C.: PDE: A Pareto-frontier Differential Evolution Approach for Multi-objective Optimization Problems. In: Proceedings of the Congress on Evolutionary Computation 2001 (CEC 2001), vol. 2, pp. 971–978. IEEE Service Center, Piscataway (2001)
13. Zhu, Z.Y., Leung, K.S.: Asynchronous Self-Adjustable Island Genetic Algorithm for Multi-Objective Optimization Problems. In: Congress on Evolutionary Computation (CEC 2002), vol. 1, pp. 837–842. IEEE Service Center, Piscataway (2002)
14. Coello Coello, C.A., Toscano Pulido, G.: A micro-genetic algorithm for multiobjective optimization. In: Zitzler, E., Deb, K., Thiele, L., Coello Coello, C.A., Corne, D.W. (eds.) EMO 2001. LNCS, vol. 1993, pp. 126–140. Springer, Heidelberg (2001)
15. Martí, L., García, J., Berlanga, A., Molina, J.M.: A Cumulative Evidential Stopping Criterion for Multiobjective Optimization Evolutionary Algorithms. In: Thierens, D. (ed.) 2007 Genetic and Evolutionary Computation Conference (GECCO 2007), vol. 1. ACM Press, London (2007)
16. Zielinski, K., Laur, R.: Adaptive Parameter Setting for a Multi-Objective Particle Swarm Optimization Algorithm. In: 2007 IEEE Congress on Evolutionary Computation (CEC 2007), pp. 3019–3026. IEEE Press, Singapore (2007)
17. Myers, R.H., Montgomery, D.C.: Response Surface Methodology-Process and Product Optimization Using Designed Experiments. John Wiley and Sons, Chichester (2002)
18. Zitzler, E., Thiele, L.: Multiobjective Evolutionary Algorithms: A Comparative Case Study and the Strength Pareto Approach. IEEE Transactions on Evolutionary Computation 3(4), 257–271 (1999)
19. Lindman, H.R.: Analysis of variance in complex experimental designs. SIAM Rev. 18(1), 134–137 (1976)
20. Zitzler, E., Deb, K., Thiele, L.: Comparison of Multiobjective Evolutionary Algorithms: Empirical Results. Evolutionary Computation 8(2), 173–195 (2000)
21. Deb, K., Thiele, L., Laumanns, M., Zitzler, E.: Scalable Test Problems for Evolutionary Multiobjective Optimization. In: Abraham, A., Jain, L., Goldberg, R. (eds.) Evolutionary Multiobjective Optimization. Theoretical Advances and Applications, pp. 105–145. Springer, USA (2005)
22. Zitzler, E., Thiele, L., Laumanns, M., Fonseca, C.M., da Fonseca, V.G.: Performance Assessment of Multiobjective Optimizers: An Analysis and Review. IEEE Transactions on Evolutionary Computation 7(2), 117–132 (2003)
23. Zitzler, E.: Evolutionary Algorithms for Multiobjective Optimization: Methods and Applications. PhD thesis, Swiss Federal Institute of Technology (ETH), Zurich, Switzerland (1999)
24. Goldberg, D.E.: Genetic Algorithms in Search, Optimization and Machine Learning. Addison-Wesley Publishing Company, Reading (1989)
25. Deb, K., Agrawal, R.B.: Simulated Binary Crossover for Continuous Search Space. Complex Systems 9, 115–148 (1995)
26. Michalewicz, Z.: Genetic Algorithms + Data Structures = Evolution Programs, 3rd edn. Springer, Heidelberg (1996)

27. Syswerda, G.: Uniform Crossover in Genetic Algorithms. In: Schaffer, J.D. (ed.) Proceedings of the Third International Conference on Genetic Algorithms, pp. 2–9. Morgan Kaufmann Publishers, San Mateo (1989)
28. Schwefel, H.P.: Evolution and Optimum Seeking. John Wiley & Sons, New York (1995)
29. Zitzler, E., Thiele, L.: Multiobjective Optimization Using Evolutionary Algorithms - A Comparative Case Study. In: Eiben, A.E., Bäck, T., Schoenauer, M., Schwefel, H.-P. (eds.) PPSN 1998. LNCS, vol. 1498, pp. 292–301. Springer, Heidelberg (1998)
30. Huband, S., Hingston, P., Barone, L., While, L.: A Review of Multiobjective Test Problems and a Scalable Test Problem Toolkit. IEEE Transactions on Evolutionary Computation 10(5), 477–506 (2006)
31. Li, H., Zhang, Q.: Multiobjective Optimization Problems With Complicated Pareto Sets, MOEA/D and NSGA-II. IEEE Transactions on Evolutionary Computation 13(2), 284–302 (2009)

Analysing the Performance of Different Population Structures for an Agent-Based Evolutionary Algorithm

J.L.J. Laredo, J.J. Merelo, C.M. Fernandes, A.M. Mora,
M.G. Arenas, P.A. Castillo, and P. Garcia-Sanchez

Department of Architecture and Computer Technology
University of Granada, Spain
juanlu@geneura.ugr.es

1 Introduction

The Evolvable Agent model is a Peer-to-Peer Evolutionary Algorithm [4] which focuses on distributed optimisation over Peer-to-Peer infrastructures [7]. The main idea of the model is that every agent (i.e. individual) is designated as a peer (i.e. network node) and adopts a decentralised population structure defined by the underlying Peer-to-Peer protocol newscast [3]. That way, the population structure acquires a small network diameter which allows a fast dissemination of the best solutions. Additionally, speed of propagation holds with scaling network sizes due to the logarithmic growth of the network diameter.

In that context, this work aims to compare performances of the approach considering two additional population structures other than newscast: a ring and a Watts-Strogatz [9] topology. Figure 1 shows snapshots for the different population structures.

Fig. 1. From left to right: ring, Watts-Strogatz and newscast population structures

Given that regular lattices represent a common approach to fine-grained Evolutionary Algorithms in the literature (e.g. see the review of Tomassini in [8]), we have chosen a ring population structure as the instance to compare the performance of regular lattices against a newscast population structure.

C.A. Coello Coello (Ed.): LION 5, LNCS 6683, pp. 582–585, 2011.

Additionally, the Watts-Strogatz method represents an easy and understandable model for creating a small-world population structure. This way, it will be possible to compare two different methods (i.e. Watts-Strogatz and newscast) for generating the same sub-type of complex network. The interest here goes a step further than in the case of the ring since there are many P2P protocols designed to work as small-world networks. Therefore, we aim to establish whether the performance of the newscast population structure lie in its small-world structure so that may be extended to other protocols implementing the same kind of topologies (e.g. any Distributed Hash Table (DHT) [5]).

2 Experiments and Results

The following experiment aims to compare the influence of the previously explained decentralised population structures on the scalability of the Evolvable Agent model when tackling a 2-trap function [1]. To that aim, optimal population sizes were estimated using the bisection method by Sastry in [6] for different instances of increasing size ($L = 12, 24, 36, 48, 60$).

Figure 2 depicts the scalability of the population size and the number of evaluations for the different population structures in the problem under study. Results show that the ring structure is able to scale better than its counterparts with respect to the population size. Nevertheless, the analysis drastically changes with respect to the computational efforts. In such case, the ring population scales worse than the small-world ones requiring therefore a larger time to converge to optimal solutions.

In addition, the comparison between the two small-world methods (Watts-Strogatz and newscast) shows that scalabilities are quite similar and there is no clear trend of an approach outperforming the other with both requiring equivalent times to solution.

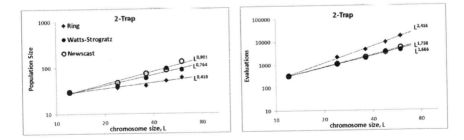

Fig. 2. Scalability of the Evolvable Agent model using a Ring, Watts-Strogatz and Newscast population structures in 2-trap function for optimal population sizes N *(left)* and the number of evaluations to solution *(right)*. Results are depicted in a *log-log* scale as a function of the length of the chromosome, L.

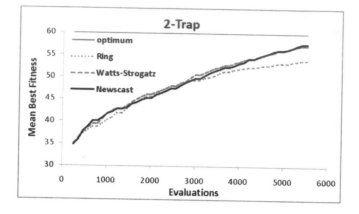

Fig. 3. Best fitness convergence on a 2-trap instance for $L = 60$ and a population size of $N = 135$. Graphs plotted represent the average of 50 independent runs.

Table 1. Wilcoxon test comparing the best fitness distribution of the Evolvable Agent model using a Ring, Watts-Strogatz and Newscast population structures. Results are obtained over 50 independent runs.

Problem Instance	Algorithm	Avg. Fitness $\pm \sigma$	Wilcoxon Test	Significantly different?
2-Trap				
L=60	Ring	53.88±1.21	W=2429 p-value=2.22e-16	yes
N=135	**Watts-Strogatz**	**57.26±1.52**	**W=1386 p-value=0.335**	no
M. Eval.= 5535	**Newscast**	**57.6±1.38**	-	-

With respect to the convergence of the algorithm, Figure 3 shows both small-world approaches having a better progress in fitness than the ring one in the larger problem instance under study (i.e. $L = 60$). In fact, either Watts-Strogatz or newscast reach the same quality in solutions at the maximum number of evaluations.

The Wilcoxon analysis [2] in Table 1 shows that differences in fitness between newscast and ring population structures are statistically significant which confirms previous results on the different convergences of the approaches. Nevertheless, such differences do not appear when comparing newscast with the Watts-Strogatz method.

3   Conclusions

This paper analyses the Evolvable Agent model using different decentralised population structures in order to assess their influence on the performance of the algorithm. A ring topology and the Watts-Strogatz method are considered for comparison against the newscast method which allows a decentralised execution of the approach in a P2P system.

Results show that the ring approach needs smaller population sizes than newscast to guarantee a reliable convergence but, in turn, it requires of a larger number of evaluations which translates into a larger times to solution. On the other hand, the Watts-Strogatz method has a similar performance and does not present statistical differences with respect to the results obtained using newscast. Therefore, the small-world population structures generated by both methods promote equivalent algorithmic performances. We find that fact promising since such a property may extend to other small-world based P2P protocols.

Acknowledgements

This work has been supported by the Junta de Andalucia projects P08-TIC-03903 and P08-TIC-03928, and FCT (ISR/IST plurianual funding), Ministério da Ciência e Tecnologia, through the POS_Conhecimento Program (SFRH / BPD / 66876 / 2009).

References

1. Ackley, D.H.: A connectionist machine for genetic hillclimbing. Kluwer Academic Publishers, Norwell (1987)
2. Garcia, S., Molina, D., Lozano, M., Herrera, F.: A study on the use of non-parametric tests for analyzing the evolutionary algorithms' behaviour: a case study on the CEC' 2005 special session on real parameter optimization. Journal of Heuristics 15(6), 617–644 (2009)
3. Jelasity, M., van Steen, M.: Large-scale newscast computing on the Internet. Technical Report IR-503, Vrije Universiteit Amsterdam, Department of Computer Science, Amsterdam, The Netherlands (October 2002)
4. Laredo, J.L.J., Eiben, A.E., van Steen, M., Merelo, J.J.: Evag: A scalable peer-to-peer evolutionary algorithm. In: Genetic Programming and Evolvable Machines (2010)
5. Ratnasamy, S., Francis, P., Handley, M., Karp, R., Shenker, S.: A scalable content addressable network. In: ACM SIGCOMM, pp. 161–172 (2001)
6. Sastry, K.: Evaluation-relaxation schemes for genetic and evolutionary algorithms. Technical Report 2002004, University of Illinois at Urbana-Champaign, Urbana, IL (2001)
7. Steinmetz, R., Wehrle, K.: What is this peer-to-peer about? In: Steinmetz, R., Wehrle, K. (eds.) Peer-to-Peer Systems and Applications. LNCS, vol. 3485, pp. 9–16. Springer, Heidelberg (2005)
8. Tomassini, M.: Spatially Structured Evolutionary Algorithms: Artificial Evolution in Space and Time. Natural Computing Series. Springer-Verlag New York, Inc., Secaucus (2005)
9. Watts, D.J., Strogatz, S.H.: Collective dynamics of "small-world" networks. Nature 393, 440–442 (1998)

EDACC - An Advanced Platform for the Experiment Design, Administration and Analysis of Empirical Algorithms

Adrian Balint, Daniel Diepold, Daniel Gall, Simon Gerber,
Gregor Kapler, and Robert Retz

Ulm University,
Institute of Theoretical Computer Science,
89069 Ulm, Germany
{adrian.balint,daniel.diepold,daniel.gall,simon.gerber,
gregor.kapler,robert.retz}@uni-ulm.de

Abstract. The design, execution and analysis of experiments using heuristic algorithms can be a very time consuming task in the development of an algorithm. There are a lot of problems that have to be solved throughout this process. To speed up this process we have designed and implemented a framework called EDACC, which supports all the tasks that arise throughout the experimentation with algorithms. A graphical user interface together with a database facilitates archiving and management of solvers and problem instances. It also enables the creation of complex experiments and the generation of the computation jobs needed to perform the experiment. The task of running the jobs on an arbitrary computer system (or computer cluster or grid) is taken by a compute client, which is designed to increase computation throughput to a maximum. Real-time monitoring of running jobs can be done with the GUI or with a web frontend, both of which provide a wide variety of descriptive statistics and statistic testing to analyze the results. The web frontend also provides all the tools needed for the organization and execution of solver competitions.

1   Introduction

Many problems that come from practical applications or from theory are known to be very hard to solve. This means that the time for solving these problems increases exponentially with the size of the input. The class of NP-complete problems is probably the most well known class of such problems. Formerly, proving that a problem was NP-complete meant that the design of a practical algorithm for this problem would be useless because of the estimated exponential time of the algorithm. The situation changed drastically with the development of heuristics, meta-heuristics and approximation algorithms for hard combinatorial problems. The size of the problems that can be solved by these kind of algorithms has increased continuously over the years.

C.A. Coello Coello (Ed.): LION 5, LNCS 6683, pp. 586–599, 2011.

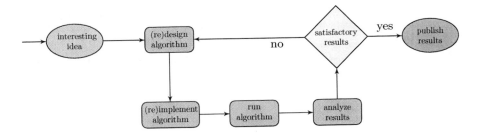

Fig. 1. A typical work flow for the development of empirical algorithms

This progress can be seen as the result of a paradigm change from "algorithms are fast if they have a theoretical good upper bound for their runtime" to "algorithms are fast if they are fast in practical experiments". This should not mean that theoretical results are not important any more, but rather that the design of algorithms has become oriented towards practical applications.

With this paradigm change methodologies have also changed a lot. A theoretical analysis of heuristics is not possible in most cases, and has been replaced by an empirical evaluation like the ones used in engineering. Most development of empirical algorithms now follows an engineering scheme like the one in figure 1.

With the use of new methodologies new problems arise. After the design and implementation phase the algorithm has to be tested and evaluated, which in most cases is a very time consuming task. The first problem that an algorithm designer encounters is the collection and selection of instances on which the solver will be evaluated. A lack of publicly available repositories can hinder this task. Dependent on the set of instances chosen for the evaluation a parameter configuration for the algorithm has to be chosen. This problem can be very often solved by automated procedures like ParamILS [5]. Having the instances and the parameters for the evaluation the user has to choose a computing system. A multi-core computer or a cluster or even a grid can speed up the computations drastically, but at the same time the problem of equally distributing the workload arises, which in most cases is solved by some home brewed scripts. After finishing the computation the results have to be gathered from the computing systems and the important information has to be extracted from the output by some parsing procedures. To find out to what extent the results are satisfactory some statistical tests have to be applied. Comparing the performance of the own algorithm with others demands further elaborated statistics.

The processes of evaluation and analysis are seldom reproducible between different researchers, because of the complexity of the process and the lack of common methods. This is probably the reason why most of the communities working on empirical algorithms periodically organize competitions. The purpose of these competitions is to provide the same evaluation and analysis environment for all the algorithms. A problem with these competitions is that the underlying evaluation system consists of scripts and databases that are not freely available.

The system EDACC (**E**xperiment **D**esign and **A**dministration for **C**omputer **C**lusters) overcomes most of these problems. The previous version of EDACC [1] was restricted to SAT-solvers and SAT-instances). EDACC is capable of managing solvers with their parameters, instances, creating experiment jobs, running them on arbitrary computing systems ranging from multi-core computers to large scale grids, collecting the results and processing them. Advanced methods for automatically extracting and archiving information from the results and from the instances are provided for users. EDACC also provides a large variety of statistical tests and descriptive statistics to analyze the results. To make the organization and execution of competitions with EDACC possible, also a competition mode that follows a widely accepted scheme is provided.

The paper is organized as follows. Chapter 2 gives an overview over the system. Chapter 3 describes the methods for extracting information from instances and from the results. The wide range of possibilities for statistically analyzing this information is presented in chapter 4. Chapter 5 describes the competition mode of the system. Some implementation details and related work is given in chapters 6 and 7. Chapter 8 concludes with some outlooks.

2   EDACC - Overview of the Main Components

A detailed description of the core functionalities of EDACC restricted to the SAT problem was given in [1]. We have considerably extended EDACC to be able to handle arbitrary solvers and instances. All further improvements such as information extraction, statistical analysis and the competition mode, are new features described in this work. To make this paper self-contained an overview of the components of EDACC is given.

Before describing the main components some entities that will be used through the rest of the paper are defined. A *solver* is an implementation of an algorithm that works on some input and has an output. The behavior of a solver is controlled by arbitrarily many *parameters*. A solver together with some fixed parameters is called a *solver configuration*. The input to a solver is called an *instance*. Any information that can be computed from an instance is called an *instance property*. A *computing system* is defined as the computer, computer cluster, or grid on which a solver is tested. When running a solver on a computing system *computational limits* can be imposed (e.g. maximum computation time or maximum memory). An *experiment* is the cartesian product of some set of algorithm configurations, a set of instances, a set of computing systems, and some computational limits. An element of an experiment is a *job*. When the computation of a job is finished it will have a *result*. Any information that is computed from a result is a *result property*.

The main components of EDACC are:

1. database (DB)
2. graphical user interface client (GUI)
3. compute client (CC)
4. web frontend (WF)

The DB is responsible for storing and archiving all the information about the entities defined above. Examples for such information are for solvers the name, version, author, binary, MD5 checksum and the source code. For instances we store the filename, the instance, and the MD5 checksum. The DB also acts as the mediator between GUI, CC, and WF.

The GUI is split into two modes: *manage DB mode* and *experiment mode*. The first mode provides all the necessary DB-operations e.g. create, remove, update and delete (CRUD) for solvers, parameters and instances. As the number of instances stored in the DB can be very large a categorization of the instances into a hierarchical class model is provided. There are two types of classes: *source classes* and *user classes*. The first one specifies the source of the instances. The second one enables the user to create its own collection of instances from different source classes. The class generation process can be done manually or automatically by using the names and the hierarchies of the directories from where the instances are imported.

The work flow of EDACC usually starts by adding solvers, specifying their parameters, and by adding instances and categorizing them into classes. When all the solvers and instances are available in the DB, the user can switch to the experiment mode. After providing some general information, e.g. a name and description of the experiment, the user can select and configure the solvers to be used in the experiment. There are a lot of solver configuration possibilities e.g. enabling or disabling parameters, automated generation of seeds for probabilistic solvers, linking seeds between solvers, for minimizing the variance, and many more.

Next, the instances to use for the created experiment have to be chosen. This operation is alleviated by the instance classes and by filters, enabling a fast selection process. To restrict the consumption of resources like cpu time or memory different limitations can be imposed on the solvers. If the tested solvers are probabilistic there is the possibility to configure the number of repetitions. After choosing a computation system (for which some basic information has to be provided), the user can generate the jobs for the experiment and the distribution package, which is an archive containing the compute client and a configuration file. The configuration file contains information about the DB connection, the experiment and the target compute system.

Copying the distribution package to the computing system and starting the CC will start the processing of the jobs. The CC consists of three programs: *launcher, watcher* and *verifier*. The launcher fetches jobs from the DB and passes them to the watcher, which monitors the use of resources and imposes the desired limitations (At the moment we use the runsolver program from the SAT Competition to achieve this [7]). When a solver finishes, the verifier is used to check the result of the solver, and upon completion the launcher writes all results back to the DB. The verifier is characteristic for each kind of instance and can be replaced or not invoked at all.

There are no limitations on how many CC's are running at the same time. If the computing system is a computer cluster or a grid, then the CC can be run on

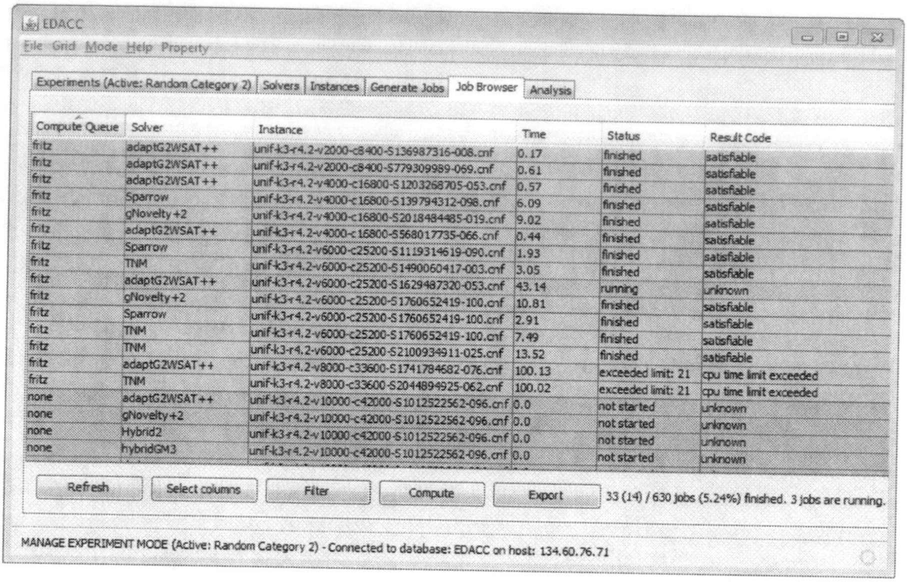

Fig. 2. A snapshot of the job browser within the experiment mode of the GUI, while monitoring the progress of an experiment

all nodes to increase throughput. If the nodes have multi-core CPU's the client can make use of this by starting multiple jobs on a node. Crashes of parts of the computing system will not affect the processing of the experiment, because failed jobs are computed by other CC's. A nice feature worth mentioning is that instances or solvers can be added and deleted during computation, without having to stop the CC's. When a CC finishes a job it writes the results (e.g. CPU time, output of solver, watcher and verifier) back to the DB and picks another job until all jobs are completed. During the computation of the jobs the job browser from the GUI or the WF enables real-time monitoring of the jobs (see Fig. 2). When all jobs of an experiment are finished, the user can extract information from the results and from the instances, and use it for descriptive statistics or statistical tests, that can be performed within the GUI or WF. These features are described in detail in the next chapters.

3   Information Extraction

To analyze the results of an experiment different kinds of statistics can be used. The more information about the experiment's results and instances are available, the more powerful these statistics can be. To make the analysis more easy for the user, EDACC supports a variety of information extraction mechanisms. All the information extracted through these mechanisms can be saved in the DB and used for statistics or can be exported.

We differentiate between two kinds of information, depending on the source. Any information that can be computed from the input instances is called an *instance property (IP)*. All other information is called a *result property (RP)*. The sources of RP's are: the parameters of the solver, and the outputs (stdout, stderr) of the solver, launcher, watcher and verifier.

Most of the information researchers are interested in is present in some output file, and can be easily extracted by a parser procedure. However there are a lot of information, e.g. the "hardness" of an instance or the "qualitiy" of a solution, that requires advanced information processing. To cover both of these scenarios, we provide two major mechanisms to extract IP's and RP's : by an internally defined parser, that can work with regular expressions, or by an external program.

Before starting to extract information, the user has to define the properties in the EDACC GUI by specifying the name, value type, description, source and the regular expression or external program. The value type of the property can be chosen from several predefined types like boolean, integer, float or string. To make the information extraction as flexible as possible, the user is also able to define further types and also to specify if the property has multiple occurrences. If the property's computation mechanism is an external program, the user has to provide a binary and a parameter line to run the program. The stdout output of the program is then interpreted as the value of the property.

Properties are stand-alone entities, and do not require the existence of instances or of results. Starting the computation of a property creates a link between the property and the instance or the result. The link contains the value of the property.

The computation of properties can take a long time, depending on the complexity and size of the input. To take advantage of current multi-core computer architectures EDACC can parallelize the computation of properties.

3.1   Instance Properties

Instance properties can be computed in the manage DB mode of EDACC and are independent of the existence of an experiment. Information about instances can be also parsed from the instance filenames. This can be very useful when the filename encodes different properties. After their computation, IP's can be displayed within the GUI, or can even be used to filter instances, according to certain values of a property. This feature can be very useful when selecting the instances for an experiment. Further, all computed IP's are available for use in the WF.

3.2   Result Properties

Result properties can be computed in the experiment mode of EDACC and assume the existence of an experiment. Most of the RP's, excepting those computed from solver parameters, can be computed only when a job is finished and the output files of solver, launcher, watcher (and verifier) are available in the DB. The computation of result properties can be started during the computation of an experiment because EDACC will take only finished jobs into consideration. Thereby preliminary analysis of the results and their properties is possible. Computed RP's can be displayed in the result browser or can be used in the WF. There are some predefined RP's within EDACC that do not have to be computed: the result time (the time it took to compute the result) and the parameters of a solver.

4   Analysis and Statistical Evaluation

Through its information extraction mechanism, EDACC provides a lot of information about an experiment. Having all this sort of information in the same DB we have extended the GUI and the WF to provide also descriptive statistics and statistical tests. This can be for example used to measure the performance of algorithms, to find out correlations between some properties of the results or to simply have a graphical representation of the results. This enables the user to directly analyze the results without having to export the data, and then process them within a statistical program.

The information that can be used for analysis is stored in the DB within IP's and RP's. We differentiate between two scenarios in which analysis is performed. Analysis of a single solver or comparison of two or more solvers. We also have to differentiate between single runs or multiple runs of a solver on the same instance. If multiple runs are available, the information used for statistics can be chosen by the user from median, mean, all runs or only a single specified run.

To improve the statistical methods the user has also the ability to select the instances used for the analysis. For example when analyzing the results of SAT solvers on random instances containing 3-SAT, 5-SAT, and 7-SAT instances, the user might be only interested in 3-SAT. This can be performed by choosing only the 3-SAT instances for the analysis. Instance selection is provided for all methods.

A RP distribution plot (see Fig. 3) and a nonparametric kernel density estimation is provided for the analysis of the results of a single solver on an arbitrary instance by means of an arbitrary RP. To analyze the results of a solver on all instances (or a selection) the user can use scatter plots. The compared information can be an IP with a RP, like for example number of variables vs. CPU time or two RP's, like memory-usage vs CPU-time. Beside the scatter plots we also compute the Spearman rank correlation coefficient and the Pearson product-moment correlation coefficient.

A scatter plot (see fig. 4 for a run time comparison) together with the two mentioned correlation tests is provided for the comparison of two solvers by means of an arbitrary result property. When the comparison is limited to one instance we also provide RP distributions comparisons together with a Kolmogorow-Smirnow two-sample test and a Mann-Whitney-U Test (Wilcoxon rank sum test). The RP distribution comparison plot can be also done for all solvers but without the tests.

A well founded comparison of the performance of two solvers can also be done with the help of a probabilistic domination test by means of an arbitrary RP.

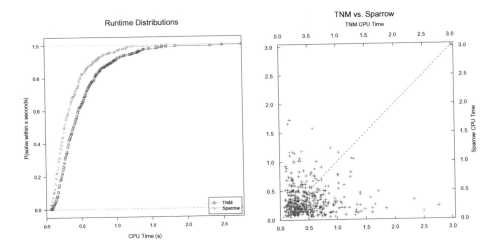

Fig. 3. Comparison of the runtime distribution of two solvers

Fig. 4. Scatter plot to compare the runtime of two solvers

Within this tests instances are split into three categories. The first category contains the instances where the first solver probabilistically dominates the second one. The second one contains the instances where the second solver probabilistically dominates the first one and the third category contains the instances where no probabilistic domination can be found because of the crossing of the RP distributions.

Analyzing one result property for one or more solvers can be done by a box plot or by a cactus plot (number of solved instances within a given amount of the RP's see Fig. 7) .

Finally EDACC can export the generated plots in a huge variety of file formats including vector graphics. To support third-party analysis tools IP's and RP's can also be exported to the widespread csv-format.

5   EDACC - Competition Mode

Solver competitions can be an incentive for researchers to implement new ideas, to improve existing solver and spark interest in the field. Recurring competitions can show the progress in the development of solvers by comparing new solvers with reference solvers from previous competitions. They can also help to identify challenging instances for state-of-the-art solvers. The results of such a competition can be used by researchers to identify the strengths and weaknesses of solvers and instances and to guide further development.

There are several competitions in the field of empirical algorithms, for example the "SAT Competition"[6][7], the "SAT-Race"[8], the "SMT-COMP"[9] or "CASC" [3]. Running such competitions is an organizational challenge and comes with the inevitable need for tools to make it possible to run dozens of solvers on a huge set of instances in a multi-computer environment and then retrieve and process the results for competition purposes. The competitions mentioned above do have such internal tools and web interfaces, but to our knowledge they are not publicly available. To make the organization of competitions to everybody possible, (who has the computational resources) we decided to extend EDACC to be able to provide all required functionalities for the organization of competitions.

We first started by analyzing the existing competition systems to find out their commonalities and to identify interesting or missing features.

From an abstract point of view all competitions have:

1. static web pages to provide information about rules and the course of events
2. user administration to control the access to the results
3. an execution system to run solvers and manage the results
4. dynamic web pages to present the results

As necessary, interesting or missing features we have identified:

1. Plausibility and verifiability of the steps taken in all competition phases by providing participants real-time access to all relevant information.
2. The results have to be reproducible, which means all required information (e.g. starting command, seeds, input files, output files) should be easily accessible through a web interface.
3. Various forms of presentation of the results with cross linking and filtering.
4. Different graphical presentations of the results, including interactive elements such as clickable points in plots that lead to detailed information.
5. All graphical presentations are exportable both as image and as numerical data.
6. Descriptive statistics and statistical tests for analysis of the results.
7. Clean encapsulation of the ranking system enabling easy implementation of new ranking systems.

We have extended the WF of EDACC to provide together with the GUI and CC all of these features. Further we have added a phase system (see Fig. 5) to specify the course of events during a competition. The phases also specify which actions should be taken by whom and control the access to the various information.

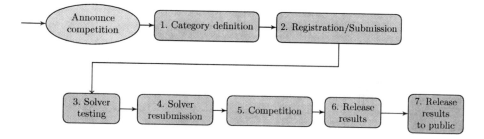

Fig. 5. The phases of a competition

Next we are going to describe the organization of a competition with the EDACC WF by describing each phase, and pointing out the interesting features that are provided. The access control to different kinds of information (e.g. own results, all results, statistics, etc.) can be configured by the organizers for each phase individually, according to their competition policies. Through the description of the phases an exemplary access control is given.

In the first phase the organizers of the competition define the competition categories (which actually can be seen as sub-competitions). A category is defined by the instances it will contain and should give the competitors an orientation where to submit their solvers. In EDACC, each category will be represented by an experiment. In this phase competitors have access only to general information, rules and the schedule. The WF provides containers for these static web pages.

In the second phase, competitors are requested to create an account for the web interface. After login they can submit their solvers (i.e. source code or binary), which are directly saved within the DB. They have to provide detailed information about their solvers like the parameters and the competition category where the solver should participate. Instances can also be submitted by specifying the origin, type and the category it would suit best. Submitted instances will be then available to organizers in the EDACC DB. During this phase competitors have no access to other competitors' solvers nor instances. The WF together with the DB provides the necessary access controls.

The solver testing phase is used to ensure that the submitted solvers are able to run on the computing system of the competition. Within the EDACC GUI organizers create test experiments, corresponding to each of the competition categories. Creating this experiments is straightforward, because solvers and instances are already in the DB. Each solver will be tested in all categories it was submitted to. The experiments are then run on the competition computing system with the help of the CC. Competitors have the possibility to real-time monitor their solvers through the WF (results of other solvers are not visible). Registration and submission of solvers or instances is no longer possible within the WF. From this phase on results are accessible in several forms[1]:

1. By solver configuration: The results for all instances computed by a solver configuration.
2. By instance: The results of all solver configurations.
3. By solver configuration and instance (if multiple runs are allowed): multiple jobs of each solver configuration on an instance are accumulated and some descriptive statistics like the minimum, maximum, median and mean runtime displayed.
4. Single result: The result of a single job, including the output of solver, launcher, watcher and verifier and also all result properties that where computed for this result.

During a solver resubmission phase, competitors have the opportunity to submit solver updates if bugs or compatibility issues with the computing system

---

[1] An example for the results of a competition can be found at
http://edacc.informatik.uni-ulm.de/

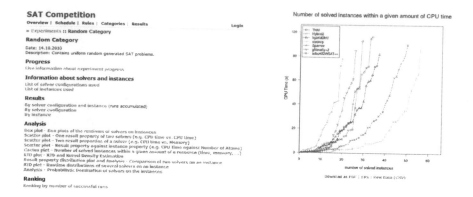

Fig. 6. View of the WF showing the result phase of a competition.

Fig. 7. Cactus plot (cumulative distribution function for run-time) of the results during the run of a competition.

occurred during the test phase. The organizers can then rerun the testing experiments with the updated solvers.

Similar to the testing phase, in the competition phase organizers create experiments based on the competition categories and choose the solvers and instances for each experiment. This task is again accomplished with the help of the GUI. The experiments are then run on the computing system and competitors have the possibility to monitor the results of their own solvers online (and of others if configured so by the organizers).

In the release phase competitors gain access to the results of all competing solvers. Before making the results available to the wide public a ranking has to be calculated. The ranking can either be calculated dynamically by the web application or simply displayed after a manual calculation. We implemented a simple, exemplary ranking using the number of correct results and breaking ties by the accumulated CPU time. Further rankings can be easily encapsulated within the application. Also available in this phase is the complete spectrum of descriptive statistics and statistical tests described in chapter 4. For pointing out interesting results or correlations the organizer have the possibility to extract instance or result properties within the GUI and make them available within the WF.

In a last phase, instances, results and possibly solver source codes and binaries are made publicly available on the web interface without requiring registration.

6   Implementation Details

The first component of EDACC, the DB, requires an user-account on a MySQL 5.1 database with read and write access. The location of the DB plays no role. The needed tables are generated by EDACC itself. The GUI of EDACC is written in Java and is independent of the operating system of the computer. It needs only the Java virtual machine version 6. For the statistical evaluation, the R programming language should also be installed on the computer.

The compute client consists of three sub programs: the launcher, the watcher and the verifier. The launcher builds a DB-connection, and is responsible for fetching the jobs and all necessary files, providing them to the watcher. The launcher is written in C and was tested only on unix-like systems. The watcher starts the solver, and monitors the consumption of resources on the computing system. If some limits are exceeded the solver will be stopped. At the moment we use the runsolver code of Olivier Roussel from the SAT Competition as a watcher. The watcher is a replaceable component in EDACC. The verifier is problem dependent and has to be provided by the user. If the results of the solver can be trusted (e.g. the solver contains a verifier procedure) the verifier can be omitted.

A MySQL proxy is provided to make the execution of the CC on computer clusters possible, where the nodes do not have Internet access, except for a login node. In such a scenario the MySQL-proxy running on the login node provides the DB-connection for the CC's. This feature was tested on several computer clusters.

The web interface for the competition mode is implemented as Python WSGI (Web Server Gateway Interface) application. The application uses a web framework and several open source libraries which are available on most platforms. All competition specific data like user accounts, instance types and the phase of the competition are stored in the central DB. To generate plots and calculate statistics it uses an interface library to the statistical computing language R

The code of EDACC components is open source and is released under the MIT License (excepting the watcher, which has an GPLv3 license). The code is available at the project site: http://sourceforge.net/projects/edacc/.

7   Related Work

We are not aware of the existence of an experimentation system for empirical algorithms that provides all the functionalities of EDACC within the same platform. Parts of EDACC's functionalities are provided by different systems or tools. GridTPT [4] for example supports the testing of SMT solvers and their distribution on computer clusters supporting a master/slave architecture. It is also able to parse information from the output and present some statistics as scatter plots.

The different competitions like [7] and the SMT Competition [9] systems have several tools similar to our WF but they lack the possibility to perform advanced analysis of the results and are not freely available nor portable to other computing systems.

8   Conclusion and Future Work

In this work we have introduced EDACC, a platform for the design, administration and analysis of experiments on empirical algorithms. EDACC consists of four major components, the database, a graphical user interface, a compute client and a web frontend. The DB is the central information storage of EDACC and provides the communication link between GUI, CC and WF. The GUI enables the user to manage solvers, their parameters and instances within the DB. It also enables the design and creation of complex experiments and their administration on different computing systems. The compute client performs the computation of the experiment jobs on arbitrary computing systems ranging from multi-core computers to large scale grids. The architecture of the compute client is designed to use the allocated resources to a maximum, increasing the computational throughput. Crashes of parts of the computational system do not affect the processing of experiment jobs, as failed jobs can be recomputed by other CC's. During the computation of an experiment the GUI and the WF provide a job browser to monitor the jobs. They also provide a wide variety of statistical analysis methods like descriptive statistics and statistical tests. For organizing solver competitions the WF provides all necessary functionalities like user administration, and different dynamic web pages for monitoring the course of events. The statistical analysis possibilities are also provided for the competition mode, enabling a fast evaluation of the results.

We think that researchers, that study empirical algorithms, can drastically speed up their experimental and analysis work by using EDACC as their experimental platform.

In the further development of EDACC we plan to integrate an automatic parameter optimizing procedure. Together with the distributed computing possibilities of EDACC, the optimization process could be sped up. We also plan to integrate different priority policies for processing the jobs within an experiment. For the competition mode of the WF an automated compilation of the source code (which is submitted by the competitors) on the computing system is planed.

Acknowledgments. We would like to thank the bwGrid [2] project for providing the test environment, and Borislav Junk and Raffael Bild for the first version of the launcher code. We would also like to thank Geoff Sutcliffe for fruitful suggestions regarding the first version of this paper.

References

1. Balint, A., Gall, D., Kapler, G., Retz, R.: Experiment design and administration for computer clusters for SAT-solvers (EDACC). JSAT 7, 77–82 (2010); system description
2. bwGRiD, member of the German D-Grid initiative, funded by the Ministry for Education and Research (Bundesministerium für Bildung und Forschung) and the Ministry for Science, Research and Arts Baden-Württemberg (Ministerium für Wissenschaft, Forschung und Kunst Baden-Württemberg), http://www.bw-grid.de
3. Sutcliffe, G.: The CADE-22 Automated Theorem Proving System Competition CASC-22. AI Communications Journal 23(1), 47–60 (2010)
4. Bouton, T., de Oliveira, D.C.B., Déharbe, D., Fontaine, P.: GridTPT: a distributed platform for Theorem Prover. In: Proc. Workshop on Practical Aspects of Automated Reasoning 2010 (2010)

5. Hutter, F., Hoos, H., Stützle, T.: Automatic Algorithm Configuration based on Local Search. In: AAAI 2007 (2007)
6. Le Berre, D., Simon, L.: The Essentials of the SAT 2003 Competition. In: Giunchiglia, E., Tacchella, A. (eds.) SAT 2003. LNCS, vol. 2919, pp. 452–467. Springer, Heidelberg (2004)
7. The SAT Competition Homepage, http://www.satcompetition.org
8. SAT-Race 2010 Homepage, http://baldur.iti.uka.de/sat-race-2010/
9. Barrett, C., De Moura, L., Stump, A.: SMT-COMP: Satisfiability Modulo Theories Competition. In: Etessami, K., Rajamani, S.K. (eds.) CAV 2005. LNCS, vol. 3576, pp. 20–23. Springer, Heidelberg (2005)
10. Homepage of the project, http://sourceforge.net/projects/edacc/

HAL: A Framework for the Automated Analysis and Design of High-Performance Algorithms

Christopher Nell, Chris Fawcett, Holger H. Hoos, and Kevin Leyton-Brown

University of British Columbia, 2366 Main Mall, Vancouver BC, V6T 1Z4, Canada
{cnell,fawcettc,hoos,kevinlb}@cs.ubc.ca

Abstract. Sophisticated empirical methods drive the development of high-performance solvers for an increasing range of problems from industry and academia. However, automated tools implementing these methods are often difficult to develop and to use. We address this issue with two contributions. First, we develop a formal description of *meta-algorithmic problems* and use it as the basis for an automated algorithm analysis and design framework called the High-performance Algorithm Laboratory. Second, we describe HAL 1.0, an implementation of the core components of this framework that provides support for distributed execution, remote monitoring, data management, and analysis of results. We demonstrate our approach by using HAL 1.0 to conduct a sequence of increasingly complex analysis and design tasks on state-of-the-art solvers for SAT and mixed-integer programming problems.

1 Introduction

Empirical techniques play a crucial role in the design, study, and application of high-performance algorithms for computationally challenging problems. Indeed, state-of-the-art solvers for prominent combinatorial problems, such as propositional satisfiability (SAT) and mixed integer programming (MIP), rely heavily on heuristic mechanisms that have been developed and calibrated based on extensive computational experimentation. Performance assessments of such solvers are also based on empirical techniques, as are comparative analyses of competing solvers for the same problem. Advanced algorithm design techniques based on empirical methods have recently led to substantial improvements in the state of the art for solving many challenging computational problems (see, *e.g.*, [1,2,3]).

Empirical analysis and design techniques are often used in an ad-hoc fashion, relying upon informal experimentation. Furthermore, despite a growing body of literature on advanced empirical methodology, the techniques used in practice are often rather elementary. We believe that this is largely due to the fact that many researchers and practitioners do not have sufficient knowledge of, or easy access to, more sophisticated techniques, and that implementations of these techniques are often difficult to use, if publicly available at all. At the same time, it is clear that much can be gained from the use of advanced empirical techniques.

To address the need for easy access to powerful empirical techniques, we developed HAL, the High-performance Algorithm Laboratory – a computational

C.A. Coello Coello (Ed.): LION 5, LNCS 6683, pp. 600–615, 2011.

environment for empirical algorithmics. HAL was conceived to support both the computer-aided design and the empirical analysis of high-performance algorithms, by means of a wide range of ready-to-use, state-of-the-art analysis and design procedures [4]. HAL was also designed to facilitate the development, dissemination, and ultimately wide adoption of novel analysis and design procedures.

By offering standardized, carefully designed procedures for a range of empirical analysis and design tasks, HAL aims to promote best practices and the correct use of advanced empirical methods. In particular, HAL was designed to support the use and development of fully automated procedures for the empirical analysis and design of high-performance algorithms. Since they operate upon algorithms, we refer to these procedures as *meta-algorithmic procedures* (or *meta-algorithms*). Example meta-algorithmic analysis procedures include the characterization of algorithm performance on a set of benchmark instances using a solution cost distribution, as well as the comparison of two algorithms' performance using the Wilcoxon signed-rank test (see, *e.g.*, [5]). Meta-algorithmic design procedures are rapidly gaining prominence and include configuration procedures, such as PARAMILS [6,7] and GGA [8], and portfolio builders like SATZILLA [9,1].

During the early stages of developing HAL, we realized that appropriately formalized notions of meta-algorithmic procedures, and of the tasks accomplished by these procedures, would provide an ideal foundation for the system. This conceptual basis promotes ease of use, by inducing a natural categorization of analysis and design procedures and by facilitating the use of multiple (or alternative) analysis or design procedures. For example, configuration procedures like PARAMILS and GGA solve the same fundamental problem, and with HAL it is easy to conduct analogous (or even parallel) experiments using either of them. Furthermore, HAL's foundation on meta-algorithmic concepts facilitates the combination of various procedures (such as configuration and algorithm selection [10]) and their sequential application (such as configuration followed by comparative performance analysis), as well as the application of analysis or design procedures to other meta-algorithmic procedures (as in the automated configuration of a configurator). Finally, meta-algorithmic concepts form a solid basis for realizing HAL in a convenient and extensible way.

HAL also offers several other features important for work in empirical algorithmics. First, to support large computational experiments, HAL uses a database to collect and manage data related to algorithms, benchmark instances, and experimental results. Second, while HAL can be used on a stand-alone computer, it also supports distributed computation on computer clusters. Third, it allows researchers to archive experiment designs into a single file, including settings, instances, and solvers if unencumbered by license restrictions. Another user can load the file into HAL and replicate exactly the same experiment.

HAL is also designed to facilitate the development and critical assessment of meta-algorithmic procedures. To this end, it is realized as an open environment that is easy to extend, and offers strong support for recurring tasks such as launching, monitoring, and analyzing individual algorithm runs. In short,

HAL allows developers to focus more on building useful and powerful meta-algorithmic procedures and less on the infrastructure required to support them. We hope that this will help to bring about methodological progress in empirical algorithmics, and specifically in the development of novel meta-algorithmic procedures, incorporating contributions from a broad community of researchers and practitioners.

HAL shares some motivation with other systems supporting the empirical study of algorithms. PAVER [11] performs automated performance analysis of optimization software through a web-based interface, but requires that input data be collected by separate invocation of a different tool, and thus is unsuitable for automated techniques that perform concurrent data collection and analysis. EDACC [12] is an experiment management framework which, like HAL, supports distributed execution on compute clusters and centralized data storage, accessed via a unified web interface; unlike HAL, EDACC is focused only on the SAT problem, and more fundamentally does not provide any support for automated meta-algorithmic design procedures. Overall, HAL is the only environment of which we are aware that is designed for the development and application of general-purpose meta-algorithmic analysis and design techniques.

The remainder of this paper is structured as follows. In Section 2, we describe in more detail our vision for HAL and the meta-algorithmic concepts underlying it. In Section 3, we explain how HAL 1.0, our initial implementation of the HAL framework, provides an extensible environment for empirical algorithmics research. We illustrate the use of HAL 1.0 with a sequence of analysis and design tasks for both SAT and MIP in Section 4: first characterizing one solver's performance, next comparing alternative solvers, and finally automating solver design using proven meta-algorithmic techniques. Finally, in Section 5 we summarize our contributions and discuss ongoing work.

2   HAL: A Framework for Meta-algorithmics

The concepts of meta-algorithmic analysis and design procedures are fundamental to HAL. In this section we formally introduce these concepts, discuss benefits we can realize from this formal understanding, and outline HAL's high-level design.

2.1   Meta-algorithmic Problems

We begin by defining a (computational) *problem* as a high-level specification of a relationship between a space of inputs and a corresponding space of outputs. An *instance* of a problem p is any set of values compatible with its input space, and a *solution* to an instance is a set of values compatible with its output space and satisfying the relationship required by p. For example, SAT can be defined as:

Input: $\langle V, \phi \rangle$, where V is a finite set of variables, and ϕ is a Boolean formula in conjunctive normal form containing only variables from V or their negations;

Output: $s = \begin{cases} \text{true} & \text{if } \exists K : V \mapsto \{\text{true}, \text{false}\} \text{ such that } \phi = \text{true under } K; \\ \text{false} & \text{otherwise.} \end{cases}$

Thus, $\langle V = \{a, b, c\}, \phi = (\neg b \vee c) \wedge (a \vee b \vee \neg c) \rangle$ is an example of a SAT instance with solution $s = \text{true}$.

An *algorithm* is any well-defined computational procedure that takes some set of inputs and produces some set of outputs. We say an algorithm A solves a problem p if it accepts any instance of p as a subset of its inputs, and a solution to that instance is identified in its outputs when executed. We observe that A may include inputs and/or outputs other than those required by p, and distinguish three types of algorithm inputs: the algorithm-independent problem instance to be solved, algorithm-specific *parameters* that qualitatively affect behaviour while solving the instance, and any other *settings* that might be required (*e.g.*, a CPU time budget or a random seed). We refer to algorithms that have parameters as *parameterized*, and to the rest as *parameterless*. Any parameterized algorithm can be made parameterless by instantiating all of its parameters with specific values. Thus, a parameterized algorithm defines a space of parameterless algorithms.

A *meta-algorithmic problem* is a problem whose instances contain one or more algorithms, and a meta-algorithm, or *meta-algorithmic procedure*, is an algorithm that solves some meta-algorithmic problem. We refer to algorithms that serve as (part of) a meta-algorithm's input as *target algorithms*, and to the problems target algorithms solve as *target problems*. An *analysis problem* is a meta-algorithmic problem whose solution must include a statement about the target algorithm(s); a *design problem* is a meta-algorithmic problem whose solutions must include one or more algorithms. Finally, we refer to an algorithm that solves an analysis problem as an *analysis procedure*, and one that solves a design problem as a *design procedure*.

Meta-algorithmic analysis problems are ubiquitous, even if they are not always solved by automated procedures. Consider the task of evaluating a solver on a benchmark instance set, using various statistics and diagnostic plots. This corresponds to the *single-algorithm analysis problem*:

Input: $\langle A, \mathcal{I}, m \rangle$, where A is a parameterless target algorithm, $\mathcal{I}$ is a distribution of target problem instances, and m is a performance metric;

Output: $\langle S, T \rangle$, where S is a list of scalars and T a list of plots; and where each $s \in S$ is a statistic describing the performance of A on $\mathcal{I}$ according to m, and each $t \in T$ is a visualization of that performance.

One meta-algorithmic procedure for solving this problem might collect runtime data for A, compute statistics including mean, standard deviation, and quantiles, and plot the solution cost distribution over the instance set [5]; other procedures might produce different plots or statistics. We can similarly define *pairwise comparison*, whose instances contain *two* parameterless algorithms, and whose output characterizes the two algorithms' relative strengths and weaknesses.

Now consider the use of PARAMILS [6,7] to optimize the performance of a SAT solver. PARAMILS is a meta-algorithmic design procedure that approximately solves the *algorithm configuration problem*:

Input: $\langle A, \mathcal{I}, m \rangle$, where A is a parameterized target algorithm, $\mathcal{I}$ is a distribution of target problem instances, and m is a performance metric;

Output: A^*, a parameterless algorithm for the target problem; where A^* corresponds to an instantiation of A's parameters to values that optimize aggregate performance on $\mathcal{I}$ according to m.

We can similarly define the per-instance *portfolio-based algorithm selection problem*, which is approximately solved by SATzilla [9,1]:

Input: $\langle \mathcal{A}, \mathcal{I}, m \rangle$, where $\mathcal{A}$ is a finite set of parameterless target algorithms, $\mathcal{I}$ is a distribution of target problem instances, and m is a performance metric;

Output: A', a parameterless algorithm for the target problem; where A' executes one $A \in \mathcal{A}$ for each input instance, optimizing performance according to m.

Other variations also fit within the framework. Since we consider a parameterized algorithm to be a space of parameterless algorithms, portfolio-based selection can be seen as a special case of the generalization of configuration sometimes referred to as *per-instance configuration*, restricted to finite sets of target algorithms. Generalizing differently, we can arrive at the *parallel portfolio scheduling problem*, which requires that A' executes multiple algorithms from $\mathcal{A}$ in parallel and returns the first solution found, allocating computational resources to optimize the expected aggregate performance on $\mathcal{I}$ according to m. Finally, one can further generalize to *per-instance parallel portfolio scheduling*, where A' executes multiple algorithms from $\mathcal{A}$ for each input instance and returns the first solution found, allocating computational resources to optimize performance according to m.

We note a parallel between meta-algorithmic problems and the idea of design patterns from software engineering, which describe recurrent problems arising frequently in a given environment, along with solutions for them [13]. Meta-algorithmic problems identify challenges that arise regularly in algorithm development and present specific solutions to those challenges. However, choosing between design patterns relies on understanding the benefits and drawbacks of each. The same holds in the meta-algorithmic context; we hope that HAL will prove useful for developing such understanding.

2.2   The High-Performance Algorithm Laboratory

HAL has been designed to align closely with the conceptual formalization from Section 2.1, thereby providing a unified environment for the empirical analysis and design of high-performance algorithms via general meta-algorithmic techniques. In particular, HAL allows explicit representation of arbitrary problems and algorithms (including input and output spaces), problem instances and distributions, and performance metrics. Meta-algorithmic problems in HAL are simply problems whose input (and perhaps output) spaces are constrained to involve algorithms; likewise, meta-algorithmic procedures are realized as a special case of algorithms. HAL presents a unified user interface that gives the user easy and uniform access to a wide range of empirical analysis and design techniques through a task-basked workflow. For example, users can design experiments simply by selecting a meta-algorithmic problem of interest (*e.g.*, configuration), a meta-algorithmic procedure (*e.g.*, PARAMILS), and additional information that specifies the meta-algorithmic problem instance to be solved (*e.g.*, a target algorithm, a distribution of target instances, and a performance metric).

This design provides the basis for five desirable characteristics of HAL. First, it allows HAL to work with arbitrary problems, algorithms and meta-algorithmic design and analysis techniques. Second, it enables HAL to automatically archive and reuse experimental data (avoiding duplication of computational effort, *e.g.*, when rerunning an experiment to fill in missing data), and to serve as a central repository for algorithms and instance distributions. Third, it makes it easy to support packaging and distribution of complete experiments (including target algorithms, instances, and other experiment settings) for independent verification, for example to accompany a publication. Fourth, it facilitates the straightforward use (and, indeed, implementation) of different meta-algorithmic procedures with compatible input spaces; in particular including procedures that solve the same meta-algorithmic problem (*e.g.*, two algorithm configuration procedures). Finally, it simplifies the construction of complex experiments consisting of sequences of distinct design and analysis phases.

To support a wide range of meta-algorithmic design and analysis procedures, HAL allows developers to contribute self-contained plug-in modules relating to specific meta-algorithmic problems and their associated procedures. A plug-in might provide a new procedure for a relatively well-studied problem, such as configuration. Alternately, it might address new problems, such as robustness analysis or algorithm simplification, and procedures for solving them drawing on concepts such as solution cost and quality distributions, runtime distributions, or parameter response curves. In the long run, the value of HAL to end users will largely derive from the availability of a library of plug-ins corresponding to cutting-edge meta-algorithmic procedures. Thus, HAL is an open platform, and we encourage members of the community to contribute new procedures.

To facilitate this collaborative approach, HAL is designed to ensure that the features offered to end users are mirrored by benefits to developers. Perhaps most importantly, the separation of experiment design from runtime details means that the execution and data management features of HAL are automatically provided to all meta-algorithmic procedures that implement the HAL API. The API also includes implementations of the fundamental objects required when building a meta-algorithm, and makes it easier for developers to implement new meta-algorithmic procedures. Adoption of this standardized API also streamlines the process of designing hybrid or higher-order procedures. For example, both HYDRA [10] and ISAC [14] solve algorithm configuration and per-instance portfolio-based selection problems; implementation using HAL would allow the underlying configuration and selection sub-procedures to be easily replaced or interchanged. Finally, as we continue to add meta-algorithmic procedures to HAL, we will compile a library of additional functionality useful for implementing design and analysis procedures. We expect this library to ultimately include components for exploring design spaces (*e.g.*, local search and continuous optimization), machine learning (*e.g.*, feature extraction and regression/classification methods), and empirical analysis (*e.g.*, hypothesis testing and plotting), adapted specifically for the instance and runtime data common in algorithm design scenarios.

3   The HAL 1.0 Core Infrastructure

The remainder of this paper describes an implementation of HAL's core functionality, HAL 1.0, which is now available online.[1] The system is essentially complete in terms of core infrastructure (*i.e.*, experiment modelling, execution management, and user interface subsystems), and includes five meta-algorithmic procedures, focused on two meta-algorithmic analysis problems—single algorithm analysis and paired comparison—and the meta-algorithmic design problem of configuration. These procedures are further described in Section 4, where we present a case study illustrating their use. As discussed above, we intend to add a variety of additional meta-algorithmic procedures to HAL in the next release, and hope that still others will be contributed by the broader community.

This section describes HAL 1.0's core infrastructure. We implemented HAL 1.0 in Java, because the language is platform independent and widely used, its object orientation is appropriate for our modular design goals, and it offers relatively high performance. The HAL 1.0 server has been tested primarily under openSUSE Linux and Mac OS X, and supports most POSIX-compliant operating systems; basic Windows support is also provided. The web-based UI can provide client access to HAL from any platform. HAL 1.0 interfaces with Gnuplot for plotting functionality, and (optionally) with R for statistical computing (otherwise, internal statistical routines are used), MySQL for data management (otherwise, an embedded database is used), and Grid Engine for cluster computing.

In the following subsections, we describe HAL 1.0's implementation in terms of the three major subsystems illustrated in Figure 1. While these details are important for prospective meta-algorithm contributors and illustrative to readers in general, one does not need to know them to make effective use of HAL 1.0.

3.1   Experiment Modelling

The components of the *experiment modelling subsystem* correspond to the concepts defined in Section 2. This subsystem includes most of the classes exposed to developers using the HAL API, including those that are extensible via plug-ins.

We will consider the running example of a user designing an experiment with HAL 1.0, which allows us to describe the Java classes in each subsystem. The user's first step is to select a meta-algorithmic problem to solve. The *Problem* class in HAL 1.0 encodes the input and output *Spaces* defined by a particular computational problem. (We hereafter indicate Java classes by capitalizing and italicizing their names.) The relationship between the inputs and outputs is not explicitly encoded, but is implicitly identified through the name of the

---

[1] `hal.cs.ubc.ca`

Problem itself. Individual variables in a *Space* are represented by named *Domains*; functionality is provided to indicate the semantics of, and conditional interdependencies between, different variables. HAL 1.0 supports a variety of *Domains*, including Boolean-, integer-, and real-valued numerical *Domains*, categorical *Domains*, and *Domains* of other HAL objects.

Once a problem is selected, the user must import an *InstanceDistribution* containing target problem *Instances* of interest. HAL 1.0 currently supports finite instance lists, but has been designed to allow other kinds of instance distributions such as instance generators. The *Instance* class provides access to problem-specific instance data, as well as to arbitrary sets of *Features* and user-provided *Tags* (used, *e.g.*, to indicate encoding formats that establish compatibility with particular *Problems* or *Algorithms*). An *Instance* of a target problem typically includes a reference to the underlying instance file; an *Instance* of a meta-algorithmic problem contains the *Algorithms*, *Instances*, and *Metrics* that define it.

The next step in experiment specification is to choose one or more target algorithms. In HAL 1.0, the *Algorithm* class encodes a description of the input and output spaces of a particular algorithm *Implementation*. For external target algorithms, the *Implementation* specifies how the underlying executable is invoked, and how outputs should be parsed; for meta-algorithmic procedures, it implements the relevant meta-algorithmic logic. Note that the base *Implementation* classes are interfaces, and meta-algorithmic procedures added via plug-ins provide concrete implementations of these. Input and output spaces are encoded using the *Space* class, and an *Algorithm* may be associated with a set of *Tags* that identify the *Problems* that the algorithm solves, and compatible *Instances* thereof. Two *Algorithm* subclasses exist: a *ParameterizedAlgorithm* includes configurable parameters in its input space, and a *ParameterlessAlgorithm* does not. Before execution, an *Algorithm* must be associated with a compatible *Instance* as well as with *Settings* mapping any other input variables to specific values.

The final component needed to model a meta-algorithmic experiment is a performance metric. A *Metric* in HAL 1.0 is capable of performing two basic actions: first, it can evaluate an *AlgorithmRun* (see Section 3.2) to produce a single real value; second, it can aggregate a collection of such values (for example, over problem instances, or over separate runs of a randomized algorithm) into a single final score. HAL 1.0 includes implementations for commonly-used performance metrics including median, average, penalized average runtime (PAR), and average solution quality, and it is straightforward to add others as required.

3.2   Execution and Data Management

The *execution subsystem* implements functionality for conducting experiments specified by the user; in HAL 1.0, it supports execution on a local system, on a remote system, or on a compute cluster. It also implements functionality for cataloguing individual resources (such as target algorithms or instance distributions) and for archiving and retrieving the results of runs from a database.

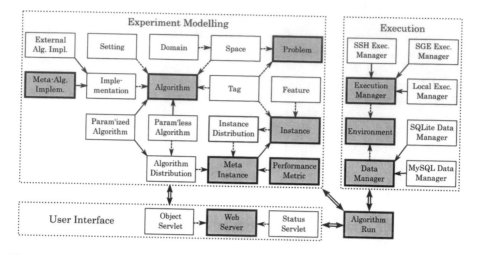

Fig. 1. Infrastructural overview of HAL 1.0. Dashed arrows indicate composition; solid arrows, inheritance. Key components are shaded. Note the distinct subsystems, with interactions between them (double arrows) typically moderated by *AlgorithmRuns*.

Once our user has completely specified an experiment, he must define the environment in which execution is to occur. An *Environment* in HAL 1.0 is defined by *ExecutionManagers* which are responsible for starting and monitoring computation and a *DataManager* which is responsible for performing archival functions. When an algorithm run request is made, the *Environment* queries the *DataManager* to see if results for the run are already available. If so, these results are fetched and returned; if not, the request is passed to an *Execution-Manager* for computation and automatic output parsing. In either case, results are returned as an *AlgorithmRun* object which allows monitoring of the run's elapsed CPU time, status, and individual output value trajectories both in real time during execution and after completion. It also exposes functionality for early termination of runs and uses this to enforce runtime caps.

HAL 1.0 includes three *ExecutionManager* implementations. The *LocalExecutionManager* performs runs using the same machine that runs HAL 1.0, and the *SSHExecutionManager* performs runs on remote machines using a secure shell connection. The *SGEClusterExecutionManager* distributes algorithm runs to nodes of a compute cluster managed by Oracle Grid Engine (formerly Sun Grid Engine). The *Environment* can be configured to use different *ExecutionManagers* in different situations. For example, for analysis of an algorithm on target problems that require a particularly long time to solve, the user might specify an *Environment* in which the parent meta-algorithm is executed on the local machine, but target algorithm runs are distributed on a cluster. Alternatively, when target algorithm runs are relatively short but require a platform different than the one running HAL 1.0, the user might specify an *Environment* in which all execution happens on a single remote host.

HAL 1.0 includes two *DataManager* implementations. By default, a subclass employing an embedded SQLite database is used. However, due to limitations of SQLite in high-concurrency applications, a MySQL-backed implementation is also provided. These *DataManagers* use a common SQL schema based on the same set of fundamental meta-algorithmic concepts to store not only experimental results, but also information sufficient to reconstruct all HAL objects used in the context of a computational experiment. We note that external problem instances and algorithms are not directly stored in the database, but instead at recorded locations on the file system, along with integrity-verifying checksums. This eliminates the need to copy potentially large data files for every run, but presently requires that all compute nodes have access to a shared file space.

3.3   User Interface

The *user interface subsystem* provides a remotely-accessible web interface to HAL 1.0, via an integrated *WebServer*. Many classes have associated *Object-Servlets* in the *WebServer*, which provide interface elements for their instantiation and modification. The *ObjectServlets* corresponding to *Problems* are used to design and execute experiments; the servlet for a given *Problem* automatically makes available all applicable meta-algorithmic procedures. Additional *ObjectServlets* allow the user to specify and examine objects such as *Algorithms*, *InstanceDistributions*, *Settings*, and *Environments*. A *StatusServlet* allows the user to monitor the progress and outputs of experiments both during and after execution, by inspecting the associated *AlgorithmRun* objects. Finally, the interface allows the user to browse and maintain all objects previously defined in HAL, as well as to export these objects for subsequent import by other users.

4   Case Study: Analysis and Design with HAL 1.0

We now demonstrate HAL 1.0 in action. Specifically, we walk through two workflow scenarios that could arise for a typical user. In this way, we also present the five meta-algorithmic procedures that are available in HAL 1.0. The outputs of these procedures are summarized in Table 1, and in the following figures (exported directly from HAL 1.0). Exports of experiment designs are available on the HAL website to facilitate independent validation of our findings.

Scenario 1: Selecting a MIP Solver. In this scenario, a user wants to select between two commercial mixed-integer program (MIP) solvers, IBM ILOG CPLEX[2] 12.1 and GUROBI[3] 3.01, on the 55-instance mixed integer linear programming (MILP) benchmark suite constructed by Hans Mittelmann.[4] Our user sets a per-target-run cutoff of 2h and uses penalized average runtime (PAR-10) as the performance metric (PAR-k counts unsuccessful runs at k times the cutoff).

Scenario 2: Adapting a SAT Solver. In this scenario, a user aims to adapt a stochastic tree search solver for SAT, version 1.2.1 of SPEAR [15], to achieve

---

[2] ibm.com/software/integration/optimization/cplex
[3] gurobi.com
[4] plato.asu.edu/ftp/milpf.html

strong performance on the 302-instance industrial software verification (SWV) benchmark training and test sets used by Hutter et al. [16]. Our user sets a per-target-run cutoff of 30s and evaluates performance by mean runtime (PAR-1).

Computational Environment. All experiments were performed on a Grid Engine cluster of 55 identical dual-processor Intel Xeon 3.2GHz nodes with 2MB cache and 4GB RAM running openSUSE Linux 11.1. Runtime data was archived using a dedicated MySQL server with the same machine specifications. Individual target algorithm runs for Scenario 1 experiments were distributed across cluster nodes, and for Scenario 2 experiments were consolidated on single nodes. Reported runtimes indicate CPU time used, as measured by HAL 1.0.

4.1   The Single-Algorithm Analysis Problem

In both scenarios, our user begins by analyzing single algorithms individually.

Analysis Procedure 1: SCD-Based Analysis. This comprehensive approach to single-algorithm analysis takes as input a single target algorithm, a set of benchmark instances, and some additional settings including a maximum number of runs per target instance, a maximum CPU time per target run, a maximum number of total target runs, and a maximum aggregate runtime budget. It collects runtime data for the target algorithm on the instance distribution (in parallel, when specified) until a stopping criterion is satisfied. Summary statistics are computed over the instance distribution, and a solution cost distribution plot (SCD; see, *e.g.*, Ch. 4 of [5]), illustrating (median) performance across all target runs on each instance, is produced.

Scenario 1(1). CPLEX is the most prominent mixed-integer programming solver. Here, our user measures its performance on the MILP instance set using the SCD-Based Analysis procedure; as CPLEX is deterministic, it is run only once per instance. The resulting summary statistics are shown in Table 1, and the SCD appears in the left pane of Figure 2.

Scenario 2(1). SPEAR was originally optimized for solving SAT instances from several applications, but was later prominently used for software verification in particular. In this phase of the case study, our user assesses the original, manually optimized version of SPEAR on the SWV test set. The summary statistics from an SCD-based analysis (performing 20 runs per instance as SPEAR is randomized) are shown in Table 1 and the SCD in the top left pane of Figure 3.

4.2   The Pairwise Comparison Problem

Now our user performs pairwise comparisons between different solvers.

Analysis Procedure 2: Comprehensive Pairwise Comparison. This procedure performs SCD-Based Analysis on two given algorithms, generates a scatter plot illustrating paired performance across the given instance set, and performs Wilcoxon signed-rank and Spearman rank correlation tests. The Wilcoxon signed-rank test determines whether the median of the paired

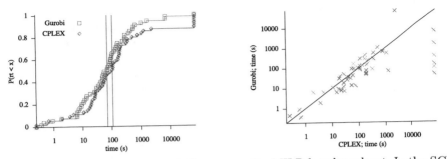

Fig. 2. Comparison of CPLEX and GUROBI on the MILP benchmark set. In the SCD, median runtimes are indicated by vertical lines.

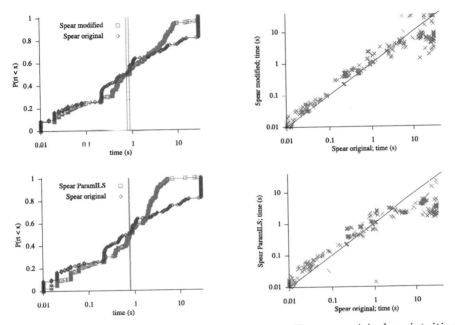

Fig. 3. Analysis of SPEAR designs on SWV test set. Top row, original vs. intuitively modified design; bottom row, original vs. best configured design (from PARAMILS).

performance differences between the two algorithms across the instance set is significantly different from zero; if so, it identifies the better-performing algorithm. The Spearman rank correlation test determines whether a significant monotonic performance correlation exists between them. Both tests are non-parametric, and so appropriate for the non-Gaussian performance data frequently seen in empirical algorithm analysis.

Scenario 1(2). Our user aims to compare CPLEX with GUROBI, a relatively recent commercial MIP solver. He uses HAL's Comprehensive Pairwise Comparison procedure on the MILP benchmark set for this task. Statistics on the performance of the two solvers are shown in Table 1. As can be seen from Figure 2, which presents the combined SCD plot and the performance correlation plot, GUROBI outperformed CPLEX on most instances; the Wilcoxon signed-rank test indicated that this performance difference was significant at $\alpha = 0.05$ ($p = 0.024$). This result is consistent with Mittelmann's observations using the MILP benchmark set. A Spearman correlation coefficient of $\rho = 0.86$ ($p = 0.0$) reflects the strong correlation seen in the scatter plot. However, the slightly better performance of CPLEX observed for a number of instances suggests a potential for modest performance gains by using automated portfolio-based algorithm selection techniques (see, e.g., [1]), which we plan to support in HAL in the near future.

Scenario 2(2). When adapting an algorithm to a new class of benchmark instances, algorithm designers often apply intuition to making important design choices; these choices are often realized by setting parameters of the algorithm to certain values. For example, Hutter et al. [16] provide an intuitive explanation of the strong performance of one particular configuration of SPEAR in solving software verification instances. Our user follows their qualitative description to manually obtain a configuration of SPEAR that he then compares against the default on the SWV test set (based on 20 runs per instance) using the Comprehensive Pairwise Comparison procedure; the results are shown in Figure 3 and Table 1. Overall, the modified configuration achieved better (PAR-1) performance than the default, as expected. However, as clearly seen from the SCDs and from the scatter plot, this was accomplished by sacrificing performance on easy instances for gains on hard instances. The Wilcoxon signed-rank test determined that if all instances were weighted equally, the median paired performance difference over the full benchmark set was *not* significantly different from zero at $\alpha = 0.05$ ($p = 0.35$). The inter-instance correlation was significant, however, with $\rho = 0.97$ ($p = 0.0$).

4.3   The Algorithm Configuration Problem

In Scenario 2(2) above, our user observed that SPEAR's performance can be improved by manually modifying its parameters. Seeking further performance gains, he turns to automatic configuration. HAL 1.0 supports three procedures for this meta-algorithmic design problem.

Design Procedure 1: Automated Configuration using ParamILS. HAL 1.0 supports the FOCUSEDILS variant of the local-search-based PARAMILS configurator [7]. The original Ruby implementation is augmented by using an adapter class to implement the plugin in HAL 1.0.

Design Procedure 2: Automated Configuration using GGA. HAL 1.0 includes a plugin that interfaces with the original implementation of GGA, which employs a gender-based genetic algorithm [8]. Unfortunately, sources for this procedure are not available, and because of copyright restrictions we are unable to further distribute the executable supplied to us by its authors.

Design Procedure 3: Automated Configuration using ROAR. HAL 1.0 also supports the Random Online Aggressive Racing (ROAR) procedure, a simple yet powerful model-free implementation of the general Sequential Model-Based

Table 1. Summary of case study results. Reported statistics are in terms of PAR-10 for CPLEX and GUROBI, and PAR-1 for SPEAR; units are CPU seconds. Only the best design in terms of training set performance is reported for each configuration procedure.

Algorithm	Training Set					Test Set				
	q25	q50	q75	mean	stddev	q25	q50	q75	mean	stddev
CPLEX						26.87	109.93	360.59	9349.1	24148.9
GUROBI						13.45	71.87	244.81	1728.8	9746.0
SPEAR default						0.13	0.80	10.78	6.78	10.62
SPEAR modified						0.19	0.89	4.35	3.40	6.31
SPEAR PARAMILS	0.22	0.80	2.63	1.72	2.54	0.19	0.80	2.21	1.56	2.22
SPEAR GGA	0.22	0.90	1.96	2.00	3.36	0.16	0.90	1.72	1.72	3.39
SPEAR ROAR	0.22	0.92	2.70	1.91	2.59	0.19	0.91	2.41	1.82	2.98

Optimization (SMBO) framework [17]. ROAR was implemented entirely within HAL 1.0, and serves as an example of developing meta-algorithmic design procedures within the HAL framework.

Unlike ROAR and GGA, PARAMILS requires sets of discrete values for all target algorithm parameters; therefore, when using PARAMILS, HAL 1.0 automatically discretizes continuous parameters. Unlike PARAMILS and ROAR, GGA requires all target runs to be performed on the same host machine, and GGA's authors recommend against the use of performance metrics other than average runtime.

Scenario 2(3) Because the three configuration procedures are easily interchangeable in HAL 1.0, our user runs all of them. He performs 10 independent runs of each configurator on the SWV training set, and sets a time budget of 3 CPU days for each run. For some of SPEAR's continuous parameters, our user indicates that a log transformation is appropriate. In these cases, HAL performs the transformations automatically when calling each configurator; it also automatically discretizes parameters for ParamILS. Our user validates the performance of each of the 30 final designs on the training set using the SCD-Based Analysis procedure with 20 runs per instance. He then compares the design with the best training performance found by each of the procedures against the default configuration using the Comprehensive Pairwise Comparison procedure on the test set, again performing 20 runs per instance. Results are shown in Figure 3 and Table 1. The best design found by each configurator was substantially better than both the default and the intuitively-modified configuration in terms of PAR-1, with PARAMILS producing slightly better results than GGA, and with GGA in turn slightly better than ROAR. In all cases, the performance difference with respect to the default was significant at $\alpha = 0.05$ according to the Wilcoxon signed rank test ($p = \{7.8, 9.7, 0.002\} \times 10^{-3}$ for PARAMILS, ROAR, and GGA respectively).

5   Conclusions and Future Work

In this work we introduced HAL, a versatile and extensible environment for empirical algorithmics, built on a novel conceptual framework that formalizes meta-algorithmic problems and procedures. HAL facilitates the application of advanced empirical methods, including computationally intensive analysis and design tasks. It also supports the development and critical assessment of novel empirical analysis and design procedures. The first implementation of our framework, HAL 1.0, can address arbitrary target problems; can run experiments on local machines, remote machines, or distributed clusters; and offers detailed experiment monitoring and control, both before, during and after execution. HAL 1.0 provides a versatile API for developing and deploying new meta-algorithmic analysis and design procedures. Using this API, we developed plugins implementing two performance analysis tasks and supporting three state-of-the-art automated algorithm configurators. We demonstrated the use of all five procedures in a case study involving prominent solvers for MIP and SAT.

Our group continues to actively develop and extend the HAL framework. We are currently working on adding support for additional meta-algorithmic design procedures, such as SATzilla [1], the Hydra instance-based portfolio-builder [10], and the Sequential Model-Based Optimization framework [17]. We are also working on adding new analysis procedures, such as comparative analysis of more than two algorithms and scaling analyses. Finally, we plan to improve HAL's support for execution of experiments on Windows platforms, and on computer clusters running Torque. Ultimately, we hope that the HAL software framework will help to promote the use of state-of-the-art methods and best practices in empirical algorithmics, and to improve the state of the art in solving challenging computational problems through the use of advanced empirical techniques.

Acknowledgements. We thank Frank Hutter for partly implementing Roar and testing HAL, and Meinolf Sellmann and Kevin Tierney for their support in using GGA. Our research has been funded by the MITACS NCE program, by individual NSERC Discovery Grants held by HH and KLB, and by an NSERC CGS M held by CN.

References

1. Xu, L., Hutter, F., Hoos, H.H., Leyton-Brown, K.: SATzilla: Portfolio-based algorithm selection for SAT. JAIR 32, 565–606 (2008)
2. Chiarandini, M., Fawcett, C., Hoos, H.H.: A modular multiphase heuristic solver for post enrollment course timetabling (extended abstract). In: PATAT (2008)
3. Hutter, F., Hoos, H.H., Leyton-Brown, K.: Automated configuration of mixed integer programming solvers. In: Lodi, A., Milano, M., Toth, P. (eds.) CPAIOR 2010. LNCS, vol. 6140, pp. 186–202. Springer, Heidelberg (2010)
4. Hoos, H.H.: Computer-aided design of high-performance algorithms. Technical Report TR-2008-16, University of British Columbia, Computer Science (2008)
5. Hoos, H.H., Stützle, T.: Stochastic Local Search—Foundations and Applications. Morgan Kaufmann Publishers, USA (2004)
6. Hutter, F., Hoos, H.H., Stützle, T.: Automatic algorithm configuration based on local search. In: AAAI (2007)
7. Hutter, F., Hoos, H.H., Leyton-Brown, K., Stützle, T.: ParamILS: An automatic algorithm configuration framework. JAIR 36, 267–306 (2009)

8. Ansótegui, C., Sellmann, M., Tierney, K.: A gender-based genetic algorithm for the automatic configuration of algorithms. In: Gent, I.P. (ed.) CP 2009. LNCS, vol. 5732, pp. 142–157. Springer, Heidelberg (2009)
9. Nudelman, E., Leyton-Brown, K., Devkar, A., Shoham, Y., Hoos, H.H.: Understanding random SAT: Beyond the clauses-to-variables ratio. In: Wallace, M. (ed.) CP 2004. LNCS, vol. 3258, pp. 438–452. Springer, Heidelberg (2004)
10. Xu, L., Hoos, H.H., Leyton-Brown, K.: Hydra: Automatically configuring algorithms for portfolio-based selection. In: AAAI (2010)
11. Mittelmann, H.D., Pruessner, A.: A server for automated performance analysis of benchmarking data. Opt. Meth. Soft. 21(1), 105–120 (2006)
12. Balint, A., Gall, D., Kapler, G., Retz, R.: Experiment design and administration for computer clusters for SAT-solvers (EDACC). JSAT 7, 77–82 (2010)
13. Gamma, E., Helm, R., Johnson, R., Vlissides, J.: Design Patterns: Elements of Reusable Object-Oriented Software. Addison-Wesley, New York (1995)
14. Kadioglu, S., Malitsky, Y., Sellmann, M., Tierney, K.: ISAC – Instance-specific algorithm configuration. In: ECAI (2010)
15. Babić, D.: Exploiting Structure for Scalable Software Verification. PhD thesis, University of British Columbia, Vancouver, Canada (2008)
16. Hutter, F., Babić, D., Hoos, H.H., Hu, A.: Boosting verification by automatic tuning of decision procedures. In: FMCAD (2007)
17. Hutter, F., Hoos, H.H., Leyton-Brown, K.: Sequential model-based optimization for general algorithm configuration (extended version). Technical Report TR-2010-10, University of British Columbia, Computer Science (2010)

HYPERION – A Recursive Hyper-Heuristic Framework

Jerry Swan, Ender Özcan, and Graham Kendall

Automated Scheduling, Optimisation and Planning (ASAP) Research Group,
School of Computer Science, University of Nottingham,
Jubilee Campus, Wollaton Road, Nottingham NG8 1BB, UK
{jps,exo,gxk}@cs.nott.ac.uk

Abstract. Hyper-heuristics are methodologies used to search the space of heuristics for solving computationally difficult problems. We describe an object-oriented domain analysis for hyper-heuristics that orthogonally decomposes the domain into generative policy components. The framework facilitates the recursive instantiation of hyper-heuristics over hyper-heuristics, allowing further exploration of the possibilities implied by the hyper-heuristic concept. We describe HYPERION, a Java™ class library implementation of this domain analysis.

1 Introduction

The idea of combining the strength of multiple (meta-)heuristics goes back to the 1960s ([1], [2]) with the term *hyper-heuristics* being introduced by Denzinger et al. [3]. There has been recent interest in using hyper-heuristics to tackle combinatorial problems. One approach is to employ heuristics as primitive operators, guided to (and hopefully beyond) local optima by a portfolio of meta-heuristics, with the choice of meta-heuristic to apply at each decision-point being determined by a hyper-heuristic. The underlying idea is that hyper-heuristic activity tends to explore the space of local (and hence hopefully global) optima by using a set of lower-level (meta-)heuristics. There are two main types of hyper-heuristics, categorised by whether they are used for *selecting* or *generating* heuristics (see [4] for the former and [5] for the latter). For further detail on hyper-heuristics the reader is referred to [6], [7], [8] and [9].

We describe an object-oriented domain analysis for hyper-heuristics that orthogonally decomposes the domain into generative policy components [10]. This decomposition yields a generative algorithm framework that facilitates rapid prototyping and allows the components that contribute to an algorithm's success to be identified in a procedural fashion. In addition, we add facilities for recursively aggregating hyper-heuristics via the hierarchical nesting of local search neighborhoods. To the knowledge of the authors, there has been no explicit investigation of the effect of instantiating hyper-heuristics to a depth greater than 2, i.e. instantiating hyper-heuristics over hyper-heuristics (perhaps recursively) rather than simply over meta-heuristics. The facility for nesting algorithms to an

C.A. Coello Coello (Ed.): LION 5, LNCS 6683, pp. 616–630, 2011.
© Springer-Verlag Berlin Heidelberg 2011

arbitrary (and possibly dynamically-determined) depth therefore allows further exploration of the possibilities implied by the hyper-heuristic concept.

2   Domain Analysis

The widespread adoption of design patterns as reusable elements of domain vocabulary has lead to the development of a number of popular local search frameworks (e.g. [11], [12],[13]). Although these offer a diversity of approaches for high-level control, the essential nature of local search is present in some elemental domain concepts (albeit appearing under different names). We present them here in the vocabulary used by Fink and Voß [11] in their generic C++ class library, HOTFRAME:

State. This type parameter represents an element of the solution-space.

ObjectiveFunction. A measure of the quality of a State.

Heuristic. This interface abstracts the mechanism for transforming an initial State into some other State of (hopefully) superior quality.

Neighbourhood. This defines some finite neighborhood of a State.

HOTFRAME also makes use of a NEIGHBOURHOODSELECTIONPOLICY, layered upon NEIGHBOURHOOD and having instances that include random neighbor, best neighbor, and best improving neighbor. Metaheuristics directly supported by HOTFRAME include iterated local search (from which random search and varieties of hillclimbing can be configured), together with varieties of simulated annealing and tabu search (the latter being configurable with a number of tabu strategies, including static and reactive tabu).

In addition to the identification of ubiquitous domain vocabulary, we were also strongly influenced in our domain decomposition by the approach of Özcan et al. [9], which achieves a highly-modular decomposition of hyper-heuristics as applied to the domain of fixed-length vectors of bits. Özcan et al. describe four separate hyper-heuristic frameworks in which primitive operations and meta-heuristics (in their case a variety of hillclimbers) are conditionally applied in turn. These four frameworks are conceptually parameterized by the choice of primitive operators, meta-heuristics and heuristic selection mechanisms. They also introduce an acceptance policy mechanism with instances that include unconditional acceptance; improving operations only; Metropolis-Hastings probabalistic acceptance of unimproving moves, and a variant of Great-Deluge.

To the knowledge of the authors, the only other *hyper-heuristic* framework is HY-FLEX [14]. In contrast to the solution-domain frameworks above, HY-FLEX is concerned with building reusable elements for common *problem domains*, and currently supports modules for SAT; one-dimensional bin-packing; permutation flow-shop and personnel scheduling. In the following sections, we describe HYPERION, a Java™ class library for the hyperheuristic solution-domain that respects the entity relationships that hold between the key domain concepts, generalizes the framework of Özcan et al. and facilitates the hierarchical nesting of meta-heuristics.

3   The HYPERION Hyper-Heuristic Framework

We employ object-oriented and generative programming methods [10] to decompose the problem domain, resulting in the key concepts (implemented either directly as classes or generatively via parameterized types) illustrated in Fig. 1-3.

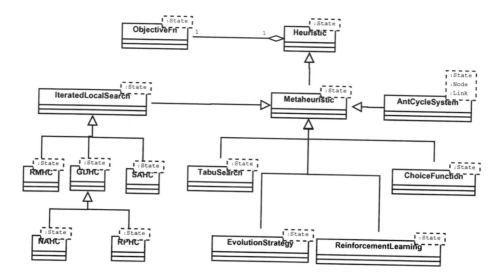

Fig. 1. Main interfaces and concrete meta-heuristics

Figure 1 depicts the heirarchy for HEURISTIC and some of its concrete specialisations. The polymorphic *update* method in the HEURISTIC class represents a single iteration of the algorithm. Formally, the method signature is:

$$update : Transition\langle State\rangle \rightarrow Transition\langle State\rangle$$

where State is a generic type, as denoted by the bracket conventions) and TRANSITION is the generically-typed 5-tuple

$$(from : State, fromValue : \mathbb{R}, operator : Operator, to : State, toValue : \mathbb{R})$$

with *operator* being a descriptor for the operation instance applied. The semantics are that the heuristic should return a result in which the *to* State represents the perturbation of the *from* State of its argument via a single application of the subclass algorithm. In [11], the existence of many-to-one relationships between state-space and objective function and state-space and neighborhood are

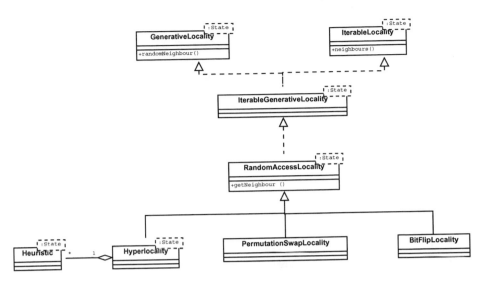

Fig. 2. Abstract and concrete localities

acknowledged, but for efficiency purposes in the implementation, the STATE concept is actually in one-to-one correspondence with its objective function. Our formulation using explicit "pass-though" of tuples representing transitions in the implied search graph (with their caching of objective values of states) allows us to achieve the desired decoupling of states, objective values and neighborhoods without loss of efficiency. Note that, in the domain of hyper-heuristics, the decoupling of states and neighborhoods is essential, since we need to interoperably consider multiple neighborhoods (perhaps operating at different hierarchical levels) over the same state representation. Figure 2 depicts the heirarchy for LOCALITY, the HYPERION term for the ubiquitous concept of local search neighbourhood. In contrast to the singular HOTFRAME neighborhood concept, the HYPERION concept is factored into three - ITERABLELOCALITY, GENERATIVELOCALITY and RANDOMACCESSLOCALITY. ITERABLELOCALITY defines some neighborhood of a state, successive elements of which are accessed via the *Iterator* design pattern [15], GENERATIVELOCALITY provides for the creation of randomly-generated neighbors and RANDOMACCESSLOCALITY allows a neighbor to be accessed via an integer index in O(1) time. The rationale for factoring out these concepts is to reduce the implementation burden for custom neighborhoods. There is explicit support within HYPERION for bit-flip and permutation-swap neighborhoods. By way of example, the interface for BitFlipLocality is given in Listing 1. HYPERION adopts a similar neighborhood selection policy approach to HOTFRAME, additionally providing stochastic tie-breaking and proportional, rank and tournament selection. We incorporate the acceptance policies of [9] as a generic parameter, and provide the following policies (depicted in Fig. 3):

```
public final class BitFlipLocality
extends RandomAccessLocality< BitVector >
{
  public BitFlipLocality( int bitVectorSize )
  {
    /* ... */
  }


  @Override
  public int neighbourhoodSize( Transition< BitVector > t )
  {
    /* ... */
  }


  @Override
  public Transition< BitVector >
  getNeighbour( Transition< BitVector > t , int i )
  {
    /* ... */
  }
}
```

Listing 1. Methods for class BitFlipLocality (implementation details omitted)

All Moves (AM). Unconditionally accepts all generated states.

Only Improving (OI). Accepts only states that improve on the objective value of the previously generated state.

Improving and Equal (IE). As OI, but states of equal objective value are also accepted.

Exponential Monte Carlo (EMC). A worsening move is accepted by this policy with the probability of $p_t = e^{-\frac{\Delta f u}{C}}$, where Δf is the change in objective value in the t-th iteration, C is a counter for successive worsening moves and u is the unit time (e.g., in minutes) that measures the duration of the heuristic execution [16].

Simulated Annealing (SA). This policy accepts unimproving states with probability $p_t = e^{-\frac{\Delta f / N}{1 - t/D}}$, where Δf is the change in objective value in the t-th iteration, D is the maximum number of iterations and N is the maximum possible fitness change [17], [18], [19].

Great Deluge (GD). A variant of the algorithm given in [20], this policy accepts states that are improving or equal relative to a dynamically-determined value that is linearly interpolated from initial to optimal (or best-known) values via the iteration count.

The hillclimbing meta-heuristics implemented in HYPERION are combinatorial generalizations of the bitwise hillclimbing variants described in [21]. Each

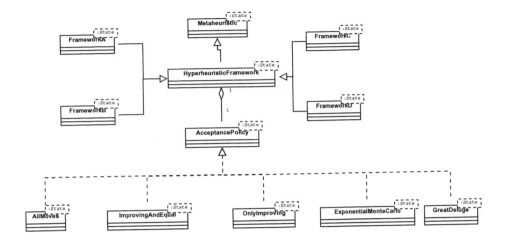

Fig. 3. Frameworks and acceptance policies

hillclimber iteratively replaces the current solution (conditional upon the acceptance policy) with a solution chosen from the current neighborhood according to a neighbor selection policy. Steepest Ascent Hillcimbing (SAHC) evaluates all neighbors and chooses the one with the best objective value. In Random Mutation Hill Climbing (RMHC), the selection policy is to choose a random neighbor. Generalized Davis Hill Climbing (GDHC) is a generalization of Davis's *random bit climber*, in which successive neighbour selections are determined by successive indices of a permutation function. Next Ascent Hillclimbing (NAHC) is then given by instantiating GDHC with the identity permutation and RPHC is GDHC with a random permutation function.

Other heuristic selection strategies implemented in [21] include Choice Function (CF) [4], Simple Random (SR) and Greedy (GR). CF is directly implemented in HYPERION, SR is equivalent to RMHC and GR may be achieved by instantiating ITERATEDLOCALSEARCH with a BESTNEIGHBOUR selection policy (optionally with stochastic tie-breaking). Other meta-heuristics implemented within HYPERION include Reinforcement Learning [22] [23], Evolutionstrategie [24], Tabu Search [25] and Ant-Cycle System [26]:

ReinforcementLearning (RL). Heuristics are ranked (ranking scores are constrained to a fixed range) with scores increasing or decreasing as a function of the heuristic's performance.

Evolutionstrategie (ES). This is a population-based approach in which the number of mutations applied to offspring is an typically a function of some aspect of parent state.

Tabu Search (TS). This restricts the local search neighbourhood by maintaining a (potentially adaptive) mechanism for identifying prohibited transitions.

Ant Cycle System (ACS). This maintains a graph of *solution components* which is repeatedly traversed by a collection of agents. Components from each traversal are assembled into a complete solution in a problem-specific manner.

Since ES is population-based, there is no entirely satisfactory way for it implement the single-solution-based *update* method. We have elected to achieve this by returning the best population member encountered so far and treating the input *from* state as a hint for conditionally reseeding the population. TS is parameterized by a TABUPOLICY in a similar manner to HOTFRAME, since design investigation of a variety of alternative tabu policy signatures revealed that the HOTFRAME approach was the most loosely-coupled of all the alternatives considered. For each of these meta-heuristics, except ACS, the neighborhood is specified via a LOCALITY parameter. In [9], hillclimbers feature as both meta-heuristics and hyper-heuristics, but are implemented separately in each case. By contrast, HYPERION facilitates the creation of hyper-heuristics from existing meta-heuristics via the HYPERLOCALITY specialization of RANDOMACCESSLO-CALITY. By adapting a sequence of heuristics into a locality, a HYPERLOCALITY (listing 2) allows the same algorithm implementation to be used in either case. Listing 3 shows the use of HYPERLOCALITY to recursively instantiate a collection of hillclimbers.

The four frameworks described by Özcan et al. are shown in Fig. 3 in the context of HYPERION and the detail of their internal operation is given in Fig. 4.

In these frameworks, primitive heuristics and hillclimbers (or more generally in HYPERION, meta- or hyper- heuristics) can be partitioned into separate groups. If we denote the application of a framework-selected primitive heuristic by h, a framework-selected higher-order (i.e. meta- or hyper-) heuristic by H and a predetermined higher-order heuristic by P, then the operation of a single invocation of the *update* method on these these frameworks can be described by the following grammar:

$$F_A ::= h|H$$
$$F_B ::= hP|H$$
$$F_C ::= hP$$
$$F_D ::= hH$$

The underlying idea is that this pattern of interaction between primitive and higher-order heuristics will promote solution diversity [9].

```java
public final class Hyperlocality< State >
extends RandomAccessLocality< State >
{
  private List< Metaheuristic< State > >  meta-heuristics;

  public Hyperlocality( List< Metaheuristic< State > > mh )
  {
    this.meta-heuristics = mh;
  }

  @Override
  public Transition< State >
  getNeighbour( Transition< State > t, int index )
  {
    return meta-heuristics.get( index ).update( t );
  }

  @Override
  public int neighbourhoodSize( Transition< State > s )
  {
    return meta-heuristics.size();
  }
}
```

Listing 2. Methods for class Hyperlocality

```
public final class HyperHillclimbers
{
  public static < State >
  List< Metaheuristic< State > >
  instantiate( RandomAccessLocality< State > locality ,
      int recursionDepth )
  {
    if( recursionDepth < 0 )
      throw new IllegalArgumentException ();
    else if( recursionDepth == 0 )
      return getHillclimbers( locality );
    else
    {
      List< Metaheuristic< State > > lm = instantiate(
          locality , recursionDepth - 1 );
      return getHillclimbers( new Hyperlocality< State >( lm )
          );
    }
  }

  //////////////////////////////////

  private static < State >
  List< Metaheuristic< State > >
  getHillclimbers( RandomAccessLocality< State > locality )
  {
    List< Metaheuristic< State >
    > result = new ArrayList< Metaheuristic< State > >();
    result.add( new SAHC< State >( locality ) );
    result.add( new RMHC< State >( locality ) );
    result.add( new NAHC< State >( locality ) );
    result.add( new RPHC< State >( locality ) );
    return result;
  }
}
```

Listing 3. Recursive instantiation of hyper-hillclimbers

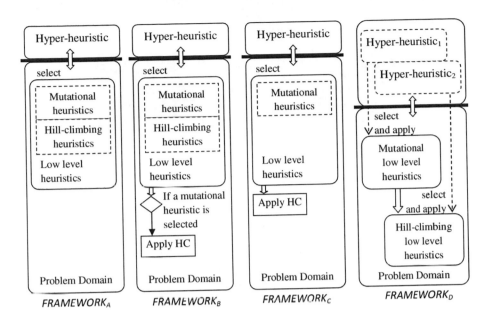

Fig. 4. Internal operation of top-level frameworks

3.1 Design-Space of Hyper-Heuristics

In [8], Burke et al. describe a design space for hyper-heuristics that has two orthogonal dimensions. The first dimension represents selection versus generation and the second the source of feedback during learning (online,offline or none). Both dimensions are further partitioned by the nature of the search space (constructive or pertubative). If we instantiate HYPERION with STATE taken to be some representation of solution state S, then this corresponds to selective hyperheuristics. If instead we take STATE to be some type representing the mapping $S \rightarrow S$, then this corresponds to generative hyper-heuristics. The only explicitly constructive heuristic implemented in HYPERION is ACS, which is additionally parameterized by NODE and LINK types, representing the vertices and edges of the graph of partial solutions traversed by the agents of the ACS. If we employ ACS as a hyperheuristic over some complete solution state, then a path in the graph of partial solutions corresponds to a sequence of lower-level heuristics and an adaptor function is used to yield the resulting complete solution state via by the sequential application of these heuristics to the *from* state. In general therefore, heuristics may be considered as constructive or perturbative as required, employing adaptors as necessary for interoperability with other solution representations. By virtue of this modularity of decomposition, HYPERION facilitates a wide variety of hyperheuristic strategies. In particular, the approaches adopted in [27], [28] and [29] may all be considered as specific configurations of HYPERION components.

Table 1. Average heuristic values obtained over 100 runs of 3-SAT instances

Problem instance	RPHC	RMHC	SAHC	SA	NAHC
uf20-01.cnf	2.96	4.36	6.85	8.42	11.54
uf20-02.cnf	3.14	4.16	6.08	7.09	9.84
uf20-03.cnf	3.37	4.99	7.56	9.24	12.35
uf20-04.cnf	3.23	5.27	8.11	10.0	13.67
uf20-05.cnf	3.84	5.75	9.13	10.97	15.32
uf20-06.cnf	3.29	5.0	7.35	9.02	12.37
uf20-07.cnf	2.91	3.93	5.55	6.78	9.07
uf20-08.cnf	3.07	4.7	6.89	8.32	11.2
uf20-09.cnf	3.07	4.71	6.84	8.41	11.88
uf20-010.cnf	2.76	4.19	6.16	7.49	10.23
uf20-011.cnf	2.9	3.76	5.84	6.81	9.88
uf20-012.cnf	2.2	2.82	4.38	4.96	7.33
uf20-013.cnf	3.42	5.24	7.82	9.79	13.17
uf20-014.cnf	2.92	4.4	6.68	8.2	11.29
uf20-015.cnf	2.67	3.71	5.49	6.49	9.2
uf20-016.cnf	2.82	4.12	6.26	7.53	10.34
uf20-017.cnf	2.55	3.94	5.75	7.2	9.76
uf20-018.cnf	3.56	5.65	8.64	10.69	14.6
uf20-019.cnf	3.28	5.26	7.85	9.67	13.17
uf20-020.cnf	3.21	4.66	7.06	8.54	12.0
percentage solved	3.7%	8.55%	10.35%	8.25%	2.9%

3.2   Application to SAT

We illustrate the use of the framework classes via application to the well-known boolean satisfiability problem (SAT). The palette of meta-heuristics is obtained from some class MyMetaheuristics, which is identical to code for the hyperhill-climbers described in listing 3, together with an instantiation of simulated annealing that has a geometric annealing schedule in which the parameters are dynamically determined by sampling the state-space [30]. The client-code for applying 'Framework A' to the SAT domain using a simple heuristic measure of the number of unsatisfied clauses is given in listing 4. Table 1 gives the average heuristic values obtained from 100 applications of this framework to the first 20 instances of the 3-SAT UF20-91 SATLIB problem set (http://www.cs.ubc.ca/~hoos/SATLIB/benchm.html) [31]. All instances have 20 variables and 91 clauses and are known to be satisfiable. RPHC can be seen to give better average performance in all cases, but if we consider the percentage of cases that are actually solved (as given in the bottom row of Table 1), we see that SAHC converges in the highest number of cases and RPHC gives the second worst performance.

```java
package hyperion.benchmarks.sat;

public final class RunSAT
{
  static final int NUM_ITERATIONS = 100000;
  static final int HYPERHEURISTIC_NESTING_LEVEL = 0;
  // ^ nesting level 0 instantiates _meta_ heuristics

  public static void main( String [] args ) throws IOException
  {
    String fileName = "resources/uf20-91/uf20-0102.cnf";
    CNF cnf = ReadCNF.readDIMACS( fileName );
    ObjectiveFn<BitVector> heuristicFn = new
        NumUnsatisfiedClauses( cnf );
    List< Metaheuristic<BitVector> > hyperheuristics =
      MyMetaheuristics.instantiate(
      new BitFlipLocality( cnf.getNumVariables() ),
      HYPERHEURISTIC_NESTING_LEVEL );

    BitVector initial = new BitVector( cnf.getNumVariables());
    AcceptancePolicy<BitVector> acceptance = new AllMoves<
        BitVector >();
    for( Metaheuristic<BitVector> alg : hyperheuristics )
    {
      FrameworkA<BitVector> framework = new FrameworkA<
          BitVector >(
          alg ,
          acceptance ,
          NUM_ITERATIONS );

      BitVector result= framework.apply(initial , heuristicFn);
      int value = heuristicFn.valueOf( result );
      System.out.println("alg:" | alg + ",value:" + value);
    }
  }
}
```

Listing 4. Client code for SAT solver

4   Conclusion and Future Work

We have presented an object-oriented analysis of the hyper-heuristic domain, incorporating generic versions of the decomposition given in [9] to produce a Java[TM] implementation (available from http://hyperion-java.sourceforge.net) that recursively aggregates local search neighborhoods to generate hyper-heuristics from meta-heuristics without the necessity for source-code duplication.

In addition, it is possible to combinatorially instantiate hyper-heuristics from collections of policy components, with the additional possibility that instantiation can recurse over available meta-heuristics to some dynamically-determined depth.

Recursion is thus of value as a facility for source code re-use. In addition, by altering the given examples of recursive instantiation to make a stochastic choice of lower-level (hyper-)heuristics, HYPERION can also be considered as a generation mechanism for *strongly-typed genetic programming* [32] in the domain of hyper-heuristics. Future work includes an investigation of the effect of recursion depth in the context of building-blocks in 'hierarchical iff' functions [33]. There are also a number of aspects of the current framework implementation that we believe could be improved upon. As discussed above, single-state and population-based meta-heuristics do not interoperate in an entirely satisfactory manner. A more loosely-coupled scheme for mediating interactions between heuristics is currently under development. Another significant improvement would be a change in the level of abstraction from that of local search neighborhoods to local search *frames*, the analogy being with stack frames in a conventional programming language. A frame encapsulates an algorithm instantiated over a locality and comes equipped with a parameter schema detailing not only the set of permissible parameter values but also other information pertinent to searching the parameter space (e.g. whether first or second derivatives exist for a parameter).

References

1. Fisher, H., Thompson, G.L.: Probabilistic learning combinations of local job-shop scheduling rules. In: Muth, J.F., Thompson, G.L. (eds.) Industrial Scheduling, pp. 225–251. Prentice-Hall, Inc., New Jersey (1963)
2. Crowston, W., Glover, F., Thompson, G., Trawick, J.: Probabilistic and parameter learning combinations of local job shop scheduling rules. In: ONR Research Memorandum. GSIA, vol. 117, Carnegie Mellon University, Pittsburgh (1963)
3. Denzinger, J., Fuchs, M., Fuchs, M.: High Performance ATP Systems by combining several AI Methods. In: Proceedings of the 4th Asia-Pacific Conference on SEAL, IJCAI, pp. 102–107 (1997)
4. Cowling, P.I., Kendall, G., Soubeiga, E.: A Hyperheuristic approach to Scheduling a Sales Summit. In: Burke, E., Erben, W. (eds.) PATAT 2000. LNCS, vol. 2079, pp. 176–190. Springer, Heidelberg (2001)
5. Burke, E.K., Hyde, M.R., Kendall, G., Ochoa, G., Özcan, E., Woodward, J.R.: Exploring Hyper-heuristic Methodologies with Genetic Programming. In: Kacprzyk, J., Jain, L.C., Mumford, C.L., Jain, L.C. (eds.) Computational Intelligence. Intelligent Systems Reference Library, vol. 1, pp. 177–201. Springer, Heidelberg (2009)
6. Ross, P.: Hyper-heuristics. In: Burke, E.K., Kendall, G. (eds.) Search Methodologies: Introductory Tutorials in Optimization and Decision Support Techniques, pp. 529–556. Springer, Heidelberg (2005)
7. Burke, E.K., Hart, E., Kendall, G., Newall, J., Ross, P., Schulenburg, S.: Hyper-heuristics: An emerging direction in modern search technology. In: Glover, F., Kochenberger, G. (eds.) Handbook of Metaheuristics, pp. 457–474. Kluwer, Dordrecht (2003)

8. Burke, E.K., Hyde, M., Kendall, G., Ochoa, G., Özcan, E., Woodward, J.R.: A classification of hyper-heuristic approaches. In: Gendreau, M., Potvin, J.Y. (eds.) Handbook of Metaheuristics. International Series in Operations Research and Management Science, vol. 146, pp. 449–468. Springer, US (2010)
9. Özcan, E., Bilgin, B., Korkmaz, E.E.: A comprehensive analysis of hyper-heuristics. Intell. Data Anal. 12, 3–23 (2008)
10. Czarnecki, K., Eisenecker, U.: Generative Programming: Methods, Tools, and Applications. Addison-Wesley Professional, Reading (2000)
11. Fink, A., Voß, S.: Hotframe: A heuristic optimization framework. In: Voß, S., Woodruff, D. (eds.) Optimization Software Class Libraries. OR/CS Interfaces Series, pp. 81–154. Kluwer Academic Publishers, Boston (2002)
12. Gaspero, L.D., Schaerf, A.: Easylocal++: An Object-oriented Framework for the flexible design of Local-Search Algorithms. Softw., Pract. Exper. 33, 733–765 (2003)
13. Voudouris, C., Dorne, R., Lesaint, D., Liret, A.: iOpt: A Software Toolkit for Heuristic Search Methods. In: Walsh, T. (ed.) CP 2001. LNCS, vol. 2239, pp. 716–729. Springer, Heidelberg (2001)
14. Burke, E.K., Curtois, T., Hyde, M., Kendall, G., Ochoa, G., Petrovic, S., Vazquez-Rodriguez, J.A.: HyFlex: A Flexible Framework for the Design and Analysis of Hyper-heuristics. In: Multidisciplinary International Scheduling Conference (MISTA 2009), Dublin, Ireland, pp. 790–797 (2009)
15. Gamma, E., Helm, R., Johnson, R.E., Vlissides, J.M.: Design patterns: Abstraction and reuse of object-oriented design. In: Wang, J. (ed.) ECOOP 1993. LNCS, vol. 707, pp. 406–431. Springer, Heidelberg (1993)
16. Ayob, M., Kendall, G.: A monte carlo hyper-heuristic to optimise component placement sequencing for multi head placement machine. In: Proceedings of the International Conference on Intelligent Technologies (InTech 2003), Chiang Mai, Thailand, pp. 132–141 (2003)
17. Kirkpatrick, S., Gelatt, C.D., Vecchi, M.P.: Optimization by simulated annealing. Science 220, 671–680 (1983)
18. Bai, R., Kendall, G.: An investigation of automated planograms using a simulated annealing based hyper-heuristics. In: Ibaraki, T., Nonobe, K., Yagiura, M. (eds.) Metaheuristics: Progress as Real Problem Solver, pp. 87–108. Springer, Heidelberg (2005)
19. Burke, E., Kendall, G., Misir, M., Özcan, E.: Monte carlo hyper-heuristics for examination timetabling. Annals of Operations Research 2, 1–18 (2010), 10.1007/s10479-010-0782-2
20. Dueck, G.: New optimization heuristics: The great deluge algorithm and the record-to record travel. Journal of Computational Physics 104, 86–92 (1993)
21. Mitchell, M., Holland, J.H.: When will a genetic algorithm outperform hill climbing? In: Proceedings of the 5th International Conference on Genetic Algorithms, vol. 647. Morgan Kaufmann Publishers Inc., San Francisco (1993)
22. Kaelbling, L.P., Littman, M.L., Moore, A.P.: Reinforcement learning: A survey. J. Artif. Intell. Res. (JAIR) 4, 237–285 (1996)
23. Özcan, E., Misir, M., Ochoa, G., Burke, E.: A reinforcement learning - great-deluge hyper-heuristic for examination timetabling. International Journal of Applied Metaheuristic Computing, 39–59 (2010)
24. Herdy, M.: Application of the evolutionsstrategie to discrete optimization problems. In: Schwefel, H.-P., Männer, R. (eds.) PPSN 1990. LNCS, vol. 496, pp. 188–192. Springer, Heidelberg (1991)
25. Glover, F.: Tabu Search - Part I. INFORMS Journal on Computing 1, 190–206 (1989)

26. Dorigo, M., Stützle, T.: Ant Colony Optimization. MIT Press, Cambridge (2004)
27. Ortiz-Bayliss, J.C., Özcan, E., Parkes, A.J., Terashima-Marin, H.: Mapping the performance of heuristics for constraint satisfaction, pp. 1–8 (2010)
28. Hyde, M., Özcan, E., Burke, E.K.: Multilevel search for evolving the acceptance criteria of a hyper-heuristic. In: Proceedings of the 4th Multidisciplinary Int. Conf. on Scheduling: Theory and Applications, pp. 798–801 (2009)
29. Ersoy, E., Özcan, E., Uyar, C.: Memetic algorithms and hyperhill-climbers. In: Baptiste, P., Kendall, G., Kordon, A.M., Sourd, F. (eds.) 3rd Multidisciplinary Int. Conf. On Scheduling: Theory and Applications, pp. 159–166 (2007)
30. White, S.: Concepts of scale in simulated annealing. In: Proc. Int'l Conf. on Computer Design, pp. 646–651 (1984)
31. Hoos, H.H., Stützle, T.: SATLIB: An online resource for research on SAT. In: Gent, I.P., Maaren, H.V., Walsh, T. (eds.) SAT 2000 (2000), SATLIB is available online at www.satlib.org
32. Montana, D.J.: Strongly typed genetic programming. Evolutionary Computation 3, 199–230 (1995)
33. Iclanzan, D., Dumitrescu, D.: Overcoming hierarchical difficulty by hill-climbing the building block structure. In: GECCO 2007: Proceedings of the 9th Annual Conference on Genetic and Evolutionary Computation, pp. 1256–1263. ACM, New York (2007)

The Cross-Domain Heuristic Search Challenge –
An International Research Competition

Edmund K. Burke[1], Michel Gendreau[2], Matthew Hyde[1], Graham Kendall[1],
Barry McCollum[3], Gabriela Ochoa[1], Andrew J. Parkes[1], and Sanja Petrovic[1]

[1] Automated Scheduling, Optimisation and Planning (ASAP) Group, School of
Computer Science, University of Nottingham, Nottingham, UK
[2] CIRRELT, University of Montreal, Canada
[3] School of Electronics, Electrical Engineering and Computer Science,
Queen's University, Belfast, UK

Abstract. The first *Cross-domain Heuristic Search Challenge* (CHeSC
2011) seeks to bring together practitioners from operational research,
computer science and artificial intelligence who are interested in developing more generally applicable search methodologies. The challenge is
to design a search algorithm that works well, not only across different instances of the same problem, but also across different problem domains.
This article overviews the main features of this challenge.

1 Introduction

The *Cross-domain Heuristic Search Challenge*[1] differs from other competitions in
search and optimisation, as it aims to measure performance over several problem
domains rather than just one. We propose a software framework (*HyFlex*) featuring a common software interface for dealing with different combinatorial optimisation problems. HyFlex provides the algorithm components that are problem
specific. In this way, we liberate algorithm designers from needing to know the
details of the problem domains and also prevent them from incorporating additional problem specific information in their algorithms. Efforts can instead be
focused on designing high-level strategies to intelligently combine the provided
problem-specific algorithmic components. The competition is organised and run
by the Automated Scheduling, Optimisation and Planning (ASAP) group at the
University of Nottingham, Nottingham, UK; with contributions from Queen's
University, Belfast, UK; Cardiff University, UK; and the Ecole Polytechnique,
Montreal, Canada. Members of these groups will not be allowed to enter the
competition.

2 The HyFlex Framework

HyFlex (Hyper-heuristics Flexible framework) [1] is a Java object oriented framework for the implementation and comparison of different iterative general-purpose

---

[1] http://www.asap.cs.nott.ac.uk/chesc2011/

C.A. Coello Coello (Ed.): LION 5, LNCS 6683, pp. 631–634, 2011.

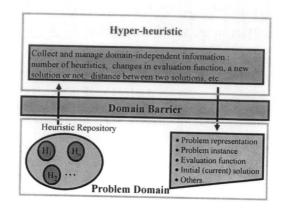

Fig. 1. Hyper-heuristic conceptual framework featuring the domain barrier [2,3]

heuristic search algorithms (also called hyper-heuristics). The framework appeals to modularity and is inspired by the notion of a domain barrier between the low-level heuristics and the hyper-heuristic [2,3] (Figure 1). HyFlex provides a software interface between the hyper-heuristic and the problem domain layers, thus enabling a clearly defined separation, and communication protocol between the domain specific and the domain independent algorithm components.

HyFlex extends the conceptual framework discussed in [2,3] (Figure 1) int that a population of solutions (instead of a single incumbent solution) is maintained in the problem layer. Also, a richer variety of low-level heuristics is provided. Another relevant antecedent to HyFlex is PISA [4], a text-based software interface for multi-objective evolutionary algorithms, which divides the implementation of an evolutionary algorithm into an application-specific part and an algorithm-specific part. HyFlex differs from PISA in that its interface is not text-based but instead given by an abstract Java class. HyFlex is not tied to evolutionary algorithms. It allows the implementation of most single-point and population-based search methods. Moreover, it provides a rich variety of combinatorial optimisation problems including real-world instance data. Each HyFlex problem domain module consists of:

1. A routine to initialise randomised solutions in the population.
2. A set of heuristics to modify solutions classified into four groups:
   mutational : makes a (randomised) modification to the current solution.
   ruin-recreate : destroys part of the solution and rebuilds it using a constructive procedure.
   local search : searches in the *neighbourhood* of the current solution for an improved solution.
   crossover : takes two solutions, combines them and returns a new solution.
3. A varied set of instances that can be easily loaded.
4. A population of one or more solutions that has to be administered.

For testing purposes, four domain modules are provided each containing around 10 low-level heuristics of the types discussed above, and 10 instances of medium to hard difficulty. The domains provided are: permutation flowshop, one dimensional bin packing, Boolean satisfiability (MAX-SAT) and personnel scheduling. Technical reports describing the details of each of these modules, are available at the competition Web site ('Documentation' section: http://www.asap.cs.nott.ac.uk/chesc2011/documentation.html).

3   Challenge Description and Scoring System

For the competition, a number instances from each of these four test domains will be considered (including both training and hidden instances). Additionally, at least two hidden domains will also form part of the competition. These additional domains will be revealed only after the competition has been completed. For each instance, a single run will be conducted, and all the competing algorithms will start from the same initial solution generated from the same random seed. The run time will be limited to 10 minutes (measured in CPU time) on a modern PC running Windows XP. This figure was selected empirically after extensive testing on our problem domains' hardest instances. A benchmarking program (for both Windows and Linux) is available from the Web site that will report the time it takes a competitor's computer to execute a set of instructions that in the competition computer takes 10 minutes (600 seconds). It is worth noting that all the competitors will be run on a standard machine therefore creating a "level playing field".

In order to compare the performance of the competing hyper-heuristics and declare the winner, we will use a scoring system inspired by Formula 1. Before 2010, the Formula 1 system had the following structure. The top eight drivers scored 10, 8, 6, 5, 4, 3, 2 and 1 points respectively, in each race. These points are added for all the events, and the winner is the driver accumulating the most points. This is adapted for the cross-domain challenge as follows. Let us assume that m instances (considering all the domains) and n competing algorithms in total are considered. For each experiment (instance) an ordinal value o_k is given representing the rank of the algorithm compared to the others ($1 \leq o_k \leq n$). The top eight ranking algorithms will receive the points as in the Formula 1 system described above, and the remaining algorithms will receive no points. The points will be added across the m instances, and the winner will be the algorithm accumulating the most points. Therefore, if for example, five problem domains are considered with five instances each, the maximum possible score is 250 points. For solving ties we will also follow Formula 1. Full details can be seen on the competition Web site ('Scoring System' section: http://www.asap.cs.nott.ac.uk/chesc2011/scoring.html).

4   Final Remarks

Extensive tests (some of them published in [5]), have confirmed that a rich set of state-of-the-art hyper-heuristics can be implemented with HyFlex. Both single point and population based search algorithms can be designed. The Java jar file implementing the framework can be downloaded from the website, which also provides a tutorial, several examples, and the relevant software and academic documentation. An additional interesting feature is the Leaderboard, a table ranking participants according to their best score on a rehearsal competition conducted every week. This rehearsal competition is based on a set of results submitted by the participants who chose to do so. Note that only the results, and not the algorithms, are required for the Leaderboard submissions.

The prize fund is 3,000 GBP to be split between the first, second and third place competitors. The winners will be announced at OR53 (UK Operational Research Society conference, to be held in Nottingham, UK in September 6 - 8, 2011) and their registration fee will be waived. Our goal is to both promote research into more general search methodologies, and also to gain a deeper understanding of the algorithm design principles and machine learning techniques that work well in practice.

Acknowledgments. We would like to thank Aptia solutions Ltd, EventMap Ltd, Staff Rostering Solutions Ltd, the PATAT steering committee and the UK Operational Research Society for sponsoring the competition and providing the prize money.

References

1. Burke, E.K., Curtois, T., Hyde, M., Kendall, G., Ochoa, G., Petrovic, S., Vazquez-Rodriguez, J.A.: HyFlex: A flexible framework for the design and analysis of hyper-heuristics. In: Multidisciplinary International Scheduling Conference (MISTA 2009), Dublin, Ireland, pp. 790–797 (August 2009)
2. Cowling, P., Kendall, G., Soubeiga, E.: A hyperheuristic approach to scheduling a sales summit. In: Burke, E., Erben, W. (eds.) PATAT 2000. LNCS, vol. 2079, pp. 176–190. Springer, Heidelberg (2001)
3. Burke, E.K., Hart, E., Kendall, G., Newall, J., Ross, P., Schulenburg, S.: Hyper-heuristics: An emerging direction in modern search technology. In: Glover, F., Kochenberger, G. (eds.) Handbook of Metaheuristics, pp. 457–474. Kluwer, Dordrecht (2003)
4. Bleuler, S., Laumanns, M., Thiele, L., Zitzler, E.: PISA – A Platform and Programming Language Independent Interface for Search Algorithms. In: Fonseca, C.M., Fleming, P.J., Zitzler, E., Deb, K., Thiele, L. (eds.) EMO 2003. LNCS, vol. 2632, pp. 494–508. Springer, Heidelberg (2003)
5. Burke, E.K., Curtois, T., Hyde, M., Kendall, G., Ochoa, G., Petrovic, S., Vazquez-Rodriguez, J.A., Gendreau, M.: Iterated local search vs. hyper-heuristics: Towards general-purpose search algorithms. In: IEEE Congress on Evolutionary Computation (CEC 2010), Barcelona, Spain, pp. 3073–3080 (July 2010)

Author Index

Abdullah, Salwani   539
Aguirre, Hernán E.   91
Akama, Kiyoshi   203
Alba, Enrique   488
Alpaydın, Ethem   1
Arbelaez, Alejandro   46
Arenas, M.G.   582
Aslan, Özlem   1
Awad, Wasan Shakr   308

Balint, Adrian   586
Barrera, Julio   226
Battiti, Roberto   336
Baumgartner, Lukas   76
Bect, Julien   176
Belgasmi, Nabil   364
Bello, Rafael   253
Benassi, Romain   170
Berlanga, Antonio   458
Blum, Christian   76
Bonnard, Nicolas   433
Bontempi, Gianluca   106
Bureerat, Sujin   379
Burke, Edmund K.   631

Campigotto, Paolo   336
Castelli, Mauro   503
Castillo, P.A.   582
Coello Coello, Carlos A.   567
Couëtoux, Adrien   433
Croitoru, Cornelius   351

Daolio, Fabio   454
Dhaenens, Clarisse   31, 116, 238
Diepold, Daniel   586
Di Gaspero, Luca   450
di Tollo, Giacomo   450
Doerner, Karl F.   61
Durillo, Juan J.   488

Fawcett, Chris   600
Fernandes, C.M.   552, 582
Fialho, Álvaro   473
Flores, Juan J.   226
Frangioni, Antonio   407

Gall, Daniel   586
García, Jesús   458
Garcia-Sanchez, P.   582
Gendreau, Michel   631
Gerber, Simon   586
Ghedira, Khaled   364
Gunawan, Aldy   278

Hamadi, Youssef   46
Hartl, Richard F.   61
Hayashi, Akira   191
Higuchi, Tetsuya   218
Hoock, Jean-Baptiste   433
Hoos, Holger H.   507, 600
Humeau, Jérémie   31
Hutter, Frank   507
Hyde, Matthew   631

Ingimundardottir, Holga   263
Iwata, Kazunori   191
Izui, Kazuhiro   161

Jourdan, Laetitia   31, 116, 238

Kampouridis, Michael   16
Kaneko, Satoshi   191
Kapler, Gregor   586
Keane, Andy J.   446
Kendall, Graham   616, 631
Kobayashi, Takumi   218
Kritzinger, Stefanie   61
Kwong, Sam   473

Laredo, J.L.J.   552, 582
Lau, Hoong Chuin   131, 278
Leyton-Brown, Kevin   507, 600
Li, Ke   473
Liefooghe, Arnaud   31, 116, 238
Lindawati,   131, 278
Lo, David   131
Lopes, Leo   524
López, Rodrigo   226

Manzoni, Luca   503
Marmion, Marie-Eléonore   238
Martí, Luis   458

Martínez, Yailen   253
McCollum, Barry   631
Melab, Nouredine   321
Merelo, J.J.   552, 582
Molina, José M.   458
Mora, A.M.   552, 582
Munawar, Asim   203
Munetomo, Masaharu   203

Nebro, Antonio J.   488
Nell, Christopher   600
Nishiwaki, Shinji   161
Nowé, Ann   253

Ochoa, Gabriela   454, 631
Otsu, Nobuyuki   218
Özcan, Ender   616

Parkes, Andrew J.   631
Passerini, Andrea   336
Perez Sanchez, Luis   407
Petrovic, Sanja   631

Raschip, Madalina   351
Retz, Robert   586
Ribeiro, Celso C.   146
Roli, Andrea   450
Rosa, A.C.   552
Rosseti, Isabel   146
Runarsson, Thomas Philip   263, 423

Said, Lamjed Ben   364
Sato, Hiroyuki   91
Saubion, Frédéric   392

Schaerf, Andrea   450
Schmid, Verena   76
Smith-Miles, Kate   524
Sokolovska, Nataliya   433
Souza, Reinaldo C.   146
Suárez, Juliett   253
Suematsu, Nobuo   191
Swan, Jerry   616

Talbi, El-Ghazali   321
Tanaka, Kiyoshi   91
Tenne, Yoel   161
Teytaud, Olivier   433
Toal, David J.J.   446
Tomassini, Marco   454
Tricoire, Fabien   61
Tsang, Edward   16
Turabieh, Hamza   539
Turco, Alessandro   293

Van Luong, Thé   321
Vanneschi, Leonardo   503
Vazquez, Emmanuel   176
Veerapen, Nadarajen   392
Verel, Sébastien   31, 116, 238, 454

Wahib, Mohamed   203

Yáñez Oropeza, Edgar G.   567
Ye, Jiaxing   218
Yıldız, Olcay Taner   1

Zapotecas Martínez, Saúl   567
Zhang, Qingfu   488